Entropy Theory in Hydrologic Science and Engineering

About the Author

Professor V. P. Singh is a distinguished professor and the inaugural holder of the Caroline and William N. Lehrer Distinguished Chair in Water Engineering in the Department of Biological and Agricultural Engineering as well as the Zachry Department of Civil Engineering at Texas A & M University. He received his BS, MS, PhD, and DSc degrees in Engineering. He is a registered professional engineer, a registered professional hydrologist, and an Honorary Diplomate of Water Resources Engineering, American Academy of Water Resources Engineers. Professor Singh has extensively published on an extraordinary range of topics in hydrology, groundwater, irrigation engineering, environmental engineering, and water resources. In addition to numerous journal articles, book chapters, conference papers, and technical reports, he has published 22 textbooks and 55 edited reference books, including the massive *Encyclopedia of Snow, Ice and Glaciers*. For his scientific contributions to the development and management of water resources and promoting the cause of their conservation and sustainable use, he has received more than 65 national and international awards and numerous honors, including the Arid Lands Hydraulic Engineering Award, Ven Te Chow Award, Richard R. Torrens Award, Norman Medal, and EWRI Lifetime Achievement Award, all given by the American Society of Civil Engineers, 2010; and Ray K. Linsley Award and Founder's Award, given by the American Institute of Hydrology, 2006. He has received three honorary doctorates. He is a fellow of ASCE, EWRI, AWRA, IWRS, ISAE, IASWC, and IE and holds membership in 16 additional professional associations. He is a fellow/member of ten international science/engineering academies. He has served as president and senior vice president of the American Institute of Hydrology (AIH).

Entropy Theory in Hydrologic Science and Engineering

Vijay P. Singh

New York Chicago San Francisco
Athens London Madrid
Mexico City Milan New Delhi
Singapore Sydney Toronto

Entropy Theory in Hydrologic Science and Engineering

1 2 3 4 5 6 7 8 9 0   QVS/QVS   19 18 17 16 15 14

ISBN 978-0-07-183546-6
MHID 0-07-183546-6

The pages within this book were printed on acid-free paper.

Sponsoring Editor
Bridget Thoreson

Copy Editors
Patti Scott

Editorial Supervisor
Donna M. Martone

Proofreader
N. Amitha Karkera

Production Supervisor
Pamela A. Pelton

Indexer
Robert Swanson

Acquisitions Coordinator
Amy Stonebraker

Art Director, Cover
Jeff Weeks

Project Manager
Yashmita Hota,
Cenveo® Publisher Services

Composition
Cenveo Publisher Services

Dedicated to My wife, Anita,
daughter, Arti, son, Vinay,
daughter-in-law, Sonali, and grandsons Ronin and Kayden

Contents

Part V Environmental Conditions

Preface

Since the pioneering work of Shannon in the late 1940s on the development of information entropy theory and the landmark contributions of Jaynes, Kullback and Leibler, and Lindley, leading, respectively, to the development of the principle of maximum entropy and theorem of concentration, the principle of minimum cross entropy, and the concept of mutual information, entropy theory has been applied to a wide spectrum of areas. These include biology, chemistry, economics, ecology, electronics and communication engineering, data acquisition and storage and retrieval, fluid mechanics, genetics, geology and geomorphology, geophysics, geography, geotechnical engineering, hydraulics, hydrology, image processing, management sciences, operations research, pattern recognition and identification, photogrammetry, psychology, physics and quantum mechanics, risk and reliability analysis, reservoir engineering, social sciences, statistical mechanics, thermodynamics, topology, transportation engineering, turbulence modeling, etc. New areas of entropy application have continued to unfold. The entropy theory is indeed versatile and its application is widespread.

In the area of hydrologic science and engineering, a range of applications of entropy have been reported during the past four decades or so, and new topics applying entropy are emerging each year. Many books on entropy have been written in statistics, communication engineering, economics, biology, and reliability analysis. However, these books have been written with different objectives in mind and to address different kinds of problems. Application of entropy concepts and techniques discussed in these books to hydrologic science and engineering problems is not always straightforward. Therefore, there is a need for a book that deals with basic concepts of entropy theory from a hydrologic engineering perspective and deals with applications of these concepts to a range of hydrologic engineering problems. Currently there is no book devoted to covering basic aspects of entropy theory and its application in hydrologic science and water engineering. This book attempts to fill this need.

The subject matter of the book is divided into 20 chapters organized in five parts. Part I, comprising five chapters, deals with preliminaries. Chapter 1 discusses the concept of entropy and entropy theory for both discrete and continuous variables. It goes on to discuss different types of entropy for both univariate and multivariate cases, the principle of maximum entropy, principle of minimum cross entropy, concentration theorem, and methodology for application of entropy theory. Chapter 1 concludes with a discussion of the types of problems and the hypothesis on the cumulative distribution function. The maximum entropy production principle constitutes the subject matter of Chap. 2. Beginning with a discussion of hydrologic systems and irreversibility, Chap. 2

goes on to discuss the laws of thermodynamics and their application, entropy production flux, and application of nonequilibrium thermodynamics to flow of water over and below the land surface. Chapter 3 discusses various types of measures employed to evaluate the performance of hydrologic models. The measures include bias, different types of errors, correlation coefficients, Nash-Sutcliffe efficiency criterion, mean square error, and related criteria. Chapter 3 concludes with a discussion of performance measures for probabilistic models.

Chapter 4 treats morphological analysis, including types of entropy, potential energy, the law of average stream fall, longitudinal channel profile and relation between entropy and Horton-Strahler order, basin magnitude, fractal dimension, scaling properties, bed profile, stream meandering, sample source uncertainty, and watershed surface description. A case study concludes the chapter. Chapter 5 addresses the evaluation and design of monitoring networks. Beginning with design considerations and correlation, it goes on to discuss entropy-based measures, isoinformation, directional information transfer index, total correlation, and maximum entropy and minimum redundancy. The chapter concludes with an application to evaluation of different types of hydrologic networks.

Part II consists of five chapters dealing with some aspects of rainfall. Chapter 6 deals with precipitation variability, including measures of variability and diversity indices, trends and persistence, and entropy measures for analysis of precipitation, rainfall disaggregation, and rainfall complexity. The chapter concludes with two case studies presenting the application of these concepts. Chapter 7 discusses rainfall frequency distributions. It argues that frequency distributions of rainfall vary with time scale, climate type, and distance from the sea. Discussing briefly the empirical distributions commonly employed for rainfall frequency analysis, it presents the entropy-based distributions using moment constraints which can accommodate bimodality in the distribution of rainfall maxima. Chapter 7 concludes with a discussion of two case studies. Chapter 8 deals with the evaluation of precipitation forecasting schemes. Starting with a presentation of a set of notations, it discusses various methods for evaluation of probabilistic forecasting, decomposition of the Briar score, as well as Kullback-Leibler divergence. Chapter 9 deals with the assessment of water resources variability, including entropy as a measure of temporal rainfall apportionment for hydrological zoning and categorization of water resources availability, and intensity and apportionment entropies as well as clustering for the assessment of water resources availability. Chapter 9 concludes with two case studies. Evaporation and transpiration are discussed in Chap. 10. The discussion includes the extremum principle for evaporation from land surfaces and its application. The chapter concludes with treatment of transpiration.

The subject matter of Part III, which comprises three chapters, encompasses subsurface flow. Chapter 11 deals with infiltration and presents the derivation of seven infiltration equations, including the equations of Horton, Kostiakov, Philip, Green and Ampt, Overton, Holtan, Singh and Yu, and a general equation. The chapter concludes with the derivation of a general infiltration equation. Chapter 12 discusses soil moisture. Providing a short introduction to soil moisture profiles and their estimation, it presents the derivation of soil moisture profiles for wetting, drying, and mixed phases; mean soil moisture content; and variation in time. Chapter 13 deals with groundwater flow. It includes groundwater movement in both confined and unconfined aquifers by employing one-constraint and two-constraint formulation, inverse problems, and linear inverse problems.

Part IV, comprising four chapters, deals with surface flow. Chapter 14 addresses rainfall-runoff modeling. It discusses runoff volume by the SCS-CN method, instantaneous unit hydrograph, volume-peak discharge relation, rational method, and linear reservoir model for computing the discharge hydrograph. Chapter 14 concludes with App. 14A, which presents the derivation of a general probability distribution. Stream flow simulation at a single site as well as at multiple sites is discussed in Chap. 15. This entails derivation of joint and conditional distributions and sampling. Hydrologic frequency analysis constitutes the subject matter of Chap. 16. Included in the discussion are methods for derivation of the frequency distribution, general probability distribution, and truncated normal distribution. The chapter concludes with the treatment of bivariate rainfall-runoff analysis and flow duration curves. Starting with a discussion of preliminaries, Chap. 17 deals with long-term stream flow forecasting. It presents maximum entropy spectral analysis using Burg entropy, configuration entropy, and minimum cross entropy as well as a methodology for the extension of the autocorrelation function. The chapter concludes with comparison with time series models.

Part V on environmental considerations contains three chapters. Chapter 18 deals with river regime classification. Introducing river regimes, it discusses methods of regime classification, grouping of regimes, and stability and instability measures. Chapter 19 addresses sediment yield. Introducing erosion processes, it discusses the application of entropy theory to deriving a sediment production equation, sediment graph, and conditional probability distribution of sediment yield. The eco-index is the subject matter of Chap. 20, containing indicators of hydrologic alteration, methodology for development of the eco-index, and computation of the nonsatisfaction eco-level and eco-index. It concludes with a case study on application of the eco-index.

Vijay P. Singh
College Station, Texas

Acknowledgments

The book draws from the works of hundreds of scientists and engineers who have dedicated their careers to the enrichment of the literature on entropy. I am indebted to them for what they did. It is because of them that we know much more about entropy today. I have tried to acknowledge them specifically. Any omission on my part has been entirely inadvertent, and I offer my apologies in advance.

For over two decades I have had the privilege of collaborating with a number of researchers from around the globe. For economy of space I will not mention all of them but would particularly like to acknowledge Prof. A. K. Rajagopal from Naval Research Laboratory, Washington, D.C.; Prof. N. B. Harmancioglu from Dokuz Elylul University, Izmir, Turkey; Prof. M. Fiorentino from University of Basilicata, Potenza, Italy; and Dr. T. Moramarco from C.N.R. Institute for Hydrogeological Protection Research, Perugia, Italy. I have learned much from these collaborators. Likewise I have had a number of graduate students, postdoctoral fellows, and visiting scholars over the years who worked with me on topics employing entropy theory. They have helped me in myriad ways, and I am grateful that they worked with me. With some of them I am still collaborating. I particularly wish to acknowledge Dr. L. Zhang from University of Akron, Ohio; Dr. G. Marini from University of Sannio, Italy; Dr. A. K. Mishra from Clemson University, Clemson, South Carolina; Dr. Lu Chen from Huazhong University of Science & Technology, China; Dr. Z. Hao from University of Oklahoma, Oklahoma City; Dr. C. Li from Stanford University, Palo Alto, California; Dr. K. Singh from University of Houston, Texas; Dr. P. F. Krstanovic from Northwest Consultants, Seattle, Washington; Ms. Luo Hao from University of Illinois, Urbana-Champaign; Zooho Kim from K-Water, South Korea; and Dr. C. P. Khedun, Mrs. H. Cui, Mrs. Deepthi Rajasekhar, and Mr. C. Sohoulande from Texas A&M University, College Station. They helped with different parts of the book, and without their support this book would not have been completed.

Finally, I would like to take this opportunity to acknowledge the support that my brothers and sisters in India and my family here in the United States have given me over the years. My brothers and sisters are most affectionate and caring and are always there to lend support. My wife, Anita, helps me in so many ways that I do not have enough words to fully describe her contribution to my life. Each day she makes sure that I am not bothered by any household chores, and I am free to do my professional work. She has always been the bedrock of support. My daughter, Arti, checks on me regularly regardless of where she is or where I am and makes sure that I am taking care of myself. I could not have asked for a more caring and affectionate daughter. My son,

Vinay, is always there to lend a helping hand. He is a reservoir of goodwill and stability. My daughter-in-law, Sonali, is what a father-in-law would wish for. I am grateful that she is my daughter-in-law. My grandsons, Ronin and Kayden, are my dreams, my hope, and my future. They are lovely to play with and make my life worthwhile. My family members always support me in whatever I elect to pursue. Therefore, I humbly dedicate this book to them, for without their support and affection this book would not have come to print.

PART I

Preliminaries

CHAPTER 1

Introduction to Entropy Theory

In 1948 Shannon formulated the concept of entropy as a measure of information or uncertainty, and Lindley (1956) developed the concept of mutual entropy as a measure of dependence. Nearly a decade later, Jaynes (1957a, 1957b, 1958, 1982, 2003) developed the principle of maximum entropy (POME) for deriving the least-biased probability distributions, subject to given information in terms of constraints, as well as the theorem of concentration for hypothesis testing. Around the same time, Kullback (1959) formulated the principle of minimum cross entropy (POMCE) which is a generalization of POME. Together these concepts constitute what can now be referred to as entropy theory. Entropy has since been extensively applied in hydrologic science and engineering, including geomorphology, hydrology, and hydrometeorology. Harmancioglu et al. (1992), Singh and Fiorentino (1992), and Singh (2010, 2011, 2013) surveyed applications of entropy in water resources, and Singh (1997) discussed the use of entropy in hydrology and water resources. New applications of entropy continue to unfold. This chapter introduces the concept of entropy and entropy theory and provides a snapshot of applications of the theory in hydrologic science and engineering.

1.1 Concept of Entropy

Entropy is regarded as a measure of disorder, chaos, uncertainty, or surprise, since these are different shades of information. Consider, e.g., a discrete random variable X that takes on N values $x_1, x_2, ..., x_N$ with probabilities $p_1, p_2, ..., p_N$, respectively; i.e., each value of X, x_i, represents an event with a corresponding probability of occurrence p_i, $i = 1, 2, ..., N$. The occurrence of an event x_i provides a measure of information about the likelihood of that probability p_i being correct (Batty, 2010). If p_i is low, say, 0.001 and if x_i actually occurred, then one would be quite surprised by the occurrence of x_i with $p_i = 0.001$, because one's anticipation of it was highly uncertain. On the other hand, if p_i is very high, say, 0.999 and if x_i did actually occur, then one would hardly be surprised by the occurrence of x_i with $p_i = 0.999$, because one's anticipation of it was quite certain. This shows that there is a connection between surprise and uncertainty or its complement, certainty. Certain and uncertain events are frequently encountered in real life.

Uncertainty about the occurrence of an event further suggests that the random variable takes on different values, and information is gained by observing it only if there is uncertainty about the event. If an event occurs with a high probability, it conveys less information and vice versa. On the other hand, more information will be needed to characterize less probable or more-uncertain events or to reduce uncertainty about the occurrence of such an event. In a similar vein, if an event is more certain to occur, its occurrence or observation conveys less information and less information will be needed to characterize it.

In a similar manner, the behavior of a system, represented by X, can be characterized by considering the set of states that the system occupies during its evolution in time. If a system is observed to occupy a state that occurs rarely, then it would be surprising to find the system in that state. This discussion shows that there is a connection between information, uncertainty, and surprise. Because entropy is a measure of information or uncertainty or surprise, the connection extends to entropy as well.

Example 1.1.   Consider an event that occurs with probability 1. What is the uncertainty about this event? What is the degree of surprise about the event? Now consider another event whose probability of occurrence was close to 0 but actually did occur. Then what can be said about the uncertainty and degree of surprise about this event? Consider another event whose probability of occurrence is almost zero that does not occur. What can be stated about the uncertainty and level of surprise of this event?

Solution.   The event occurs with probability 1; hence it is a certain event, implying that there is no uncertainty associated with it. It does not correspond to a random value, or it is not a manifestation of a random variable and its observation conveys no information. The degree of surprise about the occurrence of the event is zero. For the other event, the probability of occurrence is close to 0 but it did occur, implying a highly uncertain event and an enormously high degree of surprise. This means that the uncertainty about the occurrence of the event is enormously high, and its occurrence would provide enormously high information. On the other hand, if the probability of the event is indeed zero, then the event cannot occur and does not correspond to a random value or is not an outcome of a random variable. Here there is no uncertainty or surprise as to the non-occurrence of this event and its non-occurrence conveys no information.

Consider as an example a random variable representing instantaneous annual peak flow at a gauging station on a river. The peak discharge can take on many values. Consider an average return period of a peak discharge value as T years. If, say, $T = 500$ years, then the peak discharge has a probability of occurrence each year of $1/T = 1/500 = 0.002$. If this discharge occurred, its occurrence would be a surprise because its occurrence was not anticipated and was highly uncertain. To model such a discharge, a lot of observations or information will be needed to reduce anticipatory uncertainty. This kind of event contains greater uncertainty, and a lot more information will be needed to reduce uncertainty. This suggests that the anticipatory uncertainty of x_i prior to the observation is an increasing function of decreasing probability $p(x_i)$ of its occurrence. This seems to suggest that information varies inversely with probability p, that is, with $1/p$. Probability is an expression of the degree of plausibility of a proposition.

Now consider the case when two or more independent events occur. Let x and y denote two independent events occurring with probability p_x and p_y, respectively. The question arises: What can be said about the information on the occurrence of these two events? The probability of the joint occurrence of x and y is $p_x p_y$. It would seem logical from the discussion of single events that the information to be gained from the joint occurrence of these events would be the inverse of the probability of their occurrence, that is, $1/(p_x p_y)$. Note that this information does not equal the sum of information gained from the occurrence of event x, $1/p_x$, and the information gained from the occurrence of event y, $1/p_y$, that is,

$$\frac{1}{p_x p_y} \neq \frac{1}{p_x} + \frac{1}{p_y} \qquad (1.1a)$$

Considering a function $g(\cdot)$ of probabilities, the inequality expressed by Eq. (1.1a) can be mathematically expressed as

$$g\left(\frac{1}{p_x p_y}\right) = g\left(\frac{1}{p_x} + \frac{1}{p_y}\right) \qquad (1.1b)$$

If function $g(\cdot)$ is a logarithmic function, then Eq. (1.1b) can be expressed as

$$\log\frac{1}{p_x p_y} = \log\frac{1}{p_x} + \log\frac{1}{p_y} \qquad (1.2)$$

It seems that the logarithmic function is the only function that will allow the transition from Eq. (1.1b) to Eq. (1.2).

Thus, it can be summarized that the information gained from the occurrence of any event x with probability $p(x)$ is $s(x) = \log(1/p(x)) = -\log p(x)$. Here $p(x)$ is the probability that the system occupies state x, and $-\log p(x)$ indicates the expectation (or lack of expectation) that the system occupies state x. Since $s(x)$ is large for improbable or rare events and small for probable or frequently occurring events, Tribus (1969) regarded $s(x) = -\log p(x)$ as a measure of surprise of that event occurring or a measure of uncertainty of the event occurring with probability p. This concept can be extended to a series of N events or system states occurring with probabilities $p_1, p_2, \ldots, p_N$, which then leads to the Shannon entropy, described in what follows.

1.2 Entropy Theory

The entropy theory comprises four parts: (1) the Shannon entropy, (2) the principle of maximum entropy, (3) the principle of minimum cross entropy, and (4) the concentration theorem.

1.2.1 Shannon Entropy: Discrete Variable

Equation (1.2) can be extended by considering a discrete random variable X that takes on values $x_1, x_2, \ldots, x_N$, with each value corresponding to an event, occurring with probabilities $p_1, p_2, \ldots, p_N$, respectively, as

$$\log\frac{1}{p_1 p_2 \cdots p_N} = \log\frac{1}{p_1} + \log\frac{1}{p_2} + \cdots + \log\frac{1}{p_N} = -\sum_{i=1}^{N}\log p_i \qquad (1.3)$$

Equation (1.3) states the information gained by the joint occurrence of N independent events. One can write the average information as the expected value (or weighted average) of this series

$$H = -\sum_{i=1}^{N} p_i \log p_i \tag{1.4}$$

where H is termed the *entropy*, defined by Shannon (1948). This is frequently called the Shannon entropy. For logarithm to base 2, Eq. (1.4) gives the average number of bits needed to optimally encode independent draws $X: \{x_1, x_2, \ldots, x_N\}$ with a probability distribution $P: \{p(x_1), p(x_2), \ldots, p(x_N)\} = \{p_1, p_2, \ldots, p_N\}$. Fundamental to constructing optimal coding is $p(x_i)$, $i = 1, 2, \ldots, N$.

Equation (1.4) is a measure of uncertainty given by a set of probabilities. It is a measure of disorder in a system, or it is a measure of our ignorance about the actual state of a system. Equation (1.4) is obtained by averaging $s(x)$ over all possible states, which is a measure of the expectation of the state of the system or a measure of the information one has about the system. If the system states or behaviors are characterized by multiple parameters, then the states can be used jointly to describe the behavior of the system. In this case, the Shannon entropy will become joint entropy which can be computed by averaging over all possible states. This aspect is discussed later in the chapter.

The Shannon entropy can also be viewed as follows. Let there be a long time D for which a system is observed, let N be the total number of different things (i.e., measures of information) that occur, and let n_i be the things that occur with probability p_i, $i = 1, 2, \ldots, N$. If all p_i are equal, then no information about the system can be obtained. If a particular p_i is specified, then a certain amount of information about the system can be gained from this value. This suggests that the probability of occurrence of an event is related to information. If the probability is specified about an event that is unexpected, the specification conveys much information. The relation between information and probability can be expressed as

$$I = k \log \frac{1}{p_i}$$

where k is constant and is equal to 1 when the base of the logarithm is 2. We think of information as additive, so the use of the logarithm is justified.

Because the system is being observed for a long time, p for n_1 can be expected to have occurred with $p_1 D$, for n_2 with $p_2 D$, for n_i with $p_i D$, and so on. The total information I_{total} about the system during period D can then be expressed as

$$I_{total} = k\left(p_1 D \ln \frac{1}{p_1 D} + p_2 D \ln \frac{1}{p_2 D} + \cdots + p_N D \ln \frac{1}{p_N D} \right)$$

The average information per unit time, referred to as entropy H, is represented as

$$H = \frac{I_{total}}{D} = -k \sum_{i=1}^{N} p_i \ln p_i$$

which is the same as Eq. (1.4).

The concept of entropy is central to statistical physics. Boltzmann and then Gibbs provided statistical interpretations of H as a measure of thermodynamic entropy.

Some investigators therefore designate H as Shannon-Boltzmann-Gibbs entropy (see Papalexiou and Koutsoyiannis, 2012). An excellent account of entropy is provided by Koutsoyiannis (2013). In this text, we will simply call it the Shannon entropy. The Shannon entropy H is also written as

$$H(X) = H(P) = -K \sum_{i=1}^{N} p(x_i) \log p(x_i) \tag{1.5}$$

where $H(X)$ is the entropy of X: $\{x_1, x_2, \ldots, x_N\}$; P: $\{p_1, p_2, \ldots p_N\}$ is the probability distribution of X; N is the sample size; and K is a parameter whose value depends on the base of the logarithm used. If different units of measurement of entropy are used, then the base of the logarithm changes. For example, one uses bits for base 2, neper (N_p) for Napier or nat for base e, and decibels or dogits for base 10.

In general, K can be taken as unity, and Eq. (1.5) therefore becomes

$$H(X) = H(P) = -\sum_{i=1}^{N} p(x_i) \log [p(x_i)] \tag{1.6}$$

where $H(X)$, given by Eq. (1.6), represents the information content of random variable X or its probability distribution $P(x)$. It is a measure of the amount of uncertainty or indirectly the average amount of information content of a single value of X. Equation (1.6) satisfies a number of requirements, such as continuity, symmetry, additivity, expansibility, recursivity, and others (Shannon and Weaver, 1949).

If X is a deterministic variable, then the probability that it will take on a certain value is 1, and the probabilities of all other alternative values are 0. Then Eq. (1.6) shows that $H(x) = 0$, which can be viewed as the lower limit of the values the entropy function may assume. This corresponds to absolute certainty; i.e., there is no uncertainty and the system is completely ordered. On the other hand, when all x_i are equally likely, i.e., the variable is uniformly distributed ($p_i = 1/N$, $i = 1, 2, \ldots, N$), then Eq. (1.6) yields

$$H(X) = H_{max}(X) = \log N \tag{1.7}$$

This shows that the entropy function attains a maximum, and Eq. (1.7) thus defines the upper limit. This also reveals that the outcome has the maximum uncertainty. Equation (1.4) and in turn Eq. (1.7) show that the larger the number of events, the larger the entropy measure. This is intuitively appealing, because more information is gained from the occurrence of more events, unless, of course, events have zero probability of occurrence. The maximum entropy occurs when the uncertainty is maximized or the disorder is maximized.

One can now state that entropy of any discrete variable always assumes positive values within limits defined by

$$0 \leq H(X) \leq \log N \tag{1.8}$$

It is logical to say that many probability distributions lie between these two extremes and their entropies between these two limits.

As an example, consider a random variable X that takes on a value of 1 with a probability p and a value of 0 with a probability $q = 1 - p$. Taking different values of p, one

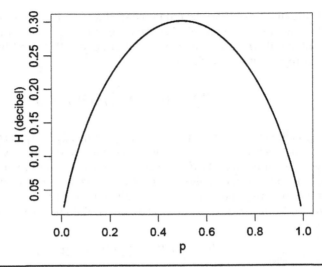

FIGURE 1.1 Entropy as a function of probability $[X: \{x_1, x_2\}; P: \{p_1, p_2 = 1 - p_1\}; p_1 + p_2 = 1]$.

can plot $H(p)$ as a function of p, as shown in Fig. 1.1. It is seen that for $p = 1/2$, $H(p) = 0.301$ dB and is the maximum. Likewise, $H(p) = 1$ bit for $p = 1/2$.

Example 1.2. Consider a random variable X taking on three values with probabilities p_1, p_2, and p_3. Using different combinations of these probabilities as shown in Table 1.1, compute the entropy and determine the combination for which the entropy will be maximum. Tabulate the entropy values for different combinations of probabilities.

Solution

$$H(X) = H(P) = -\sum_{i=1}^{3} p(x_i) \log\ p(x_i)$$

The computed entropy values are listed in the last column of Table 1.1. It is clear from the table that entropy is maximum at 0.477 dB when $p_1 = p_2 = p_3 = 0.333$. For all other combinations of p, the entropy value is less than this value. In this case, $0 \leq H(p) \leq 0.477$ dB.

p_1	p_2	p_3	H(X) (dB)	p_1	p_2	p_3	H(X) (dB)
0.05	0.05	0.9	0.171	0.333	0.333	0.333	0.477
0.1	0.1	0.8	0.278	0.4	0.3	0.3	0.473
0.1	0.2	0.7	0.348	0.5	0.3	0.2	0.447
0.1	0.3	0.6	0.390	0.6	0.2	0.2	0.413
0.2	0.2	0.6	0.413	0.7	0.2	0.1	0.348
0.2	0.3	0.5	0.447	0.8	0.1	0.1	0.278
0.3	0.3	0.4	0.473	0.9	0.05	0.05	0.171

TABLE 1.1 Probabilities p_1, p_2, and p_3 of Values x_1, x_2, and x_3 of a Random Variable and Corresponding Entropy Values.

1.2.2 Shannon Entropy: Continuous Variable

Now the question arises: What happens to entropy if the random variable is continuous? Let X be a continuous random variable within a certain range with a probability density function $f(x)$. Then the range within which the continuous variable assumes values can be divided into N intervals of width Δx. One can then express the probability that a value of X is within the nth interval as

$$p_n = P\left(x_n - \frac{\Delta x}{2} \le X \le x_n + \frac{\Delta x}{2}\right) = \int\limits_{x_n - \Delta x/2}^{x_n + \Delta x/2} f(x)\,dx \qquad (1.9)$$

For a relatively small value of Δx, the probability p_n can be approximated as

$$p_n \cong f(x_n)\Delta x \qquad (1.10)$$

The marginal entropy of X expressed by Eq. (1.6) for a given class interval Δx can be written as

$$H(X; \Delta x) \cong -\sum_{n=1}^{N} p_n \log p_n = -\sum_{n=1}^{N} f(x_n)\log[f(x_n)\,\Delta x]\,\Delta x \qquad (1.11)$$

This approximation would have an error whose sign would depend on the form of the function $-f(x) \log \Delta x$. To reduce this approximation error, the Δx interval can be chosen as small as possible. Let $p_i = p(x_i)\Delta x$, and let the interval size Δx tend to zero; here p_i is the probability and $p(x_i)$ is the probability density. Then Eq. (1.11) can be expressed as

$$H(X; \Delta x) = -\lim_{\Delta x \to 0}\sum_{i=1}^{N} p(x_i)\,\Delta x \,\log[p(x_i)\,\Delta x] \qquad (1.12)$$

Equation (1.12) can be expanded as

$$H(X; \Delta x) = -\lim_{\Delta x \to 0}\sum_{i=1}^{N} p(x_i)\,\Delta x \log p(x_i) - \lim_{\Delta x \to 0}\sum_{i=1}^{N} p(x_i)\,\ln(\Delta x)\,\Delta x \qquad (1.13)$$

Equation (1.13) can also be extended to the case where Δx_i varies with i, and it shows that the discrete entropy of Eq. (1.13) increases without bound. Equation (1.13) can also be written as

$$H(X; \Delta x) = -\sum_{i=1}^{N} p(x_i)\,\Delta x \log p(x_i) - \sum_{i=1}^{N} p(x_i)\,\ln(\Delta x)\,\Delta x \qquad (1.14)$$

Equation (1.13) converges to

$$H(X; \Delta x) = -\int\limits_{0}^{\infty} f(x)\,\ln f(x)\,dx - \lim_{\Delta x \to 0}\sum_{i=1}^{N} p(x_i)\,\ln(\Delta x)\,\Delta x \qquad (1.15)$$

Equation (1.15) yields

$$H(X;\Delta x) = -\int_0^\infty f(x)\ln f(x)\,dx - \lim_{\Delta x \to 0}\ln(\Delta x) \qquad (1.16)$$

Equation (1.16) is also written as

$$H(X;\Delta x) = -\int_0^\infty f(x)\log f(x)\,dx - \log(\Delta x) \qquad (1.17)$$

Combining $-\log(\Delta x)$ with the right side, Eq. (1.17) can be written as

$$H(X;\Delta x) \cong -\sum_{n=1}^N p_n \log \frac{p_n}{\Delta x} = -\sum_{n=1}^N f(x_n)\log f(x_n)\,\Delta x \qquad (1.18)$$

The right side of Eq. (1.18) can be written as

$$H(X) = -\int_0^\infty f(x)\log f(x)\,dx = -\int_0^\infty \log f(x)\,dF(x) = E[-\log f(x)] \qquad (1.19)$$

Equation (1.18) is also referred to as the *spatial entropy* if x is a space dimension and Eq. (1.19) is the commonly used expression for continuous Shannon entropy. Here $F(x)$ is the cumulative (probability) distribution function (CDF) of X, $E[\cdot]$ is the expectation of $[\cdot]$, and $H(X)$ is the entropy and is a measure of the uncertainty of random variable X of the system. It can also be understood as a measure of the amount of information required, on average, to describe the random variable. Thus, entropy is a measure of the amount of uncertainty represented by the probability distribution or of the lack of information about a system represented by the probability distribution. Sometimes it is referred to as a measure of the amount of *chaos* characterized by the random variable. If complete information is available, entropy = 0, that is, there is no uncertainty; otherwise, it is greater than zero. Thus, the uncertainty can be quantified using entropy by taking into account all different kinds of available information.

A more general form of the Shannon entropy for a continuous case is often written as

$$H(X) = -K\int_0^\infty f(x)\log \frac{f(x)}{m(x)}\,dx \qquad (1.20)$$

where $m(x)$ is an invariance measure which guarantees the invariance of entropy $H(X)$ under any allowable change of variable and provides an origin of measurement of $H(X)$; K is an arbitrary constant or scale factor depending on the choice of measurement units of entropy. Equation (1.20) is the same as Eq. (1.19) with $m(x) = 1$ and $K = 1$.

Example 1.3.   Consider a random variable $X \in [0,\infty)$ that is described by a gamma distribution whose probability density function (PDF) can be expressed as

$$f(x) = \frac{1}{b\Gamma(a)}\left(\frac{x}{b}\right)^{a-1}\exp\left(-\frac{x}{b}\right)$$

where a and b are shape and scale parameters and $\Gamma(a) = (a-1)!$ For illustrative purposes, take $a = 5$, $b = 1$, and $X = [0, 20]$. Shrinkage of the support from $[0, \infty)$ to $[0, 20]$ is reasonable since the PDF decays very close to 0 when $X = 20$. The entropy theory shows (Singh, 1998, 2013) that the gamma distribution can be determined by noting that $ab = \bar{x} = E[x]$ and $\Psi(a) - \ln a = \overline{\ln x} = E[\ln x]$, where $\Psi(a)$ is the digamma function, which is defined as the derivative of the logarithm of the gamma function

$$\Psi(a) = \frac{d}{da} \log \Gamma(a) = \frac{\Gamma'(a)}{\Gamma(a)}$$

and can be approximated as

$$\Psi(a) = \log a - \frac{1}{2a} - \frac{1}{12a^2} + \frac{1}{120a^4} - \frac{1}{252a^6} + O\left(\frac{1}{a^8}\right) \tag{1.21}$$

Select an interval size for discrete approximation and compute the entropy, using the discrete approximation as well as the continuous form. Then use different interval sizes, repeat the calculations, and determine the impact of the choice of interval size.

Solution. First, consider the continuous form. Then substituting the gamma PDF in the continuous form of entropy [Eq. (1.19)], one obtains

$$H(X) = a \ln b + \ln \Gamma(a) + (1-a)\overline{\ln x} + \bar{x}/b$$

By substituting $\bar{x} = ab$ and $\overline{\ln x} = \Psi(a) - \ln(a)$, the entropy equation becomes

$$H(X) = a + a \ln b + \ln \Gamma(a) + (1-a)[\Psi(a) - \ln a]$$
$$= 5 + 0 + \ln(24) - (1-5)(1.506 - 1.609) = 7.766 \text{ Np}$$

(The unit neper, abbreviated Np, honored Napier.)

Now consider discrete entropy, with interval Δx. The continuous entropy can be written as

$$H(X) = -\int_0^\infty f(x)\log f(x)dx = -\sum_{i=1}^N f(x_i) \Delta x \ln\left[f(x_i)\frac{\Delta x}{\Delta x}\right] = -\sum_{i=1}^N p_i \ln \frac{p_i}{\Delta x}$$

which can also be written as

$$H(x) = -\sum_{i=1}^N p_i \ln p_i + \ln \Delta x$$

where p_i can be computed with Eq. (1.9) as

$$p_i = F\left(x_i + \frac{\Delta x}{2}\right) - F\left(x_i - \frac{\Delta x}{2}\right)$$

where F denotes the cumulative distribution function (CDF) of the gamma distribution. Taking $\Delta x = 2$ as an example, it is assumed that one wants to compute p_{6-8}, which is actually the shaded area shown in Fig. 1.2 and can be computed as

$$p_{6-8} = F(8) - F(6) = 0.185$$

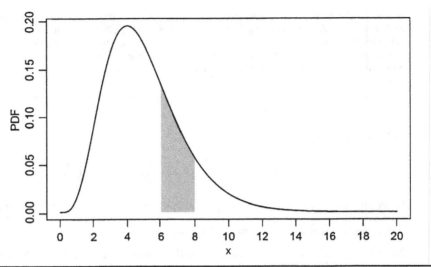

FIGURE 1.2   Illustration for the computation of p_4.

In the same way, the other p_i values can be computed, as listed in Table 1.2. The continuous entropy corresponding to an interval size of 2 is

$$H(x) = -(0.053 \times \ln 0.053 + 0.319 \times \ln 0.319 + \cdots) + \ln 2 = 2.2041 \text{ Np}$$

In the same way, entropy is computed for different interval sizes as shown in Table 1.3.
   Table 1.3 shows that the entropy values significantly depend on the interval size.

Interval	0–2	2–4	4–6	6–8	8–10	10–12	12–14	14–16	16–18	18–20
p_i	0.053	0.319	0.344	0.185	0.070	0.022	0.006	0.001	0.0003	0.00006

TABLE 1.2   Values of p_i when Interval Size is 2

Δx	N	H(x) (Np)
2	10	2.046
1	20	2.167
0.4	50	2.156
0.2	100	2.154
0.02	1,000	2.153
0.01	2,000	2.153
0.002	10,000	2.153

TABLE 1.3   Values of Entropy for Different Interval Sizes

1.3   Types of Entropies

1.3.1   Marginal Entropy and Information

The entropy of a probability distribution can be regarded as a measure of information or a measure of uncertainty. The amount of information obtained when observing the result of an experiment can be considered numerically equal to the amount of uncertainty as regards the outcome of the experiment before it is conducted. There are different types of entropy or measures of information: marginal entropy, approximation entropy, sample entropy, conditional entropy, joint entropy, transinformation, and interaction information. *Marginal entropy* is the entropy of a single variable and is defined by Eq. (1.19) if the variable is continuous or by Eq. (1.6) if the variable is discrete. Other types of entropies are defined when more than one variable is considered.

Entropy $H(X)$ permits one to measure information, and for that reason it is also referred to as informational entropy. Intuitively, information reduces uncertainty which is a measure of surprise. Consider an input-output channel or transmission conduit with H_I as the entropy (or uncertainty) of input and H_O is the entropy (or uncertainty) of output. For an input-output system, the information I is a reduction in uncertainty $H(X)$ about the system due to the available information and can be defined as

$$I = H_I - H_O \tag{1.22}$$

Were there no noise in the channel, the output would be certain as soon as the input was known. This means that the uncertainty in output H_O would be 0 and Eq. (1.22) would yield $I = H_I$, meaning no change in information.

1.3.2   Relative Redundancy

By comparing H with H_{max}, a measure of information can be constructed as

$$I = H_{max} - H = \log N + \sum_{i=1}^{N} p_i \log p_i \tag{1.23}$$

Normalizing I by H_{max}, Eq. (1.23) becomes

$$R = \frac{I}{H_{max}} = 1 - \frac{H}{H_{max}} \tag{1.24}$$

where R is the relative redundancy and varies between 0 and 1.

If we recall that H_{max} is obtained when all probabilities are of the same value, i.e., all events occur with the same probability, Eq. (1.23) can be written as

$$I = \sum_{i=1}^{N} p_i \log \frac{p_i}{1/N} = \sum_{i=1}^{N} p_i \log \frac{p_i}{q_i} \tag{1.25}$$

where $q_i = 1/N$. In Eq. (1.25), $\{q_i\}$ can be considered as a prior distribution and $\{p_i\}$ as a posterior distribution. Then I can be referred to as *cross entropy*; this is discussed in greater detail later.

1.3.3 Multivariate Entropy

Now consider two random variables X and Y that are not necessarily independent. Then the marginal entropy $H(X)$, given by Eq. (1.6), can be defined as the potential information of variable X; this is also the information of its underlying probability distribution. For two variables, the joint entropy $H(X, Y)$ is the total information content in both X and Y; i.e., it is the sum of the marginal entropy of one of the variables and the uncertainty that remains in the other variable when a certain amount of information that it can convey is already present in the first variable. This leads to the definition of the *joint entropy* of two independent variables as

$$H(X, Y) = H(X) + H(Y) \tag{1.26}$$

Clearly $H(X)$ depends on the probability distribution of X, and the same is true for Y. In Eq. (1.26), $H(X, Y)$ indicates the total amount of uncertainty that X and Y entail or the total amount of information they convey. Consider, e.g., the flood process characterized by flood peak X and flood volume Y if they are independent. A practical implication is that more observations of X and Y would reduce the uncertainty more than more observations of only X or only Y.

Let X take on values $x_1, x_2, x_3, \ldots, x_N$ with corresponding probabilities $P = \{p_1, p_2, p_3, \ldots, p_N\}$; and let Y take on values $y_1, y_2, y_3, \ldots, y_M$ with corresponding probabilities $Q = \{q_1, q_2, q_3, \ldots, q_M\}$ such that

$$\sum_{i=1}^{N} p_i = 1 \qquad p_i \geq 0 \qquad \sum_{j=1}^{N} q_j = 1 \qquad q_j \geq 0 \tag{1.27}$$

Note that each value of a random variable represents an event. The joint probability of x_i and y_j can be denoted as $p(x_i, y_j) = p_{ij}$; that is, $p(x_i, y_j)$ is the joint probability of $X = x_i$ and $Y = y_j$; N is the number of values X takes on; and M is the number of values Y takes on. Then the joint entropy of X and Y can be defined as

$$H(X, Y) = -\sum_{i=1}^{N}\sum_{j=1}^{M} p(x_i, y_j) \log p(x_i, y_j) \tag{1.28}$$

Then for the case of the flood process, the uncertainty would be represented by Eq. (1.28). If X and Y are stochastically independent, then $p(x_i, y_j) = p(x_i)\, p(y_j)$ or $p_{ij} = p_i p_j$, and the joint probability can be written as

$$H(X, Y) = -\sum_{i=1}^{N}\sum_{j=1}^{M} p(x_i)p(y_j)[\log p(x_i) + \log y_j]$$

$$\tag{1.29}$$

$$= -\sum_{i=1}^{N} p(x_i) \log p(x_i) \sum_{j=1}^{M} p(y_j) - \sum_{i=1}^{N} p(x_i) \sum_{j=1}^{M} p(x_j) \log p(x_j) = H(X) + H(Y)$$

Equation (1.28) can be generalized to any number of variables as

$$H(X_1, X_2, ..., X_n) = -\sum_{i=1}^{n_1} \sum_{j=1}^{n_2} ... \sum_{m=1}^{n_n} p(p_{1j}, p_{2j}, ..., p_{nm}) \log p(p_{1i}, p_{2j}, ..., p_{nm}) \qquad (1.30)$$

where n_i is the number of values X_i takes on.

1.3.4 Conditional Entropy

Consider two simultaneous outcomes or events A and B occurring with probabilities P_A and P_B, respectively. If events A and B are correlated, then the occurrence of event A gives some indication about the occurrence of event B. The probability of occurrence of B given the occurrence of A is $P(B|A)$, which is different from $P(B)$. This explains the statistical dependence. The conditional entropy of $B|A$ is $H(B|A)$, and the global entropy of the two outcomes is

$$H(AB) = H(A) + H(B|A) \qquad (1.31a)$$

Since $H(AB) = H(BA)$, one can write

$$H(A) + H(B|A) = H(B) + H(A|B) \qquad (1.31b)$$

Now consider the conditional entropy for two variables denoted as $H(X|Y)$. This is a measure of the information content of X that is not contained in Y, or entropy of X given the knowledge of Y or the amount of information that still remains in X even if Y is known. If X is input and Y is the resulting output, then note that conditional entropy $H(X|Y)$ represents the amount of information loss during transmission, meaning the part of X that never reaches Y. On the other hand, $H(Y|X)$ represents the amount of information received as noise; i.e., this part was never sent by X but was received by Y. Clearly, both of these quantities must be positive.

The conditional entropy $H(X|Y)$ gives the amount of uncertainty still remaining in X after Y becomes known or has been observed, and it is expressed in terms of conditional probabilities. It is the expected uncertainty in X given that Y is known and can be computed using the joint PDF $f(x, y)$. With the knowledge of the values of Y, say, y_j ($j = 1, 2,..., M$), the uncertainty in X will be reduced. In other words, the knowledge of Y will convey information about X. Thus it can be stated that conditional entropy $H(X|Y)$ is a measure of the amount of uncertainty remaining in X even with the knowledge of Y; the same amount of information can be gained by observing Y. The amount of reduction in uncertainty in X equals the amount of information gained by observing Y. It then follows that the conditional entropy of one variable, say, X, with respect to the other, say, Y, will be less than the marginal entropy of X:

$$H(X|Y) \le H(X) \qquad (1.32a)$$

Likewise,

$$H(Y|X) \le H(Y) \qquad (1.32b)$$

wherein equality holds if X and Y are independent of each other. Note that

$$H(Y\,|\,X) \neq H(X\,|\,Y) \qquad\qquad (1.33a)$$

but

$$H(X) - H(X\,|\,Y) = H(Y) - H(Y\,|\,X) \qquad\qquad (1.33b)$$

When X and Y are dependent, as may frequently be the case, e.g., flood peak and flood volume, then their joint (or total) entropy equals the sum of the marginal entropy of the first variable (say, X) and the entropy of the second variable (say, Y) conditioned on the first variable (X) (i.e., the uncertainty remaining in Y when a certain amount of information it conveys is already present in X):

$$H(X, Y) = H(X) + H(Y\,|\,X) \qquad\qquad (1.34a)$$

or

$$H(X, Y) = H(Y) + H(X\,|\,Y) \qquad\qquad (1.34b)$$

Noting Eqs. (1.34a) and (1.34b), we see that the total entropy $H(X, Y)$ of two dependent variables X and Y will be less than the sum of the marginal entropies of X and Y:

$$H(X, Y) \leq H(X) + H(Y) \qquad\qquad (1.34c)$$

If $p(x_i, y_j)$, $i = 1, 2, 3, \ldots, N$, $j = 1, 2, \ldots, M$, are the joint probabilities and $p(x_i\,|\,y_j)$ and $p(y_j\,|\,x_i)$ are conditional probabilities, then the conditional entropies $H(X\,|\,Y)$ and $H(Y\,|\,X)$ can be expressed mathematically as

$$H(X\,|\,Y) = -\sum_{i=1}^{N}\sum_{j=1}^{M} p(x_i, y_j)\log p(x_i\,|\,y_i) \qquad\qquad (1.35a)$$

$$H(Y\,|\,X) = \sum_{i,j} p(x_i, y_j)\log p(y_j\,|\,x_i) \qquad\qquad (1.35b)$$

Equation (1.35a) or (1.35b) can be easily generalized to any number of variables. Consider n variables denoted as $(X_1, \ldots, X_n)$. Then the conditional entropy can be written as

$$H[(X_1, X_2, \ldots, X_{n-1})\,|\,X_n] = -\sum_{1_i}^{N_1}\cdots\sum_{n_{i-1}}^{N_{n-1}}\sum_{n_i}^{N_n} p(x_{1i}, x_{2i}, \ldots, x_{(n-1)_i}, x_{n_i})\log[p(x_{1_i}, x_{2_i}, \ldots, x_{(n-1)_i})\,|\,x_n)]$$

$$(1.36a)$$

or

$$H[(X_1, X_2, \ldots, X_{n-1})|X_n] = H(X_1, X_2, \ldots, X_{n-1}, X_n) - H(X_n) \qquad (1.36b)$$

where N_i is the number of values X_i, $x_{i1}, x_{i2}, \ldots, x_{iN_i}$, takes on.

The conditional entropy is also an asymmetric quantity, as seen below:

$$H(X|Y) = -\sum_{i=1}^{N} \sum_{j=1}^{M} p(x_i, y_j) \log p(x_i|y_j) = H(X, Y) - H(Y) \qquad (1.37a)$$

$$H(Y|X) = -\sum_{i=1}^{N} \sum_{j=1}^{M} p(x_i, y_j) \log p(y_j|x_i) = H(X, Y) - H(X) \qquad (1.37b)$$

However, since

$$H(X|Y) - H(Y|X) = H(X) - H(Y) \qquad (1.37c)$$

the conditional entropy is asymmetric due to different individual entropies, not due to information flow.

1.3.5 Transinformation

Transinformation of different events A and B is the quantity of information common to the two events and is also called *mutual information* (MI), $I(A, B)$, and can be expressed as

$$T(A, B) = I(A, B) = H(A) - H(A|B) \qquad (1.38a)$$

Likewise,

$$T(B, A) = I(B, A) = H(B) - H(B|A) \qquad (1.38b)$$

Mutual information is symmetric as noted below:

$$T(B, A) = H(A) + H(B|A) - H(A|B) - H(B|A) = H(A) - H(A|B) = T(A, B) = I(A, B) \qquad (1.38c)$$

Mutual information can be extended to any number of events. To illustrate, consider three events A, B, and C. Then one can write

$$T(AB, C) = H(AB) - H(AB|C) \qquad (1.39)$$

The concept of mutual information can be used in a variety of problems, such as network design, flow regime classification, soil moisture clustering, watershed regionalization, and so on. In the network design the objective is to minimize mutual information between gauges to avoid repetition of observed information. The bigger the mutual information between two gauges A and B, the less useful the two are together. It can be said that the role of gauge B, used after gauge A, must be proportional to the new information it actually conveys after deduction of the mutual information already accounted for in A, i.e., proportional to $H(B) - T(A, B) = H(B|A)$.

Now the discussion is extended to two subsystems characterized by discrete random variables X and Y, as shown in Fig. 1.3. Transinformation represents the amount of information common to both X and Y or repeated in both or shared between X and Y, and it is denoted as $T(X, Y)$. Since $H(X)$ represents the uncertainty about the system input X before observing the system output Y and the conditional entropy $H(X|Y)$ represents the uncertainty about the system input after observing the system output, the difference between $H(X)$ and $H(X|Y)$ must represent the uncertainty about the system input that is reduced by observing the system output. This difference is often called the *average mutual information* between X and Y, or *transinformation T* of X and Y. Transinformation is also referred to as mutual information and is a measure of the dependence between X and Y and is always nonnegative. It is a measure of the amount of information random variable X contains about random variable Y; that is, if something is known about X, then transinformation indicates the amount of information known about Y and vice versa. In other words, MI quantifies or measures the relevance of knowledge about one subsystem to knowledge about another subsystem. This then suggests that MI can be used for identifying and selecting a set of relevant hydrologic variables that can help in the prediction of another hydrologic variable. Furthermore, $T(X, Y)$ defines the amount of uncertainty reduced in X when Y is known. It equals the difference between the sum of two marginal entropies and the total entropy:

$$T(X,Y) = H(X) + H(Y) - H(X, Y) \tag{1.40}$$

Equation (1.40) shows that MI is the difference between the amount of information one has when the system is considered jointly and the amount of information one has when considering the system as two individual subsystems. When X and Y are independent, $T(X, Y) = 0$. The mutual information measures the distance between the joint distribution $f(x, y)$ and one for which X and Y are independent, that is, $f(x, y) = f(x)f(y)$. It represents the degree of dependency between random variables. MI is a measure of

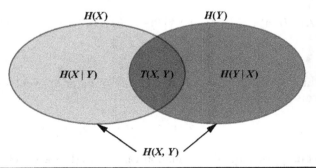

FIGURE 1.3 $T(X, Y)$ is information common to X and Y; $H(X|Y)$ is information only in X; $H(Y|X)$ is information only in Y; and $H(X, Y)$ is total information only in X and Y together.

true statistical independence, whereas concepts such as decorrelation describe independence up to second order. Note that two variables or systems can be uncorrelated, yet still not independent. In principle, *uncorrelated* means that the covariance matrix is of diagonal form.

By taking into account Eqs. (1.37a) and (1.37b), Eq. (1.40) can be written as

$$T(Y, X) = H(Y) - H(Y \mid X) \tag{1.41a}$$

$$T(X, Y) = H(X) - H(X \mid Y) \tag{1.41b}$$

The marginal entropy can be shown to be a special case of transinformation or mutual information, because

$$H(X) = T(X, X) \tag{1.42a}$$

$$H(Y) = T(Y, Y) \tag{1.42b}$$

Transinformation, denoted by $T(X,Y)$—also called *mutual entropy* (information) between X and Y—is interpreted as the reduction in the original uncertainty of X, due to the knowledge of Y. It can also be defined as the information content of X that is contained in Y. In other words, it is the difference between the total entropy of X and Y and the sum of the entropy of X and the entropy of Y. This is the information repeated in both X and Y, and it defines the amount of uncertainty that can be reduced in one of the variables when the other variable is known as shown in Fig. 1.3.

Transinformation $T(X, Y)$ is symmetric, that is, $T(X, Y) = T(Y, X)$, and is nonnegative. A zero value occurs when two variables are statistically independent so that no information is mutually transferred; that is, $T(X, Y) = 0$ if X and Y are independent. When two variables are functionally dependent, the information in one variable can be fully transmitted to another variable with no loss of information at all. Subsequently, $T(X, Y) = T(X) = T(Y)$. For any other case, $0 \le T(X, Y) \le \min(H(X), H(Y))$. Larger values of T correspond to greater amounts of information transferred. Thus, T is an indicator of the capability of information transmission and the degree of dependency of two variables. Transinformation or mutual information measures information transferred among information emitters (predictor variables) and information receivers (predicted variables). This means that the information contained in different variables can be inferred, to some extent, from the information in other variables. Mutual information is used for measuring the inferred information or equivalently information transmission.

Entropy and mutual information have advantages over other measures of information, for they provide a quantitative measure of (1) the information in a variable, (2) the information transferred and information lost during transmission, and (3) the relationship among variables based on their information transmission characteristics. For an input-output system, $H(X, Y)$ represents the total amount of information one has when both input information and output information are known. One can say that $H(X, Y)$ is the sum of the lost information, transmitted information, and added information or noise:

$$H(X, Y) = H(X \mid Y) + T(X,Y) + H(Y \mid X) \tag{1.43}$$
$$= \text{equivocation} + \text{transmission} + \text{noise}$$

The information transmitted from variable X to variable Y is represented by the mutual information $T(X, Y)$ and is given (Lathi, 1969) as

$$T(X,Y) = \sum_i \sum_j p(x_i, y_j) \log \frac{p(x_i \mid y_j)}{p(x_i)} \qquad (1.44a)$$

or

$$T(X,Y) = \sum_i \sum_j p(x_i, y_j) \log \frac{p(x_i, y_j)}{p(x_i)p(y_j)} \qquad (1.44b)$$

or

$$T(X,Y) = H(X,Y) - H(X \mid Y) \qquad (1.45)$$

Equation (1.45) can be generalized as

$$T[(X_1, X_2, \ldots, X_{n-1}); X_n] = H(X_1, X_2, \ldots, X_n) - H[(X_1, X_2, \ldots, X_{n-1}) \mid X_n] \qquad (1.46)$$

Note that although transinformation is a measure of association between two variables, it is not equivalent to correlation. If event A is determined by event B, then $p(A \mid B) = 1$ and $B \subset A$. The information is $(AB) = B$ and is measured by $H(B)$, which may be small. The common information carried by two partially correlated variables may be greater than the information of a group of two variables where one completely governs the other.

The above discussion shows that information or entropy theory permits computation of the amount of information that knowledge of one variable yields about another variable, regardless of whether the relationship between the variables is linear or nonlinear. The computation depends on higher-order statistics, such as probability density functions. There are other approaches, such as correlation analysis, empirical orthogonal functions (EOFs), principal-component analysis (PCA), and Granger causality (Granger, 1969), that also permit computation of information, but they are limited by their dependence on second-order statistics, which is tantamount to approximating with the use of the normal distribution. A further advantage of information theory is that information quantities are connected to the concept of relevance (Knuth, 2004, 2005), thus providing a basis for identifying relevant variables and their interactions with one another.

Example 1.4.   Consider data on flow discharge in cubic feet per second (cfs) at three stations (say, A, B, and C) for a river in Texas. The data are given in Table 1.4. Compute the marginal entropies of stations A, B, and C. Then compute conditional entropies $H(A \mid B)$, $H(B \mid C)$, and $H(A \mid C)$ and joint entropies $H(A, B)$, $H(A, C)$, and $H(A, C)$. Also, compute transinformation $T(A, B)$, $T(B, C)$, and $T(A, C)$.

Solution
    Computation of Marginal Entropy. Take station A as an example. By dividing the range of stream flow into five equal-sized intervals with an interval size of 68.78 cfs, the contingency table can be constructued as follows:

	1	2	3	4	5	6	7	8	9	10	11	12
Station A												
2000	61.2	40.6	122.7	97.5	179.8	110.5	12.5	4.1	62.2	56.7	347.9	137.2
2001	120.1	104.9	175.0	14.5	89.2	336.6	41.7	117.4	168.7	103.9	72.4	144.5
2002	49.8	57.9	82.6	59.4	101.9	104.7	90.7	25.7	29.7	140.7	99.8	125.2
2003	18.5	189.2	51.6	30.5	56.6	123.9	111.3	39.4	90.2	93.5	116.3	59.9
2004	110.2	156.9	75.9	131.1	105.2	212.3	54.1	105.7	50.0	179.1	249.7	66.8
Station B												
2000	2.5	8.1	22.1	18.8	23.9	120.7	12.5	0.5	16.5	109.5	84.1	14.7
2001	30.5	36.8	33.3	14.7	60.2	5.8	12.5	70.1	52.3	21.8	77.9	4.8
2002	10.7	30.2	41.91	9.4	32	41.2	73.2	16	32.8	124.7	13.2	26.2
2003	4.1	42.2	35.3	11.7	38.9	148.3	16	59.2	68.8	91.7	20.6	0
2004	35.8	48.5	48.3	61.2	22.9	105.4	62.2	119.4	37.1	150.4	159	10.4
Station C												
2000	6.9	45.9	15.8	26.2	80.5	109.2	7.9	7.1	48	185.7	140.4	34.54
2001	64.8	30.7	59.4	39.6	72.4	22.6	25.2	92.7	67.3	43.4	126.5	34.8
2002	6.6	6.1	27.7	60.5	50.6	50.3	416.6	17.3	144.8	198.1	23.9	37.3
2003	19.1	45.2	41.2	4.1	23.6	112.3	182.1	46.2	119.4	66.3	23.1	1.5
2004	57.9	44.7	95.3	176.5	31.2	232.2	32.3	59.4	76.2	77.7	142.2	5.8

TABLE 1.4   Stream Flow Observations (cfs)

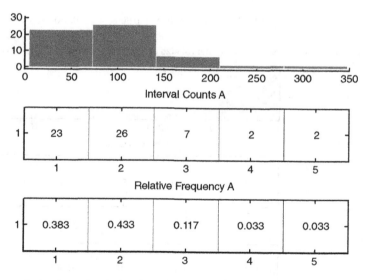

Then the marginal entropy for station A can be computed as

$$H(A) = -\sum_{i=1}^{N} p(x_i) \log_2 [p(x_i)]$$

$$= -0.383 \log_2 0.383 - 0.433 \log_2 0.433 - \cdots - 0.033 \log_2 0.033 = 1.7418 \text{ bits}$$

Similarly, for station B the contingency table with an interval size of 31.80 cfs can be constructed as follows:

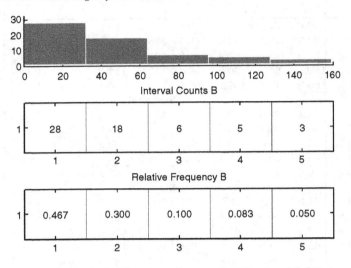

Interval Counts B

1	28	18	6	5	3
	1	2	3	4	5

Relative Frequency B

1	0.467	0.300	0.100	0.083	0.050
	1	2	3	4	5

The marginal entropy for station B can be obtained by

$$H(B) = - \sum_{i=1}^{N} p(x_i) \log_2 [p(x_i)]$$

$$= -0.467 \log_2 0.467 - 0.300 \log_2 0.300 - \cdots - 0.050 \log_2 0.050 = 1.8812 \text{ bits}$$

The contingency table with an interval size of 83.01 cfs for station C is given as follows:

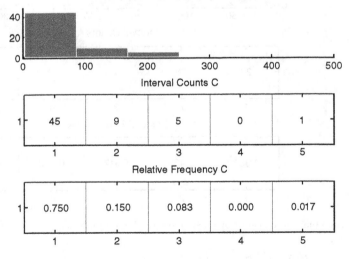

Interval Counts C

1	45	9	5	0	1
	1	2	3	4	5

Relative Frequency C

1	0.750	0.150	0.083	0.000	0.017
	1	2	3	4	5

From the contingency, the marginal entropy for station C is easily obtained by

$$H(C) = -\sum_{i=1}^{N} p(x_i) \log_2 [p(x_i)]$$

$$= -0.750 \log_2 0.750 - 0.150 \log_2 0.150 - \cdots - 0.017 \log_2 0.017 = 1.1190 \text{ bits}$$

Computation of Conditional Entropies: Take $H(A \mid B)$ as an example. First, the joint contingency table is constructed as follows:

Counts Table

	1	2	3	4	5
1	15	6	1	1	0
2	10	8	4	3	1
A 3	2	4	0	0	1
4	0	0	0	1	1
5	1	0	1	0	0

B

Joint Prob. Table

	1	2	3	4	5
1	0.250	0.100	0.017	0.017	0.000
2	0.167	0.133	0.067	0.050	0.017
A 3	0.033	0.067	0.000	0.000	0.017
4	0.000	0.000	0.000	0.017	0.017
5	0.017	0.000	0.017	0.000	0.000

B

Station A: 1 [4.06, 72.84); 2 [72.84, 141.62); 3 [141.62, 210.40); 4 [210.40, 279.18); 5 [279.18, 347.96)

Station B: 1 [0.00, 31.80); 2 [31.80, 63.60); 3 [63.60, 95.40); 4 [95.40, 127.20); 5 [127.20, 159.00)

From the definition of conditional entropy,

$$H(A \mid B) = -\sum_{i=1}^{N}\sum_{j=1}^{M} p(A_i, B_j)\log_2 \frac{p(A_i, B_j)}{p(B_j)}$$

One can see that the marginal probability distribution of B is required. The marginal distribution of B can be obtained by marginalizing out A from the bivariate contingency table. The results are as follows:

Joint Prob. + Margin. Y Table

	1	2	3	4	5
1	0.250	0.100	0.017	0.017	0.000
2	0.167	0.133	0.067	0.050	0.017
A 3	0.033	0.067	0.000	0.000	0.017
4	0.000	0.000	0.000	0.017	0.017
5	0.017	0.000	0.017	0.000	0.000
	0.467	0.300	0.100	0.083	0.050

B

The shaded row corresponds to the marginal distribution of B. The first entry in the shaded row is taken as an example. It is obtained by summing all elements of the first column in the bivariate contingency table, i.e.,

$$0.467 = 0.250 + 0.167 + 0.033 + 0.000 + 0.017$$

Using the definition of conditional entropy, $H(A \mid B)$ can be computed as

$$H(A \mid B) = -\sum_{i=1}^{N}\sum_{j=1}^{M} p(A_i, B_j) \log_2 \frac{p(A_i, B_j)}{p(B_j)}$$

$$= -0.250 \log_2 \frac{0.250}{0.467} - 0.167 \log_2 \frac{0.167}{0.467} - \cdots - 0.017 \log_2 \frac{0.017}{0.467}$$

$$\cdots - 0.000 \log_2 \frac{0.000}{0.050} - 0.017 \log_2 \frac{0.017}{0.050} - \cdots - 0.000 \log_2 \frac{0.000}{0.050}$$

$$= 1.4575 \text{ bits}$$

Similarly, the joint contingency table for B and C is as follows:

Counts Table

	1	2	3	4	5
1	27	0	1	0	0
2	15	2	1	0	0
B 3	1	4	0	0	1
4	1	1	3	0	0
5	1	2	0	0	0

C

Joint Prob. Table

	1	2	3	4	5
1	0.450	0.000	0.017	0.000	0.000
2	0.250	0.033	0.017	0.000	0.000
B 3	0.017	0.067	0.000	0.000	0.017
4	0.017	0.017	0.050	0.000	0.000
5	0.017	0.033	0.000	0.000	0.000

C

Station B: 1 [0.00, 31.80); 2 [31.80, 63.60); 3 [63.60, 95.40); 4 [95.40, 127.20); 5 [127.20, 159.00)

Station C: 1 [1.52, 84.53); 2 [84.53, 167.54); 3 [167.54, 250.55); 4 [250.55, 333.56); 5 [333.56, 416.57)

The marginal distribution of C can be obtained by marginalizing out B. The results are presented in the shaded row.

Joint Prob. + Margin. Z Table

	1	2	3	4	5
1	0.450	0.000	0.017	0.000	0.000
2	0.250	0.033	0.017	0.000	0.000
B 3	0.017	0.067	0.000	0.000	0.017
4	0.017	0.017	0.050	0.000	0.000
5	0.017	0.033	0.000	0.000	0.000
	0.750	0.150	0.083	0.000	0.017

C

The conditional entropy $H(B\,|\,C)$ can be computed as follows:

$$H(B\,|\,C) = -\sum_{i=1}^{N}\sum_{j=1}^{M} p(B_i, C_j)\log_2 \frac{p(B_i, C_j)}{p(C_j)}$$

$$= -\,0.450\,\log_2 \frac{0.450}{0.750} - 0.250\,\log_2 \frac{0.250}{0.750} - \cdots - 0.017\,\log_2 \frac{0.017}{0.750}$$

$$\cdots$$

$$-\,0.000\,\log_2 \frac{0.000}{0.017} - 0.000\,\log_2 \frac{0.000}{0.017} - \cdots - 0.000\,\log_2 \frac{0.000}{0.017} = 1.3922 \text{ bits}$$

The joint contingency table for A and C is shown here:

Counts Table

	1	2	3	4	5
1	20	17	7	0	1
2	2	5	7	1	1
C 3	1	3	0	1	0
4	0	0	7	0	0
5	0	1	0	0	0

A

Joint Prob. Table

	1	2	3	4	5
1	0.333	0.283	0.117	0.000	0.017
2	0.033	0.083	0.000	0.017	0.017
C 3	0.017	0.050	0.000	0.017	0.000
4	0.000	0.000	0.000	0.000	0.000
5	0.000	0.017	0.000	0.000	0.000

A

Station A: 1 [4.06, 72.84); 2 [72.84, 141.62); 3 [141.62, 210.40); 4 [210.40, 279.18); 5 [279.18, 347.96)

Station C: 1 [1.52, 84.53); 2 [84.53, 167.54); 3 [167.54, 250.55); 4 [250.55, 333.56); 5 [333.56, 416.57)

The marginal distribution of C can be obtained as shown here:

Joint Prob. + Margin. X Table

	1	2	3	4	5
1	0.333	0.283	0.017	0.000	0.017
2	0.033	0.083	0.000	0.017	0.017
C 3	0.017	0.050	0.000	0.017	0.000
4	0.000	0.000	0.000	0.000	0.000
5	0.000	0.017	0.000	0.000	0.000
	0.383	0.433	0.017	0.033	0.033

A

Similarly,

$$H(C \mid A) = -\sum_{i=1}^{N}\sum_{j=1}^{M} p(C_i, A_j) \log_2 \frac{p(C_i, A_j)}{p(A_j)}$$

$$= -0.333 \log_2 \frac{0.333}{0.383} - 0.250 \log_2 \frac{0.033}{0.383} - \cdots - 0.000 \log_2 \frac{0.000}{0.383}$$

$$\cdots$$

$$- 0.017 \log_2 \frac{0.017}{0.033} - 0.017 \log_2 \frac{0.017}{0.033} - \cdots - 0.000 \log_2 \frac{0.000}{0.033}$$

$$= 0.9327 \text{ bits}$$

Computation of Joint Entropies. From the joint contingency table of A and B, the joint entropy $H(A, B)$ can be computed thus:

$$H(A, B) = -\sum_{i=1}^{N}\sum_{j=1}^{M} p(A_i, B_j) \log_2 p(A_i, B_j)$$

$$= -0.250 \log_2 0.250 - 0.167 \log_2 0.167 - \cdots - 0.017 \log_2 0.017$$

$$\cdots$$

$$- 0.000 \log_2 0.000 - 0.017 \log_2 0.017 - \cdots - 0.000 \log_2 0.000$$

$$= 3.3388 \text{ bits}$$

From the joint contingency table of B and C, the joint entropy $H(B, C)$ can be computed as

$$H(B, C) = -\sum_{i=1}^{N}\sum_{j=1}^{M} p(B_i, C_j) \log_2 p(B_i, C_j)$$

$$= -0.450 \log_2 0.450 - 0.250 \log_2 0.250 - \cdots - 0.017 \log_2 0.017$$

$$\cdots$$

$$- 0.000 \log_2 0.000 - 0.000 \log_2 0.000 - \cdots - 0.000 \log_2 0.000$$

$$= 2.5112 \text{ bits}$$

Similarly from the joint contingency table of A and C, the joint entropy $H(A, C)$ can be computed as

$$H(A, C) = -\sum_{i=1}^{N}\sum_{j=1}^{M} p(A_i, C_j) \log_2 p(A_i, C_j)$$

$$= -0.333 \log_2 0.333 - 0.283 \log_2 0.283 - \cdots - 0.017 \log_2 0.017$$

$$\cdots$$

$$- 0.000 \log_2 0.000 - 0.017 \log_2 0.017 - \cdots - 0.000 \log_2 0.000$$

$$= 2.6745 \text{ bits}$$

Computation of Transinformation. There are three different approaches for computing the transinformation.

Approach 1:

$$T(A,B) = H(A) - H(A|B) = 1.7418 - 1.4575 = 0.2843 \text{ bit}$$

$$T(B,C) = H(B) - H(BC) = 1.8812 - 1.3922 = 0.4890 \text{ bit}$$

$$T(A,C) = H(C) - H(C|A) = 1.1190 - 0.9327 = 0.1863 \text{ bit}$$

Approach 2:

$$T(A,B) = H(A) + H(B) - H(A,B) = 1.7418 + 1.8812 - 3.3388 = 0.2843 \text{ bit}$$

$$T(B,C) = H(B) + H(C) - H(B,C) = 1.8812 + 1.1190 - 2.5112 = 0.4890 \text{ bit}$$

$$T(A,C) = H(A) + H(C) - H(A,C) = 1.7418 + 1.1190 - 2.6745 = 0.1863 \text{ bit}$$

Approach 3:

The third method is to compute the transinformation directly from its definition rather than use shortcut formulas as in approach 1 and approach 2. Now compute the transinformation between A and B first. The bivariate contingency table between A and B was already given when computing their joint entropy. From the joint contingency table, one can compute the marginal distribution of A and B by marginalizing out one of the variables. The results are as follows:

Joint + Margin. Prob. Table

1	0.250	0.100	0.017	0.017	0.000	0.383
2	0.167	0.133	0.067	0.050	0.017	0.433
A 3	0.033	0.067	0.000	0.000	0.017	0.117
4	0.000	0.000	0.000	0.017	0.017	0.033
5	0.017	0.000	0.017	0.000	0.000	0.033
	0.467	0.300	0.100	0.083	0.050	
	1	2	3	4	5	
			B			

The marginal distributions of A and B are shown in the shaded column and row of the above contingency table. From the definition of transinformation, one has

$$T(A, B) = \sum_i \sum_j p(A_i, B_j) \log \frac{p(A_i, B_j)}{p(A_i)p(B_j)}$$

$$= 0.250 \log_2 \frac{0.250}{0.467 \times 0.383} + 0.167 \log_2 \frac{0.167}{0.467 \times 0.433} + \cdots + 0.017 \log_2 \frac{0.017}{0.467 \times 0.117}$$

$$\cdots$$

$$+ 0.000 \log_2 \frac{0.000}{0.050 \times 0.383} + 0.017 \log_2 \frac{0.017}{0.050 \times 0.433} + \cdots + 0.000 \log_2 \frac{0.000}{0.050 \times 0.033}$$

$$= 0.2843 \text{ bits}$$

Similarly, the transinformation between B and C, denoted by $H(B, C)$, can be computed. From their joint contingency table, one can compute the marginal distributions of B and C by marginalizing out one of the variables. The results are as shown:

Joint + Margin. Prob. Table

	1	2	3	4	5	
1	0.450	0.000	0.017	0.000	0.000	0.467
2	0.250	0.033	0.017	0.000	0.000	0.300
B 3	0.017	0.067	0.000	0.000	0.017	0.100
4	0.017	0.017	0.050	0.000	0.000	0.083
5	0.017	0.033	0.000	0.000	0.000	0.050
	0.750	0.150	0.083	0.000	0.017	
	1	2	3	4	5	

C

The marginal distributions of B and C are shown in the shaded column and row of the above contingency table. From the definition of transinformation, one has

$$T(B, C) = \sum_i \sum_j p(B_i, C_j) \log \frac{p(B_i, C_j)}{p(B_i)p(C_j)}$$

$$= 0.450 \log_2 \frac{0.450}{0.750 \times 0.467} + 0.250 \log_2 \frac{0.250}{0.750 \times 0.300} + \cdots + 0.017 \log_2 \frac{0.017}{0.750 \times 0.050}$$

$$\cdots$$

$$+ 0.000 \log_2 \frac{0.000}{0.017 \times 0.467} + 0.000 \log_2 \frac{0.000}{0.017 \times 0.300} + \cdots + 0.000 \log_2 \frac{0.000}{0.017 \times 0.050}$$

$$= 0.4890 \text{ bit}$$

Similarly, the transinformation between A and C, $H(A, C)$, can be computed. From their joint contingency table, the marginal distributions of A and C can be computed by marginalizing out one of the variables. The results are shown.

Joint + Margin. Prob. Table

	1	2	3	4	5	
1	0.333	0.033	0.017	0.000	0.000	0.383
2	0.283	0.083	0.050	0.000	0.017	0.433
A 3	0.117	0.000	0.000	0.000	0.000	0.117
4	0.000	0.017	0.017	0.000	0.000	0.033
5	0.017	0.017	0.000	0.000	0.000	0.033
	0.750	0.150	0.083	0.000	0.017	
	1	2	3	4	5	

C

The marginal distributions of A and C are shown in the shaded column and row of the above contingency table. From the definition of transinformation, one has

$$T(A, C) = \sum_i \sum_j p(A_i, C_j) \log \frac{p(A_i, C_j)}{p(A_i)p(C_j)}$$

$$= 0.333 \log_2 \frac{0.333}{0.750 \times 0.383} + 0.283 \log_2 \frac{0.283}{0.750 \times 0.433} + \cdots + 0.017 \log_2 \frac{0.017}{0.750 \times 0.033}$$

$$\cdots$$

$$+ 0.000 \log_2 \frac{0.000}{0.017 \times 0.383} + 0.017 \log_2 \frac{0.017}{0.017 \times 0.433} + \cdots + 0.000 \log_2 \frac{0.000}{0.017 \times 0.033}$$

$$= 0.1863 \text{ bit}$$

1.3.6   Calculation of Transinformation and Interaction Entropy

From the point of view of calculating T, consider a discrete input variable X (say, rainfall) taking on values $i = 1, 2, 3, \ldots, N$ and a discrete output variable Y (say, flood) taking on values $j = 1, 2, 3, \ldots, M$. It is assumed that when i occurs, j is caused. Thus, one can think of a joint input-output event (i, j) having a probability $p(i, j)$. Of course, it is true that

$$\sum_{i=1}^{N} p(i) = \sum_{j=1}^{M} p(i) = \sum_{i,j} p(i, j) = 1 \tag{1.47a}$$

The amount of information transmitted in bits per value (or signal) can be written from Eq. (1.40) as

$$T(X, Y) = H(X) + H(Y) - H(X, Y)$$

where

$$H(X) = \sum_{i=1}^{N} p(i) \log_2 p(i) \qquad H(Y) = -\sum_{j=1}^{M} p(j) \log_2 H(Y) \qquad -\sum_{i,,j} p(i, j) \log_2 p(i, j) \tag{1.47b}$$

One bit equals $-\log_2 (1/2)$ and denotes the information transmitted by a choice between two equally probable alternatives.

If there exists a relation between X and Y, then $H(X) + H(Y) > H(X, Y)$ and the size of the inequality defines $T(X, Y)$, which is a bivariate positive quantity measuring the association between X and Y. Now suppose there are n observations of events (i, j) and n_{ij} denotes the number of times i occurred and j was caused. In other words,

$$n_i = \sum_j n_{ij} \qquad n_j = \sum_i n_{ij} \qquad n = \sum_{i,j} n_{ij} \tag{1.48}$$

where n_i denotes the number of times i occurred, n_j denotes the number of times j was caused, and n denotes the total number of observations. For doing calculations it is easier to use contingency tables where XY would represent cells and n_{ij} would be entries.

In that case, the probabilities would be $p(i) = n_i/n$, $p(j) = n_j/n$, and $p(i, j) = n_{ij}/n$; these actually are relative frequencies. Rather than use relative frequencies, one can also use a simpler notation for entropies in terms of absolute frequencies

$$s_{ij} = \frac{1}{n}\sum_{i,j} n_{ij} \log_2 n_{ij} \qquad s_i = \frac{1}{n}\sum_i n_i \log_2 n_i \qquad s_j = \frac{1}{n}\sum_j n_j \log_2 n_j \qquad s = \log_2 n \qquad (1.49a)$$

Then it can be shown that

$$T(X, Y) = s - s_i - s_j + s_{ij} \qquad (1.49b)$$

The two-dimensional case of the amount of information transmitted can be extended to three-dimensional or greater cases (McGill, 1953, 1954), as shown in Figs. 1.4 and 1.5. Consider three random variables U, V, Y, where U and V constitute sources and Y is the effect. In this case, X of the two-dimensional case has been replaced by U and V. Then, as shown in Fig. 1.4, one can write

$$T(U, V; Y) = H(U, V) + H(Y) - H(U, V, Y) \qquad (1.50)$$

Here X is divided into two classes U and V, with values of U as $k = 1, 2, 3, \ldots, K$ and values of V as $w = 1, 2, 3, \ldots, W$. The subdivision of X is made such that the ranges of values of U and V jointly constitute the values of X, with the implication that the input event i can be replaced by the joint event (k, w). This means that $n_i = n_{kw}$.

Here $T(U, V; Y)$ measures the amount of information transmission that U and V transmit to Y. It can be shown that the direction of transmission is irrelevant, because

$$T(U, V; Y) = T(Y; U, V) \qquad (1.51)$$

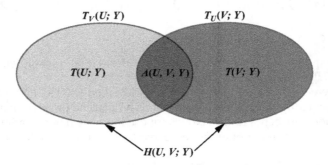

Figure 1.4   Schematic of the components of three-dimensional transmitted information. Three-dimensional information = part of bivariate transmissions plus an interaction term. $H(U,V;Y)$ is the joint entropy between (U, V) and Y; $T(U; Y)$ is the transinformation between U and Y; $T(V; Y)$ is the transinformation between V and Y; $A(U, V, Y)$ is the interaction information between U, V and Y; $T_V(U; Y)$ is the averaged transinformation between U and Y conditional upon each possible value of V; $T_U(V; Y)$ is the averaged transinformation between V and Y conditional upon each possible value of U.

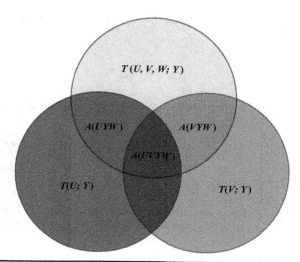

FIGURE 1.5 Schematic of the components of four-dimensional transmitted information with three sources and a single output. $T(U; Y)$ is the transinformation between U and Y; $T(V; Y)$ is the transinformation between V and Y; $A(UYW)$ is the interaction information between U, Y, and W; $A(VYW)$ is the interaction information between V, Y, and W; $A(UVYW)$ is the interaction information among U, V, Y, and W; $T(U,V,W; Y)$ is the transinformation between the joint effect of U, V, and W and that of Y.

This also implies that distinguishing sources from effects or transmitters from receivers does not gain anything, because the amount of information transmitted measures the association between variables; hence the direction in which information travels is immaterial. However, it is important to note that symbols cannot be permuted at will, because

$$T(U, Y; V) = H(U, Y) + H(V) - H(U, Y; V) \tag{1.52}$$

and it is not necessarily equal to $T(U, V; Y)$.

For computing $T(U, V; Y)$, note that it can be expressed as a function of bivariate transmission between U and Y, and V and Y. Observations of the joint event (k, w, j) can be organized into a three-dimensional contingency table with UVY cells and n_{kwj} entries. Then one can compute the terms of

$$T(U,V;Y) = H(U,V) + H(Y) - H(U,V,Y) = s - s_j - s_{kw} + s_{kwj} \tag{1.53}$$

where

$$s_{kwj} = \frac{1}{n} \sum_{k,w,j} n_{kwj} \log_2 n_{kwj} \tag{1.54}$$

Other terms can be expressed in a similar manner.

One can investigate transmission between U and Y. This would involve eliminating V, which can be done in two ways. First, the three-dimensional contingency table can be reduced to a two-dimensional table by summing over V, resulting in the entries of the reduced table

$$n_{kj} = \sum_w n_{kwj} \tag{1.55}$$

The transmitted information between U and Y can be expressed as

$$T(U;Y) = H(U) + H(Y) - H(U,Y) = s - s_k - s_j + s_{kj} \tag{1.56}$$

The second way of eliminating V is to compute the transmission between U and Y separately for each value of V and then average the transmitted values together. Designating $T_w(U;Y)$ as the information transmitted between U and Y for a single value of V, namely, w, one can write the transmitted information $T_v(U;Y)$ as

$$T_v(U;Y) = \sum_w \frac{n_w}{n}[T_w(U;Y)] \tag{1.57}$$

It can be shown that

$$T_v(U;Y) = s_w - s_{kw} - s_{wj} + s_{kwj} \tag{1.58}$$

In a three-dimensional contingency table, three different pairs of variables occur. For transmission between V and Y, one can write

$$T(V;Y) = s - s_w - s_j + s_{wj} \tag{1.59}$$

$$T_u(V;Y) = s_k - s_{kw} - s_{kj} + s_{kwj} \tag{1.60}$$

The transmission between U and V can be expressed as

$$T(U;V) = s - s_k - s_w + s_{kw} \tag{1.61}$$

$$T_y(U;V) = s_j - s_{kj} - s_{wj} + s_{kwj} \tag{1.62}$$

Now, the information transmitted between U and Y can be reconsidered. If V affects the transmission between U and Y, that is, U and V are related, then $T_V(U;Y) \neq T(U;Y)$. This effect can be measured as

$$A(UVY) = T_V(U;Y) - T(U;Y) \tag{1.63}$$

$$A(UVY) = -s + s_k + s_w + s_j - s_{kw} - s_{wj} - s_{kj} + s_{kwj} \tag{1.64}$$

A little algebra shows

$$A(UVY) = T_v(U;Y) - T(U;Y) = T_u(V;Y) - T(V;Y) = T_y(U;V) - T(U;V) \tag{1.65}$$

Keeping this symmetry in mind, $A(UVY)$ can be regarded as the $U \cdot V \cdot Y$ interaction information and is the gain (or loss) in information transmitted between any two of the variables due to the knowledge of the third variable.

Now the three-dimensional information transmitted from U, V to Y, that is, $T(U, V; Y)$, can be expressed as a function of its bivariate components:

$$T(U, V; Y) = T(U; Y) + T(V; Y) + A(UVY) \qquad (1.66)$$

$$T(U, V; Y) = T_v(U; Y) + T_u(V; Y) - A(UVY) \qquad (1.67)$$

Following these two equations, taken together, $T(U, V; Y)$ can be shown as in Fig. 1.4 with overlapping circles. This figure assumes that there is a positive interaction between $U, V,$ and Y, meaning that when one of the interacting variables is held constant, the amount of association between the remaining two increases, that is, $T_v(U; Y) > T(U; Y)$ and $T_u(V; Y) > T(V; Y)$.

For the three-dimensional case, one can write

$$H(Y) = H_{uv}(Y) + T(U; Y) + T(V; Y) + A(UVY) \qquad (1.68)$$

where $H(Y) = s - s_j$ and $H_{uv}(Y) = s_{kwj} - s_{kw}$. This shows that the marginal information is partitioned into an error term and a set of correlation terms due to input variables. The error term is $H_{uv}(Y)$ and denotes the unexplained or residual variance in the output Y after the information due to inputs U and V has been removed. For the two-dimensional case, one can write

$$H(Y) = H_u(Y) + T(U; Y) \qquad (1.69)$$

In this case H_u is the error term, because there is only one input variable U. Shannon (1948) has shown that

$$H_u(Y) \geq H_{uv}(Y) \qquad (1.70)$$

This shows that if only U is controlled, the error term cannot be increased if V is also controlled. It can be shown that

$$H_u(Y) = H_{uv}(y) + T_{uv}(V; Y) \qquad (1.71)$$

Now the issue of independence in three-dimensional transmission is considered. If the output is independent of the joint input, then $T(U, V; Y) = 0$, that is,

$$n_{kwj} = \frac{n_{kw} n_j}{n} \qquad (1.72)$$

Then it can be shown that

$$s_{kwj} = s_{kw} + s_j - s \qquad (1.73)$$

This equation can be used to show that $T(U, V; Y) = 0$.

Now assume that $T(U, V; Y) > 0$, but V and Y are independent, i.e.,

$$n_{wj} = \frac{n_w n_j}{n} \qquad (1.74)$$

This results in

$$s_{wj} = s_w + s_j - s \tag{1.75}$$

If s_{wj} from Eq. (1.59) is used in Eq. (1.75), then $T(V; Y) = 0$. Equation (1.75) does not lead to a unique condition for independence between V and Y.

If the input variables are correlated, then the question arises: How is the transmitted information affected? The three-dimensional transmitted information $T(U, V; Y)$ would account for only part of the total association in a three-dimensional contingency table. Let $C(U, V; Y)$ be the correlated information. Then one can write

$$C(U, V; Y) = H(U) + H(V) + H(Y) - H(U, V; Y) \tag{1.76a}$$

Adding to and subtracting from Eq. (1.76a) $H(U, V)$, one obtains

$$C(U, V; Y) = T(U; V) + T(U, V; Y) \tag{1.76b}$$

$$C(U, V; Y) = T(U; Y) + T(U; Y) + T(V; Y) + A(UVY) \tag{1.76c}$$

It is seen that $C(U, V; Y)$ can be employed to generate all possible components of the three correlated sources of information U, V, and Y. And $T(U, V, Y)$ is the common information between the joint effect of U and V and that of Y.

Example 1.5.   Illustrate the calculation of transinformation and interaction information, using the data from Example 1.4.

Solution
   Transinformation Calculation. First, $T(A, B)$ is computed. From Eq. (1.49b) it is known that s, s_i, s_j, and s_{ij} need to be computed. All the components can be obtained from Eq. (1.49a). In the following, by taking the transinformation $T(A, B)$ as an example, all the components are computed one by one.

(a) Computation of s: By dividing the ranges of variables A and B into five equal-size intervals and counting the number of occurrences in all combinations of these subintervals, the contingency table can be constructed as shown:

n Table

	1	2	3	4	5
1	15	6	1	1	0
2	10	8	4	3	1
A 3	2	4	0	0	1
4	0	0	0	1	1
5	1	0	1	0	0
	1	2	3	4	5
			B		

Station A: 1 [4.06, 72.84); 2 [72.84, 141.62); 3 [141.62, 210.40); 4 [210.40, 279.18); 5 [279.18, 347.96)

Station B: 1 [0.00, 31.80); 2 [31.80, 63.60); 3 [63.60, 95.40); 4 [95.40, 127.20); 5 [127.20, 159.00)

From Eq. (1.48), one has

$$n = \sum_{i,j} n_{ij} = 15 + 10 + 2 + 0 + 1 + 6 + 8 + 4 + 0 + 0 + \cdots + 0 + 1 + 1 + 1 + 0 = 60$$

From Eq. (1.49a), one has

$$s = \log_2 n = \log_2 60 = 5.9069$$

(b) Computation of s_i and s_j: By marginalizing out one of the two variables, the contingency of A and B can be obtained from the above contingency table. From Eq. (1.48), we have

$$n_i(1) = \sum_j n_{1j} = 15 + 6 + 1 + 1 + 1 = 23$$

$$n_i(2) = \sum_j n_{2j} = 10 + 8 + 4 + 3 + 1 = 26$$

$$\vdots$$

$$n_i(5) = \sum_j n_{5j} = 1 + 0 + 1 + 0 + 0 = 2$$

The results (contingency table for A) are shown in the shaded column in the following table.

n + Margin. n Table

		1	2	3	4	5	
	1	14	6	1	1	0	23
	2	10	8	4	3	1	26
A	3	2	4	0	0	1	7
	4	0	0	0	1	1	2
	5	1	0	1	0	0	2
		28	18	6	5	3	
		1	2	3	4	5	

B

Similarly, one has

$$n_j(1) = \sum_i n_{i1} = 15 + 10 + 2 + 0 + 1 = 28$$

$$n_j(2) = \sum_i n_{i2} = 6 + 8 + 4 + 0 + 0 = 18$$

$$\vdots$$

$$n_j(5) = \sum_i n_{i5} = 0 + 1 + 1 + 1 + 0 = 3$$

Finally, from Eq. (1.49a) one has

$$s_i = \frac{1}{n} \sum_i n_i \log_2 n_i = \frac{1}{60}(23 \log_2 23 + 26 \log_2 26 + 7 \log_2 7 + 2 \log_2 2 + 2 \log_2 2)$$

$$= 4.1651$$

Also, one has

$$s_j = \frac{1}{n}\sum_j n_j \log_2 n_j = \frac{1}{60}(28\log_2 28 + 18\log_2 18 + 6\log_2 6 + 5\log_2 5 + 3\log_2 3) = 4.0256$$

(c) Computation of s_{ij}: From the joint contingency table and Eq. (1.49a) one can also compute s_{ij}:

$$s_{ij} = \frac{1}{n}\sum_{i,j} n_{ij}\log_2 n_{ij}$$

$$= \frac{1}{60}\left(15\log_2 15 + 10\log_2 10 + 2\log_2 2 + \cdots + 0\log_2 0 + 1\log_2 1 + \cdots + 0\log_2 0\right)$$

$$= 2.5681$$

Finally,

$$T(A,B) = s - s_i - s_j + s_{ij} = 5.9069 - 4.1651 - 4.0256 + 2.5681 = 0.2843$$

Similarly, $T(B, C)$ and $T(C, A)$ can be computed.

$$T(B,C) = 0.4890 \qquad T(C,A) = 0.1863$$

Interaction Information Calculation. By dividing the ranges of variables A, B, and C into five equal-size intervals, the trivariate contingency table can be created. The resulting contingency table is shown.

Counts, C: [1.50, 84.53)

A \ B	1	2	3	4	5
1	14	5	1	0	0
2	7	6	1	2	0
3	3	4	0	0	1
4	0	0	0	0	0
5	1	0	0	0	0

Counts, C: [84.53, 167.53)

A \ B	1	2	3	4	5
1	2	2	0	0	0
2	0	1	0	1	1
3	0	0	0	0	0
4	0	0	0	0	1
5	0	0	1	0	0

Counts, C: [167.53, 250.54)

A \ B	1	2	3	4	5
1	0	0	0	0	1
2	1	0	0	0	0
3	0	0	1	0	1
4	0	0	0	0	0
5	0	0	0	0	1

Counts, C: [250.54, 333.55)

A \ B	1	2	3	4	5
1	0	0	0	0	0
2	0	0	0	0	0
3	0	0	0	0	0
4	0	0	0	0	0
5	0	0	0	0	0

Counts, C: [333.55, 416)

	1	2	3	4	5
1	0	0	0	0	0
2	0	0	0	0	0
A 3	0	0	1	0	0
4	0	0	0	0	0
5	0	0	0	0	0

B

Station A: 1 [4.06, 72.84); 2 [72.84, 141.62); 3 [141.62, 210.40); 4 [210.40, 279.18); 5 [279.18, 347.96)

Station B: 1 [0.00, 31.80); 2 [31.80, 63.60); 3 [63.60, 95.40); 4 [95.40, 127.20); 5 [127.20, 159.00)

Summing all the elements in the above trivariate joint contingency table, one has $n = 60$. Therefore,

$$s = \log_2 60 = 5.9069$$

The marginal contingency table can be obtained in the following way. The marginal contingency table of A and B given $C \in (1.5, 84.53]$ is as follows:

Counts, C: [1.5, 84.53)

	1	2	3	4	5	
1	14	5	1	0	0	20
2	7	6	1	2	0	16
A 3	3	4	0	0	1	8
4	0	0	0	0	0	0
5	1	0	0	0	0	1
	25	15	2	2	1	

B

The marginal contingency table of A and B given $C \in (84.53, 167.53]$ is shown next.

Counts, C: [84.53, 167.53)

	1	2	3	4	5	
1	2	2	0	0	0	4
2	0	1	0	1	1	3
A 3	0	0	0	0	0	0
4	0	0	0	0	1	1
5	0	0	1	0	0	1
	2	3	1	1	2	

B

The marginal contingency table of A and B given $C \in (167.53, 250.54]$ is as follows:

Counts, C: [167.53, 250.54)

A \ B	1	2	3	4	5	
1	0	0	0	0	1	1
2	1	0	0	0	0	1
3	0	0	1	0	1	2
4	0	0	0	0	0	0
5	0	0	0	0	1	1
	1	0	1	0	3	

The marginal contingency table of A and B given $C \in (250.54, 333.55]$ is shown here:

Counts, C: [250.54, 333.55)

A \ B	1	2	3	4	5	
1	0	0	0	0	0	0
2	0	0	0	0	0	0
3	0	0	0	0	0	0
4	0	0	0	0	0	0
5	0	0	0	0	0	0
	0	0	0	0	0	

The marginal contingency table of A and B given $C \in (250.54, 333.55]$ is

Counts, C: [333.55, 416)

A \ B	1	2	3	4	5	
1	0	0	0	0	0	0
2	0	0	0	0	0	0
3	0	0	1	0	0	1
4	0	0	0	0	0	0
5	0	0	0	0	0	0
	0	0	1	0	0	

Then the marginal contingency table for A can be obtained by summing over the last shaded columns on the right in the above five tables:

$$n_k(1) = \sum_{w,j} n_{1wj} = 20 + 4 + 1 + 0 + 0 = 25$$

$$n_k(2) = \sum_{w,j} n_{2wj} = 16 + 3 + 1 + 0 + 0 = 20$$

$$n_k(3) = \sum_{w,j} n_{3wj} = 8 + 0 + 2 + 0 + 1 = 11$$

$$n_k(4) = \sum_{w,j} n_{4wj} = 0 + 1 + 0 + 0 + 0 = 1$$

$$n_k(5) = \sum_{w,j} n_{5wj} = 1 + 1 + 1 + 0 + 0 = 3$$

The results are tabulated as follows:

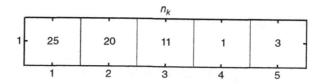

Similarly, the marginal contingency table for B can be obtained by summing over the last shaded rows on the right in the above five tables:

$$n_j(1) = \sum_{k,w} n_{kw1} = 25 + 2 + 1 + 0 + 0 = 28$$

$$n_j(2) = \sum_{k,w} n_{kw2} = 15 + 3 + 0 + 0 + 0 = 18$$

$$n_j(3) = \sum_{k,w} n_{kw3} = 2 + 1 + 1 + 0 + 1 = 5$$

$$n_j(4) = \sum_{k,w} n_{kw4} = 2 + 1 + 0 + 0 + 0 = 3$$

$$n_j(4) = \sum_{k,w} n_{kw4} = 1 + 2 + 3 + 0 + 0 = 6$$

The results are tabulated thus:

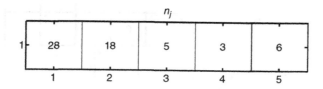

The marginal contingency table for C can be computed by summing, respectively, over all the unshaded elements on the right in the above five tables:

$$n_w(1) = 25 + 15 + 2 + 2 + 1 = 20 + 16 + 8 + 0 + 1 = 45$$

$$n_w(2) = 2 + 3 + 1 + 1 + 2 = 4 + 3 + 0 + 1 + 1 = 9$$

$$n_w(3) = 1 + 0 + 1 + 0 + 3 = 1 + 1 + 2 + 0 + 1 = 5$$

$$n_w(4) = 0 + 0 + 0 + 0 + 0 = 0 + 0 + 0 + 0 + 0 = 0$$

$$n_w(5) = 0 + 0 + 1 + 0 + 0 = 0 + 0 + 1 + 0 + 0 = 1$$

The results are tabulated thus:

$$n_w$$

1	45	9	5	0	1
	1	2	3	4	5

Therefore, one has

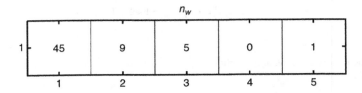

$$s_k = \frac{1}{60}(25 \log_2 25 + 20 \log_2 20 + 11 \log_2 11 + 1 \log_2 1 + 3 \log_2 3) = 4.0891$$

$$s_j = \frac{1}{60}(28 \log_2 28 + 18 \log_2 18 + 5 \log_2 5 + 3 \log_2 3 + 6 \log_2 6) = 4.0256$$

$$s_w = \frac{1}{60}(45 \log_2 45 + 9 \log_2 9 + 5 \log_2 5 + 0 \log_2 0 + 1 \log_2 1) = 4.7879$$

Now s_{kw}, s_{wj}, and s_{kj} are computed. The bivariate contingency table can also be obtained from the trivariate contingency by marginalizing out one of the three variables. Using Eq. (1.55), one can have the bivariate joint contingency table of A and B:

$$n_{kj}$$

	1	2	3	4	5
1	16	7	1	0	1
2	8	7	1	3	1
A 3	3	4	2	0	2
4	0	0	0	0	1
5	1	0	1	0	1
	1	2	3	4	5
			B		

Therefore, s_{kj} can be computed as

$$s_{kj} = \frac{1}{60}(16 \log_2 16 + 8 \log_2 8 + 3 \log_2 3 + \cdots + 1 \log_2 1 + 1 \log_2 1) = 2.4802$$

Similarly, the bivariate joint contingency table of B and C is as shown:

$$n_{wj}$$

	1	2	3	4	5
1	25	2	1	0	0
2	15	3	0	0	0
B 3	2	1	1	0	1
4	2	1	0	0	0
5	1	2	3	0	0

C

Now s_{wj} can be computed as

$$s_{wj} = \frac{1}{60}(25 \log_2 25 + 15 \log_2 15 + 2 \log_2 2 + \cdots + 0 \log_2 0 + 0 \log_2 0) = 3.2035$$

This is the bivariate joint contingency table of A and C:

$$n_{kw}$$

	1	2	3	4	5
1	20	4	1	0	0
2	16	3	1	0	0
A 3	8	0	2	0	1
4	0	1	0	0	0
5	1	1	1	0	0

C

and s_{kw} is computed as

$$s_{kw} = \frac{1}{60}(20 \log_2 20 + 16 \log_2 16 + 8 \log_2 8 + \cdots + 0 \log_2 0 + 0 \log_2 0) = 3.1532$$

By using the trivariate contingency table and Eq. (1.54), s_{kwj} can be computed as

$$s_{kwj} = \frac{1}{n} \sum_{k,w,j} n_{kwj} \log_2 n_{kwj}$$

$$= \frac{1}{60} (14 \log_2 14 + 7 \log_2 7 + 3 \log_2 3 + \cdots + 0 \log_2 0 + 0 \log_2 0$$

$$+ 2 \log_2 2 + 0 \log_2 0 + 0 \log_2 0 + \cdots + 1 \log_2 1 + 0 \log_2 0$$

$$+ 0 \log_2 0 + 1 \log_2 1 + 0 \log_2 0 + \cdots + 0 \log_2 0 + 1 \log_2 1$$

$$+ 0 \log_2 0 + 0 \log_2 0 + 0 \log_2 0 + \cdots + 0 \log_2 0 + 0 \log_2 0$$

$$+ 0 \log_2 0 + 0 \log_2 0 + 0 \log_2 0 + \cdots + 0 \log_2 0 + 0 \log_2 0)$$

$$= 1.9805$$

Thus, Eq. (1.64) can be used to compute the interaction information:

$$A(A, B, C) = -s + s_k + s_w + s_j - s_{kw} - s_{wj} - s_{kj} + s_{kwj}$$

$$= -5.9069 + 4.0891 + 4.7879 + 4.0256 - 3.1532 - 3.2035 - 2.4802 + 1.9805$$

$$= 0.1393$$

1.3.7 Transfer Information

Mutual information (MI) is important for identifying relationships between variables, but it says nothing about the causality of their relationships. To illustrate this point, consider two variables or subsystems X and Y that are interactive. Note that MI is symmetric with respect to the interchange of X and Y, but their causal interactions are not. Schreiber (2000) introduced the concept of transfer entropy (TE) for measuring causal interactions. *TE* is an asymmetric quantity. It shares some of the desired properties of MI and takes into account the dynamics of information transport. Thus, TE can quantify the exchange of information between two variables or subsystems separately for both directions, i.e., from X to Y and Y to X, and also conditional to common input, if needed. Thus, it is possible to introduce direction dependency into mutual information by introducing a time lag τ in either one of the two variables

$$T_{XY}(\tau) = \sum p(x_n, y_{n-\tau}) \log \frac{p(x_n, y_{n-\tau})}{p(x_n)p(y_{n-\tau})} \qquad (1.77)$$

The dynamic structure can be introduced by considering transition probabilities in place of static probabilities. Consider that a system is approximated by a stationary Markov process of order k. This means that the conditional probability of determining the state x_{n+1} of system X at time $n + 1$ is independent of the state $n - k$. Expressed algebraically, $p(x_{n+1}|x_n, x_{n-1}, ..., x_{n-k+1}) = p(x_{n+1}|x_n, x_{n-1}, ..., x_{n-k+1}, x_{n-k})$.

Consider two subsystems or variables X and Y with data as time series of measurements: $X: \{x_1, x_2, ..., x_t, x_{t+1}, ..., x_n\}$ and $Y: \{y_1, y_2, ..., y_s, y_{s+1}, ..., y_n\}$ with $t = s + l$ or $s = t - l$, where l is the lag time and the two consecutive observations are Δt time apart. The transfer entropy can be written as

$$TE(X_{t+1} | X_t, Y_s) = I(X_{t+1}, Y_s) - I(X_t, X_{t+1}, Y_s) \qquad (1.78)$$

where $I(X_{t+1}, Y_s)$ is the rank 2 co-information (mutual information), and $I(X_t, X_{t+1}, Y_s)$ is the rank 3 co-information that denotes the information that the three variables share. Equation (1.80) states that the transfer information denotes the information shared by Y and future values of X minus the information shared by Y and X and future values of X. In this manner it captures the predictive information Y has about X, and in this sense it is a measure of a possible causal interaction between X and Y. As suggested by Schreiber (2000), TE can also be expressed as

$$TE(X_{t+1} \mid X_t, Y_s) = -H(X_t) + H(X_t, Y_s) + H(X_t, X_{t+1}) - H(X_t, X_{t+1}, Y_s) \qquad (1.79)$$

where $H(X_t, X_{t+1}, Y_s)$ denotes the joint entropy between X and Y and time-shifted version of X, namely, X_{t+1}.

To illustrate the point that TE is not symmetric with respect to the interchange of X and Y, note that

$$TE(Y_{s+1} \mid X_t, Y_s) = -H(Y_s) + H(X_t, Y_s) + H(Y_s, Y_{s+1}) - H(X_t, Y_s, Y_{s+1}) \qquad (1.80)$$

Equations (1.79) and (1.80) are not the same. This asymmetric property of TE is fundamental to identifying causal interactions.

Example 1.6. Suppose we have a bivariate time series: monthly precipitation and monthly mean temperature from years 1950 to 2010 at a weather station in Texas, as shown in Fig. 1.6. For simplicity, we denote the precipitation and temperature time series as X and Y, respectively, and assume both the series are a first-order Markov process. Compute the transfer entropy $TE(X_{t+1} \mid X_t, Y_t)$ and $TE(Y_{t+1} \mid Y_t, X_t)$. Note that here we simply assume $l = 1$.

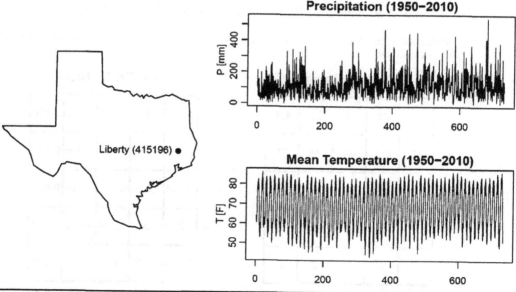

FIGURE 1.6   Bivariate time series of monthly precipitation and mean temperature of station TX415196 from 1950 to 2010.

Solution. We first compute the transfer entropy $TE(X_{t+1} \mid X_t, Y_t)$. To that end, we need to create three time series, namely, X_{t+1}, X_t, and Y_t. This procedure is the same as before, so it is not described here. Based on Eq. (1.79), next we need to compute $H(X_t)$, $H(X_t, Y_t)$, $H(X_t, X_{t+1})$, and $H(X_{t+1}, X_t, Y_t)$. From Example 1.5, we already know how to compute these entropy terms. For ease of understanding, we repeat the step-by-step procedure again. We arbitrarily use five equal-size intervals to discretize these time series. Note that instead of using 5 as the interval number, one can also search some value such that the value is optimal for obtaining the underlying density estimates of the variable of interest.

Computation of $H(X_t)$. By discretizing the continuous time series X_t into five intervals, the contingency table of X_t can be constructed as follows:

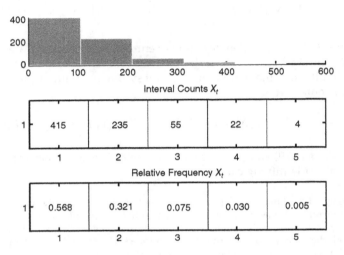

From this table, the marginal entropy $H(X_t)$ can be computed as

$$H(X_t) = -0.568 \log_2 0.568 - 0.321 \log_2 0.321 - \cdots - 0.005 \log_2 0.005 = 1.4641 \text{ bits}$$

Computation of $H(X_t, Y_t)$. Using the same discretization scheme, the joint contingency table of X_t and Y_t can be constructed as shown.

Counts Table

X_t	1	2	3	4	5
1	30	95	88	72	130
2	16	52	44	48	75
3	2	15	9	14	15
4	0	2	3	5	12
5	0	0	0	2	2

Y_t

Joint Prob Table

X_t	1	2	3	4	5
1	0.041	0.130	0.120	0.098	0.178
2	0.022	0.071	0.060	0.066	0.103
3	0.003	0.021	0.012	0.019	0.021
4	0.000	0.003	0.004	0.007	0.016
5	0.000	0.000	0.000	0.003	0.003

Y_t

Then the joint entropy $H(X_t, Y_t)$ can be computed as

$$H(X_t, Y_t) = -0.041 \log_2 0.041 - 0.022 \log_2 0.022 - 0.003 \log_2 (0.003) - \cdots - 0.003 \log_2 0.003$$
$$= 3.6324 \text{ bits}$$

Computation of $H(X_t, X_{t+1})$. Similarly, the joint contingency table of X_t and X_{t+1} is created as follows:

Counts Table

X_t	1	2	3	4	5
1	236	137	32	7	3
2	139	69	15	11	1
3	26	20	7	2	0
4	12	8	1	1	0
5	2	1	0	1	0

X_{t+1}

Joint Prob Table

X_t	1	2	3	4	5
1	0.323	0.187	0.044	0.010	0.004
2	0.190	0.094	0.021	0.015	0.001
3	0.036	0.027	0.010	0.003	0.000
4	0.016	0.011	0.001	0.001	0.000
5	0.003	0.001	0.000	0.001	0.000

X_{t+1}

The joint entropy is

$$H(X_t, X_{t+1}) = -0.323 \log_2 0.323 - 0.190 \log_2 0.190 - \cdots - 0.001 \log_2 0.001$$
$$= 2.9144 \text{ bits}$$

Computation of $H(X_t, X_{t+1}, Y_t)$. The joint contingency table of X_t, X_{t+1}, and Y_t is shown here.

Counts Table, Y_t: [42, 51)

X_t	1	2	3	4	5
1	6	6	5	0	0
2	4	4	4	1	1
3	2	5	1	1	1
4	0	4	0	0	0
5	0	0	2	0	1

X_{t+1}

Prob. Table, Y_t: [42, 51)

X_t	1	2	3	4	5
1	0.008	0.008	0.007	0.000	0.000
2	0.005	0.005	0.005	0.001	0.001
3	0.003	0.007	0.001	0.001	0.001
4	0.000	0.005	0.000	0.000	0.000
5	0.000	0.000	0.003	0.000	0.001

X_{t+1}

Counts Table, Y_t: [51, 60)

X_t		X_{t+1} 1	2	3	4	5
	1	24	22	10	2	0
	2	25	15	9	4	0
	3	11	13	5	3	0
	4	7	5	1	4	1
	5	3	0	0	0	0

Prob. Table, Y_t: [51, 60)

X_t		X_{t+1} 1	2	3	4	5
	1	0.033	0.030	0.014	0.003	0.000
	2	0.034	0.021	0.012	0.005	0.000
	3	0.015	0.018	0.007	0.004	0.000
	4	0.010	0.007	0.001	0.005	0.001
	5	0.004	0.000	0.000	0.000	0.000

Counts Table, Y_t: [60, 68)

X_t		X_{t+1} 1	2	3	4	5
	1	30	22	10	1	1
	2	16	15	6	2	0
	3	6	10	8	1	0
	4	4	6	1	0	0
	5	2	2	0	1	0

Prob. Table, Y_t: [60, 68)

X_t		X_{t+1} 1	2	3	4	5
	1	0.041	0.030	0.014	0.001	0.001
	2	0.022	0.021	0.008	0.003	0.000
	3	0.008	0.014	0.011	0.001	0.000
	4	0.005	0.008	0.001	0.000	0.001
	5	0.003	0.003	0.000	0.001	0.000

Counts Table, Y_t: [68, 77)

X_t		X_{t+1} 1	2	3	4	5
	1	36	26	8	4	2
	2	21	17	4	2	1
	3	3	9	1	0	0
	4	3	2	0	0	0
	5	1	0	0	1	0

Prob. Table, Y_t: [68, 77)

X_t		X_{t+1} 1	2	3	4	5
	1	0.049	0.036	0.011	0.005	0.003
	2	0.029	0.023	0.005	0.003	0.001
	3	0.004	0.012	0.001	0.000	0.000
	4	0.004	0.003	0.000	0.000	0.000
	5	0.001	0.000	0.000	0.001	0.000

Counts Table, Y_t: [77, 86)

	1	2	3	4	5
1	49	40	14	6	1
2	40	25	7	4	4
X_t 3	11	8	5	1	2
4	6	5	0	0	0
5	0	3	1	1	1
	1	2	3	4	5

X_{t+1}

Prob. Table, Y_t: [77, 86)

	1	2	3	4	5
1	0.067	0.055	0.019	0.008	0.001
2	0.055	0.034	0.010	0.005	0.005
X_t 3	0.015	0.011	0.007	0.001	0.003
4	0.008	0.007	0.000	0.000	0.000
5	0.000	0.004	0.001	0.001	0.001
	1	2	3	4	5

X_{t+1}

The joint entropy is computed thus:

$$H(X_t,\, X_{t+1},\, Y_t) = -0.008\log_2 0.008 - 0.005\log_2 0.005 - \cdots - 0.001\log_2 0.001 = 5.6925 \text{ bits}$$

From Eq. (1.79), the transfer entropy can be computed.

$$TE(X_{t+1}\,|\,X_t, Y_t) = -H(X_t) + H(X_t, Y_t) + H(X_t, X_{t+1}) - H(X_t, X_{t+1}, Y_t)$$

$$= -1.4641 + 3.6324 + 2.9144 - 5.6925$$

$$= -0.6098 \text{ bit}$$

In a similar vein, we can compute the transfer entropy $TE(Y_{t+1}\,|\,Y_t, X_t)$. Because of space limitations, we do not present the detailed procedure here, but directly give the results.

$$H(Y_t) = 2.1847 \text{ bits} \qquad H(Y_t, Y_{t+1}) = 3.7163 \text{ bits}$$

$$H(X_t, Y_t) = 3.6324 \text{ bits} \qquad H(X_t, Y_t, Y_{t+1}) = 5.1348 \text{ bits}$$

$$TE(Y_{t+1}|X_t, Y_t) = -H(Y_t) + H(X_t, Y_t) + H(Y_t, Y_{t+1}) - H(X_t, Y_t, Y_{t+1})$$

$$= -2.1847 + 3.6324 + 3.7163 - 5.1348$$

$$= 0.0292 \text{ bit}$$

1.3.8 Entropy Rate

Consider a system that can occupy a finite number of states whose probabilities are known. Then the Shannon entropy yields the average number of bits needed to enclose a given number of states. Let the system be described by a Markov process of order k, and let $x_n^{(k)} = (x_n,\, x_{n-1},\, \ldots,\, x_{n-k+1})$, where x_n denotes the state at time n and x_{n-k+1} denotes the state at time $n - k + 1$. The entropy rate (Schreiber, 2000) is the average number of bits needed to encode one additional state of the system if all previous states are known, and it can be expressed as

$$h(X) = -\sum p\left(x_{n+1}, x_n^{(k)}\right) \log\left[p\left(x_{n+1} \mid x_n^{(k)}\right)\right] \tag{1.81}$$

Equation (1.81) denotes the difference between the Shannon entropies of the processes given by $(k + 1)$ and k-dimensional delay vectors obtained from X:

$$h(X) = H(X^{(k+1)}) - H(X^{(k)}) \tag{1.82}$$

Equation (1.82) stems from the fact that

$$p\left(x_{n+1} \mid x_n^{(k)}\right) = \frac{p\left(x_{n+1}^{(k+1)}\right)}{p\left(x_n^{(k)}\right)} \tag{1.83}$$

The dynamics of shared information among processes can be investigated by generalizing the entropy rate in place of the Shannon entropy. The reason is that the dynamics of information lie in the transition probabilities.

Consider two processes X and Y. The mutual information rate can now be defined by maximizing the deviation from independence. Using the generalized Markov property,

$$p\left(x_{n+1} \mid x_n^{(k)}\right) = p\left(x_{n+1} \mid x_n^{(k)}, y_n^{(l)}\right) \tag{1.84}$$

If there is no flow of information from Y to X, the state (value) of Y has no influence on system X. Then the transfer entropy can be defined as

$$TE(Y \rightarrow X) = \sum p\left(x_{n+1}, x_n^{(k)}, y_n^{(l)}\right) \log \frac{p\left(x_{n+1} \mid x_n^{(k)}, y_n^{(l)}\right)}{p\left(x_{n+1} \mid x_n^{(k)}\right)} \tag{1.85}$$

Often $l = k$ or 1. And $T(Y \rightarrow X)$ measures the degree of dependence of X on Y, not vice versa, and is hence asymmetric.

Now consider a Markov process of order k with time lag discretized with an interval of Δt. The probability distribution at any particular spatial point depends on the values at this particular location for the previous k times. For an order $k = 1$ Markov process, the transfer entropy can be defined as

$$TE(Y \rightarrow X, t) = \iiint f(x(t + \Delta t), x(t), y(t)) \log \frac{f(x(t + \Delta t), x(t), y(t))}{f(x(t + \Delta t) \mid x(t))} \, dx\,(t + \Delta t) dx(t)\, dy(t) \tag{1.86}$$

Equation (1.86) denotes the expected change in the probability function at a particular spatial location due to the knowledge of a random variable at another location and earlier time that would result from the knowledge of the first variable at the earlier time. The transfer entropy can be written in terms of conditional probabilities as

$$TE(Y \rightarrow X, t) = H(X(t + \Delta t) \| X(t)) - H(X(t + \Delta t) \mid X(t), Y(t)) \tag{1.87}$$

Equations (1.86) and (1.87) can be generalized for $k > 1$.

The *time-lagged mutual information* (TLMI) of random variables $I(X(t_1); Y(t_2))$ can now be defined as

$$I(X(t_1); Y(t_2)) = H(X(t_1)) - H(X(t_1) \mid Y(t_2))$$ (1.88)

The time-lagged mutual information is a bivariate function, whereas the transfer entropy is a function of order $k + 1$.

1.3.9 Informational Correlation Coefficient

The information correlation coefficient R_0 is a measure of transferable information; it measures the mutual dependence between random variables X and Y and does not assume any type of relationship between them. It is a dimensional quantity and is expressed in terms of transinformation as

$$R_0 = \sqrt{1 - \exp(-2T_0)}$$ (1.89)

where T_0 is transinformation or mutual information, representing the upper limit of transferable information between two variables X and Y. If the values of probabilities are computed from the corresponding sample frequencies (Harmancioglu et al., 1999), then the transinformation so obtained represents the upper limit of transferable information between the two variables. When X and Y are normally distributed and linearly correlated, R_0 reduces to the classical Pearson correlation coefficient between X and Y, denoted r_{xy}:

$$r_0 = r_{XY} = \frac{\mathrm{Cov}(X, Y)}{\sigma_X \, \sigma_Y}$$ (1.90)

where $\mathrm{Cov}(X, Y)$ is the covariance between X and Y, σ_X is the standard deviation of X, and σ_Y is the standard deviation of Y. These quantities can be computed from sample data as follows:

$$\mathrm{Cov}(X, Y) = \frac{1}{N} \sum_{i=1}^{N} (x_i - \bar{x})(y_i - \bar{y})$$ (1.91)

where $\bar{x}$ is the sample mean of X; $\bar{y}$ is the sample mean of Y; and N is the sample size. The standard deviations are computed as

$$\sigma_x = \sqrt{\frac{1}{N-1} \sum_{i=1}^{N} (X_i - \bar{X})^2} \qquad \sigma_y = \sqrt{\frac{1}{N-1} \sum_{i=1}^{N} (y_i - \bar{X})^2}$$ (1.92)

Example 1.7. Using data from Example 1.4 in Table 1.4, compute the Pearson correlation coefficient between stations A and B, stations B and C, and stations A and C. Now compute the information coefficient between stations A and B, stations B and C, and stations A and C. Compare the two sets of correlation coefficients.

Solution. By using the data in Table 1.4 and Eq. (1.90), the Pearson correlation coefficient between stations A and B is found to be 0.3668. Also the Pearson correlation coefficient between stations B, and

C is 0.5564, and that between stations A and C is 0.2580. To compute the information coefficient, first the transinformation between stations is required. The second step is to compute the transinformation. From Example 1.4, it is already known that the values of transinformation between stations A and B, between B and C, and between A and C are 0.2843, 0.4890 and 0.1863 bit, respectively. From Eq.(1.89), the information coefficient between stations A and B can be computed:

$$R_{AB} = \sqrt{1 - \exp(-2T_{AB})} = \sqrt{1 - \exp(-2 \times 0.2843)} = 0.6585$$

Similarly, the information coefficient between stations B and C and between stations A and C can be computed.

$$R_{BC} = \sqrt{1 - \exp(-2T_{BC})} = \sqrt{1 - \exp(-2 \times 0.4890)} = 0.7899$$

$$R_{AC} = \sqrt{1 - \exp(-2T_{AC})} = \sqrt{1 - \exp(-2 \times 0.1863)} = 0.5577$$

1.4 Approximate Entropy and Sample Entropy

Approximate entropy (ApEn) is a measure for characterizing the complexity of a time series (Pincus, 1991). This complexity measure has a strong anti-interference ability for short time series, and it can be applied to a variety of time series. However, ApEn suffers from bias and hence leads to inconsistent results. To overcome the shortcomings of ApEn, a related complexity measure, called *sample entropy* (SampEn), has been developed (Richman and Moorman, 2000). It is a measure of orderly structure in a time series and can determine if there are repeated patterns of various lengths. SampEn has the same physical meaning as ApEn, but does not depend on the length of the series and can obtain accurate estimates even for short time series and is hence preferable. SampEn can also be used for mixed signals composed of random and deterministic components, and it can extract more practically meaningful features than can known simple statistical parameters, such as the mean and standard deviation. When a time series is more complex, it has lower self-similarity, but it will have a larger value of SampEn; and vice versa, the higher self-similarity of the series, the smaller the value of its SampEn.

SampEn is the exact value of the negative average natural logarithm of the conditional probability. For a time series of length N, denoted, $Y = \{y(i), i = 1, 2, \ldots, N\}$, it can be computed as follows.

1. Construct m-dimensional vectors $X(i) = \{x(i), x(i+1), \ldots, x(i+m-1)\}, i = 1, 2, \ldots, N-m+1$, that comprise parts of the time series. These vectors are in an embedding space R^m, and term m corresponds to the embedding dimension. The proper minimum value of m is obtained by analyzing the behavior of nearest neighbors when the embedding dimension is changed from m to $m+1$ (Kennell et al., 1992). In practice, m can be regarded as the length of series to be compared.

2. Define the m vector, using the method of delay, as

$$X(i) = \{y(i), y(i+1), \ldots, y(i+m-1)\} \qquad 1 \le i \le N-m+1 \qquad (1.93)$$

3. Define the euclidean distance between two vectors $X(i)$ and $X(j)$ as the maximum absolute difference between their corresponding scalar elements, i.e.,

$$d[X(i), X(j)] = \max_{k=1, 2, \ldots, m} \left(|y(i+k-1) - y(j+k-1)| \right) \qquad (1.94)$$

4. Define a tolerance criterion r as a positive real number that can be regarded as a criterion of similarity.

5. Count the number of distances $d[X(i), X(j)]$ smaller than the threshold r. That is, for given $x_m(i)$, count the number of $X(j)$ ($j = 1$ to $N - m + 1$, $j \neq i$) for distances being less than r. These distances can be denoted as $C_i^m(r)$, which is the number of sequences in the time series that match (without self-matching) the template with length m within the tolerance criterion r.

6. Compute the ratio of the number in step 5 to the total number $N - m$, which is the average for all i, as

$$C^m(r) = \frac{\sum\limits_{i=1}^{N-m+1} C_i^m(r)}{N - m + 1} \tag{1.95}$$

7. Similarly, for given $x_{m+1}(i)$, count the number of $X(j)$ ($j = 1$ to $N - m$, $j \neq i$)) for which distances $d[X(i), X(j)] < r$. These distances can be denoted as $A_i^m(r)$. Then for $i = 1$ to $N - m$,

$$A^m(r) = \frac{\sum\limits_{i=1}^{N-m} A_i^m(r)}{N - m} \tag{1.96}$$

8. Define theoretically SampEn as

$$\text{SampEn}(m, r) = \lim_{N \to \infty} \left\{ -\ln \frac{A^m(r)}{C^m(r)} \right\} \tag{1.97}$$

In reality data points are finite. Therefore, the above steps yield an approximate value or an estimate of

$$\text{SampEn}(m, r, N) = \left\{ -\ln \frac{A^m(r)}{C^m(r)} \right\} \tag{1.98}$$

SampEn (m, r, N) is the negative natural log of the conditional probability that two sequences similar within a tolerance r for m points remain similar at the next point, where N is the total number of points and self-matches are not included. A low value of SampEn (m, r, N) indicates greater regularity or order in the data series.

The threshold factor r, also called the filter, is an important factor. In principle, if N is almost infinity, r will tend to approach 0. For a finite amount of data (N is finite), or with measurement error, Pincus (1991) suggested an r value between 10 and 20 percent of the time series standard deviation.

For the ease of understanding, we use a simple schematic example to show how to compute the sample entropy of a given time series, as seen in Fig. 1.7. Suppose the length of the series N is 6, and the embedding dimension m is 2; see Fig. 1.7a. First, one

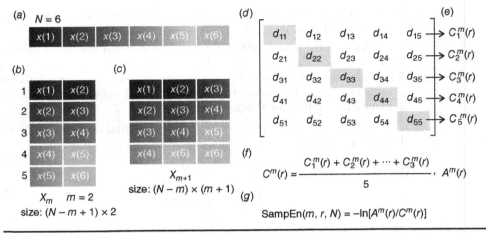

FIGURE 1.7   Schematic illustration for computing sample entropy.

needs to construct two-dimensional embedding vectors (m-dimensional vectors) and three-dimensional embedding vectors (($m + 1$)-dimensional vectors) from the original time series. One can save these vectors in a matrix, with each row representing one vector, as illustrated in Fig. 1.7b and c, respectively. The size of the matrix formed by the m-dimensional vectors (here $m = 2$) and that of the matrix formed by the ($m + 1$)-dimensional vectors are ($N - m + 1$) × m and ($N - m$) × ($m + 1$), respectively. Note that each row in the m-dimensional embedding vector matrix can be viewed as a point.

The next step is to compute the distances of each point in the m-dimensional embedding vector matrix to the other points, following Eq. (1.94). The distance values thus obtained are saved in a matrix, as shown in Fig. 1.7d. For a given threshold r and a given embedding vector $x_m(i)$, which corresponds to the ith row of the embedding vector matrix, one needs to count the numbers of points whose distance to the point $x_m(i)$ is less than r. For simplicity of presentation, we denote the obtained number as n_i. Then as illustrated in Fig. 1.7e, $C_i^m(r)$ can be computed as

$$C_i^m(r) = \frac{n_i}{N - m} \qquad (1.99)$$

Note that the denominator of Eq. (1.99) is equal to the number of columns of the distance matrix minus 1. Then by averaging the obtained $C_i^m(r)$ ($i = 1, 2,..., N - m + 1$), as illustrated in Fig. 1.7f, one can obtain the value of $C^m(r)$, namely,

$$C^m(r) = \frac{\sum_{i=1}^{N-m+1} C_i^m(r)}{N - m + 1} \qquad (1.100)$$

By setting m to $m+1$ and through the same procedure, one can obtain $A^m(r)$. Then the sample entropy can be computed by using Eq. (1.98).

Month	1	2	3	4	5	6	7	8	9	10	11	12
Precipitation (in.)	4.19	2.60	11.42	7.74	6.90	3.51	2.70	0.56	3.31	4.45	4.27	0.77

TABLE **1.5**   Monthly Precipitation Observations (in.) of year 1980 at Station LA169806

Example 1.8.   Suppose we have monthly precipitation records for the period from 1980 to 2010 from 18 stations in Louisiana. Compute the sample entropy for each station and plot it. Take r as 0.2 standard deviation of the time series and $m = 2, 6, 12, 24$.

Solution.   In an attempt to show how the calculation of sample entropy is done, we choose station LA169806 as an example. For economy of space, we do not explicitly show all the observations, only those for the year 1980, as given in Table 1.5. Calculations can be done in a similar fashion for other years and other stations.

Suppose the embedding dimension m is 2. The first step is to embed vectors. The 2-dimensional embedding vector is constructed as shown:

4.19	2.60
2.60	11.42
11.42	7.74
7.74	6.90
6.90	3.51
3.51	2.70
...	...

Here the actual size of the two-dimensional embedding vector is $(N - m + 1) \times 2 = (372 - 2 + 1) \times 2 = 371 \times 2$.

Then we calculate the distance between each point in this vector to others. For example, for the first point [4.19, 2.60], its distance to the second point is

$$d_{12} = \max(|\,[4.19, 2.60] - [2.60, 11.42]\,|) = \max([1.59, 8.82]) = 8.82$$

Distances between point [4.19, 2.60] and other points can be computed in the same way. Therefore, the distance matrix corresponding to the two-dimensional embedding vector is as follows:

	8.82	7.23	4.30	2.71	0.68	...
8.82		8.82	5.14	7.91	8.72	...
7.23	8.82		3.68	4.52	7.91	...
4.30	5.14	3.68		3.39	4.23	...
2.71	7.91	4.52	3.39		3.39	...
0.68	8.72	7.91	4.23	3.39		...
...	...	...	...	...	...	

Note that the actual size of the above matrix is 371 × 371. Now we choose $r = 0.2 \times \text{std}(X)$, where $\text{std}(X)$ represents the sample standard deviation of observations, which is 3.2245. Thus the value of r

is 0.6449. Then compare each element in the above matrix with r. If the element is smaller than r, the result will be 1; otherwise, it will be 0. Therefore, we can have another new matrix:

	0.00	0.00	0.00	0.00	0.00	...
0.00		0.00	0.00	0.00	0.00	...
0.00	0.00		0.00	0.00	0.00	...
0.00	0.00	0.00		0.00	0.00	...
0.00	0.00	0.00	0.00		0.00	...
0.00	0.00	0.00	0.00	0.00		...
...	...	...	...	...	...	

It should be acknowledged that the above matrix represents only a part of the whole matrix. Therefore, one should not be confused as to why all the elements are 0. Next, we compute

$$C_i^m(r) = \frac{n_i}{N-m}$$

where n_i is the number of 1s in the ith row of the above matrix; $N = 372$; and $m = 2$. The computed C_i^m vector is

0.0270
0.0000
0.0000
0.0000
0.0135
0.0432
...

Then C^m is simply the average of the elements of the above matrix, or $C^m = 0.0187$.
 Similarly, the three-dimensional embedding vector can be constructed and is presented here:

4.19	2.60	11.42
2.60	11.42	7.74
11.42	7.74	6.90
7.74	6.90	3.51
6.90	3.51	2.70
3.51	2.70	0.56
...	...	...

Then, in exactly the same vein, one can compute A^m, which is 0.0024 here. Finally, the sample entropy can be computed as

$$H = -\ln(A^m/C^m) = 2.032$$

One can compute the sample entropy for other stations with the use of the same procedure. The results are depicted in Fig. 1.8.

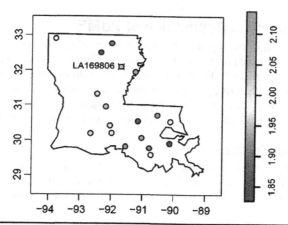

FIGURE 1.8   Sample entropy of the monthly precipitation series from stations in Louisiana for 1980 to 2000. The numbers on the horizontal axis indicate the longitude and those on the vertical axis are indicate the latitude. The negative longitude is in lieu of west. Texas spans from latitude 25°50′ N to 36°30′ N and longitude 93°31′ W to 106°39′ W or (−93°31′ to −106°39′).

1.5   Principle of Maximum Entropy

It is common that some information is available on the random variable X. The question arises: What should be the probability density function of X that is consistent with the given information.[2] The chosen probability distribution should then be consistent with the given information. Laplace's principle of insufficient reason says that all outcomes of an experiment on the random variable are equally likely unless there is information to the contrary. The principle of maximum entropy (POME) states that the probability distribution should be selected in such a way that it maximizes entropy subject to the given information; i.e., POME takes into account all the given information and at the same time avoids consideration of any information that is not given. This is consistent with Laplace's principle. In other words, for given information the best possible distribution that fits the data is the one with the maximum entropy, since this contains the most reliable assignment of probabilities. Since the POME-based distribution is favored over those with less entropy among those that satisfy the given constraints, according to the Shannon entropy as an information measure, entropy defines a kind of measure on the space of probability distributions. Intuitively, distributions of higher entropy represent greater disorder, are smoother, are more probable, are less predictable, or assume less. The POME-based distribution is maximally noncommittal with regard to the missing information and is hence least biased.

Constraints encode the given information. POME leads to the distribution that is most conservative and hence most uninformative. If a distribution with lower entropy were chosen, we would be assuming information that was not available, while a distribution with higher entropy would violate the known constraints. The maximum entropy leads to a probability distribution of a particular macrostate occurring among all possible arrangements (or microstates) of the events under consideration.

1.6   Methodology for Application of POME

The methodology for application of POME involves the following steps: (1) specification of constraints; (2) maximization of entropy, which can be accomplished by using the method of Lagrange multipliers; (3) determination of the entropy-based PDF; (4) determination of the Lagrange multipliers in terms of constraints; (5) determination of entropy in terms of constraints; and (6) derivation of the desired relation. These parts are now briefly discussed for both discrete and continuous variable cases in what follows.

1.6.1   Discrete Case

Let there be a discrete random variable $X = \{x_i, i = 1, 2, ..., N\}$ with probability distribution $P = \{p_i, i = 1, 2, ..., N\}$. The Shannon entropy is defined by Eq. (1.6). The constraints on the random variable $G = \{G_0, G_1, G_2, ..., G_m\}$ can be expressed as

$$G_0 = \sum_{i=1}^{N} p_i = 1 \tag{1.101}$$

$$G_j = \sum_{i=1}^{N} g_j(x_i)p_i = \overline{g_j}, \qquad j = 1, 2, ..., m < N \tag{1.102}$$

For determining $P = \{p_i, i = 1, 2, ..., N\}$, the lagrangian function L can be constructed as

$$L = -\sum_{i=1}^{N} p_i \ln p_i - (\lambda_0 - 1)\left(\sum_{i=1}^{N} p_i - 1\right) - \sum_{j=1}^{m} \lambda_j \left[\sum_{i=1}^{N} p_i g_j(x_i) - \overline{g_j}\right] \tag{1.103}$$

where λ_i, $i = 0, 1, ..., m$, are the Lagrange multipliers. Differentiation of L with respect to p_i gives

$$\frac{\partial L}{\partial p_i} = -\sum_{i=1}^{N} (\ln p_i + 1) - (\lambda_0 - 1)\sum_{i=1}^{N} 1 - \sum_{j=1}^{m} \lambda_j \left[\sum_{i=1}^{N} g_j(x_i)\right] = 0 \tag{1.104}$$

Equation (1.104) yields the probability distribution

$$p_i = p(x_i) = \exp\left[-\lambda_0 - \sum_{j=1}^{m} \lambda_j g_j(x_i)\right] \tag{1.105}$$

Now the Lagrange multipliers are determined. Substituting Eq. (1.105) in Eq. (1.101), one gets the partition function denoted as $Z(\lambda)$:

$$\sum_{i=1}^{N} p_i = 1 = \sum_{i=1}^{N} \exp\left[-\lambda_0 - \sum_{j=1}^{m} \lambda_j g_j(x_i)\right]$$

$$\exp(\lambda_0) = Z(\lambda) = \sum_{i=1}^{N} \exp\left[-\sum_{j=1}^{m} \lambda_j g_j(x_i)\right] \qquad \lambda_0 = \lambda_0(\lambda_1, \lambda_2, ..., \lambda_m) = \lambda \tag{1.106}$$

Taking the logarithm of Eq. (1.106) gives

$$\lambda_0 = \ln Z(\lambda) = \ln \sum_{i=1}^{N} \exp\left[-\sum_{j=1}^{m} \lambda_j g_j(x_i)\right] \tag{1.107}$$

Then Eq. (1.105) can be expressed as

$$p_i = \frac{1}{Z(\lambda)} \exp\left[-\lambda_0 - \sum_{j=1}^{m} \lambda_j g_j(x_i)\right] = p(i \,|\, GC) \tag{1.108}$$

where C is the prior information.

Differentiating Eq. (1.108) with respect to λ_j, one obtains

$$\frac{\partial \lambda_0}{\partial \lambda_i} = -\frac{\sum_{i=1}^{N} \exp\left[-\sum_{j=1}^{m} \lambda_j g_j(x_i)\right] g_j(x_i)}{\sum_{i=1}^{N} \exp\left[-\sum_{j=1}^{m} \lambda_j g_j(x_i)\right]} = -\frac{\sum_{i=1}^{N} g_j(x_i)\exp\left[-\lambda_0 - \sum_{j=1}^{m} \lambda_j g_j(x_i)\right]}{\sum_{i=1}^{N} \exp\left[-\lambda_0 - \sum_{j=1}^{m} \lambda_j g_j(x_i)\right]} = -\overline{g_j} \tag{1.109}$$

or

$$\frac{\partial \lambda_0}{\partial \lambda_1} = -\overline{g_1} = -G_1 \qquad \frac{\partial \lambda_0}{\partial \lambda_2} = -\overline{g_2} = -G_2, \ldots, \frac{\partial \lambda_0}{\partial \lambda_m} = -\overline{g_m} = -G_m \tag{1.110}$$

One can also write

$$\frac{\partial \ln Z}{\partial \lambda_j} = -\frac{\overline{g_j}}{1} = -G_j, \qquad 1 \le j \le m \tag{1.111}$$

$$\frac{\partial^2 \lambda_0}{\partial \lambda_j \partial \lambda_k} = \frac{\partial}{\partial \lambda_k}\left\{-\sum_{i=1}^{N} g_j(x_i)\exp\left[-\lambda_0 - \sum_{j=1}^{m} \lambda_j g_j(x_i)\right]\right\} \tag{1.112}$$

$$= -\sum_{i=1}^{N} g_j(x_i)\exp\left[-\lambda_0 - \sum_{j=1}^{m} \lambda_j g_j(x_i)\right]\left[-\frac{\partial \lambda_0}{\partial \lambda_k} - g_k(x_i)\right] = B_{jk} \tag{1.113}$$

$$\frac{\partial^2 \lambda_0}{\partial \lambda_j \partial \lambda_k} = -\sum_{i=1}^{N} g_j(x_i)\exp\left[-\lambda_0 - \sum_{j=1}^{m} \lambda_j g_j(x_i)\right]\left(-\frac{\partial \lambda_0}{\partial \lambda_k}\right)$$

$$-\sum_{i=1}^{N} g_j(x_i)\exp\left[-\lambda_0 - \sum_{j=1}^{m} \lambda_j g_j(x_i)\right][-g_k(x_i)]$$

$$= -\overline{g_j(x_i)g_k(x_i)} + \overline{g_j(x_i)}\,\overline{g_k(x_i)}$$

$$= \overline{g_j(x_i)}\,\overline{g_k(x_i)} - \overline{g_j(x_i)g_k(x_i)} \qquad 1 \le j, k \le m \tag{1.114}$$

Also,

$$\frac{\partial \lambda_0}{\partial \lambda_j} = \frac{\partial \lambda_0}{\partial \lambda_k} \quad \text{and} \quad \frac{\partial^2 \lambda_0}{\partial \lambda_j \, \partial \lambda_k} = \frac{\partial^2 \ln Z(\lambda)}{\partial \lambda_j \, \partial \lambda_k} = \frac{\partial G_k}{\partial \lambda_j} = \frac{\partial G_j}{\partial \lambda_k} \quad 1 \le j, k \le m \tag{1.115}$$

The maximum entropy can now be expressed as

$$H = -\sum_{i=1}^{N} p_i \ln p_i = -\sum_{i=1}^{N} p_i \ln \left\{ \frac{1}{Z(\lambda)} \exp \left[-\sum_{j=1}^{m} \lambda_j g_j(x_i) \right] \right\}$$

$$= -\sum_{i=1}^{N} p_i \left[-\ln Z(\lambda) - \sum_{j=1}^{m} \lambda_j g_j(x_i) \right] = \ln Z(\lambda) + \sum_{j=1}^{m} \lambda_j G_j \tag{1.116}$$

Recall that

$$G_k = \overline{g_k(x_i)} = \frac{\partial \ln Z(\lambda)}{\partial \lambda_k} \quad 1 \le k \le m \tag{1.117}$$

Therefore,

$$H = \ln Z(\lambda) + \sum_{j=1}^{m} \lambda_j \frac{\partial \ln Z(\lambda)}{\partial \lambda_j} \tag{1.118}$$

Let that H, as a function of constraints G, be denoted by $S(G)$. Then (1.119)

$$S(G) = \ln Z(\lambda) + \sum_{j=1}^{m} \lambda_j G_j = \ln \sum_{i=1}^{N} \exp \left[-\sum_{j=1}^{m} \lambda_j g_j(x_i) \right] + \sum_{j=1}^{m} \lambda_j G_j$$

$$= \ln \sum_{i=1}^{N} \exp \left[-\lambda_0 - \sum_{j=1}^{m} \lambda_j g_j(x_i) \right] \frac{1}{\exp(-\lambda_0)} + \sum_{j=1}^{m} \lambda_j G_j \tag{1.120}$$

$$= \ln \left[\frac{1}{\exp(-\lambda_0)} \right] + \sum_{j=1}^{m} \lambda_j G_j = \lambda_0 + \sum_{j=1}^{m} \lambda_j G_j$$

Differentiation of S with respect to constraint G_j can be written as

$$\frac{\partial S(G)}{\partial G_j} = \lambda_j \quad 1 \le j \le m \tag{1.121}$$

Equation (1.121) expresses a general orthogonality property of the POME-based distribution.

$$\frac{\partial^2 S}{\partial G_j \, \partial G_k} = -\frac{\partial \lambda_j}{\partial G_k} = -\frac{\partial \lambda_k}{\partial G_j} = A_{jk} \tag{1.122}$$

Matrices A_{jk} and B_{jk} are symmetric, positive, definite matrices. Matrix B_{jk} is the covariance matrix, and matrix A_{jk} is the entropy curvature matrix, provided the constraints are linearly independent, that is, $\partial G_j / \partial G_k = \delta_{jk}$.

1.6.2 Continuous Case

Let X be a continuous variable having a probability density function (PDF) as $f(x)$ and cumulative distribution function (CDF) as $F(x)$.

Choice and Specification of Constraints

Tsallis et al. (1998) have discussed the role of constraints in the context of the Tsallis entropy formulas. Papalexiou and Koutsoyiannis (2012) have provided an excellent discussion of the rationale for choosing different types of constraints in hydrology. Entropy maximization shows that there is a unique correspondence between a probability distribution and the constraints that lead to it. Therefore, choosing appropriate constraints is extremely important in the entropy formalism.

Constraints encode the information or summarize the knowledge that can be garnered from empirical observations or theoretical considerations. To start, it is foremost to determine or at least have a good idea as to the type of probability distribution that will best represent the empirical hydrologic data, such as rainfall, runoff, or temperature. Once an idea about the distribution shape is formed, the issue of constraints is addressed.

Clearly there can be a large, if not an infinite, number of constraints that can perhaps summarize the information on the random variable. The question then arises: How should the appropriate constraints be chosen? First, constraints should be simple, but simplicity is subjective and therefore quantitative criteria are needed. Second, constraints should be as few as absolutely needed.

When empirical data suggest, without any consideration of entropy, a particular shape of the probability distribution, it is important to keep in mind that these data represent only a small part of the past. Therefore, any inference on the random variable characteristics may vary in the future. This suggests that constraints should be defined such that they are more or less preserved in the future. To be more specific, some constraints, especially lower-order moments such as mean and variance, are less susceptible to change than are others, especially higher-order moments, such as kurtosis.

From a physical point of view, it would be desirable to specify constraints in terms of the laws of mathematical physics—mass conservation, momentum conservation, and energy conservation—or constitutive laws. Then these three concepts can link statistical analysis based on entropy with physical concepts in hydrologic science and engineering. However, water resources and hydrologic processes are complicated and do not often allow for defining constraints in this manner except for very simple cases. For example, in case of groundwater flow, it is possible to express constraints in terms of mass conservation and Darcy's law when one is solving very simple problems. Likewise, for movement of soil moisture, infiltration, velocity distribution, and suspended-sediment concentration, mass conservation can be used as a constraint. However, analysis becomes unwieldly for a majority of cases, and therefore constraints need to be defined while keeping the empirical evidence in mind.

If the observed data show that the random variable can be described by a bell-shaped distribution, then all that is needed is the mean and variance. That means these constitute the constraints. Likewise, if the distribution is heavy-tailed or light-tailed, then appropriate constraints need to be specified accordingly. However, only a few hydrologic processes follow a normal distribution. Of course, most hydrologic variables are nonnegative, and with the nonnegative condition imposed, the resulting distribution would be truncated normal. The Tsallis entropy maximization leads to a symmetric bell-shaped curve with power-type tails. If the mean is zero, then this becomes the Tsallis distribution. For nonzero mean this is the Pearson type VII distribution. A majority of hydrologic and environmental processes exhibit a rich variety of asymmetries. Examples are rainfall extremes corresponding to small time intervals, flow maxima, flow minima, extreme temperatures, and extreme winds among others. It is therefore more convenient to express the information in terms of moments, such as mean, variance, covariance, cross-covariance, or linear combinations of these statistics. The constraints encode relevant information.

Mean—a measure of central tendency—is one of the most frequently specified constraints:

$$E[X] = \mu = \int_0^\infty x f(x)\, dx \tag{1.123a}$$

which is approximated by the sample mean, denoted by $\bar{x}$ where the overbar denotes the sample average. If the random variable takes on only nonnegative values, then the geometric mean, denoted by, μ_G, which is smaller than the arithmetic mean, is another useful constraint for hydrologic processes. For a sample size N, an estimate of μ_G can be defined as

$$\mu_G = \left(\sum_{i=1}^N x_i \right)^{1/N} = \exp\left[\ln\left(\sum_{i=1}^N x_i \right)^{1/N} \right] = \exp\left[\frac{1}{N} \ln\left(\sum_{i=1}^N \right)^{1/N} \right]$$

$$= \exp\left[\frac{1}{N} \sum_{i=1}^N \ln x_i \right] = \exp[\overline{\ln x}] \tag{1.123b}$$

Equation (1.123b) suggests that a constraint for entropy maximization can be expressed as

$$E[\ln X] = \ln \mu_G \tag{1.123c}$$

This constraint would be useful when empirical data are positively skewed or even heavy-tailed. The logarithm of the geometric mean, because of its logarithmic character, is likely to be preserved in the future.

The second moment or variance, a measure of dispersion about the central tendency or mean, is also expressed as a constraint:

$$\sigma^2 = \int_0^\infty x^2 f(x)\, dx \tag{1.124}$$

where x is about μ. Papalexiou and Koutsoyiannis (2012) reasoned that if the second moment is preserved, then its square root, the standard deviation, is even more likely to be preserved and more robust to outliers. This reasoning can be extended to lower-order fractional moments.

Recall that for the logarithmic function one can write

$$\lim_{a \to 0} \frac{x^a - 1}{a} = \ln x \tag{1.125}$$

For small values of a, it can be argued that x^a would behave as $\ln x$. Therefore, it may be deemed logical to allow the order of the moment to remain unspecified so that even fractional values of order are accommodated. Thus, any moment of order r, denoted M_r, as a constraint can be expressed by

$$M_r = E[X^r] = \int_0^\infty x^r f(x) dx \tag{1.126a}$$

Further, recalling the limiting definition of the exponential function

$$\exp x^a = \lim_{b \to 0} \frac{1 + bx^a}{b} \tag{1.126b}$$

one can define

$$x_b^a = \frac{\ln(1 + bx^a)}{b} \tag{1.127a}$$

which becomes a power x^a for $b = 0$:

$$x_0^a = \lim_{b \to 0} \frac{\ln(1 + bx^a)}{b} = x^a \tag{1.127b}$$

Now the regular moments, designated as p-moments, can be generalized as

$$M_p' = E[X_p'] = \int_0^\infty \ln(1 + px^r) f(x) \, dx \tag{1.128}$$

This is one generalization among many that can be constructed. Papalexiou and Koutsoyiannis (2012) provided the following rationale for this generalization: (1) The p moments are simple, and for $p = 0$ they reduce to regular moments. (2) Their basis is the x_p' function that has the desirable properties of the $\ln x$ function and therefore are appropriate for positively skewed random variables. (3) The p moments lead to flexible power-type distributions, including the Pareto (for $m = 1$) and Tsallis ($m = 2$) distributions. (4) These moments are no more arbitrary than generalized entropy measures. (5) These moments lead to distributions that represent many hydrologic processes as well.

If observations are available, then, for simplicity of analysis, let the information on the random variable $X \varepsilon (a, b$, be expressed in terms of constraints $C_r, r = 0, 1, 2,..., n$, defined as

$$C_0 = \int_a^b f(x)\,dx = 1 \tag{1.129a}$$

$$C_r = \int_a^b g_r(x) f(x)\,dx = \overline{g_r(x)}, \qquad r = 1, 2,..., n \tag{1.129b}$$

where $g_r(x), r = 1, 2,..., n$, represent some functions of x; n denotes the number of constraints; and $\overline{g_r(x)}$ is the expectation of $g_r(x)$, with $g_0(x) = 1$. Equation (1.129a) states that the PDF must satisfy the total probability. Other constraints, defined by Eq. (1.129b), have physical meaning. For example, if $r = 1$ and $g_1(x) = x$, Eq. (1.131b) would correspond to the mean $\overline{x}$; likewise, for $r = 2$ and $g_2(x) = (x - \overline{x})^2$, it would denote the variance of X. For many hydrologic engineering problems, more than two or three constraints are not needed.

Maximization of Entropy

To obtain the least-biased $f(x)$, the entropy given by Eq. (1.19) is maximized, subject to Eqs. (1.129a) and (1.129b). One simple way to achieve the maximization is through the use of the method of Lagrange multipliers. To that end, the Lagrangian function L can be constructed:

$$L = -\int_a^b f(x) \ln f(x)\,dx - (\lambda_0 - 1)\left[\int_a^b f(x)\,dx - C_0\right] - \sum_{r=1}^n \lambda_r \left[\int_a^b f(x) g_r(x)\,dx - C_r\right] \tag{1.130a}$$

where $\lambda_1, \lambda_2,..., \lambda_n$ are the Lagrange multipliers. To obtain $f(x)$ which maximizes L, one may recall the Euler-Lagrange calculus of variation, and therefore one differentiates L with respect to $f(x)$ (noting X as the parameter and f as the variable), equates the derivative to zero, and obtains

$$\frac{\partial L}{\partial f} = 0 \Rightarrow -[1 + \ln f(x)] - (\lambda_0 - 1) - \sum_{r=1}^n \lambda_r g_r(x) = 0 \tag{1.130b}$$

Probability Distribution

Equation (1.130b) leads to the probability density function of X in terms of the given constraints:

$$f(x) = \exp\left[-\lambda_0 - \sum_{r=1}^n \lambda_r g_r(x)\right] \tag{1.130c}$$

where the Lagrange multipliers $\lambda_r, r = 0, 1, 2, \ldots, n,$ can be determined with the use of Eqs.(1.129a) and (1.129b). Equation (1.130c) is also written as

$$f(x) = \frac{1}{Z(\lambda_1, \lambda_2, \ldots, \lambda_n)} \exp\left[-\sum_{r=1}^{n} \lambda_r g_r(x)\right] \tag{1.130d}$$

where $Z = \exp \lambda_0$ is called the *partition function*. Integration of Eq. (1.130d) leads to the cumulative distribution function or simply the probability distribution of X as

$$F(x) = \int_a^x \exp\left[-\lambda_0 - \sum_{r=1}^{n} \lambda_r g_r(x)\right] dx \tag{1.131}$$

Substitution of Eq. (1.132d) in Eq. (1.19) results in the maximum entropy of $f(x)$ or X:

$$H = \lambda_0 + \sum_{r=1}^{n} \lambda_r \overline{g_r(x)} = \lambda_0 + \sum_{r=1}^{n} \lambda_1 E[g_r(x)] = \lambda_0 + \sum_{r=1}^{n} \lambda_r C_r \tag{1.132}$$

where $E[g(x)]$ is the expectation of $g(x)$. Equation (1.129a) shows that the entropy of the probability distribution of X depends only on the specified constraints, since the Lagrange multipliers themselves can be expressed in terms of the specified constraints. Equations (1.19), (1.129a), (1.129b), (1.130a), and (1.130d) constitute the building blocks for the application of entropy theory. Entropy maximization results in the probability distribution that is most conservative and hence most informative. If a distribution with lower entropy were chosen, that would imply the assumption of information that is not available. On the other hand, a distribution with higher entropy would violate the known constraints. Thus, it can be stated that the maximum entropy leads to a probability distribution of a particular macrostate among all possible configurations of the states (or events) under consideration.

Determination of Lagrange Multipliers

Equation (1.130d) is the POME-based probability distribution containing Lagrange multipliers $\lambda_0, \lambda_1, \lambda_2, \ldots, \lambda_n$ as parameters which can be determined by inserting Eq. (1.130d) in Eqs. (1.129a) and (1.129b):

$$\exp(\lambda_0) = \int_a^b \exp\left[-\sum_{r=1}^{n} \lambda_r g_r(x)\right] dx \qquad r = 1, 2, \ldots, n \tag{1.133a}$$

and

$$C_r = \int_a^b g_r(x) \exp\left[-\sum_{r=1}^{n} \lambda_r g_r(x)\right] dx \qquad r = 1, 2, \ldots, n \tag{1.133b}$$

Equation (1.133a) can be written for the zeroth Lagrange multiplier as

$$\lambda_0 = \ln \int_a^b \exp\left[-\sum_{r=1}^{n} \lambda_r g_r(x)\right] dx \qquad r = 1, 2, \ldots, n \tag{1.133c}$$

Equation (1.133c) shows that λ_0 is a function of $\lambda_1, \lambda_2, \ldots, \lambda_n$ and expresses the partition function Z as

$$Z(\lambda_1, \lambda_2, \ldots, \lambda_m) = \int_a^b \exp\left[-\sum_{r=1}^m \lambda_r g_r(x)\right] dx \qquad (1.134)$$

Differentiating Eq. (1.133c) with respect to λ_r, one gets

$$\frac{\partial \lambda_0}{\partial \lambda_r} = -\frac{\displaystyle\int_a^b g_r(x) \exp\left[-\sum_{r=1}^n \lambda_r g_r(x)\right] dx}{\displaystyle\int_a^b \exp\left[-\sum_{r=1}^n \lambda_r g_r(x)\right] dx} \qquad r = 1, 2, \ldots, n \qquad (1.135)$$

Multiplying the numerator as well as the denominator of Eq. (1.135) by $\exp(-\lambda_0)$, one obtains

$$\frac{\partial \lambda_0}{\partial \lambda_r} = -\frac{\displaystyle\int_a^b g_r(x) \exp\left[-\lambda_0 - \sum_{r=1}^n \lambda_r g_r(x)\right] dx}{\displaystyle\int_a^b \exp\left[-\lambda_0 - \sum_{r=1}^n \lambda_r g_r(x)\right] dx} \qquad r = 1, 2, \ldots, n \qquad (1.136)$$

The denominator in Eq. (1.136) equals unity by virtue of Eq. (1.130c). Therefore, Eq. (1.136) becomes

$$\frac{\partial \lambda_0}{\partial \lambda_r} = -\int_a^b g_r(x) \exp\left[-\lambda_0 - \sum_{r=1}^n \lambda_r g_r(x)\right] dx \qquad r = 1, 2, \ldots, n \qquad (1.137)$$

Note that substitution of Eq. (1.137) in Eq. (1.123b) yield

$$C_r = \int_a^b g_r(x) \exp\left[-\lambda_0 - \sum_{r=1}^n \lambda_r g_r(x)\right] dx \qquad r = 1, 2, \ldots, n \qquad (1.138)$$

Therefore, Eqs. (1.137) and (1.138) yield

$$\frac{\partial \lambda_0}{\partial \lambda_r} = -C_r \qquad r = 1, 2, \ldots, n \qquad (1.139)$$

Likewise, Eq. (1.133a) can be written analytically as

$$\lambda_0 = \lambda_0(\lambda_1, \lambda_2, \ldots, \lambda_n) \qquad (1.140)$$

Equation (1.140) can be differentiated with respect to λ_r, $r = 1, 2, \ldots, n$, and each derivative can be equated to the corresponding derivative in Eq. (1.139). This would lead to a system of $n - 1$ equations with $n - 1$ unknowns, whose solution would lead to the expression of Lagrange multipliers in terms of constraints.

Example 1.9.   Assume a discrete random variable X that takes on values $x_1, x_2, x_3, \ldots, x_N$ with probabilities $p_1, p_2, p_3, \ldots, p_N$. The expected value or the average value of the variable is known from observations. What should be its probability distribution?

Solution.   The expected value of the variable is known:

$$p_1 x_1 + p_2 x_2 + \cdots + p_N x_N = \sum_{i=1}^{N} x_i p_i = \overline{x} \tag{1.141}$$

Of course, the total probability law holds:

$$\sum_{i=1}^{N} p_i = 1 \qquad p_i \geq 0 \qquad i = 1, 2, \ldots, N \tag{1.142}$$

The objective is to derive the POME-based distribution $p = \{p_1, p_2, \ldots, p_N\}$, subject to Eqs. (1.141) and (1.142). In other words, one maximizes the Shannon entropy given by Eq. (1.18), subject to Eqs. (1.141) and (1.142).

Following the POME formalism, one constructs the lagrangian L:

$$L = -\sum_{i=1}^{N} p_i \ln p_i - (\lambda_0 - 1)\left(\sum_{i=1}^{N} p_i - 1\right) - \lambda_1\left(\sum_{i=1}^{N} p_i x_i - \overline{x}\right) \tag{1.143}$$

Differentiating Eq. (1.143) with respect to p_i, $i = 1, 2, \ldots, N$, and equating each derivative to zero, one obtains

$$\frac{\partial L}{\partial p_i} = 0 \Rightarrow -\ln p_i - \lambda_0 - \lambda_1 x_i = 0 \qquad i = 1, 2, \ldots, N \tag{1.144}$$

Equation (1.144) yields

$$p_i = \exp\left(-\lambda_0 - \lambda_1 x_i\right) \qquad i = 1, 2, \ldots, N \tag{1.145}$$

Equation (1.145) contains parameters λ_0 and λ_1 that are determined with the use of Eqs. (1.141) and (1.142). Inserting Eq. (1.145) in Eq. (1.142), one gets

$$\exp\left(-\lambda_0\right) = \left[\sum_{i=1}^{N} \exp\left(-\lambda_1 x_i\right)\right]^{-1} \tag{1.146}$$

When Eq. (1.146) is substituted in Eq. (1.145), the result is

$$p_i = \frac{\exp(-\lambda_1 x_i)}{\sum\limits_{i=1}^{N} \exp(-\lambda_1 x_i)} \tag{1.147}$$

Equation (1.147) is called the *Maxwell-Boltzmann* (M-B) *distribution* used in statistical mechanics.

Now its parameter λ_1 must be determined in terms of constraint $\bar{x}$. Inserting Eq. (1.145) in Eq. (1.141), one gets

$$\sum_{i=1}^{N} x_i \exp(-\lambda_0 - \lambda_1 x_i) = \bar{x} \tag{1.148}$$

Taking advantage of Eq. (1.146), Eq. (1.148) yields

$$\frac{\sum_{i=1}^{N} x_i \exp(-\lambda_1 x_i)}{\sum_{i=1}^{N} \exp(-\lambda_1 x_i)} = \bar{x} \tag{1.149}$$

Equation (1.147) permits the estimation of λ_1 in terms of $\bar{x}$. Note that if $\lambda_1 = 0$, clearly Eq. (1.145) would be a rectangular distribution with $p_i = 1/N$. If λ_1 is negative, then the probability increases as x_i increases; and if $\lambda_1 > 0$, then the probability decreases as x_i increases. In physics the M-B distribution has been employed to derive the microstates of a system on the basis of the mean of the macroscopic data. For example, if a system had a large number of particles, each with an energy level, then the M-B distribution would be employed to determine the probability distribution of the energy levels of the particles, provided the expected energy of the system was somehow known. Fiorentino et al. (1993) employed the M-B distribution to describe the probability distribution of elevations of links in a river basin if the mean basin elevation was known.

Example 1.10.   Consider a random variable X varying over a semi-infinite interval $(0, \infty)$ and having a probability density function (PDF) $f(x)$. From observations, the expected value of X, denoted $E[x]$, is known. Derive the PDF $f(x)$ of X.

Solution.   In this case the constraint equation is given as

$$\int_0^{\infty} x f(x)\, dx = E[x] = \bar{x} = k \tag{1.150}$$

The Shannon entropy is given by Eq. (1.19), where

$$\int_0^{\infty} x f(x)\, dx = 1 \tag{1.151}$$

The least-biased $f(x)$ is determined by maximizing Eq. (1.19), subject to Eqs. (1.150) and (1.151). To that end, the lagrangian L is constructed as

$$L = -\int_0^{\infty} f(x) \ln f(x)\, dx - (\lambda_0 - 1)\left[\int_0^{\infty} f(x)\, dx - 1\right] - \lambda_1\left[\int_0^{\infty} x f(x)\, dx - k\right] \tag{1.152}$$

Taking the derivative of L with respect to $f(x)$ and equating it to 0, one obtains

$$\frac{L}{f(x)} = 0 \Rightarrow -[1 + \ln f(x)] - (\lambda_0 - 1) - \lambda_1 x = 0 \tag{1.153}$$

Therefore,

$$f(x) = \exp(-\lambda_0 - \lambda_1 x)$$

(1.154)

Equation (1.154) is the POME-based distribution with λ_0 and λ_1 as parameters. Substituting Eq. (1.154) in Eq. (1.151), one obtains

$$\int_0^\infty \exp(-\lambda_0 - \lambda_1 x)\, dx = \lambda_1 \exp \lambda_0 = 1$$

(1.155)

Substituting Eq. (1.155) in Eq. (1.154), one gets

$$f(x) = \lambda_1 \exp(-\lambda_1 x)$$

(1.156)

Inserting Eq. (1.156) in Eq. (1.150), one gets

$$\int_0^\infty \lambda_1 x \exp(-\lambda_1 x)\, dx = k \qquad \text{or} \qquad \lambda_1 = \frac{1}{k}$$

(1.157)

Thus Eq. (1.154) becomes

$$f(x) = \frac{1}{k} \exp\left(-\frac{x}{k}\right), \qquad k = \bar{x}$$

(1.158)

which is the exponential distribution.

Maximum Entropy

Substitution of Eq. (1.127) into Eq. (1.19) yields the maximum entropy

$$H(X) = \lambda_0 + \sum_{r=1}^{n} \lambda_r \overline{g_r(x)}]$$

(1.159)

The maximum entropy is a function of the Lagrange multipliers and constraints. Because the Lagrange multipliers can be expressed in terms of constraints, the entropy can be expressed in terms of constraints alone. It can be shown that $H(X)$ is a concave function of constraints.

1.7   Concentration Theorem

The *concentration theorem* (CT), formulated by Jaynes (1958), has two aspects. First, it shows that the POME-based probability distribution best represents our knowledge about the state of the system by showing the spread of lower entropies around the maximum entropy value. Second, POME is the preferred method to obtain this distribution. The concentration theorem can be employed to measure the bias in this preference. The basis for these two aspects is contained in the Shannon inequality and the relation between entropy and the chi-square test.

The Shannon inequality says that the probability distribution given by POME has greater entropy than any other distribution. Let $P = \{p_i, i = 1,2,...,N\}$ be the distribution given by the maximization of the Shannon entropy. The entropy corresponding to this distribution obtained by the method of Lagrange multipliers is H_{max}. Let there be another distribution $Q = \{q_i, i = 1,2,...,N\}$ satisfying the same constraints as P. Its entropy is H. For these two probability distributions, the Shannon inequality can be expressed as

$$\sum_{i=1}^{N} q_i \log \frac{q_i}{p_i} \geq 0 \tag{1.160}$$

where the equality holds only if $q_i = p_i, i = 1,2,...,N$. Thus,

$$H_{max} - H \geq 0 \tag{1.161}$$

where $H_{max} - H = 0$ only if $Q = P$. If Q satisfied the same constraints as P, it can be shown that

$$H_{max} - H = \sum_{i=1}^{N} q_i \log \frac{q_i}{p_i} \tag{1.162}$$

Following Keshavan and Kapur (1992), an interesting result is obtained by letting $q_i = p_i(1 + \omega_i)$, where

$$\sum_{i=1}^{N} p_i \omega_i = 0 \tag{1.163}$$

Then, neglecting summation terms involving third or higher powers of ω, one can approximate Eq. (1.162) as

$$\Delta H = H_{max} - H = \sum_{i=1}^{N} q_i \log \frac{q_i}{p_i} = \sum_{i=1}^{N} p_i(1 + \omega_i) \log (1 + \omega_i) \tag{1.164}$$

By expanding the logarithmic term inside the summation, Eq. (1.164) can be approximated as

$$\Delta H \approx \sum_{i=1}^{N} p_i(1 + \omega_i)\left[\omega_i - \frac{\omega_i^2}{2} + \frac{\omega_i^3}{3} - \cdots \right] \approx \frac{1}{2} \sum_{i=1}^{N} p_i \omega_i^2$$

$$= \frac{1}{2} \sum_{i=1}^{N} p_i \frac{(q_i - p_i)^2}{p_i^2} = \frac{1}{2} \sum_{i=1}^{N} \frac{(q_i - p_i)^2}{p_i} = \frac{1}{2M} \sum_{i=1}^{N} \frac{(Mq_i - Mp_i)^2}{Mp_i} \tag{1.165}$$

where M is the total number of observations. Then Mq_i, for $i = 1, 2, ..., N$, can be interpreted as the observed frequencies; N is the number of class intervals; and Mp_i, $i = 1, 2, ..., N$, is the frequencies computed from the POME-based distribution P which has the same moments as Q. Let $O_i = Mq_i$ and $C_i = Mq_i$. Then, Eq. (1.165) can be written as

$$2M\,\Delta H = \sum_{i=1}^{N} \frac{(Mq_i - Mp_i)^2}{Mp_i} = \sum_{i=1}^{N} \frac{(O_i - C_i)^2}{C_i} = \chi^2 \qquad (1.166)$$

where χ^2 is chi-square with n degrees of freedom defined by

$$n = N - m - 1 \qquad (1.167)$$

where $(m+1)$ is the number of constraints. Because the chi-square distribution is known, one can determine the probability that the computed chi-square is less than the chi-square at the 95 or 99 percent significance level:

$$P[(\chi^2) \le \chi_n^2(0.95)] = 0.95 \qquad P[(\chi^2) \le \chi_n^2(0.99)] = 0.99 \qquad (1.168)$$

For different degrees of freedom n, the values of $\chi_n^2(0.95)$ and $\chi_n^2(0.99)$ are found in standard statistical textbooks or can be computed from the chi-square distribution as given in Table 1.6.

To determine the 95 and 99 percent entropy intervals, Eq. (1.168) with the use of Eq. (1.166) can be expressed as

$$P\left[H_{max} - H \le \frac{\chi_n^2(0.95)}{2M}\right] = 0.95 \qquad (1.169)$$

$$P\left[H_{max} - H \le \frac{\chi_n^2(0.99)}{2M}\right] = 0.99 \qquad (1.170)$$

Equations (1.169) and (1.170) can be expressed as

$$P\left[H_{max} \ge H \ge H_{max} - \frac{\chi_n^2(0.95)}{2M}\right] = 0.95 \qquad (1.171)$$

$$P\left[H_{max} \ge H \ge H_{max} - \frac{\chi_n^2(0.99)}{2M}\right] = 0.99 \qquad (1.172)$$

N	2	3	4	5	6	8	10	15	20	25	30
$\chi_n^2(0.95)$	5.99	7.80	9.49	11.07	12.59	15.51	18.31	25.00	31.40	37.66	43.78
$\chi_n^2(0.99)$	9.21	11.35	12.28	15.09	16.81	20.09	25.21	30.58	37.67	44.31	50.89

TABLE 1.6 Chi-Square Distribution

Thus, the 95 and 99 percent entropy intervals, respectively, are

$$\left[H_{max} - \frac{\chi_n^2(0.95)}{2M}, H_{max} \right] \quad \left[H_{max} - \frac{\chi_n^2(0.95)}{2M}, H_{max} \right] \qquad (1.173)$$

For a random variable X having a probability distribution that satisfies the same constraints as the POME-based distribution, its entropy will be greater than $\left[H_{max} - \chi_n^2(0.95)/2M \right]$ with a 95 percent probability. Likewise, its entropy with 99 percent probability will be greater than $H_{max} - \chi_n^2(0.99)/2M$. If M is large, entropies of most probability distributions that satisfy a given set of constraints will be concentrated near the maximum entropy values. This means that the POME-based distribution is the best choice. Also note that the length of the entropy interval increases with N and decreases with m, increases with confidence interval, and decreases rapidly with M. Thus, the concentration theorem states that for large M, $2M(H_{max} - H)$ is distributed as chi-square with $N - m - 1$ degrees of freedom.

The concentration theorem can measure the bias in the decision of selecting a model. According to the concentration theorem, a certain fraction F of all outcomes in class C of outcomes will have entropy in the range

$$H_{max} - \Delta H \le H(f_1, f_2, \ldots, f_M) \le H_{max} \qquad (1.174)$$

Here H_{max} is obtained by POME subject to the specified constraints. The CT shows that $2M\Delta H$ is distributed in C as χ_k^2 with $n = N - m - 1$ degrees of freedom, $m =$ the number of constraints, and $N =$ the number of possible outcomes. Denoting χ_n^2 at the $100p$ percent significance level by $\chi_n^2(p)$, one gets

$$2M\Delta H = \chi_n^2(1 - F) \qquad (1.177)$$

Equation (1.175) determines the percentage chance that the entropy of the observed frequency distribution will be outside the interval computed by Eq. (1.174). If N is large, the entropy of the overwhelming majority of all possible distributions will be close to H_{max}. Since POME probabilities are mean frequencies in reproducible experiment, for large N, the POME probability and observed probability will have the same value.

Example 1.11.   A six-face die that is thrown 100 times, and a mean value of 3.5 is obtained. Compute 95 and 99 percent entropy intervals. Compare these intervals if there were no constraints.

Solution.   Here $N = 6$ and $M = 100$. The constraints here are $\sum_{i=1}^{6} f_i = 1$ and $\sum_{i=1}^{6} n_i f_i = 3.5$. The entropy reaches a maximum value at $H_{max} = 12.969$. If $F = 95$ percent and $1 - F = 0.05$, then by the CT, $N - 1 - 1 = 6 - 1 - 1 = 4$ degrees of freedom. From Table 1.6, one gets $\chi_4^2(0.05) = 9.49$. This means that a 95 percent fraction of all possible outcomes will have entropy in the range of $2M\Delta H = 9.49$ or $\Delta H = \chi_n^2(1 - F)/2M = 0.047$ and $12.922 \le H \le 12.969$. This shows that virtually all possible outcomes that satisfy the constraints have frequencies close to the POME-based distribution.

If $F = 99$ percent and $1 - F = 0.01$, then $\chi_4^2(0.01) = 12.28$ from Table 1.6. It becomes $\Delta H = \chi_n^2(1 - F)/2M = 0.061$. Thus, 99 percent of all possible outcomes will have entropy in the range of $12.908 \le H \le 12.969$.

1.8   Principle of Minimum Cross Entropy

On the basis of intuition, experience, or theory, one may guess an a priori probability distribution of the random variable. Then the Shannon entropy is maximum when the probability distribution of the random variable is as close to the a priori distribution as possible. This *principle of minimum cross entropy* (POMCE) was formulated by Kullback and Leibler (1951) and is described in full by Kullback (1959). Sometimes it is also referred to as the Kullback-Leibler (K-L) principle. POMCE is also referred to as the principle of minimum discrimination information, principle of minimum directed divergence, principle of minimum distance, or principle of minimum relative entropy. To explain POMCE, let one guess, based on intuition, experience, or theory, a probability distribution $Q = \{q_1, q_2, ..., q_N\}$ for a random variable X that takes on N values. If Q is a uniform distribution, then this would represent the maximum uncertainty. The guessed distribution constitutes the prior information in the form of a priori distribution. To verify one's guess, one takes some observations and computes some moments of the distribution, using these observations, and expresses them in the form of constraints. If the prior or guessed distribution satisfies the constraints, then the guessed probability distribution $Q = \{q_1, q_2, ..., q_N\}$ is the desired distribution of X. If that is not the case, which is often true, then another probability distribution is to be sought. To derive the distribution $P = \{p_1, p_2, ..., p_N\}$ of X, one takes all the given information and makes the distribution as near to Q (based on one's intuition and experience) as possible; i.e., minimize the distance between P and Q. This means that the closer P is to Q, the greater will be its uncertainty. Thus, POMCE is expressed as

$$D(P, Q) = \sum_{i=1}^{N} p_i \ln \frac{p_i}{q_i} \tag{1.176}$$

where $D(P, Q)$ is the crossentropy or distance or discrimination information and the objective is to minimize D, the cross entropy. In Eq. (1.176) we often use the convention that $0 \log(0/q) = 0$ and $p \log(p/0) = \infty$. If no a priori distribution is available in the form of constraints and Q is chosen to be a uniform distribution $Q = \{q_i = 1/N, i = 1, 2, ..., N\}$, then Eq. (1.176) takes the form

$$D(P, Q) = \sum_{i=1}^{N} p_i \ln \frac{p_i}{1/N} = \ln N + \sum_{i=1}^{N} p_i \ln p_i = \ln N - H \tag{1.177}$$

where H is the Shannon entropy. Equation (1.177) shows that minimizing $D(P, Q)$ is equivalent to maximizing the Shannon entropy. Because D is a convex function, its local minimum is its global minimum. Thus, a posterior distribution P is obtained by combining a priori Q with specified constraints. The distribution P minimizes the cross (or relative) entropy with respect to Q, defined by Eq. (1.176), where the entropy of Q is defined or given. Minimization of cross entropy results asymptotically from Bayes' theorem.

Note that POMCE involves two concepts: (1) a priori probability distribution and (2) a measure of distance. First, as stated earlier, $Q = \{q_1, q_2, ..., q_N\}$ is chosen based on some knowledge about X, but this Q does not satisfy the prescribed constraints. Second, D is a measure of distance that has certain desirable properties, such as

$D(P, Q) \geq 0$ and $D(P, Q) = 0$ if and only if $P = Q$. In essence, POMCE is a measure between two probability distributions, one of which is related to the system to be characterized or the source to be estimated and assumed to be unknown and the other (called prior) to the model chosen to describe the system. The system is characterized by a set of moments or by the mean and any symmetric part of the covariance matrix of the system called constraints. The POMCE measure is obtained by minimizing the discrimination information with respect to the given prior distribution over all probabilistic descriptions of the system that concur with the given constraints.

If a different distribution $q(x_i)$ was used, then the excess number of bits that will be needed is described by the Kullback-Leibler entropy or cross entropy:

$$D(X) = \sum p(x_i) \log \frac{p(x_i)}{q(x_i)} \tag{1.178}$$

The Kullback-Leibler entropy can be used for conditional probabilities $p(x_i | y_i)$. For a single state or signal y_j,

$$D(X) = \sum p(x_i | y_j) \log \frac{p(x_i | y_j)}{q(x_i | y_j)} \tag{1.179}$$

Summing over all j with respect to $p(j)$ leads to

$$D(X | Y) = \sum p(x_i, y_j) \log \frac{p(x_i | y_j)}{q(x_i | y_j)} \tag{1.180}$$

The mutual information of two processes or subsystems X and Y with joint probability $p(x_i, y_j)$ can be interpreted as the excess amount of code resulting from the erroneous assumption that the two systems or processes are indeed independent, i.e., using $q(x_i, y_j) = p(x_i)p(x_j)$ instead of $p(x_i, y_j)$. The corresponding cross entropy is

$$D = \sum p(x_i, y_j) \log \frac{p(x_i, y_j)}{p(x_i)p(y_j)} \tag{1.181}$$

which is the transinformation of X and Y, a measure of the deviation from independence of the two processes:

$$T(X, Y) = H(X) + H(Y) - H(X, Y) \geq 0 \tag{1.182}$$

1.9 Application of POMCE to Continuous Variables

For a continuous random variable X in the range of 0 to infinity, let $q(x)$ be the prior probability density function (PDF) and $p(x)$ be the posterior PDF. Note that $q(x)$ is a prior estimate of $p(x)$. One can also generalize by taking X as a random variable vector

$X^T = (X_1, X_2, ...)$. Then $p(x)$ will be a multivariate PDF. For simplicity, we will deal with the univariate case and will make remarks as and when pertinent about the multivariate case. The objective is to determine $p(x)$ subject to specified constraints and given the prior PDF. Then the relative entropy or cross entropy of $p(x)$ relative to $q(x)$ is expressed as

$$D(p, q) = D(x) = \int_0^\infty p(x) \ln \frac{p(x)}{q(x)} dx \qquad (1.183)$$

where both the prior and the posterior PDFs satisfy

$$\int_0^\infty q(x)\, dx = 1 \qquad (1.184)$$

$$\int_0^\infty p(x)\, dx = 1 \qquad (1.185)$$

To derive $p(x)$ by applying POMCE, the following constraints can be specified:

$$\int_0^\infty g_r(x)p(x)\, dx = \overline{g_r} = C_r, \qquad r = 1, 2, ..., m \qquad (1.186)$$

The problem is to determine $p(x)$, subject to the information given by Eqs. (1.184), (1.185), and (1.186). Equation (1.183) does not lead to a unique $p(x)$ but does restrict the permissible densities that could be plausible. To obtain $p(x)$ uniquely, $D(p, q)$ is minimized, subject to Eqs. (1.184) to (1.186), and minimization can be achieved by using the method of Lagrange multipliers where the lagrangian function can be expressed as

$$L = \int_0^\infty p(x) \ln \frac{p(x)}{q(x)} dx + (\lambda_0 - 1)\left[\int_0^\infty p(x)dx - 1 \right] + \sum_{r=1}^m \lambda_r \left[\int_0^\infty g_r(x)p(x)dx - C_r \right] \qquad (1.187)$$

where $\lambda_r, r = 1, 2, ..., m$, are the Lagrange multipliers. Through minimization of the variation in L with respect to $p(x)$ or differentiating Eq. (1.187) with respect to $p(x)$ while recalling the calculus of variation and equating the derivative to zero, the following is obtained:

$$\int_0^\infty \delta p \left[\ln \frac{p(x)}{q(x)} + 1 + (\lambda_0 - 1) + \sum_{r=1}^m \lambda_r g_r(x) \right] dx = 0 \qquad (1.188)$$

Since δp is arbitrary, terms inside the square bracket must vanish. Therefore,

$$\ln\frac{p(x)}{q(x)}+1+(\lambda_0-1)+\sum_{r=1}^{m}\lambda_r g_r(x)=0 \qquad (1.189)$$

Equation (1.189) leads to the posterior PDF $p(x)$:

$$p(x)=q(x)\exp\left[-\lambda_0-\sum_{r=1}^{m}\lambda_r g_r(x)\right] \qquad (1.190)$$

The Lagrange multipliers are determined by using Eqs. (1.185) and (1.186).
    Let $\exp(-\lambda_0)=C$. Then Eq. (1.190) can be written as

$$p(x)=q(x)C\exp\left[-\sum_{r=1}^{m}\lambda_r g_r(x)\right] \qquad (1.191)$$

which can be written in a product or multiplicative form:

$$p(x)=Cq(x)\prod_{r=1}^{m}\exp[-\lambda_r g_r(x)] \qquad (1.192)$$

Equation (1.190) is also conveniently written as

$$p(x)=\frac{1}{Z(\lambda_0)}q(x)\exp\left[-\sum_{r=1}^{m}\lambda_r g_r(x)\right] \qquad (1.193)$$

where Z is called the *partition function* obtained by substituting Eq. (1.190) in Eq. (1.185):

$$Z=\exp\lambda_0=\int_{0}^{\infty}q(x)\exp\left[-\sum_{r=1}^{m}\lambda_r g_r(x)\right]dx \qquad (1.194)$$

The Lagrange multipliers are determined using Eq. (1.186) with known values of constraints:

$$-\frac{1}{Z}\frac{\partial Z}{\partial\lambda_r}=-\frac{\partial\ln Z}{\partial\lambda_r}=\overline{g_r(x)}=C_r \qquad (1.195)$$

Equation (1.195) does not lend itself to an analytical solution except for simple cases, but numerical solution is not difficult. Equation (1.190) shows that specific forms of $p(x)$ depend on the specification of $q(x)$ and $g_r(x)$. Here two simple cases of the prior $q(x)$ are dealt with.

Example 1.12. It is assumed that the domain of X is bounded and a prior density is uniform as

$$q(x) = \frac{1}{b-a} \qquad (1.196)$$

where a and b are limits of X for q. The mean constraint is defined as

$$\int_0^\infty x p(x) = \bar{x} \qquad (1.197)$$

Determine the PDF of X.

Solution. The PDF $p(x)$ is determined, subject to Eqs. (1.185) and (1.197) and given the prior as Eq. (1.196). In this case Eq. (1.190) yields

$$p(x) = \frac{1}{b-a} \exp(-\lambda_0 - \lambda_1 x) \qquad (1.198)$$

Substitution of Eq. (1.198) in Eqs. (1.185) and (1.197) yields

$$\int_0^\infty \frac{1}{b-a} \exp(-\lambda_0 - \lambda_1 x)\, dx = 1 \qquad (1.199)$$

$$\int_0^\infty \frac{1}{b-a} x \exp(-\lambda_0 - \lambda_1 x)\, dx = \bar{x} \qquad (1.200)$$

Solution of Eqs. (1.199) and (1.200) yields

$$(b-a)\exp\lambda_0 = \frac{1}{\lambda_1} \qquad (1.201)$$

and

$$(b-a)\,\bar{x}\,(\exp\lambda_0) = \frac{1}{\lambda_1^2} \qquad (1.202)$$

Equations (1.201) and (1.202) give

$$\frac{1}{\lambda_1} = \bar{x} \qquad (1.203)$$

Substitution of Eqs. (1.203) and (1.201) in Eq. (1.198) leads to

$$p(x) = \frac{1}{(b-a)\bar{x}} \exp(-x/\bar{x}) \qquad (1.204)$$

which is of exponential type.

Example 1.13.   The prior PDF is exponential, given as

$$q(x) = \frac{1}{k}\exp(-x/k) \tag{1.205}$$

and the constraints are the mean and mean log specified as

$$\int_0^\infty \ln x p(x)\, dx = \overline{\ln x} \tag{1.206}$$

Equation (1.190) in light of Eqs. (1.205), (1.197), and (1.204) becomes

$$p(x) = \frac{1}{k}\exp(-x/k)\exp(-\lambda_0 - \lambda_1 x - \lambda_2 \ln x) \tag{1.207}$$

which can be expressed as

$$p(x) = \frac{x^{-\lambda_2}}{k}\exp(-x/k)\exp(-\lambda_0 - \lambda_1 x) \tag{1.208}$$

Substitution of Eq. (1.208) in Eqs. (1.185), (1.197), and (1.206) yields, respectively,

$$\int_0^\infty \frac{x^{-\lambda_2}}{k}\exp\left(\frac{-x}{k}\right)\exp(-\lambda_0 - \lambda_1 x)\, dx = 1 \tag{1.209}$$

$$\int_0^\infty \frac{x^{1-\lambda_2}}{k}\exp\left(\frac{-x}{k}\right)\exp(-\lambda_0 - \lambda_1 x)\, dx = \overline{x} \tag{1.210}$$

$$\int_0^\infty \frac{x^{1-\lambda_2}}{k}\ln x \exp(-x/k)\exp(-\lambda_0 - \lambda_1 x)\, dx = \overline{\ln x} \tag{1.211}$$

Equations (1.209) to (1.211) can be solved numerically for the Lagrange multipliers.

1.10   Application of POMCE with Fractile Constraints

Let there be a random variable X with a specific value denoted as x. The domain or region of X can be finite or infinite $R: [x_0, x_{r+1}]$ of real numbers which can be partitioned into $r + 1$ subintervals $R_0 = [x_0, x_1), R_1 = [x_1, x_2), \ldots, R_r = [x_r, x_{r+1})$. Each subinterval is like a class interval. It is assumed that a prior PDF of X is given or assumed: $q(x)$, and its CDF is $Q(x)$. Note that $Q(x)$ is positive everywhere in domain R.

$$Q_i = \int_{R_i} q(x)\, dx \tag{1.212}$$

The objective is to determine the posterior PDF of X, denoted as $p(x)$, and CDF $P(x)$. If observations on the random variable are available, then information can be obtained in the form of fractile pairs $(x_i, P(x_i))$, $i = 1, 2, \ldots, r$. That is, the observed values of the random variable X can be arranged as a sequence of $r + 1$ elements, each of which is a realization of an *independent identically distributed* (IID) random variable with distribution $P(x)$ such that over interval or partition i

$$P(x_i) = P_i = \int_{R_i} p(x)\, dx = -\frac{i}{r+1} \qquad i = 1, 2, \ldots, r+1 \tag{1.213}$$

Equation (1.213) defines fractile constraints, which are analogous to plotting positions and can be denoted as

$$C_i: P(x_i) = P_i, \qquad P_i \in (0, 1) \qquad i \in n = \{0, 1, 2, \ldots, r\} \tag{1.214}$$

Now the constraints for individual intervals or interval constraints can be defined as

$$C_i = \int_{R_i} p(x)\, dx = P_i \qquad i \in n, \qquad \sum_{i=1}^{n} P_i = 1 \tag{1.215}$$

By using the indicator function I, Eq. (1.215) can be written in terms of expectation over the whole domain as

$$C_i = \int_R I_i(x) p(x)\, dx - P_i = 0 \qquad i \in n \tag{1.216}$$

where, I_i is the indicator function for the ith interval, defined as

$$I_i(x) = \begin{cases} 1 & \text{if } x \in R_i \\ 0 & \text{otherwise} \end{cases} \qquad \text{or} \qquad x_i \le x < x_{i+1} \tag{1.217}$$

The objective is to seek a posterior PDF $p(x)$ with CDF $P(x)$ which minimizes the cross entropy function D

$$D(p, q) = \int_R q(x) \log \frac{p(x)}{q(x)}\, dx = \int_R p(x) \log p(x)\, dx - \int_R p(x) \log q(x)\, dx \tag{1.218}$$

which satisfies the fractile constraints given by Eq. (1.212), (1.213), or (1.216). Equation (1.218) can be minimized, subject to the constraints defined by Eq. (1.216), using the method of Lagrange multipliers. To that end, one writes the lagrangian function L as

$$L = D(p, q) + \sum_{i=0}^{n} \lambda_i C_i = \int_R \{p(x)[\log p(x) - \log q(x)]\} + \sum_{i=1}^{N} \lambda_i [I_i(x) p(x) - P_i] \tag{1.219}$$

Recalling the Euler-Lagrange calculus of variation, differentiating Eq. (1.219) with respect to $p(x)$, and equating the derivative to zero, one gets

$$\frac{\partial L}{\partial w(x)} = 0 = \log w(x) - \log g(x) + 1 + \sum_{i=1}^{n} \lambda_i I_i(x) \tag{1.220}$$

Therefore,

$$p(x) = q(x)\exp\left[-1 - \sum_{i=1}^{n} \lambda_i I_i(x)\right] \qquad (1.221)$$

Let $U(x) = \exp\left[-1 - \sum_{i=1}^{n} \lambda_i I_i(x)\right]$, which is a step function that has a constant u_i in interval I_i

$$u_i = \exp(-1 - \lambda_i) \qquad i \in n \qquad (1.222)$$

Therefore,

$$p(x) = q(x)U(x) \qquad (1.223)$$

Equation (1.223) is the posterior PDF of X in terms of the prior PDF $q(x)$ and the Lagrange multipliers embedded in the step function $U(x)$. Substitution of Eq. (1.223) in Eq. (1.218) yields the minimum cross entropy value as

$$D_{\min} = \int_R p(x)\log U(x)dx = \int_R q(x)U(x)\log U(x)dx \qquad (1.224)$$

The minimum cross entropy can be interpreted as the directional distance between the prior PDF and the sample, suggesting that the prior PDF should be chosen so as to minimize the cross entropy.

Now the constraint Eq. (1.194) can be written with the use of Eqs. (1.223) and (1.222) as

$$\int_R I_i(x)q(x)U(x)\,dx = u_i\int_{R_i = I_i} q(x)\,dx = u_i Q_i \qquad (1.225)$$

where Q_i is the prior probability of $x \in I_i$, that is,

$$P_i = P(x_{i+1}) - P(x_i) \qquad (1.226)$$

The constraints given by Eq. (1.192) are therefore satisfied when

$$u_i = \frac{P_i}{Q_i} \qquad i \in n \qquad (1.227)$$

$$p(x) = q(x)\frac{P_i}{Q_i} \qquad i \in n \qquad (1.228)$$

Then with the use of Eqs. (1.227), (1.223), and (1.214) the minimum cross entropy becomes

$$D_{\min} = \sum_{i=1}^{n} P_i \log\frac{P_i}{Q_i} \qquad (1.229)$$

$$D_{\min} = \frac{1}{n+1}H(x;w) - \log(n+1) \qquad (1.230)$$

The minimum cross entropy–based PDF $p(x)$ given by Eq. (1.222) retains the piecewise form of the prior PDF $q(x)$ scaled over each interval I_i by the constant factor $u_i = P_i / Q_i$, given by Eq. (1.226). The minimum cross entropy is a function of the r prior distribution values Q_i and the fractile constraints P_i. It is independent of the values assumed by the prior distribution at any other point, and it is not directly dependent on the prior distribution $q(x)$.

The minimum value of the cross entropy given by Eq. (1.229) is equivalent to that for discrete distributions—corresponding to a set of $r + 1$ possible events with prior probabilities Q_i and minimum cross entropy probabilities P_i. With the use of Eqs. (1.218), (1.219), and (1.226) in Eq. (1.222) and integration

$$P(x) = P_i + u_i[Q(x) - Q_i] \qquad x \in I_i \qquad i \in n \tag{1.231}$$

Note that the posterior PDF $w(x)$ given by Eq. (1.223) is the prior PDF scaled over each interval that is, $p(x)$ has the piecewise form of $q(x)$ scaled by $U(x)$ over each interval I_i as given by Eq. (1.227). The prior PDF $q(x)$ does not influence the posterior PDF $p(x)$ at the fractiles.

Now the question of selecting a prior PDF arises. Since $q(x)$ serves as an interpolation function between fractiles and more importantly an extrapolation function outside the fractile range $[x_1, x_r]$, it should be selected judiciously. The selection of a prior PDF can be made with the use of Shannon entropy. Of course, the selection may vary from one problem to another.

1.11   Comparison between POME and POMCE

The die problem is considered to illustrate the comparison between POME and POMCE. A die has six faces, and it is assumed that nothing is known about the appearance of the faces. The objective is to determine the frequency distribution $\{f_1, f_2, \ldots, f_6\}$ that satisfies

$$\sum_{n=1}^{6} f_n = 1 \tag{1.232}$$

The entropy

$$H = -\sum_{n=1}^{6} f_n \ln f_n \tag{1.233}$$

is maximized, subject to Eq. (1.232). The solution in this case is

$$f_n = \frac{1}{6} \qquad n = 1, 2, \ldots, 6 \tag{1.234}$$

Now consider the case where another piece of information is known:

$$\sum_{n=1}^{6} n f_n = 4.5 \tag{1.235}$$

This is the expected value constraint. In this case the POME-based probability distribution is

$$f_n = \frac{\exp(-\lambda_1 n)}{\sum\limits_{i=1}^{6}\exp(-\lambda_1 n)} \tag{1.236}$$

To determine λ_1, substitution of Eq. (1.236) in Eq. (1.235) yields

$$\sum_{i=1}^{6} n\left[\frac{\exp(-\lambda_1 n)}{\sum\limits_{i=1}^{6}\exp(-\lambda_1 n)}\right] = 4.5 \tag{1.237}$$

Solution of Eq. (1.237) yields λ_1, and the probability distribution is obtained from Eq. (1.236). The solution is

$$f = (0.0543, 0.0788, 0.1142, 0.1654, 0.2398, 0.3475)$$

with $H = 1.6136$ bits.

Now this same die problem is addressed using POMCE. In the first case, the die is thrown N times. Let n_i be number of times the ith face shows up in N throws. Then

$$\sum_{i=1}^{6} n_i = N \tag{1.238}$$

Let the average number of spots be denoted by a:

$$\frac{n_1 + 2n_2 + 3n_3 + 4n_4 + 5n_5 + 6n_6}{N} = a = \frac{1}{N}\sum_{i=1}^{6} in \tag{1.239}$$

or

$$a = \frac{n_1}{N} + \frac{2n_2}{N} + \frac{3n_3}{N} + \frac{4n_4}{N} + \frac{5n_5}{N} + \frac{6n_6}{N}$$
$$= m_1 + 2m_2 + 3m_3 + 4m_4 + 5m_5 + 6m_6 \tag{1.240}$$

where m_i are the relative frequencies of faces in the experiment that are not known. Also

$$\sum_{i=1}^{6}\frac{n_i}{N} = \sum_{i=1}^{6} m_i = 1 \tag{1.241}$$

Thus, there are two constraints expressed as above.

First, the prior distribution is assumed to be uniform: $q_i = \frac{1}{6}$, $i = 1, 2, \ldots, 6$. Second, the only constraint is the total probability:

$$\sum_{i=1}^{6} p_i = 1 \tag{1.242}$$

and $p = p_i, i = 1, 2, \ldots, 6$, needs to be determined using POMCE. Now

$$1 = \int_M p(m) \left(\sum_{i=1}^{6} g_i m_i \right) dm = \int p(m) \left(\sum_{i=1}^{6} m_i \right) dm = 1 \tag{1.243}$$

From POMCE,

$$D = \int_M p(m) \ln \frac{p(m)}{q(m)} \, dm = \int_M p(m) \ln \frac{p(m)}{\frac{1}{6}} \, dm$$
$$= \int p(m) \ln p(m) \, dm - \int p(m) \ln 6 \, dm \tag{1.244}$$

This will yield $p = q = \left(\frac{1}{6}, \frac{1}{6}, \ldots, \frac{1}{6} \right)$ which is the same as the POME-based distribution.

Now the mean constraint is also specified:

$$\overline{C_1} = \overline{a} = \int_M p(m) \left(\sum_{i=1}^{6} g_{1,i} m_i \right) dm \tag{1.245}$$

in which $g_{1,i} = (1,2,3,4,5,6)$ and

$$\overline{C_2} = 1 = \int_M p(m) \left(\sum_{i=1}^{6} g_{2,i} m_i \right) dm \tag{1.246}$$

where $g_{2,i} = (1,1,1,1,1,1)$.

Two problems can be formulated: (1) to obtain a posterior PDF $p(m)$ and (2) to obtain the average values of the frequencies $\overline{m}$ that describe the experiment. Now $\overline{m}$ is the expected value of $p(m)$ and is the most probable frequency. Given are only two data points and six unknown model parameters, which are frequencies in the experiment, and hence the problem is ill posed.

For $\overline{m} = \overline{a} = 4.5$, lower limit as 0 and upper limit as 1, the POMCE-based distribution is

$$\overline{m} = (0.0672, 0.08260, 0.10554, 0.14533, 0.27322, 0.37549)$$

with entropy $H = 1.609$. In this case, the solution does not equal the POME-based solution.

1.12   Entropy Scaling

Let there be a continuous time series of random variable X and a time lag between two observations denoted by τ. If the time lag τ autocorrelation for the time series is denoted by $\rho(\tau)$, then under stationarity the scale of fluctuation (Vanmarcke, 1983) can be defined as

$$\theta = \int_{-\infty}^{\infty} \rho_X(\tau)d\tau \tag{1.247}$$

where $\rho(\tau)$ is the lag τ autocorrelation coefficient of the series (θ is now known as twice the integral time scale). For a discrete case, the scale of fluctuation can be represented as

$$\theta(k) = \left(1 + 2\sum_{\tau=1}^{k} \rho_\tau\right)\tau \qquad k \to \infty \tag{1.248}$$

For $\theta < \infty$, the process represented by the time series $X(t)$ is said to have a short-term memory (finite). This means the autocorrelation function rapidly converges to 0 (that is, exponentially). If $\theta = \infty$, the autocorrelation decays as a power law: $\rho(\tau) \sim \tau^{-\alpha}$ with $\alpha < 1$ when $\tau \to \infty$.

Let T be the time of aggregation of some characteristic of the process. For example, in the case of rainfall, the rainfall intensity (say, x) can be integrated over different durations to produce corresponding rainfall depths (say, y). Under stationary conditions, the variance of the aggregated characteristic y, denoted $\sigma^2(t)$, is directly related to the autocorrelation function of the original characteristic x, or $\rho_x(\tau)$, as

$$\sigma^2(T) = 2\sigma_x^{2}(t)\int_{0}^{T}(T-\tau)\rho_x(\tau)d\tau \tag{1.249}$$

For short-term memory, $\sigma^2(T) \sim T^1$ for $T \to \infty$. For long-term memory, $\sigma^2(T) \sim T^\delta$, with $\delta = 2 - \alpha (1 < \delta < 2)$. Near the origin $(T \to 0)$, a short-term memory process behaves as $\sigma^2(T) \sim T^2$, whereas a long-term memory process behaves as $\sigma^2 \sim T^\delta$, $1 < \delta < 2$. Taking into account these characteristics, Marini (2003, 2005) proposed three different regimes for $\sigma^2(T)$ of rainfall: (1) an inner region for $t = 10$ to 15 min where $\sigma^2(T) \sim T^2$; (2) a transition regime up 20 to 80 h where there exists a non-power-law scaling; and (3) a scaling regime onward where $\sigma^2(T) \sim T^\delta$ with $1 < \delta < 2$. Expressed another way, $\sigma^2(T)/T \sim T$ for small T, $\sigma^2(T)/T \sim T^0$ (constant) as $T \to \infty$ for the short-term memory process, and $\sigma^2(T)/T \sim T^{1-\alpha}$ as $T \to \infty$ for the long-term memory process.

For any τ, $I(\tau)$ denotes the value of τ autoinformation. Following Poveda (2011), the *scale of information* can be defined as

$$\psi = \int_{-\infty}^{\infty} I(\tau)\,d\tau \tag{1.250}$$

Similarly, for the discrete case at constant times τ,

$$\psi(k) = H + 2\sum_{\tau=1}^{k} I_\tau \qquad k \to \infty \tag{1.251}$$

where H is the Shannon entropy and I_τ denotes the value of lag-τ autocorrelation. Note that the scale of information is defined in a manner analogous to the scale of fluctuation.

If there are two series $X(t)$ and $Y(t)$, sampled at fixed τ, the lag τ mutual information (MI) can be defined as

$$I_{X,Y}(\tau) = I(X_t, Y_{t-\tau}) \tag{1.252}$$

Dividing Eq. (1.252) by the product of the standard deviations of X and Y yields the normalized mutual information

$$I^*_{X,Y}(\tau) = \frac{I_{X,Y}(\tau)}{\sqrt{S(X)}\sqrt{S(Y)}} \in (0, 1) \tag{1.253}$$

where $S(X)$ is the variance of X and $S(Y)$ is the variance of Y. If $X = Y$, $I_{X,Y}(\tau = 0) = S(X)$, then Eq. (1.250) becomes

$$I^*_{X,X}(\tau) = \frac{I_{X,Y}(\tau)}{S_X} \tag{1.254}$$

The scale of information for the continuous case can now be expressed as

$$\psi = \int_{-\infty}^{\infty} I^*_{X,X}(\tau)d\tau \tag{1.255}$$

and for the discrete case

$$\psi(k) = \left[1 + 2\sum_{\tau=1}^{k} I^*_\tau\right]\tau \qquad k \to \infty \tag{1.256}$$

In case of rainfall and stream flow, Poveda (2011) found that the Shannon entropy computed for different time scales seems to follow a power law

$$H(T) = cT^b \tag{1.257}$$

where b can take on different values for different locations. The scaling regime seems to hold up to a certain scale $T_{max\ ent}$ at which time entropy reaches a maximum value and then levels off ($b = 0$). The values of $T_{max\ ent}$ and entropy may vary. The question arises: why does the maximum entropy rise and why does it level off? First, in case of rainfall, larger time scale corresponds to more nonzero values. Poveda (2011) also found that the Shannon entropy and rescaled range seem to behave similarly at increasing aggregation times. The autoinformation function and sample autoinformation function exhibit an exponential decay at small lags and a power law at large lags.

Koutsoyiannis (2005) noted that the scaling exponent can be quantified using a simple indicator, such as the coefficient of variation. The state scaling behavior is different from normal distribution behavior and is characterized by an asymptotic J-shaped density function $f(x)$ and a scaling property of its survival function $F^*(x)$:

$$F^*(lx) = l^{-1/k} F^*(x) \tag{1.258}$$

for any $l > 0$ and a specified parameter $k > 0$. This yields

$$F^*(x) = \left(\frac{\lambda}{kx}\right)^{1/k} \qquad f(x) = \frac{1}{\lambda}\left(\frac{\lambda}{kx}\right)^{1/k} \qquad x = \frac{\lambda}{k}\left(\frac{T}{\delta}\right)^k \qquad (1.259)$$

where T = return period, δ = interarrival time, and λ = a parameter.

1.13   Partial and Constrained Entropy

For a given system, both the Shannon entropy and the Kullback-Leibler cross entropy can be expressed in fully constrained form. That is, they can incorporate constraints, regardless of their form or number, that are imposed on the system. On the other hand, the constrained entropy functions can be decomposed into partial or local constrained forms that are local maxima or minima. Thus, it permits one to evaluate the probability of any outcome or system state with respect to its corresponding maximum or minimum. Niven (2004) has discussed these entropy forms, and this section draws from his work for discussing these constrained entropy forms.

1.13.1   Partial Shannon Entropy

The *partial Shannon entropy* can be defined as

$$H_i = H(p_i) = -p_i \ln p_i \qquad (1.260)$$

where

$$H = \sum_{i=1}^{N} H_i = \sum_{i=1}^{N} H(p_i) = -\sum_{i=1}^{N} p_i \ln p_i \qquad (1.261)$$

where $\{p_i, i = 1, 2, \ldots, N\}$ is the probability distribution. The partial entropy can be employed to analyze the individual behavior of each system state i (Young, 1971; Arndt, 2001; Niven, 2004). Figure 1.9 shows the partial entropy. Note that $0 \ln 0 = \ln 0^0 = \ln 1 = 0$.

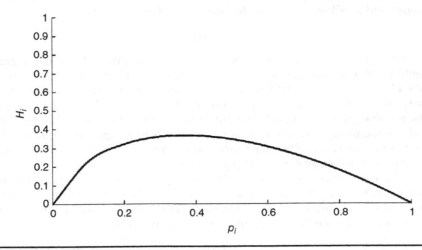

FIGURE **1.9**   Partial Shannon entropy

H_i is minimum at $p_i = 0$ as well as $p_i = 1$, that is, $H_i(1) = 0$ and is hence concave. The maximum of H_i occurs at $p_i = e^{-1}$, that is, $H_i(e^{-1}) = e^{-1}$. The maximum is constant. The partial entropy function does not incorporate the effect of constraints and is hence not constrained.

1.13.2 Constrained Shannon Entropy

Niven (2004) formulated the concept of the constrained entropy function which has a number of interesting attributes. To define the concept, we recall the constraints defined by the total probability (also called normalizing condition or natural constraint) given as

$$\sum_{i=1}^{N} p_i = \sum_{i=1}^{N} p(x_i) = 1 \tag{1.262}$$

and moment constraints given as

$$\sum_{i=1}^{N} g_j(x_i) p_i = \sum_{i=1}^{N} g_{ji} p_i = \overline{g_j(x)} = \overline{g_j} \tag{1.263}$$

where $\overline{g_j}$ is the expectation of function $g_j(x_i) = g_{ji}$, which can be chosen appropriately. For maximization of Eq. (1.262), the lagrangian function L can be constructed as

$$L = -\sum_{i=1}^{N} p_i \ln p_i - (\lambda_0 - 1)\left(\sum_{i=1}^{N} p_i - 1\right) - \sum_{j=1}^{m} \lambda_j \left[\sum_{i=1}^{N} p_i g_{ji} - \overline{g_j}\right] \tag{1.264}$$

and $\lambda_j, j = 1, \ldots, m$, are the Lagrange multipliers. For the purposes of entropy maximization, the lagrangian function can simply be written as

$$L = \sum_{i=1}^{N} p_i \ln p_i - (\lambda_0 - 1)\sum_{i=1}^{N} p_i - \sum_{j=1}^{m} \lambda_j \sum_{i=1}^{N} p_i g_{ji} \tag{1.265}$$

Differentiation of Eq. (1.265) with respect to p_i yields the maximum entropy-based probability distribution

$$p_i^* = \exp\left(-\lambda_0 - \sum_{j}^{m} \lambda_j g_{ji}\right) \qquad i = 1, 2, \ldots, N \tag{1.266}$$

where the superscript asterick (*) denotes that the probability distribution $P^* = \{p_i^*, 1 = 1, 2, \ldots, N\}$ is the probability distribution obtained by maximizing entropy. It can be considered as the equilibrium distribution.

Applying the total probability constraint Eq. (1.263) to Eq. (1.266), gives the generalized Maxwell-Boltzmann distribution, often employed in statistical thermodynamics and physics (Tribus, 1961):

$$P^* = \frac{\exp\left(-\sum_{j=1}^{m} \lambda_j g_{ji}\right)}{Z} \qquad i = 1, 2, \ldots, N \tag{1.267}$$

where

$$Z = \sum_{i=1}^{N} \exp\left(-\sum_{j=1}^{m} \lambda_j g_{ji}\right) \tag{1.268}$$

Function Z is called the *generalized partition function*. The Lagrange multipliers λ_j, $j = 0, 1, \ldots, m$, are obtained with the use of Eq. (1.264), albeit implicitly.

In a closed system subject to fluctuations between states, i.e., where p_i are able to change, $P = \{p_i, i = 1, 2, \ldots, N\}$ will tend toward a single unique, stationary equilibrium distribution $P^* = \{p_i^*, i = 1, 2, \ldots, N\}$. At the equilibrium position the Lagrange multipliers and constraints will become constant in time. By taking the logarithm of Eq. (1.268), the zeroth Lagrange multiplier can be expressed as

$$\lambda_0 = -\ln p_i^* - \sum_{j=1}^{m} \lambda_j g_{ji} \tag{1.269}$$

Substituting Eq. (1.269) in Eq. (1.265), the lagrangian L is obtained as

$$L = \sum_{i=1}^{N} (-p_i \ln p_i - p_i \ln p_i^* + p_i) \tag{1.270}$$

which can be cast as

$$H^c(P \mid P^*) = \sum_{i=1}^{N} -P_i \ln \frac{p_i}{p_i^*} + p_i \tag{1.271}$$

Function $H^c(P \mid P^*)$ is called the *constrained entropy function* with superscript c indicating constrained. Niven (2004) noted that Eq. (1.271) is a generic time-dependent form of the lagrangian function and therefore of the Shannon entropy function. Maximization of Eq. (1.260) yields

$$P = P^* = \{p_i = p_i^*, i = 1, 2, \ldots, N\} \tag{1.272}$$

Here Eq. (1.271) is fully constrained and can be used in place of the Shannon entropy. Probabilities p_i are fully constrained and may be considered independent. From Eq. (1.271), the constrained partial entropy or local entropy function can be expressed as

$$H_i^c(p_i \mid p_i^*) = -p_i \ln \frac{p_i}{p_i^*} + p_i \qquad i = 1, 2, \ldots, N \tag{1.273}$$

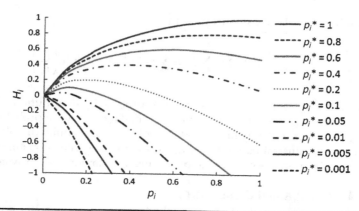

FIGURE 1.10 Plot of constrained partial entropy versus p_i for various values of p_i^*.

The constrained entropy H_i^c attains its maximum where $p_i = p_i^*$ for each local state i. Further, the maximum of $L = H^c$ is the same as that of all H_i^c.

The constrained partial entropy function as a function of p_i for various values of p_i^* is plotted in Fig. 1.10. It is seen that $H_i^c(0) = 0$ at $p_i = 0$ and $H_i^c(1) = 1 + \ln p_i^*$ at $p_i = 1$, and hence it is concave. In contrast with the Shannon entropy, the maximum of H_i^c occurs at $p_i = p_i^*$ as $H_i^c(p_i^*) = p_i^*$. This shows that H_i^c incorporates the effect of constraints on each state i. If $p_i^* = e^{-1}$, then H_i^c reduces to the partial Shannon entropy. Recalling that $\sum_{i=1}^{N} p_i = 1$, it can be shown that for each state $p_i^* < 1/N$ there can be one or several $p_i^* > 1/N$, $i \neq j$.

The constrained entropy H_i^c can take on positive or negative values and is 0 at $H_i^c(0) = 0$ and $H_i^c(p_i^*e) = 0$. That is, H_i^c and hence H^c do not follow the nonnegativity condition. Fortunately, this characteristic exerts no effect on the relative values of H_i^c and H^c, and hence does not influence comparisons between realizations of different entropies. The maximum of H^c is always nonnegative. The total constrained entropy $H^c = \sum_{i=1}^{N} H_i$ can be negative for systems away from equilibrium but is always positive.

Niven (2004) showed that the partial Shannon entropy constitutes the divide between the negative and positive values of H_i^c. Function $H_i^c(p_i \mid p_i^* < e^{-1})$ becomes negative over the interval $p_i^*e < p_i < 1$ for which the global minimum is $H_i^c(1)$. Function $H_i^c(p_i \mid p_i^* > e)$ is positive over the entire range of p_i for which the global minimum is $H_i^c(0)$.

1.13.3 Partial Cross Entropy

In a manner similar to the partial Shannon entropy, the partial Kullback-Leibler (K-L) cross entropy can be defined as

$$D_i = D(p_i \mid q_i) = p_i \ln \frac{p_i}{q_i} \tag{1.274}$$

The Kullback-Leibler cross entropy is then

$$D(P|Q) = \sum_{i=1}^{N} D_i = \sum_{i=1}^{N} D_i(p_i|q_i) = \sum_{i=1}^{N} p_i \ln\left(\frac{p_i}{q_i}\right) \qquad (2.275)$$

in which $Q = \{q_i, i = 1, 2, \ldots, N\}$ is the prior probability distribution. The cross entropy can be used as a probabilistic measure of the distance between probability distributions P and Q. The cross entropy minimization yields the probability distribution P that satisfies the given constraints while being closest to the given probability distribution Q. And D_i can be used to evaluate the probabilistic behavior of each system state i.

1.13.4   Constrained Cross Entropy

Following the same method as before, the lagrangian function for minimization of the K-L cross entropy, considering constraints given by Eqs. (1.262) and (1.263), can be expressed as

$$L = \sum_{i=1}^{N} p_i \ln\frac{p_i}{q_i} - (\lambda_0 - 1)\sum_{I=1}^{N} p_i - \sum_{j=1}^{m}\lambda_j \sum_{i=1}^{N} p_i g_{ji} \qquad (1.276)$$

Differentiation of Eq. (1.276) yields the minimum cross entropy–based probability distribution as

$$p_i^* = q_i \exp\left(-\lambda_0 - \sum_{j=1}^{m}\lambda_j g_{ji}\right) \qquad i = 1, 2, \ldots, N \qquad (1.277)$$

With the use of the natural constraint [Eq. (1.262)], Eq. (1.277) becomes

$$p_i^* = \frac{q_i \exp\left(-\lambda_0 - \sum_{j}^{m}\lambda_j g_{ji}\right)}{Z}, \; i = 1, 2, \ldots, N \qquad (1.278)$$

where

$$Z = \sum_{i=1}^{N} q_i \exp\left(\sum_{j=1}^{m}\lambda_j g_{ji}\right) \qquad (1.279)$$

Equation (1.278) is the Maxwell-Boltzmann cross entropy distribution, and Eq. (1.279) specifies the generalized cross entropy partition function.

The constrained cross entropy function can now be defined in a similar manner. It is assumed that the a priori probabilities q_i remain unaltered. Taking logarithm of Eq. (1.277), we obtain

$$-\ln q_i + \lambda_0 + \sum_{j=1}^{m}\lambda_j g_{ji} = -\ln p_i^* \qquad i = 1, 2, \ldots, N \qquad (1.280)$$

Substitution of Eq. (1.280) in Eq. (1.276) yields

$$L = D^c(P \mid P^*) = -H(P \mid P^*) = \sum_{i=1}^{N} p_i \ln \frac{p_i}{p_i^*} - p_i \qquad (1.281)$$

Equation (1.281) specifies the constrained cross entropy function which is the same as the constrained entropy function with the sign reversed. The partial form of this function becomes

$$D_i^c(p_i \mid p_i^*) = -H(p_i \mid p_i^*) = p_i \ln \frac{p_i}{p_i^*} - p_i \qquad (1.282)$$

Equation (1.282) shows that the a priori probabilities do not have any influence on the constrained cross entropy function or partial constrained cross entropy function.

Figure 1.11 shows a plot of the partial K-L cross entropy function D_i for various values of q_i. Similarly Fig. 1.12 shows a plot of the constrained partial K-L cross entropy function D_i^c. The minimum of D_i is attained when $D_i(p_i = e^{-1}q_i) = -e^{-1}q_i$. The negative cross entropy occurs in the triangular region between $p_i = 0$ and $p_i = q_i$, whereas the minimum of D_i^c occurs when $D_i^c(p_i = p_i^c) = -p_i^c$. The negative D_i^c occurs in the region between $p_i = 0$ and $p_i = ep_i^*$. The minimum D_i^c reflects the local equilibrium position.

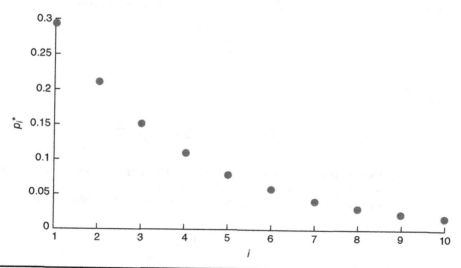

FIGURE 1.11 Plot of equilibrium probability.

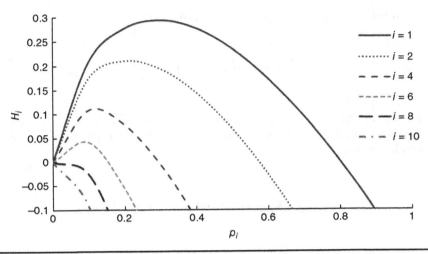

FIGURE **1.12**   Plot of constrained partial entropy versus p_i.

1.13.5   Continuous Constrained Entropy

For an n-dimensional system, the continuous form of the Shannon entropy can be expressed as

$$H = H(P) = -\iint \ldots \int p \ln p \; d^n x \tag{1.283}$$

where $p = p(x)$ is the probability density function of $X = \{x_1, x_2, \ldots, x_n\}$, and $d^n x = dx_1 \, dx_2 \ldots dx_n$. Integrations are done over the entire range of p in each coordinate. The natural and moment constraints for the continuous case can be written as

$$\iint \cdots \int p \; d^n x = 1 \tag{1.284}$$

and

$$\iint \cdots \int g_j(x) p \; d^n x = E[g_j(x)] = \overline{g_j} \tag{1.285}$$

Although the continuous case does not have a partial entropy function equivalent of the discrete case, $p \ln p$ from Eq. (1.283) is often employed.

Maximization of Eq. (1.283), subject to Eqs. (1.284) and (1.285), yields

$$p^*(x) = \frac{\exp\left(-\sum_{j=1}^{m} \lambda_j g_{ji}\right)}{Z} \tag{1.286}$$

where

$$Z = \iint \cdots \int \exp\left(-\sum_{j=1}^{m} \lambda_j g_{ji}\right) d^n x \tag{1.287}$$

in which Z is the continuous partition function and $p^*(x)$ is the POME-based PDF of X—the generalized Maxwell-Boltzmann equation.

Using Eq. (1.283), the continuous constrained entropy can be expressed as

$$L = H^c(p\,|\,p^*) = \iint \cdots \int \left(-p\ln\frac{p}{p^*} + p\right) d^n x \tag{1.288}$$

Following the discrete analogy, the partial entropy can be written as

$$H^c_{\text{local}}(p\,|\,p^*) = -p\ln\frac{p}{p^*} + p \tag{1.289}$$

1.13.6 Constrained Continuous Cross Entropy

The continuous form of the K-L cross entropy for an n-dimensional system can be written as

$$D = D(p\,|\,q) = \iint \cdots \int p\ln\frac{p}{q}\, d^n x \tag{1.290}$$

where p is the posterior distribution and q is the prior distribution. In a similar manner the K-L cross entropy, subject to Eqs. (1.284) and (1.285), yields the POMCE-based probability distribution as

$$p^*(x) = \frac{q\exp\left(-\sum_{j=1}^{m} \lambda_j g_{ji}\right)}{Z_q} \tag{1.291}$$

where

$$Z = \iint \cdots \int q\exp\left(-\sum_{j=1}^{m} \lambda_j g_{ji}\right) d^n x \tag{1.292}$$

Equation (1.292) is the cross entropy partition function. Using the generic constraint, the continuous constrained cross entropy can be written as

$$L_D = D^c(p\,|\,p^*) = -H^c(p\,|\,p^*) = \iint \cdots \int \left(p\ln\frac{p}{p^*} - p\right) d^n x \tag{1.293}$$

The partial constrained cross entropy is given as

$$D_{\text{local}}^c = -H_{\text{local}}^c(p \mid p^*) = p \ln \frac{p}{p^*} - p \qquad (1.294)$$

1.14 Entropy under Transformation of Variables

Let there be two random variables X and Y with probability density functions (PDFs) $f_X(x)$ and $f_Y(y)$, respectively, and cumulative distribution functions (CDFs) $F_X(x) = P(X \le x)$ and $F_Y(y) = P(Y \le y)$, respectively. Let $Y = (X - b)/a$ or $X = aY + b$. Then one can write

$$F_Y(y) = P(X \le ay + b) = F_X(ay + b) \quad \text{and} \quad f_Y(y) = af_X(ay + b) \qquad (1.295)$$

Following Saeb (2013), the Shannon entropy can be written as

$$H(Y) = -\int_{-\infty}^{\infty} f_Y(y) \log f_Y(y)\, dy = -\int_{-\infty}^{\infty} af_X(ay + b) \log[af_X(ay + b)]\, dy$$

$$= -\int_{-\infty}^{\infty} af_X(ay + b)[\log a + \log f_X(ay + b)]\, dy = -\log a - \int_{-\infty}^{\infty} f_X(w) \log f_X(w)\, dw \qquad (1.296)$$

$$= -\log a + H(X)$$

Likewise, let $Y = (X^c - b)/a$, where X is a positive random variable, that is, $x \ge 0$. Then, $X = (aY + b)^{1/c}$. In a similar fashion as above,

$$F_Y(y) = P(X \le (ay + b)^{1/c}) = F_X((ay + b)^{1/c})$$

and

$$f_Y(y) = \frac{a}{c}(ay + b)^{(1-c)/c} f_X((ay + b)^{1/c}) \qquad y > \frac{-b}{a} \qquad (1.297)$$

Now the Shannon entropy can be written as

$$H(Y) = -\int_{-\infty}^{\infty} f_Y(y) \log f_Y(y)\, dy$$

$$= -\int_{-b/a}^{\infty} \frac{a}{c}(ay + b)^{(1-c)/c} f_X((ay + b)^{1/c}) \log\left[\frac{a}{c}(ay + b)^{1/c} f_X((ay + b)^{1/c})\right] dy \qquad (1.298)$$

Let $w = (ay + b)^{1/c}$. Then Eq. (1.298) can be simplified as

$$H(Y) = -\int_{0}^{\infty} f_X(w)\left[\log \frac{a}{c} z^{1-c} f_X(w)\right] dw$$

$$= \log \frac{a}{c} + (c - 1)E[\log X] + H(X) \qquad (1.299)$$

Similarly, if X is negative, that is, $X \leq 0$, and $Y = (-X)^c - b/a$, then following the same method as above, the entropy of Y can be expressed as

$$H(Y) = -\log \frac{a}{c} + (c-1)E[\log(-X)] + H(X) \tag{1.300}$$

1.15   Types of Hydrologic Problems

The entropy concept has been applied to a wide range of problems in hydrology. These problems can be classified into three groups: (1) statistical, (2) physical, and (3) mixed. In the first group, problems are entirely statistical and do not invoke any physical laws. Examples of such problems are the derivation of frequency distributions for given constraints, estimation of frequency distribution parameters in terms of given constraints, evaluation of monitoring networks in space or/and time, flow forecasting, spatial and inverse spatial analysis, complexity analysis, clustering analysis, and so on.

In the second group, problems involve a kind of physical law in the form of a flux-concentration relation and a hypothesis on the cumulative probability distribution of either flux or concentration, depending on the problem at hand. Examples of such problems are infiltration in watersheds, soil moisture, unit hydrograph, stage-discharge relation, volume-peak discharge relation, and erosion and sediment yield.

The third group involves problems wherein relations are derived between entropy and design variables, and then relations between design variables and system characteristics are established. Examples of such problems include geomorphologic relations for elevation, slope, and fall, and reliability of water distribution systems. In hydrology the entropy theory has been applied to address problems from all three groups (Singh, 2010, 2011, 2013).

1.16   Hypothesis on Cumulative Distribution Function

For deriving physical relations using entropy theory, one deals with the formulation of a hypothesis on the cumulative distribution function of flux in terms of concentration, such as the CDF of infiltration of water in soils in terms of depth of infiltrated water or cumulative infiltration, the CDF of soil moisture in terms of depth, the CDF of discharge or runoff in terms of soil wetness, and so on. Now depending on the problem at hand, the CDF of X, say, flux, can be hypothesized as

$$F(x) = a + b \left(\frac{h}{h_{max}} \right)^c \tag{1.301}$$

where h_{max} is the maximum value of h, say, concentration, and a, b, and c are constants. A frequently used form of Eq. (1.301) is such that $a = 0$, $b = c = 1$ or $a = 1$, $b = -1$ and $c = 1$ or $a = 0$, $b = -1$ and $c = 1$. In the former case, Eq. (1.301) expresses that the probability of flux X being less than or equal to a given value x is equal to the concentration h at $X = x$

divided by the maximum concentration h_{max}. From a sampling standpoint, all values of h are equally likely to be sampled; in other words h follows a uniform distribution. This is a simple hypothesis and may not be entirely true.

If flux is a design variable, then time-averaged or space-averaged flux is assumed to be a random variable. Let X be flux and h be the associated concentration. In many problems, the time-averaged flux can be considered a random variable. For example, in open channel flow, time-averaged velocity at a given cross section can be considered a random variable. If X is space averaged, it can be considered a random variable. For example, the space-averaged infiltration capacity rate can be considered a random variable. Likewise, space-averaged soil moisture can be considered a random variable. This assumption is fundamental to the derivation of entropy-based relationships. How valid this assumption is may depend on the problem at hand. What is interesting, however, is that the final result does not seem to be greatly affected by the perfect or less-than-perfect validity of this assumption—a marvelous attribute of the entropy theory.

Example 1.13.   Using field measurements of soil moisture given below, plot Eq. (1.301) and show values of parameters a, b, and c.

Depth (cm)	Θ	Depth (cm)	Θ
15	0.86	15	0.894
25	0.803	20	0.86
35	0.509	30	0.675
45	0	45	0. 79

Soil Moisture Data for Wetting Phase

Solution.   For given soil moisture data, parameters of Eq. (1.301) were obtained: $a = 0.681$, $b = 1$, and $c = 1$. Thus, $F(\theta) = 1 - (Z/L)^a$. Values of computed F and observed F are given below and are plotted in Fig. 1.13.

Depth (cm)	Observed F(θ)	Estimated F(θ)
15	0.722	1
25	0.555	0.527
35	0.222	0.241
45	0.111	0
15	0.889	1
20	0.722	0.705
30	0.333	0.377
45	0.167	0

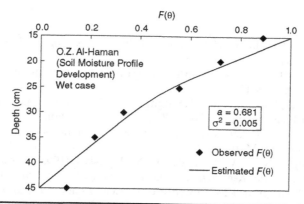

FIGURE 1.13   Comparison of computed and observed CDF for soil moisture in wet case.

Questions

1. An event occurs with a probability $p_1 = 0.1$. Compute its entropy. Likewise, six different events occur with probabilities, respectively, of $p_2 = 0.15$, $p_3 = 0.2$, $p_4 = 0.25$, $p_5 = 0.3$, $p_6 = 0.35$, and $p_7 = 0.4$. Compute the entropy for each value of p. What can be said about the uncertainty, degree of surprise, and information with respect to each event?

2. Follow the discharge values given in Table 1.3 for two stations A and B for 5 yr. Compute the value of information reduction.

3. Compute the joint entropy, using the discharge values in Question 2.

4. Compute the conditional entropy, using the data in Question 2.

5. Compute the transinformation, using the data in Question 2.

6. Select stage data for a gauging station. Compute the CDF for discharge as a function of stage or flow depth.

7. Using entropy theory, compute the probability density function of a random variable whose mean and variance are known. Compute the Lagrange multipliers and the partition function and provide their physical interpretations.

8. Using entropy theory, compute the probability density function of a random variable whose mean and the mean of logarithmic values of the variable are known. Compute the Lagrange multipliers and the partition function and provide their physical interpretations.

9. Using entropy theory, compute the probability density function of a random variable whose mean and variance are known. However, the variable takes on values over the semi-infinite domain, that is, 0 to infinity. Compute the Lagrange multipliers and the partition function and provide their physical interpretations.

10. Using the entropy theory, compute the probability density function of a random variable where nothing is known about the variable. The variable takes on values over a finite interval. Compute the Lagrange multipliers and the partition function, and provide their physical interpretations.

References

Arndt, C. (2001). *Information Measures: Information and Its Description in Science and Engineering*. Springer, Berlin.

Batty, M. (2010). Space, scale, and scaling in entropy maximizing. *Geographical Analysis*, vol. 42, pp. 395–421.

Fiorentino, M., P. Claps, and V. P. Singh, (1993). An entropy-based morphological analysis of river-basin networks. *Water Resources Research*, vol. 29, no. 4, pp. 1215–1224.

Granger, C. W. J. (1969). Investigating Causal Relations by Econometric Models and Cross-Spectral Methods. *Econometrica*, vol. 37, no. 3, pp. 424–438.

Harmancioglu, N. B., V. P. Singh, and N. Alpaslan, (1992). *Versatile Uses of the Entropy Concept in Water Resources. Entropy and Energy Dissipation in Water Resources*, edited by V. P. Singh and M. Fiorentino. Kluwer Academic Publishers, Dordrecht, The Netherlands, pp. 91–118.

Harmancioglu, N. B., Alpaslan, M. N., Singh, V. P., Fistikoglu, O., and Ozkul, S. D. (1999). Water Quality Monitoring Network Design. Kluwer Academic Publishers, Boston, pp. 290.

Jaynes, E. T. (1957a). Information theory and statistical mechanics, I. *Physical Reviews*, vol. 106, pp. 620–630.

Jaynes, E. T. (1957b). Information theory and statistical mechanics, II. *Physical Reviews*, vol. 108, pp. 171–190.

Jaynes, E. T. (1958). Probability theory in science and engineering. *Colloquium Lectures in Pure and Applied Science*, no. 4. Socony Mobil Oil Company, Dallas, Tex.

Jaynes, E. T. (1982). On the rationale of maximum entropy methods. *Proceedings of IEEE*, vol. 70, pp. 939–952.

Jaynes, E. T., (2003). *Probability Theory*. Cambridge University Press, Cambridge, England.

Kapur, J. N., and Kesavan, H. K. (1992). Entropy Optimization Principles and Their Applications. Academic Press, San Diego.

Kennell, M. B., Brown, R., and Abarbanel, H. D. I. (1992). Determining embedding dimension for phase-space reconstruction using a geometrical construction. *Physical Review A*, vol. 45, pp. 3403–3411.

Knuth, K. H. (2004). Measuring questions: relevance and its relation to entropy. In: Bayesian Inference and Maximum Entropy Methods in Science and Engineering, eds. R. Fischer, V. Dose, American Institute of Physics, Melville, New York.

Knuth, K. H. (2005). Optimal data-based binning for histograms. 30 pages and 6 figures. http://arxiv.org/pdf/physics/0605197. Accessed 19 August 2014.

Koutsoyiannis, D. (2005). Uncertainty, entropy, scaling and hydrological stochastics. 2. Time dependence of hydrological processes and time scaling. *Hydrological Sciences Journal*, vol. 50, no. 3, pp. 405–426.

Koutsoyiannis, D. (2013). Physics of uncertainty, the Gibbs paradox and indistinguishable particles. *Studies in History and Philosophy of Modern Physics*, vol. 44, pp. 480–489.

Kullback, S. (1959). *Information Theory and Statistics*. Wiley, New York.

Kullback, S., and Leibler, R. A. (1951). On Information and Sufficiency. *Annals of Mathematical Statistics*, vol. 22, pp. 79–86.

Lindley, D. V. (1956). On a measure of information provided by an experiment. *Annals of Mathematics*, vol. 27, pp. 986–1005.

Marini, M. (2003). On the correlation structure of point rainfall. *Water Resources Research*, vol. 39, no. 5, pp. 1128–1135.

Marini, M. (2005). Non-power-law scale properties of rainfall in space and time. *Water Resources Research*, vol. 41, W08413, doi: 10.1029/2004WR003822.

McGill, W. J. (1953). Multivariate transmission of information and its relation to analysis of variance. Report No. 32, Human Factors Operations Research Laboratories, MIT, Cambridge, Massachusetts.

McGill, W. J. (1954). Multivariate information transmission. *Psychometrica*, vol. 19, no. 2, pp. 97–116.

Niven, R. K. (2004). The constrained entropy and cross-entropy functions. *Physica A*, vol. 334, pp. 444–458.

Papalexiou, S. M., and Koutsoyiannis, D. (2012). Entropy based derivation of probability distributions: A case study to daily rainfall. *Advances in Water Resources*, vol. 25, pp. 51–57.

Pincus, S. M. (1991). Approximate entropy as a measure of system complexity. *Proceedings of National Academy of Science (USA)*, vol. 88, no. 6, pp. 2297–2301.

Poveda, G. (2011). Mixed memory, (non) Hurst effect, and maximum entropy of rainfall in the tropical Andes. *Advances in Water Resources*, vol. 34, pp. 243–256.

Richman, J. S., and Moorman, J. R. (2000). Physiological time-series analysis using approximate entropy and sample entropy. *American Journal of Physiology: Heart and Circulatory Physiology*, vol. 278, pp. 2039–2049.

Schreiber, T. (2000). Measuring information transfer. *Physical Review Letters*, vol. 85, no. 2, pp. 461–464.

Shannon, C. E. (1948). The mathematical theory of communications, I and II. *Bell System Technical Journal*, vol. 27, pp. 379–423.

Singh, V. P. (1997). The use of entropy in hydrology and water resources. *Hydrological processes*, vol. 11, pp. 587–626.

Singh, V. P. (2010). Entropy theory for earth science modeling, *Indian Geological Congress Journal*, vol. 2, no. 2, pp. 5–40.

Singh, V. P. (2011). Hydrologic synthesis using entropy theory: A review. *Journal of Hydrologic Engineering, ASCE*, vol. 16, no. 5, pp. 421–433.

Singh, V. P. (2013). *Entropy Theory in Environmental and Water Engineering*. Wiley, Sussex, United Kingdom.

Singh, V. P., and Fiorentino, M. (1992). A historical perspective of entropy applications in water resources. In: *Entropy and Energy Dissipation in Water Resources*, edited by V. P. Singh and M. Fiorentino, Kluwer Academic Publishers, Dordrecht, The Netherlands, pp. 21–61.

Tribus, M. (1961). *Thermostatics and Thermodynamics*. D. Van Nostrand, Princeton, N. J.

Tribus, M. (1969). *Rational Descriptions, Decisions and Designs*. Pergamon Press, New York.

Tsallis, C., Mendes, R. S., and Plastino, A. R. (1998). The role of constraints within non-extensive statistics. *Physica A*, vol. 261, pp. 534–554.

Vanmarcke, E. (1983). *Random Fields: Analysis and Synthesis*. MIT Press, Cambridge, Mass. Available from <http:www.princeton.edu/wilding/rbs/vanmEZ01htm>.

Young, J. F. (1971). *Information Theory*. Butterworths, London.

Further Reading

Abramowitz, M., and I. A. Stegun, (1970). *Handbook of Mathematical Functions*. Dover, New York.

Akaike, H., 1973. Information theory and an extension of the maximum likelihood principle. *Proceedings, 2nd International Symposium on Information Theory*, B. N. Petrov and F. Csaki, eds., Publishing House of the Hungarian Academy of Sciences, Budapest.

Amorocho, J. and Espildora, B. (1973). Entropy in the assessment of uncertainty in hydrologic systems and models. *Water Resources Research*, vol. 9, no. 6, pp. 1511–1522.

Awumah, K., Goulter, I., and Bhatt, S. K. (1991). Entropy-based redundancy measures in water distribution networks. *Journal of Hydraulic Engineering, ASCE*, vol. 117, no. 5, pp. 595–614.

Barbé, D. E., Cruise, J. F., and Singh, V. P. (1991). Solution of three-constraint entropy-based velocity distribution. *Journal of Hydraulic Engineering, ASCE*, vol. 117, no. 10, pp. 1389–1396.

Burg, J. P. (1975). Maximum entropy spectral analysis. Unpublished Ph.D. thesis, Stanford University, Palo Alto, Calif.

Burg, J. P. (1977). Maximum entropy spectral analysis. Paper presented at the 37th Annual Meeting of Society of Exploration Geophysics, Oklahoma City, Okla.

Chapman, T. G. (1980). *Equations for Entropy in Hydrologic Time Series*. Department of Civil Engineering, Faculty of Military Studies, University of New South Wales, Duntroon, N.S.W., Australia.

Chapman, T. G. (1986). Entropy as a measure of hydrological data uncertainty and model performance. *Journal of Hydrology*, vol. 85, pp. 111–126.

Chen, Y. C., and Chiu, C. L. (2004). A fast method of flood discharge estimation. *Hydrological Processes*, vol. 18, pp. 1671–1684.

Chen, Y. C., Hsu, Y. C., and Kuo, K. T. (2013). Uncertainties in the methods of flood discharge measurement. *Water Resources Management*, vol. 27, pp. 153–167.

Chiu, C. L. (1987). Entropy and probability concepts in hydraulics. *Journal of Hydraulic Engineering, ASCE*, vol. 113, no. 5, pp. 583–600.

Chiu, C. L. (1988). Entropy and 2-D velocity distribution in open channels. *Journal of Hydraulic Engineering, ASCE*, vol. 114, no. 7, pp. 738–756.

Chiu, C. L. (1989). Velocity distribution in open channel flow. *Journal of Hydraulic Engineering, ASCE*, vol. 115, no. 5, pp. 576–594.

Chiu, C. L. (1991). Application of entropy concept in open channel flow study. *Journal of Hydraulic Engineering, ASCE*, vol. 117, no. 5, pp. 615–628.

Chiu, C. L., and Chen, Y. C. (2003). An efficient method of discharge estimation based on probability concept. *Journal of Hydraulic Research, IAHR*, vol. 41, no. 6, pp. 589–596.

Chiu, C. L., and Chiou, J. D. (1986). Structure of 3-D flow in rectangular open channels. *Journal of Hydraulic Engineering*, vol. 112, no. 11, pp. 1050–1068.

Chiu, C. L., Hsu, S. M., and Tung, N. C. (2005). Efficient methods of discharge measurements in rivers and streams based on the probability concept. *Hydrological Processes*, vol. 19, pp. 3935–3946.

Chiu, C. L., Jin, W., and Chen, Y. C. (2000). Mathematical models of distribution of sediment concentration. *Journal of Hydraulic Engineering, ASCE*, vol. 126, no. 1, pp. 16–23.

Chiu, C. L., and Lin, G. F. (1983). Computation of 3-D flow and shear in open channels. *Journal of Hydraulic Engineering*, vol. 109, no. 11, pp. 1424–1440.

Chiu, C. L., and Murray, D. W. (1992). Variation of velocity distribution along non-uniform open channel flow. *Journal of Hydraulic Engineering, ASCE*, vol. 118, no. 7, pp. 989–1001.

Chiu, C. L., and Said, C. A. A. (1995). Maximum and mean velocities and entropy in open-channel flow. *Journal of Hydraulic Engineering*, vol. 121, no. 1, pp. 26–35.

Chiu, C. L., and Tung, N.C. (2002). Maximum velocity and regularities in open channel flow. *Journal of Hydraulic Engineering, ASCE*, vol. 128, no. 4, pp. 390–398.

Choo, T. H. (2000). An efficient method of the suspended sediment-discharge measurement using entropy concept. *Water Engineering Research*, vol. 1, no. 2, pp. 95–105.

Chou, C. M. (2012). Applying multiscale entropy to the complexity analysis of rainfall-runoff relationships. *Entropy*, vol. 14, pp. 945–957.

Dalezios, N. R., and Tyraskis, P. A. (1989). Maximum entropy spectra for regional precipitation analysis and forecasting. *Journal of Hydrology*, vol. 109, pp. 25–42.

Denbigh, K. G. (1989). Note on entropy, disorder and disorganization. *British Journal of Philosophy and Science*, vol. 40, pp. 323–332.

Denbigh, K. G., and Denbigh, J. S. (1985). *Entropy in Relation to Incomplete Knowledge*. Cambridge University Press, Cambridge United Kingdom.

Fass, D. M. (2006). *Human Sensitivity to Mutual Information*. Rutgers, The State University of New Jersey, New Brunswick.

Fast, J. D. (1968). *Entropy: The Significance of the Concept of Entropy and Its Application in Science and Technology*, Gordon and Breach, Science Publishers Inc., New York.

Fiorentino, M., Claps, P., and Singh, V. P. (1993). An entropy-based morphological analysis of river basin networks. *Water Resources Research*, vol. 29, no. 4, pp. 1215–1224.

Fougere, P. F., Zawalick, E. J., and Radoski, H. R. (1976). Spontaneous life splitting in maximum entropy power spectrum analysis. *Physics of the Earth and Planetary Interiors*, vol. 12, pp. 201–207.

Gallager, R. G. (1968). *Information Theory and Reliable Communication*. Wiley, New York.

Goda, T., Tai, S., and Yamane, A. T. (1981). Evaluation of thermodynamic efficiency of reverse osmosis process using entropy. *Water Research*, vol. 15, pp 1305–1311.

Harmancioglu, N. B. (1981). Measuring the information content of hydrological processes by the entropy concept. *Journal of Civil Engineering Faculty of Ege University*, Special Issue: *Centennial of Ataturk's Birth, Izmir*, Turkey, pp. 13–40.

Harmancioglu, N. B. (1997). The need for integrated approaches to environmental data management. In: *Integrated Approaches to Environmental Data Management Systems*, edited by N. B. Harmancioglu, M. N. Alsplan, S. D. Ozkul and V. P. Singh. Kluwer Academic Publishers, NATO ASI Series, 2, Environment, vol. 31, pp. 3–14.

Harmancioglu, N. B., and Alpaslan, N. (1992). Water quality monitoring network design. *Water Resources Bulletin*, vol. 28, no. 1, pp. 179–192.

Harmancioglu, N. B., Alpaslan, N., and Singh, V. P. (1992). *Application of the Entropy Concept in Design of Water Quality Monitoring Networks. Entropy and Energy Dissipation in Water Resources*, edited by V. P. Singh and M. Fiorentino. Kluwer Academic Publishers, Dordrecht, The Netherlands, pp. 283–302.

Harmancioglu, N. B., Alpaslan, N., and Singh, V. P. (1994). *Assessment of Entropy Principle as Applied to Water Quality Monitoring Network Design. Time Series Analysis in Hydrology and Environmental Engineering*, vol. 3, edited by K. W. Hipel, A. I. McLeopd, U. S. Panu, and V. P. Singh. Kluwer Academic Publishers, Dordrecht, The Netherlands, pp. 135–148.

Harmancioglu, N. B., Alpaslan, N. M., and Singh, V. P. (1998). Needs for environmental data management. In: *Environmental Data Management*, edited by N. B., Harmancioglu, V. P. Singh, and M. N. Alpaslan. Kluwer Academic Publishers, Dordrecht, The Netherlands, Chap. 1, pp. 1–2.

Harmancioglu, N. B., Alkan, A., Singh, V. P., and Alpaslan, N. (1996). Entropy-based approaches to assessment of monitoring networks. *Proceedings, IAHR International Symposium on Stochastic Hydraulics*, edited by Tickle, I. C., Goulter, C. C., Xu, S. A., Wasimi, and F. Bouchart, pp. 183–190.

Harmancioglu, N. B., Alpaslan, M. N., Whitfield, P., Singh, V. P., Literathy, P., Mikhailov, N., and Fiorentino, M. (1998). *Assessment of Water Quality Monitoring Networks—Design and Redesign.* Final Report to NATO, Brussels, Belgium.

Harmancioglu, N. B., Alpaslan, N., Ozkul, S. D., and Singh, V. P. (1997). *Integrated Approach to Environmental Data Management Systems.* NATO ASI Series, vol. 31. Kluwer Academic Publishers, Dordrecht, The Netherlands.

Harmancioglu, N. B., Alpaslan, M. N., Singh, V. P., Fistikoglu, O., and Ozkul, S. D. (1999). *Water Quality Monitoring Network Design.* Kluwer Academic Publishers, Boston.

Harmancioglu, N. B., Fistikoglu, O., and Singh, V. P. (1998). Modeling of environmental processes. In: *Environmental Data Management,* edited by N. B., Harmancioglu, V. P., Singh, and M. N. Alpaslan, Kluwer Academic Publishers, Dordrecht, The Netherlands, Chap. 9, pp. 213–242.

Harmancioglu, N. B., and Singh, V. P. (1990). *Design of Water Quality Networks.* Technical Report WRR14, Water Resources Program, Department of Civil Engineering, Louisiana State University, Baton Rouge.

Harmancioglu, N. B., and Singh, V. P. (1991). An information based approach to monitoring and evaluation of water quality data. In: *Advances in Water Resources Technology,* edited by G. Tsakiris, A. A. Balkema, Rotterdam, The Netherlands, pp. 377–386.

Harmancioglu, N. B., and Singh, V. P. (1998). Entropy in environmental and water resources. In: *Encyclopedia of Hydrology and Water Resources,* edited by D. R. Herschy. Kluwer Acdemic Publishers, Dordrecht, The Netherlands, pp. 225–241.

Harmancioglu, N. B., and Singh, V. P. (1999). On redesign of water quality networks. In: *Environmental Modeling,* edited by V. P. Singh, I. L. Seo, and J. H. Sonu, Water Resources Publications, Littleton, Colo., pp. 47–60.

Harmancioglu, N. B., and Singh, V. P. (2002). Data accuracy and validation. In: *Encyclopedia of Life Support Systems,* edited by A. Sydow. EOLSS Publishers Co., Ltd., Oxford, United Kingdom.

Harmancioglu, N. B., Singh, V. P., and Alpaslan, N. (1992). Design of water quality monitoring networks. In: *Geomechanics and Water Engineering in Environmental Management,* edited by R. N. Chowdhury. A. A. Balkema Publishers, Rotterdam, The Netherlands, Chap. 8, pp. 267–296.

Harmancioglu, N. B., Singh, V. P., and Alpaslan, N., eds. (1997). *Environmental Data Management.* Kluwer Academic Publishers, Dordrecht, The Netherlands.

Harmancioglu, N. B., and Singh, V. P. (1998). Entropy in environmental and water resources. *Encyclopedia of Hydrology and Water Resources,* edited by R. W. Hershey and R. W. Fairbridge, Kluwer Academic Publishers, Boston, pp. 225–241.

Harmancioglu, N. B., Yevjevich, V., and Obeysekara, J. T. B. (1986). Measures of Information transfer between variables. In: *Proceedings of the Fourth International Hydrology Symposium on Multivariate Analysis of Hydrologic Processes,* edited by H. W. Shen, Fort Collins, Colorado State University, July 1985, pp. 481–499.

Huang, F., Xia, Z., Zhang, N., Zhang, Y., and Li, J. (2011). Flow-complexity analysis of the upper reaches of the yangtze river, China. *Journal of Hydrologic Engineering,* vol. 16, no. 11, pp. 914–919. doi: 10.1061/(ASCE)HE.1943-5584.0000392

Jakulin, A., and Bratko, I. (2003). Quantifying and visualizing attribute interactions. Arxiv preprint cs.AI/0308002.

Jakulin, A., and Bratko, I. (2004). Testing the significance of attribute interactions. ICML 2004 Proceedings of the Twenty-First International Conference on Machine Learning, 52 pages, ACM, New York. doi: 10.1145/1015330.1015377.

Jaynes, E. T. (1957). Information theory and statistical mechanics, I. *Physical Review*, vol. 106, pp. 620–630.

Jaynes, E. T. (1979). Concentration of distributions at entropy maxima. Paper presented at the 19th NBER-NSF Seminar on Bayesian Statistics, Montreal, October 1979. In: E. T. Jaynes, *Papers on Probability, Statistics and Statistical Physics*, edited by R. D. Rosenkratz, D. Reidel Publishing Co., Boston, 1983, pp. 315–336.

Jaynes, E. T. (1982). On the rationale of maximum entropy methods. *Proceedings of the IEEE*, vol. 70, pp. 939–952.

Kaplan, S., and Garrick, B. J. (1981). On the quantitative definition of risk. *Risk Analysis*, vol. 1, no. 1, pp. 11–27.

Kawachi, T., Maruyama, T., Singh, V. P. (2001). Rainfall entropy for delineation of water resources zones in Japan. *Journal of Hydrology*, vol. 246, pp. 36–44.

Klir, G. J. (2006). *Uncertainty and Information: Foundations of Generalized Information Theory*, Wiley, New York.

Koutsoyiannis, D. (2005). Uncertainty, entropy, scaling and hydrological stochastics. 1. Marginal distributional properties of hydrological processes and state scaling. *Hydrological Sciences Journal*, vol. 50, no. 3, pp. 381–404.

Koutsoyiannis, D. (2011). Hurst-Kolmogorov dynamics as a result of extremal entropy production. *Physica A.*, vol. 390, no. 8, pp. 1424–1432.

Krasovskaia, I. (1997). Entropy-based grouping of river flow regimes. *Journal of Hydrology*, vol. 202, pp. 173–1191.

Krstanovic, P. F. and Singh, V. P. (1991a). A univariate model for long-term streamflow forecasting: 1. Development. *Stochastic Hydrology and Hydraulics*, vol. 5, pp. 173–188.

Krstanovic, P. F., and Singh, V. P. (1991b). A univariate model for long-term streamflow forecasting: 2. Application. *Stochastic Hydrology and Hydraulics*, vol. 5, pp. 189–205.

Krstanovic, P. F., and Singh, V. P. (1992a). Evaluation of rainfall networks using entropy: 1. Theoretical development. *Water Resources Management*, vol. 6, pp. 279–293.

Krstanovic, P. F., and Singh, V. P. (1992b). Evaluation of rainfall networks using entropy: II. Application. *Water Resources Management*, vol. 6, pp. 295–314.

Krstanovic, P. F., and Singh, V. P. (1993a). A real-time flood forecasting model based on maximum entropy spectral analysis: 1. Development. *Water Resources Management*, vol. 7, pp. 109–129.

Krstanovic, P. F., and Singh, V. P. (1993b). A real-time flood forecasting model based on maximum entropy spectral analysis: 2. Application. *Water Resources Management*, vol. 7, pp. 131–151.

Kullback, S., and Leibler, R. A. (1951). On information and sufficiency. *Annals of Mathematical Statistics*, vol. 22, pp. 79–86.

Kumar, P. T. K., and Sekimoto, H. (2009). Reduction of systematic uncertainty in the transmission measurement of iron using entropy based mutual information. *Radiation Measurement*, vol. 44, no. 2, pp. 168–172.

Lake, D. E., Richman, J. S., Griffin, M. P., and Moorman, J. R. (2002). Sample entropy analysis of neonatal heart rate variability. *American Journal of Physiology: Regulatory, Integrative and Comparative Physiology*, vol. 283, pp. 789–797.

Lathi, B. P. (1969). *An Introduction to Random Signals and Communication Theory*. International Textbook Company, Scanton, Penn.

Levine, R. D., and Tribus, M., eds. (1978). *The Maximum Entropy Formalism*, MIT Press, Cambridge, Mass.

Li, S. C., Zhou, Q. F., Wu, S. H., and Dai, E. (2009). Measurement of climate complexity using sample entropy. *International Journal of Climatology*, vol. 26, no. 15, pp. 2131–2139.

Li, A., and Zhang, Y. K. (2012). Multiscale entropy analysis of Mississippi River Flow. *Stochastic Environmental Research and Risk Analysis*, vol. 22, pp. 507–512.

Maruyama, T., and Kawachi, T. (1998). Evaluation of rainfall characteristics using entropy. *Journal of Rainwater Catchment Systems*, vol. 4, no. 1, pp. 7–10.

Maruyama, T., Kawachi, T., and Singh, V. P. (2005). Entropy-based assessment and clustering of potential water resources availability. *Journal of Hydrology*, vol. 309, pp. 104–113.

Miller, G. A. (1953). What is information measurement? *American Psychologist*, vol. 8, pp. 3–11.

Mogheir, Y., and Singh, V. P. (2002). Application of information theory to groundwater quality monitoring networks. *Water Resources Management*, vol. 16, no. 1, pp. 37–49.

Mogheir, Y., de Lima, J. L. M. P., and Singh, V. P. (2003). Assessment of spatial structure of groundwater quality variables based on the entropy theory. *Hydrology and Earth System Sciences*, vol. 7, no. 5, pp. 707–721.

Mogheir, Y., de Lima, J. L. M. P., and Singh, V. P. (2004a). Characterizing the spatial variability of groundwater quality using the entropy theory: 1, Synthetic data. *Hydrological Processes*, vol. 18, pp. 2165–2179.

Mogheir, Y., de Lima, J. L. M. P., and Singh, V. P. (2004b). Characterizing the spatial variability of groundwater quality using the entropy theory: 2, Case study from Gaza Strip. *Hydrological Processes*, vol. 18, pp. 2579–2590.

Mogheir, Y., de Lima, J. L. M. P., and Singh, V. P. (2005). Assessment of informativeness of groundwater monitoring in developing regions (*Gaza Strip Case Study*). *Water Resources Management*, vol. 19, pp. 737–757.

Mogheir, Y., Singh, V. P., and de Lima, J. L. M. P. (2006). Spatial assessment and redesign of groundwater quality network using the entropy theory. *Hydrogeology Journal*, vol. 14, pp. 700–712.

Moramarco, T., Barbetta, S., Melone, F., and Singh, V. P. (2005). Relating local stage and remote discharge with significant lateral inflow. *Journal of Hydrologic Engineering, ASCE*, vol. 10, no. 1, pp. 58–69.

Moramarco, T., Saltalippi, C., and Singh, V. P. (2002). Estimating the cross-sectional mean velocity in natural channels by the entropy approach. In: *Water Resources Planning and Management*, edited by M. Al-Rashid, V. P. Singh, and M. M. Sherif. A. A. Balkema, Rotterdam, The Netherlands, pp. 435–449.

Moramarco, T., Saltalippi, C., and Singh, V. P. (2004). Estimating the cross-sectional mean velocity in natural channels using Chiu's velocity distribution. *Journal of Hydrologic Engineering, ASCE*, vol. 9, no. 1, pp. 42–50.

Moramarco, T., Saltalippi, C., and Singh, V. P. (2009).Velocity profiles assessment in natural channels during high floods. In: *Water, Environment, Energy and Society*, vol. 2: *Statistical and Systems Analysis Techniques*, Proceedings of WEES-09, edited by S. K. Jain, V. P. Singh, V. Kumar, R. Kumar, R. D. Singh, and K. D. Sharma, Allied Publishers, New Delhi, India, pp. 780–786.

Moramarco, T., and Singh, V. P. (2001). Simple method for relating local stage and remote discharge. *Journal of Hydrologic Engineering, ASCE*, vol. 6, no. 1, pp. 78–81.

Moran, P. A. P. (1969). Statistical inference with bivariate gamma distributions. *Biometrika*, vol. 56, no. 3, pp. 627–634.

Padmanabhan, G., and Rao, A. R. (1986). Maximum entropy spectra of some rainfall and river flow time series from southern and central India. *Theoretical and Applied Climatology.* vol. 37, pp. 63–73.

Padmanabhan, G., and Rao, A. R. (1988). Maximum entropy spectral analysis of hydrologic data. *Water Resources Research,* vol. 24, no. 9, pp. 1519–1533.

Pal, N. R., and Pal, S. K. (1991a). Entropy: A new definition and its applications. *IEEE: Transactions on Systems, Man, and Cybernetics,* vol. 21, no. 5, pp. 1260–1270.

Pal, N. R., and Pal, S. K. (1991b). Image model, Poisson distribution and object extraction. *International Journal of Pattern Recognition and Artificial Intelligence,* vol. 5, no. 3, pp. 459–483.

Pal, N. R., and Pal, S. K. (1992). Some properties of the exponential entropy. Informational Sciences, vol. 66, pp. 119–137.

Pal, N. R., and Bezdek, J. C. (1994). Measuring fuzzy uncertainty. IEEE Transactions on Fuzzy Systems, vol. 2, no. 2, pp. 107–118.

Pincus, S. M. (1995). Approximate entropy (ApEn) as a complexity measure. *Chaos,* vol. 5, no. 1, pp. 110–117.

Pincus, S. M., and Viscarello, R. R. (1992). Approximate entropy: A regulatory measure for fetal heart rate analysis. *Obstetrics and Gynecology,* vol. 79, pp. 249–255.

Prigogine, I. (1989). What is entropy? *Naturwissenschaften,* vol. 76, pp. 1–8.

Rao, A. R., Padmanabhan, G., and Kashyap, R. L. (1980). Comparison of recently developed methods of spectral analysis. *Proceedings, Third International Symposium on Stochastic Hydraulics,* Tokyo, Japan, pp. 165–175.

Saeb, A. (2013). On extreme value theory and information theory. Unpublished Ph.D. thesis, University of Mysore, Mysore, India.

Shannon, C. E. (1948). A mathematical theory of communications, I and II. *Bell System Technical Journal,* vol. 27, pp. 379–443.

Shannon, C. E., and Weaver, W. (1949). *The Mathematical Theory of Communication.* University of Illinois Press, Urbana.

Shore, J. E. (1979). Minimum cross-entropy spectral analysis. NRL Memorandum Report 3921, Naval Research Laboratory, Washington, D.C.

Singh, V. P. (1998). The use of entropy in hydrology and water resources. *Hydrological Processes,* vol. 11, pp. 587–626.

Singh, V. P. (1998). *Entropy-Based Parameter Estimation in Hydrology.* Kluwer Academic Publishers, Boston.

Singh, V. P., and Fiorentino, M., eds. (1992). *Entropy and Energy Dissipation in Water Resources.* Kluwer Academic Publishers, Dordrecht, The Netherlands.

Srinivas, S., (2008). A review of multivariate mutual information. University of Notre Dame, unpublished manuscript.

Tai, S., and Goda, T. (1980). Water quality assessment using the theory of entropy. In: *River Pollution Control,* edited by M. J. Stiff. Ellis Horwood Limited, Chichester, United Kingdom.

Tai, S., and Goda, T. (1985). Entropy analysis of water and wastewater treatment processes. International Journal of Environmental Studies, vol. 25, pp. 13–21.

Tribus, M. (1969). *Rational Description: Decision and Designs.* Pergamon Press, New York.

Tsallis, C. (1988). Possible generalization of Boltzmann-Gibbs statistics. *Journal of Statistical Physics,* vol. 32, no. ½, pp. 479–487.

Ulrych, T., and Clayton, R. W. (1976). Time series modeling and maximum entropy. Physics of the Earth and Planetary Interiors, vol. 12, no. 2–3, pp. 188–200.

Xia, R. (1997). Relation between mean and maximum velocities in a natural river. *Journal of Hydraulic Engineering*, vol. 123, no. 8, pp. 720–723.

Yan, K., Cai, H., and Song, S. (2011). A measure of hydrological system complexity based on sample entropy. *IEEE*, vol. 1, pp. 470–473.

Zhou, Y., Zhang, Q., Li, K., and Chen, X. (2012). Hydrological effects of water reservoirs on hydrological processes in the East River (China) Basin: Complexity evaluation based on the multi-scale entropy analysis. *Hydrological Processes*, vol. 26, pp. 3253–3262, doi: 10.1002/hyp.8406.

CHAPTER 2
Maximum Entropy Production Principle

In recent years, optimality-based principles have been proposed for estimating hydrologic model parameters and modeling hydrologic processes. Examples of such principles are the principle of maximum energy dissipation (Zehe et al., 2010), principle of minimum water stress (Caylor et al., 2009; Porporato et al., 2001; Rodriguez-Iturbe et al., 1999), principle of maximum transpiration and minimum water stress and oxygen stress (Brolsma and Bierkens, 2007), principle of net carbon profit (Schymanski et al., 2009), principle of maximum entropy production (MEP) (McDonnell et al., 2007; Kleidon and Schymanski, 2008; Kleidon, 2009, 2010a, b; Zehe and Sivapalan, 2009; Schaefli et al., 2011), principle of maximum power production (Westhoff and Zehe, 2013), principle of minimum energy expenditure (Rodriguez-Iturbe et al., 1992a, b; Rinaldo et al., 1992), principle of minimum energy dissipation rate (Singh et al., 2003a, b; Yang, 1971), and principle of minimum frictional dissipation (West et al., 1997). On the other hand, maximization principles state that systems organize themselves such that they maximize power dissipation or entropy production. The optimality principles and their application provide independent criteria for parameter selection and their need for model calibration. Different systems optimize their behavior or response, depending on different objective functions, such as maximizing water storage (Rodriguez-Iturbe et al., 1999; Porporato et al., 2001), maximizing transpiration or minimizing water stress (Brolsma and Bierkens, 2007), and maximizing net carbon profit (Schymanski et al., 2009).

Power represents the rate at which work is performed through time and the generation of free energy. In steady state, the free energy is dissipated into heat, producing entropy. This maximization is also connected to minimization. For example, when the dissipation of the energy of a moving fluid is minimized, the ability of the fluid to transport matter along a gradient is maximized. On one hand, gradients give the power to derive the fluid dynamics, but the same dynamics dissipate the gradients.

The basic concept underlying optimality principles is that a natural system, under given external constraints, organizes itself in such a manner that its response is optimal during steady-state conditions. The objective of this chapter is to discuss the principle of maximum entropy production (MEP) and its application to hydrologic modeling.

2.1   Hydrologic Systems

Systems can be isolated, closed, or open. At the global scale, the hydrologic cycle can be regarded as a closed system. At small scales, such as a watershed, it is an open system. According to the second law of thermodynamics, any flow or process increases the entropy of the universe. For an isolated system the entropy increases as the system tends toward a state of maximum entropy. In this state, the entropy production vanishes. However, most systems are open, not isolated, and hence an exchange of entropy occurs between these systems and their surroundings. The open systems can achieve a steady state when their entropy production equals the net entropy exchange.

2.2   Irreversibility of System Processes and Entropy Production

Geophysical systems in general and hydrologic systems in particular are dominated by irreversible processes and produce entropy (Piexoto et al., 1991; Goody, 2000; Kleidon and Lorenz, 2005). Some examples follow. The hydrologic cycle is intimately linked with the irreversible nature of these processes. Solar radiation, once absorbed at the surface, cannot be re-emitted at the same wavelength but is emitted as longwave radiation at much cooler temperatures of the earth's surface in comparison with the hot emission temperature of the sun. Hence, absorption of solar radiation at prevailing temperatures of the earth's surface is irreversible. The process of evaporation from the earth's surface into the earth's atmospheric boundary layer cannot be reversed unless the evaporated moisture is cooled to saturation, usually through lifting by atmospheric motion. Atmospheric motion is driven by the gradient in heating, and the associated generation of kinetic energy is dissipated by friction into heat. It is not possible to convert the dissipated heat back to kinetic energy. The entry of water into the soil at its surface, called *infiltration*, is an irreversible process. Once the water enters the soil, it cannot get back to the surface, except for a small quantity through exfiltration or as water vapor by evaporation. Likewise, once snow melts, it cannot become snow again. Of course, if there is a sudden drop in temperature, the melt water can become ice but cannot assume the same form or shape. As water moves as overland flow or channel flow down the slope, it does not get back to where it started. The movement of water in this manner is an irreversible process. If a certain amount of pollutant is injected into channel flow, it will be transported by the flow velocity and diffusion, and the pollutant transport is an irreversible process. A nonequilibrium thermodynamic framework can be employed to describe these processes.

Now consider the state of thermodynamic equilibrium (TE) as a reference state. The atmospheric water vapor attains thermodynamic equilibrium with a water surface, when relative humidity reaches 100 percent. At this stage there is no net exchange of moisture between the atmospheric water vapor and the water surface, because the net evaporation balances out net condensation. On the other hand, evaporation of water into unsaturated atmosphere is irreversible and produces entropy, making the atmosphere closer to TE. The unsaturated atmosphere is maintained by atmospheric motion which acts as a dehumidifier (Pauluis, 2005). The motion lifts the air which cools and its vapor gets closer to saturation. The condensation of supersaturated vapor is irreversible and produces entropy, and so does diffusion of water vapor. This shows that the irreversibility of hydrologic cycle processes is closely connected to the power of atmospheric circulation.

The irreversibility of earth system processes is associated with the production of entropy in steady state. When a system is maintained away from thermodynamic equilibrium, processes occur in order to bring the system back to thermodynamic equilibrium, and they produce entropy in doing so. The magnitude of entropy production characterizes the extent to which systems are maintained away from thermodynamic equilibrium. To maintain this system state, the conditions at the system boundary allow the entropy produced by the processes within the system to be exported to the surroundings. The steady state is characterized by a comparatively low entropy production within the system and equivalently by a net export of entropy to the surroundings.

2.3 Laws of Thermodynamics

The first law of thermodynamics is the energy conservation law, which states that the sum of all changes of energy within the system equals the energy exchanges with the surroundings. Let U represent the total energy of the system and dU the change in U; Q represent external heating and dQ the change in Q; and W represent the work done by the system and dW the change in W. The first law of thermodynamics can be expressed as

$$dU = dQ - dW \qquad (2.1)$$

From the perspective of the total energy of the system, dW is transformed into another form of energy. For example, when there is a change in potential energy, the amount of work done to produce motion is dW, and the motion increases the kinetic energy by the amount dW at the expense of potential, which is reduced by $-dW$. In this process of conversion, the total energy of the system does not change, but the form of energy is changed. When contributions of different forms of energy to the total energy of the system are considered, the first law restricts the energy conversions within the system and dW indicates the conversion of heat to some other form of energy. Specifically, dW is the work performed to generate a gradient in another variable under the laws of conservation of mass, momentum, and energy. For example, when work is performed to accelerate mass, motion is generated, that is, a velocity gradient is created at the expense of the geopotential gradient. Extending this reasoning further, we can state that the dynamics within the system essentially deal with the conversion of gradients of another form of energy; that is, a buildup and a depletion of gradients of different types occur. These gradients are associated with energy that is available to perform work or free energy; in other words, work is derived from these gradients. Since a gradient associated with a certain form of energy can be used to generate another gradient, the power P associated with this conversion can be denoted as

$$\frac{dW}{dt} = P \qquad (2.2)$$

The first law then shows that the sum of all energy conversions between different forms of energy within a system must equal the energy exchanges with the surroundings.

The second law of thermodynamics states that entropy cannot be consumed, only produced. This holds during irreversible processes that cannot be reversed in time. Following Kleidon et al. (2013), the change in internal energy dU can be expressed as

$$dU = d(TS) - d(pV) + \sum_i d(M_i \phi_i) \tag{2.3}$$

in which U is the internal energy (J), T is the absolute temperature (K), S is the entropy (J K^{-1}), p is the pressure (N m^{-2}), V is the volume (m^3), M_i is the mass of matter within the system (kg), and ϕ_i is the corresponding chemical potential or energy level (m^2 s^{-2}). Under steady state, $dU = 0$, that is, the internal energy is constant; $dT = 0$ (also temperature gradients do not generate water fluxes), $dp = 0$, $dV = 0$, and $d\phi_i = 0$, and incoming fluxes balance outgoing fluxes. Then Eq. (2.3) can be simplified as

$$T \, dS + \sum_i \phi_i \, dM_i = 0 \tag{2.4}$$

Dividing Eq. (2.4) by $T \, dt$, we obtain

$$\frac{dS}{dt} = \frac{1}{T} \sum_i (-\phi_i) \frac{dM_i}{dt} \tag{2.5}$$

The quantity $\phi_i \, (dM_i/dt)$ is the flux times gradient and equals power (kg s^{-1} m^2 s^{-2}). Equation (2.5) states that the rate of entropy production is equal to the power divided by temperature. This implies that maximization of one is the same as maximization of the other.

For an isolated system in which there is no exchange of mass and energy with the surroundings, the second law of thermodynamics can be expressed as

$$\frac{dS}{dt} = w \tag{2.6}$$

where S is the entropy of the system (ML^2 T^{-2} θ^{-1}, θ = temperature), w is the rate of entropy production or entropy production flux (ML^2 T^{-3} θ^{-1}), t is time, and dS/dt expresses the change in entropy with time. The entropy production is associated with irreversible processes, such as diffusion, that occur within the system. As time increases, $dS/dt \rightarrow 0$, w tends to zero, and S will reach a maximum value and the system will tend toward a state of maximum entropy—steady state.

For an open system, by taking into account exchanges of entropy the second law of thermodynamics can be expressed as (Kondepudi and Prigogine, 1998)

$$\frac{dS}{dt} = w - w_{\text{net}} \tag{2.7}$$

in which w_{net} denotes the net entropy exchange flux (or rate) associated with exchanges of mass and energy (ML^2 T^{-3} θ^{-1}). At steady state, $dS/dt = 0$ and $w = w_{\text{net}}$; that is, the internally produced entropy is balanced by the net export of entropy to the surroundings.

The second law points to the direction in which the processes evolve and is reflected in the depletion of gradients. In other words, natural processes occur in such a way that they dissipate their driving gradients. For example, velocity gradients are dissipated by friction, thermal gradients are dissipated by heat conduction, vapor pressure gradients are dissipated by evaporation, concentration gradients are dissipated by dispersion, and so on.

2.4   Principle of Maximum Entropy Production

The principle of maximum entropy production (MEP) states that complex systems far from thermodynamic equilibrium organize in such a way that the rate of entropy production is maximized in steady state. The MEP principle provides a mechanism to understand optimality in a wide range of geophysical systems. Paik and Kumar (2010) state that MEP has a physical basis, although Dewar (2009) states that it is a statistical concept and the system state corresponding to it is just the most probable one. The MEP principle entails a flux and a gradient where the gradient drives the flux but the same flux dissipates the gradient.

Entropy production is caused by several irreversible processes. For example, in the atmospheric system these processes may include evaporation, condensation of supersaturated vapor, water vapor, diffusion and expansion, as well as reevaporation and frictional dissipation of raindrops (Pauluis et al., 2000; Pauluis and Held, 2002; Kleidon, 2008; Goody, 2000). Kleidon (2008) reported that of the total entropy production of $w = 23 \ mW \ m^{-2} \ T^{-1}$, the entropy production due to evaporation into an unsaturated atmosphere contributes about $8 \ mW \ m^{-2} \ K^{-1}$, subject to strong geographic and seasonal variations. The diffusion in w is due to other processes. Hydrologic fluxes maximize entropy production. Wang et al. (2004, 2007) showed that evaporation is maximized under given environmental constraints in the field.

2.5   Application of Maximum Entropy Production Principle

When the maximum entropy production principle is applied in hydrology, certain issues need to be addressed. First, MEP considers entropy production in steady state, but the time interval for achieving such a state in hydrology is often difficult to determine or is beyond the realm of most practical applications. Second, the variability of exchange fluxes and gradients is not taken into account. As shown by Porporato et al. (2004), there is evidence that different variability regimes can lead to different hydrologic flux partitioning.

Third, hydrologic fluxes interact with other dissipative processes. For example, hydrologic fluxes of flow on the land surface are strongly coupled with those in the vadose zone. Similarly, hydrologic processes on land and vegetation processes are strongly coupled. Evapotranspiration is controlled by vegetation type, density, root biomass and depth, leaf area, and stomatal condensation. Carbon uptake by photosynthesis, which transforms low-entropy solar radiation into chemical free energy associated with carbohydrates, and respiration of carbohydrates into heat are also dissipative processes but are of a biogeochemical nature. When MEP is applied to the vegetation activity, the implication is the maximization of the mean gross carbon uptake (Kleidon, 2004a,b).

Since hydrologic fluxes are affected by vegetation activity, which, in turn, depends on water availability, this might lead to conflicting predictions by MEP of hydrologic and biogeochemical fluxes. However, hydrological and biogeochemical fluxes have different time scales and hence should be optimized at these time scales. For example, the redistribution of soil moisture occurs at a time scale of days, whereas the maximization of vegetation activity through adaptation processes in ecohydrological functioning occurs most likely at a much greater scale of years or longer. The interaction of quick hydrological fluxes and slow ecohydrological processes needs to be better understood and validated.

The above issues notwithstanding, MEP has advantages: (1) It is based on physics; (2) it is applicable to physical exchange processes of heat and matter as well as chemical and biogeochemical processes, since they are thermodynamic in nature; (3) MEP is objective with respect to choosing an objective function and subsequent optimization; (4) MEP helps describe large-scale behavior as the system becomes more complex; and (5) prediction of hydrologic fluxes can be improved without the need for scaling up small-scale heterogeneous processes to the large scale.

2.6   Entropy Production Flux

In an open or nonisolated system, there will be exchanges of mass and energy between the system and its surroundings. As a result, there will be exchanges of entropy. The rate of entropy production is a measure of irreversibility. In open systems, organized structures persist; at steady state, the entropy production is maximum, and this corresponds to a state of minimum total energy (Kleidon, 2010a). Were there no external forcing, the system would reach thermodynamic equilibrium as quickly as possible.

The entropy production flux can be expressed as a product of thermodynamic force and causative flux (Kleidon and Schymanski, 2008). The entropy production flux due to mass flux F_{mass} can be expressed as

$$w = F_{mass} \frac{z_{high} - z_{low}}{T} \tag{2.8}$$

in which the mass flux $F_{mass} = \rho g Q$, ρ is the density of water (ML^{-3}), g is the acceleration of gravity ($L\,T^{-2}$), Q is the flow or discharge ($L^3\,T^{-1}$), T is the absolute temperature, and z is the chemical potential or energy level ($ML^2\,T^{-2}$). In Eq. (2.8), $(z_{high} - z_{low})/T$ denotes the thermodynamic force.

Consider a system of two boxes A and B of the same size and mass. Box A is warm with temperature denoted by T_{warm}, and box B is cold with temperature denoted by T_{cold}. It is assumed that the system is isolated. Since the boxes are connected, the changes in the box temperatures are related to the exchange of heat between the two boxes. The entropy production (EP) flux due to the heat flux F_{heat} from the reservoir of low temperature T_{cold} can be written as

$$w = F_{heat} \left(\frac{1}{T_{cold}} - \frac{1}{T_{warm}} \right) \tag{2.9}$$

where the heat flux is expressed as $F_{heat} = k(T_{cold} - T_{warm})$ with a certain conductivity k. Then entropy production can be computed from Eq. (2.9).

The change in entropy S of the total system (boxes A and B) with time is then

$$S(t) = c \ln[T_{warm}(t)] + c \ln[T_{cold}(t)] = S_0 + \int_0^t w(x)\, dx \tag{2.10}$$

where $S_0 = c \ln[T_{warm}(0)] + c \ln[T_{cold}(0)]$ is the initial entropy of the system.

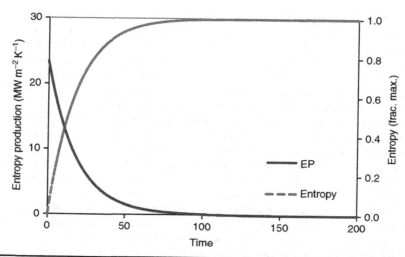

FIGURE 2.1 Entropy production and entropy fraction as a function of time (after Kleidon, 2009).

Example 2.1. The values used in simulations are $T_{warm}(0) = 38.7°C$, $T_{cold}(0) = -5.5 °C$, $c = 2 \times 10^8$ J K^{-1} m^{-2}, and $k = 1$ W m^{-2} K^{-1}. Time is expressed in months. All fluxes are expressed per unit area. Compute the entropy production and entropy, and plot them as a function of time.

Solutions. The quantity F_{heat} in each interval is calculated from $F_{heat} = k\ (T_{warm} - T_{cold})$. Then entropy production is obtained from Eq. (2.9) following Kleidon (2009), and the time evolution of entropy is shown in Fig. 2.1.

For a typical hydrologic system or under isothermal conditions, there may not be a significant temperature differential and w may vanish. Therefore Eq. (2.9) is not used. Under isothermal conditions, MEP corresponds to maximizing power; then maximum power and maximum entropy production can be used interchangeably. By definition, power equals flux times the driving gradient; the flux is the driving gradient divided by resistance. The maximum power can be caused by maximum flux and maximizing the driving gradient in which resistance is considered to be the degree of freedom of the system.

2.7   Application of Nonequilibrium Thermodynamics

The state of thermodynamic equilibrium of atmospheric moisture over a wet surface represents the situation where the atmosphere is saturated with water vapor. In this equilibrium state the evaporation and condensation would be reversible, because in the saturated atmosphere these phase transitions would occur at the same temperature. The unsaturated atmosphere (i.e., relative humidity less than 100 percent) indicates a state away from thermodynamic equilibrium. In this state the process of evaporation becomes irreversible, and it produces entropy by the mixing of saturated air from the surface where evaporation occurs with the unsaturated air from the boundary layer. Evaporation increases the moisture content of the atmospheric boundary layer closer to the thermodynamic equilibrium. The atmospheric circulation is the main driver for maintaining the state away from thermodynamic equilibrium.

Peixoto et al. (1991) estimated the amount of entropy production by the hydrologic cycle, using global averages of latent heat flux of $LH = 79$ W m^{-2}, and the mean surface temperature of $T_s = 288$ K. The average estimate $EP = LH[(1/T_c) - (1/T_s)] = 23$ mW m^{-2} K^{-1}. Determination of entropy production helps characterize the thermodynamic nature that the entropy balance is able to only partially capture, and the rate of entropy is maximized according to the MEP principle (Ozawa et al., 2003; Kleidon and Lorenz, 2005; Martyushev and Seleznev, 2006). This deduces that the hydrologic cycle is maintained farthest away from thermodynamic equilibrium at a comparatively low average relative humidity, and that is likely associated with maximum precipitation and evapotranspiration rates. Since the terrestrial biota substantially affects the exchange fluxes of energy and water, terrestrial vegetation plays a critical role in maintaining the thermodynamic state away from equilibrium. The atmosphere-biosphere interactions and dissipative nature of land surface processes influence the state.

Entropy production greatly varies among regions, from values of near zero in the polar regions to rates as high as 16 mW m^{-2} K^{-1} and more in the midlatitude regions over oceans. The peak values of entropy production are dependent on not only the patterns of ET alone but also the variations in relative humidity. Lower relative humidity should correspond to higher entropy production rates, but also regions of higher water availability tend to have a higher relative humidity.

For present-day conditions, the global mean entropy production of evaporation is 8.4 mW m^{-2} K^{-1}, which is about one-third of the estimated entropy production of the entire hydrologic cycle. On average, ocean surfaces generally produce twice the entropy produced on land surfaces. On land, high rates of entropy production of up to 16 mW m^{-2} K^{-1} are found in regions of high evapotranspiration. Land averaged evapotranspiration decreases from 2.4 to 1.4 mm/d, while entropy production is reduced comparatively less from 4.2 to 3.1 mW m^{-2} K^{-1}. This is related to the reduction in relative humidity as a compensating effect.

When nonequilibrium thermodynamics are applied to hydrologic modeling, two issues need to be considered: (1) the watershed is an open system and (2) the system is far from equilibrium. For the first issue, exchanges of energy and mass of different entropies between the watershed and its surroundings need to be incorporated and expressed in terms of their chemical potential. For the second issue, since the system is far from equilibrium, the principle corresponding to MEP in equilibrium thermodynamics needs to be formulated. The MEP principle has been proposed to overcome this problem (Ozawa et al., 2003; Kleidon and Lorenz, 2005; Martyushev and Selenznev, 2006). The MEP principle may provide the key to the fundamental understanding of optimality and macroscopic behavior in a wide spectrum of systems away from thermodynamic equilibrium.

2.7.1   Potential of Water Vapor

In the atmosphere, the potential of water vapor can be expressed as the sum of its chemical potential and gravitational potential (Kleidon and Schymanski, 2008). The chemical potential is equal to $R_v T_{air} \ln(\text{RH})$, where R_v is the gas constant, T_{air} is the air temperature, and RH is the relative humidity. Similarly the gravitational potential is equal to the product of the height above mean sea level z and the acceleration due to gravity. Therefore,

$$z_{atm} = R_v T_{air} \ln(\text{RH}) + gz \qquad (2.11)$$

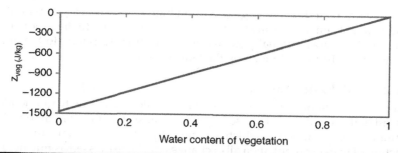

FIGURE 2.2 Relationship between the relative water content of vegetation and potential (after Porada et al., 2011).

Let the height of the soil surface be z_s. The potential of rain z_{rain} at the soil surface is equal to the gravitational potential at z_s. The potential of free water in river channel $z_{channel}$ is equal to the true gravitational potential at height z_c of the channel.

The potential of water in the vegetation z_{veg} is expressed as

$$z_{veg} = [\theta_{veg} - \max(1.0, \theta_{veg})]\psi_{pwp}g \qquad (2.12a)$$

or

$$z_{veg} = (\theta_{veg} - 1)\psi_{pwp} \qquad (2.12b)$$

where ψ_{pwp} is the permanent wilting point (pwp) and θ_{veg} is the relative content of vegetation (see Fig. 2.2). Here z_{veg} decreases linearly with plant available water content from 0 to the minimum possible root water potential of the wilting point (Roderick and Canny, 2005; Schymanski, 2007).

Example 2.2. If the saturated air at the surface is mixed with the unsaturated air of the atmospheric boundary layer (with a temperature T_a and a relative humidity RH_a), the entropy production associated with this mixing can be directly calculated by Eq. (2.12). Let the annual evapotranspiration rate be 100 mm, the relative humidity RH_a be 60 percent, and the gas constant of water vapor $R_v = 461.5$ J kg^{-1} K^{-1}. Calculate the entropy production.

Solution. The entropy production of evaporation can be written as

$$\sigma_{evap} = \rho R_v E[\ln(RH_a) - \ln(RH_s)]$$

Since RH_s is near saturation, $\ln(RH_s) \approx 0$, and the entropy production of evaporation can be calculated as

$$\sigma_{evap} = -\rho R_v E[\ln(RH_a)]$$

$$= -1000 \times 461.5 \times 0.1 \div 365 \div 3600 \div 24 \times \ln 0.6 \times 1000 = 0.75 \text{ mW}$$

2.7.2 Potential of Water

Flow of water is caused by the gradient in the gravitational or/and chemical potential. For example, soil water moves as a result of the gradient of matric potential of soil water between two locations. Water moves from higher potential to lower potential, from wet to dry soil. On the land surface, water moves as a result of a gravitational potential gradient due to topography from high to low, moving downhill. The entropy production of runoff over the land surface is proportional to the product of the flow of water and the gravitational potential gradient. It corresponds to the amount of heat generated by the flow, divided by the temperature.

2.7.3   Soil Water Potential

Soil is a nonequilibrium open system where water flows due to gradients in the water potential. If a small part of the soil hydrological system far from thermodynamic equilibrium can be assumed in which a chemical potential of water can be computed as a function of water content, then all exchanges can be expressed as functions of gradients in the combined chemical and gravitational potential of water. The combined chemical potential and gravitational water potential can be denoted as z_{soil}. It is then possible to write the entropy budget of soil hydrological processes.

The modified matric potential ψ_m at height z depends on the relative soil water content $\theta_{soil}(z)$ at that height z. In unsaturated soils, the relationship between $\psi_m(z)$ and $\theta_{soil}(z)$ is expressed as a characteristic curve, such as the van Genuchten soil water retention curve (van Genuchten, 1980; Maulem, 1976). The value of $\psi_m(z)$ is negative

$$\psi_m(z) = -\frac{g}{a_{vg}}\left\{\left[\frac{\theta_{soil}(z)}{\theta_{soil,max}}\right]^{1/b_{vg}} - 1\right\}^{1/c_{vg}} \tag{2.13}$$

where $\theta_{soil,max}(z)$ is the relative soil water content of saturation and a_{vg}, b_{vg}, and c_{vg} are parameters of the van Genuchten soil water curve and they depend on the soil type. Under unsaturated conditions $\psi_m(z)$ can be replaced by the hydraulic head (Atkins, 1998).

To obtain z_{soil} for the whole soil column, the water is assumed to reach an equilibrium distribution in each time step of the model. In that case, the soil water potential is constant across the soil profile, or $z_{soil}(z) = $ constant. This suggests a nonuniform distribution of water in the soil column. Each value of $z_{soil}(z) = $ constant is associated with a different vertical equilibrium distribution of water. Let $\theta_{soil}(z)$ be the relative soil water content. The equilibrium soil moisture distribution whose integral is equal to the value of $\theta_{soil}(z)$ is computed. The relationship of $z_{soil}(z)$ is shown in Fig. 2.3.

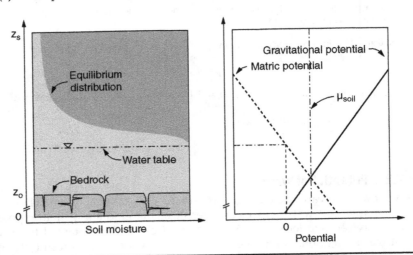

FIGURE 2.3   Soil moisture and potential (after Porada et al., 2011).

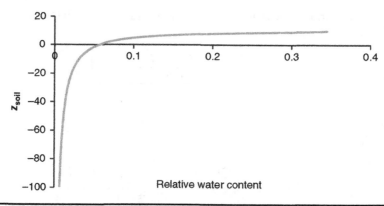

FIGURE 2.4   Relationship between soil water content and potential.

Example 2.3.   Consider values of $\theta_{soil, max}$ and θ_{soil} corresponding to sandy loam soil. Parameters of the van Genuchten (vg) soil water retention curve are a_{vg}, b_{vg}, and c_{vg}, and their values for sandy loam can be taken as $a_{vg} = 7.5$, $b_{vg} = 1.89$, $c_{vg} = 0.47$, $\theta_{soil, max} = 0.345$, $z = 1$ m, and θ_{soil} ranges from 0 to 0.345. Calculate z_{soil}.

Solution.   The values of soil water content corresponding to different values of z_{soil} are computed. Then the relationship of soil water content θ_{soil} and z_{soil} is plotted as shown in Fig. 2.4. Take $\theta_{soil} = 0.1$, for example.

$$\psi_m(z) = -\frac{g}{a_{vg}}\left\{\left[\frac{\theta_{soil}(z)}{\theta_{soil, max}}\right]^{1/-b_{vg}} - 1\right\}^{1/c_{vg}}$$

$$= -\frac{9.81}{7.5}\times\left[\left(\frac{0.1}{0.345}\right)^{1/-0.47} - 1\right]^{1/1.89} = -1.316 \times 12.94^{0.53} = -5.11$$

$$z_{soil} = \psi_M(Z) + gz = -5.11 + 9.81 = 4.70$$

2.7.4   Entropy Production in Water Flows

Evaporation and Transpiration

Bare soil evaporation q_{evap} and transpiration q_{trans} are computed from the atmospheric demand q_{epot} and the amount v of water available in the soil and vegetation, respectively, as

$$q_{evap} = \min\left(q_{epot}\frac{\theta_{soil} - \Delta_s}{dt}\right) \tag{2.14}$$

$$q_{trans} = \min\left(q_{epot}\frac{\theta_{veg}\Delta_v}{dt} + q_{root}\right) \tag{2.15}$$

where Δ_s is the bucket depth of soil, Δ_v is the bucket depth of vegetation, and dt is the model time step (say, day).

The demand is calculated with the equilibrium approach (McNaughton and Jarvis, 1983).

$$q_{epot} = \frac{\left(\frac{ds\,dT}{ds\,dT + \gamma} f\,net\,\Gamma \right)}{\lambda} \tag{2.16}$$

with

$$ds\,dT = \frac{\exp\left[p_{vpt}\left(\frac{zT}{p_{vpt} + zT} \right) \right] p_{vp1}\, p_{vp2}\, p_{vp3}}{(p_{vp2} + zT)^2} \rho \tag{2.17}$$

where zT is the surface temperature in Kelvins (melting temperature of water), $fnet\Gamma$ is the net radiation, ds/dT is the slope of saturation vapor pressure–temperature relationship, λ is the latent heat of vaporization, and p_{vp1}, p_{vp2}, and p_{vp3} are parameters to calculate the vapor pressure. At lower soil water contents, the hydraulic conductivity decreases and evaporation from bare soil occurs as long as the difference between the maximum relative water content and the actual one is lower than 0.01. For this value the decrease in hydraulic conductivity at the top of the soil column is approximately two orders of magnitude (van Genuchten, 1980). The bare soil evaporation is relatively small, and it is limited by the fraction of soil in each grid cell.

The entropy production of bare soil evaporation and transpiration can be expressed as

$$w_{evap} = q_{evap}\, \rho \frac{z_{soil} - z_{atmos}}{T_{surf}} \tag{2.18}$$

$$w_{trans} = q_{trans}\, \rho \frac{z_{veg} - z_{atmos}}{T_{surf}} \tag{2.19}$$

where z_{atmos} is the atmospheric water vapor potential and T_{surf} is the surface temperature.

Example 2.3. If $q_{evap} = 5 \times 10^{-8}\,\text{m/s}$, $q_{tran} = 5 \times 10^{-8}\,\text{m/s}$, $\rho = 1000\,\text{kg/m}^3$, $z_{soil} = -15\,\text{J/kg}$, $z_{veg} = -100\,\text{J/kg}$, $R_v = 461.5\,\text{J kg}^{-1}\,\text{K}^{-1}$, RH = 60 percent, $T_{surf} = 280\,\text{K}$, $T_{air} = 280\,\text{K}$, $z_a = 1$, and $g = 9.81\,\text{m/s}^2$, calculate the entropy production of bare soil evaporation and transpiration.

Solution

$$z_{atmos} = R_v \cdot T_{air} \cdot \ln(\text{RH}) + gz_a = 461.5 \times 280 \times \ln 0.6 + 9.81 \times 1 = -66{,}018\,\text{J/kg}$$

$$w_{evap} = q_{evap}\, \rho \frac{z_{soil} - z_{atmos}}{T_{surf}}$$

$$= 5 \times 10^{-8} \times 1000 \times (-15 + 66{,}018)/280 = 11.78\ \text{mW m}^{-2}\,\text{K}^{-1}$$

$$w_{trans} = q_{trans}\, \rho \frac{z_{veg} - z_{atmos}}{T_{surf}}$$

$$= 5 \times 10^{-8} \times 1000 \times (-100 + 66{,}018)/280 = 11.77\ \text{mW m}^{-2}\,\text{K}^{-1}$$

Rootwater Uptake. The root water uptake is a function of the gradient in water potential between the soil and vegetation as

$$q_{root} = C_{root}(z_{soil} - z_{veg})$$ (2.20)

where z_{soil} is the soil water potential, z_{veg} is the potential of water in the vegetation, and C_{root} is the effective conductivity at the soil root interface. The entropy production of the root water uptake is expressed as

$$w_{root} = q_{root}\,\rho\frac{z_{soil} - z_{veg}}{T_{soil}}$$ (2.21)

where T_{soil} is the soil temperature, and ρ is the density of water in order to express the entropy production in units of W m^{-2} K^{-1}.

Example 2.4. Let $c_{root} = 3.5 \times 10^{-11}$, $z_{soil} = -15$ J/kg, $z_{veg} = -100$ J/kg, and $T_{soil} = 280$ K. Calculate the entropy production of the root water uptake.

Solution

$$q_{root} = c_{root}(z_{soil} - z_{veg})$$

$$= 3.5 \times 10^{-11} \times (-15 + 100) = 2.97 \times 10^{-9}\ \text{m/s}$$

$$w_{root} = q_{root}\,\rho\frac{z_{soil} - z_{veg}}{T_{soil}}$$

$$= 2.97 \times 10^{-9} \times 1000 \times 85/280 = 9.01 \times 10^{-4}\ \text{mW m}^{-2}\ \text{K}^{-1}$$

Surface Runoff The saturation runoff is the surface excess flow and is controlled by the bucket size. The entropy production of surface runoff is expressed as

$$w_{surf} = q_{surf}\,\rho\frac{z_{rain} - z_{channel}}{T_{surf}}$$ (2.22)

where z_{rain} and $z_{channel}$ occur because water flows from the soil surface to the nearest channel.

Example 2.5. If $q_{surf} = 1.2 \times 10^{-7}$ m/s, $\rho = 1000$ kg m^{-3}, $z_r = 5$ m, $z_c = 1$ m, $T_{surf} = 280$ K, and $g = 9.81$ m/s^2, calculate the entropy production of surface runoff.

Solution

$$w_{surf} = q_{surf}\,\rho\frac{z_{rain} - z_{channel}}{T_{surf}}$$

$$= 1.2 \times 10^{-7} \times 1000 \times 4 \times 9.81/280 = 4 \times 10^{-6}\ \text{mW m}^{-2}\ \text{K}^{-1}$$

River Discharge Entropy production of river discharge q_{river} into the seas, which is comprised of surface runoff and base flow, can be written as

$$w_{river} = (q_{surf} + q_{base})\,\rho\frac{z_{channel} - z_{msl}}{T_{surf}}$$ (2.23)

where z_{msl} is the potential of free water at the mean sea level set to 0. Because gradients z_{rain} and $z_{channel}$ are constant, both w_{surf} and w_{river} vary with only the flow rate.

Example 2.6.   If $q_{surf} = 1.2 \times 10^{-7}$ m/s, $q_{base} = 1.2 \times 10^{-7}$ m/s, $\rho = 1000$ kg m^{-3}, $z_r = 5$ m, and $T_{surf} = 280$ K, calculate the entropy production of the river discharge.

Solution

$$w_{river} = (q_{surf} + q_{base})\rho \frac{z_{channel} - z_{msl}}{T_{surf}}$$

$$= 2.4 \times 10^{-7} \times 1000 \times 5 \times 9.81/280 = 4.2 \times 10^{-2} \text{ mW m}^{-2} \text{ K}^{-1}$$

Infiltration   Entropy is also produced by the infiltration (inf) of water into the soil and can be expressed as

$$w_{inf} = (q_{rain} - q_{surf})\rho \frac{z_{rain} - z_{soil}}{T_{soil}} \tag{2.24}$$

The hydrologic cycle is a system that is far from equilibrium and is characterized by its rate of entropy production. Over land, the cycle is strongly affected by the presence of terrestrial vegetation.

Example 2.7.   If rainfall is 200 mm/month, 120 mm of rainfall generated surface flow, $z_{soil} = -15$ J/kg, $T_{surf} = 280$ K, $T_{soil} = 280$ K, $z_s = 5$ m, and $g = 9.81$ m/s^2, calculate the entropy production of infiltrated water.

Solution

$$q_{rain} = 200/1000/30/24/3600 = 7.71 \times 10^{-8} \text{ m/s}$$

$$q_{surf} = 120/1000/30/24/3600 = 4.62 \times 10^{-8} \text{ m/s}$$

$$w_{inf} = (q_{rain} - q_{surf})\rho \frac{z_{surf} - z_{soil}}{T_{soil}}$$

$$= (7.71 - 4.62) \times 10^{-8} \times 1000 \times (5 \times 9.81 + 15)/280 = 7 \times 10^{-3} \text{ mW m}^{-2} \text{K}^{-1}$$

Base Flow   Base flow is expressed as

$$q_{base} = C_{base}(z_{soil} - z_{channel}) \tag{2.25}$$

where $z_{channel}$ is the potential of the water in the stream channel and C_{base} is the effective conductivity of the interface between the soil and the channel. The entropy production of base flow can be expressed as

$$w_{base} = q_{base}\,\rho \frac{z_{soil} - z_{channel}}{T_{soil}} \tag{2.26}$$

Example 2.8. Let $C_{base} = 8.6 \times 10^{-9}$, $z_{soil} = 35$ J/kg, $z_{channel} = 9.8$ J/kg, and $T_{soil} = 280$ K. Calculate the entropy production of the root water uptake.

Solution

$$q_{base} = C_{base} (z_{soil} - z_{channel})$$

$$= 8.6 \times 10^{-9} \times (35 - 9.8) = 2.16 \times 10^{-7} \text{ m/s}$$

$$w_{base} = q_{base} \, \rho \frac{z_{soil} - z_{channel}}{T_{soil}}$$

$$= 2.16 \times 10^{-7} \times 1000 \times (35 - 9.8)/280 = 1.94 \times 10^{-2} \text{ mW m}^{-2} \text{K}^{-1}$$

Questions

1. Let $T_{warm}(0) = 40°C$, $T_{cold}(0) = -10°C$, $c = 2 \times 10^8$ J K^{-1}m^{-2}, and $k = 1$ W m^{-2}K^{-1}. Time is expressed in months, and all fluxes are expressed per unit area. Compute entropy production and entropy, and plot them as a function of time.

2. If the saturated air at the surface is mixed with the unsaturated air of the atmospheric boundary layer (with a temperature T_a and a relative humidity RH_a), the entropy production associated with this mixing can be directly calculated by Eq. (2.12). Let the annual evapotranspiration rate be 150 mm, the relative humidity RH_a be 70 percent, and the gas constant of water vapor $R_v = 461.5$ J kg^{-1}K^{-1}. Calculate the entropy production.

3. Consider values of $\theta_{soil,max}$ and θ_{soil} corresponding to the sandy loam soil. Parameters of the van Genuchten soil water retention curve are a_{vg}, b_{vg}, and c_{vg}, and their values for sandy loam can be taken as $a_{vg} = 7.5$, $b_{vg} = 0.47$, $c_{vg} = 1.89$, $\theta_{soil,max} = 0.345$, $z = 2$ m, and θ_{soil} ranges from 0 to 0.345. Calculate z_{soil}.

4. If $q_{evap} = 3 \times 10^{-8}$ m/s, $q_{tran} = 3 \times 10^{-8}$ m/s, $\rho = 1000$ kg/m^3, $z_{soil} = -10$ J/kg, $z_{veg} = -90$ J/kg, $R_v = 461.5$ J kg^{-1} K^{-1}, $RH = 50$ percent, $T_{surf} = 270$ K, $T_{air} = 270$ K, $z_a = 2$, and $g = 9.81$ m/s^2, calculate the entropy production of bare soil evaporation and transpiration.

5. Let $c_{root} = 3.0 \times 10^{-11}$, $z_{soil} = -10$ J/kg, $z_{veg} = -90$ J/kg, and $T_{soil} = 280$ K. Calculate the entropy production of the root water uptake.

6. If $q_{surf} = 1.0 \times 10^{-7}$ m/s, $\rho = 1000$ kg m^{-3}, $z_r = 10$ m, $z_c = 5$ m, $T_{surf} = 280$ K, and $g = 9.81$ m/s^2, calculate the entropy production of surface runoff.

7. If $q_{surf} = 1.0 \times 10^{-7}$ m/s, $q_{base} = 1.1 \times 10^{-7}$ m/s, $\rho = 1000$ kg m^{-3}, $z_r = 5$ m, and $T_{surf} = 280$ K, calculate the entropy production of river discharge.

8. If rainfall is 150 mm/month, 100 mm of rainfall generated surface flow, $z_{soil} = -15$ J/kg, $T_{surf} = 270$ K, $T_{soil} = 270$ K, $z_s = 5$ m, and $g = 9.81$ m/s^2, calculate the entropy production of infiltrated water.

9. Let $C_{base} = 5 \times 10^{-9}$, $z_{soil} = 35$ J/kg, $z_{channel} = 9.8$ J/kg, and $T_{soil} = 280$ K. Calculate the entropy production of the root water uptake.

References

Atkins, P. W. (1998). *Physical Chemistry*. 6th edition, Oxford University Press, Oxford, pp. 111.

Brolsma, R. J., and M. F. P. Bierkens (2007). Groundwater-soil water-vegetation dynamics in a temperate forest ecosystem along a slope. *Water Resources Research*, vol. 43, no. 1, W01, 414, doi: 10.1029/2005WR004696.

Caylor, K., T. M. Scanlon, and I. Rodriguez-Iturbe (2009). Ecohydrological optimization of pattern and processes in water limited ecosystems: a trade-off-based hypothesis. *Water resources Research*, vol. 45, no. 8, W08, 4-07, doi: 10.1029/2008WR007230.

Dewar, R. C. (2009). Maximum entropy production as an algorithm that translates physical assumptions into microscopic predictions: Do not shoot the messenger. *Entropy*, vol. 11, pp. 931–944, doi: 10.3390/e11040931.

Goody, R. (2000). Sources and sinks of climate entropy. *Quarterly Journal of Royal Meteorological Society*, vol. 126, pp. 1953–1970.

Kleidon, A. (2004a). Beyond Gaia: Thermodynamics of life and Earth system functioning. *Climate Change*, vol. 66, pp. 271–319.

Kleidon, A. (2004b). Optimized stomatal conductance of vegetated land surfaces and its effects on simulated productivity and climate. *Geophysical Research Letters*, vol. 31, L21203, doi: 10.1029/2004GL020769.

Kleidon, A. (2008). Entropy production by evapotranspiration and its geographic variation. *Soil Water Research*, vol. 3, pp. S89–S94.

Kleidon, A. (2009). Nonequilibrium thermodynamics and maximum entropy production in the earth system. *Naturwissenshaften*, vol. 96, pp. 653–677, doi: 10.1007/s00114-009-0509-x.

Kleidon, A. (2010a). A basic introduction to the thermodynamics of the earth system far from equilibrium and maximum entropy production. *Philosophical Transactions, Royal Society*, B, vol. 365, pp. 1303–1315, doi: 10.1098/rstb.2009.0310.

Kleidon, A. (2010b). Life, hierarchy, and the thermodynamics machinery of planet earth. *Physics Life Review*, vol. 7, no. 4, pp. 424–460, doi: 10.1016/j.plrev.2010.10.002.

Kleidon, A. and R. D. Lorenz (2005). *Non-equilibrium Thermodynamics and the Production of Entropy: Life, Earth and Beyond*, Springer Verlag, Heidelberg, Berlin.

Kleidon, A. and S. Schymanski (2008). Thermodynamics and optimality of the water budget on land: A review. Geophysical Research Letters, vol. 35, L20404, doi: 10.1029/2008GL035393.

Kleidon, A., E. Zehe, U. Ehrer, and U. Scherer (2013). Thermodynamics, maximum power, and dynamic of preferential river flow structures at the continental scale. *Hydrology and Earth System Sciences*, vol. 17, pp. 225–251.

Kondepudi, D., and I. Prigogine (1998). *Modern Thermodynamics: From Heat Engines to Dissipative Structures*, Wiley, Chichester, United Kingdom.

Martyushev, L. M. and V. D. Seleznev (2006). Maximum entropy production principle in physics, chemistry, and biology. *Physical Reports*, vol. 426, pp. 1–45.

McDonnell, J., M. Sivapalan, K. Vache, S. Dunn, G. Grant, R. Haggerty, C. Hinz, et al. (2007). Moving beyond heterogeneity and process complexity: a new vision for watershed hydrology. *Water Resources Research*, vol. 43, W07301, doi: 10.1029/2006WR005467.

McNaughton, K. G. and P. G. Jarvis (1983). Predicting effects of vegetation changes on transpiration and evaporation. In: *Water Deficits and Plant Growth*, edited by T. L. Kozlowski, Academic Press, New York, pp. 1–47.

Mualem, Y. (1976). A new model for predicting the hydraulic conductivity of unsaturated porous media. *Water Resources Research*, vol. 12, pp. 513–522.

Ozawa, H., A. Ohmura, R. D. Lorenz, and T. Pujol (2003). The second law of thermodynamics and the global climate system: A review of the maximum entropy production principle. *Reviews of Geophysics*, vol. 41, no. 4, 1018, doi: 10.1029/2002RG000113.

Paik, K. and P. Kumar (2010). Optimality approaches to describe characteristic fluvial patterns on landscapes. *Philosophical Transactions, Royal Society*, B, vol. 365, pp. 1387–1395.

Pauluis, O. M. (2005). Water vapor and entropy production in the earth's atmosphere. In: *Non-equilibrium Thermodynamics and the Production of Entropy: Life, Earth and Beyond*, by A. Kleidon, and R. D., Lorenz, Springer Verlag, Heidelberg, Berlin.

Pauluis, O. M., V. Balaji, and I. M. Held (2000). Frictional dissipation in a precipitating atmosphere. *Journal of Atmospheric Science*, vol. 57, pp. 987–994.

Pauluis, O. M., and I. M. Held (2002). Entropy budget of an atmosphere in radiative convective equilibrium. Part II. Latent heat transport and moist processes. *Journal of Atmospheric Science*, vol. 59, pp. 140–149.

Peixoto, J. P., A. H. Oort, M. de Almeida, and A Tome (1991). Energy budget of the atmosphere. *Journal of Geophysical Research*, vol. 96, pp. 10981–10988.

Porada, P., A. Kleidon, and S. J. Schymanski (2011). Entropy production of soil hydrological processes and its maximization. *Earth System Dynamics Discussions*, vol. 2, pp. 105–2011.

Porporato, A., F. Laio, L. Ridolfi, and I. Rodriguez-Iturbe (2001). Plants in water-controlled ecosystems: Active role in hydrologic processes and response to water stress: III. Vegetation water stress. *Advances in Water Resources*, vol. 24, no. 7, pp. 725–744, doi: 10.1016/S0309-1078(01)00006-9.

Porporato, A., E. Daly, and I. Rodriguez-Iturbe (2004). Soil water balance and ecosystem response to climate change. *American Nature*, vol. 164, pp. 625–632.

Rinaldo, A., I., R. Rodriguez-Iturbe, R. L. Rigon, E. Bras, Ijjasz-Vasquez, and A. Marani (1992). Minimum energy and fractal structures of drainage networks. *Water Resources Research*, vol. 28, pp. 2183–2195.

Roderick, M. L., and M. J. Canny (2005). A mechanical interpretation of pressure chamber measurements—what does the strength of the squeeze tell us? *Plant Physiology and Biochemistry*, vol. 43, pp. 323–336, doi: 10.1016/j.plaphy.2005.02.014.

Rodriguez-Iturbe, I., A. Rinaldo, R. Rigon, R. L. Bras, A. Marani, and E. Ijjasz-Vasquez (1992a). Energy dissipation. Runoff production, and the three dimensional structure of river basins. *Water Resources Research*, vol. 28, pp. 1095–1103.

Rodriguez-Iturbe, I., A. Rinaldo, R. Rigon, R. L. Bras, E. Ijjasz-Vasquez, and A. Marani (1992b). Fractal structures as least energy patterns: The case of river networks. *Geophysical Research Letters*, vol. 9, pp. 889–892.

Rodriguez-Iturbe, I., P. D'Odorico, L. Porporato, and A. Ridolfi (1999). On the spatial and temporal links between vegetation, climate and soil moisture. *Water Resources Research*, vol. 35, No. 12, pp. 3709–3722, doi: 10.1029/1999WR900255.

Schaefli, B., C. J. Harman, M., Sivapalan, and S. J. Schymanski (2011). HESS opinions: Hydrologic predictions in a changing environment: Behavioral modeling. *Hydrology and Earth System Sciences*, vol. 15, pp. 635–646, doi: 10.5194/hess-15-635-2100.

Schymanski, S. J. (2007). "Transpiration as the Leak in the Carbon Factory: A Model of Self-optimizing Vegetation." Unpublished Ph.D. thesis, The University of Western Australia, Perth.

Schymanski, S. J., M. Sivapalan, M. L. Roderick, L. B. Hutley, and J. Beringer (2009). An optimality-based model of the dynamic feedbacks between natural vegetation and the water balance. *Water Resources Research,* vol. 45, W01412, doi: 10.1029/2008WR006841.

Singh, V. P., C. T. Yang, and Z. Q. Deng (2003a). Downstream hydraulic geometry relations: 1. Theoretical development. *Water Resources Research,* vol. 39, no. 12, pp. SWC2-1 to SWC2-15.

Singh, V. P., C. T. Yang, and Z. Q. Deng (2003b). Downstream hydraulic geometry relations: 2. Calibration and testing. *Water Resources Research,* vol. 39, no. 12, pp. SWC3-1 to SWC3-10.

Van Genuchten, M. T. (1980). A closed form equation for predicting the hydraulic conductivity of unsaturated soils. *Soil Science Society of America Journal,* vol. 44, no. 5, pp. 892–898.

Wang, J., G. D. Salvucci, and R. L. Bras (2004). An extremum principle of evaporation. *Water Resources Research,* vol. 40, W09303, doi: 10.1029/2004WR0003087.

Wang, J., R. L. Bras, M. Lerdau, and G. D. Salvucci (2007). A maximum hypothesis of transpiration. *Journal of Geophysical Research,* vol. 112, G03010, doi: 10.1029/2006JG000255.

West, G. B., J. H. Brown, and B. J. Enquist (1997). A general model for the origin of allometric scaling laws in biology. *Science,* vol. 276, pp. 122–126.

Westhoff, M. C., and E. Zehe (2013). Maximum entropy production: Can it be used to constrain conceptual hydrological models? *Hydrology and Earth System Sciences,* vol. 17, pp. 3141–3157.

Yang, C. T. (1971). Potential energy and stream morphology. *Water Resources Research,* vol. 7, No. 2, pp. 311–324.

Zehe, E., T. Blume, and G. Bloschl (2010). The principle of "maximum energy dissipation': A novel thermodynamic perspective on rapid water flow in connected soil structures. *Philosophical Transactions, Royal Society, B,* vol. 365, pp. 1377–1386.

Zehe, E., and M. Sivapalan (2009). Threshold behavior in hydrological systems as (human) geo-ecosystems: Manifestations, controls, implications. *Hydrology and Earth System Sciences,* vol. 13, pp. 1273–1297, doi: 10.5194/hess-13-1273-2009.

Further Reading

Dewar, R. C. (2005). Maximum entropy production and fluctuation theorem. *Journal of Physics A: Mathematical Sciences and General,* vol. 38, pp. L371–L381.

Dewar, R. C. (2010). Maximum entropy production and plant optimization theories. *Philosophical Transactions, Royal Society, B,* vol. 365, pp. 1429–1435, doi: 10.1098/rstb.2009.0293.

Evans, D. J. (2002). The fluctuation theorem. *Advances in Physics,* vol. 51, pp. 1529–1585.

Gonzales, R. G., A. Verhoef, P. L. Vidale, and I. Braud (2012). Incorporation of water vapor transfer in the JULES land surface model: Implications for the key soil variables and land surface fluxes. *Water Resources Research,* vol. 48, W05538, doi: 10.1029/2011WR011811.

Goody, R. (2007). Maximum entropy production in climate theory. *Journal of the Atmospheric Sciences,* vol. 64, pp. 2735–2739.

Grinstein, G. and R. Linsker (2007). Comments on a derivation and application of the "maximum entropy production" principle. *Journal of Physics A: Mathematical Sciences and General,* vol. 40, pp. 9717–9720.

Holdaway, R. J., A. D. Sparro, and D. A. Coomes (2010). Trends in entropy production during ecosystem development in the Amazon basin. *Philosophical Transactions, Royal Society B*, vol. 365, pp. 1437–1447, doi: 10.1098/rstb.2009.0298.

Jaynes, E. T. (1980). The minimum entropy production principle. *Annual Reviews in Physics and Chemistry*, vol. 31, pp. 579–601.

Juretic, D. and Zupanovic, P. (2003). Photosynthetic models with maximum entropy production in irreversible charge transfer steps. *Computational Biology and Chemistry*, vol. 27, pp. 541–553.

Katul, G. G. (1994). A model for sensible heat flux probability density function for near-neutral and slightly-stable atmospheric flows. *Boundary-Layer Meteorology*, vol. 71, pp. 1–20.

Kleidon, A. (2010c). Non-equilibrium thermodynamics, maximum entropy production and Earth-system evolution. *Philosophical Transactions, Royal Society, A*, vol. 368, pp. 181–196, doi: 10.1098/rsta.2009.0188.

Kleidon, A., K. Fraedrich, T. Kunz, and F. Lunkeit (2003). The atmospheric circulation and states of maximum entropy production. *Geophysical Research Letters*, vol. 30, No. 23, pp. CLM 9-1—CLM 9–4 2223, doi: 10.1029/2003GL018363.

Kleidon, A., K. Fraedrich, E. Kirk, and F. Lunkeit (2006). Maximum entropy production and the strength of boundary layer exchange in an atmospheric general circulation model. *Geophysical Research Letters*, vol. 33, L06706, doi: 10.1029/2005GL025373.

Kleidon, A. and S. Schymanski (2008). Thermodynamics and optimality of the water budget on land: A review. *Geophysical Research Letters*, vol. 35, L20404, doi: 10.1029/2008GL035393.

Kleidon, A., S. Schymanski, and M. Stieglitz (2009). Thermodynamics, irreversibility, and optimality in land surface hydrology. In: *Bioclimatologist and Natural Hazards*, edited by K. Strelcova et al., Springer, pp. 107–118, doi: 10.1007/978-1-4020-8876-6-9.

Li, J. C. M. (2011). Thermodynamics for nonequilibrium systems: The principle of macroscopic separability and the thermokinetic potential. *Journal of Applied Physics*, vol. 33, no. 2, pp. 616–624.

Martyushev, L. M. (2010). The maximum entropy production principle: Two basic questions. *Philosophical Transactions, Royal Society, B*, vol. 365, pp. 1333–1334.

Martyushev, L. M. (2013). Entropy and entropy production: Old misconceptions and new breakthroughs. *Entropy*, vol. 13, pp. 1152–1170.

Muller, I. (2008). Entropy and energy—a universal competition. *Entropy*, vol. 10, pp. 462–476.

Paltridge, G. W. (1975). Global dynamics and climate—a system of minimum entropy exchange. *Quarterly Journal of Royal Meteorological Society*, vol. 101, pp. 47–484.

Schymanski, S. J., A. Kleidon, M. Stieglitz, and J. Narula (2010). Maximum entropy production allows a simple representation of heterogeneity in semiarid ecosystems. *Philosophical Transactions, Royal Society, B*, vol. 365, pp. 1449–1455.

Volk, T., and O. Pauluis (2010). It is not the entropy you produce, rather, how you produce it. *Philosophical Transactions, Royal Society, B*, vol. 365, pp. 1317–1322.

Wang, J., and R. L. Bras (2010). An extremum solution of the Monin-Obukhov similarity equations. *Journal of the Atmospheric Sciences*, vol. 67, pp. 485–499.

Wu, W., and Y. Liu (2010). Radiation entropy flux and entropy production of the earth system. *Reviews of Geophysics*, vol. 48, RG2003/2010, pp. 1–27, paper no. 2008RG000275.

CHAPTER 3

Performance Measures

$\mathbf{C}$alibration and validation of hydrologic models, whether deterministic, statistical, or stochastic, are done using one or more quantitative criteria or performance measures. The ASCE Watershed Management Task Committee (ASCE, 1993) defined measures for evaluation of watershed models, and a number of measures have since been developed. Legates and McCabe (1999) reviewed some of these measures. There is no one best measure that will be satisfactory for all purposes. Therefore, sometimes two or more measures are used jointly. However, if these measures are not independent, then interdependence among them and the resulting consequences for the goodness-of-fit results must be enumerated. Schaefli and Gupta (2007) emphasized the need for describing and interpreting the performance measures in the proper context. The objective of this chapter is to discuss some of these measures.

3.1 Symbols and Definitions

Before the discussion of the performance measures or criteria, it is useful to define the symbols and terms that will be used throughout the chapter. A random variable is denoted X, and it takes on N values. The ith observed value of the variable is denoted x_i, and the ith predicted or computed value is y_i. The *arithmetic mean* of observed ($\bar{x}$) and computed ($\bar{y}$) values is defined as

$$\bar{x} = \frac{1}{N}\sum_{i=1}^{N} x_i \qquad (3.1a)$$

$$\bar{y} = \frac{1}{N}\sum_{i=1}^{N} y_i \qquad (3.1b)$$

The standard deviation of observed (S_x) and of computed (S_y) values is defined as

$$S_x = \sqrt{\frac{1}{N-1}\sum_{i=1}^{N}(x_i - \bar{x})^2} \qquad (3.2a)$$

$$S_y = \sqrt{\frac{1}{N-1}\sum_{i=1}^{N}(y_i - \bar{y})^2} \qquad (3.2b)$$

The *coefficient of variation* CV of observed [CV(x)] and of computed [CV(y)] values is defined as

$$CV(x) = \frac{S_x}{\bar{x}} \tag{3.3a}$$

$$CV(y) = \frac{S_y}{\bar{y}} \tag{3.3b}$$

The covariance of observed and computed values, denoted Cov(x, y), is defined as

$$Cov(x, y) = \frac{1}{N} \sum_{i=1}^{N} (x_i - \bar{x})(y_i - \bar{y}) \tag{3.4}$$

Example 3.1. For a number of rainfall events for the Pee Dee River, the watershed values of rainfall amounts and runoff peak are given in Table 3.1. Calculate the mean, standard deviation, and coefficient of variation of both rainfall amounts and runoff peaks. Now plot runoff peak versus rainfall amount, and fit a curve, using regression analysis, and express it mathematically. The regression equation now becomes the model. Compute peak runoff values for rainfall amounts, and compute the above statistics for the model-computed runoff peak values.

Solution.   For rainfall values,

$$\bar{P} = \frac{1}{N} \sum_{i=1}^{N} P_i = \frac{1}{69}(2.33 + 2.06 + 2.19 + \cdots + 0.1) = 0.67 \text{ in.}$$

$$S_P = \sqrt{\frac{1}{N-1} \sum_{i=1}^{N} (P_i - \bar{P})^2} = \sqrt{\frac{1}{68}[(2.33 - 0.67)^2 + (2.06 - 0.67)^2 + \cdots + (0.1 - 0.67)^2]} = 0.522 \text{ in.}$$

$$CV(P) = \frac{S_P}{\bar{P}} = \frac{0.522}{0.67} = 0.785$$

For runoff peak

$$\bar{Q} = \frac{1}{N} \sum_{i=1}^{N} Q_i = \frac{1}{69}(12 + 2.28 + 7.2 + \cdots + 0.48) = 3.04 \text{ in.}$$

$$S_Q = \sqrt{\frac{1}{N-1} \sum_{i=1}^{N} (Q_i - \bar{Q})^2} = \sqrt{\frac{1}{68}[(12 - 3.04)^2 + (2.28 - 3.04)^2 + \cdots + (0.48 - 3.04)^2]} = 3.856 \text{ in.}$$

$$CV(Q) = \frac{S_Q}{\bar{Q}} = \frac{3.856}{3.04} = 1.268$$

Based on the regression analysis, shown in Fig. 3.1, the rainfall-runoff model becomes

$$Q = 2.8741 P^{0.8834}$$

Thus, the peak runoff value for maximum rainfall is

$$(Q_c)_{max} = 2.8741 P_{max}^{0.8834} = 2.8741 \times 2.33^{0.8834} = 6.07 \text{ in.}$$

$$\bar{Q}_C = \frac{1}{N} \sum_{i=1}^{N} Q_{ci} = \frac{1}{69}(6.068 + 5.442 + \cdots + 0.376) = 1.95 \text{ in.}$$

P (in.)	Q₀ (in.)	Q_c (in.)	P (in.)	Q₀ (in.)	Q_c (in.)
2.33	12	6.068	0.54	0.12	1.668
2.06	2.28	5.442	0.52	0.84	1.613
1.9	7.2	5.067	0.5	1.56	1.558
1.76	11.28	4.736	0.49	3.6	1.530
1.64	13.32	4.449	0.45	6.12	1.420
1.62	3.6	4.401	0.44	0.6	1.392
1.62	7.32	4.401	0.43	1.32	1.364
1.41	8.4	3.893	0.38	0.84	1.223
1.33	9.48	3.698	0.38	0.72	1.223
1.25	17.28	3.500	0.37	0.24	1.194
1.15	3.84	3.252	0.35	0.6	1.137
1.14	0.6	3.227	0.35	0.12	1.137
1.13	16.68	3.202	0.3	0.48	0.992
1	2.16	2.874	0.29	1.32	0.963
1	3	2.874	0.28	4.08	0.934
0.93	1.2	2.696	0.25	0.24	0.845
0.92	3.12	2.670	0.24	0.84	0.815
0.9	6.72	2.619	0.24	1.2	0.815
0.9	1.68	2.619	0.22	0.84	0.754
0.84	1.08	2.464	0.21	0.72	0.724
0.84	6.6	2.464	0.19	0.72	0.663
0.74	1.56	2.203	0.18	0.12	0.632
0.73	1.44	2.177	0.18	2.28	0.632
0.7	1.56	2.097	0.18	1.8	0.632
0.69	2.16	2.071	0.18	2.04	0.632
0.65	1.44	1.964	0.17	0.72	0.601
0.64	4.2	1.938	0.17	0.96	0.601
0.63	0.48	1.911	0.15	0.48	0.538
0.62	1.68	1.884	0.14	2.64	0.506
0.61	0.12	1.857	0.13	0.84	0.474
0.6	1.32	1.830	0.13	0.72	0.474
0.6	3.96	1.830	0.13	2.04	0.474
0.6	6.84	1.830	0.11	0.12	0.409
0.55	0.6	1.695	0.1	0.48	0.376
0.55	1.32	1.695			

TABLE 3.1   Rainfall and Runoff Peaks Observed on Pee Dee River Watershed

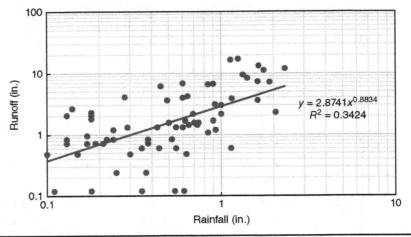

FIGURE **3.1**   Regression of runoff and rainfall.

$$S_{Q_c} = \sqrt{\frac{1}{N-1}\sum_{i=1}^{N}(Q_{ci} - \bar{Q}_C)^2} = \sqrt{\frac{1}{68}[(6.068 - 1.95)^2 + \cdots + (0.376 - 1.95)^2]} = 1.358 \text{ in.}$$

$$CV(Q_c) = \frac{S_{Q_m}}{\bar{Q}_m} = \frac{1.358}{6.07} = 0.696$$

3.2   Performance Measures for Deterministic Models

3.2.1   Bias

Modeled or predicted values always differ from observed values, and the differences or errors reflect the prediction accuracy or inaccuracy of the model. These errors are not all of the same magnitude, but vary from one predicted value to another. The error variation in the predicted values can be systematic or nonsystematic, depending on its cause. Bias corresponds to the systematic error variation. If the model error is denoted Δy, then one can write

$$y = x + \Delta y \tag{3.5}$$

The model is perfect when

$$\Delta y = 0 \tag{3.6a}$$

that is,

$$y = x \tag{3.6b}$$

Bias is a measure of discrepancy between the predicted value and the observed value, and it is estimated using the average error. It is a dimensional quantity and has the same dimensions as the measured or predicted value. Bias is determined using the average error as

$$B = \frac{1}{N}\sum_{i=1}^{N}(y_i - x_i) = \bar{y} - \bar{x} \qquad (3.7)$$

Dividing by the mean of observed values, bias is expressed, for easier interpretation, as a ratio B_r, often referred to as *relative bias*:

$$B_r = \frac{B}{\bar{x}} \qquad (3.8)$$

In some cases, a 5 percent or higher value of relative bias B_r may be considered significant. Bias is also normalized as

$$B_* = \frac{B}{S_x} \qquad (3.9a)$$

and

$$B_*^2 = \frac{B^2}{S_x^2} \qquad (3.9b)$$

It is used as a nondimensional measure of unconditional bias of the model (Murphy, 1988; Murphy and Epstein, 1989).

The purpose of model calibration is to minimize the error variation. However, calibrated models are not always unbiased. McCuen et al. (1990) have shown that power models are often biased when logarithms are used for calibration. Frequently the model is biased both conditionally and unconditionally. The regression line of x versus y does not match the one : one line; i.e., the regression coefficient $(a_{x|y})$ is different from unity. As the difference $y - \bar{y}$ increases, so does the absolute value of the model prediction error Δy. This explains the reason why the vanishing of conditional bias of a model, represented by the equality $a_{x|y} = 1$, is a desirable property.

3.2.2 Maximum Absolute Error

The *maximum absolute* error (MxAE) can be defined as

$$\text{MxAE} = \max|\Delta y| = \max|y_i - x_i| \qquad (3.10)$$

3.2.3 Relative Error

The *relative error* (RE) can be defined as

$$\text{RE} = \frac{y_i - x_i}{x_i} \qquad (3.11)$$

3.2.4 Mean Absolute Error

The *mean absolute error* (MAE) can be defined as

$$\text{MAE} = \overline{|\Delta y|} = \frac{1}{N}\sum_{i=1}^{N}|y_i - x_i| \qquad (3.12)$$

3.2.5 Mean Square Error

The *mean square error* (MSE) can be expressed as

$$\text{MSE} = \frac{1}{N}\sum_{i=1}^{N}(y_i - x_i)^2 \tag{3.13}$$

If any one of MxAE, MAE, RE, and MSE vanishes, then the remaining three will also vanish.

The bias measure is different from the measure given by Eq. (3.13) in the sense that it does not vanish when the other does; but the reverse is true, as shown below.

$$\text{Var}(y - x) = \frac{1}{N}\sum_{i=1}^{N}[(y_i - x_i) - \overline{(y - x)}]^2 \tag{3.14}$$

$$\text{Var}(\Delta y) = \frac{1}{N}\sum_{i=1}^{N}(y_i - x_i)^2 + \frac{1}{N}\sum_{i=1}^{N}[\overline{(y - x)}]^2 - 2\frac{1}{N}\sum_{i=1}^{N}(y_i - x_i)\overline{(y - x)} \tag{3.15}$$

Note that

$$\overline{y - x} = \frac{1}{N}\sum_{i=1}^{N}(y_i - x_i) = \frac{1}{N}\sum_{i=1}^{N}y_i - \frac{1}{N}\sum_{i=1}^{N}x_i = \overline{y} - \overline{x} \tag{3.16}$$

Equation (3.15) can be written as

$$\text{Var}(\Delta y) = \overline{(\Delta y)^2} + B^2 - 2B^2 \tag{3.17}$$

Therefore,

$$\text{MSE} = \overline{(\Delta y)^2} = \text{Var}(\Delta y) + B^2 \tag{3.18}$$

Equation (3.17) shows that if $\text{Var}(\Delta y)$ is zero, the bias is also zero.

3.2.6 Integral Square Error

The *integral square error* (ISE) can be expressed as

$$\text{ISE} = \frac{\sqrt{\overline{(\Delta y)^2}}}{\overline{\overline{x}}} = \frac{\text{MSE}}{\overline{\overline{x}}} \tag{3.19}$$

Equation (3.19) shows that ISE can be referred to as the relative root mean square error. If ISE = 0, the fit is perfect, i.e., MSE = 0.

3.2.7 Coefficient of Correlation

The coefficient of correlation, r, also called the Pearson product-moment correlation coefficient, is one of the most commonly used criteria to measure the goodness of fit for model calibration and/or validation. It can be expressed as

$$r = \frac{\text{Cov}(x, y)}{S_x S_y} = \overline{\left(\frac{x - \overline{x}}{S_x}\right)\left(\frac{y - \overline{y}}{S_y}\right)} = \frac{\overline{xy} - (\overline{x})(\overline{y})}{S_x S_y} \tag{3.20}$$

It is a dimensionless measure of the covariation of the observed and computed values and can vary from −1 to +1.

The correlation coefficient is applicable to any linear model that includes the intercept; that is, r estimates the concentration of the (x, y) points along an arbitrary line, not along the 1:1 line which is the item of interest. This implies that the correlation coefficient is insensitive to the bias of the model. The correlation coefficient assumes that the model being validated is unbiased; in other words, the sum of errors equals zero (McCuen et al., 1990). Therefore, the correlation coefficient is not a good measure of the goodness of fit, even for a commonly used power model $y = ax^b$ (where a and b are parameters).

To illustrate the sensitivity to the model bias, consider a set of observed (x_i) and predicted (y_i) values, $i = 1, 2, ..., N$. The correlation coefficient is not sensitive to the scale. Let $y_* = ay + b$ in which a and b are parameters. The value of r is the same in both cases, shown analytically as

$$\bar{y}_* = \frac{1}{N}\sum_{i=1}^{N}(ay_i + b) = a\bar{y} + b \tag{3.21a}$$

$$S_{y_*} = \sqrt{\frac{1}{N}\sum_{i=1}^{N}[(ay_i + b) - (a\bar{y} + b)]^2} = \sqrt{\frac{a^2}{N}\sum_{i=1}^{N}(y_i - \bar{y})^2} \tag{3.21b}$$

$$r = \frac{\frac{1}{N}\sum_{i=1}^{N}(x_i - \bar{x})(y_{*i} - \bar{y}_*)}{S_x S_{y_*}} = \frac{\frac{1}{N}\sum_{i=1}^{N}(x_i - \bar{x})(y_i - \bar{y})}{S_x S_y} \tag{3.21c}$$

Both model values y and y_* are equally good when judged by r, even though in the case of y_* the model is biased and is therefore inferior to the model in the previous case.

The slopes of regression lines and the coefficient of correlation are related. Since different values of regression coefficient a are obtained for the MSE criterion, the implication of regression lines is

$$r^2 = k_y k_x \tag{3.22}$$

$$k_y = \frac{\text{Cov}_{yx}}{S_y^2} = \frac{r}{a} \tag{3.23a}$$

$$k_x = \frac{\text{Cov}_{yx}}{S_x^2} = ra \tag{3.23b}$$

$$r = \frac{\text{Cov}_{yx}}{S_y S_x} \tag{3.24}$$

where Cov_{yx} is the covariance between model-predicted (y) and observed (x) values, k_y is the slope of the regression line between observed and predicted values, and k_x is the slope of the regression line between predicted and observed values.

The square of r, or r^2, is called the *coefficient of determination* and is a measure of the amount of variance explained by the model. The total variation (TV) can be partitioned into two parts: (1) the explained variation (EV) and (2) the unexplained variation (UV) by the model:

$$TV = EV + UV \tag{3.25a}$$

Dividing Eq. (3.25a) by TV gives

$$1 = \frac{EV}{TV} + \frac{UV}{TV} \tag{3.25b}$$

For a perfect model, UV = 0 and

$$1 = \frac{EV}{TV} = r^2 \tag{3.25c}$$

Thus, the coefficient of determination equals the ratio of explained variation to the total variation. Willmott (1984) has shown that r^2 is insensitive to additive and proportional differences between computed and observed values.

Correlation-based measures are more sensitive to outliers than to observed values near the mean (Legates and Davis, 1997; Moore, 1991), resulting in bias toward extreme values. Recognizing these deficiencies, McCuen and Snyder (1975) proposed an adjustment factor of

$$\left[\sum_{i=1}^{N}(x_i - \bar{x})^2 \sum_{i=1}^{N}(y_i - \bar{y})^2 \right]^{-0.5}$$

which is multiplied by the coefficient of correlation between observed and computed values to account for the differences in standard deviations of observed and computed values. This factor is based on the assumption that the observed variance is less than the computed variance. Legates and McCabe (1999) pointed out that this adjustment factor does not consider differences in the mean values of observed and computed values. Further, if the computed variance were larger than the observed variance, then the factor would increase the correlation and might even cause it to exceed 1.0 in extreme cases; hence this factor is not recommended.

3.2.8   Special Correlation Coefficient

The *special correlation coefficient* r_s can be defined as

$$r_s = \sqrt{\frac{2\overline{xy} - \overline{y^2}}{\overline{x^2}}} \tag{3.26a}$$

Equation (3.26a) can be cast in clearer form as

$$r_s^2 = 1 - \frac{\overline{(y - x)^2}}{\overline{x^2}} = 1 - \frac{\overline{(\Delta y)^2}}{S_x^2 + (\bar{x})^2} = 1 - \frac{MSE}{S_x^2 + (\bar{x})^2} \tag{3.26b}$$

Example 3.2.   For the observed and computed runoff peak values in Example 3.1, show by computation the relation between the special correlation coefficient and the mean square error.

Solution

$$r_s = \sqrt{\frac{2\overline{xy} - \overline{y}^2}{\overline{x}^2}} = \sqrt{\frac{2 \times 9.308 - 5.618}{23.908}} = 0.737$$

Thus, $r_s^2 = 0.737^2 = 0.544$ and

$$MSE = \frac{1}{N}\sum_{i=1}^{N}(y_i - x_i)^2 = \frac{1}{69}[(6.068 - 12)^2 + (5.442 - 2.28)^2 + \cdots] = 10.911 \text{ in.}^2$$

Thus, r_s^2 from Eq. (3.26b) becomes

$$r_s^2 = 1 - \frac{MSE}{S_x^2 + (\overline{x})^2} = 1 - \frac{10.991}{3.856^2 + 3.042^2} = 0.544$$

3.2.9   Nash-Sutcliffe Efficiency and Related Criteria

Nash and Sutcliffe (1970) proposed a goodness of fit measure, often called the *Nash-Sutcliffe efficiency* (NSE) index or coefficient. This is one of the most popular transformations of MSE. Like MSE, NSE is one of the most widely used criteria employed for calibrating and validating hydrologic models with observed data. The MSE value is dimensional and varies from 0 to ∞, and NSE is dimensionless and varies from −∞ to 1. Gupta et al. (2009) noted that the NSE value can be obtained by dividing MSE by the variance of observed values and subtracting that ratio from 1. The NSE is defined as

$$NSE = 1 - \frac{\sum_{i=1}^{N}(x_i - y_i)^2}{\sum_{i=1}^{N}(x_i - \overline{x})^2} = 1 - \frac{MSE}{S_x^2} \tag{3.27}$$

The NSE can be interpreted as a classic skill score (Murphy, 1988) where skill is considered the comparative ability of a model, with regard to a baseline model, which in the case of NSE is used as the mean of observations (i.e., if NSE ≤ 0, the model is no better than using the observed mean as predictor). If MSE is zero, then NSE is 1, indicating that the model is perfect because it simulates the target values or observations perfectly. If MSE = S_x^2, then NSE = 0, indicating that the model is as good as the mean target value as prediction; and if MSE < S_x^2, then the model is worse than mean prediction and its choice is questionable. However, Criss and Winston (2008) note that a negative value does not necessarily mean the model is abysmally poor, but may mean that observed values are steady or otherwise closely match the benchmark values. The preferred value of NSE should be greater than zero, tending toward 1. The NSE is sensitive to differences in observed and computed values of the mean and variance.

Equation (3.27) shows that the NSE compares the mean square error of a model to the variance of the observed values which are the model targets. In other words, it is a kind of noise-to-signal ratio, comparing the average size or variability of model residuals to the size or variability of the target output (Schaefli and Gupta, 2007). The NSE does not measure the goodness of fit of a model in absolute terms. In NSE, the mean observed value is used as a reference or benchmark which can be a poor predictor. Garrick et al. (1978) discussed a benchmark model and Schaefli et al. (2005) used this model. Following Legates and McCabe (1999) and Seibert (2001), Schaefli and Gupta (2007) defined a normalized benchmark efficiency (BE), analogous to NSE, for measuring the model performance over the benchmark model as

$$BE = 1 - \frac{\sum_{i=1}^{N} (y_i - x_i)^2}{\sum_{i=1}^{N} (x_i - x_{bi})^2} \tag{3.28}$$

where x_{bi} is ith benchmark model value. With the use of Eq. (3.28), the explanatory power of the model over the benchmark model can be evaluated.

For unbiased predictions by a linear model, NSE lies in the interval from 0 to +1. For biased predictions, NSE may be negative. For nonlinear hydrologic models, negative values of NSE may occur even if the model is unbiased. NSE has an advantage in that it can be applied to a range of model types, reflecting its flexibility as a goodness of fit measure. Examples of its application to different types of models include intercomparison of hydrologic models (Martinec and Rango, 1989), continuous soil moisture accounting models (Birikundavyi et al., 2002; Johnson et al., 2003; Downer and Ogdon, 2004), storm event models (Kalin et al., 2003), calibration and verification of catchment model parameters (Merz and Bloschl, 2004), nonlinear regression models (Erpul et al., 2003), and rating curves (Jain and Sudheer, 2008) among others. ASCE Watershed Management Committee (ASCE, 1993) recommended NSE for evaluation of continuous moisture accounting models. One way to remedy the bias is to use a linear transformation of the model values which maximizes NSE, as shown by Krzysztofowicz and Watada (1986).

McCuen et al. (2006) investigated the characteristics of NSE as a goodness of fit measure and noted its values are not easily interpreted because its sampling distribution is not known. They presented an approximate sampling distribution. Jain and Sudheer (2008) showed that this measure alone is not adequate for measuring the performance of a hydrologic model, for a relatively poor model can have a high value of NSE and vice versa. Several factors affect a sample value of NSE, such as sample size and sample variance. Looking at Eq. (3.27), one notes that NSE can be high if the variance of observed values is quite high. Values of NSE depend on the sample size; that means that interpretation of a good versus weak model depends on the sample size.

NSE entails three quantities, including measured values, the mean of measured values, and predicted or modeled values, which are employed to constitute two terms, as shown in Eq. (3.27). The denominator in Eq. (3.27) is the variance of observed values, reflecting the total variance of observed values about the mean. The model should be able to explain this variability. The second term, i.e., the numerator of Eq. (3.27), is the mean squared error, reflecting the variation in the data that is not explained by the model. Dividing the numerator by the degrees of freedom and taking the square root would yield the standard error of the estimate.

For a linear model,

$$NSE = 1 - \frac{UV}{TV} \tag{3.29a}$$

and NSE is directly related to r as

$$NSE = r^2 \qquad (3.29b)$$

The values of NSE that are computed are sample values and may differ from the true values which are usually not known.

NSE has an underlying probability distribution that depends on the sample size and the underlying population value NSE_0. The exact distribution of NSE is not known, but an approximate distribution can be obtained. McCuen et al. (2006) tested the null hypothesis

$$H_0: e - e_0 \qquad (3.30)$$

whether or not a sample estimate of NSE, or e, is likely to have been drawn from a population based on a true value e_0. One can also formulate an alternative hypothesis for a one-tailed upper, one-tailed lower, or a two-tailed test.

To test the hypothesis in Eq. (3.30), one can use a standard normal transform

$$z = \frac{e - \bar{e}}{S_e} \qquad (3.31)$$

where $\bar{e}$ is the average sample value of NSE, and S_e is the standard deviation of sample NSE. For moderate-size samples, test statistic z has an approximately normal distribution (McCuen et al., 2006):

$$e = 0.5 \ln \frac{1 + NSE^{0.5}}{1 - NSE^{0.5}} \qquad (3.32a)$$

$$\bar{e} = 0.5 \ln \frac{1 + e_0^{0.5}}{1 - e_0^{0.5}} \qquad (3.32b)$$

$$S_e = (N - 3)^{-0.5} \qquad (3.32c)$$

Here N is the sample size. The value of z computed by using Eq. (3.31) is compared to a critical value obtained from a standard normal distribution table for a probability significance level of α or $\alpha/2$.

For population values of $e_0 = 0.5, 0.7$, and 0.9 for $N = 10, 25, 50$, and 100, the probability distribution of NSE is plotted as shown in Fig. 3.2. It is seen that the distribution becomes more skewed with the long tail on the lower side of e_0 as e_0 tends toward 1. With increasing sample size, the spread of the distribution decreases.

Now the confidence intervals on NSE can be constructed, based on the sampling distribution developed by Fisher (1928)

$$\left[\frac{\exp(x) - 1}{\exp(x) + 1} \right]^2 \qquad (3.33a)$$

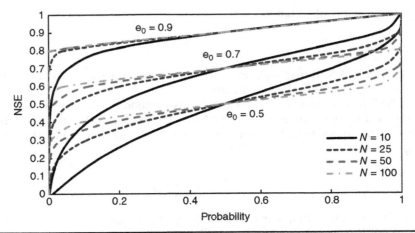

FIGURE 3.2 Probability distributions of the Nash-Sutcliffe efficiency index for population values of $e_o = 0.5$ (bottom set), 0.7 (medium set), and 0.9 (top set).

where

$$x = \ln \frac{1+\text{NSE}}{1-\text{NSE}} + \frac{2z}{(N-3)^{0.5}} \qquad (3.33b)$$

in which z is the standard normal variate and x is e in this case. The value of z can be negative or positive. For the one-sided lower γ percent ($= 100 - \alpha\%$) confidence interval, z of Eq. (3.33b) will be negative for α percent of the standard normal distribution in the lower tail. For the one-sided upper γ confidence interval, z will be positive for an α in the upper tail. For two-sided confidence intervals, one would use z values for $\alpha/2$ percent from each tail. The shape of the confidence intervals will not be symmetric but will depend on both NSE and N.

McCuen et al. (2006) commented on the use and misuse of NSE in assessing model accuracy with reference to regression equations and single-event hydrograph analyses. Considering first the effect of bias on NSE, a biased predicted value can be expressed as a sum of an unbiased estimate y_u and bias B. Then the computed NSE for a biased model, NSE_b, can be expressed as

$$\text{NSE}_b = 1 - \frac{\sum_{i=1}^{N}(y_{ui} - x_i)^2 + NB^2}{\sum_{i=1}^{N}(x_i - \bar{x})^2} \qquad (3.34)$$

where y_{ui} is the i-th unbiased computed value and B is the bias. Equation (3.34) shows that if B is negative, the numerator will be greater for a biased model than for an unbiased model, causing NSE to be smaller than that for the unbiased model. It may therefore be desirable to report the bias and relative bias along with NSE. The precise influence of bias on NSE will depend on the magnitude of two terms in Eq. (3.34), but it is greatly influenced by the model bias (at least in the case of the gamma distribution).

The squared form, Eq. (3.27), shows that NSE gives greatest weight to large values than to other values, as noted by Legates and McCabe (1999), Krause et al. (2005), and Criss and Winston (2008). To overcome this problem, Legates and McCabe (1999) proposed a general form of Eq. (3.27) (GE) as

$$GE = 1 - \frac{\sum_{i=1}^{N} |(y_i - x_i)|^n}{\sum_{i=1}^{N} |(x_i - x_{bi})|^n} \qquad (3.35a)$$

where x_{bi} is the objective benchmark value, n is a power, and a value of $n = 1$ was found to be the best.

Krause et al. (2005) defined relative NSE as

$$NSE_r = 1 - \frac{\sum_{i=1}^{N} \left[\frac{x_i - y_i}{x_i} \right]^2}{\sum_{i=1}^{N} \left[\frac{x_i - \bar{x}}{\bar{x}} \right]^2} \qquad (3.35b)$$

To overcome the insensitivity of correlation-based measures to differences in the observed and computed mean and variance values, Willmott (1981) developed an index of agreement d, representing 1 minus the ratio of mean square error to the potential error (PE), expressed as

$$d = 1 - \frac{\sum_{i=1}^{N} (x_i - y_i)^2}{\sum_{i=1}^{N} \left(|y_i - \bar{x}| + |x_i - \bar{x}| \right)^2} = 1 - N \frac{MSE}{PE} \qquad PE = \sum_{i=1}^{N} \left[|x_i - \bar{x}| + |y_i - \bar{x}| \right]^2 \quad (3.36a)$$

The value of d varies from 0.0 to 1.0 with larger values indicating better agreement between computed and observed values. The potential error is the sum of squared absolute values of the distances from y_i to $\bar{x}$ to x_i. It represents the largest value that $(x_i - y_i)^2$ can obtain for each pair of observed and computed values. The index of agreement is also sensitive to extreme values.

Willmott (1981) generalized the index of agreement as

$$d_g = 1 - \frac{\sum_{i=1}^{N} (x_i - y_i)^n}{\sum_{i=1}^{N} \left[|y_i - \bar{x}| + |x_i - \bar{x}| \right]^n} = 1 - N \frac{MSE}{PE} \qquad PE = \sum_{i=1}^{N} \left[|x_i - \bar{x}| + |y_i - \bar{x}| \right]^n \quad (3.36b)$$

where n is a power. In this index, errors and differences can be appropriately weighted.

Krause et al. (2005) defined a relative index of agreement as

$$d_r = 1 - \frac{\sum_{i=1}^{N} \left(\frac{x_i - y_i}{x_i} \right)^2}{\sum_{i=1}^{N} \left[\frac{|y_i - \bar{x}| + |x_i - \bar{x}|}{\bar{x}} \right]^2} \qquad (3.37)$$

This index significantly reduces the influence of absolute high differences, on one hand, and increases the influence of lower differences, on the other hand. This relative index d_r is expected to be more sensitive to systematic over- or underprediction, especially of low values.

Example 3.3. For the observed and computed runoff peak values in Example 3.1, calculate the maximum absolute error, mean square error, mean absolute error, integral square error, special correlation coefficient, Nash-Sutcliffe efficiency index, generalized efficiency index, and index of disagreement for relation between runoff peak and rainfall amount.

Solution

$$\text{MxAE} = \max|\Delta y| = \max|y_i - x_i| = \max\left(\left|6.068 - 12\right|, \left|5.442 - 2.28\right|, \cdots\right) = 13.780 \text{ in.}$$

$$\text{MSE} = \frac{1}{N}\sum_{i=1}^{N}(y_i - x_i)^2 = \frac{1}{69}[(6.068 - 12)^2 + (5.442 - 2.28)^2 + \cdots] = 10.911 \text{ in.}^2$$

$$\text{MAE} = \left|\overline{\Delta y}\right| = \frac{1}{N}\sum_{i=1}^{N}|y_i - x_i| = \frac{1}{69}\left(\left|6.068 - 12\right| + \left|5.442 - 2.28\right| + \cdots\right) = 1.887 \text{ in.}$$

$$\text{ISE} = \frac{\sqrt{\overline{(\Delta y)^2}}}{\overline{x}} = \frac{\text{MSE}}{\overline{x}} = \frac{10.911}{3.042} = 3.587$$

$$r_s = \sqrt{\frac{2\overline{xy} - \overline{y^2}}{\overline{x^2}}} = \sqrt{\frac{2 \times 9.308 - 5.618}{23.908}} = 0.737$$

$$\text{NSE} = 1 - \frac{\sum_{i=1}^{N}(x_i - y_i)^2}{\sum_{i=1}^{N}(x_i - \overline{x})^2} = 1 - \frac{\text{MSE}}{S_x^2} = 1 - \frac{10.911}{3.856^2} = 0.266$$

$$\text{GE} = 1 - \frac{\sum_{i=1}^{N}\left|(y_i - x_i)\right|^n}{\sum_{i=1}^{N}\left|(x_i - x_{bi})\right|^n} = 1 - \frac{\left|6.068 - 12\right| + \left|5.442 - 2.28\right|}{\left|6.068 - 3.042\right| + \cdots} = 0.313 \qquad \text{with } n = 1$$

$$\text{PE} = \sum_{i=1}^{N}\left[\left|x_i - \overline{x}\right| + \left|y_i - \overline{x}\right|\right]^2 = \left(\left|12 - 3.042\right| + \left|6.068 - 3.042\right|\right)^2$$

$$+ \left(\left|2.28 - 3.042\right| + \left|5.442 - 3.042\right|\right)^2 + \cdots = 1768.59$$

$$d = 1 - \frac{\sum_{i=1}^{N}(x_i - y_i)^2}{\sum_{i=1}^{N}\left[\left|y_i - \overline{x}\right| + \left|x_i - \overline{x}\right|\right]^2} = 1 - N\frac{\text{MSE}}{\text{PE}} = 1 - 69\left(\frac{10.911}{1768.59}\right) = 0.574$$

Example 3.4. What do the values of different measures computed in Example 3.2 say about the model in Example 3.1?

Solution. The above statistics suggest that the performance of the computed model is not perfect but is acceptable. The coefficient of correlation is not low, and the Nash-Sutcliffe efficiency index is greater than 0. The MSE is larger than S_x, which means it is better than mean prediction.

3.2.10 Relation between Mean Square Error and Correlation Coefficient

The variance of model discrepancy can be expressed as

$$\mathrm{Var}(\Delta y) = \mathrm{Var}(y - x) = \mathrm{Var}(y) - 2\mathrm{Cov}(x, y) + \mathrm{Var}(x)$$
$$= S_y^2 - 2rS_xS_y + S_x^2 \tag{3.38}$$

Combining Eqs. (3.38) and (3.18) and adding and subtracting $r^2S_x^2$ give (Murphy, 1988; Weglarczyk, 1998)

$$\mathrm{MSE} = B^2 + S_y^2 - 2rS_xS_y + S_x^2 = S_x^2\left[(1 - r^2) + \left(\frac{S_y}{S_x} - r\right)^2 + \frac{B^2}{S_x^2}\right] \tag{3.39}$$

Equation (3.39) is a combination of an observation-model covariability measure r, conditional bias, and an unconditional bias. The latter two are undesirable model characteristics.

The second term within brackets on the right side of Eq. (3.39) is a dimensionless measure of conditional bias

$$B_c^2 = \left(\frac{S_y}{S_x} - r\right)^2 \tag{3.40}$$

It is a measure of covariability between model error Δy and model prediction y. It vanishes when the model covariance vanishes.

$$\mathrm{Cov}(\Delta y, y) = \mathrm{Cov}(y - x, y) = \mathrm{Cov}(y, y) - \mathrm{Cov}(x, y) = S_x^2 - rS_xS_y \tag{3.41}$$

If $\mathrm{Cov}(\Delta y, y) = 0$, then Eq. (3.41) reduces to

$$r = \frac{S_y}{S_x} \tag{3.42}$$

and the correlation between Δy and y is zero, implying independence between model output and model discrepancy Δy. This characteristic may help improve the model.

The third term within brackets on the right side of Eq. (3.39) is a measure of unconditional bias. If this term can be eliminated, then

$$\mathrm{MSE} = S_x^2(1 - r^2) \tag{3.43}$$

Consider the case where the model output is not correlated with the model error. In the regression plot of y versus x, the slope of regression of x versus y can be expressed as

$$a_{x|y} = \frac{\mathrm{Cov}(x, y)}{\mathrm{Var}(y)} = \frac{S_y^2}{S_y^2} = 1 \tag{3.44}$$

Likewise, the coefficient of regression of y versus x can be written as

$$a_{y|x} = \frac{\text{Cov}(x, y)}{\text{Var}(x)} = \frac{S_y^2}{S_x^2} = r^2 \tag{3.45}$$

In this case the regression will coincide with the 1:1 line, and the slope of the line, y versus x, $a_{y|x}$, is r^2. The model efficiency is NSE $= r^2$. This is an infrequent situation.

Example 3.5.   For the observed and computed runoff peak values in Example 3.1, show by computation the relation between the mean square error and the coefficient of correlation.

Solution

$$r = \frac{\text{Cov}(x,y)}{S_x S_y} = \frac{3.426}{3.856 \times 1.358} = 0.654$$

$$\text{MSE} = \frac{1}{N} \sum_{i=1}^{N} (y_i - x_i)^2 = \frac{1}{69}[(6.068 - 12)^2 + (5.442 - 2.28)^2 + \cdots] = 10.911$$

$$B = \frac{1}{N} \sum_{i=1}^{N} (y_i - x_i) = \bar{y} - \bar{x} = 1.950 - 3.042 = -1.092$$

$$\text{MSE} = B^2 + S_y^2 - 2rS_x S_y + S_x^2 = S_x^2 \left[(1 - r^2) + \left(\frac{S_y}{S_x} - r \right)^2 + \frac{B^2}{S_x^2} \right]$$

$$= 1.092^2 + 1.358^2 - 2 \times 0.654 \times 3.856 \times 1.358 + 3.856^2 = 11.053$$

which is comparable to 10.911 computed from definition.

Example 3.6.   For the observed and computed runoff peak values in Example 3.1, compute the nondimensional measure of conditional bias as well as of unconditional bias.

Solution

$$B_c^2 = \left(\frac{S_y}{S_x} - r \right)^2 = \left(\frac{1.358}{3.856} - 0.654 \right)^2 = 0.09$$

$$C = \frac{B^2}{S_x^2} = \left(\frac{-1.092}{3.856} \right)^2 = 0.080$$

Example 3.7.   Considering computed runoff peak as a dependent variable and observed runoff peak as an independent variable, fit a regression line and express the regression relation. Give the values of the regression coefficient and intercept. Also draw the 1:1 line on the graph. Compute the coefficient of correlation. Compute the bias in computed values of runoff peak, and show that the regression relation is biased.

Solution.   The values of computed of runoff (y) are regressed versus the observed values of runoff (x), as shown in Fig. 3.3. The regression equation is

$$y = 0.2304x + 1.249$$

where y represents the computed runoff and x represents the observed runoff. The correlation coefficient between x and y is

$$r = \frac{\text{Cov}(x,y)}{S_x S_y} = \frac{3.426}{3.856 \times 1.358} = 0.654$$

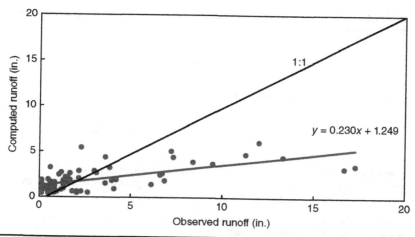

FIGURE 3.3 Regression of computed versus observed runoff values.

The bias of the regression equation is

$$B = \frac{1}{N}\sum_{i=1}^{N}(y_i - x_i) = 1.950 - 3.042 = -1.092$$

Example 3.8. For the observed and computed runoff peak values in Example 3.1, take the observed runoff peak as the dependent variable and the computed peak runoff as the independent variable. Then do the same calculations as in Example 3.7. Now compare the values of the coefficient of correlation and the regression coefficients.

Solution. The computed values of runoff (y) are regressed versus the observed values of runoff (x), as shown in Fig. 3.4. The regression equation is

$$y = 1.8588x - 0.5826$$

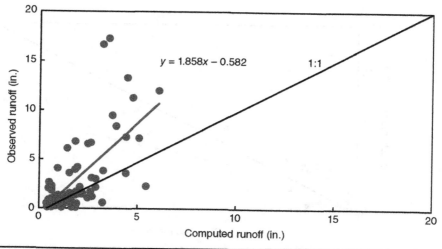

FIGURE 3.4 Regression of computed versus observed runoff values.

where y represents the observed runoff and x represents the computed runoff. The correlation coefficient between the x and y values is

$$r = \frac{\text{Cov}(x,y)}{S_x S_y} = \frac{3.426}{1.358 \times 3.856} = 0.654$$

The bias of the regression equation is

$$B = \frac{1}{N} \sum_{i=1}^{N} (y_i - x_i) = 3.042 - 1.950 = 1.092$$

Example 3.9.   Now transform the computed values of runoff peak by using the regression equation, and then plot the transformed values versus observed runoff peak values. Finally, do the same calculations as in Example 3.7. What do you conclude?

Solution.   The computed values of runoff (y) are regressed versus the observed values of runoff (x), as shown in Fig. 3.5. The regression equation is

$$y = 0.4283x + 1.739$$

where y represents the transformed runoff and x represents the observed runoff. The correlation coefficient between x and y is

$$r = \frac{\text{Cov}(x,y)}{S_x S_y} = \frac{6.369}{3.856 \times 2.524} = 0.654$$

The bias of the regression equation is

$$B = \frac{1}{N} \sum_{i=1}^{N} (y_i - x_i) = 3.042 - 3.042 = 0$$

The bias has been reduced, but there is no change in the coefficient of correlation.

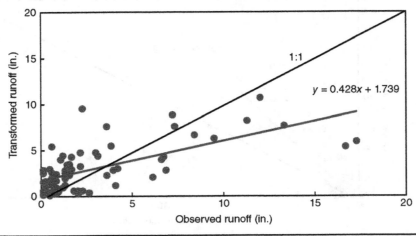

FIGURE 3.5   Regression of computed versus observed runoff values.

3.2.11   Decomposition of Mean Square Error and NSE

The MSE decomposition can be expressed as

$$\text{MSE} = 2S_y S_x(1-r) + (S_y - S_x)^2 + (\bar{y} - \bar{x})^2 \tag{3.46}$$

It can be shown that many different combinations of the three components can lead to the same overall value of MSE or NSE, potentially yielding considerable ambiguity in the comparative evaluation of alternative models and parameter sets. The relative contribution of each of these components to the overall MSE value can be computed as

$$w_i = \frac{W_i}{\sum_{i=1}^{3} F_i} \tag{3.47}$$

$$W_1 = 2S_y S_x(1-r) \tag{3.48}$$

$$W_2 = (S_y - S_x)^2 \tag{3.49}$$

$$W_3 = (\bar{y} - \bar{x})^2 \tag{3.50}$$

Each of the measures can be employed to assess the goodness of fit or model quality. However, it is preferable to have relative or dimensionless measures that indicate the model is biased when the measure is zero, is perfect when the measure is 1, and has other possibilities between these two limits of 0 and 1. A relative measure is obtained by dividing the measure by a simple statistical characteristic of observations, such as mean or standard deviation, or by comparing to a reference quantity. Further, each of these measures can be employed to define derivative measures. However, note that the derivative measures derived from the same basic measure are mutually dependent, and care should be exercised when using them together.

Example 3.10.   For the observed and computed runoff peak values in Example 3.1, show by computation the partition of the mean square error into three components. Compute the relative contribution of each component to the error.

Solution

$$W_1 = 2S_y S_x(1-r) = 2 \times 1.358 \times 3.856(1-0.654) = 3.618$$

$$W_2 = (S_y - S_x)^2 = (1.358 - 3.856)^2 = 6.243$$

$$W_3 = (\bar{y} - \bar{x})^2 = (-1.092)^2 = 1.193$$

$$\text{MSE} = W_1 + W_2 + W_3 = 11.054$$

Murphy (1988) and Weglarczek (1988) have shown that

$$\text{NSE} = A - B - C \tag{3.51}$$

where

$$A = r^2 \tag{3.52}$$

$$B = \left(r - \frac{S_y}{S_x}\right)^2 \tag{3.53}$$

$$C = \left(\frac{\bar{y} - \bar{x}}{S_x} \right)^2 \tag{3.54}$$

where r is the coefficient of correlation between predicted and observed values and S_x is the standard deviation of observed values. Here A measures the strength of the linear relationship between predicted and observed values, B measures the conditional bias, and C measures the unconditional bias. NSE is also called the *quadratic score* (Krzysztofowicz, 1992). Unlike MSE, its value increases with increasing model quality. Its form better indicates both desirable and undesirable features of the model.

Example 3.11.   For the observed and computed runoff peak values in Example 3.1, show by computation the partition of the Nash-Sutcliffe efficiency criterion into three components. Compute the relative contribution of each component to the criterion.

Solution

$$A = r^2 = 0.654^2 = 0.428$$

$$B = \left(r - \frac{S_y}{S_x} \right)^2 = \left(0.654 - \frac{1.358}{3.856} \right)^2 = 0.091$$

$$C = \left(\frac{\bar{y} - \bar{x}}{S_x} \right)^2 = \left(\frac{-1.092}{3.856} \right)^2 = 0.080$$

$$\text{NSE} = A - B - C = 0.257$$

Gupta et al. (2009) decomposed NSE as

$$\text{NSE} = 2ar - a^2 - b_N^2 \tag{3.55}$$

where

$$a = \frac{S_y}{S_x} \tag{3.56}$$

$$b_N = \frac{\bar{y} - \bar{x}}{S_x} \tag{3.57}$$

Here a is a measure of the relative variability of the predicted and observed values, and b_N is the bias normalized by the standard deviation of the observed values. Note that $b_N = \sqrt{C}$.

Two of the three components of NSE relate to the ability of the model to reproduce the first and second moments of the distribution of observed values (i.e., mean and standard deviation), whereas the third component relates to the ability of the model to reproduce the timing and shape measured by the correlation coefficient. The ideal values of the three components are $r = 1$, $a = 1$, and $b_N = 0$. When NSE is used as a criterion for optimization of model performance, this means searching for optimal values of the three components leading to a balanced solution. This is analogous to a multicriteria approach. In hydrologic modeling, the aim is to match the overall volume of flow, time distribution of flow, timing, and shape of the hydrograph or probability density function.

In using NSE, two things must be noted. First, the bias $(\bar{y} - \bar{x})$ appears in a normalized form, scaled by the standard deviation of observed values. This means that if the variability (standard deviation) is high, the bias component will have a smaller contribution to NSE, and this might result in large volume balance errors. Equivalently, this is like using a weighted objective function with a low weight given to the bias component.

Second, a appears twice in Eq. (3.55). Taking the first derivative of NSE with respect to a shows that the maximum of NSE occurs when $a = r$. Since r is always smaller than unity, during maximization the tendency is to select a that underestimates the variability in flows, i.e., one will tend to prefer models and parameters that underestimate variability.

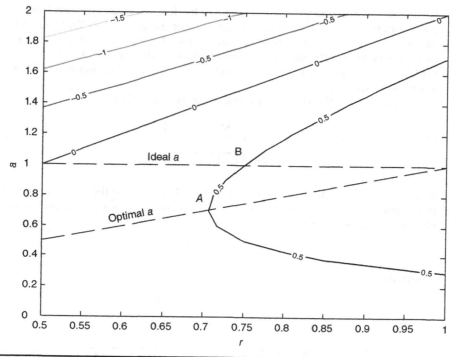

FIGURE 3.6   Relationship of NSE with a and r.

Taken together, when $b_N = 0$ and $a = r$, NSE is equivalent to the coefficient of determination r^2. The implication is that r^2 can be interpreted as a maximum or potential value of NSE, if the other two components achieve their maximum values.

Assuming $b_N = 0$ (b_N is an additive term), the relationship between NSE and r and a is shown in Fig. 3.6. For specified r, the optimal values of a for maximizing NSE lie on the 1:1 line, even the ideal value of a is on horizontal line 1:0. The interplay between a and r is likely to be of importance for any type of hydrologic model.

3.2.12   Alternative Performance Criteria

To overcome the problem of likely underestimation of variability in flows due to the interaction between a and r, Gupta et al. (2009) suggested inflating the variability while preserving the mean and linear correlation of observations as

$$x_* = c_0(x - \bar{x}) + \bar{x} \tag{3.58}$$

where c_0 is a correction factor for inflating the variability in observed flows. By substituting Eq. (3.58) in Eq. (3.55), the corrected version of NSE becomes

$$\text{NSE}_{\text{corr}} = \frac{1}{c_0} 2ar - \frac{1}{c_0^2} a^2 - \frac{1}{c_0^2} b_N^2 \tag{3.59}$$

If $c_0 = 1/r$, then a value of $a = 1$ will maximize NSE_{corr} (as opposed to $a = r$ maximizing NSE).

Alternatively, since MSE and NSE can be decomposed into correlation, variability error, and bias error, as separate criteria to be optimized, the calibration and validation can be viewed from a multiobjective perspective. Further, it is meaningful to express the bias component by the ratio of mean values of predicted and observed flow values (b) for further analysis as opposed to using b_N. Now all three components have their ideal values at unity.

Gupta et al. (2009) defined an efficiency criterion (KGE) as

$$KGE = 1 - ED \tag{3.60}$$

$$ED = \sqrt{(r-1)^2 + (a-1)^2 + (b-1)^2} \tag{3.61}$$

where ED is the euclidean distance from the ideal point in the scaled space and b is the bias defined by the ratio of the mean of predicted flows or values to the mean of observed values (flows):

$$b = \frac{\bar{y}}{\bar{x}} \tag{3.62}$$

Alternatively,

$$KGE_s = 1 - ED_s \tag{3.63}$$

where

$$ED_s = \sqrt{[s_r(r-1)]^2 + [s_a(a-1)]^2 + [s_b(b-1)]^2} \tag{3.64}$$

Here s_r, s_a, and s_b are the scaling factors, and they can also be used to adjust the relative contributions of different components.

The relative contributions of the three components of KGE can be expressed as

$$g_i = \frac{G_i}{\sum\limits_{i=1}^{3} G_i} \tag{3.65}$$

where

$$G_1 = (r-1)^2 \tag{3.66}$$

$$G_2 = (a-1)^2 \tag{3.67}$$

$$G_3 = (b-1)^2 \tag{3.68}$$

3.2.13 Multidimensional Measures

When two more goodness-of-fit measures are employed and are related, then wrong conclusions can be drawn if care is not exercised. Weglarcyzk (1998) addressed this issue by using the criteria employed by Sarma et al. (1973). These criteria were the integral square error ISE and the special correlation coefficient r_s, both based on the mean square error of a model, and the correlation coefficient between x and y.

Equation (3.26) shows that models with $MSE > S_x^2 + (\bar{x})^2$ cannot be evaluated with r_s, for $r_s^2 < 0$ and no real number can be assigned to r_s. Comparing Eqs. (3.19) and (3.26), one sees that r_s tends to 1 much quicker than ISE does to 0 and r_s is less conservative in assessing the goodness of fit. This is easily seen by combining Eqs. (3.19) and (3.26) as

$$\frac{ISE^2}{1-r_s^2} = 1 + CV^2 > 1 \tag{3.69}$$

where CV is the coefficient of variation of observed values x: $CV = S_x/\bar{x}$.

Equation (3.69) shows the relation between ISE and r_s, since they both depend on MSE and clearly they are not independent. Ranking models by ISE, r_s, and r may produce contradictory results. Figure 3.7 plots r_s versus ISE for various values of CV and the ranking of models following Sarma et al. (1973). A model ranked excellent by r_s may be ranked differently by ISE as excellent, very good, good, fair, or even poor. If r_s ranks a model as very good, then ISE cannot rank the model as very good. The same holds for other ratings, such as good and fair. This discussion suggests that when rating a model, an additional explanatory note should be provided, indicating the criteria for rating. Or else r_s and ISE should not be used at the same time.

NSE also entails MSE and is therefore related to both ISE and r_s. Analogous to Eq. (3.69), combining Eq. (3.28) and the first equality in Eq. (3.19) yields

$$\frac{ISE^2}{1-NSE} = CV^2 \tag{3.70}$$

Since CV is fixed for observations, only one of ISE, r_s, and NSE, as an independent measure, should be used as a goodness-of-fit measure. The second measure can be computed from Eq. (3.69) and CV. Combining Eqs. (3.69) and (3.70), one obtains

$$\frac{1}{ISE^2} + \frac{1}{1-NSE} = \frac{1}{1-r_s^2} \tag{3.71}$$

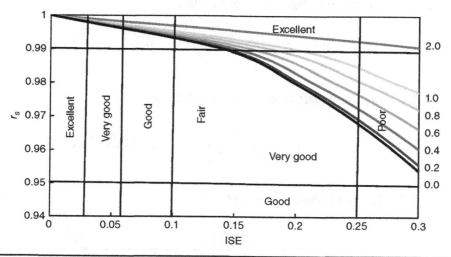

FIGURE 3.7 Special correlation coefficient r_s versus integral square error ISE for certain values of coefficient of variation CV.

Thus, if two measures are known, then the third measure is uniquely determined from Eq. (3.71). When a model is unbiased and its predicted values are independent of prediction error, then $MSE = S_x^2 - S_y^2$, $NSE = r^2$, and r is related to r_s through Eq. (3.71). If the same values of r and r_s are considered, then an unacceptable results occurs: $1/ISE^2 = 0$.

3.3   Goodness-of-Fit Measures for Probabilistic Models

To measure the goodness-of-fit of frequency distributions to empirical data, several goodness-of-fit statistics have been employed. These include the Kolmogorov-Smirnov (K-S) statistic D_N, Cramer-von Mises (C-M) statistic W_N^2, Aderson-Darling (A-D) statistic A_N^2, modified weighted Watson statistic U_N^2, and Liao and Shimokawa statistic L_N. These test statistics measure the distance between an empirical distribution function and a theoretical continuous distribution function, and they are defined as follows (Park and Kim, 1992; Shimokawa and Liao, 1999; Amal, 2005).

3.3.1   Kolmogorov-Smirnov (K-S) Statistic D_N

$$D_N = \max_{1 \le i \le N}(\hat{\delta}_i) \qquad \hat{\delta}_i = \max\left[\frac{i}{N} - F_0(x_i;\hat{\theta}), F_0(x_i;\hat{\theta}) - \frac{i-1}{N}\right] \qquad (3.72)$$

where $\hat{\theta}$ is a parameter vector and F_0 is the cumulative probability distribution function.

3.3.2   Cramer-von Mises (C-M) Statistic W_N^2

$$W_N^2 = \frac{1}{12N} + \sum_{i=1}^{N}\left[F_0(x_i,\hat{\theta}) - \frac{2i-1}{2N}\right]^2 \qquad (3.73)$$

3.3.3   Anderson-Darling (A-D) Statistic A_N^2

$$A_N^2 = -N - \frac{1}{N}\sum_{i=1}^{N}(2i-1)\ln\{F_0(x_i,\hat{\theta})[1 - F_0(x_{n+1-i},\hat{\theta})]\} \qquad (3.74)$$

3.3.4   Modified Weighted Watson Statistic U_N^2

$$U_n^2 = N^2\sum_{i=1}^{N}d_i^2 - N\left(\sum_{i=1}^{N}d_i\right)^2 \qquad d_i = \frac{F_0(x_i,\hat{\theta}) - i/N + 1}{\sqrt{i(N-i+1)}} \qquad (3.75)$$

3.3.5   Liao and Shimokawa Statistic L_N

$$L_N = \frac{1}{\sqrt{N}} + \sum_{i=1}^{n}\frac{\max\left[\frac{i}{N} - F_0(x_i,\hat{\theta}), F_0(x_i,\hat{\theta}) - \frac{i-1}{N}\right]}{\sqrt{F_0(x_i,\hat{\theta})\left[1 - F_0(x_i,\hat{\theta})\right]}} \qquad (3.76)$$

To apply these goodness-of-fit statistics to a sample of size N whose population is defined by a continuous cumulative probability distribution function $F(x)$, let $F_0(x_i, \theta)$ be a specified distribution that contains a set of parameters θ, and let $x_1 < x_2 < \cdots < x_N$ be order statistics. Then the steps for testing the null hypothesis H_0—that the theoretical distribution mimics F_0 with parameters θ—are as follows (Amal, 2005):

1. Sort original observations $x_1 < x_2 < \cdots < x_N$ in increasing order.
2. Estimate parameters $\hat{\theta}$ of a given distribution $F_0(x; \theta)$.
3. Calculate statistics D_N, W_N^2, A_N^2, U_N^2 and L_N.
4. Generate a fixed sample $u_1 < u_2 < \cdots < u_N$ of N-order statistics from a uniform distribution.
5. Transform these uniform distribution values to the given distribution with parameters $\hat{\theta}$, $x_i^* = F^{-1}(u_i)$, $i = 1, 2, \cdots, N$; that is, generate independent identically distributed (IID) observations $x_1^* < x_2^* < \cdots < x_N^*$ with parameters $\hat{\theta}$.
6. Estimate parameters $\hat{\theta}$ of the given distribution $F_0(x; \theta)$.
7. Calculate the probability distribution $F_0(x_i^*; \hat{\theta})$ with parameters $\hat{\theta}$.
8. Calculate statistics $D_N^*, W_N^{*2}, A_N^{*2}, U_N^{*2}$ and L_N^*.
9. Repeat steps 4 to 8 M times, with M being the number of bootstrap repetitions. For a given significance level α, the desired critical values for the K-S test, C-M test, A-D test, modified weighted Watson test, and Liao-Shimokawa test are determined as the $(1-\alpha)$ quantile of the values $D_N^{*(1)}, D_N^{*(2)}, \cdots, D_N^{*(M)}$; $W_N^{*2(1)}, W_N^{*2(2)}, \cdots, W_N^{*2(M)}$; $A_N^{*2(1)}, A_N^{*2(2)}, \cdots, A_N^{*2(M)}$; $U_N^{*2(1)}, U_N^{*2(2)}, \cdots, U_N^{*2(M)}$, and $L_N^{*(1)}, L_N^{*(2)}, \cdots, L_N^{*(M)}$, respectively.
10. The null hypothesis that the true distribution is F_0 with assumed parameters is then rejected (at significance level α, for sample size N) if these test statistic values are greater than their corresponding Critical Values.

The theoretical distribution can be tested by the chi-square goodness-of-fit test. The test statistic is a chi-square random variable χ^2 defined as

$$\chi_{K-m-1}^2 = \sum_{i=0}^{K} \frac{(o_i - e_i)^2}{e_i} \tag{3.77}$$

where o_i is the observed frequency count for the ith level of a variable; e_i is the expected frequency count for the ith level of the variable; K is the number of levels of the variable; $K - m - 1$ is the degrees of freedom; and m is the number of parameters of the distribution.

Example 3.12. Using observed annual peak stream flow given in Table 3.2, compute the goodness-of-fit measures with the use of K-S, C-M, A-D, modified weighted Watson, Liao and Shimokawa, and chi-square tests. Take the gamma distribution as a test probability distribution.

Solution. The gamma distribution is given as

$$f(x; \alpha, \beta) = \frac{1}{\beta^\alpha \Gamma(\alpha)} x^{\alpha-1} \exp\left(-\frac{x}{\beta}\right)$$

No.	Peak (cfs)	No.	Peak (cfs)	No.	Peak (cfs)	No.	Peak (cfs)
1	2,300	13	4,980	26	4,730	39	1,520
2	3,390	14	2,840	27	1,060	40	29,800
3	1,710	15	3,220	28	3,290	41	2,740
4	9,780	16	2,440	29	7,880	42	1,740
5	10,500	17	1,320	30	13,800	43	557
6	13,700	18	16,000	31	10,500	44	5,350
7	6,500	19	16,100	32	7,150	45	11,200
8	3,710	20	1,180	33	1,030	46	4,930
9	536	21	5,440	34	13,100	47	3,490
10	17,000	22	2,420	35	2,920	48	2,990
11	6,630	23	9,140	36	5,210	49	6,160
12	1,220	24	6,700	37	4,460	50	1,480
		25	912	38	3,100	51	496

TABLE 3.2   Observed Annual Peak Stream Flow

Goodness-of-Fit for K-S, C-M, A-D, modified Weighted Watson, and Liao and Shimokawa Tests
Per the test procedures discussed above, the following steps are needed to evaluate the goodness-of-fit study.

Step 1: Order the peak flow values (assumed random) in increasing order, and estimate the parameters for the probability distribution, as shown in Table 3.3. The parameters of the gamma

Order	Peak (cfs)	Order	Peak (cfs)	Order	Peak (cfs)	Order	Peak (cfs)
1	496	13	1,740	26	3,710	39	7,880
2	536	14	2,300	27	4,460	40	9,140
3	557	15	2,420	28	4,730	41	9,780
4	912	16	2,440	29	4,930	42	10,500
5	1,030	17	2,740	30	4,980	43	10,500
6	1,060	18	2,840	31	5,210	44	11,200
7	1,180	19	2,920	32	5,350	45	13,100
8	1,220	20	2,990	33	5,440	46	13,700
9	1,320	21	3,100	34	6,160	47	13,800
10	1,480	22	3,220	35	6,500	48	16,000
11	1,520	23	3,290	36	6,630	49	16,100
12	1,710	24	3,390	37	6,700	50	17,000
		25	3,490	38	7,150	51	29,800
		Parameters: $\alpha = 1.3164$, $\beta = 4.4737 \times 10^3$					

TABLE 3.3   Ordered Annual Peak Stream Flow and Parameters Estimated with MLE

distribution are estimated using the maximum likelihood estimation (MLE), but can also be estimated using entopy.

Step 2: Compute the corresponding test statistics.

1. Table 3.4 lists the cumulative probability distribution function (CDF) computed from increasingly ordered annual peak stream flow data for the fitted gamma distribution.

2. Compute the test statistics. The computation example uses $Q_{(1)} = 496$ cfs for a sample size of $N = 51$ as an example. The full list of computed values is given in Table 3.4.

- K-S test (D_N) [Eq. (3.72)]:

Order	Peak (cfs)	CDF	Test Statistics				
			K-S (δ_i)	C-M (CM d_i)	A-D (AD d_i)	U_N^2 (d_i)	L_N (Ld_i)
1	496	0.0441	0.0441	0.0012	−9.0297	0.0035	0.2147
2	536	0.0486	0.0290	0.0004	−18.6652	0.0010	0.1347
3	557	0.0510	0.0117	3.73E-06	−29.9324	−0.0006	0.0534
4	912	0.0933	0.0345	0.0006	−37.5253	0.0012	0.1185
5	1,030	0.1079	0.0295	0.0004	−42.8437	0.0008	0.0950
6	1,060	0.1117	0.0136	1.46E-05	−51.7633	−0.0002	0.0433
7	1,180	0.1267	0.0105	5.38E-07	−57.9304	−0.0004	0.0317
8	1,220	0.1318	0.0251	0.0002	−60.4520	−0.0012	0.0742
9	1,320	0.1444	0.0321	0.0005	−64.5542	−0.0015	0.0913
10	1,480	0.1646	0.0315	0.0005	−69.6576	−0.0014	0.0849
11	1,520	0.1697	0.0460	0.0013	−73.3189	−0.0020	0.1226
12	1,710	0.1936	0.0417	0.0010	−74.3267	−0.0017	0.1055
13	1,740	0.1974	0.0575	0.0023	−74.0839	−0.0023	0.1445
14	2,300	0.2665	0.0116	3.37E-06	−68.0529	−0.0001	0.0263
15	2,420	0.2810	0.0131	1.11E-05	−69.0385	−0.0003	0.0292
16	2,440	0.2834	0.0304	0.0004	−73.1189	−0.0010	0.0674
17	2,740	0.3187	0.0146	2.34E-05	−73.1350	−0.0003	0.0314
18	2,840	0.3302	0.0227	0.0002	−74.0442	−0.0006	0.0483
19	2,920	0.3393	0.0332	0.0005	−72.2151	−0.0010	0.0701
20	2,990	0.3473	0.0449	0.0012	−74.5592	−0.0015	0.0943
21	3,100	0.3596	0.0522	0.0018	−75.8782	−0.0017	0.1087
22	3,220	0.3728	0.0585	0.0024	−76.1786	−0.0020	0.1211
23	3,290	0.3805	0.0705	0.0037	−78.3911	−0.0024	0.1452
24	3,390	0.3913	0.0793	0.0048	−78.8187	−0.0027	0.1626
25	3,490	0.4019	0.0883	0.0062	−78.4213	−0.0030	0.1801

TABLE 3.4   CDF and Corresponding Quantities Computed for Ordered Annual Peak Stream Flow Data.

Order	Peak (cfs)	CDF	Test Statistics				
			K-S (δ_i)	C-M (CM d_i)	A-D (AD d_i)	U_N^2 (d_i)	L_N (Ld_i)
26	3,710	0.4248	0.0850	0.0056	−71.8666	−0.0029	0.1719
27	4,460	0.4979	0.0315	0.0005	−64.2055	−0.0008	0.0631
28	4,730	0.5222	0.0268	0.0003	−63.0316	−0.0006	0.0537
29	4,930	0.5396	0.0290	0.0004	−62.4556	−0.0007	0.0582
30	4,980	0.5439	0.0444	0.0012	−63.4603	−0.0013	0.0891
31	5,210	0.5630	0.0448	0.0012	−62.2242	−0.0013	0.0904
32	5,350	0.5743	0.0531	0.0019	−61.8107	−0.0016	0.1074
33	5,440	0.5815	0.0656	0.0031	−62.1859	−0.0021	0.1329
34	6,160	0.6349	0.0318	0.0005	−57.2926	−0.0008	0.0660
35	6,500	0.6579	0.0284	0.0003	−55.3675	−0.0006	0.0598
36	6,630	0.6664	0.0395	0.0009	−52.4782	−0.0011	0.0838
37	6,700	0.6708	0.0547	0.0020	−53.2252	−0.0017	0.1163
38	7,150	0.6983	0.0468	0.0014	−50.1841	−0.0014	0.1020
39	7,880	0.7383	0.0264	0.0003	−40.2878	−0.0005	0.0600
40	9,140	0.7960	0.0313	0.0005	−35.0231	0.0012	0.0777
41	9,780	0.8205	0.0362	0.0007	−31.0876	0.0015	0.0942
42	10,500	0.8446	0.0407	0.0010	−28.9422	0.0018	0.1124
43	10,500	0.8446	0.0211	0.0001	−27.6059	0.0009	0.0583
44	11,200	0.8651	0.0220	0.0001	−24.8973	0.0010	0.0643
45	13,100	0.9084	0.0457	0.0013	−20.6088	0.0024	0.1583
46	13,700	0.9190	0.0367	0.0007	−18.4602	0.0021	0.1344
47	13,800	0.9207	0.0187	7.92E-05	−18.3082	0.0011	0.0692
48	16,000	0.9497	0.0281	0.0003	−14.2122	0.0019	0.1284
49	16,100	0.9507	0.0101	8.56E-08	−9.9778	0.0007	0.0466
50	17,000	0.9591	0.0213	0.0001	−9.0618	−0.0002	0.1075
51	29,800	0.9973	0.0169	5.02E-05	−4.8272	0.0023	0.3244

TABLE 3.4 CDF and Corresponding Quantities Computed for Ordered Annual Peak Stream Flow Data. (*Continued*)

$$\hat{\delta}_1 = \max\left[\frac{1}{N} - F(x_{(1)}), F(x_{(1)}) - \frac{1-1}{N}\right] = \max\left[\frac{1}{59} - 0.0441, 0.0441 - \frac{1-1}{51}\right] = 0.0441$$

- C-M test (W_N^2) [Eq. (3.73)]: The quantity inside the summation (that is, CMd_i) for $i = 1$ is computed as

$$CMd_1 = \left[F(x_{(1)}) - \frac{2(1)-1}{2N}\right]^2 = \left[0.0441 - \frac{1}{2(51)}\right]^2 = 0.0012$$

- AD test (A_N^2) [Eq. (3.74)]: The quantity inside the summation (that is, ADd_i) for $i = 1$ is computed as

$$ADd_1 = [2(1) - 1]\ln[F(x_{(1)}) \{1 - F(x_{(1)})\}] = \ln[0.0441(1 - 0.0441)] = -3.1670$$

- Modified weighted Watson test (U_N^2) [Eq. (3.75)]: The quantity inside the summation for $i = 1$ is computed as

$$d_1 = \frac{F(x_{(1)}) - \dfrac{1}{N+1}}{\sqrt{1(N-1+1)}} = \frac{0.0441 - \dfrac{1}{51+1}}{\sqrt{1(51-1+1)}} = 0.0035$$

- Liao and Shimokawa test (L_N) [Eq. (3.76)]: The quantity inside the summation (that is, Ld_i) for $i = 1$ is computed as

$$Ld_1 = \frac{\max\left[\dfrac{1}{N} - F(x_{(1)}), F(x_{(1)}) - \dfrac{1-1}{N}\right]}{\sqrt{F(x_{(1)})[1 - F(x_{(1)})]}} = \frac{\max\left[\dfrac{1}{51} - 0.0441, 0.0441\right]}{\sqrt{0.0441(1 - 0.0441)}} = 0.2147$$

Now, substituting the quantities computed in Table 3.4 back into Eqs. (3.72) to (3.76), we calculate the final test statistics for each goodness-of-fit measure as follows:

KS test: $D_N = 0.0883 \cdot$

CM test: $W_N^2 = 0.0558$

AD test: $A_N^2 = 0.3534$

Modified weighted Watson test: $U_N^2 = 0.2993$

Liao and Shimokawa test: $L_N = 5.1695$

3. Apply the parametric bootstrap method M times to approximate the P-value with given significance level α. Here we choose $M = 1000$ and $\alpha = 0.05$. To illustrate the procedure, one parametric bootstrap simulation is used as an example.
   - Generate IID random values from the fitted gamma distribution (with parameters given in Table 3.3 for sample size $N = 51$), and sort the simulated random values in increasing order (Table 3.5).
   - Reestimate the parameters of the gamma distribution, and calculate the CDF and the corresponding test statistics using the simulated random variable. Since it has already been discussed how to compute test statistics (steps 1 and 2), here only the final results are presented.
   (a) Estimated parameters: $\alpha_1^* = 1.3241$, $\beta_1^* = 4.8206 \times 10^3$.
   (b) Test statistics computed from simulated random values with re-estimated parameters are

$$D_{N1}^* = 0.1400; W_{N1}^{2*} = 0.1496; A_{N1}^{2*} = 0.8237; U_{N1}^{2*} = 0.6595; \text{ and } L_{N1}^* = 6.8445$$

   - Repeat the parametric bootstrap simulation 1000 times. One can approximate the P value and the corresponding critical value using the K-S test as an example:

$$P \text{ value} = \frac{\sum_{i=1}^{M} 1(D_{Ni}^* > D_N)}{M}$$

Generated Values				Sorted Values			
No.	Generated	No.	Generated	Order	Sorted	Order	Sorted
1	8,683.20	26	12,416.01	1	51.56	26	6,351.30
2	921.76	27	872.26	2	127.24	27	6,756.52
3	7,874.64	28	4,003.93	3	574.63	28	7,025.33
4	10,470.50	29	3,752.08	4	766.02	29	7,581.81
5	3,019.36	30	6,756.52	5	872.26	30	7,719.17
6	5,625.04	31	12,419.87	6	921.76	31	7,789.85
7	1,548.26	32	9,953.94	7	1,317.86	32	7,874.64
8	7,719.17	33	10,547.60	8	1,411.29	33	8,067.79
9	15,787.45	34	4,895.35	9	1,548.26	34	8,329.23
10	1,592.99	35	13,512.85	10	1,592.99	35	8,385.82
11	19,530.55	36	2,193.47	11	2,007.60	36	8,683.20
12	12,160.63	37	51.56	12	2,193.47	37	8,872.19
13	1,411.29	38	7,025.33	13	2,194.08	38	8,948.61
14	13,026.83	39	574.63	14	2,431.96	39	9,190.72
15	8,385.82	40	8,329.23	15	2,801.57	40	9,953.94
16	3,906.03	41	4,407.35	16	3,019.36	41	10,131.30
17	9,190.72	42	7,581.81	17	3,282.55	42	10,470.50
18	8,067.79	43	1,27.24	18	3,643.24	43	10,547.60
19	8,948.61	44	2,801.57	19	3,752.08	44	11,060.80
20	11,060.80	45	7,789.85	20	3,906.03	45	12,160.63
21	2,431.96	46	2,007.60	21	4,003.93	46	12,416.01
22	1,317.86	47	766.02	22	4,407.35	47	12,419.87
23	2,194.08	48	10,131.30	23	4,895.35	48	13,026.83
24	5,589.25	49	6,351.30	24	5,589.25	49	13,512.85
25	3,643.24	50	8,872.19	25	5,625.04	50	15,787.45
		51	3,282.55			51	19,530.55

TABLE 3.5 Generating Gamma Distributed Random Values and Sorting Them in Increasing Order

The critical value can be approximated by interpolation from computed D_{Ni}^*, $i = 1, \ldots, M$, and its empirical distribution.

K-S test final result:

$$D_N = 0.0883, \; P = 0.222, \; \text{Critical value} = 0.1156$$

C-M test final results:

$$W_N^2 = 0.0558 \qquad P = 0.456 \qquad \text{Critical Value} = 0.1327$$

A-D test final results:

Modified weighted Watson test final results:

$$U_N^2 = 0.2993 \qquad P = 0.532 \qquad \text{Critical Value} = 0.6821$$

Liao and Shimokawa test final results:

$$L_N = 5.1695 \qquad P = 0.438 \qquad \text{Critical Value} = 6.9574$$

Chi-Square Goodness-of-Fit Test

Step 1: To apply the chi-square goodness-of-fit test, one should first examine the frequency histogram. To determine the number of bins needed, one can choose the commonly applied Sturges formula $k = [1 + \log_2 N]$.

For $N = 51$, one has $k = [1 + \log_2 51] = 7$. The observed relative frequency is shown in Fig. 3.8 and Table 3.6.

Step 2: Compute the relative frequencies from the fitted gamma distribution (parameters listed in Table 3.3) for the corresponding data interval, as shown in Table 3.6. Using the data interval of [496, 4682.2857], e.g., one has

$$e_1 = F(4682.2857; 1.3164, \ 4.4737 \times 10^3) - F(496; 1.3164, \ 4.4737 \times 10^3) = 0.4739$$

The rest of the results are listed in Table 3.6.

Step 3: Computing the test statistic using Eq. (3.77), one has

$$\chi^2_{K-m-1} = 0.1867$$

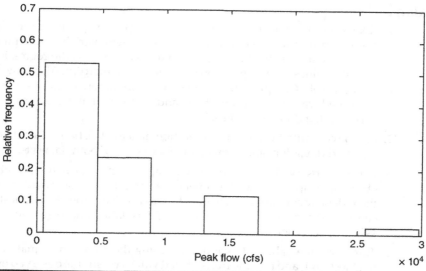

FIGURE **3.8**   Relative frequency plot.

Relative Frequency (Observed)	Data Interval	Estimated Frequency Computed from the Fitted Gamma Distribution
0.5294	[496, 4682.2857]	0.4739
0.2353	[4682.2857, 8868.5714]	0.2667
0.0980	[8868.5714, 13,054.8571]	0.1228
0.1176	[13,054.8571, 17,241.1429]	0.0536
0	[17,241.1429, 21,427.4286]	0.0227
0	[21,427.4286, 25,613.7343]	0.0095
0.0196	[25,613.7143, 29,800]	0.0039

TABLE 3.6   Relative Frequencies and Corresponding Class Intervals

From the chi-square goodness-of-fit test it is known that the test statistic should follow the chi-square distribution with the degrees of freedom $(DOF) = K - m - 1 = 7 - 2 - 1 = 4$
    Choosing the significance level $\alpha = 0.05$, one can calculate the corresponding critical value as

$$\chi_4^{2(-1)} (0.95) = 9.4877$$

All the tests applied in the example show that gamma distribution may be applied to model the annual peak stream flow data given in Table 3.2. For K-S, C-M, A-D, modified weighted Watson, and the Liao and Shimokawa tests, the parametric bootstrap method is applied. The approximated P values are all greater than 0.05. For the chi-square goodness-of-fit test, the limiting distribution is the chi-square distribution of a given degree of freedom. The test results shows that the computed test statistic is less than the critical value, i.e., the P value is greater than 0.05.

Questions

1. Take a set of values on yearly rainfall and pan evaporation for a given area. Then, considering evaporation as a dependent variable and rainfall as an independent variable, fit a regression line and express the regression relation. Give the values of the regression coefficient and intercept. Also draw the 1:1 line on the graph. Compute the coefficient of correlation. Compute the bias in the computed values of evaporation, and show that the regression relation is biased. What do you conclude?

2. For the computed values of evaporation data in Question 1, calculate the mean, standard deviation, and coefficient of variation of both data sets.

3. For the observed and computed evaporation values in Question 1, take observed evaporation as the dependent variable and computed evaporation as the independent variable. Then do the same calculations as in Question 1. Now compare the values of the coefficient of correlation and regression coefficients. What do you conclude?

4. Transform the values of evaporation using the regression equation computed in Question 1, and plot the transformed values versus the observed evaporation. Then, do the same calculations. What do you conclude?

5. For the observed and computed evaporation values in Question 1, calculate the maximum absolute error, mean square error, mean absolute error, integral square error, special correlation coefficient, Nash-Sutcliffe efficiency index, generalized efficiency index, and index of disagreement.

6. What do the values of the different measures computed in Question 5 say about the regression model?

7. For the observed and computed evaporation values in Question 1, show their relationship by computing the mean square error and coefficient of correlation.

8. For the observed and computed evaporation values in Question 1, compute the nondimensional measure of conditional bias as well as of unconditional bias.

9. For the observed and computed evaporation values in Question 1, show the partition by computation of the Nash-Sutcliffe efficiency criterion into three components. Compute the relative contribution of each component to the criterion.

10. For the observed and computed evaporation values in Question 1, show the relationship by computing the special correlation coefficient and mean square error.

11. For the observed and computed evaporation values in Question 1, show the partition by computing of the mean square error into three components. Compute the relative contribution of each component to the error.

References

Amal, S. H. (2005). Goodness-of-fit for the generalized exponential distribution. http://interstat.statjournals.net/YEAR/2005/articles/0507001.pdf

ASCE (1993). Criteria for evaluation of watershed models. *Journal of Irrigation and Drainage Engineering*, vol. 119, no. 3, pp. 429–442. (American Society of Civil Engineers Task Committee on Definition of Criteria for Evaluation of the Watershed Management, Irrigation, and Drainage Division).

Birikundavyi, S., R. Labib, H. T. Trung, and J. Rousselle (2002). Performance of neural networks in daily streamflow forecasting. *Journal of Hydrologic Engineering*, vol. 7, no. 5, pp. 392–398.

Criss, R. E., and W. E. Winston (2008). Do Nash values have value? Discussion and alternative proposals. *Hydrological Processes*, vol. 22, pp. 2723–2725.

Downer, C. W., and F. L. Ogden (2004). GSSHA: Model to simulate diverse stream flow producing processes. *Journal of Hydrologic Engineering*, vol. 9, no. 3, pp. 161–173.

Erpul, G., L. D. Norton, and D. Gabriels (2003). Sediment transport from interrill areas under wind-driven rain. *Journal of Hydrology*, vol. 276, pp. 1984–1987.

Fisher, R. A. (1928). The general sampling distribution of the multiple correlation coefficient. *Proceedings, Royal Society, London*, vol. 121, pp. 654–673.

Garrick, M., C. Cunnane, and J. E. Nash (1978). A criterion for efficiency of rainfall-runoff models. *Journal of Hydrology*, vol. 36, pp. 375-381.

Gupta, H. V., H. Kling, K. Yilmaz, and G. F. Martinez (2009). Decomposition of the mean squared error and NSE performance criteria: Implications for improving hydrological modeling. *Journal of Hydrology*, vol. 377, pp. 80–91.

Jain, S. K., and K. P. Sudheer (2008). Fitting of hydrologic models: A close look at the Nash-Sutcliffe index. *Journal of Hydrologic Engineering*, vol. 13, no. 10, pp. 981–986.

Johnson, M. S., W. Coon, V. Mehta, T. Steehuis, E. Brooke, and J. Boll (2003). Applications of two hydrologic models with different runoff mechanisms to a hillslope dominated watershed in the northeastenr U.S.: A comparison of HSPF and SMR. *Journal of Hydrology*, vol. 284, pp. 57–83.

Kalin, L., R. S. Govindaraju, and M. M. Hantush (2003). Effect of geomorphological resolution on modeling of runoff hydrograph and sedimentograph over small watersheds. *Journal of Hydrology*, vol. 276, pp. 89–111.

Krause, P., D. Boyle, and F. Base (2005). Comparison of different efficiency criteria for hydrological model assessment. *Advances in Geosciences*, vol. 5, pp. 89–97.

Krzysztofowicz, R., and L. M. Watada (1986). Stochastic model of seasonal runoff forecasts. *Water Resources Research*, vol. 22, no. 3, pp. 296–302.

Legates, S. R., and R. E. Davis (1997). The continuing search for an anthropogenic climate change signal: Limitations of correlation-based approaches. *Geophysical Research Letters*, vol. 24, pp. 2319–2322.

Legates, D. R., and G. J. McCabe (1999). Evaluating the use of 'goodness-of-fit' measures in hydrologic and hydroclimatic model validation. *Water Resources Research*, vol. 35, no. 1, pp. 233–241.

Martinec, J., and A. Rango (1989). Merits of statistical criteria for the performance of hydrologic models. *Water Resources Bulletin*, vol. 25, no. 2, pp. 421–432.

McCuen, R. H., R. B. Leahy, and P. A. Johnson (1990). Problems with logarithmic transformations in regression. *Journal of Hydraulic Engineering*, vol. 116, no. 3, pp. 414–428.

McCuen, R. H., Z. Knight, and A. G. Cutter, (2006). Evaluation of the Nash-Sutcliffe efficiency index. *Journal of Hydrologic Engineering*, vol. 11, no. 6, pp. 597–602.

McCuen, R. H., and W. M. Snyder (1975). A proposed index for computing hydrographs. *Water Resources Research*, vol. 11, pp. 1021–1023.

Merz, R., and G. Bloschl (2004). Regionalization of catchment model parameters. *Journal of Hydrology*, vol. 287, pp. 95–123.

Moore, D. S. (1991). *Statistics: Concepts and Controversies*, 3d ed., W. H. Freeman, New York.

Murphy, A. H. (1988). Skill scores based on the mean square error and their relationships to the correlation coefficient. *Monthly Weather Review*, vol. 116, pp. 2417–2423.

Murphy, A. H., and E. S. Epstein (1989). Skill scores and correlation coefficients in model verification. *Monthly Weather Review*, vol. 117, pp. 572–581.

Park, W. J., and Y. G. Kim (1992). Goodness-of-fit tests for the power-law process. IEEE *Transactions on Reliability*, vol. 41, no. 1, pp. 107–111.

Sarma, P. B. S., J. W. Delleur, and A. R. Rao (1973). Comparison of rainfall-runoff models for urban areas. *Journal of Hydrology*, vol. 18, no. 3/4, pp. 329–347.

Schaefli, B., and H. V. Gupta (2007). Do Nash values have value? *Hydrological Processes*, vol. 21, pp. 2075–2080.

Schaefli, B., B. Hingray, M. Niggli, and A. Musy (2005). A conceptual glacio-hydrologic model for high mountainous catchments. *Hydrology and Earth System Sciences*, vol. 9, pp. 95–109.

Seibert, J., (2001). On the need for benchmarks in hydrological modeling. *Hydrological Processes*, vol. 15, pp. 1063–1064.

Shimokawa, T., and M. Liao (1999). Goodness-of-fit tests for type-I extreme-value and 2-parameter Weibull distribution. *IEEE Transactions on Reliability*, vol. 48, no. 1, pp. 79–86.

Weglarczyk, S. (1998). The interdependence and applicability of some statistical quality measures of hydrological models. *Journal of Hydrology*, vol. 206, pp. 98–103.

Willmott, C. J. (1981). On the validation of models. *Physical Geography,* vol. 2, pp. 184–193.

Willmott, C. J. (1984). On the evaluation of model performance in physical geography. In: *Spatial Statistics and Models,* edited by G. L. Gaile and C. J. Willmott D. Reidel, Dordrecht, The Netherlands, pp. 443–460.

Further Reading

Aitkens, A. P. (1973). Assessing systematic errors in rainfall-runoff models. *Journal of Hydrology,* vol. 20, pp. 121–136.

Beldring, S. (2002). Multi-criteria validation of a precipitation-runoff model. *Journal of Hydrology,* vol. 257, pp. 189–211.

Cavadias, G., and G. Morin (1988). Approximate confidence intervals for verification criteria of the WMO intercomparison of snowmelt runoff models. *Hydrological Sciences Journal,* vol. 33, no. 4, pp. 69–77.

Green, I. R. A. and D. Stephenson (1986). Criteria for comparison of single event models. *Hydrological Sciences Journal,* vol. 31, no. 3, pp. 395–411.

Gupta, V. K., and S. Sorooshian (1998). Toward improved calibration of hydrologic models: Multiple and noncommensurable measures of information. *Water Resources Research,* vol. 34, no. 4, pp. 751–763.

Gupta, V. K., T. Thorsten, and Y. Liu (2008). Reconciling theory with observations: Elements of a diagnostic approach to model evaluation. *Hydrological Processes,* vol. 22, pp. 3802–3813.

Krzysztofowicz, R. (1992). Bayesian correlation score: A utilitarian measure of forecast skill. *Monthly Weather Review,* vol. 120, pp. 208–381.

Martinec, J., and A. Rango (1989). Merits of statistical criteria for the performance of hydrological models. *Water Resources Bulletin,* vol. 25, pp. 421–432.

McCuen, R. H., and W. M. Snyder (1975). A proposed index for comparing hydrographs. *Water Resources Research,* vol. 11, no. 6, pp. 1021–1023.

Murphy, A. H. (1974). A sample skill score for probability forecasts. *Monthly Weather Review,* vol. 102, pp. 48–55.

Murphy, A. H., and E. S. Epstein (1970). Skill scores and correlation coefficients in model verification. *Monthly Weather review,* vol. 117, pp. 572–581.

Nash, J. E. and J. V. Sutcliffe (1970). River flow forecasting through conceptual models. Part 1: A discussion of principles. *Journal of Hydrology,* vol. 10, pp. 282–290.

Willmott, C. J., S. G. Ackleson, R. E. Davis, J. J. Feddema, D. R. Klink, J. O'Donnell, and C. M. Rowe (1985). Statistics for the evaluation and comparison of models. *Journal of Geophysical Research,* vol. 90, pp. 8995–9005.

Xu, C.-Y. (2001). Statistical analysis of parameters and residuals of a conceptual water balance model-Methodology and case study. *Water Resources Management,* vol. 15, no. 2, pp. 75–92.

Yates, J. F. (1982). External correspondence: decomposition of the mean probability score. *Organizational Behavior and Human Performance,* vol. 30, pp. 132–156.

CHAPTER 4

Morphological Analysis

Watershed surface characteristics influence a large spectrum of agricultural, hydrologic, and environmental areas. The surfaces are often complex, varying with both location and direction, meaning that these surfaces are inhomogeneous and anisotropic. River basins exhibit certain characteristics that help one understand the hydrologic and hydraulic behavior of the basins. Many of these river basin characteristics are so universally found that they have been formulated as morphological laws. Examples of such laws include Horton's laws of stream number, stream length, and stream area; Schumm's law of stream slope; and Yang's law of average stream fall, among others. In this chapter we explore connections between entropy and certain basin characteristics under simplified assumptions.

4.1 Basic Terminology

Before we undertake morphological analysis using entropy, it will be appropriate to define certain terms that are commonly used. A basin may consist of a network of channels, as shown in Fig. 4.1. This network is also ordered following the Horton-Strahler ordering scheme. The network can be simplified in terms of links and nodes, as shown in Fig. 4.2.

Link: A link can be a segment of a channel or a portion thereof or a channel itself; a channel may be formed by more than one segment, called a link. Channels can be idealized as single lines and links as channel segments.

Node: A new link is formed when two links join, and the point where they meet is called a *node*.

Source: Sources are the points farther upstream in the channel network. Upstream of these points there can be no links or channels. From these points originate links or channels.

Magnitude of a link: The magnitude of a link is the number of sources upstream draining into that link.

Magnitude of a basin n: The magnitude of a basin or channel network is the magnitude of the outlet link and equals the number of first-order streams in the basin, as shown in Fig. 4.2.

Topological distance d: The topological distance d of a node from the outlet is the number of links forming the path between that node and the outlet. It would be more appropriate to designate it as topological number.

Topological level: The topological level of a link that originates from a node and goes downstream is equal to the topological distance.

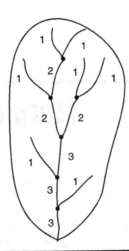

FIGURE 4.1   A hypothetical small third-order watershed based on the Horton-Strahler ordering scheme.

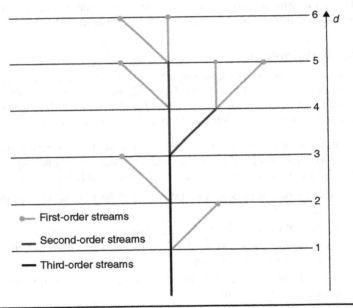

FIGURE 4.2   Simplified drainage network formed by links and nodes.

Topological diameter: The maximum of topological distances within the channel network is called the topological diameter D, as shown in Fig. 4.2.

Width function: The width function can be defined as the number of channel segments or network links at a given distance upstream from the basin outlet (Rodriguez-Iturbe and Rinaldo, 1997). This suggests that the density of stream reaches in a drainage network can be described by the width function. The width function describes the network geomorphology by counting all stream links or segments located at the same distance from the outlet without considering differences in flow conveyance. This averaging, therefore, prevents complete and accurate description of the spatial variability of hydrodynamic parameters. The width function has been included in various studies dealing with the effect of changes in watersheds on floods.

Example 4.1. Consider the Brazos River basin in Texas. Order the river basin following the Horton-Strahler scheme. Indicate the basin order.

Solution. The Brazos River basin map is shown in Fig. 4.3. The basin is ordered following the Horton-Strahler ordering scheme.

Example 4.2. Represent the Brazos River basin network by using links and nodes. Then draw a tree diagram of the basin. Indicate the basin magnitude and diameter.

Solution. The tree diagram is shown in Fig. 4.4. The basin magnitude and diameter are also depicted there. The basin magnitude is 20 (first-order channels). The topological diameter is 11.

Example 4.3. A small watershed of 6.4 km^2 area in South Korea is given, as shown in Fig. 4.5. It is ordered following the Horton-Strahler scheme. Figure 4.6 shows stream segment numbers, and Fig. 4.7 shows lengths and distances of stream segments from the outlet. Construct the width function.

Solution. The maximum distance is 5051 m, and the minimum distance is 0. The number of stream segments is 49. By choosing a class interval size of 600 m, a contingency table of link distance is prepared. Table 4.1 gives the distances of stream lengths, and Table 4.2 shows the contingency table. The width function is plotted in Fig. 4.8.

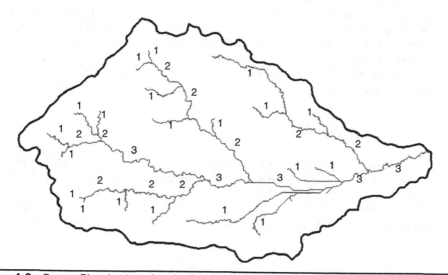

FIGURE **4.3** Brazos River basin ordered using the Horton-Strahler ordering scheme.

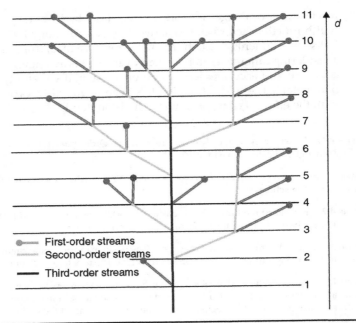

First-order streams
Second-order streams
Third-order streams

FIGURE 4.4   Brazos River basin represented as tree diagram by lines and nodes.

4.2   Distribution of Stream Elevations

We consider a drainage basin of magnitude n. The basin would have the number of links M in the basin as $M = 2n - 1$. Let elevation E be the state variable. Consider a link i having an average elevation E_i, $i = 1, 2, \ldots, m$, $m \leq M$. Note that any link can have one of m average elevations. With this representation of the drainage basin, the Shannon informational entropy of the drainage basin can be defined.

Let p_i be the probability of a link having an elevation E_i, $i = 1, 2, \ldots, m$, for $m =$ number of links. Thus, one can write $\{E_1, E_2, \ldots, E_m\}$ having probabilities $\{p_1, p_2, \ldots, p_m\}$. The informational entropy S (Shannon, 1948) of the drainage basin can be expressed as

$$S = -\sum_{i=1}^{m} p_i \ln p_i \tag{4.1}$$

By maximizing S following the principle of maximum entropy (POME) (Jaynes, 1957), subject to the specified constraints on link elevations, the probability distribution $P = \{p_1, p_2, \ldots, p_m\}$ can be determined.

Let us assume that the only constraint or the only information to be specified is the mean elevation of all the links $(\bar{E})$ in the drainage basin, expressed as

$$\bar{E} = \sum_{i=1}^{m} p_i E_i \tag{4.2}$$

Yuchunchun (Area: 6.4 km²)

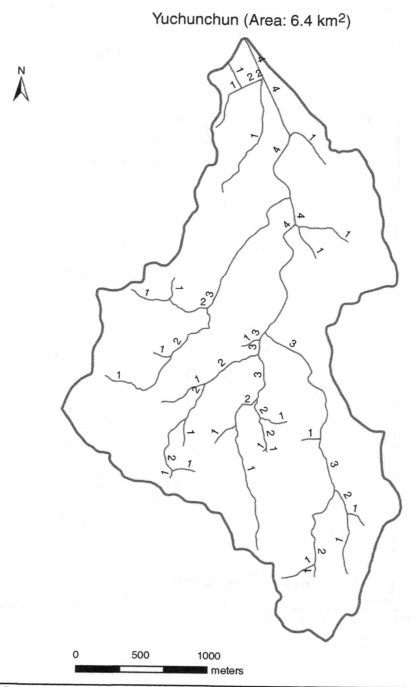

FIGURE 4.5 Yuchunchun watershed in South Korea ordered following the Horton-Strahler ordering scheme.

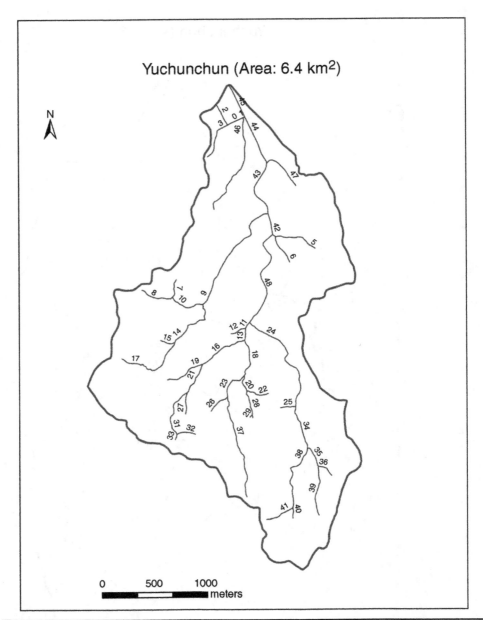

FIGURE 4.6 Numbering of stream segments.

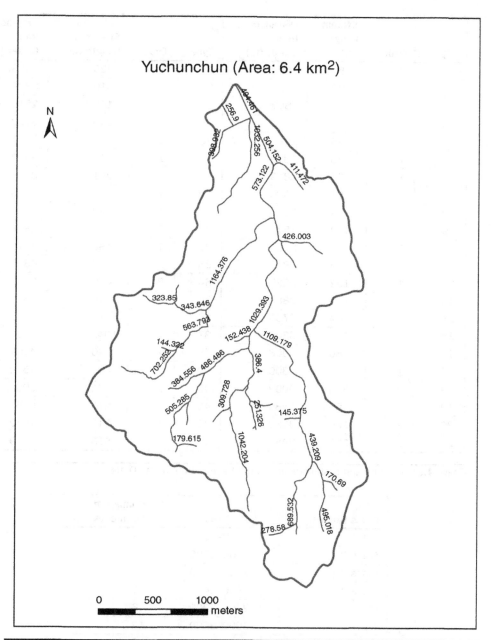

FIGURE 4.7 Distances of stream segments from the outlet of the Yuchunchun watershed.

Type	Order	Stream Length (m)	Distance from Outlet (m)	Type	Order	Stream Length (m)	Distance from Outlet (m)
0	2	159	653	22	1	220	3558
1	2	23	494	23	2	310	3389
2	1	257	812	24	3	1109	2813
3	1	399	812	25	1	145	3922
4	2	35	1783	26	1	221	3699
5	1	426	1819	27	1	215	3813
6	1	294	1819	28	2	251	3558
7	1	187	3080	29	1	33	3810
8	1	324	3080	30	1	29	3810
9	3	1164	1572	31	2	505	3813
10	2	344	2736	32	1	180	4318
11	3	73	2813	33	1	57	4318
12	1	152	2886	34	3	439	3922
13	3	117	2886	35	2	216	4361
14	2	564	2736	36	1	171	4578
15	1	144	3300	37	1	1042	3699
16	2	486	3003	38	2	690	4361
17	1	702	3300	39	1	495	4578
18	3	386	3003	40	1	101	5051
19	1	385	3489	41	1	279	5051
20	2	169	3389	42	4	212	1572
21	2	323	3489	43	4	573	999

TABLE 4.1   Lengths and Distances of Stream Segments from the Outlet

Class Number	Bin Size (m)	Number of Segments
Class 1	0–600	3
Class 2	600–1200	6
Class 3	1200–1800	4
Class 4	1800–2400	2
Class 5	2400–3000	6
Class 6	3000–3600	12
Class 7	3600–4200	8
Class 8	4200–4800	6
Class 9	4800–5400	2

TABLE 4.2   Arrangement of Stream Segment Distances in Class Intervals

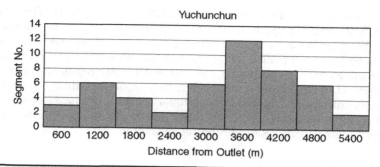

FIGURE 4.8 Width function relating the number of segments to the distance from the outlet.

and, of course, the p_i satisfy

$$\sum_{i=1}^{m} p_i = 1 \tag{4.3}$$

Without undue error it can be assumed that $\bar{E}$ is approximately equal to the mean basin elevation. There may be an infinite number of probability distributions satisfying Eqs. (4.2) and (4.3), for we have two equations and m (> 2) quantities to be determined. Thus, following Jaynes (1957), the least-biased P is the one that maximizes S defined by Eq. (4.1), subject to Eqs. (4.2) and (4.3). Accordingly, the invocation of POME yields

$$p_i = \frac{\exp(-\lambda_1 E_i)}{\sum_{i=1}^{m} \exp(-\lambda_1 E_i)} \tag{4.4}$$

where λ_1 is the Lagrange multiplier, which can be determined with the use of Eq. (4.2) as

$$\frac{\sum_{i=1}^{m} E_i \exp(-\lambda_1 E_i)}{\sum_{i=1}^{m} \exp(-\lambda_1 E_i)} = \bar{E} \tag{4.5}$$

The probability distribution given by Eq. (4.4) is incidentally the Maxwell-Boltzmann distribution. The maximum entropy $S_{\max}$ is given by the substitution of Eq. (4.4) in Eq. (4.1):

$$S_{\max} = \lambda_1 \bar{E} + \ln \sum_{i=1}^{m} \exp(-\lambda_1 E_i) \tag{4.6}$$

So far nothing has been stated about the average link elevation E_i. One way to obtain the E_i that has a hydrological appeal is to obtain the E_i as the mean elevations of nodes having the same topological distance from the basin outlet. In this manner, the p_i are the proportions of the links at the ith topological level, which also define the topological width function (Troutman and Karlinger, 1984) of the network. Then, Eqs. (4.1) to (4.6) hold, with the topological distance D being equal to the possible number of states m.

Following the discussion in Chap. 1, two types of entropies may be defined for a drainage basin: (1) primary entropy and (2) secondary entropy. Primary entropy is obtained where no constraints are imposed on the drainage basin, and secondary entropy when constraints are imposed. The secondary entropy is always less than the primary entropy. For computing the primary entropy of a basin as the value of information entropy under no constraints other than Eq. (4.3), maximization of S in Eq. (4.1), subject to Eq. (4.3), leads to a uniform distribution

$$p_i = \frac{1}{D}$$ (4.7a)

which yields the information entropy as

$$S = \ln D$$ (4.7b)

The entropy in Eq. (4.7b) can be referred to as *primary entropy* of the drainage basin, i.e., the entropy without any constraints.

When constraints are imposed on the drainage basin, such as the mean basin elevation, and so on, its maximum entropy will be reduced. This reduced entropy can be referred to as the *secondary entropy* which, e.g., is given by Eq. (4.6) for one constraint. The difference between the primary entropy and the secondary entropy quantifies the reduction in uncertainty due to the knowledge expressed as constraints. Figure 4.9 shows primary entropy [Eq. (4.7b)] and computed information entropy [Eq. (4.1)] for subnetworks of the main channel of a basin (Fiorentino et al., 1993). The maximum unconstrained value of entropy (primary) corresponds to a basin with a uniform topological width function. Although far from reality, the entropy of this simplified basin configuration may provide a first-order approximation to the value of information entropy given by Eq. (4.6).

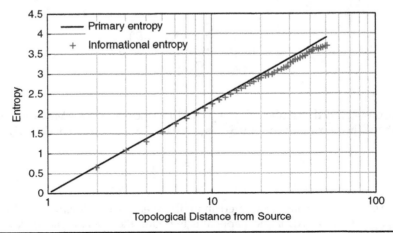

FIGURE 4.9 Primary entropy and information entropy as a function of topological distance from the source (after Fiorentino et al., 1993).

4.3   Entropy and Potential Energy

In basin morphology, potential energy, represented by elevation, plays a critical role. This energy can be specified by the elevation of the network nodes above a datum, say, the basin outlet. For establishing a relation between entropy and potential energy, we consider the mean node elevation y_d at the topological distance d from the outlet. The value of y_d is obtained by averaging the elevations of the nodes at distance d from the outlet. We consider $\bar{y}$ as the mean elevation of all the nodes and consider it approximately as the mean basin elevation; $\bar{y}$ can be regarded as a measure of the average potential energy of the drainage basin. The topological distance d can vary from 1 to D, where D is the network diameter. Physically speaking, the drainage basin at any time can be surmised as the result of its history dating back to infinity. It may then be useful to evaluate the entropy of the basin in its most probable state under the constraint specified by the mean basin elevation. Thus, we can write

$$\sum_{d=1}^{D} p_d y_d = \bar{y} \tag{4.8}$$

and

$$\sum_{d=1}^{D} p_d = 1 \tag{4.9}$$

where p_d is the probability of m_d number of nodes with topological distance d having the mean elevation y_d. The probability distribution $P = \{p_1, p_2, ..., p_D\} = \{p_d, d = 1, 2, ..., D\}$ is given by Eq. (4.4), with y_d replacing E_i and $1/DT$ replacing λ_1, as

$$p_d = \frac{\exp[-y_d/(DT)]}{\sum_{d=1}^{D} \exp[-y_d/(DT)]} \tag{4.10}$$

where T is a parameter.

By reparameterization, Eq. (4.10) can be written as

$$y_d = -\alpha \ln \beta p_d \tag{4.11}$$

where

$$\alpha = DT \tag{4.12}$$

and

$$\beta = \sum_{d=1}^{D} \exp\left(-\frac{y_d}{\alpha}\right) \tag{4.13}$$

Substitution of Eq. (4.11) in Eq. (4.8) yields

$$\bar{y} = -\alpha \sum_{d=1}^{D} p_d \ln \beta p_d \tag{4.14}$$

Expanding Eq. (4.14) gives

$$\bar{y} = -\alpha \sum_{d=1}^{D} p_d \ln \beta - \alpha \sum_{d=1}^{D} p_d \ln p_d \tag{4.15a}$$

Using Eqs. (4.1) and (4.9), Eq. (4.15a) can be written as

$$\bar{y} = -\alpha \ln \beta + \alpha S \tag{4.15b}$$

Equations (4.15) yield the relation between the potential energy of the drainage basin and entropy under the constraint given by Eq. (4.8). This relation should also hold for any drainage subnetwork within a larger basin.

For a drainage basin the distribution of potential energy can be assumed to be controlled by two fundamental quantities D and T measured for the entire basin. Based on an analogy with a thermodynamic system where T can be thought of as proportional to the energy content of the system, α and β may be assumed to be constant as a first-order approximation. Then α and β can be theoretically derived as follows.

The primary entropy for a drainage network of diameter D is given by Eq. (4.7b). Similarly, for a drainage subnetwork of diameter δ the primary entropy is given as

$$S_\delta = \ln \delta \tag{4.16}$$

Let $\bar{y}_\delta$ be the mean basin elevation of the drainage subnetwork of diameter δ whose outlet lies on the channel of diameter D. Let $\bar{y}$ be the mean elevation of the entire network of diameter D. At the basin source $\bar{y}_\delta = 0$ and $S = 0$. In Eq. (4.15b), $\ln \beta = 0$, which yields $\beta = 1$. Equation (4.15b) therefore becomes

$$\bar{y} = \alpha S \tag{4.17}$$

Application of Eq. (4.7b) to Eq. (4.17) yields

$$\alpha = \frac{\bar{y}}{S} = \frac{\bar{y}}{\ln D} \tag{4.18}$$

Since Eq. (4.17) applies to any drainage subnetwork of diameter δ, it can be expressed as

$$\bar{y}_\delta = \alpha S_\delta \tag{4.19}$$

Substitution of Eq. (4.18) in Eq. (4.19) yields

$$\overline{y_\delta} = \overline{y}\,\frac{\ln\delta}{D\ln D} \qquad (4.20)$$

Equation (4.20) expresses the relation between the mean elevation of a drainage subnetwork and its diameter δ. Using Eqs. (4.12) and (4.18), parameter T can be found as

$$T = \frac{\overline{y}}{D\ln D} \qquad (4.21)$$

Equation (4.21) defines parameter T, which is proportional to the potential energy of the basin and inversely proportional to a coefficient that monotonically increases with the diameter of the drainage network. Equation (4.21) provides a quantitative measure of the macroscopic thermodynamic property of the drainage system.

Fiorentino et al. (1993) verified Eq. (4.15) for three drainage basins in Italy. They computed the average elevation and the entropy of the subnetwork of each link which is part of the mainstream. The values of entropy were obtained using Eq. (4.1). The average elevation of the subnetwork was plotted versus its entropy, as shown in Fig. 4.10 for the basins considered. Also plotted in the figure is the least-squares line obtained by treating $-\alpha\ln\beta$ and α as regression constants. For all the basins Eq. (4.15b) explained more than 74 percent of the variability in the average elevation.

Fiorentino et al. (1993) evaluated the adequacy of Eq. (4.20). Figure 4.11 shows the relation between the average elevation of a subnetwork and its topological diameter for three basins. For each basin the amount of variance explained in the average elevation of the subnetwork was greater than 74 percent. Note that part of the unexplained

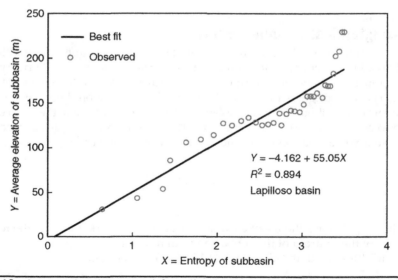

FIGURE 4.10   Average elevation versus informational entropy of subnetworks whose outlets lie on the main channel.

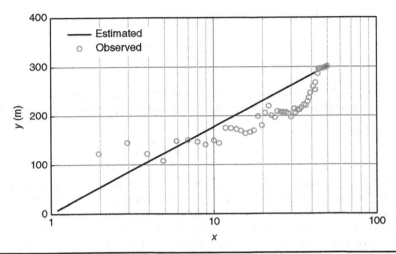

FIGURE 4.11 Average elevation y versus topological diameter x of subnetworks whose outlets lie on the main channel.

variance may be ascribed to the use of topological distances instead of actual distances. The observations that link lengths tend to increase in the downstream direction, suggesting that graphs of Fig. 4.11 would spread in the x direction if the real distances were considered. With the use of basin planimetric configuration, Eq. (4.21) can be reliably employed for the estimation of subnetwork mean elevation when elevations are used for investigating such network characteristics as channel profiles and scaling properties of slopes.

4.4 Longitudinal Channel Profile

For considering the longitudinal bed profile of a channel, a uniform width function provides a first-order approximation for the p_i as being equal to $1/D$. Furthermore, invoking the principle of equal energy expenditure, which is analogous to the concepts used by Yang (1971) and Rodriguez-Iturbe et al. (1992), the basin elevation drop may be assumed to be uniformly distributed between topological levels. This means that the mean elevation drops between consecutive levels are the same. Therefore, following Eq. (4.8), the total fall H_δ from the source to the outlet of a channel of subnetwork diameter δ is given by

$$H_\delta = \frac{\delta \overline{y}_\delta}{\overline{d}_\delta} \qquad (4.22)$$

where $\overline{y}_\delta$ denotes the mean elevation of the subnetwork and $\overline{d}_\delta$ is the topological distance of the centroid of the subbasin width function.

Multiplying both sides of Eq. (4.15b) by $\delta / \overline{d}_\delta$, one gets

$$H = -\alpha_\infty \ln \beta + \alpha_\infty S \qquad (4.23)$$

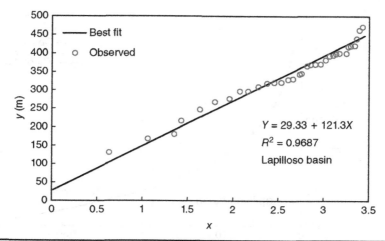

FIGURE **4.12**   Elevation drop y from the source versus entropy x for the main channel.

where $\alpha_* = \alpha\delta / \overline{d_\delta}$; α is defined by Eq. (4.12); and H is the fall in elevation from the source to the outlet of the main channel of the drainage basin whose entropy is S. Treating $\alpha_* \ln\beta$ and α_* as regression constants, Fiorentino et al. (1993) obtained the best-fit lines for the plots of elevation drop versus entropy for each link of the mainstream, as shown in Fig. 4.12. For all basins Eq. (4.23) explained more than 90 percent of the variance in the fall in elevation.

Parameters α_* and β can be theoretically evaluated as was done for Eq. (4.15b). Accordingly, Eq. (4.23) provides

$$H_\delta = \alpha_* S_\delta = \frac{H}{\ln D} S_\delta \qquad (4.24)$$

Let y_0 be the elevation of the source of the channel, and y_δ be the elevation of the downstream node at a distance δ from the source. Then with the substitution of Eq. (4.16) in Eq. (4.24), the bed elevation profile of a channel of topological length D is given by

$$y_\delta = y_0 - H\frac{\ln\delta}{\ln D} \qquad (4.25)$$

Equation (4.25) expresses the relation between elevation at a point and its topological distance. Fiorentino et al. (1993) plotted this relation for three drainage basins and found that for each basin more than 95 percent of the variability in the nodal elevation was explained, as shown in Fig. 4.13.

4.5   Law of Average Stream Fall

Using entropy and energy considerations, Yang (1971) derived the law of average stream fall. Consider a drainage basin whose channel network is ordered following the Horton-Strahler ordering scheme. Let y_i be the average fall of the ith-order stream

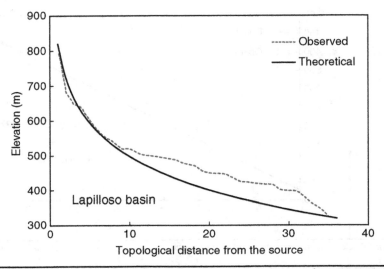

FIGURE **4.13** Relation between nodal elevation and topological distance from the source for the main channel.

from the beginning to the end. The average potential energy e_{pi} per unit mass of water in an ith-order stream can be considered as proportional to the average fall for that stream, i.e.,

$$e_{pi} = ky_i \qquad (4.26)$$

where k is a factor of conversion between energy and fall. The total fall Y_m from the start of the first-order stream to the mth-order stream is

$$Y_m = \sum_{i=1}^{m} y_i = \text{const} \qquad (4.27a)$$

where const stands for constant. Similarly the total potential energy (loss) per unit mass of water from the beginning of the first-order stream to the end of the mth-order stream (or highest order), E_m, is

$$E_{pm} = \sum_{i=1}^{m} e_{pi} = \sum_{i=1}^{m} ky_i = \text{const} \qquad (4.27b)$$

where const is the constant of integration.

Analogous to Eq. (4.26), the relation between E_m and Y_m can be written as

$$E_{pm} = kY_m \qquad (4.28)$$

The energy available for a unit mass of water at the start of the first-order stream is its potential energy. When this unit mass of water flows down the slope in the stream, it will convert its potential energy into kinetic energy and friction loss or heat energy. The kinetic energy thus generated is the only useful energy that the flow can employ to carve its own channel and form its stream network. Thus, the sum of the kinetic energy change and friction loss in any reach must equal the total potential energy loss in that reach.

In thermodynamics, entropy ϕ is defined as

$$\phi = \int \frac{dH}{T} \tag{4.29}$$

where H is the thermal energy per unit mass and T is the absolute temperature. Now one can assume that the temperature in a heat system is analogous to elevation E in a stream system, and thermal energy in a heat system can be analogous to potential energy E_p in a stream system. With this analogy in mind, the entropy of a stream system can be expressed, with the use of $E_p = kY$, as

$$\phi = \int \frac{dE_p}{E} = k \int \frac{dY}{Y} \tag{4.30a}$$

where Y is the total fall of the stream system and E is the total elevation of the remotest point from the outlet. Likewise, for an mth-order stream path, one can write

$$\phi_m = \int \frac{dE_{pm}}{E_m} = k \int \frac{dY_m}{Y_m} \tag{4.30b}$$

where E_{pm} is the total elevation of the mth-order stream path and Y_m is the total fall of the mth-order stream path.

Likewise, the entropy of an ith-order stream can be written as

$$\phi_i = \int \frac{de_{pi}}{E_i} = k \int \frac{dy_i}{Y_i} \tag{4.31}$$

Now the probability that a particular amount of potential energy will be lost in the ith-order stream can be expressed as

$$p_i = \frac{e_{pi}}{E} = \frac{ky_i}{kY} = \frac{y_i}{Y} \tag{4.32}$$

Substituting Eq. (4.32) in Eq. (4.31), one obtains

$$\phi_i = k \int \frac{dp_i}{p_i} = k \ln p_i + \text{const} \tag{4.33}$$

where const is the constant of integration. There are many streams of ith-order—the smaller the order, the larger the number. Thus it may be more appropriate to take the average value for the ith-order streams.

Following Prigogine (1967), the total entropy of a stream system is equal to the sum of the entropies of all order streams:

$$\phi = \sum_{i=1}^{m} \phi_i = -k \sum_{i=1}^{m} \ln p_i + \text{const} \tag{4.34}$$

A more robust expression would be in the form of the Shannon entropy as

$$\phi = -k \sum_{i=1}^{m} p_i \ln p_i + \text{const} \tag{4.35}$$

where const is the constant of integration. According to Lewis and Randall (1961), the maximum entropy leads to the most probable distribution of energy. Therefore,

$$-\sum_{i=1}^{m} p_i \ln p_i \rightarrow \text{maximum} \tag{4.36}$$

Note that

$$\sum_{i=1}^{m} p_i = 1 \qquad 0 \le p_i \le 1$$

It can then be shown that the maximum entropy results when

$$p_1 = p_2 = \cdots = p_m \tag{4.37}$$

Then, with the use of Eq. (4.32), one obtains

$$y_1 = y_2 = \cdots = y_m \tag{4.38}$$

Equation (4.38) shows that when a stream system has reached dynamic equilibrium, each order stream must exhibit the same amount of fall as the next-higher- or next-lower-order stream. The morphological behavior is such that a stream will increase its order after an equal amount of fall has occurred. Equation (4.38) can be generalized as

$$\frac{y_i}{y_{i+1}} = \frac{y_i}{y_{i+2}} = \cdots = \frac{y_i}{y_m} = 1 \tag{4.39}$$

Equation (4.39) is the law of average stream fall and states that for any river basin in dynamic equilibrium, the ratio of the average fall between any two different-order streams in the same basin is unity.

4.6 Entropy, Horton-Strahler Order, Magnitude, and Fractal Dimension

Now we consider a drainage network of Horton-Strahler order Ω. In the Horton-Strahler ordering scheme, a channel of order ω, $2 \leq \omega \leq \Omega$, is defined as the succession of links of order ω whose last link drains into a link of order $\omega^* > \omega$ or into the outlet. The drainage basin has a bifurcation ratio R_B and stream length ratio R_L defined as

$$R_B = \frac{N_{\omega-1}}{N_\omega} \tag{4.40}$$

and

$$R_L = \frac{L_\omega}{L_{\omega-1}} \tag{4.41}$$

where N_ω is the number of streams of order ω in the drainage network of order Ω, and L_ω is the average length of streams of order ω.

In a Horton network the topological diameter D can be expressed in terms of the stream length ratio as

$$D = \frac{R_L^\Omega - 1}{R_L - 1} \tag{4.42}$$

Taking the logarithm of Eq. (4.42), one gets

$$\ln D = \ln \frac{R_L^\Omega - 1}{R_L - 1} \tag{4.43}$$

By virtue of Eq. (4.7b), Eq. (4.43) expresses the entropy of the Horton-Strahler network of order Ω. Therefore,

$$S = \ln(R_L^\Omega - 1) - \ln(R_L - 1) \tag{4.44}$$

Equation (4.44) can be simplified further. Note that $R_L > 1$ and if Ω is much greater than 1, then $R_L^\Omega - 1 \approx R_L^\Omega$. Under this condition, Eq. (4.44) simplifies to

$$S = \Omega \ln R_L - \ln(R_L - 1) \tag{4.45}$$

Equation (4.45) is a linear relation between entropy and the drainage basin order, wherein the Horton-Strahler order can be thought of as a measure of network complexity (Wang and Waymire, 1991).

The magnitude of a drainage basin network is one of its most important characteristics, for it provides a surrogate for the drainage area. In terms of the bifurcation ratio, the magnitude can be expressed as

$$n = R_B^{\Omega-1} \quad \text{or} \quad \Omega = 1 + \frac{\ln n}{\ln R_B} \tag{4.46}$$

Introducing Eq. (4.46) in Eq. (4.45) and eliminating Ω, one obtains

$$S = \frac{\ln R_L}{\ln R_B} \ln n + \ln \frac{R_L}{R_L - 1} \tag{4.47}$$

or

$$S = (\Omega - 1)\ln R_L + \ln \frac{R_L}{R_L - 1} \tag{4.48}$$

Equations (4.47) and (4.48) express the relation between entropy S, the magnitude n, and the Horton order Ω of a drainage basin.

Fiorentino et al. (1993) plotted Eq. (4.47) for the average entropy for fixed magnitude n versus the logarithm of n, and they found a strong relation for all the basins, with a coefficient of determination greater than 95 percent. Figure 4.14 shows such a plot. When the average entropy for fixed Horton order was plotted versus the order, a strong linear relation was found, explaining the variance of more than 99 percent for all the basins.

To further explore the entropic properties of drainage networks, the cumulative entropy was computed as the sum of information entropies of Horton streams having the same order for each of the three basins, and was plotted versus the Horton order. The plot is shown in Fig. 4.15. There is a strong relation between the logarithm of cumulative entropy and the Horton order. The figure suggests a relation between cumulative entropy and the Horton order as

$$S_\Sigma = (2n - 1)\exp(-\omega) = M\exp(-\omega) \tag{4.49}$$

where S_Σ is the cumulative entropy. This shows that $\omega = 0$, $S_\Sigma = 2n - 1 = M = $ total number of links in the basin. For two larger basins (fifth order) the differences between actual and estimated values of M were significantly small.

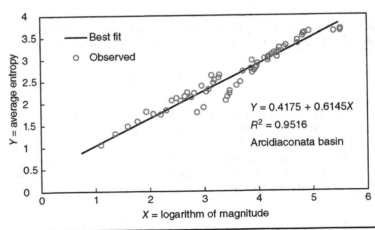

FIGURE 4.14 Relation between average entropy and logarithm of the basin magnitude.

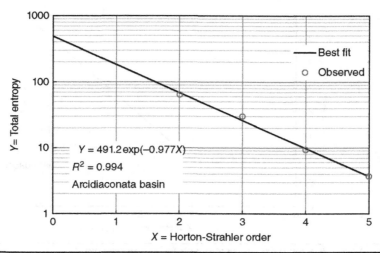

Figure 4.15   Relation between cumulative entropy and Horton order.

4.7   Scaling Properties

A well-known empirical relation between the length of the main channel and the drainage area is expressed as

$$L \propto A^{\Theta} \tag{4.50}$$

in which Θ has been found to be, on average, 0.568 by Gray (1961) and 0.554 by Mueller (1973). Mandelbrot (1983) used Eq. (4.50) to derive the fractal dimension $D_l = 2\Theta \approx 1.14$. La Barbera and Rosso (1989) defined the fractal dimension of stream networks as

$$F = \frac{\ln R_B}{\ln R_L} \tag{4.51}$$

Tarboton et al. (1990) recognized F to be the fractal dimension due to the branching process, with the product FD_l being the total fractal dimension of the stream network. Thus, according to Mandelbrot's value of D_l, F should be equal to 1.75 for space-filling river systems. A relation analogous to Eq. (4.50) can be derived by equating Eq. (4.7b) to Eq. (4.48):

$$D = Cn^{1/F} \tag{4.52}$$

where $C = R_L/(R_L - 1)$. Equation (4.52) has an obvious analogy with Eq. (4.50) and sheds more light on the meaning of Θ, which turns out to equal $1/F$. For $F = 1.75$, Eq. (4.52) leads to $\Theta = 0.57$, which agrees well with Gray (1961).

Substitution of Eq. (4.52) in Eq. (4.47) produces

$$S = \ln\left(\frac{R_L}{R_L - 1} n^{1/F}\right) \tag{4.53}$$

which expresses the link between entropy, magnitude, and fractal dimension of networks.

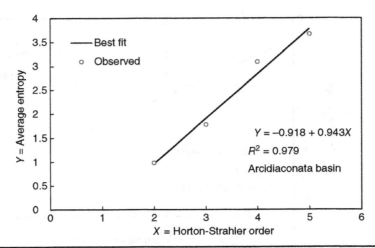

FIGURE **4.16**   Relation between average entropy and Horton order.

Fiorentino et al. (1993) plotted values of S versus the values of logarithms of magnitude n for various values of the stream length ratio and bifurcation ratio, as shown in Fig. 4.16. The S was obtained by averaging the values of all the subbasins of the same magnitude. Given the variance in R_L and R_B found for a drainage basin, the agreement between observed and computed values of S versus n was quite good.

Another important relation that has been a subject of a number of investigations (e.g., Gupta and Waymire, 1989; Willgoose et al., 1991) is one between channel slope s and drainage area A:

$$s \propto A^{-\gamma} \tag{4.54}$$

with γ varying from 0.37 to 0.83, with an average value of 0.6 (Flint, 1974). The entropy-based river profile derived earlier provides an explanation of Eq. (4.54). To that end, differentiating Eq. (4.7b), we get

$$s \propto \frac{1}{D} \tag{4.55}$$

and by virtue of Eq. (4.52),

$$s \propto n^{-1/F} \tag{4.56}$$

which, by analogy between n and A, leads to $\gamma = 1/F$. For $F = 1.75$, $\gamma = 0.57$, which is close to the average value obtained by Flint (1974).

4.8 Horton-Strahler Equations

Using the Horton-Strahler ordering scheme (Horton, 1945; Strahler, 1957), Horton's laws of geomorphology can be expressed. Denoting stream order by w, number of streams of order w by N_w, the average length of w-order streams by L_w, the average slope of w-order streams by S_w, and the average drainage area of w-order streams by A_w, these laws can be written as follows:

For stream numbers:

$$N_w = \exp(a - bw) \quad \text{or} \quad \ln N_w = a - bw \qquad (4.57)$$

For average stream length:

$$L_w = \exp(a_l - b_l w) \quad \text{or} \quad \ln L_w = a_l - b_l w \qquad (4.58)$$

Average stream slope:

$$S_w = \exp(a_s - b_s w) \quad \text{or} \quad \ln S_w = a_s - b_s w \qquad (4.59)$$

Average drainage area:

$$A_w = \exp(a_a - b_a w) \quad \text{or} \quad \ln A_w = a_a - b_a w \qquad (4.60)$$

where a, b, a_l, b_l, a_s, b_s, a_a, and b_a are constants. When Eqs. (4.57) to (4.60) are plotted on semilogarithmic paper, straight lines result and their slopes yield values of $\exp b$, $\exp b_l$, $\exp b_s$, and $\exp b_a$ (Yang, 1971).

Equations (4.57) to (4.60) lead to the following insightful geomorphological parameters:

Bifurcation ratio of stream number:

$$\frac{N_w}{N_{w+1}} = \exp b \qquad (4.61)$$

Stream length ratio:

$$\frac{L_w}{L_{w+1}} = \exp b_l \qquad (4.62)$$

Stream concavity:

$$\frac{S_w}{S_{w+1}} = \frac{Y_w / L_w}{Y_{w+1} / L_{w+1}} = \exp b_s \qquad (4.63)$$

Drainage area ratio:

$$\frac{A_w}{A_{w+1}} = \exp b_a \tag{4.64}$$

The shape factor for a drainage basin can be computed from its length and area. Let α denote the shape factor. Then

$$\alpha = \frac{\alpha_w}{\alpha_{w+1}} = \frac{L_w^2 / A_w}{L_{w+1}^2 / A_{w+1}} \tag{4.65}$$

where α_w is the shape factor of w-order drainage area.

By rearranging Eqs. (4.61) to (4.64), additional interesting geomorphological parameters can be derived following Yang (1971):

Stream fall ratio:

$$\frac{Y_w}{Y_{w+1}} = \exp b_l \exp b_s \tag{4.66}$$

Stream frequency ratio:

$$\frac{N_w / A_w}{N_{w+1} / A_{w+1}} = \exp b \exp - b_a \tag{4.67}$$

An index of the stream frequency ratio can be expressed if the drainage area ratio is unknown:

$$\frac{1}{\alpha} \frac{N_w / A_w}{N_{w+1} / A_{w+1}} = \frac{\exp b}{\exp 2b_l} \tag{4.68}$$

As discussed before, under dynamic equilibrium, the value of the stream fall ratio must be unity:

$$\frac{Y_w}{Y_{w+1}} = \exp b_l \exp b_s = 1.0 \tag{4.69}$$

Using data from 14 river basins collected by Stall and Fok (1968) and Stall and Yang (1970), Yang (1971) showed observed fall ratios ($\exp b_l \exp b_s$) were close to 1.0.

4.9 Streambed Profile

First, for deriving the streambed profile, the average fall of the w-order stream can be expressed, with the use of the definition of average slope and Eqs. (4.58) and (4.59), as

$$Y_w = \exp[(a_l + a_s) - (b_l + b_s)w] \tag{4.70}$$

A similar result was obtained by Fok (1969). The total fall from the beginning of the first-order stream to the end of the mth-order stream can be written as

$$Z_m = \sum_{w=1}^{m} Y_w = \exp(a_l + a_s) \sum_{w=1}^{m} \exp[-(b_l + b_s)w]$$ (4.71)

The total horizontal length from the beginning of the first-order stream to the end of the mth-order stream can be obtained from the law of average stream length, given by Eq. (4.58) as

$$X_m = \sum_{w=1}^{m} L_w = \exp(a_l) \sum_{w=1}^{m} \exp(-wb_l)$$ (4.72)

Equations (4.71) and (4.72) permit determination of a theoretical longitudinal profile with stream order as the parameter. This profile can be considered as a representative profile. Under the condition of dynamic equilibrium, each stream order should have an average fall as given by Eq. (4.69). Then Eq. (4.70) can be written as

$$Y_w^* = \exp(a_l + a_s)$$ (4.73)

and the total fall as

$$Z_w^* = m \exp(a_l + a_s)$$ (4.74)

One can compute the difference between the longitudinal profile given by Eq. (4.73) or (4.74) and the theoretical profile obtained from Eq. (4.71) or (4.72), or the observed profile can be considered as a measure of the maturity of the river system. The theoretical profile not only represents general characteristics of the stream system, but also can reflect the maturity of the system, local geological constraints on the stream, and the possibility of historical changes in stream geomorphology.

4.10 Stream Meandering

During the basin evolution toward the dynamic equilibrium state, the rate of production of entropy per unit mass should be minimum, subject to specified constraints. Therefore, from Eqs. (4.32), (4.34), and (4.37) and the total probability, the rate of entropy production can be expressed as

$$\frac{d\phi}{dt} = k\frac{d}{dt}(\ln p_i) = km\frac{d}{dt}(\ln e_i - \ln E) = \min$$ (4.75)

or

$$\frac{de_i}{dt} = \min$$ (4.76)

where m is constant and min is minimum.

Equation (4.76) can be generalized as

$$\frac{\Delta E}{\Delta t} = k\frac{\Delta y}{\Delta t} = \min \tag{4.77}$$

where ΔE is the potential energy loss per unit mass of water in a reach along the stream length L and fall y, min is the minimum, and Δt is the average time required for a unit mass of water to travel through the reach. Equation (4.77) expresses the least rate of energy expenditure.

Stream meandering can be explained by using Eq. (4.77). To that end, let there be points in a river basin with an elevation difference of Y. Between any two different points, there can be many reaches or courses with different horizontal distances that a stream can choose. It is assumed that discharge and bed roughness are the same for all these courses. For simplicity, let us consider a course of length L of slope Y/L, and let the stream cross section be a segment of a semicircle with radius r and angle θ, as shown in Fig. 4.17. The cross-sectional area can be written as

$$A_c = \frac{1}{2}r^2(\theta - \sin\theta) \tag{4.78}$$

If Q is the discharge passing through the cross section, then the average velocity can be expressed as

$$V = \frac{2Q}{r^2(\theta - \sin\theta)} \tag{4.79}$$

The average cross-sectional velocity can also be computed by using Manning's formula:

$$V = \frac{1.49}{n}\left[\frac{r^2(\theta - \sin\theta)}{2r\theta}\right]^{2/3}\left(\frac{Y}{L}\right)^{1/2} \tag{4.80}$$

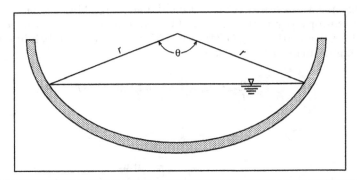

Figure 4.17 Segment of a semicircle with radius r and angle θ.

where the wetted perimeter is $r\theta$ and n is the Manning coefficient of roughness. From Eqs. (4.79) and (4.80), the relation between L and θ can be derived as

$$\frac{(\theta - \sin\theta)^{5/3}}{(2\theta)^{2/3}} = \left(\frac{2Qn}{1.49r^{8/3}Y^{1/2}}\right)L^{1/2} = \beta L^{1/2} \qquad \beta = \frac{2Qn}{1.49r^{8/3}Y^{1/2}} \qquad (4.81)$$

where $\beta^2 L$ can be considered as a stream length factor. The time required for a unit mass of water to travel through this reach of length L can be determined as

$$\Delta t = \frac{L}{V} = \frac{r^2}{2\beta^2 Q}[(\beta^2 L)(\theta - \sin\theta)] \qquad (4.82)$$

Now the potential energy spent by water to travel through the river course can be expressed as

$$\frac{\Delta H}{\Delta t} = \frac{kY}{\Delta t} = \left(\frac{2\beta^2 QkY}{r^2}\right)\left[\frac{1}{(\beta^2 L)(\theta - \sin\theta)}\right] = \frac{1}{\xi}\frac{1}{(\beta^2 L)(\theta - \sin\theta)} \qquad \xi = \frac{2\beta^2 QkY}{r^2} \qquad (4.83)$$

Here $\xi\Delta H/\Delta t$ can be regarded as the rate of the potential energy expenditure factor. If the energy expenditure factor is plotted versus the stream length factor on a log-log paper, a straight line will result, as shown in Fig. 4.18. Yang (1971) showed the straight-line plot as

$$\xi\left(\frac{\Delta H}{\Delta t}\right) = \frac{1}{(\theta - \sin\theta)(\beta^2 L)} = 0.52(\beta^2 L)^{-1.354} \qquad (4.84)$$

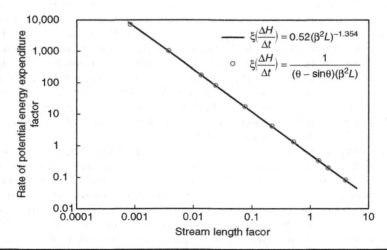

Figure 4.18 Rate of potential energy expenditure factor versus stream length factor.

Equations (4.83) and (4.84) show that if the rate of potential energy expenditure is to remain constant for a particular fall Y, then the stream must satisfy one of two conditions: either increase its time of travel or increase the distance. This is rendered possible by meandering. This conclusion, of course, can also be shown for other cross sections. This may explain why straight rivers seldom exist in nature, as observed by Leopold and Wolman (1962).

Now, taking the derivative of Eq. (4.84) with respect to stream length L (considering it as an independent variable), one gets

$$\xi \frac{d(\Delta H/\Delta t)}{dL} = -0.704\beta^{-2.708}L^{-2.354} \tag{4.85}$$

Equation (4.85) shows that the change in potential energy expenditure per unit mass of water with respect to the change in stream length increases with increasing stream length, i.e., decreasing stream slope. The implication is that streams tend to meander in areas of mild or gentle slopes rather than in areas of steep slopes. This means that streams in plain areas will have higher sinuosity than those in mountainous areas. In mountainous areas river valley widths are narrower, valley slopes are steeper, and bedrocks are harder, reflecting stronger geological controls. These constraints are not conducive to meandering.

In comparing streams of different sizes, note that in general smaller streams (1) have steeper slopes and shallower depths than do larger streams and (2) have larger rates of potential energy expenditure per unit mass of water along their courses of flow. To reduce the rate of energy expenditure, small streams tend to join other smaller steams and form large streams. The joining occurs either during floods or by chance. However, large streams will seldom divide themselves into smaller streams that will flow parallel to one another.

Streams flowing downslope tend to increase in both width and depth, and hence in size. Empirical evidence suggests that width increases more quickly than does depth. During the course of increasing width, a point is reached when a further increase in width would lead to a decrease in depth. This would then lead to a decrease in velocity, with the result that the stream would be unable to carry the sediment load. To maintain the balance between sediment inflow and outflow, the stream, at this juncture, would bifurcate into smaller but steeper channels; i.e., the steam would likely braid. In that case the channel width would exceed that needed to carry the given discharge. Also note that braided reaches are both wider than undivided reaches and steeper. This explanation is in agreement with Leopold and Wolman (1962) and Hydraulics Research Station (1967).

To satisfy the law of average stream fall and the law of least rate of energy expenditure, smaller streams should join to form larger streams after an equal amount of fall has occurred. This process of joining together leads to the type of stream networks that are observed in nature. A stream, along its flow path, always tends to minimize its rate of potential energy expenditure per unit mass of water. Of course, this is subject to constraints. As long as characteristics, such as width or slope, are insufficient for the stream to transport the inflow of water and sediment, changes will occur to maintain a condition of dynamic equilibrium.

Example 4.4. Plot the stream length ratio versus stream concavity, using the data from Table 4.3. Fit Eq. (4.69) to the data.

Solution. The stream length ratio is plotted versus concavity in Fig. 4.19, and the relationship is clearly quite strong. As stream concavity increases, the length ratio decreases.

			Parameters				
River Basin	Bifurcation Ratio $\exp b$	Stream Length Ratio $\exp b_l$	Stream Concavity $\exp b_s$	Area Ratio $\exp b_a$	Fall Ratio $\exp b_l$ $\exp b_s$	Frequency Ratio $\exp b/$ $\exp b_a$	Index of Frequency Ratio $\exp b/$ $\exp 2b_l$
Big Muddy, IL	3.287	0.478	2.065	0.344	0.987	9.555	14.387
Big Sandy, KY	3.931	0.414	2.423	0.203	1.003	19.365	14.387
Colorado, CO	4.104	0.467	2.115	0.199	0.988	20.623	18.818
Kaskaskia, IL	4.055	0.417	2.094	0.229	0.986	17.707	18.278
Mackinaw, IL	3.877	0.505	1.929	0.261	0.974	14.854	15.202
Merrimack, NH	4.468	0.391	2.612	0.285	1.021	15.677	29.225
Neches, TX	4.787	0.406	2.455	0.266	0.997	17.996	29.042
Roanoke, NC	4.306	0.476	2.100	0.194	1.000	22.196	19.004
Rogue, OR	3.721	0.484	2.123	0.309	1.028	12.042	15.884
Sangamon, IL	3.736	0.540	1.866	0.304	1.008	12.289	12.813
Spoon, IL	3.999	0.474	1.904	0.291	0.902	14.742	17.800
Susquehanna, NY	3.987	0.427	2.532	0.241	1.081	16.544	21.867
Vermillion, IL	3.967	0.525	1.809	0.431	0.950	9.204	14.392
White, IN	4.513	0.425	2.349	0.241	0.998	18.726	24.986

TABLE 4.3 Stream Morphology Parameters (after Yang, 1971)

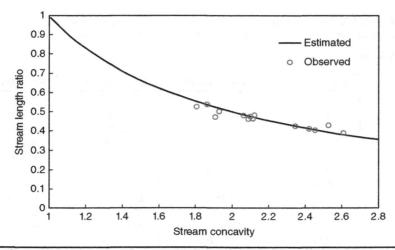

FIGURE 4.19 Stream length ratio versus stream concavity.

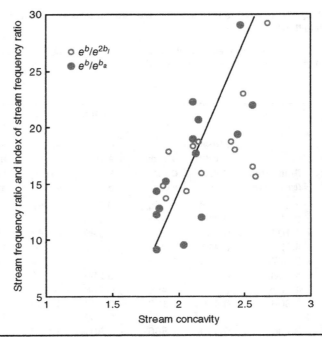

Figure 4.20   Stream frequency ratio and index of stream frequency ratio versus stream concavity.

Example 4.5.   Plot the stream frequency ratio and the index of the stream frequency ratio versus stream concavity, using data from Table 4.3, and fit a line.

Solution.   The stream frequency ratio and index of stream frequency ratio are plotted versus stream concavity, in Fig. 4.20. There is a strong direct relation.

Example 4.6.   Data on vertical fall are given for three rivers (Big Sandy River, Big Muddy River, and Rogue River) in Tables 4.4 to 4.6. Plot the vertical fall versus the horizontal distance.

Solution

Distance (mi)	Fall (100 ft)	Distance (mi)	Fall (100 ft)
0.0	1.583	24.458	11.957
0.436	2.591	29.156	12.356
1.515	4.035	38.967	14.456
3.752	6.077	54.825	14.917
5.333	7.002	61.777	14.256
6.922	7.953	78.883	14.576
7.664	8.252	104.873	15.070
8.798	8.768	146.534	15.450
11.606	9.417	191.073	15.607
15.803	10.337	247.559	16.340
18.308	11.068	274.410	16.422

Table 4.4   Profile of Big Sandy River

Distance (mi)	Fall (100 ft)	Distance (mi)	Fall (100 ft)
14.867	40.696	67.818	67.114
15.185	42.638	82.044	69.530
16.334	45.428	84.138	71.376
18.145	48.234	94.990	72.995
20.848	50.924	108.215	75.411
24.596	52.924	128.202	77.690
31.407	55.323	138.000	80.137
35.471	57.493	152.350	80.622
42.875	59.329	157.740	82.824
51.499	62.852	167.821	84.310
60.175	65.548		

TABLE 4.5 Profile of Upper Colorado River

Distance (mi)	Fall (100 ft)	Distance (mi)	Fall (100 ft)
0.358	2.751	30.050	27.874
0.884	5.090	31.664	28.576
0.935	8.249	35.026	30.067
1.491	10.173	39.477	31.833
2.088	11.517	40.388	32.685
2.608	14.939	40.419	34.182
4.098	16.383	40.896	35.273
4.129	17.880	41.523	36.202
6.009	18.755	42.488	38.219
8.695	19.981	44.229	39.090
10.775	21.941	45.434	39.700
12.690	22.317	54.722	42.137
15.990	22.727	57.023	42.546
24.634	24.483	64.967	44.977
24.977	25.096	71.965	45.054
27.393	26.233	80.781	45.924
28.586	27.009	89.725	46.963

TABLE 4.6 Profile of Rogue River

4.11 Stream Order as a Measure of Sample Source Uncertainty

In a drainage basin, the water, sediment, or pollutant that appears at the basin outlet is supplied by different parts or sources of the basin. These different sources are located at different distances from the outlet. That is the reason why runoff hydrograph, sediment graph, or pollutant graph due to a given rainfall has a rising limb and a falling limb, and its shape is a skewed distribution. The question arises: What is the uncertainty about the source of any constituent, sediment or pollutant, sampled at the outlet? In a similar manner, one can ask related questions as posed by Sharp (1970). Suppose one observed a speck of precious metal (say, gold) in her pan at a downstream location. The question then is: How much searching will be needed to determine the source of the precious metal? If a pollutant is being observed in runoff at the outlet, then one may ask the question: How many sampling stations should be installed to locate the sources of pollution? Sharp (1970) employed the Horton stream order at the outlet of the basin to measure the uncertainty of the source supplying any discrete quantity of pollutant at the outlet. This suggests that Horton's stream order can be used to design efficient methods for basin sampling. The discussion in this section closely follows the work of Sharp (1970).

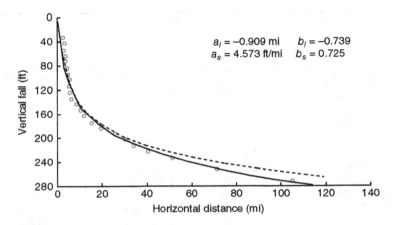

(a) Big Muddy River, Illinois

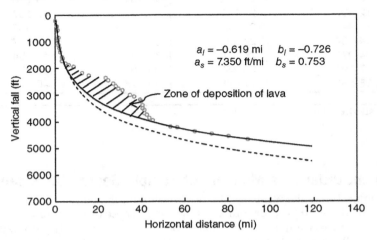

(b) Rogue River, Oregon

Figure 4.21 Vertical fall versus horizontal distance. Solid lines are for Eqs. (4.71) and (4.72), and dashed lines are for Eqs. (4.72) and (4.74).

4.11.1 Stream Order and Numbers

Following the Horton-Strahler ordering scheme, let N_i be the number of streams of order i. Then the law of stream numbers (Horton, 1945) can be expressed as

$$N_i = R_b N_{i+1} \tag{4.86}$$

where R_b is the bifurcation ratio and $N_i{+}_1$ is the number of streams of order $i + 1$. Let the basin order be defined by M. Then the total number of streams of order i can be expressed as

$$N_i = R_b{}^{M-i} \tag{4.87}$$

Note that $N_M = 1$.

Interestingly, Eq. (4.87) can be related to the population growth model (MacArthur and Connell, 1966) as follows (Sharp, 1970). If the initial population is assumed to be unity, then N_i will correspond to the population size after $M - i$ generations. The bifurcation ratio will correspond to the net population growth or reproductive rate, and $M - i$ will correspond to the number of generations.

4.11.2 Uncertainty of Source Location

Let it be assumed that there is only one source supplying the constituent that is being observed at the basin outlet. The question is one of finding a way to measure the uncertainty about the source or which tributary is the source. To answer this question, it is assumed that segments of ith-order streams constitute random numbers. Then the source lies in the drainage area of a randomly chosen ith-order stream segment. If j is the jth stream segment, then $j = 1, 2, ..., N_i$, and N_i is the total number of ith-order stream segments. If all ith-order streams are equally likely to contribute to the pollutant being observed at the basin outlet, i.e., their probabilities are the same, then

$$p_{ij} = \frac{1}{N_i} \tag{4.88}$$

where p_{ij} is the probability of the source lying in the drainage area of the jth segment of the ith-order streams.

Now p_{ij} can be expressed in terms of the bifurcation ratio by substituting Eq. (4.87) in Eq. (4.88) as

$$p_{ij} = \frac{1}{R_b{}^{M-i}} \tag{4.89}$$

Note that if $i = M$, then

$$p_{ij} = \frac{1}{R_b{}^{M-M}} = \frac{1}{R_b{}^{0}} = 1 \tag{4.90}$$

Equation (4.90) suggests that to detect pollutant at the outlet, there must be a source upstream. In a similar manner, probabilities for decreasing values of i as $i = M - 1$, $M - 2, ..., 3, 2, 1$ can be written as $R_b^{-1}, R_b^{-2}, ..., R_b^{-(M-2)}, R_b^{-(M-1)}$ which decrease with each generation or step and are related in a geometric sequence.

To determine the source uncertainty, it would be desirable to do minimum sampling within the constraints of the stream network. It can be shown that the minimum number of sampling will equal the number of generations in the basin. It is therefore hypothesized that if there is only one source and only one sample for each generation, then the minimum number of sequential samples to locate the source is equal to number of generations separating the outlet from the source

$$U = M - i \qquad (4.91)$$

where U is the minimum number of samples that can be employed to denote the source uncertainty.

The source uncertainty can also be related to the Shannon entropy (Shannon, 1948) as

$$U = -c \sum_{j=1}^{N_i} p_{ij} \log p_{ij} \qquad (4.92)$$

where c is a scale factor or unit of measurement, and p_{ij} is the probability of the jth source of the ith-order streams. Substituting Eq. (4.88) in Eq. (4.92) yields

$$U = -c \sum_{j=1}^{N_i} \frac{1}{N_i} \log \frac{1}{N_i} = -c \log N_i \qquad (4.93)$$

Inserting Eq. (4.87) in Eq. (4.93), one obtains

$$U = c(M - i) \log R_b \qquad (4.94)$$

Since the value of c is arbitrary, it can be expressed in any convenient units, and a logical choice for a stream network would be

$$c = \frac{1}{\log R_b} \qquad (4.95)$$

With this choice, Eq. (4.94) reduces to Eq. (4.91). Note that the smallest order in the Horton-Strahler ordering scheme is 1 ($i = 1$). In this case, the entropy or uncertainty will be maximum: $U = M - 1$. If $i = 0$, then $U = M$, exactly equal to the basin order.

4.11.3   Search for Contaminant Source

First, consider an example where a stream nework has a bifurcation ratio of 3 and a maximum order of 6 on a map scale of of 1:25,000. The postulate in Eq. (4.91) states that the minimum number of sequential samples needed to locate the source starting at the outlet would be $U = M - i = M - 1 = 6 - 1 = 5$ samples, under the condition that there is only one source and one sample needed for each generation. In practice, the number of these sequential samples will depend on the effectiveness of the sampling method in distinguishing various tributaries of each generation. The number of samples to track down a source will be

$$S = U \times n \tag{4.96}$$

where n is the average number of samples needed for each generation and will depend on the sampling method and bifurcation nature, S is the total number of samples, and U is the number of generations or source uncertainty. For illustrative purposes the following examples are worked out.

Consider a stream network such that streams of order i meet at a single place to form streams of order $i + 1$. In this case, the bifurcation ratio is 2, and one sample will be needed for each generation. If there exists only one source, then the ideal sampling efficiency will be one sample per generation.

For any constant integer value of bifurcation ratio, Sharp (1970) showed that per generation the minimum number of samples would be 1 and the maximum would be $R_b - 1$ and derived the average number of samples as

$$n = \frac{(R_b + 2)(R_b - 1)}{2R_b} \tag{4.97}$$

For large values of R_b, n would approach $R_b/4$. Equation (4.97) shows that the sampling effectiveness decreases with increasing R_b. For the previous case of stream, $U = 5$, $R_b = 3$, and the average number of samples $= (3 + 2)(3 - 1)/6 = 10/6 = 1.667$, or approximately 2. The total number of samples $S = 5 \times 1.667 = 8.37$, or approximately 9. The minimum number of samples would be 5.

Assume that there is an unknown number of multiple sources. Then for each contaminated segment of order i, an average of R_b segments of order $i - 1$ would have to be sampled. If the number of multiple sources is small, then the maximum number of samples required to locate these sources can be approximated as

$$S \approx U(R_b - 1)N \tag{4.98}$$

where N is the number of contaminated sources.

The maximum number of samples required to locate all possible sources would be the number of forks in the basin. Recall that the number of forks plus the outlet equals the number of ith order streams (Shreve, 1966). Then the number of samples S to search a basin is

$$U \leq S \leq 2N_i \tag{4.99}$$

where N_i is given by Eq. (4.87).

In natural basins, there are streams that do not lead to forming new higher-order streams; i.e., a stream order i may be absorbed in a stream of order $i + 1$. This type of stream network can be handled using the Fibonacci tree (Hoggatt, 1968), as shown by Sharp (1970). The Fibonacci method is not discussed here. However, the methodology can be extended to the stream ordering method proposed by Shreve (1967), called *link magnitude*. The magnitude of any link in a stream network equals the total number of sources which become tributary to it. Again, it is assumed that there is only one source of contamination, and the stream network is represented in terms of link magnitude.

Let u be the link magnitude at the outlet of a stream. The question is: How many samples are needed to determine a contamination source located on the source link? The number of samples required to find the location of a source on the first-order link is given by Sharp (1970) as

$$S = 1 + \log_2 u = \log_2(2u) \qquad (4.100)$$

The search process is conducted as follows. One traverses from the basin outlet upstream until the link magnitude of the mainstream is halved. Should the sample exhibit the presence of a contaminant, the search is continued by traversing upstream until the link magnitude is halved again. Thus, the search is continued by successively halving the link magnitude. It is now recalled that

$$u = N_1 = R_b{}^{M-1} \qquad (4.101)$$

Substitution of Eq. (4.101) in Eq. (4.100) yields

$$S = 1 + \log_2 R_b{}^{M-1} = 1 + (M-1)\log_2 R_b \qquad (4.102)$$

Consequently, the average number of samples per generation is given by

$$n = \frac{1 + (M-1)\log_2 R_b}{M - 1} \qquad (4.103)$$

For the river considered before where $M = 6$ and $R_b = 3$, one gets $S = 1 + (6-1)\log_2 3 = 8.92$. This yields 9 samples if searching is based on link magnitudes.

In natural stream networks, there may be many confluences between streams differing in order by 2 or more. Considering that the bifurcation ratio reflects these confluences, the expected number of samples may be given by the product $U \times R_b$. Based on testing on six South Carolina rivers, Sharp (1970) suggested a better number

$$S = U(R_b - 1) = (M-1)(R_b - 1) \qquad (4.104)$$

For the river considered, the expected average number of sequential samples needed to determine the first-order source is $(6-1)(3-1) = 10$.

In practice, sequential searches are not routinely done. Instead, permanent monitoring stations are situated that will perform some fraction of a sequential search. If the sequential search is reduced to one-third, then this is equivalent to determining $U/3$ or $(M - i)/3, i = 1$. If $M = 6$, then $U/3$ will equal 2 when $i = 1$. The order of a river with $U = 2$ is found to be $2 = M - 1$ or $M = 3$. This means that a series of stations covering all third-order rivers will reduce the sequential search by one-half. The number of stations needed would be $N_3 = R_b^{6-3} = 3^3 = 9$ stations which is close to the actual number of stations needed. One can then chose to monitor all fifth-order streams. This will correspond to a reduction of sequential search by $(U'/U = (5 - 1)/(6 - 1))4/6$; one would need $N_5 = R_b^{6-5} = 3$ stations.

4.12 Watershed Surface Description

Watershed surfaces impact the economic use of land, agricultural activities and development, forest management, urban development, road development, pollutant movement, and design of drainage systems. These surfaces can be classified into macro- and microsurfaces. Microsurfaces or surfaces with a few macro contour lines are desirable for agricultural operations. Indeed machine operations are easier on microsurfaces, and these surfaces can be used to quantify the effect of tillage operations and tools. In many instances, tools are designed considering surface configuration. Microsurfaces permit the development of sheet/overland flow when rain falls on them.

Macrosurfaces relate to the geomorphological development of an area and reflect the runoff and erosion potential of the area, which, in turn, can be used to design soil and water conservation measures. These surfaces can be described using topographic maps with a contour interval of, say, 1 ft and can be employed to delineate the channel network and hence describe the channel hydrology.

Surface irregularities contribute to deep percolation loss and reduce water application efficiency. That is one reason that land leveling is often practiced for improving agricultural irrigation by flooding. These irregularities also affect cropping patterns and methods of irrigation to be developed.

4.12.1 Methods of Description

There are myriad ways to describe a watershed surface, such as (1) topographic or contour mapping, (2) dimensional analysis, and (3) geomorphologic (e.g., basin relief, relief ratio, drainage density, distribution of slopes, etc.), among others. One of the problems with these methods is that they are designed primarily for macrosurfaces and therefore are limited in their application. Because of the continuing changes occurring in the watershed surface, a mathematical description becomes difficult. Further, most surfaces are inherently nonuniform, and this nonuniformity makes it difficult to develop a method that would apply to all areas of interest or all surfaces. Hence, a new way of describing surfaces is desired.

For the purposes of description, surface nonuniformity can be characterized by elevation. If observations of elevation for a surface are available, then an average elevation can be determined and the nonuniformity can be expressed by the deviations from the average or mean elevation. The deviation values will depend on the areal extent and the scale of nonuniformity. Observations reveal that surfaces are nonuniform in both directions; e.g., the deviations are location- and direction-dependent. Therefore, it desirable to define the nonuniformity in terms of heterogeneity and isotropy according to location and direction, respectively. To define these terms, consider two traces that are perpendicular to each other. A surface is isotropic if the statistics of the surface estimated by measurements along trace 1 are identical to those obtained by measurements along trace 2. In a similar manner, at two different locations if the two traces at one location are identical to those at the other location, then the traces are homogeneous. Thus, a surface can be classified as (1) homogeneous and isotropic, (2) homogeneous and anisotropic, (3) heterogeneous and isotropic, and (4) heterogeneous and anisotropic. Most large basins are heterogeneous and anisotropic. A plowed field is an example of an isotropic and homogeneorus surface. Some agricultural fields are homogeneous but anisotropic, and others are heterogeneous but isotropic.

Since elevation varies in two dimensions and randomly, it would be logical to consider a probabilistic decription of surface nonuniformity. Merva et al. (1970) presented such a description using entropy, and we follow their work here. To that end, let the elevation from some mean plane (deviation) be a random variable denoted by Z. The objective is to employ the entropy theory to derive the probability distribution of the elevation deviations. From the definition of entropy, one can then write

$$S(Z) = -\int_{-\infty}^{\infty} f(z) \log f(z) dz \qquad (4.105)$$

where $f(z)$ is the probability density function (PDF) of Z and z is a particular value of Z. In accordance with the second law of thermodynamics, the entropy of a system (in this watershed surface) will attain a maximum value when the system tends to reach equilibrium. Since a watershed surface is formed as a result of natural proecsses, it is reasonable to assume that the surface is in a state of equilibrium or quasi-equilibrium. This means that $S(Z)$ must be maximized, and following the discussion in Chap. 1, the $f(z)$ that will correspond to the maxmium entropy must be determined. Of course, entropy maximization will depend on the constraints that the surface must satisfy. Recall that for a given variance the continuous distribution having the largest entropy is the normal distribution. If the elevation deviations are considered discrete then a discrete distribution corresponding to the maximum entropy will need to be determined.

Assuming that the surface is near equilibrium and deviations are continuous, the deviations may be approximated by the normal distribution

$$F(Z \leq z) = \frac{1}{\sqrt{2\pi\sigma^2}} \int_{-\infty}^{z} \exp\left(\frac{z^2}{2\sigma^2}\right) dz \qquad (4.106)$$

where $F(Z \leq z)$ is the probability that Z is equal to or less than a given value of deviation z and σ^2 is the variance of Z. Equation (4.106) involves specification of the mean plane, variance, and PDF. However, in practice many surfaces having the same variance can be very different from one another. Therefore, additional information is needed to probabilistically describe the surface uniquely. To that end, one can consider the correlation of deviations between any two points on the surface. Let w denote the distance between the two points. Then one can compute the coefficient of correlation between pairs of points on the surface. The correlation will be a function of the distance w between points and can be represented as a correlation function $R(w)$. If the normal distribution is assumed to probabilistically represent the surface, then the surface is described completely by its mean plane, variance, and correlation function $R(w)$. In this case, the second-order joint probability is specified such that a second point separated by a distance w from the first point will have a certain deviation.

Let the mean plane be described by the x and y coordinates. For any point (x, y) on the mean plane, the corresponding deviation will be $z(x, y)$. The deviations z are measured in the r direction. The probability that the deviation will lie between z and $z + \Delta z$ is given by the normal two-parameter probability distribution

$$F(r \leq Z(x,y) \leq r + \Delta r) = \frac{1}{\sqrt{2\pi\sigma^2}} \int_{r}^{r+\Delta r} \exp\left\{-\frac{[z(xy)]^2}{\sigma^2}\right\} dz(x,y) \tag{4.107}$$

If the surface is isotropic, then the correlation distance between two points (x_1, y_1) and (x_2, y_2) is given as

$$w = [(x_2 - x_1)^2 + (y_2 - y_1)^2]^{1/2} \tag{4.108}$$

For an anisotropic surface, the orientation of the line separating the two points should be taken into account so that the correlation function can be expressed as $R[(x_1, y_1), (x_2, y_2)]$; that is, both the separation distance and the direction of the second point from the first point are important.

4.12.2 Spectral Density Function of a Surface

Another function that is useful for describing the surface is the spectral density function. It is known that the power spectrum $P(f)$ can be expressed in terms of the Fourier transform of the correlation function $R(w)$. Defining frequency f in terms of cycles per unit length, $P(f)$ can be expressed as

$$P(f) = 4\int_{0}^{\infty} R(w)\cos(2\pi fw)\,dw \tag{4.109}$$

For illustration, consider a cornfield. There would be significant surface irregularities occurring about 40 in. after the tillage operation. This means $f = 0.3$ cycle/ft. Of course, this will depend on the row spacing used and the degree of the tillage operation. For defining the spectral density function, it is usual practice to normalize the correlation function by dividing it by the value of the function for $w = 0$. This value of correlation is the same as the variance associated with the first-order probability distribution. The normalized correlation function is referred to as the *autocorrelation function* and can be denoted as

$$\rho(w) = \frac{R(w)}{R(0)} = \frac{R(w)}{\sigma^2} \tag{4.110}$$

Similarly, the spectral density function is obtained by dividing the power spectrum by the variance:

$$p(f) = \frac{P(f)}{\sigma^2} = 4\int_0^\infty \rho(w)\cos(2\pi f w)\,dw \tag{4.111}$$

The advantage of using the spectral density function for describing random surafces is that it can be construed as the frequency composition of the variance of the surface. From the Fourier series theory it is known that the trace of any function can be expressed as a trace of a series of sine and cosine functions of all frequencies having amplitudes such that the series sums to the desired function. The power spectrum can be viewed as the relative significance of a given frequency in contributing to the nonuniformity of a surface trace.

The spectral density can be computed as follows: (1) Lay out two traverses perpendicular to each other. (2) Measure elevations at fixed intervals along each traverse. (3) Determine the correlation values between elevation pairs and the correlation function. (4) Determine the frequency. (5) Find the power spectrum for each frequency. (6) Calculate the variance of the elevations. (7) Determine the spectral density function. (8) Plot the spectral density as a function of frequency. (8) Determine the Nyquist frequency f_n, which is the frequency beyond which no spectral density information can be obtained because of the sampling interval selected. Here,

$$f_n = \frac{1}{2\Delta w} \tag{4.112}$$

where Δw is the sampling distance.

4.13   Case Study

A watershed named Bogue Chitto with a drainage area over 3142 km^2 or 1213 mi^2 is selected. It has a flow discharge measuring gauge with identification number USGS 02490500 which is located at latitude 30°37'45", longitude 89°53'50" NAD27, and the USGS Hydrological Unit Code (HUC) for the watershed is 03180005. The watershed is located in the lower Mississippi River basin and is shared by the states of Louisiana and Mississippi. Using DEM with a 30-m by 30-m spatial resolution, the watershed is subdivided into 29 subwatersheds, numbered from 1 to 29, as shown in Fig. 4.22. The watershed is ordered following the Horton-Strahler ordering scheme, as shown in Fig. 4.23. The watershed is also represented as a tree diagram consisting of links and nodes, as shown in Fig. 4.24. Then the lengths, elevations, slopes, and distances of the stream segments to the outlet are computed, as shown in Table 4.7. Distances of stream segments from the outlet are arranged in class intervals, as in Table 4.8. The width function of Bogue Chitto watershed relating the number of segments to distance from the

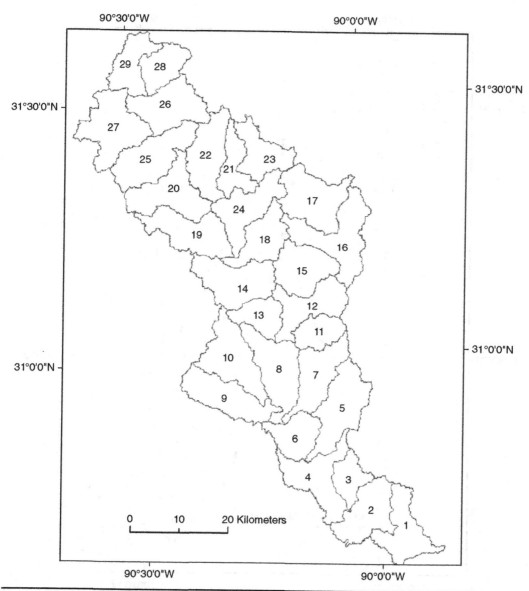

FIGURE 4.22   Subbasins in the Bogue Chitto watershed.

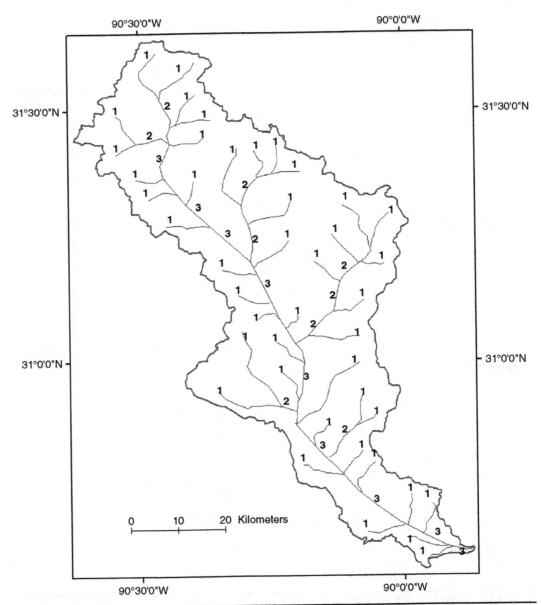

FIGURE 4.23   Bogue Chitto watershed ordered using the Horton-Strahler ordering scheme.

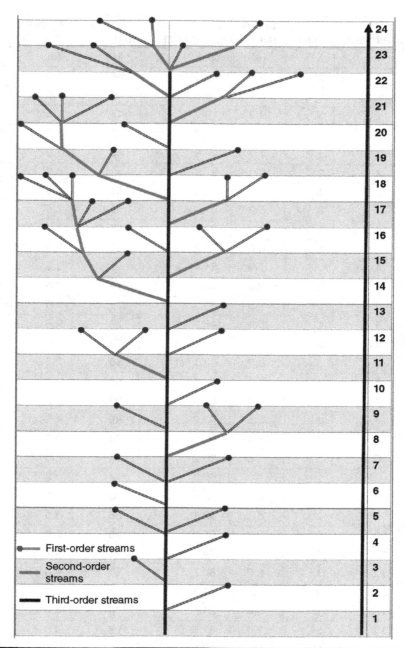

FIGURE 4.24   Simplified drainage network formed by links and nodes for Bogue Chitto catchment.

Link No.	Subbasin	Order	Length (km)	Distance Outlet (km)	Elevation (m)	Terrain Slope (%)	Stream Slope (%)
1	1	1	8.5	12.3	44	2.3	0.36
2	1	1	5.5	9.3	44	2.3	0.36
3	1	1	6.4	6.4	44	2.3	0.36
4	1	3	10.2	10.2	44	2.3	0.36
5	2	1	7.7	21.7	42	3.0	0.29
6	2	1	7.2	20.3	42	3.0	0.29
7	2	3	11.0	21.1	42	3.0	0.29
8	3	1	8.1	32.3	51	2.6	0.31
9	3	3	5.9	27.1	51	2.6	0.31
10	4	1	7.9	37.4	51	3.3	0.32
11	4	1	6.3	36.2	51	3.3	0.32
12	4	3	7.3	34.4	51	3.3	0.32
13	5	2	8.6	42.8	79	2.6	0.27
14	5	1	7.3	50.3	79	2.6	0.27
15	5	1	5.4	48.3	79	2.6	0.27
16	5	3	1.1	43.9	79	2.6	0.27
17	6	3	8.9	43.1	55	2.2	0.33
18	6	1	6.2	43.4	55	2.2	0.33
19	7	1	18.9	62.9	89	4.2	0.24
20	7	3	1.1	43.9	89	4.2	0.24
21	8	3	15.4	61.5	74	3.0	0.42
22	8	1	6.8	56.2	74	3.0	0.42
23	8	1	7.6	66.2	74	3.0	0.42
24	9	1	12.7	62.5	85	3.5	0.32
25	9	2	3.1	49.1	85	3.5	0.32
26	10	1	19.2	67.9	95	3.6	0.22
27	11	1	9.4	74.0	99	4.7	0.41
28	12	1	6.7	80.2	97	4.0	0.64
29	12	2	10.1	71.7	97	4.0	0.64
30	15	1	6.4	84.5	108	4.0	0.54
31	15	2	12.0	82.6	108	4.0	0.54
32	16	1	8.6	97.2	116	3.5	0.35
33	16	1	4.1	88.7	116	3.5	0.35
34	16	1	8.6	92.1	116	3.5	0.35

TABLE 4.7 Lengths, Elevations, Slopes, and Distances to Outlet of the Stream Segments in Bogue Chitto Watershed

Link No.	Subbasin	Order	Length (km)	Distance Outlet (km)	Elevation (m)	Terrain Slope (%)	Stream Slope (%)
35	16	2	3.6	87.4	116	3.5	0.35
36	17	1	11.3	98.7	123	3.1	0.25
37	13	1	3.9	69.8	88	4.2	0.75
38	13	1	4.5	72.0	88	4.2	0.75
39	13	3	10.4	75.9	88	4.2	0.75
40	14	1	6.3	85.2	106	4.5	0.53
41	14	1	6.3	78.6	106	4.5	0.53
42	14	3	6.5	79.0	106	4.5	0.53
43	18	1	8.8	89.3	115	4.3	0.45
44	18	3	3.1	81.3	115	4.3	0.45
45	19	1	8.0	100.5	113	4.5	0.49
46	19	3	10.5	91.9	113	4.5	0.49
47	20	1	6.0	107.1	119	4.1	0.54
48	20	1	5.5	104.6	119	4.1	0.54
49	20	3	13.4	105.8	119	4.1	0.54
50	24	1	9.7	100.0	116	4.5	0.41
51	24	2	12.6	94.0	116	4.5	0.41
52	22	1	14.3	108.1	133	3.3	0.25
53	21	1	6.6	107.3	127	4.0	0.50
54	21	2	6.8	101.5	127	4.0	0.50
55	23	1	6.6	109.9	133	3.3	0.45
56	23	1	5.2	108.5	133	3.3	0.45
57	23	2	2.7	103.4	133	3.3	0.45
58	25	1	6.6	119.9	128	4.1	0.42
59	25	1	6.2	111.3	128	4.1	0.42
60	25	3	8.4	112.9	128	4.1	0.42
61	26	1	8.4	125.0	134	3.7	0.29
62	26	1	7.6	125.4	134	3.7	0.29
63	26	2	9.6	123.7	134	3.7	0.29
64	28	1	12.8	136.4	142	2.8	0.22
65	29	1	12.1	135.7	141	2.7	0.21
66	27	1	7.3	126.9	136	3.3	0.20
67	27	1	4.5	124.1	136	3.3	0.20
68	27	2	6.7	119.6	136	3.3	0.20

TABLE 4.7   Lengths, Elevations, Slopes, and Distances to Outlet of the Stream Segments in Bogue Chitto Watershed (*Continued*)

Classification	Bin Size (km)	Number of Segments
Class 1	0–20	4
Class 2	20–40	8
Class 3	40–60	9
Class 4	60–80	12
Class 5	80–100	13
Class 6	100–120	15
Class 7	120–140	7

TABLE 4.8 Arrangement of Stream Segment Distances in Class Intervals

outlet is shown in Fig. 4.25. The primary entropy and the secondary entropy (Shannon) are computed for the Bogue Chitto watershed, as shown in Fig. 4.26. Then the entropy for subnetworks discharging in the main channel of the Bogue Chitto watershed is computed and is plotted versus average elevation, as seen in Fig. 4.27. The average elevation versus topological diameter of subnetworks whose outlets lie on the main channel is plotted in Fig. 4.28. The average elevation versus the entropy of the subnetwork is plotted in Fig. 4.29, and the relation between nodal elevation and the topological distance from the source of the main channel is plotted in Fig. 4.30. The relation

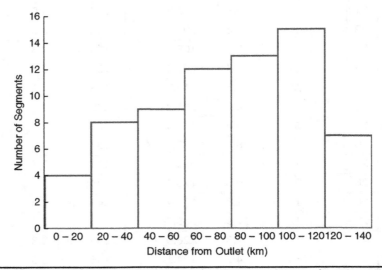

FIGURE 4.25 Width function relating the number of segments to distance from the outlet (Bogue Chitto watershed).

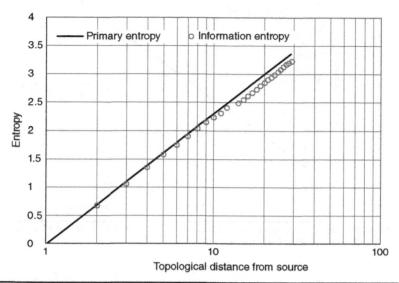

FIGURE 4.26 Primary entropy and information entropy as a function of distance from the outlet (Bogue Chitto watershed).

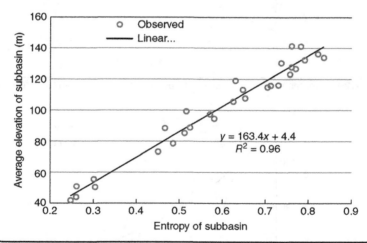

FIGURE 4.27 Average elevation versus information entropy of subnetworks with outlets lying in the mainstream of the Bogue Chitto watershed.

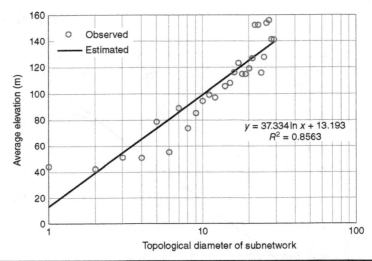

FIGURE 4.28 Average elevation versus topological diameter of subnetworks whose outlets lie on the main channel (Bogue Chitto watershed).

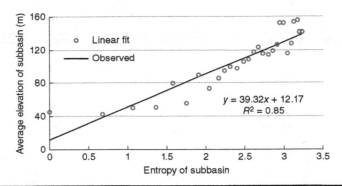

FIGURE 4.29 Average elevation versus entropy of subnetwork (Bogue Chitto watershed).

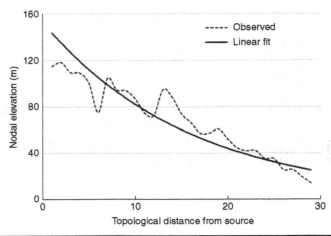

FIGURE 4.30 Relation between nodal elevation and topological distance from the source of the main channel (Bogue Chitto watershed).

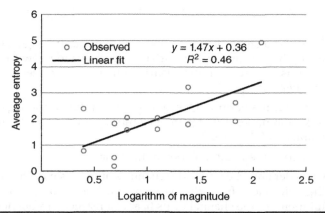

FIGURE 4.31 Relation between average entropy and the basin magnitude (Bogue Chitto watershed).

between average entropy and the basin magnitude is plotted in Fig. 4.31, and the relation between the cumulative entropy and Horton-Strahler order is plotted in Fig. 4.32. A simplified drainage network formed by links and nodes is similar to what is shown in Fig. 4.3, and the Brazos River basin represented by lines and nodes is shown in Fig. 4.4.

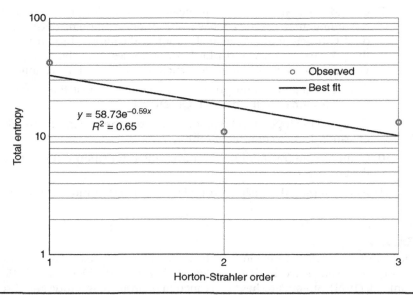

FIGURE 4.32 Relation between cumulative entropy and Horton-Strahler order (Bogue Chitto watershed).

Questions

1. Consider a small (less than 100 km^2) natural drainage basin. Order the basin using the Horton-Strahler ordering scheme and show the ordering on a map. What is the order of this basin? Tabulate the channels of all orders and their number.

2. Show for the basin in Question 1 the simplified drainage network formed by links and nodes. What is the magnitude of this basin? What is the topological diameter of the basin?

3. Construct the width function of the basin.

4. Consider only the first-order channels of the basin in Question 1. Then compute their distances from the basin outlet. The distance can be from either the midpoint of the channel or its head. Construct a frequency histogram of these distances. Does it have any relation to the width function?

5. Compute the primary information entropy of the basin in Question 1. Assuming that the only information available is the mean topological distance, derive its probability density function and then its entropy. How does this entropy compare with the primary entropy? Plot the two entropies.

6. Consider elevations of links in Question 1 and assume that the mean elevation of links in the basin is known. Derive the probability density function of link elevation and compute its entropy. How does this entropy value compare with that obtained in Question 5? Also compute the primary entropy. Plot the two entropies.

7. Consider elevations and their mean corresponding to a topological distance. Assuming that the mean basin elevation is known, derive the probability density function of the mean elevation and compute its entropy. Also compute the primary entropy and plot both entropies. Plot the mean basin elevation versus its subnetwork diameter whose outlet lies on the channel diameter.

8. Construct the elevation profile of a channel against the topological distance from the source for the basin in Question 1.

9. Compute the topological diameter from the stream length ratio for the basin in Question 1. Also compute its entropy.

10. Compute the fractal dimension from the bifurcation ratio and length ratio for the basin in

11. Also compute the basin entropy from the length ratio and fractal dimension.

References

Fiorentino, M., P. Claps, and Singh, V. P. (1993). An entropy-based morphological analysis of river basin networks. *Water Resources Research*, vol. 29, no. pp. 1215–1224.

Flint, J. J. (1974). Stream gradient as a function of order, magnitude and discharge. *Water Resources Research*, vol. 10, no. 5, pp. 969–974.

Fok, Y. S. (1969). The law of stream elevations in stream morphology. Paper presented at the Fall Meeting of the American Geophysical Union, San Francisco.

Gray, D. M. (1961). Interrelationships of water characteristics. *Journal of Geophysical Research*, vol. 66, pp. 1215–1224.

Gupta, V. K., and Waymire E. (1989). Statistical self-similarity in river networks parameterized by elevation. *Water Resources Research*, vol. 25, no. 3, pp. 463–476.

Hoggatt, V. E., Jr. (1968). Generalized rabbits for generalized Fibonacci numbers. *Fibonacci Quarterly*, vol. 6, pp. 105–114.

Horton, R. E. (1945). Erosional development of streams and their drainage basins: Hydrophysical approach to quantitative geomorphology. *Bulletin of Geological Society of America*, vol. 56, no. 3, pp. 275–370.

Hydraulics Research Station (1967). Hydraulics research report, report of the Hydraulics Research Station Steering Committee with the report of the Director of Hydraulics Research, Wallingford, Berkshire, England.

Jaynes, E. T. (1957). Information theory and statistical mechanics, I. *Physical Reviews*, vol. 106, pp. 620–630.

La Barbera, P., and R. Rosso (1989). On the fractal dimension of stream networks. *Water Resources Research*, vol. 25, no. 4, pp. 735–741.

Leopold, L. B., and Wolman M. G. (1962). The concept of entropy in landscape evolution. U.S. Geological Survey, Professional Paper 500-A, Washington, D.C.

Lewis, G. N., and Randall, M. (1961). *Thermodynamics*. 2d ed., revised by K. S. Pitzer and L. Brewer, McGraw-Hill, New York, pp. 91–92.

Mandelbrot, B. B. (1983). *The Fractal Geometry of Nature*. W. H. Freeman, New York.

MacArthur, R. H., and Connell J. H. (1966). *The Biology of Populations*. Wiley, New York.

Merva, G. E., Brazee, R. D., Schwab, G. O., and Curry R. B. (1970). Theoretical considerations of watershed surface description. *Transactions of the ASAE*, vol. 13, pp. 462–465.

Mueller, J. E. (1973). Re-evaluation of the relationship of master streams and drainage basins: Reply. *Geological Society of America Bulletin*, vol. 84, pp. 3127–3130.

Prigogine, I. (1967). *Introduction to Thermodynamics of Irreversible Processes*, 3d ed., Wiley, New York.

Rodriguez-Iturbe, I., and Rinaldo A. (1997). *Fractal River Basins*. Cambridge University Press, Cambridge, United Kingdom.

Rodriguez-Iturbe, I., Rinaldo, A., Rigon, R., Bras, R. L., Marani, A., and Ijjasz-Vasquez E. (1992). Energy, dissipation, runoff production and the 3-dimension structure of river basins. *Water Resources Research*, vol. 28, no. 4, pp. 1095–1104.

Shannon, C. E. (1948). The mathematical theory of communication, I and II. *Bell System Technical Journal*, vol. 27, pp. 379–424.

Sharp, W. E. (1970). Stream order as a measure of sample source uncertainty. *Water Resources Research*, vol. 6, no. 3, pp. 919–926.

Shreve, R. L. (1966). Statistical law of stream numbers. *Journal of Geology*, vol. 74, pp. 17–37.

Shreve, R. L. (1967). Infinite topologically random channel networks. *Journal of Geology*, vol. 75, pp. 178–186.

Stall, J. B., and Fok Y. S. (1968). Hydraulic geometry of Illinois streams. *Water Resources Center Research Report 15*, University of Illinois, Urbana.

Stall, J. B., and Yang T. T. (1970). Hydraulic geometry of 12 selected stream systems of the United States. Research Report No. 32, Illinois State Water Survey, University of Illinois, Urbana-Champaign.

Strahler, A. N. (1957). Quantitative analysis of watershed geomorphology. *Transactions of the American Geophysical Union*, vol. 38, no. 6, pp. 913–920.

Tarboton, D. G., Bras, R. L., and Rodriguez-Iturbe I. (1990). Comment on "On the fractal dimension of stream networks," by P. La Babera and R. Rosso. *Water Resources Research*, vol. 26, no. 9, pp. 2243–2244.

Troutman, B. M., and Karlinger, M. R. (1984). On the expected width function for topologically random channel networks. *Journal of Applied Probability*, vol. 21, pp. 836–844.

Wang, S. X., and Waymire E. C. (1991). A large deviation rate and central limit theorem for Horton ratios. *SIAM Journal of Discrete Mathematics*, vol. 4, no. 4, pp. 575–588.

Willgoose, G., Bras, R. L., and Rodriguez-Iturbe I. (1991). A physical explanation of observed link area-slope relationship. *Water Resources Research*, vol. 27, no. 7, pp. 1697–1704.

Yang, C. T. (1971). Potential energy and stream morphology. *Water Resources Research*, vol. 7, no. 2, pp. 311–324.

Further Reading

Broscoe, A. J. (1959). Quantitative analysis of longitudinal stream profiles of small watersheds. Office of Naval Research Technical Report 18, Columbia University, New York.

Claps, P., and Oliveto G. (1994). Fractal structure, entropy and energy dissipation in river networks. *IAH Hydrology Journal*, vol. 17, no. 3, pp. 38–51.

Claps, P., Fiorentino, M. and Oliveto G. (1996). Informational entropy of fractal river networks. *Journal of Hydrology*, vol. 187, pp. 145–156.

Davy, B. W., and Davies T. R. H. (1979). Entropy concepts in fluvial geomorphology: A reevaluation. *Water Resources Research*, vol. 15, no. 1, pp. 103–106.

Gray, D. M. (1961). Interrelationships of watershed characteristics. *Journal of Geophysical Research*, vol. 66, pp. 1215–1224.

Mock, S. J. (1971). A classification of channel links in stream networks. *Water Resources Research*, vol. 7, pp. 1558–1566.

Rodriguez-Iturbe, I., Rinaldo, A., Rigon, R., Bras, R. L. Marini, A., and Ijjasz-Vasquez R. (1992). Energy dissipation, runoff production and the 3-dimensional structure of river basins. *Water Resources Research*, vol. 28, no. 4, pp. 1095–1104.

Scheidegger, A. E. (1962). A thermodynamic analogy for meander systems. *Water Resources Research*, vol. 3, no. 4, pp. 1041–1046.

Scheidegger, A. E. (1968). Horton's law of stream numbers. *Water Resources Research*, vol. 4, no. 3, pp. 655–658.

Sharp, W. E. (1979). Stream order as a measure of sample source uncertainty. *Water Resources Research*, vol. 6, no. 3, pp. 919–926.

Shreve, R. L. (1966). Statistical law of stream numbers. *Journal of Geology*, vol. 74, pp. 17–34.

Singh, V. P., and Fiorentino M. (1992). *Entropy and Energy Dissipation in Water Resources*. Kluwer Academic Publishers (now Springer), Dordrecht, The Netherlands.

Smart, J. S. (1969). Topological properties of channel networks. *Geological Society of America Bulletin*, vol. 80, pp. 1754–1774.

Strahler, A. N. (1952). Hypsometric (area-altitude) analysis of erosional topography. *Bulletin of Geological Society of America*, vol. 63, pp. 1117–1144.

Scheidegger, A. E. (1964). Some implications of statistical mechanics in geomorphology. *International Association of Scientific Hydrology Bulletin*, vol. 9, no. 1, pp. 12–16.

Werner, C., and Smart J. S. (1973). Some new methods of topologic classification of channel networks. *Geographical Analysis*, vol. 5, pp. 271–295.

CHAPTER 5

Evaluation and Design of Sampling and Measurement Networks

Hydrometric data are required for efficient planning, design, operation, and management of virtually all water resources systems, including water supply reservoirs, recreation and fisheries facilities, flood control structures, and hydroelectric plants, to name but a few. Many studies have applied entropy theory to assess and optimize data collection networks (e.g., water quality, rainfall, stream flow, hydrometric, groundwater, elevation data, landscape). Decision making in water resources project design and evaluation is closely linked to the amount of hydrologic information available. If enough accurate and relevant information is available, the likelihood of an underdesign or overdesign is reduced. Thus, economic losses can be minimized, resulting in an overall increase in the benefit/cost ratio. However, it is not always easy to quantitatively define the optimum level of information needed for planning, design, and development of a specific project in a watershed. This is largely due to the difficulty of developing cost and benefit functions of hydrologic information. This then leads to the difficulty of achieving an optimum balance between the economic risk due to inadequate information and the cost of a network capable of providing the required information. The objective of this chapter is to present basic entropy-related considerations needed for evaluation and design of monitoring networks.

5.1   Design Considerations

A methodology for data collection network design must take into account the information of each gauging station or potential gauging station in the network. A station with higher information content would generally be given a higher priority over other stations that have lower information contents. The information content of a station must, however, be balanced with site-specific uses and users of the data collected at the station. For example, a station that is used by one user might be given a lower priority than a station that has diverse uses. Burn and Goulter (1991) developed a data collection network design framework considering such issues.

In general, a framework for network design or evaluation considers a range of factors, including: (1) objectives of sampling, (2) variables to be sampled, (3) locations of measurement stations, (4) frequency of sampling, (5) duration of sampling, (6) uses and users of data, and (7) socioeconomic considerations. Effective monitoring is also related to these factors. Evaluation of a network has two modes: (1) number of gauges and their locations (space evaluation) and (2) time interval for measurement (time evaluation). The information in one mode may be supplemented by the other with appropriate transfer mechanisms and by cross-correlation structure (space-time tradeoff). Space-time evaluation of networks should not be considered as fixed but should periodically be subject to revision and is indeed evolutionary. Uslu and Tanriover (1979) analyzed the entropy concept for the delineation of optimum sampling intervals for data collection systems in both space and time. Harmancioglu et al. (1992a,c; 1997; 1999), Harmancioglu and Singh (1990, 1991, 1998, 2002) and Ozkul et al. (1996; 2000a,b) presented methodologies using entropy theory for assessing water quality monitoring networks, following the work by Harmancioglu and Alpaslan (1992).

All designs, whether of the network or the monitoring program, must be cost-effective in gathering data and cost-efficient in obtaining information from data. These two requirements call for evaluating the performance of a network. Such an evaluation must consider the benefits of monitoring with respect to the objectives of monitoring and the cost, both marginal and average, of obtaining those benefits. Sometimes it is the budget that controls the design of the network and monitoring program. Then the problem reduces to one of obtaining the greatest benefit (most information) for the available budget.

5.2   Correlation

A measure of information is cross-correlation among records from nearby sites. Cross-correlation helps with space-time tradeoff and regional data collection. The correlation coefficient r_{xy} is calculated for each pair of stations (X, Y) as

$$r_{xy} = \frac{\text{Cov}_{xy}}{S_x S_y} \tag{5.1a}$$

where Cov_{xy} is the covariance between random variables X and Y measured by the two stations, S_x is the standard deviation of variable X, and S_y is the standard deviation of variable Y. And Cov_{xy} can be computed as

$$\text{Cov}_{xy} = \frac{\sum_{i=1}^{n}(x_i - \overline{x})(y_i - \overline{y})}{n-1} \tag{5.1b}$$

where $\overline{x}$ is the mean of variable X; $\overline{y}$ is the mean of variable Y; and n is the number of observations or values. In this chapter, the station and the variable being measured there will be used interchangeably.

5.3   Entropy-Based Measures

Since entropy is a measure of information or uncertainty associated with a random variable or its probability distribution, entropy can be used for measuring the information content of observations of the random variable at a gauging station. The measures of information are marginal entropy, conditional entropy, joint entropy, and transinformation, which are discussed in Chap. 1.

Let the data being collected at a station correspond to a random variable X. Then the *marginal entropy* $H(X)$ can be defined as the potential information of the variable; this is also the information of that gauging station.

$$H(X) = -\sum_{i=1}^{n} p(x_i) \ln p_i = -\sum_{i=1}^{n} p_i \ln p_i \qquad (5.2a)$$

where random variable X takes on values x_i, $i = 1, 2, \ldots, n$, with probabilities p_i, $i = 1, 2, \ldots, n$. In a similar manner, the marginal entropy of random variable Y can be defined.

The *joint entropy* $H(X, Y)$ is the total information content contained in both X and Y and can be defined as

$$H(X, Y) = -\sum_{i=1}^{N} \sum_{j=1}^{M} p(x_i, y_j) \log p(x_i, y_j) \qquad (5.2b)$$

where $p(x_i, y_j)$, $i = 1, 2, 3, \ldots, N$ and $j = 1, 2, \ldots, M$, are the joint probabilities. It can be shown that the joint entropy is the sum of the marginal entropies of X and Y, provided that they are independent. Otherwise, it is the sum of the marginal entropy of one of the stations and the uncertainty that remains in the other station when a certain amount of information that it can convey is already present in the first station.

For two gauging stations X and Y, the conditional entropy $H(X \mid Y)$ is a measure of the information content of X that is not contained in Y, or the entropy of X given the knowledge of Y or the amount of information that still remains in X even if Y is known. If $p(x_i \mid y_j)$ and $p(y_j \mid x_i)$ are the conditional probabilities, then the conditional entropies $H(X \mid Y)$ and $H(Y \mid X)$ can be expressed mathematically as

$$H(X \mid Y) = -\sum_{i=1}^{N} \sum_{j=1}^{M} p(x_i, y_j) \log p(x_i \mid y_j) \qquad H(Y \mid X) = -\sum_{i,j} p(x_i, y_j) \log p(y_i \mid x_j) \qquad (5.2c)$$

A series of observations about an uncertain event contains more information about the event than do observations of a less certain event. The information observed at different sites (gauging stations) can be inferred, to some extent, from the observations at other sites. The information transferred among information emitters (predictor stations) and the information receivers (predicted stations) can be measured by transinformation or mutual information. The information transmitted from variable X to variable Y can be represented as

$$T(X, Y) = \sum_i \sum_j p(x_i, y_j) \log \frac{p(x_i \mid y_j)}{p(x_i)} \qquad T(X, Y) = \sum_i \sum_j p(x_i, y_j) \log \frac{p(x_i, y_j)}{p(x_i)p(y_j)} \qquad (5.2d)$$

Mutual information is used for measuring the inferred information or, equivalently, information transmission. Entropy and mutual information have advantages over other measures of information, for they provide a quantitative measure of (1) the information at a station, (2) the information transferred and information lost during transmission, and (3) a description of the relationship among stations based on their information transmission characteristics.

Thus, the joint entropy is equal to the sum of marginal entropies minus the common entropy or transinformation. The mutual entropy (information) between X and Y, also called *transinformation* $T(X, Y)$, is interpreted as the reduction in the original uncertainty of X, due to the knowledge of Y. It can also be defined as the information content of X that is contained in Y. In other words, it is the difference between the total entropy and the sum of the entropies of two variables or stations. This is the information repeated in both X and Y, and it defines the amount of uncertainty that can be reduced in one of the stations when the other station is known. The information transmitted from station X to station Y is represented by the mutual information $T(X, Y)$ (Lathi, 1969). The conditional entropy can be interpreted as the amount of lost information.

The multidimensional joint entropy for n gauges in a watershed represents the common uncertainty of their measured data sets. The amount of uncertainty left in the central gauge when the records of all other gauges are known is expressed by the multivariate conditional entropy of the central gauge conditioned on all other records. Similarly, the uncertainty left in the group of gauges $(i_1, ..., i_n)$ when any new gauge is added (i.e., the expansion of the existing gauge network) can be defined as

$$H[(X_1, X_2, ..., X_n) \mid X_{n+1}] = -\sum_{i_1} \cdots \sum_{i_{n+1}}^{N_1} \sum_{i_{n+1}}^{N_{n+1}} p(i_1, i_2, ..., i_n, i_{n+1}) \log[p(i_1, i_2, ..., i_n) \mid i_{n+1}] \qquad (5.3a)$$

or

$$H[(X_1, X_2, ..., X_n) \mid X_{n+1}] = H(X_1, X_2, ..., X_n, X_{n+1}) - H(X_{n+1}) \qquad (5.3b)$$

where N_i is the number of values of X_i. Corresponding to Eqs. (5.3a) and (5.3b), the multidimensional transinformation between the n existing gauges and the new (added) gauge $(n + 1)$ can be defined as

$$T[(X_1, X_2, ..., X_n); X_{n+1}] = H(X_1, X_2, ..., X_n, X_{n+1}) - H[(X_1, X_2, ..., X_n) \mid X_{n+1}] \qquad (5.4)$$

The transinformaton $T(X, Y)$ is symmetric, that is, $T(X, Y) = T(Y, X)$ and is nonnegative. A zero value occurs when two stations are statistically independent so that no information is mutually transferred; that is, $T(X, Y) = 0$ if X and Y are independent. When two stations are functionally dependent, the information at one site can be fully transmitted to another site with no loss of information at all. Subsequently, $T(X, Y) = H(X) = H(Y)$. For any other case, $0 \leq T(X, Y) \leq \min(H(X), H(Y))$. Larger values of T correspond to greater amounts of information transferred. Thus, T is an indicator of the capability of information transmission and the degree of dependency of stations.

5.4   Isoinformation

To determine the spatial distribution of information, the coefficient of nontransferred information, which is a measure of the nontransferred information, can be defined. This coefficient is the percentage of the total nontransferable information (Harmancioglu and Yevjevich, 1985, 1987). To illustrate, it is assumed that X_1 is associated with the rainfall record of the central rain gauge in the region, and $X_2 = X_i (i = 2, ..., n)$ associated with rainfall record of any other rain gauge. Let T_0 be the total transferable information (not necessarily achieved by the monitoring network) and T_1 be the measured trans-information between X_1 and X_2. Then

$$T_1 = T(X_1, X_i) = H(X_1) - H(X_1 | X_i) \qquad i = 2, \ldots, n \qquad (5.5)$$

The value of T_0 can be expressed as

$$T_0 = H(X_1) \qquad (5.6)$$

Then the coefficient of nontransferred information is

$$t_1 = \frac{H(X_1 | X_i)}{H(X_1)} \qquad 0 \leq t_1 \leq 1 \qquad (5.7)$$

and the coefficient of transferred information (or transferred information), $1 - t_1$, is defined as the fraction of the total transferable information: $1 - t_1 = T_1/T_0$.

By computing the coefficient of transferred information for all gauges, the isoinformation contours can be constructed. These contours will be the lines of equal transfer of information in the region (Krstanovic and Singh, 1988). The isoinformation contour of a bivariate record is the line of equal common information between any gauge in the watershed and the other existing gauges located in the watershed. In the selection process, the first chosen gauge has the highest information content and this can be designated as the central gauge. Thus, it is convenient to choose that gauge as the reference point in the construction of all isoinformation contours when choosing gauges in order of descending importance. It is true that the isoinformation contour can be constructed between any two gauges, but it will not benefit the selection process.

Example 5.1.   Monthly precipitation records from two stations in Texas are available, as presented in Table 5.1. Compute the coefficients of transferred and nontransferred information of station 1 with respect to station 2.

Station 1											
Month	**1**	**2**	**3**	**4**	**5**	**6**	**7**	**8**	**9**	**10**	**11**	**12**
2006	0.17	0.09	0.81	0.25	0.91	0.47	0.5	0.31	5.17	0.67	0.02	1.02
2007	3.8	0.05	2.95	0.92	5.33	4.55	3.19	1.11	2.53	0.34	0.3	0.21
2008	0.45	0.05	0.56	0.15	1.63	0.09	0.09	4.77	0.99	0.61	0.07	0.2
2009	0.07	0.05	0.96	0.1	1.76	0.23	0.02	0.47	2.79	2.84	1.34	0.94
2010	2.23	1.83	0.74	0.01	2.39	1.92	0.83	0.22	3.54	0.03	0.38	0.15
Station 2											
2006	0.08	0.26	0.2	0.02	4.51	0.09	2.22	0.01	2.52	0.68	0.01	1.64
2007	2.98	0.32	2.58	0.73	4.38	1.77	11.11	0.18	4.02	0.24	0.7	0.77
2008	0.24	0.05	0.55	3.19	3.43	0.31	3.97	10.35	3.32	1.02	0	0.35
2009	0.04	0.07	0.09	0.34	1.74	0	0.42	0.68	1.8	0.26	0.52	0.71
2010	4.98	3.63	0.65	2.74	0.41	0.01	3.08	0.26	5.62	0.06	0.01	0.08

TABLE 5.1   Monthly Precipitation (in.) of Two Stations in Texas

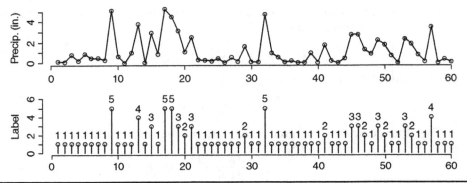

FIGURE **5.1** Illustration for continuous data discretization.

Solution. First, the marginal entropy of station 1 is computed. By dividing the observations into 5 equal sized intervals, discretization of the continuous precipitation sequence can be explicitly illustrated as in Fig. 5.1.

Following the method illustrated in Fig. 5.1, the continuous time series can be discretized and labeled, as tabulated in Table 5.2.

From Table 5.2, the univariate contingency table for station 1 can be constructed as follows:

	0–1.064	1.065–2.128	2.129–3.192	3.193–4.256	4.256–5.320
Counts	41	6	7	2	4
Frequency	0.683	0.100	0.117	0.033	0.067

	1	**2**	**3**	**4**	**5**	**6**	**7**	**8**	**9**	**10**	**11**	**12**
						Station 1						
2006	1	1	1	1	1	1	1	1	5	1	1	1
2007	4	1	3	1	5	5	3	2	3	1	1	1
2008	1	1	1	1	2	1	1	5	1	1	1	1
2009	1	1	1	1	2	1	1	1	3	3	2	1
2010	3	2	1	1	3	2	1	1	4	1	1	1

1: 0–1.064; **2:** 1.065–2.128; **3:** 2.129–3.192; **4:** 3.193–4.256; **5:** 4.256–5.320

	1	**2**	**3**	**4**	**5**	**6**	**7**	**8**	**9**	**10**	**11**	**12**
						Station 2						
2006	1	1	1	1	3	1	1	1	2	1	1	1
2007	2	1	2	1	2	1	5	1	2	1	1	1
2008	1	1	1	2	2	1	2	5	2	1	1	1
2009	1	1	1	1	1	1	1	1	1	1	1	1
2010	3	2	1	2	1	1	2	1	3	1	1	1

1: 0–2.222; **2:** 2.223–4.444; **3:** 4.445–6.666; **4:** 6.667–8.888; **5:** 8.889–11.110

TABLE **5.2** Discretization

Now the marginal entropy of station 1 can be computed.

$$H(X_1) = -\sum_{i=1}^{5} p_i \log_2 p_i = -0.683 \log_2 0.683 - 0.100 \log_2 0.100 - \cdots - 0.067 \log_2 0.067$$

$$= 1.4932 \text{ bits}$$

Second, the conditional entropy of precipitation of station 1 given that of station 2 is computed. Following the same method illustrated in Fig. 5.1, dividing the precipitation series of station 2 into five equal-size intervals, the joint contingency table for station 1 and station 2 can be constructed as Table 5.3.

From the joint contingency table, Table 5.3, the univariate contingency table of station 2 can be obtained by marginalizing station 1 out. The marginal contingency table for station 2 so obtained is as follows:

	0–2.222	2.223–4.444	4.445–6.666	6.667–8.888	8.889–11.110
Counts	43	12	3	0	2
Frequency	0.717	0.200	0.050	0.000	0.033

Then the conditional entropy of station 1 given station 2 can be computed:

$$H(X_1 \mid X_2) = -\sum_{i=1}^{5} p_{ij} \log_2 \frac{p_{ij}}{p_j}$$

$$= -0.583 \log_2 \frac{0.583}{0.717} - 0.083 \log_2 \frac{0.083}{0.200} - 0.017 \log_2 \frac{0.017}{0.050} - \cdots - 0.017 \log_2 \frac{0.017}{0.033}$$

$$= 1.2202 \text{ bits}$$

Counts Table					
		Station 2			
		0–2.222	2.223–4.444	4.445–6.666	6.667–8.888	8.889–11.110
Station 1	0–1.064	35	5	1	0	0
	1.065–2.128	4	2	0	0	0
	2.129–3.192	3	2	1	0	1
	3.193–4.256	0	1	1	0	0
	4.256–5.320	1	2	0	0	1
Frequency Table					
		Station 2			
		0–2.222	2.223–4.444	4.445–6.666	6.667–8.888	8.889–11.110
Station 1	0–1.064	0.583	0.083	0.017	0.000	0.000
	1.065–2.128	0.067	0.033	0.000	0.000	0.000
	2.129–3.192	0.050	0.033	0.017	0.000	0.017
	3.193–4.256	0.000	0.017	0.017	0.000	0.000
	4.256–5.320	0.017	0.033	0.000	0.000	0.017

TABLE 5.3 Contingency Table for Stations 1 and 2

Then the nontransferred information of station 1 with respect to station 2 is equal to the conditional entropy of station 1 given station 2. Therefore, the coefficient of nontransferred information is

$$t_1 = \frac{H(X_1 \mid X_2)}{H(X_1)} = \frac{1.2202}{1.4932} = 0.8172$$

The coefficient of transferred information between station 1 and station 2 can be obtained as

$$1 - t_1 = 1 - 0.8172 = 0.1828$$

5.5   Directional Information Transfer Index

Although transinformation indicates the dependence of two variables, it is not a good index of dependence, since its upper bound varies from site to site (it varies from 0 to marginal entropy H). Therefore, the original definition of T can be normalized by dividing by the marginal entropy (Yang and Burn, 1994) as follows

$$\frac{T}{H} = \mathrm{DIT} = \frac{(H - H_{\mathrm{Lost}})}{H} = 1 - \frac{H_{\mathrm{Lost}}}{H} \tag{5.8}$$

where H_{Lost} is the amount of information lost. The ratio of T by H is called the *directional information transfer* (DIT) index. Mogheir and Singh (2002) called it the *information transfer index* (ITI). The physical meaning of DIT is the fraction of information transferred from one site to another. DIT varies from 0 to 1 when T varies from 0 to H. The zero value of DIT corresponds to the case where sites are independent and therefore no information is transmitted. A value of unity for DIT corresponds to the case where sites are fully dependent and no information is lost. Any other value of DIT between 0 and 1 corresponds to a case between fully dependent and fully independent.

DIT is not symmetric, since $\mathrm{DIT}_{XY} = T/H(X)$ in general will not be equal to $\mathrm{DIT}_{YX} = T/H(Y)$. DIT_{XY} describes the fractional information inferred by station X about station Y, whereas DIT_{YX} describes the fractional information inferred by station Y about station X. Between two stations, the station with the higher DIT should be given higher priority because of its higher capability of inferring (predicting) the information at other sites.

DIT can be applied to the regionalization of the network or watersheds. If both DIT_{XY} and DIT_{YX} are high, the two related stations can be arranged in the same group, since they are strongly dependent and information can be mutually transferred between them. If neither of the DIT values is high, they should be kept in separate groups. If DIT_{XY} is high, station Y whose information can be predicted by X can join station X, if station Y does not belong to another group; otherwise it stays in its own group. The predictor station X cannot join station Y's group, since if it were discontinued, the information at that site would be lost.

DIT can be both a measure of information transmission capability and an indicator of the dependency of a station. This is an indicator of the information connection. In the station selection process, a predicted station should be removed first, because its information is recovered efficiently and by itself it does not predict information of other stations efficiently. When all remaining stations in the group have strong mutual connections with one another (i.e., both DIT_{XY} and DIT_{YX} are high), they can be further selected based on a criterion, designated as SDIT_i, defined as

$$\mathrm{SDIT}_i = \sum_{j=1, j \neq i}^{m} \mathrm{DIT}_{ij} \tag{5.9}$$

where DIT_{ij} is the information inferred by station i about station j, and $m =$ the number of stations in the group. The station in each group with the highest value of SDIT, in comparison with members in the group, should be retained in the network.

To be able to use transinformation and DIT, the probability density function of the variable being sampled must be determined. To that end, nonparametric estimation methods such as the histogram and kernel density estimator can be employed. Nonparametric estimation does not describe a probability density function by a formula and parameters, but rather by a set of point values everywhere in the domain. If the values of the density function are known everywhere, then the function is known numerically. This method of describing the density function is known as the *nonparametric method*, and it can be described as follows:

For a random variable X, let $x_1, x_2, ..., x_N$ be a sample of observations, independently and identically distributed with PDF $f(x)$. If $f(x)$ is the derivative of $F(x)$ and is continuous, then at a given point x, it can be estimated by a kernel estimator as

$$f_N(x) = \int_{-\infty}^{\infty} \frac{1}{h} K\left(\frac{x - x_j}{h}\right) dF_N\left(\frac{x - x_j}{N}\right) \approx \frac{1}{Nh} \sum_{j=1}^{N} K\left(\frac{x - x_j}{h}\right) \tag{5.10}$$

where $f_N(x)$ is the kernel estimator of $f(x)$; function $K(\cdot)$ is called the *kernel*; h is a positive number and is a smoothing factor for the kernel; and $F_N(x)$ is an estimate of the CDF of X. Parzen (1962) has shown that $f_N(x)$ is an unbiased estimate of $f(x)$ under certain conditions.

For a multidimensional case where $X = X(x_1, x_2, ..., x_p)$, Cacoullos (1966) has shown that at a given point $X^O = X^O(x_1, x_2, ..., x_p)$ the kernel estimator has the form

$$f_N(X^O) = \frac{1}{N} \frac{1}{h_1 h_2 \cdots h_p} \sum_{j=1}^{N} K\left(\frac{x_1 - x_{j1}}{h_1}, \frac{x_2 - x_{j2}}{h_2}, ..., \frac{x_p - x_{jp}}{h_p}\right) \tag{5.11}$$

where $X^j = X^j(x_{j1}, x_{j2}, ..., x_{jp})$ is the jth observation of X which is a p-dimensional variable.

The components of $X : \{X_1, X_2, ..., X_p\}$ can be mutually dependent or independent, but observations of each component are still assumed to be independent and identically distributed. A simple approximation of this equation is the product function of the form

$$K\left(\frac{x_1 - x_{j1}}{h_1}, \frac{x_2 - x_{j2}}{h_2}, ..., \frac{x_p - x_{jp}}{h_p}\right) = \prod_{i=1}^{p} K_i\left(\frac{x_i - x_{ji}}{h_i}\right) \tag{5.12}$$

For practical purposes, it can be assumed that

$$h_1 = h_2 = \cdots = h_p = h \tag{5.13}$$

According to Adamowski (1989), the choice of a kernel is not crucial for the estimation of the probability distribution. However, the selection of h is crucial, because it affects both the bias and the variance of the estimator. Some common forms of kernels are presented by Parzen (1962) and Wertz (1978) and are as follows: h is frequently expressed as a function of $f(x)$ which is not known. A widely used method is cross-validation maximum likelihood (Scott, 1992), which minimizes the integral mean square error of $f_N(x)$. Then h is selected to satisfy

$$\max L(h) = \prod_{i=1}^{N} f_N(x_i, h) = \prod_{i=1}^{N} \frac{1}{h^p} \sum_{j=1, i \neq j}^{p} K(x_{ij}, h) \tag{5.14}$$

where $f_N(x_j, h)$ is the estimated density value at x_j but with x_j removed.

In practice, hydrologic variables are nonnegative. Therefore, the original variables are logarithmically transformed, and a Gaussian kernel is chosen to form the estimator:

$$K(x, y) = K_1(x) K_2(y) = \frac{1}{2\pi} \left[\exp\left(-\frac{x^2}{2}\right) \right] \left[\exp\left(-\frac{y^2}{2}\right) \right]$$

$$= \frac{1}{2\pi} \left[\exp\left(-\frac{x^2 + y^2}{2}\right) \right] \tag{5.15}$$

where $x = (x^* - x_j)/h$; $y = (y^* - y_j)/h$; (x_j, y_j) is an observation point; and (x^*, y^*) is the coordinate of any point in the field. All x^* and y^* (or x_j and y_j) are logarithmically transformed. The optimal h values are derived from Eq. (5.12). Then $f_N(x, y)$ becomes

$$f_N(x, y) = \frac{1}{Nh^2} \sum_{i=1}^{N} \frac{1}{2\pi} \exp\left[-\frac{(x - x_i)^2 + (y - y_i)^2}{2} \right] \tag{5.16}$$

Example 5.2. Suppose there is a random sample of size 5: {13, 14.5, 15, 18, 19}. Estimate the probability density that is most likely to have generated this sample, using the kernel estimator.

Solution. Before explaining how to estimate the density, it is necessary to note here that just five data points are used purely for demonstrating the working machinery of the kernel estimator, as shown in Fig. 5.2.

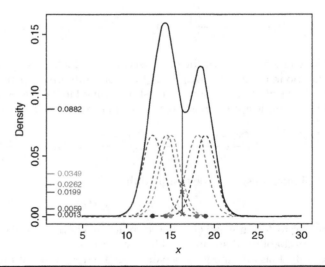

FIGURE 5.2 Kernel density estimate based on five observations with a smoothing factor $h = 1.2$.

Suppose the kernel function is chosen as Gaussian. The first step is to determine an appropriate smoothing factor. Again for simplicity, the smoothing factor h used here is 1.2, which is selected by eye rather than by other sophisticated, data-based smoothing factor selectors. The kernel estimate is constructed by centering a scaled kernel at each observation (point), as shown by the bell-shaped dashed curves with one related to each observation. The value of the kernel estimate at a query point x is simply the summation of the n (here 5) scaled kernel ordinates at that point. To be more precise, suppose one wants to estimate the density value at $x = 16.36$, as illustrated by the gray vertical line in Fig. 5.2. Now evaluate the scaled kernel centered at the first observation $X_1 = 13$ as

$$y_1 = \frac{1}{n}\frac{1}{\sqrt{2\pi}\,h}e^{-(x-X_1)^2/2h^2} = \frac{1}{5}\times\frac{1}{1.2\times\sqrt{2\pi}}e^{-(16.36-13)^2/(2\times1.2^2)} = 0.0013$$

which represents the contribution of the kernel centered at the first observation to the density estimate for point x. Following the same method, one can evaluate the scaled kernels centered at other observations: $y_2 = 0.0199$, $y_3 = 0.0349$, $y_4 = 0.0262$, $y_5 = 0.0059$, as marked on the left side of the figure. Then the kernel estimate at point $x = 16.36$ is

$$y = y_1 + y_2 + y_3 + y_4 + y_5 = 0.0013 + 0.0199 + 0.0349 + 0.262 + 0.0059 = 0.0882$$

One can also examine the effect of the smoothing factor h on the density estimate. When $h = 0.5$, as in the left plot of Fig. 5.3, it is observed that the kernel spreads narrowly, which means that only relatively few observations contribute to the estimation, resulting in a rough density estimate. This estimate pays too much attention to the particular data set at hand. Such an estimate is said to be *undersmoothed*. On the other hand, when $h = 2.5$, this results in a much smoother estimate. However, it seems to be too smooth, since the dimodality structure of the data is masked.

Example 5.3.   Monthly precipitation records from two stations, denoted as X and Y, respectively, in Texas are presented in Table 5.1. Compute the directional information transfer indices DIT_{XY} and DIT_{YX}.

Solution.   It is assumed that ranges of the values of X and Y can be divided into five equal-size intervals, and then their distributions can be estimated. First, compute the marginal entropy of X and Y (i.e., station 1 and station 2). From Example 3.1, we have already calculated the marginal entropy

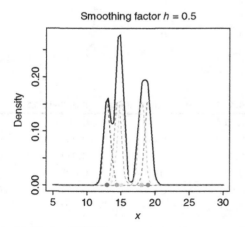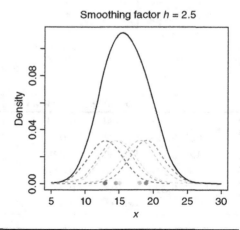

FIGURE **5.3**   Kernel density estimates using different smoothing factors.

of X, which has a value of 1.4932 bits. Also from Example 3.1, we have summarized the marginal contingency table for Y (i.e., station 2), from which the marginal entropy of Y is calculated:

$$H(Y) = -\sum_{i=1}^{5} p_i \log_2 p_i = -0.717 \log_2 0.717 - 0.200 \log_2 0.200 - \cdots - 0.033 \log_2 0.033$$

$$= 1.1885 \text{ bits}$$

Second, the transinformation between X and Y is computed. To improve the readability, here we present the joint contingency table and marginal contingency tables together. The joint contingency table of X and Y and the corresponding marginal contingency tables can be constructed as Table 5.4, following the same method used in Example 5.1.

Using the definition of transinformation, the transinformation between X and Y can be computed.

$$T(X,Y) = \sum_{i=1}^{5}\sum_{j=1}^{5} p_{ij} \log_2 \frac{p_{ij}}{p_i p_j}$$

$$= 0.583 \log_2 \frac{0.583}{0.683 \times 0.717} + 0.083 \log_2 \frac{0.083}{0.683 \times 0.200} + \cdots + 0.017 \log_2 \frac{0.017}{0.067 \times 0.033}$$

$$= 0.2730 \text{ bit}$$

Then the directional information transfer indices are

$$\text{DIT}_{XY} = \frac{T(X,Y)}{H(X)} = \frac{0.2730}{1.4932} = 0.1828$$

$$\text{DIT}_{YX} = \frac{T(X,Y)}{H(Y)} = \frac{0.2730}{1.1885} = 0.2297$$

Counts Table						
		Y				
		0–2.222	2.223–4.444	4.445–6.666	6.667–8.888	8.889–11.110
X	0–1.064	35	5	1	0	0	41
	1.065–2.128	4	2	0	0	0	6
	2.129–3.192	3	2	1	0	1	7
	3.193–4.256	0	1	1	0	0	2
	4.256–5.320	1	2	0	0	1	4
		43	12	3	0	2	
Frequency Table						
		Y				
		0–2.222	2.223–4.444	4.445–6.666	6.667–8.888	8.889–11.110
X	0–1.064	0.583	0.083	0.017	0.000	0.000	0.683
	1.065–2.128	0.067	0.033	0.000	0.000	0.000	0.100
	2.129–3.192	0.050	0.033	0.017	0.000	0.017	0.117
	3.193–4.256	0.000	0.017	0.017	0.000	0.000	0.033
	4.256–5.320	0.017	0.033	0.000	0.000	0.017	0.067
		0.717	0.200	0.050	0.000	0.033	

TABLE 5.4 Contingency Tables

5.6   Information Correlation Coefficient

Transferability of information between hydrologic variables depends on the degree and the structure of the interdependence of the variables. The most likely measure of association between variables is the correlation coefficient. Its use is valid under the assumption of normality of variables and linearity of relationship between them. If variables are nonlinearly related, then either the variables have to be transformed to linearize the regression function or nonlinear regression has to be developed. For both linear and nonlinear types of interdependence, the correlation coefficient measures the amount of information that is actually transferred by the assumed regression relationship. If the correlation coefficient is zero, it does not necessarily mean the absence of association between variables, but may also be due to the inappropriate choice of regression relation. The information correlation coefficient and transinformation represent the extent of transferable information without assuming any particular type of interdependence. They also provide a means of judging the amount of information actually transferred by regression.

When the marginal and joint probability distributions of stations X and Y are approximated by their relative frequency distributions or when no particular probability distribution is assumed for the stations, then T_0 represents the upper limit of transferable information between stations. It can also be computed as the difference between joint entropy and conditional entropy. The information correlation coefficient R_0 (Linfoot, 1957) is a dimensionless measure of stochastic interdependence, varies between 0 and 1, and is expressed in terms of transinformation as

$$R_0 = \sqrt{1 - \exp(-2T_0)} \tag{5.17}$$

Thus, R_0 or R_0^2 measures the transferable information, whereas r or r^2 measures transferred information via regression. Equation (5.17) does not assume normality or any type of functional relationship between stations and therefore has an advantage over the ordinary correlation coefficient. It reduces to the classical correlation coefficient when the normality and linearity assumptions are satisfied. Although R_0 and T_0 do not provide any functional relationship between stations to transfer information, they serve as criteria for checking the validity of assumed types of dependence and probability distributions of the stations.

The amount of actually transferred information under the assumptions made can also be defined as

$$t_1 = \frac{T_0 - T_1}{T_0} = 1 - \frac{T_1}{T_0} \tag{5.18}$$

where

$$T_1 = -\frac{\ln(1 - r^2)}{2}$$

Here T_1 represents the transinformation between variables, but T_0 represents the upper limit of transferable information of one station with respect to the other. In Eq. (5.18), t_1 measures the portion $T_0 - T_1$ of the nontransferred information, and $1 - t_1$ measures the amount of transferred information of the total transferable information. Likewise R_0 and t_1 can be used as criteria to judge the validity of assumptions made about the dependence between stations. Entropy or transinformation does not provide any means of transferring information but provides a means for measuring transferable information. Thus, it can help improve the transfer of information by regression analysis.

If the actual amount of information transferred by regression analysis is far below the transferable information defined by entropy measures, then one can attempt to improve the regression analysis. This can be accomplished by (1) selecting marginal and joint distributions that better fit the data, (2) searching for better relationships between stations, (3) analyzing the effect of autocorrelation of each station upon interdependence and regression, and (4) analyzing the effect of lag cross-correlation upon interdependence and information transfer.

Example 5.4. Compute the information correlation coefficient between station 1 and station 2 in Example 5.1.

Solution.   To calculate the information correlation coefficient, we need to calculate the transinformation between the two stations. In Example 5.3, we have already calculated the transinformation, which has a value of 0.2730 bit. Therefore, in terms of Eq. (5.18) the information correlation coefficient is

$$R_0 = \sqrt{1 - \exp(-2T_1)} = \sqrt{1 - \exp(-2 \times 0.2730)} = 0.6486$$

5.7   Total Correlation

McGill (1954) and Watanabe (1960), among others, have defined *total correlation* as

$$C(X_1, X_2, \ldots, X_n) = \sum_{i=1}^{n} H(X_i) - H(X_1, X_2, \ldots, X_n) \tag{5.19}$$

The total correlation can be used to quantify the amount of information shared by gauges (or variables) at the same time and hence provides an alternative way of examining multivariate dependency. The total correlation is always positive, because the sum of marginal entropies of n variables will be greater than their joint entropy. It is symmetric with respect to its arguments. If these n variables are independent, then C will be zero. A large value of C may imply either a strong dependency among a few variables or a relatively small dependency among all of them. If $n = 2$, Eq. (5.19) will reduce to the usual transinformation T.

Equation (5.19) involves two components: marginal entropies and joint entropy. The total correlation can be computed without resorting to the computation of multivariate entropy or multivariate probabilities. This can be accomplished by recalling the grouping property of total correlation and accordingly a systematic grouping of bivariate entropies. According to this property, if new variables are formed by the union of two or more variables such that the marginal entropy of a new variable is equal to the joint entropy of the two forming variables, then the total correlation of the original variables is obtained by summing the marginal entropies of the new variables (Kraskov, 2003). If the number of variables is 2, then transinformation becomes a special case of total correlation. Thus, total correlation is a kind of multivariate extension of bivariate transinformation. From Eq. (5.19), the multivariate joint entropy can be expressed as

$$H(X_1, X_2, \ldots, X_n) = \sum_{i=1}^{n} H(X_i) - C(X_1, X_2, \ldots, X_n) \tag{5.20}$$

Although the total correlation concept has been widely used in medicine, neurology, psychology, clustering, significant feature selection, and genetics (Jakulin and Bratko, 2004; Fass, 2006), there appears to have been limited applications of total correlation in hydrology and water resources. Krstanovic and Singh (1992a, b) used it for evaluating multivariate ($N > 2$) dependence, where it was assumed that the random variables followed a normal or lognormal distribution, which is not always the case for hydrologic variables, such as discharge and rainfall intensity. In network design, most analyses about the dependence of time series belonging to different gauging stations have been restricted up to bivariate analysis in terms of transinformation or directional information transfer index rather than multivariate analysis, partly because of the difficulty of estimating multivariate probability distributions and the limited availability of data. However, often there is a need to evaluate the total amount of information/ uncertainty duplicated by several stations under consideration. Recently Alfonso (2010) used total correlation to evaluate the performance of different optimal water level monitoring stations.

The total correlation can be computed directly using the grouping property. For trivariate total correlation $C(X_1, X_2, X_3)$, the grouping property can be expressed as (Kraskov et al., 2005)

$$C(X_1, X_2, X_3) = C(X_1, X_2) + C(X_{1:2}, X_3) \tag{5.21}$$

where $X_{1:2}$ denotes the grouped variable formed by grouping X_1 and X_2. Using this property sequentially, the multivariate total correlation can be computed recursively as

$$
\begin{aligned}
C(X_1, X_2, X_3, \cdots, X_N) &= C(X_1, X_2, X_{3 \to N}) \\
&= C(X_1, X_2) + C(X_{1:2}, X_{3 \to N}) \\
&= C(X_1, X_2) + C(X_{1:2}, X_3, X_{4 \to N}) \\
&= C(X_1, X_2) + C(X_{1:2}, X_3) + C(X_{1:3}, X_{4 \to N}) \\
&\vdots \\
&= C(X_1, X_2) + C(X_{1:2}, X_3) + C(X_{1:3}, X_4) + \cdots + C(X_{1:N-1}, X_N) \\
&= \sum_{i=1}^{N-1} C(X_{1:i}, X_{i+1})
\end{aligned}
\tag{5.22}
$$

The notation $X_{1:i}$ represents the merging of variables $X_1, X_2, \ldots, X_i$. Equation (5.22) shows that the total correlation is finally factored as a summation of bivariate total correlation which is just the transinformation. In other words, the grouping property can reduce the dimension of multivariate total correlation, and thus the estimation of multivariate probability distribution can be avoided.

To apply the grouping property, the key is to merge two discrete random samples into a single one such that the marginal entropy of the merged sample is equal to the joint entropy of the two original samples. In other words, the amount of information will be invariant before and after agglomeration. Two discrete or categorical random samples X_1 and X_2 can be merged in such a way as placing a unique value for every combination of the corresponding records in X_1 and X_2 (Kraskov, 2003; Alfonso et al., 2010). For instance, if $X_1 = [1\ 2\ 1\ 2\ 1\ 3\ 3]$ and $X_2 = [1\ 2\ 2\ 2\ 1\ 3\ 2]$, then one of the options to merge X_1 and X_2 is by putting all the corresponding digits of

X_1 and X_2 together, that is, [11 22 12 22 11 33 32]. However, this option has a serious defect in that it will cause the problem of "out of memory" as the number of merged samples increases especially when dealing with large samples. Alfonso (2010) suggested an alternative which can also obtain the same objective. Figure 5.4 provides a step-by-step schematic illustration of this data merging procedure.

First, a new sample is created following the previously described method, illustrated by step 1 in Fig. 5.4. Then, as illustrated by step 2 in the same figure, pick out the unique values and rank them in ascending order to obtain a new sample, called *unique ranked series*, i.e., [11 12 22 32 33]. Finally, as shown by step 3, access the location index for each element of the previously merged sample in the sorted series, and assign the index so obtained as the new label for the corresponding element in the direct merged series. Results of computation in Fig. 5.4 show that this merging method keeps the amount of information retained by the composite samples invariant. However, this conclusion will not hold for continuous random samples. This data merging method can also be used for more than two random samples. Further it is symmetric with respect to its composites, which means one can merge several samples in any sequence one wants.

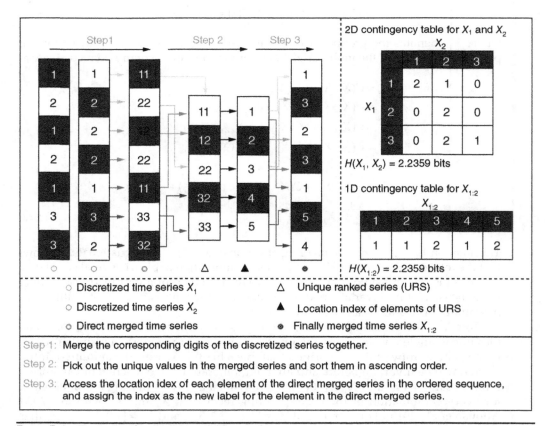

FIGURE 5.4 Illustration for discrete variable merging.

Example 5.5. Considering the two records of monthly precipitation presented in Example 5.1, create a merged series such that entropy of the merged series is equal to the joint entropy of the original two series.

Solution. For the ease of understanding the step-by-step procedure of discrete variable merging, we start with the original continuous time series by discretizing them using the method illustrated in Example 5.1. After discretization, the resulting time series are shown in Table 5.5.

Recall the illustration in Fig. 5.4. The first step in variable merging is to put all the corresponding digits of the two discretized series together. The direct merged series is obtained, as shown in Table 5.6.

Following the illustration in step 2 in Fig. 5.4, pick out the unique values in the merged series and then sort them in ascending order. This leads to the following:

Sorted Unique Value Sequence													
1	2	3	4	5	6	7	8	9	10	11	12	13	14
11	12	13	21	22	31	32	33	35	42	43	51	52	55

	Station 1											
Month	1	2	3	4	5	6	7	8	9	10	11	12
2006	1	1	1	1	1	1	1	1	5	1	1	1
2007	4	1	3	1	5	5	3	2	3	1	1	1
2008	1	1	1	1	2	1	1	5	1	1	1	1
2009	1	1	1	1	2	1	1	1	3	3	2	1
2010	3	2	1	1	3	2	1	1	4	1	1	1

1: 0–1.064; **2:** 1.065–2.128; **3:** 2.129–3.192; **4:** 3.193–4.256; **5:** 4.257–5.320

	Station 2											
2006	1	1	1	1	3	1	1	1	2	1	1	1
2007	2	1	2	1	2	1	5	1	2	1	1	1
2008	1	1	1	2	2	1	2	5	2	1	1	1
2009	1	1	1	1	1	1	1	1	1	1	1	1
2010	3	2	1	2	1	1	2	1	3	1	1	1

1: 0–2.222; **2:** 2.223–4.444; **3:** 4.445–6.666; **4:** 6.667–8.888; **5:** 8.889–11.110

TABLE 5.5   Contingency Tables for Station 1 and Station 2

	Station 1 + Station 2											
Month	1	2	3	4	5	6	7	8	9	10	11	12
2006	11	11	11	11	13	11	11	11	52	11	11	11
2007	42	11	32	11	52	51	35	21	32	11	11	11
2008	11	11	11	12	22	11	12	55	12	11	11	11
2009	11	11	11	11	21	11	11	11	31	31	21	11
2010	33	22	11	12	31	21	12	11	43	11	11	11

TABLE 5.6   Direct Welding of the Digits of Original Series

The next step, in line with step 3 in Fig. 5.4, is to access the location index of each element of the direct merged series in the unique ranked series and to assign the location index as the new label for the corresponding element in the direct merged series. For example, the fifth element of the direct merged sample is 13, in the unique ranked series the location index of 13 is 3, then 3 is the new label for the fifth element of the direct merged sample. Following this method, the new merged series is shown in Table 5.7.

Then the marginal contingency table for the merged series can be summarized as in Table 5.8.

From the contingency in Table 5.8, the marginal entropy of the merged series can be computed as follows:

$$H(X_{12}) = -0.5833\log_2 0.5833 - 0.0833\log_2 0.0833 - 0.0167\log_2 0.0167 - \cdots - 0.0167\log_2 0.0167$$

$$- 0.0167\log_2 0.0167 - 0.0333\log_2 0.0333 - 0.0167\log_2 0.0167$$

$$= 2.4087 \text{ bits}$$

On the other hand, from the joint contingency table in Example 5.2, one can compute the joint entropy of station 1 and station 2:

$$H(X_1, X_2) = -0.583\log_2 0.583 - 0.083\log_2 0.083 - 0.017\log_2 0.017 - \cdots - 0.017\log_2 0.017$$

$$= 2.4087 \text{ bits}$$

Obviously, one can see that the information content retained by the two stations keeps invariant before and after the variable merging.

Station 1 + Station 2												
Month	**1**	**2**	**3**	**4**	**5**	**6**	**7**	**8**	**9**	**10**	**11**	**12**
2006	1	1	1	1	3	1	1	1	13	1	1	1
2007	10	1	7	1	13	12	9	4	7	1	1	1
2008	1	1	1	2	5	1	2	14	2	1	1	1
2009	1	1	1	1	4	1	1	1	6	6	4	1
2010	8	5	1	2	6	4	2	1	11	1	1	1

TABLE 5.7   Merged Series

Counts						
Label	**1**	**2**	**3**	**4**	**5**	**6**	**7**
	35	5	1	4	2	3	2
Label	**8**	**9**	**10**	**11**	**12**	**13**	**14**
	1	1	1	1	1	2	1
Frequency						
Label	**1**	**2**	**3**	**4**	**5**	**6**	**7**
	0.5833	0.0833	0.0167	0.0667	0.0333	0.0500	0.0333
Label	**8**	**9**	**10**	**11**	**12**	**13**	**14**
	0.0167	0.0167	0.0167	0.0167	0.0167	0.0333	0.0167

TABLE 5.8   Marginal Contingency of Merged Variables

Example 5.6. Considering the monthly precipitation records from two stations in Texas in Example 5.1 and the same records for another station as in Table 5.9, compute the total correlation among the three stations.

Solution. Using the same method as in Example 5.1 to discretize the continuous time series from station 3, the result is tabulated in Table 5.10.

There are two different ways to compute the total correlation. One is to use the definition of total correlation, as in Eq. (5.20). The other one is to use the grouping property and variable merging. In the following, the two methods are illustrated.

Calculation of Total Correlation Using Its Definition   First, we need to calculate the marginal entropy of each station. From Examples 5.1 and 5.2 we have already calculated the marginal entropy of stations 1 and 2, respectively.

$$H(X_1) = 1.4932 \text{ bits}$$

$$H(X_2) = 1.1885 \text{ bits}$$

From Table 5.10, we can summarize the marginal contingency table of station 3.

	0–1.88	1.89–3.76	3.77–5.64	5.65–7.52	7.53–9.20
Counts	35	13	7	3	2
Frequency	0.5833	0.2167	0.1167	0.0500	0.0333

The marginal entropy for station 3 is

$$H(X_3) = -0.5833 \log_2 0.5833 - 0.2167 \log_2 0.2167 - \cdots - 0.0333 \log_2 0.0333$$

$$= 1.6729 \text{ bits}$$

Month	Station 3											
Month	**1**	**2**	**3**	**4**	**5**	**6**	**7**	**8**	**9**	**10**	**11**	**12**
2006	0.35	0.62	1.36	1.4	3.8	1.63	1.41	0.03	4.11	3.44	0.75	2.44
2007	4.33	0.09	7.24	2.07	5.39	4.21	9.43	4.41	1.82	0.77	0.78	0.32
2008	0.6	0.39	1.44	0.62	1.26	0.29	1.9	3.94	0.37	0.36	0.16	0.25
2009	0.33	0.42	2.18	3.26	1.73	0.43	0.54	0.82	6.47	7.72	3.73	1.69
2010	3.52	2.63	1.7	2.11	3.49	2.99	2.19	0.41	7.01	0.04	0.99	0.85

Table 5.9   Monthly Precipitation (in.) of a Station in Texas

Month	Station 3											
Month	**1**	**2**	**3**	**4**	**5**	**6**	**7**	**8**	**9**	**10**	**11**	**12**
2006	1	1	1	1	3	1	1	1	3	2	1	2
2007	3	1	4	2	3	3	5	3	1	1	1	1
2008	1	1	1	1	1	1	2	3	1	1	1	1
2009	1	1	2	2	1	1	1	1	4	5	2	1
2010	2	2	1	2	2	2	2	1	4	1	1	1
1: 0–1.88; **2:** 1.89–3.76; **3:** 3.77–5.64; **4:** 5.65–7.52; **5:** 7.53–9.20												

Table 5.10   Discretization of Monthly Precipitation Series in Table 5.9

Then we compute the joint entropy in terms of discrete variable merging. In Example 5.4, we have already obtained the merged series of station 1 and station 2 (see Table 5.7). For the ease of understanding, we tabulate the table once more as follows:

Month	Station 1 + Station 2											
	1	**2**	**3**	**4**	**5**	**6**	**7**	**8**	**9**	**10**	**11**	**12**
2006	1	1	1	1	3	1	1	1	13	1	1	1
2007	10	1	7	1	13	12	9	4	7	1	1	1
2008	1	1	1	2	5	1	2	14	2	1	1	1
2009	1	1	1	1	4	1	1	1	6	6	4	1
2010	8	5	1	2	6	4	2	1	11	1	1	1

The joint contingency table of the merged series and the discretized series of station 3 can be summarized, as shown in Table 5.11. The joint entropy of the three stations can then be computed as

$$H(X_1, X_2, X_3) = -0.5000 \log_2 0.5000 - 0.0333 \log_2 0.0333 - \cdots - 0.0167 \log_2 0.0167$$

$$= 3.0807 \text{ bits}$$

Then the total correlation from Eq. (5.20) can be computed.

$$C(X_1, X_2, X_3) = H(X_1) + H(X_2) + H(X_3) - H(X_1, X_2, X_3)$$

$$= 1.4932 + 1.1885 + 1.6729 - 3.0807$$

$$= 1.2739 \text{ bits}$$

Calculation of Total Correlation Using Grouping Property　To compute the total correlation of the three stations from Eq. (5.21), the transinformation between station 1 and station 2 must be computed first. From Example 5.2, we have already computed the transinformation between station 1 and station 2, which is

$$C(X_1, X_2) = T(X_1, X_2) = 0.2730 \text{ bit}$$

Then we need to compute the transinformation between the merged series of station 1 and station 2 and the series of station 3. Tabulating the joint contingency table and the corresponding marginal contingency tables, one can obtain Table 5.12.

The transinformation can be calculated from the contingency table, Table 5.12.

$$C(X_{1:2}, X_3) = T(X_{1:2}, X_3) = 0.5000 \log_2 \frac{0.5000}{0.5834 \times 0.5883} + 0.0833 \log_2 \frac{0.0833}{0.2167 \times 0.0833} + \cdots$$

$$+ 0.0167 \log_2 \frac{0.0167}{0.1001 \times 0.0167}$$

$$= 1.0010 \text{ bits}$$

Using the grouping property of total correlation, it can be computed as

$$C(X_1, X_2, X_3) = C(X_1, X_2) + C(X_{1:2}, X_3)$$

$$= 0.2730 + 1.0010$$

$$= 1.2740 \text{ bits}$$

which is the same as the value computed using the definition of total correlation, neglecting the round-off error.

Counts Table						
		Station 3				
		1	2	3	4	5
Merged Station 1 + Station 2	1	30	5	0	0	0
	2	2	3	0	0	0
	3	0	0	1	0	0
	4	1	2	1	0	0
	5	1	1	0	0	0
	6	0	1	0	1	1
	7	1	0	0	1	0
	8	0	1	0	0	0
	9	0	0	0	0	1
	10	0	0	1	0	0
	11	0	0	0	1	0
	12	0	0	1	0	0
	13	0	0	2	0	0
	14	0	0	1	0	0

Frequency Table						
		Station 3				
		1	2	3	4	5
Merged Station 1 + Station 2	1	0.5000	0.0833	0	0	0
	2	0.0333	0.0500	0	0	0
	3	0	0	0.0167	0	0
	4	0.0167	0.0333	0.0167	0	0
	5	0.0167	0.0167	0	0	0
	6	0	0.0167	0	0.0167	0.0167
	7	0.0167	0	0	0.0167	0
	8	0	0.0167	0	0	0
	9	0	0	0	0	0.0167
	10	0	0	0.0167	0	0
	11	0	0	0	0.0167	0
	12	0	0	0.0167	0	0
	13	0	0	0.0333	0	0
	14	0	0	0.0167	0	0

TABLE 5.11   Contingency Table of Merged Series and Discretized Series of Stations

Frequency Table							
		Station 3					
		1	**2**	**3**	**4**	**5**	
Station 1 + Station 2	1	0.5000	0.0833	0	0	0	0.5833
	2	0.0333	0.0500	0	0	0	0.0833
	3	0	0	0.0167	0	0	0.0167
	4	0.0167	0.0333	0.0167	0	0	0.0667
	5	0.0167	0.0167	0	0	0	0.0334
	6	0	0.0167	0	0.0167	0.0167	0.0501
	7	0.0167	0	0	0.0167	0	0.0334
	8	0	0.0167	0	0	0	0.0167
	9	0	0	0	0	0.0167	0.0167
	10	0	0	0.0167	0	0	0.0167
	11	0	0	0	0.0167	0	0.0167
	12	0	0	0.0167	0	0	0.0167
	13	0	0	0.0333	0	0	0.0333
	14	0	0	0.0167	0	0	0.0167
		0.5834	0.2167	0.1001	0.0501	0.0334	

TABLE 5.12   Joint Contingency Table and Corresponding Marginal Contingency Tables

5.8   Maximum Information Minimum Redundancy

This method, as the name suggests, is based on maximizing the amount of effective information retained by the selected gauging stations and minimizing the amount of redundant information due to the dependence among the selected stations. This method has two other advantages. First, a mechanism that reflects the preference of different decision makers can be built in the selection criterion by introducing two weights—effective information weight and redundancy weight. Second, it is easy to extend the criterion to cover more design considerations, such as cost and benefit of hydraulic information obtained from a network.

Let there be N potential candidate measurement stations located in the area of interest, denoted by $X_1, X_2, X_3, ..., X_N$; for example, the area may be a watershed, river, canal, estuary, or pipeline. It is assumed that for each candidate station there are some years of records about the hydrologic variable of interest, denoted by X, such as discharge, rainfall amount, sediment discharge, or water quality constituent. Let S denote the set of gauging stations already selected for the network and its elements are denoted by $X_{S_1}, X_{S_2}, X_{S_3}, ..., X_{S_k}$, where S_i can be 1, 2, ..., k. Similarly, let F denote the set of candidate stations to be selected, and similarly its elements are denoted by $X_{F_1}, X_{F_2}, X_{F_3}, ..., X_{F_m}$, where F_i can be 1, 2, ..., m. The summation of k and m is equal to N, the total number of potential candidate stations. The amount of effective information retained by S can be modeled in terms of joint entropy and transinformation as

$$H(X_{S_1}, X_{S_2}, ..., X_{S_k}) + \sum_{i=1}^{m} T(X_{S_1:S_k} ; X_{F_i})$$ (5.23)

where $X_{S_1:S_k}$ denotes the merged time series of $X_{S_1}, X_{S_2}, X_{S_3}, ..., X_{S_k}$ such that its marginal entropy is the same as the multivariate joint entropy of $X_{S_1}, X_{S_2}, X_{S_3}, ..., X_{S_k}$. In other words, the merged variable $X_{S_1:S_k}$ contains the same amount of information as that retained by all its individual members $X_{S_1}, X_{S_2}, X_{S_3}, ..., X_{S_k}$. The same notation will be used to denote merged variables; for example, $X_{A:B}$ denotes the merged variable of those variables whose subscripts are from A to B.

The effective information contains two parts. The first part is the joint entropy of the selected stations, measuring the total but not duplicated amount of information which can be obtained from the selected stations. The second part is the summation of trans-ferred information from the group of already selected stations to each individual station which is still in the candidate set, respectively. To illustrate, consider $T(X_{S_1:S_k}, X_{F_i})$ which is the transinformation of $X_{S_1:S_k}$ and X_{F_i} that quantifies the common information shared by these two variables. When the network is designed, the major function of a network is to monitor the variables of interest and to make predictions. Therefore, the predictive ability of the network should not be neglected in the design. Transformation $T(X_{S_1:S_k}, X_{F_i})$ is a quantitative measure of the amount of information about the unselected station X_{F_i} which can be inferred from the selected stations. In other words, it is a measure of the predictive ability of the selected stations.

Husain (1987) and Al-Zahrani and Husain (1998) considered the predictive ability of an optimal network. However, the effective information, Eq. (5.19), differs from the one they used in two respects. The first is that the predictive ability as measured in Eq. (5.19) considers the selected stations as a whole group containing the same amount of information as that of all its elements rather than treating them separately. This pre-dictive ability measure can successfully filter the duplicated information (redundancy) of the selected stations. Second, the multivariate joint entropy is used to quantify the total information rather than using the summation of marginal entropies in which the duplicated information is summed again and again.

The effective information can also be expressed as

$$H(X_{S_1}, X_{S_2}, ..., X_{S_k}) + T(X_{S_1:S_k}; X_{F_1:F_m}) \qquad (5.24)$$

It contains two parts: the total effective information part and the predictive ability part. In this definition the unselected stations are also treated as a whole group. Transinfor-mation $T(X_{S_1:S_k}; X_{F_1:F_m})$ is the amount of information about the unselected group that can be inferred from the selected group.

Another key point worthy of consideration in network design is the redundant information among selected stations. Such redundancy means the selected stations are not fully and effectively used, since a lot of information obtained from the network may be overlapping. In other words, some of the stations are not necessary, and therefore the network is not an economical or an optimal one. Even worse, the redundancy may deteriorate the predictive ability of the network, even though the same amount of infor-mation can be obtained from a redundant network as that obtained from a minimum redundant one, considering that no redundancy is impossible in real practice.

The total correlation of the selected stations can measure the redundancy among them, that is, $C(X_{S_1}, X_{S_2}, ..., X_{S_k})$. The total correlation of already selected stations mea-sures the common information shared by any combination of these stations, unlike the interaction which only measures the information shared by all these stations. Interaction information is sensitive to the newly added stations; in other words, it may change significantly from positive to negative or from a large value to a small value, if a new station is added to the network. To understand it, there may be a large amount of duplicated information between two stations; however, when a new station is added, these three stations may have no simultaneous common information. In this sense the total correlation is a more reliable measure of redundancy than interaction information.

5.8.1 Optimization I

A network should provide as much information as possible and at the same time restrict the redundant information as much as possible. This kind of maximum information and minimum redundancy network can be determined as follows:

$$
\begin{cases}
\text{Max}: H(X_{S_1}, X_{S_2}, \ldots, X_{S_k}) + \sum_{i=1}^{m} T(X_{S_1:S_k}; X_{F_i}) \\
\text{Min}: C(X_{S_1}, X_{S_2}, \ldots, X_{S_k})
\end{cases}
\tag{5.25}
$$

This constitutes a multiobjective optimization problem which can be reduced to a single-objective optimization problem by recalling that both the effective information part and the redundancy part have the same unit. The two objectives can therefore be unified as

$$
\text{Max}: w_1 \left[H(X_{S_1}, X_{S_2}, \ldots, X_{S_k}) + \sum_{i=1}^{m} T(X_{S_1:S_k}; X_{F_i}) \right] - w_2 C(X_{S_1}, X_{S_2}, \ldots, X_{S_k})
\tag{5.26}
$$

where w_1 and w_2, whose summation is 1, are the information weight and redundancy weight, respectively, since sometimes the decision maker needs a tradeoff between the informativeness and redundancy of the hydrometric network.

One can also unify the information and redundancy objectives as

$$
\text{Max}: \frac{H(X_{S_1}, X_{S_2}, \ldots, X_{S_k}) + \sum_{i=11}^{m} T(X_{S_1:S_k}; X_{F_i})}{w C(X_{S_1}, X_{S_2}, \ldots, X_{S_k})}
\tag{5.27}
$$

where w is the coefficient that makes a tradeoff between information ability and redundancy of the network.

5.8.2 Optimization II

The maximum information minimum redundancy (MIMR) network can also be determined as follows:

$$
\begin{cases}
\text{Max}: H(X_{S_1}, X_{S_2}, \ldots, X_{S_k}) + \sum_{i=1}^{m} T(X_{S_1:S_k}; X_{F_i:Fm}) \\
\text{Min}: C(X_{S_1}, X_{S_2}, \ldots, X_{S_k})
\end{cases}
\tag{5.28}
$$

This multiobjective optimization problem can be reduced to a single-objective optimization problem by recalling that both the effective information part and the redundancy part have the same unit:

$$
\text{Max}: w_1[H(X_{S_1}, X_{S_2}, \ldots, X_{S_k}) + T(X_{S_1:S_k}; X_{F_i:F_m})] - w_2 C(X_{S_1}, X_{S_2}, \ldots, X_{S_k})
\tag{5.29}
$$

One can also unify the information and redundancy objectives as

$$\text{Max:} \frac{H(X_{S_1}, X_{S_2}, ..., X_{S_k}) + T(X_{S_1:S_k}; X_{F_1:F_m})}{wC(X_{S_1}, X_{S_2}, ..., X_{S_k})} \quad (5.30)$$

Any of the two methods unifying information and redundancy objectives can be adopted in the network design.

5.8.3   Selection Procedure

Using the MIMR criterion, the selection procedure for a network design entails the following steps:

1. Collect the hydrologic data for variable(s) of interest, e.g., hourly, daily, weekly, or monthly discharge or rainfall from each of the measuring stations. If the record is continuous, then discretize the continuous time series in such a way that each of the records is labeled 1, 2, 3, ..., b, where b is the number of bins used in the histogram. Therefore, the continuous time series will become a discrete time series.

2. Calculate marginal entropies for all candidate stations.

3. Identify the station having the maximum marginal entropy, and designate it as the central station.

4. Update set S in which stations already selected are saved and set F in which all unselected candidate stations are saved.

5. Select the next station from set F by the MIMR criterion. In this step scan all the unselected stations, and locate the station that can maximize the unified objective function [Eq. (5.26) or Eq. (5.29)]. To compare, the multivariate joint entropy $H(X_{S_1}, X_{S_2}, ..., X_{S_k})$, the mutual information between the grouped variable and another single variable $T(X_{S_1:S_k}, X_{F_i})$ or between two grouped variables $T(X_{S_1:S_k}, X_{F_i:F_m})$, and the total correlation $C(X_{S_1}, X_{S_2}, ..., X_{S_k})$ should be computed. Note that all these terms involve multiple variables as soon as the second important station is selected.

6. Repeat steps 4 and 5 recursively until the expected number of stations has been selected.

The convergence of the selection can be determined by the ratio of joint entropy of selected stations to that of all potential candidate stations. If the ratio is over a threshold, such as 0.95, or the ratio will not change significantly, then no additional station still in the candidate set can provide, significant amount of new information. These steps show that if no convergence threshold is provided, then all the potential candidate stations will be ranked in descending order. This may be helpful when determining the station with the least importance or area or stations with the least degree of importance. These steps illustrate a forward selection procedure. A pseudocode displaying the forward MIMR selection procedure is presented in Table 5.13. This algorithm is designed for ranking all the potential candidate stations such that it can provide the decision maker with as much information as possible. That is why the convergence judgment appears outside of the selection loop.

1: $F \leftarrow$ candidate set including all candidate stations $S \leftarrow$ empty set	Initialize candidate set F and empty set S.
2: Discretize the continuous time series.	
3: $tInfo \leftarrow H(F) \leftarrow$ Equation (5.20)	Compute the total information ($tInfo$) of all the potential stations.
4: For $i = 1 : N$ $H(X_i) \leftarrow$ Equation (5.2a) End	Compute the marginal entropy of each potential station.
5: $X_i \leftarrow \underset{X_i}{\arg\max}[H(X_i)]$	Select the first center station.
6: $F \leftarrow F - X_i$ $S \leftarrow S + X_i$	Update F and S for the first time.

7: For $i = 2 : N$
   $m \leftarrow$ size (F)
   $n \leftarrow$ size (S)
   For $k = 1: m$
   $MIMR_{S+X_{F_k}} \leftarrow$ Equation (5.26) or (5.29)
   End
   $X_{F_k} = \underset{X_{F_k}}{\arg\max}(MIMR_{S+X_{F_k}})$
   $F \leftarrow F - X_{F_k}$
   $S \leftarrow S + X_{F_k}$
End
	Sequentially select station from the updated candidate set according to MIMR criterion.
	
	Update the candidate set and already selected set successively.

8: For $i = 1 : N$
   $sInfor \leftarrow H(X_{S_1}, X_{S_2}, ..., X_{S_i})$
   $pct \leftarrow \dfrac{sInfo}{tInfo}$
   If $|pct - threshold| < eps$
      $S_{final} \leftarrow \{X_{S_1}, X_{S_2}, ..., X_{S_i}\}$
      return
   End
End
	Determine the final optimal set S_{final} according to the information fraction of the selected set to the total information. Joint entropy of selected stations is denoted by $sInfo$.

TABLE 5.13   Pseudocode for the MIMR-Based Greedy Selection Algorithm

A network can also be optimized in a backward manner, in which the criterion should be changed to minimum reduced information and maximum reduced redundancy. This criterion is also based on the principle of MIMR. The reduced information and redundancy can be quantified by the difference between joint entropies, total correlations of stations before and after one station is deleted, respectively. At the same time it should guarantee that the information of the station to be deleted can be inferred from the left-out station set as much as possible.

In the selection procedure, one may use the simple histogram method to estimate the probability distribution of the hydrologic variable of interest. One can also use more sophisticated methods for probability density estimation, such as kernel density estimation, cross-entropy density estimation, and so on. However, the histogram method is sufficient if a proper bin width or number of bins is used by which data-based choice for the optimal bin width, h_{opt}, can be expressed (Scott, 1979) as

$$h_{opt}^* = 3.49sn^{-1/3} \tag{5.31}$$

where s is the sample variance of the sample data and n is the sample size. Scott (1979) used Eq. (5.31) to estimate the probability density for several heavy-tailed non-Gaussian distributions and concluded that it produced satisfactory results. Compared to a kernel estimator, the histogram method, due to its slower convergence speed to the underlying density, is less sensitive to the choice of the smoothing parameter, such as the kernel band width and the bin width.

Given a continuous time series, one can compute the optimal bin width, the number of bins, and the endpoints for each discretization interval. For example, if the optimal number of bins is b, then the empirical i/b quantiles can be computed, where $i = 0, 1, 2, ..., b$. These different i/b quantiles $q_{i/b}$ are just the thresholds used to relabel the continuous time series following the procedure that all the records of the time series falling in the interval $[q_{k-1}, q_{k/b}]$, where $k = 1, 2, ..., b$, will be labeled k, and this is repeated for each interval.

Using the variable agglomeration method, the joint entropy $H(X_{S_1}, X_{S_2}, ..., X_{S_k})$, transinformation between a grouped variable and a single variable $T(X_{S_1:S_k}, X_{F_i})$, and the transinformation between two grouped variables $T(X_{S_1:S_k}, X_{F_1:F_m})$ can be computed. This requires merging multiple variables sequentially to form a new variable without changing the information retained by them first; thus, all computations can be reduced to a univariate or bivariate case.

Example 5.7.   The monthly precipitation observations from 2007 through 2011 at each station are shown in Fig. 5.5. Precipitation data are available at United States Historical Climatology Network (http://cdiac.ornl.gov/epubs/ndp/ushcn/ushcn_map_interface.html). Rank these stations in descending priority using MIMR with Eq. (5.26) as the objective function.

Solution

Step 1. Continuous time series discretization: Taking stations 1 and 2 as an example, we explain how to discretize a continuous time series using the histogram method. Using Eq. (5.31), one can obtain the optimal bin width, with which one can construct the histogram, as illustrated in the left panel in Fig. 5.6. Then we label observations falling in the first bin as 1, and those in the second as 2, and so on. Following this method, one can obtain discretized series as shown in the right panel in Fig. 5.6.

The discretized precipitation time series of the selected stations are shown in Table 5.14.

Step 2. Candidate set and selected set initialization: At the initial state, the candidate set and selected set are

$$F = \{\text{station 1, station 2, station 3, station 4, station 5}\}$$

$$S = \{\} \qquad \text{empty set}$$

Step 3. Central station identification: The central station is identified as one with the maximum marginal entropy. To do that, we need to compute the marginal entropy of each candidate station. To illustrate the computation of marginal entropy, we choose station 1 as an example. From the discretized time series of station 1, one can compute the contingency table, as given below.

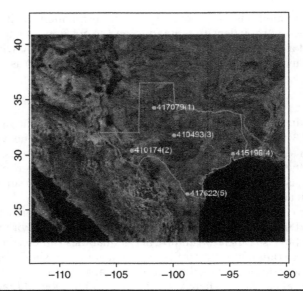

FIGURE **5.5**   Location of the selected precipitation stations in Texas.

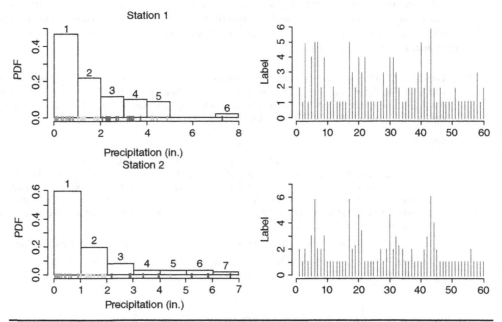

FIGURE **5.6**   Illustration for continuous time series discretization using the histogram method.

Station 1												
Month	1	2	3	4	5	6	7	8	9	10	11	12
2007	2	1	5	1	4	5	5	2	4	1	1	2
2008	1	1	1	1	5	3	2	4	3	4	1	1
2009	1	1	1	3	2	4	4	3	2	1	1	2
2010	2	2	3	5	2	3	6	2	1	2	1	1
2011	1	1	2	1	1	1	1	1	1	3	1	2

1: 0–1; **2:** 1–2; **3:** 2–3; **4:** 3–4; **5:** 4–5; **6:** 7–8.

Station 2												
Month	1	2	3	4	5	6	7	8	9	10	11	12
2007	2	1	2	1	3	7	2	2	3	1	1	1
2008	1	1	1	1	2	2	4	6	5	1	1	1
2009	1	1	2	1	2	5	2	3	3	2	1	1
2010	2	1	1	1	2	3	6	4	2	1	1	1
2011	1	1	1	1	1	1	1	2	1	1	1	1

1: 0–1; **2:** 1–2; **3:** 2–3; **4:** 3–4; **5:** 4–5; **5:** 5–6; **6:** 6–7.

Station 3												
Month	1	2	3	4	5	6	7	8	9	10	11	12
2007	1	1	3	1	4	4	1	2	1	1	1	1
2008	1	1	3	1	1	1	1	3	2	1	1	1
2009	1	1	1	1	2	1	2	1	2	2	1	1
2010	2	1	1	2	1	1	2	1	2	1	1	1
2011	1	1	1	1	2	1	1	2	1	2	1	1

1: 0–2; **2:** 2–4; **3:** 4–6; **4:** 8–10.

Station 4												
Month	1	2	3	4	5	6	7	8	9	10	11	12
2007	4	2	2	4	2	2	6	1	3	2	3	1
2008	2	2	2	2	3	2	3	4	7	2	5	4
2009	1	2	3	6	2	1	4	3	2	7	1	4
2010	2	4	2	1	2	1	6	1	2	1	3	2
2011	2	1	1	1	1	2	3	1	1	1	2	4

1: 0–2; **2:** 2–4; **3:** 4–6; **4:** 6–8; **5:** 8–10; **6:** 10–12; **7:** 14–16.

Station 5												
Month	1	2	3	4	5	6	7	8	9	10	11	12
2007	2	1	1	1	2	4	6	3	2	2	1	1
2008	1	1	1	1	2	1	4	6	3	1	1	1
2009	1	1	1	1	1	1	1	1	5	1	2	2
2010	1	1	1	1	1	3	2	1	4	1	1	1
2011	1	1	1	1	1	2	1	1	1	1	1	2

1: 0–2; **2:** 2–4; **3:** 4–6; **4:** 6–8; **5:** 8–10; **6:** 10–12.

TABLE 5.14 Discretized Precipitation Time Series

Counts Table						
Label	1	2	3	4	5	6
Count	27	14	7	6	5	1
Frequency Table						
Label	1	2	3	4	5	6
Frequency	0.4500	0.2333	0.1167	0.1000	0.0833	0.0167

From this contingency table, one can compute the marginal entropy as follows:

$$H(\text{station 1}) = -0.4500 \log_2 0.4500 - \cdots - 0.0167 \log_2 0.0167 = 2.0993 \text{ bits}$$

In the same way, the marginal entropies for other stations are computed.

	Station 1	Station 2	Station 3	Station 4	Station 5
$H(x)$	2.0993	1.8196	1.2179	2.2925	1.5004

Since station 4 has the highest marginal entropy, it is identified as the central station, that is, $s_1 = $ station 5.
Then the candidate and selected sets can be updated.

$$F = \{\text{station 1, station 2, station 3, station 5}\}$$
$$S = \{\text{station 4}\}$$

Step 4. Second station selection: According to MIMR, first we need to specify the information and redundancy weights. For simplicity, we use equal weights for information and redundancy, that is, $\lambda_1 = \lambda_2 = 1$.

Computation of Joint Entropy of Set S with Each Station in F   Taking the joint entropy of station 4 and station 1 as an example, the joint contingency table is calculated as shown in Table 5.15.
Using Eq. (5.2*b*), the joint entropy is computed as

$$H(\text{station 4, station 1}) = -0.1167 \log_2 0.1167 - 0.2000 \log_2 0.2000 - \cdots - 0.0167 \log_2 0.0167$$
$$= 4.0689 \text{ bits}$$

Similarly,

$$H(\text{station 4, station 2}) = 3.7522 \text{ bits}$$
$$H(\text{station 4, station 3}) = 3.2656 \text{ bits}$$
$$H(\text{station 4, station 5}) = 3.4531 \text{ bits}$$

Computation of Transinformation between (Stations in S and Each Station in F) and (the Other Stations in F)   Here we need to merge variables such that information retained in the merged variable is the same as that in the original variables. For merging variables, one follows the procedure in Example 5.2. Assume station 1 is selected in this step. Then the predictive ability of S is evaluated.

$$T(\text{station 4 + station 1, station 2}) = 4.6638 \text{ bits}$$
$$T(\text{station 4 + station 1, station 3}) = 4.5805 \text{ bits}$$
$$T(\text{station 4 + station 1, station 5}) = 4.6477 \text{ bits}$$

Assume station 2 is selected in this step. Then the predictive ability of S is evaluated.

$$T(\text{station 4 + station 2, station 1}) = 4.6638 \text{ bits}$$
$$T(\text{station 4 + station 2, station 3}) = 4.2046 \text{ bits}$$
$$T(\text{station 4 + station 2, station 5}) = 4.3003 \text{ bits}$$

Count Table							
		Station 1					
		1	**2**	**3**	**4**	**5**	**6**
Station 4	1	7	5	2	1	1	0
	2	12	4	2	2	2	0
	3	4	1	1	1	1	0
	4	2	4	0	2	0	0
	5	1	0	0	0	0	0
	6	0	0	1	0	1	1
	7	1	0	1	0	0	0
Frequency Table							
		Station 1					
		1	**2**	**3**	**4**	**5**	**6**
Station 4	1	0.1167	0.0833	0.0333	0.0167	0.0167	0.0000
	2	0.2000	0.0667	0.0333	0.0333	0.0333	0.0000
	3	0.0667	0.0167	0.0167	0.0167	0.0167	0.0000
	4	0.0333	0.0667	0.0000	0.0000	0.0000	0.0000
	5	0.0167	0.0000	0.0000	0.0000	0.0000	0.0000
	6	0.0000	0.0000	0.0167	0.0167	0.0167	0.0167
	7	0.0167	0.0000	0.0167	0.0000	0.0000	0.0000

TABLE 5.15   Joint Contingency Table of Station 4 and Station 1

Assume station 3 is selected in this step. Then the predictive ability of S is evaluated.

$$T(\text{station } 4 + \text{station } 3, \text{station } 1) = 4.5805 \text{ bits}$$
$$T(\text{station } 4 + \text{station } 3, \text{station } 2) = 4.2046 \text{ bits}$$
$$T(\text{station } 4 + \text{station } 3, \text{station } 5) = 4.0958 \text{ bits}$$

Assume station 5 is selected in this step. Then the predictive ability of S is evaluated.

$$T(\text{station } 4 + \text{station } 5, \text{station} 1) = 4.6477 \text{ bits}$$
$$T(\text{station } 4 + \text{station } 5, \text{station } 2) = 4.3003 \text{ bits}$$
$$T(\text{station } 4 + \text{station } 5, \text{station } 3) = 4.0958 \text{ bits}$$

Computation of Total Correlation of All in S and Each Gauge in F   Since there is only one station in set S, the total correlation is reduced to transinformation. Therefore,

$$C(\text{station } 4 \text{ and station } 1) = 0.3229 \text{ bit}$$
$$C(\text{station } 4 \text{ and station } 2) = 0.3598 \text{ bit}$$
$$C(\text{station } 4 \text{ and station } 3) = 0.2448 \text{ bit}$$
$$C(\text{station } 4 \text{ and station } 5) = 0.3398 \text{ bit}$$

Computation of Final Score for Each Gauge in F   From Eq. (5.26), we have the following:
Station 1:

$$H(\text{station } 4 \text{ and station } 1) + T(\text{station } 4 + \text{station } 1, \text{station } 2) + T(\text{station } 4 + \text{station } 1, \text{station } 3)$$
$$+ T(\text{station } 4 + \text{station } 1, \text{station } 5)$$
$$- C(\text{station } 4, \text{station } 1)$$

$$= 4.0689 + 4.6638 + 4.5805 + 4.6477 - 0.3229$$

$$= 17.6380 \text{ bits}$$

Station 2:

H(station 4, station 2) + T(station 4 + station 2, station 1)
  + T(station 4 + station 2, station 3) + T(station 4 + station 2, station 5)

  − C(station 4, station 2)

  $= 3.7522 + 4.6638 + 4.2046 + 4.3003 - 0.3598$

  $= 16.5611 \text{ bits}$

Station 3:

H(station 4, station 3) + T(station 4 + station 3, station 1)
  + T(station 4 + station 3, station 2) + T(station 4 + station 3, station 5)

  − C(station 4, station 3)

  $= 3.2656 + 4.5805 + 4.2046 + 4.0958 - 0.2448$

  $= 15.9017 \text{ bits}$

Station 5:

H(station 4, station 5) + T(station 4, station 5, station 1)
  + T(station 4, station 5, station 2) + T(station 4, station 5, station 3)

  − C(station 4, station 5)

  $= 3.4531 + 4.6477 + 4.3003 + 4.0958 - 0.3398$

  $= 16.1571 \text{ bits}$

Clearly, station 1 has the highest score. Therefore, it is selected in set S. Then set F and set S can be updated as

$$F = \{\text{station 2, station 3, station 5}\}$$

$$S = \{\text{station 4, station 1}\}$$

Step 5. Third gauge selection: Following the same procedure as in step 4, we can select the third gauge by using the MIMR criterion.

Computation of Joint Entropy of Set S with Each Gauge in F   Multivariate joint entropy can be computed using the shortcut formula between marginal entropies and total correlation. Assume station 1 is selected in this step. Then, using the computation procedure, the joint entropies are

$$H(\text{station 4, station 1, station 2}) = 4.6638 \text{ bits}$$

Similarly,

$$H(\text{station 4, station 1, station 3}) = 4.5805 \text{ bits}$$
$$H(\text{station 4, station 1, station 5}) = 4.6477 \text{ bits}$$

Computation of Transinformation between Stations in S and Each Station in F and the Other Stations in F   Assume station 2 is selected in this step. Then the predictive ability of S is evaluated.

$$T(\text{station 4 + station 1 + station 2, station 3}) = 1.0264 \text{ bits}$$
$$T(\text{station 4 + station 1 + station 2, station 5}) = 1.2641 \text{ bits}$$

Assume station 3 is selected in this step. Then the predictive ability of S is evaluated.

$$T(\text{station 4} + \text{station 1} + \text{station 3, station 2}) = 2.0993 \text{ bits}$$
$$T(\text{station 4} + \text{station 1} + \text{station 3, station 5}) = 1.2199 \text{ bits}$$

Assume station 5 is selected in this step. Then the predictive ability of S is evaluated.

$$T(\text{station 4} + \text{station 1} + \text{station 5, station 2}) = 1.5671 \text{ bits}$$
$$T(\text{station 4} + \text{station 1} + \text{station 5, station 3}) = 1.0047 \text{ bits}$$

Computation of Total Correlation of All Stations in S and Each Station in F   Assume station 2 is selected in this step. Then

$$C(\text{station 4, station 1, station 2}) = 1.5475 \text{ bits}$$

Assume station is selected in this step. Then

$$C(\text{station 4, station 1, station 3}) = 1.0292 \text{ bits}$$

Assume station 5 is selected in this step. Then

$$C(\text{station 4, station 1, station 5}) = 1.2445 \text{ bits}$$

Computation of Final Score for Each Gauge in F

Station 2:

$H(\text{station 4, station 1, station 2}) + T(\text{station 4} + \text{station 1} + \text{station 2, station 3})$
  $+ T(\text{station 4} + \text{station 1} + \text{station 2, station 5}) - C(\text{station 4, station 1, station 2})$

$= 4.6638 + 1.0264 + 1.2641 - 1.5475$

$= 5.4068 \text{ bits}$

Station 3:

$H(\text{station 4, station 1, station 3}) + T(\text{station 4} + \text{station 1} + \text{station 3, station 2})$
  $+ T(\text{station 4} + \text{station 1} + \text{station 3, station 5}) - C(\text{station 4, station 1, station 3})$

$= 4.5805 + 2.0993 + 1.2199 - 1.0292$

$= 6.8705 \text{ bits}$

Station 5:

$H(\text{station 4, station 1, station 5}) + T(\text{station 4} + \text{station 1} + \text{station 5, station 2})$
  $+ T(\text{station 4} + \text{station 1} + \text{station 4, station 3}) - C(\text{station 4, station 1, station 5})$

$= 4.6477 + 1.5671 + 1.0047 - 1.2445$

$= 5.9750 \text{ bits}$

Obviously, station 3 has the highest score. Therefore, it is selected in set S. Then sets F and S can be updated as

$$F = \{\text{station 2, station 5}\}$$
$$S = \{\text{station 4, station 1, station 3}\}$$

Step 6: Select the fourth gauge 3

Computation of Joint Entropy of Set S with Each Gauge in F

$$H(\text{station 4, station 1, station 3, station 2}) = 4.8553 \text{ bits}$$
$$H(\text{station 4, station 1, station 3, station 5}) = 4.8610 \text{ bits}$$

Computation of Transinformation between Stations in S and Each Station in F and the Other Stations in F   Assume station 2 is selected in this step. Then the predictive ability of S is evaluated.

$$T(\text{station 4} + \text{station 1} + \text{station 3} + \text{station 2, station 5}) = 1.2740 \text{ bits}$$

Assume station 5 is selected in this step. Then the predictive ability of S is evaluated.

$$T(\text{station 4} + \text{station 1} + \text{station 3} + \text{station 5, station 2}) = 1.5988 \text{ bits}$$

Computation of Total Correlation of All in S and Each Gauge in F   Assume station 2 is selected in this step. Then

$$C(\text{station 4, station 1, station 3, station 2}) = 2.5740 \text{ bits}$$

Assume gauge 3 is selected in this step. Then

$$C(\text{station 4, station 1, station 3, station 5}) = 2.2490 \text{ bits}$$

Computation of Final Score for Each Gauge in F
Station 2:

$H(\text{station 4, station 1, station 3, station 2}) + T(\text{station 4} + \text{station 1} + \text{station 3} + \text{station 2, station 5}) - C(\text{station 4, station 1, station 3, station 2})$

$= 4.8553 + 1.2740 - 2.5740$

$= 3.5553 \text{ bits}$

Station 5:

$H(\text{station 4, station 1, station 3, station 5}) + T(\text{station 4} + \text{station 1} + \text{station 3} + \text{station 5, station 2}) - C(\text{station 4, station 1, station 3, station 5})$

$= 4.8610 + 1.5988 - 2.2490$

$= 4.2108 \text{ bits}$

Station 5 has a larger final score. Therefore, station 5 is selected in this step. Since there are only 5 stations, among which 4 stations have been ranked, the final rank for these 5 stations is

$$\{\text{station 4, station 1, station 3, station 5, station 2}\}$$

Example 5.8.   Using the same precipitation data as in Example 5.7, rank these stations in descending priority, using MIMR with Eq. (5.29) as the objective function.

Solution.   Ranking stations using Eq. (5.29) is more or less the same as that using Eq. (5.26). For the ease of understanding, we explain the procedure again step by step here.

Step 1: Continuous time series discretization

The step 1 here is the same as the one in Example 5.7. For simplicity, we omit the results.

Step 2: Candidate set and selected set initialization

This is the same as in Example 5.7, and the candidate set and selected set are

$F = \{\text{station 1, station 2, station 3, station 4, station 5}\}$

$S = \{\ \}$ empty set

Step 3: Central station identification

Similarly, the central station is identified as one with the maximum marginal entropy. From Example 5.7, we have already learned that station 4 has the highest marginal entropy. It is therefore identified as the central station, that is, $s_1 = \text{station 5}$. Then the candidate and selected sets can be updated:

$F = \{\text{station 1, station 2, station 3, station 5}\}$

$S = \{\text{station 4}\}$

Step 4: The second station selection

Also for simplicity, we use equal weights for information and redundancy, that is, $\lambda_1 = \lambda_2 = 1$.

Computation of Joint Entropy of Set S with Each Station in F Suppose now that station 1 is selected in this step. From Example 5.7, we know that the joint entropy between station 4 and station 1 is

$$H(\text{station 4, station 1}) = 4.0689 \text{ bits}$$

Similarly, suppose that each of the other stations in set F is the one to be selected in this step. One can compute the joint entropies as follows:

$$H(\text{station 4, station 2}) = 3.7522 \text{ bits}$$
$$H(\text{station 4, station 3}) = 3.2656 \text{ bits}$$
$$H(\text{station 4, station 5}) = 3.4531 \text{ bits}$$

Computation of Transinformation between Stations in S and Each Station in F and the Other Stations in F after Grouping Assume station 1 is selected in this step. Then by following Eq. (5.29) one can evaluate the predictive ability of S.

$$T(\text{station 4 + station 1, station 2 + station 3 + station 5}) = 2.3486 \text{ bits}$$

Assume station 2 is selected in this step. Then the predictive ability of S is evaluated.

$$T(\text{station 4 + station 2, station 1 + station 3 + station 5}) = 2.5360 \text{ bits}$$

Assume station 3 is selected in this step. Then the predictive ability of S is evaluated.

$$T(\text{station 4 + station 3, station 1 + station 2 + station 5}) = 2.1331 \text{ bits}$$

Assume station 5 is selected in this step. Then the predictive ability of S is evaluated.

$$T(\text{station 4 + station 5, station 1 + station 2 + station 3}) = 2.1907 \text{ bits}$$

Computation of Total Correlation of All Stations in S and Each Gauge in F Suppose station 4 is selected in this step. The total correlation between station 4 and other stations, which has been calculated in Example 5.7, is

$$C(\text{station 4, station 1}) = 0.3229 \text{ bit}$$
$$C(\text{station 4, station 2}) = 0.3598 \text{ bit}$$
$$C(\text{station 4, station 3}) = 0.2448 \text{ bit}$$
$$C(\text{station 4, station 5}) = 0.3398 \text{ bit}$$

Computation of Final Score for Each Gauge in F From Eq. (5.29), we have

Station 1:

$H(\text{station 4, station 1}) + T(\text{station 4 + station 1, station 2 + station 3 + station 5})$
  $- C(\text{station 4, station 1}) = 4.0689 + 2.3486 - 0.3229$

  $= 6.0946 \text{ bits}$

Station 2:

$H(\text{station 4, station 2}) + T(\text{station 4 + station 2, station 1 + station 3 + station 5})$
  $- C(\text{station 4, station 2}) = 3.7522 + 2.5360 - 0.3598$

  $= 5.9284 \text{ bits}$

Station 3:

$H(\text{station 4, station 3}) + T(\text{station 4 + station 3, station 1 + station 2 + station 5})$
  $- C(\text{station 4, station 3}) = 3.2656 + 2.1331 - 0.2448$

  $= 5.1539 \text{ bits}$

Station 5:

H(station 4, station 5) + T(station 4 + station 5, station 1 + station 2 + station 3)
  $-$ C(station 4, station 5) = 3.4531 + 2.1907 $-$ 0.3398

  = 5.3040 bits

Obviously, station 1 has the highest score. Therefore, it is selected in set S. Then set F and set S can be updated as

$$F = \{\text{station 2, station 3, station 5}\}$$

$$S = \{\text{station 4, station 1}\}$$

Step 5: The third gauge selection

Following the same procedure as in step 4, we can select the third gauge using the MIMR criterion with Eq. (5.29) as the objective function.

Computation of Joint Entropy of Set S with Each Gauge in F   By following the same method as in Example 5.7 and assuming station 1 to be selected in this step, the joint entropies are

$$H(\text{station 4, station 1, station 2}) = 4.6638 \text{ bits}$$

Similarly,

$$H(\text{station 4, station 1, station 3}) = 4.5805 \text{ bits}$$

$$H(\text{station 4, station 1, station 5}) = 4.6477 \text{ bits}$$

Computation of Transinformation between Stations in S and Each Station in F and the Other Stations in F after Grouping
Assume station 2 is selected in this step. Then the predictive ability of S is evaluated.

$$T(\text{station 4 + station 1 + station 2, station 3 + station 5}) = 2.0627 \text{ bits}$$

Assume station 3 is selected in this step. Then the predictive ability of S is evaluated.

$$T(\text{station 4 + station 1 + station 3, station 2 + station 5}) = 2.2903 \text{ bits}$$

Assume station 5 is selected in this step. Then the predictive ability of S is evaluated.

$$T(\text{station 4 + station 1 + station 5, station 2 + station 3}) = 2.2004 \text{ bits}$$

Computation of Total Correlation of All Stations in S and Each Station in F   Assume station 2 is selected in this step. Then

$$C(\text{station 4, station 1, station 2}) = 1.5475 \text{ bits}$$

Assume station 3 is selected in this step. Then

$$C(\text{station 4, station 1, station 3}) = 1.0292 \text{ bits}$$

Assume station 5 is selected in this step. Then

$$C(\text{station 4, station 1, station 5}) = 1.2445 \text{ bits}$$

Computation of Final Score for Each Gauge in F

Station 2:

H(station 4, station 1, station 2) + T(station 4 + station 1 + station 2, station 3 + station 5)
  $-$ C(station 4, station 1, station 2)

  = 4.6638 + 2.0627 $-$ 1.5475

  = 5.1790 bits

Station 3:

H(station 4, station 1, station 3) + T(station 4 + station 1 + station 3, station 2 + station 5)
  − C(station 4, station 1, station 3)

= 4.5805 + 2.2903 − 1.0292

= 5.8416 bits

Station 5:

H(station 4, station 1, station 5) + T(station 4 + station 1 + station 5, station 2 + station 3)
  − C(station 4, station 1, station 5)

= 4.6477 + 2.2004 − 1.2445

= 5.6036 bits

Station 3 has the highest score. Therefore it is selected into set S. Then sets F and S can be updated as

$$F = \{\text{station 2, station 5}\}$$
$$S = \{\text{station 4, station 1, station 3}\}$$

Step 6: Select the fourth gauge

Computation of Joint Entropy of Set S with Each Gauge in F Suppose station 2 is selected. Then the joint entropy of stations in S and station 2 is

$$H(\text{station 4, station 1, station 3, station 2}) = 4.8553 \text{ bits}$$

Similarly, suppose station 5 is to be selected. Then the joint entropy of stations in S and station 5 is

$$H(\text{station 4, station 1, station 3, station 5}) = 4.8610 \text{ bits}$$

Computation of Transinformation between (Stations in S and Each Station in F) and (the Other Stations in F after Grouping)
Assume station 2 is selected in this step. Then the predictive ability of S is evaluated:

$$T(\text{station 4 + station 1 + station 3 + station 2, station 5}) = 1.2740 \text{ bits}$$

Assume station 5 is selected in this step. Then the predictive ability of S is evaluated.

$$T(\text{station 4 + station 1 + station 3 + station 5, station 2}) = 1.5988 \text{ bits}$$

Computation of Total Correlation of All in S and Each Gauge in F Assume station 2 is selected in this step. Then

$$C(\text{station 4, station 1, station 3, station 2}) = 2.5740 \text{ bits}$$

Assume gauge 3 is selected in this step. Then

$$C(\text{station 4, station 1, station 3, station 5}) = 2.2490 \text{ bits}$$

Computation of Final Score for Each Gauge in F

Station 2:

H(station 4, station 1, station 3, station 2) + T(station 4 + station 1 + station 3 + station 2, station 5)
  − C(station 4, station 1, station 3, station 2)

= 4.8553 + 1.2740 − 2.5740

= 3.5553 bits

Station 5:

H(station 4, station 1, station 3, station 5) + T(station 4 + station 1 + station 3 + station 5, station 2)
  − C(station 4, station 1, station 3, station 5)

= 4.8610 + 1.5988 − 2.2490

= 4.2108 bits

Station 5 has a larger final score. Therefore, station 5 is selected in this step. Since there are only 5 stations, among which 4 stations have been ranked, the final rank for these 5 stations is {station 4, station 1, station 3, station 5, station 2}. Note that the station ranks obtained with the use of Eq. (5.29) as the objective function are the same as those obtained by using Eq. (5.26) as the objective function. However, note also that this observation may not necessarily hold in other situations.

5.9   Application to Monitoring, Evaluation, and Design

5.9.1   Use of Normal Distribution

Calculation of entropy measures requires that the probability distribution of the variable to be sampled be known. When applying entropy for bivariate and multivariate cases, it is simpler to use bivariate and multivariate normal distributions than other distributions, such as the gamma, Pearson type, or Weibull. The normal distribution is chosen because of the complexity involved in the application of entropy with other distributions. The normal distribution requires (Guiasu, 1977) replacement of discrete probabilities by a continuous probability distribution function (PDF); replacement of a system of events by a function space, where PDFs are defined; and replacement of the summation in the domain by integration throughout the domain. Before the use of multivariate normal distributions, data must be normalized. The normalization can be accomplished by employing the Box-Cox transformation (Bras and Rodriguez-Iturbe, 1985). The data are transformed for every sampling interval.

5.9.2   Rainfall Networks

The main objective of a rain gauge network design is to select an optimum number of rain gauges and their locations such that the desired amount of information on rainfall is obtained. Other considerations of importance in the network design may include the obtaining of an adequate record length prior to utilization of the data, the development of a mechanism for transferring information from gauged to ungauged locations whenever needed, and the estimation of the probable magnitude of error arising from network density, distribution, and record length.

Krastanovic and Singh (1992) used the marginal entropy measure to draw contour maps of the rainfall network in Louisiana and evaluated the network according to the entropy map. Lee and Ellis (1997) compared kriging and the maximum entropy estimator for spatial interpolation and their subsequent use in optimizing monitoring networks. Husain (1989) and Bueso et al. (1999) used entropy to illustrate a framework for spatial sampling of a monitoring network.

Let n be the number of rain gauges, each with X_1 and record length N_i ($i = 1, ..., n$) and S_c the cross-correlation matrix of $n \times n$ dimensions. The use of the multivariate normal PDF yields an expression for multivariate entropy (Harmancioglu and Yevjevich, 1985; Krstanovic and Singh, 1988). The essential condition for the existence of $H(X)$ is the positive definiteness of the cross-correlation matrix S_c. This enables nonsingular determinant $|S_c|$ and the existence of S_c^{-1}.

It is convenient that all rain gauges have a common record length, that is, $N_1 = N_2 = \cdots = N_n$. This enables easy computation of different covariance matrices S_c. Records associated with X_{j1} ($j = 1, ..., n$) $= \bar{X}_1$ for each rain gauge are constructed first, then $\bar{X}_2 = (X_i, X_j)$ ($i \neq j$) for all rain gauge pairs, $\bar{X}_3 = (X_i, X_j, X_k)$ for all rain gauge triples, etc., until $\bar{X}_{n-1} = (X_i, ..., X_{n-1})$ and $\bar{X}_n = (X_1, ..., X_n)$. For every vector, the corresponding joint entropy $(H(\bar{X}_1), H(\bar{X}_1), ..., H(\bar{X}_1))$ is computed, with S_c varying from S_{c1} of 1×1 dimensions to

S_{cn-1} of $(n-1) \times (n-1)$ dimensions. At each stage, for every combination of rain gauges, the corresponding conditional entropies and transinformation are computed. Then, depending on these entropy values, either an additional rain gauge is selected or the procedure is terminated.

To summarize, rain gauge evaluation involves the following steps:

1. Compute the marginal entropies of all rain gauges. From the computed entropy values, the rain gauge (S_1) with the highest uncertainty or entropy [that is, $H(X_1)$] is found. This is regarded as the most important rain gauge and is designated as the central rain gauge.

2. Compute the conditional entropy of rain gauge S_1 with respect to all other rain gauges (S_2 to S_n).

3. Find the rain gauge that gives the lowest reduction in uncertainty or transinformation, or find $\min\{H(X_1) - H(X_1 \mid X_2)\} = \min\{T(X_1, X_2)\}$. The rain gauge that has the least common information with the first (central) rain gauge is the rain gauge of the second-highest importance. Keep rain gauges S_1 and S_2; compute the conditional entropy of the rain gauges with respect to all other rain gauges (S_3 to S_n); and find the rain gauge that gives the minimum reduction in uncertainty, or $\min\{H(X_1, X_2) - H((X_1, X_2) \mid X_3)\} = \min\{T((X_1, X_2), X_3)\}$. Rain gauge S_3 is the third most important rain gauge.

Repeat the procedure such that the jth important rain gauge is the one that gives

$$\min\{H(X_1, \dots, X_{j-1}) - H((X_1, \dots, X_{j-1}) \mid X_j)\} = \min\{T((X_1, \dots, X_{j-1}), X_j)\}$$

For a multivariate normal distribution, the transinformation can be expressed as $T[(X_1, X_2, \dots, X_{j-1}), X_j] = -\frac{1}{2}\ln(1 - R^2)$, where R is the multiple correlation coefficient. Thus, the rain gauge having the smaller multiple correlation than the other rain gauges is always chosen.

To determine the number of essential rain gauges, the coefficient of nontransferred information t_i is computed. If, at step i, $t_{i+1} \geq t_i$ occurs, then the nontransferred information of the $i + 1$ rain gauges is greater than the nontransferred information of the ith rain gauge. Then a rain gauge is added. If the value of keeping i rain gauges is greater than the value of $i + 1$ rain gauges, then the other nonchosen rain gauges are discontinued. The discontinuation depends on the measurement and the sampling interval of the rainfall record. If rain gauges collect daily data, then daily evaluation should be most useful. If rain gauges measure only long-term rainfall depths (monthly or yearly), then monthly and yearly evaluations should be performed.

Two other alternatives may be employed to choose rain gauges: (1) the maximum joint entropy $H(X_1, \dots, X_k)$ and (2) the maximum weighted entropy

$$\frac{H(X_1 \mid X_k) + H(X_2 \mid X_k) + \cdots + H(X_{k-1} \mid X_k)}{k}$$

Both alternatives were examined by Krstanovic and Singh (1988) and produced results comparable to those by the maximum conditional entropy.

For evaluation of rainfall networks, isocorrelation lines of rain gauges with respect to a central rain gauge can be plotted. In a similar vein, isoinformation contours or the lines of the coefficient of nontransferred information t_i can be plotted, where t_i is computed. Thus, t_i measures only the information transferred from the central rain gauge to any other rain gauge. Contours of isoinformation encompass central rain gauges, and higher concentrations of isoinformation contours surrounding the central rain gauges apparently require more rain gauges in the watershed to transfer that information.

Let $X_1, X_2, ..., X_n$ denote the rainfall depths corresponding to different rainfall data sets $S_1, S_2, ..., S_n$; $p(X_1, X_2, ..., X_n) = p_{i1}, ..., p_{in}$, the joint probability; $p(x_m | x_1, ..., x_n) = p(i_m | i_1, ..., i_n)$, the conditional probability record of the mth rain gauge when all other rainfall depth data sets are known; and $p[(x_1, ..., x_n) | x_{n+1}] = p[(i_1, ..., i_n) | i_{n+1}]$, the conditional probability of the n rainfall data sets (of n rain gauges) when the rainfall data set of the $(n+1)$st rain gauge is added. The multidimensional joint entropy for n rain gauges in a watershed represents the uncertainty of their measured rainfall data sets. The uncertainty left in the central rain gauge when the records of all other rain gauges are known is expressed by the multivariate conditional entropy of the central rain gauge conditioned on all other records. Similarly, the uncertainty left in the group of rain gauges $(i_1, ..., i_n)$ when any new rain gauge is added (i.e., the expansion of the existing rainfall network) can be defined. Correspondingly, the multidimensional transinformation between the n existing rain gauges and the new (added) rain gauge $(n+1)$ can be defined by Eq. (5.4).

Now we construct a network by consecutively adding rain gauges one by one for a specified watershed and sampling interval. At any step, this addition can be terminated, depending upon the consecutive coefficients of transferred information $1 - t_i$. At the first step, only the central rain gauge is chosen and

$$H(X_1) = H(X_{\text{central}}) \tag{5.32}$$

In choosing the second rain gauge,

$$t_1 = \frac{H(X_1 | X_j)}{H(X_1)} \qquad \text{where } j \neq 1 \tag{5.33}$$

is computed. In choosing the ith rain gauge,

$$t_{i-1} = \frac{H((X_1, ..., X_{i-1}) | X_i)}{H(X_1, ..., X_{i-1})} \tag{5.34}$$

is computed.

The maximum possible transformation T_0 varies from one step to the other, and it is always equal to the joint entropy of the rain gauges already chosen before that step. For example, in the second step $T_0 = H(X_1)$, and in the ith step $T_0 = H(X_1, ..., X_{i-1})$. The rain gauges are worth adding as long as the transferred information decreases significantly, i.e.,

$$1 \geq t_1 > t_2 > \cdots > t_i \tag{5.35}$$

If, at any step, $t_i \geq t_{i-1}$, then the new rain gauge being added at the ith step has repetitive information and is not worth adding.

If $t_{i-1} > t_i$, the new ith rain gauge contains new information. The magnitude of the difference between t_{i-1} and t_i is always in the $[0, 1]$ range. The higher the magnitude, the more new information is added; the lower the magnitude, the less new information is added. For the latter, other considerations (economics, rain gauge access, etc.) might terminate the addition of new rain gauges before the $t_i = t_{i-1}$ point is reached. The significance of the magnitude corresponds to the significance in a multiple correlation coefficient among i and $i - 1$ variables (r_i and r_{i-1}), assuming a certain functional relationship among these variables. However, it also depends on the number of existing rain gauges. For example, for 10 rain gauges in a watershed, the added information content is significant if $t_{i-1} - t_i \geq 0.10$ (or the average expected information increases from the first rain gauge to the tenth rain gauge). Comparison of multiple correlation coefficient with transferred information was made by Harmancioglu and Yevjevich (1985).

At each step during the rain gauge selection process, the coefficient of nontransferred information t_i between the already chosen rain gauge and each of the remaining rain gauges j ($j = i, ..., n$) is computed. Then isoinformation contours can be constructed. The isoinformation contour of the multivariate rainfall record (at the ith step) is the line of equal common information between already chosen (i) rain gauges and any not-chosen existing rain gauge (j) located in the region.

Example 5.9. Suppose there are five candidate precipitation stations in a field. Construct a monitoring network by consecutively adding stations one by one, using the coefficient of nontransferred information. The candidate stations are those in Example 5.7.

Solution. As the first step, we need to compute the marginal entropy of each candidate station. The first station, also called the central station, is the one with maximum marginal entropy. From Example 5.7 we already learned that station 4 has an entropy of 2.2925 bits, which is the maximum among the 5 stations. Therefore, it is selected as the central station. Then we need to compute the conditional entropy of station 4, given each of the other stations. For illustration, suppose now that we want to compute H(station 4 | station 1). To do that, we need to construct the joint contingency table of station 4 and station 1. Following the same discretization scheme used in Example 5.7, we can construct the contingency table as follows:

		\multicolumn{7}{c}{Count Table}	
		\multicolumn{6}{c}{Station 1}	
		1	2	3	4	5	6	
Station 4	1	7	5	2	1	1	0	16
	2	12	4	2	2	2	0	22
	3	4	1	1	1	1	0	8
	4	2	4	0	2	0	0	8
	5	1	0	0	0	0	0	1
	6	0	0	1	0	1	1	3
	7	1	0	1	0	0	0	2

		\multicolumn{7}{c}{Frequency Table}	
		\multicolumn{6}{c}{Station 1}	
		1	2	3	4	5	6	
Station 4	1	0.1167	0.0833	0.0333	0.0167	0.0167	0.0000	0.2667
	2	0.2000	0.0667	0.0333	0.0333	0.0333	0.0000	0,3667
	3	0.0667	0.0167	0.0167	0.0167	0.0167	0.0000	0.1333
	4	0.0333	0.0667	0.0000	0.0000	0.0000	0.0000	0.1333
	5	0.0167	0.0000	0.0000	0.0000	0.0000	0.0000	0.0167
	6	0.0000	0.0000	0.0167	0.0167	0.0167	0.0167	0.0500
	7	0.0167	0.0000	0.0167	0.0000	0.0000	0.0000	0.0333

From the definition of conditional entropy, we have

$$H(\text{station 4} \mid \text{station 1}) = -\sum_{i=1}^{7}\sum_{j=1}^{6} p_{ij} \log_2 \frac{p_{ij}}{p_{\cdot j}}$$

$$= -0.1167 \log_2 \frac{0.1167}{0.2667} - 0.2000 \log_2 \frac{0.2000}{0.3667} - 0.0667 \log_2 \frac{0.0667}{0.1333} - \cdots - 0.0167 \log_2 \frac{0.0167}{0.0333}$$

$$= 1.9696 \text{ bits}$$

In the same vein, we can compute the other conditional entropies:

$$H(\text{station 4} \mid \text{station 2}) = 1.9327 \text{ bits}$$
$$H(\text{station 4} \mid \text{station 3}) = 2.0477 \text{ bits}$$
$$H(\text{station 4} \mid \text{station 5}) = 1.9527 \text{ bits}$$

Then, we can compute t_1.

$$t_1 = \left\{ \frac{H(\text{station 4} \mid \text{station 1})}{H(\text{station 4})}, \frac{H(\text{station 4} \mid \text{station 2})}{H(\text{station 4})}, \frac{H(\text{station 4} \mid \text{station 3})}{H(\text{station 4})}, \frac{H(\text{station 4} \mid \text{station 5})}{H(\text{station 4})} \right\}$$

$$= \left\{ \frac{1.9696}{2.2925}, \frac{1.9327}{2.2925}, \frac{2.0477}{2.2925}, \frac{1.9527}{2.2925} \right\}$$

$$= \{0.8591, 0.8431, 0.8932, 0.8518\}$$

It can be seen that station 3 has the maximum t_1 value, which means that station 3 has the minimum common information with station 5. In other words, if station 3 is selected into the network, it will provide more useful information. It is therefore selected in the network at this step.

Next, since station 4 and station 3 are already in the network, we need to first compute the joint entropy of these two stations. Previously, we have repeatedly shown how to compute the joint entropy. We therefore omit the detailed procedure here. The joint entropy is

$$H(\text{station 4, station 3}) = 3.2656 \text{ bits}$$

We then need to compute the conditional entropy of $H(\text{station 4, station 3} \mid \text{station } i)$, in which i is 1, 2, or 5. We start with $i = 1$. For simplicity, we first merge the time series of station 4 and station 3 such that the merged series contains the same information amount as that of the original time series. Following the procedure discussed in Sec. 5.7, we can merge two discrete time series. Here, for simplicity, we omit the results. The conditional entropy of the merged time series given that of station 1 can be calculated following the method illustrated above. The conditional entropy is

$$H(\text{station 4, station 3} \mid \text{station 1}) = 2.4812 \text{ bits}$$
$$H(\text{station 4, station 3} \mid \text{station 2}) = 2.3850 \text{ bits}$$
$$H(\text{station 4, station 3} \mid \text{station 5}) = 2.5953 \text{ bits}$$

Then we can compute t_2.

$$t_2 = \left\{ \frac{H(\text{station 4, station 3} \mid \text{station 1})}{H(\text{station 4, station 3})}, \frac{H(\text{station 4, station 3} \mid \text{station 2})}{H(\text{station 4, station 3})}, \frac{H(\text{station 4, station 3} \mid \text{station 5})}{H(\text{station 4, station 3})} \right\}$$

$$= \left\{ \frac{2.4812}{3.2656}, \frac{2.3850}{3.2656}, \frac{2.5953}{3.2656} \right\}$$

$$= \{0.7598, 0.7303, 0.7947\}$$

Obviously, station 5 has the maximum value of t_2. Therefore, station 5 is selected into the network at this step.

Similarly, we now have to compute the joint entropy of station 4, station 3, and station 5:

$$H(\text{station 4, station 3, station 5}) = 5.0958 \text{ bits}$$

We then need to compute the conditional entropies:

$$H(\text{station 4, station 3, station 5} \mid \text{station 1}) = 2.7617 \text{ bits}$$

$$H(\text{station 4, station 3, station 5} \mid \text{station 2}) = 2.8351 \text{ bits}$$

At this step, t_3 is

$$t_3 = \left\{ \frac{H(\text{station 4, station 3, station 5} \mid \text{station 1})}{H(\text{station 4, station 3, station 5})}, \frac{H(\text{station 4, station 3, station 5} \mid \text{station 2})}{H(\text{station 4, station 3, station 5})} \right\}$$

$$= \left\{ \frac{2.7617}{4.0958}, \frac{2.8351}{4.0958} \right\}$$

$$= \{0.6743, 0.6922\}$$

One can see that station 2 has the maximum value of t_3. Therefore, at this step station 2 is selected in the network. Now only one station—station 1—is left in the candidate set. It should be the last station to be selected.

5.9.3 Stream Flow Networks

A stream flow network for a river can be designed based on the transfer of information between observations of stream gauging stations on the river. The same concept can be applied to the observations of stream gauges in the river basin. Harmancioglu (1981) and Harmancioglu and Yevjevich (1987) employed the entropy coefficient to measure transferable information between river gauging stations. Yang and Burn (1994) described an analytical comparison between the correlation among gauging stations and their joint entropy. The information transmission from station X to station Y is represented by the mutual information $T(X, Y)$ (Lathi, 1969) and the maximum transferable information by the resulting entropy coefficient. The correlation coefficient measures the transfer of information.

For a river basin, two cases can be identified for the transfer of information along rivers. In the first case, flows from upstream stations pass through downstream stations. These upstream flows can be considered as through flow which, when combined with intermediate flows from tributaries, constitutes the downstream flow. In the second case, the upstream flow and the intermediate flow of the downstream stations are correlated. Referring to Fig. 5.7, information transfer among river points can occur in three ways (Harmancioglu and Yevjevich, 1987): (1) Information transfer can occur among river points located on the same courses, such as transfer between points 1 and 2, 2 and 3, or 4 and 3. It is immaterial if one point is located on a tributary and the other on the main stem after the confluence of the tributary. The flow of the upstream point that passes the downstream point is designated as through flow, whereas intermediate flows represent the flows from tributaries that go into the river between the two points. (2) Information transfer can occur between points that have different drainage areas, such as points 2 and 4 or 1 and 4. In this case, there is no through flow. The correlation between these points is due to climatic factors. (3) Information transfer can occur among three or four points, such as points 2 and 1, or 3 and 4. Through flows may or may not be involved. This calls for multivariate analysis.

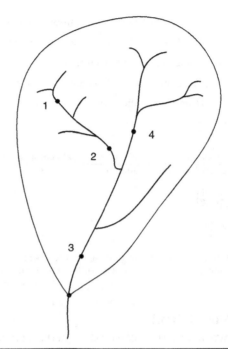

FIGURE **5.7** Scheme for information transfer among stream gauges in a watershed.

It is possible to investigate the effect of (1) trends, periodicities, stochastic dependence, and stochastic independent noise on the transferable and/or transferred information; (2) through flow on the correlation between stream flows at different points and comparison with the case having no through flow; and (3) inclusion of zero as well as nonzero lag correlation between stream flows at different stations. Since stream flow is a continuous time series, it is important that a proper class interval be selected for discretizing the series, for the number of class intervals has a material influence on the values of transferable information used in Eq. (5.18) and the entropy coefficient given by Eq. (5.17).

Caution is required when establishing regression (linear or nonlinear) between stream flows of different stations for transfer of information. Let streamflow at an upstream station be designated as X, at a downstream station as Z, and at an intermediate station as Y. Then it is clear that $Z = X + Y$. For correlation between X and Z, one can write $Z = f(X)$ or $X = g(Z)$. Then, $X + Y = f(X)$ or $X = g(X + Y)$. For random variables X and Y, this illustrates the case of spurious correlation or redundancy in information, for through flow X appears on both sides. A larger value of X in comparison with Y will lead to a larger value of the correlation coefficient. This suggests that the effect of through flow must be carefully taken into account when doing regression.

Example 5.10. Suppose one has monthly stream flow observations from 2001 to 2005 at six stream gauges on the main stem of the Colorado River, as shown in Fig. 5.8. The data set can be downloaded from http://www.esrl.noaa.gov/psd/data/reanalysis/reanalysis.shtml. Rank these gauges in descending order of priority using MIMR with Eq. (5.26) as the objective function.

FIGURE 5.8   Stream gauges on the main stream of the Colorado River.

Solution

Step 1. Continuous time series discretization: Using the Scott algorithm [Eq. (5.31)] to determine the number of bins, the original stream flow time series are discretized and plotted in Figure 5.9.

Step 2. Candidate set and selected set initialization: At the initial state, the candidate set and selected set obviously are

$$F = \{\text{gauge 1, gauge 2, gauge 3, gauge 4, gauge 5}\}$$
$$S = \{\} \qquad \text{empty set}$$

Step 3. Central gauge identification: The central gauge is identified as one with the maximum marginal entropy. The marginal entropies for the candidate gauges are found and are listed:

	Gauge 1	Gauge 2	Gauge 3	Gauge 4	Gauge 5	Gauge 6
$H(x)$	2.1640	2.1946	2.1185	1.3298	1.4566	1.3659

Since gauge 2 has the highest marginal entropy, it is identified as the central station, that is, $s_1 = $ gauge 2. Then the candidate and selected sets can be updated as

$$F = \{\text{gauge 1, gauge 3, gauge 4, gauge 5, gauge 6}\}$$
$$S = \{\text{gauge 2}\}$$

Step 4. The second gauge selection: Here we arbitrarily use equal weights for information and redundancy, that is, $\lambda_1 = \lambda_2 = 1$.

Computation of Joint Entropy of Set S with Each Station in F   The computation of joint entropy has been explained in detail in Examples 5.7 and 5.8. We therefore omit the step-by-step calculation. The joint entropy of gauge 2 and each gauge in the candidate set F is as follows.

$$H(\text{gauge 2, gauge 1}) = 2.4157 \text{ bits}$$
$$H(\text{gauge 2, gauge 3}) = 2.2413 \text{ bits}$$
$$H(\text{gauge 2, gauge 4}) = 2.4096 \text{ bits}$$
$$H(\text{gauge 2, gauge 5}) = 2.5713 \text{ bits}$$
$$H(\text{gauge 2, gauge 6}) = 2.4624 \text{ bits}$$

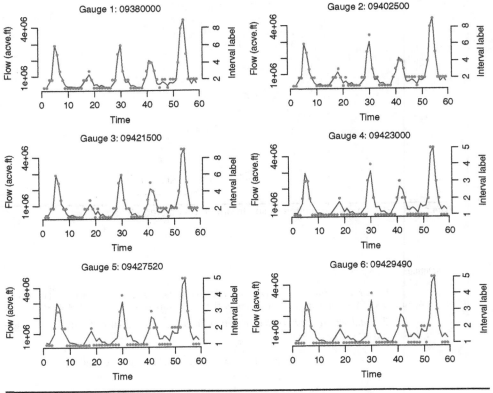

FIGURE 5.9 Discretization of time series.

Computation of Transinformation between (Stations in S and Each Station in F) and (the Other Stations in F) Again in this step we need to merge variables such that information retained in the merged variable is the same as that in the original variables. Assume gauge 1 is selected in this step. Then evaluate the predictive ability of S.

$$T(\text{gauge } 2 + \text{gauge } 1, \text{gauge } 3) = 2.0718 \text{ bits}$$
$$T(\text{gauge } 2 + \text{gauge } 1, \text{gauge } 4) = 1.1216 \text{ bits}$$
$$T(\text{gauge } 2 + \text{gauge } 1, \text{gauge } 5) = 1.1027 \text{ bits}$$
$$T(\text{gauge } 2 + \text{gauge } 1, \text{gauge } 6) = 1.1086 \text{ bits}$$

Assume gauge 3 is selected in this step. Then the predictive ability of S is evaluated.

$$T(\text{gauge } 2 + \text{gauge } 3, \text{gauge } 1) = 1.9429 \text{ bits}$$
$$T(\text{gauge } 2 + \text{gauge } 3, \text{gauge } 4) = 1.1615 \text{ bits}$$
$$T(\text{gauge } 2 + \text{gauge } 3, \text{gauge } 5) = 1.1266 \text{ bits}$$
$$T(\text{gauge } 2 + \text{gauge } 3, \text{gauge } 6) = 1.1448 \text{ bits}$$

Assume gauge 4 is selected in this step. Then the predictive ability of S is evaluated.

$$T(\text{gauge } 2 + \text{gauge } 4, \text{gauge } 1) = 1.9497 \text{ bits}$$
$$T(\text{gauge } 2 + \text{gauge } 4, \text{gauge } 3) = 2.1185 \text{ bits}$$
$$T(\text{gauge } 2 + \text{gauge } 4, \text{gauge } 5) = 1.2015 \text{ bits}$$
$$T(\text{gauge } 2 + \text{gauge } 4, \text{gauge } 6) = 1.1681 \text{ bits}$$

Assume gauge 5 is selected in this step. Then the predictive ability of S is evaluated.

$$T(\text{gauge } 2 + \text{gauge } 5, \text{gauge } 1) = 1.9657 \text{ bits}$$
$$T(\text{gauge } 2 + \text{gauge } 5, \text{gauge } 3) = 2.1185 \text{ bits}$$
$$T(\text{gauge } 2 + \text{gauge } 5, \text{gauge } 4) = 1.2364 \text{ bits}$$
$$T(\text{gauge } 2 + \text{gauge } 5, \text{gauge } 6) = 1.2642 \text{ bits}$$

Assume gauge 6 is selected in this step. Then the predictive ability of S is evaluated.

$$T(\text{gauge } 2 + \text{gauge } 6, \text{gauge } 1) = 1.9534 \text{ bits}$$
$$T(\text{gauge } 2 + \text{gauge } 6, \text{gauge } 3) = 2.1185 \text{ bits}$$
$$T(\text{gauge } 2 + \text{gauge } 6, \text{gauge } 4) = 1.1848 \text{ bits}$$
$$T(\text{gauge } 2 + \text{gauge } 6, \text{gauge } 5) = 1.2460 \text{ bits}$$

Computation of Total Correlation of All in S and Each Gauge in F   Since there is only one station in set S, the total correlation is reduced to transinformation. Therefore,

$$C(\text{gauge } 2 \text{ and gauge } 1) = 1.9429 \text{ bits}$$
$$C(\text{gauge } 2 \text{ and gauge } 3) = 2.0718 \text{ bits}$$
$$C(\text{gauge } 2 \text{ and gauge } 4) = 1.1148 \text{ bits}$$
$$C(\text{gauge } 2 \text{ and gauge } 5) = 1.0799 \text{ bits}$$
$$C(\text{gauge } 2 \text{ and gauge } 6) = 1.0981 \text{ bits}$$

Computation of Final Score for Each Gauge in F   From Eq. (5.26), we have

Gauge 1:

$H(\text{gauge } 2 \text{ and gauge } 1) + T(\text{gauge } 2 + \text{gauge } 1, \text{gauge } 3) + T(\text{gauge } 2 + \text{gauge } 1, \text{gauge } 4)$
$\quad + T(\text{gauge } 2 + \text{gauge } 1, \text{gauge } 5) + T(\text{gauge } 2 + \text{gauge } 1, \text{gauge } 6)$

$\quad - C(\text{gauge } 2, \text{gauge } 1)$

$\quad = 2.4157 + 2.0718 + 1.1216 + 1.1027 + 1.1086 - 1.9429$

$\quad = 5.8775 \text{ bits}$

Gauge 3:

$H(\text{gauge } 2 \text{ and gauge } 3) + T(\text{gauge } 2 + \text{gauge } 3, \text{gauge } 1) + T(\text{gauge } 2 + \text{gauge } 3, \text{gauge } 4)$
$\quad + T(\text{gauge } 2 + \text{gauge } 3, \text{gauge } 5) + T(\text{gauge } 2 + \text{gauge } 3, \text{gauge } 6)$

$\quad - C(\text{gauge } 2, \text{gauge } 3)$

$\quad = 2.2413 + 1.9429 + 1.1615 + 1.1266 + 1.1448 - 2.0718$

$\quad = 5.5453 \text{ bits}$

Gauge 4:

$H(\text{gauge } 2 \text{ and gauge } 4) + T(\text{gauge } 2 + \text{gauge } 4, \text{gauge } 1) + T(\text{gauge } 2 + \text{gauge } 4, \text{gauge } 3)$
$\quad + T(\text{gauge } 2 + \text{gauge } 4, \text{gauge } 5) + T(\text{gauge } 2 + \text{gauge } 4, \text{gauge } 6)$

$\quad - C(\text{gauge } 2, \text{gauge } 4)$

$\quad = 2.4096 + 1.9497 + 2.1185 + 1.2015 + 1.1681 - 1.1148$

$\quad = 7.7326 \text{ bits}$

Gauge 5:

$H(\text{gauge } 2 \text{ and gauge } 5) + T(\text{gauge } 2 + \text{gauge } 5, \text{gauge } 1) + T(\text{gauge } 2 + \text{gauge } 5, \text{gauge } 3)$
$\quad + T(\text{gauge } 2 + \text{gauge } 5, \text{gauge } 4) + T(\text{gauge } 2 + \text{gauge } 5, \text{gauge } 6)$

$\quad - C(\text{gauge } 2, \text{gauge } 5)$

$\quad = 2.5713 + 1.9657 + 2.1185 + 1.2364 + 1.2642 - 1.0799$

$\quad = 8.0762 \text{ bits}$

Gauge 6:

H(gauge 2 and gauge 6) + T(gauge 2 + gauge 6, gauge 1) + T(gauge 2 + gauge 6, gauge 3)
+ T(gauge 2 + gauge 6, gauge 4) + T(gauge 2 + gauge 6, gauge 5)

$- C$(gauge 2, gauge 6)

$= 2.4624 + 1.9534 + 2.1185 + 1.1848 + 1.2460 - 1.0981$

$= 7.8670$ bits

Clearly, gauge 5 has the highest score. Therefore, it is selected in set S. Then set F and set S can be updated as

$$F = \{\text{gauge 1, gauge 3, gauge 4, gauge 6}\}$$
$$S = \{\text{gauge 2, gauge 5}\}$$

Step 5. The third gauge selection: Following the same procedure as in step 4, we can select the third gauge using the MIMR criterion.

Computation of Joint Entropy of Set S with Each Gauge in F   Multivariate joint entropy can be computed using the shortcut formula between marginal entropies and total correlation.

$$H(\text{gauge 2, gauge 5, gauge 1}) = 2.7696 \text{ bits}$$
$$H(\text{gauge 2, gauge 5, gauge 3}) = 2.5713 \text{ bits}$$
$$H(\text{gauge 2, gauge 5, gauge 4}) = 2.6646 \text{ bits}$$
$$H(\text{gauge 2, gauge 5, gauge 6}) = 2.6730 \text{ bits}$$

Computation of Transinformation between Stations in S and Each Station in F and the Other Stations in F   Assume gauge 1 is selected in this step. Then the predictive ability of S is evaluated.

$$T(\text{gauge 2 + gauge 5 + gauge 1, gauge 3}) = 2.1185 \text{ bits}$$
$$T(\text{gauge 2 + gauge 5 + gauge 1, gauge 4}) = 1.2364 \text{ bits}$$
$$T(\text{gauge 2 + gauge 5 + gauge 1, gauge 6}) = 1.2642 \text{ bits}$$

Assume gauge 3 is selected in this step. Then the predictive ability of S is evaluated.

$$T(\text{gauge 2 + gauge 5 + gauge 3, gauge 1}) = 1.9657 \text{ bits}$$
$$T(\text{gauge 2 + gauge 5 + gauge 3, gauge 4}) = 1.2364 \text{ bits}$$
$$T(\text{gauge 2 + gauge 5 + gauge 3, gauge 6}) = 1.2642 \text{ bits}$$

Assume gauge 4 is selected in this step. Then the predictive ability of S is evaluated.

$$T(\text{gauge 2 + gauge 5 + gauge 4, gauge 1}) = 1.9657 \text{ bits}$$
$$T(\text{gauge 2 + gauge 5 + gauge 4, gauge 3}) = 2.1185 \text{ bits}$$
$$T(\text{gauge 2 + gauge 5 + gauge 4, gauge 6}) = 1.2642 \text{ bits}$$

Assume gauge 6 is selected in this step. Then the predictive ability of S is evaluated.

$$T(\text{gauge 2 + gauge 5 + gauge 6, gauge 1}) = 1.9657 \text{ bits}$$
$$T(\text{gauge 2 + gauge 5 + gauge 6, gauge 3}) = 2.1185 \text{ bits}$$
$$T(\text{gauge 2 + gauge 5 + gauge 6, gauge 4}) = 1.2364 \text{ bits}$$

Computation of Total Correlation of All Stations in S and Each Station in F   Assume gauge 1 is selected in this step. Then

$$C(\text{gauge 2, gauge 5, gauge 1}) = 3.0456 \text{ bits}$$

Assume gauge 3 is selected in this step. Then

$$C(\text{gauge 2, gauge 5, gauge 3}) = 3.1983 \text{ bits}$$

Assume gauge 4 is selected in this step. Then

$$C(\text{gauge 2, gauge 5, gauge 4}) = 2.3163 \text{ bits}$$

Assume gauge 6 is selected in this step. Then

$$C(\text{gauge 2, gauge 5, gauge 6}) = 2.3441 \text{ bits}$$

Computation of Final Score for Each Gauge in *F*

Gauge 1:

$H(\text{gauge 2, gauge 5, gauge 1}) + T(\text{gauge 2 + gauge 5 + gauge 1, gauge 3})$
  $+ T(\text{gauge 2 + gauge 5 + gauge 1, gauge 4}) + T(\text{gauge 2 + gauge 5 + gauge 1, gauge 6})$
  $- C(\text{gauge 2, gauge 5, gauge 1})$

  $= 2.7696 + 2.1185 + 1.1264 + 1.2642 - 3.0456$

  $= 4.2331 \text{ bits}$

Gauge 3:

$H(\text{gauge 2, gauge 5, gauge 3}) + T(\text{gauge 2 + gauge 5 + gauge 3, gauge 1})$
  $+ T(\text{gauge 2 + gauge 5 + gauge 3, gauge 4}) + T(\text{gauge 2 + gauge 5 + gauge 3, gauge 6})$
  $- C(\text{gauge 2, gauge 5, gauge 3})$

  $= 2.5713 + 1.9657 + 1.2364 + 1.2642 - 3.1983$

  $= 3.8393 \text{ bits}$

Gauge 4:

$H(\text{gauge 2, gauge 5, gauge 4}) + T(\text{gauge 2 + gauge 5 + gauge 4, gauge 1})$
  $+ T(\text{gauge 2 + gauge 5 + gauge 4, gauge 3}) + T(\text{gauge 2 + gauge 5 + gauge 4, gauge 6})$
  $- C(\text{gauge 2, gauge 5, gauge 4})$

  $= 2.6646 + 1.9657 + 2.1185 + 1.2642 - 2.3163$

  $= 5.6967 \text{ bits}$

Gauge 6:

$H(\text{gauge 2, gauge 5, gauge 6}) + T(\text{gauge 2 + gauge 5 + gauge 6, gauge 1})$
  $+ T(\text{gauge 2 + gauge 5 + gauge 6, gauge 3}) + T(\text{gauge 2 + gauge 5 + gauge 6, gauge 4})$
  $- C(\text{gauge 2, gauge 5, gauge 6})$

  $= 2.6730 + 1.9657 + 2.1185 + 1.2364 - 2.3441$

  $= 5.6495 \text{ bits}$

Obviously, gauge 4 has the highest score. Therefore, it is selected in set *S*. Then sets *F* and *S* can be updated as

$$F = \{\text{gauge 2, gauge 5, gauge 4}\}$$
$$S = \{\text{gauge 1, gauge 3, gauge 6}\}$$

Step 6. Select the fourth gauge.

Computation of Joint Entropy of Set *S* with Each Gauge in *F*

$$H(\text{gauge 2, gauge 5, gauge 4, gauge 1}) = 2.8630 \text{ bits}$$
$$H(\text{gauge 2, gauge 5, gauge 4, gauge 3}) = 2.6646 \text{ bits}$$
$$H(\text{gauge 2, gauge 5, gauge 4, gauge 6}) = 2.7663 \text{ bits}$$

Computation of Transinformation between Gauges in S and Each Gauge in F and the Other Gauges in F   Assume gauge 1 is selected in this step. Then the predictive ability of S is evaluated.

$$T(\text{gauge 2} + \text{gauge 5} + \text{gauge 4} + \text{gauge 1}, \text{gauge 3}) = 2.1185 \text{ bits}$$
$$T(\text{gauge 2} + \text{gauge 5} + \text{gauge 4} + \text{gauge 1}, \text{gauge 6}) = 1.2642 \text{ bits}$$

Assume gauge 3 is selected in this step. Then the predictive ability of S is evaluated.

$$T(\text{gauge 2} + \text{gauge 5} + \text{gauge 4} + \text{gauge 3}, \text{gauge 1}) = 1.9657 \text{ bits}$$
$$T(\text{gauge 2} + \text{gauge 5} + \text{gauge 4} + \text{gauge 3}, \text{gauge 6}) = 1.2642 \text{ bits}$$

Assume gauge 6 is selected in this step. Then the predictive ability of S is evaluated.

$$T(\text{gauge 2} + \text{gauge 5} + \text{gauge 4} + \text{gauge 6}, \text{gauge 1}) = 1.9657 \text{ bits}$$
$$T(\text{gauge 2} + \text{gauge 5} + \text{gauge 4} + \text{gauge 6}, \text{gauge 3}) = 2.1185 \text{ bits}$$

Computation of Total Correlation of All in S and Each Gauge in F   Assume gauge 1 is selected in this step. Then

$$C(\text{gauge 2}, \text{gauge 5}, \text{gauge 4}, \text{gauge 1}) = 4.2820 \text{ bits}$$

Assume gauge 3 is selected in this step. Then

$$C(\text{gauge 2}, \text{gauge 5}, \text{gauge 4}, \text{gauge 3}) = 4.4348 \text{ bits}$$

Assume gauge 6 is selected in this step. Then

$$C(\text{gauge 2}, \text{gauge 5}, \text{gauge 4}, \text{gauge 6}) = 3.5805 \text{ bits}$$

Computation of Final Score for Each Gauge in F

Gauge 1:

$H(\text{gauge 2}, \text{gauge 5}, \text{gauge 4}, \text{gauge 1}) + T(\text{gauge 2} + \text{gauge 5} + \text{gauge 4} + \text{gauge 1}, \text{gauge 3})$
  $+ T(\text{gauge 2} + \text{gauge 5} + \text{gauge 4} + \text{gauge 1}, \text{gauge 6}) - C(\text{gauge 2}, \text{gauge 5}, \text{gauge 4}, \text{gauge 1})$

$= 2.8630 + 2.1185 + 1.2642 - 4.2820$

$= 1.9637 \text{ bits}$

Gauge 3:

$H(\text{gauge 2}, \text{gauge 5}, \text{gauge 4}, \text{gauge 3}) + T(\text{gauge 2} + \text{gauge 5} + \text{gauge 4} + \text{gauge 3}, \text{gauge 1})$
  $+ T(\text{gauge 2} + \text{gauge 5} + \text{gauge 4} + \text{gauge 3}, \text{gauge 6}) - C(\text{gauge 2}, \text{gauge 5}, \text{gauge 4}, \text{gauge 3})$

$= 2.6646 + 1.9657 + 1.2642 - 4.4348$

$= 1.4597 \text{ bits}$

Gauge 6:

$H(\text{gauge 2}, \text{gauge 5}, \text{gauge 4}, \text{gauge 6}) + T(\text{gauge 2} + \text{gauge 5} + \text{gauge 4} + \text{gauge 6}, \text{gauge 1})$
  $+ T(\text{gauge 2} + \text{gauge 5} + \text{gauge 4} + \text{gauge 6}, \text{gauge 3}) - C(\text{gauge 2}, \text{gauge 5}, \text{gauge 4}, \text{gauge 6})$

$= 2.7663 + 1.9657 + 2.1185 - 3.5805$

$= 3.2700 \text{ bits}$

Gauge 6 has a larger final score. Therefore, gauge 6 is selected in this step. Then sets F and S can be updated as

$$F = \{\text{gauge 2, gauge 5, gauge 4, gauge 6}\}$$
$$S = \{\text{gauge 1, gauge 3}\}$$

Step 6. Select the fifth gauge.

Computation of Joint Entropy of Set *S* with Each Gauge in *F*

H(gauge 2, gauge 5, gauge 4, gauge 6, gauge 1) = 2.9647 bits

H(gauge 2, gauge 5, gauge 4, gauge 6, gauge 3) = 2.7663 bits

Computation of Transinformation between Gauges in *S* and Each Gauge in *F* and the Other Gauges in *F*   Assume gauge 1 is selected in this step. Then the predictive ability of *S* is evaluated.

$$T(\text{gauge 2} + \text{gauge 5} + \text{gauge 4} + \text{gauge 6} + \text{gauge 1}, \text{gauge 3}) = 2.1185 \text{ bits}$$

Assume gauge 3 is selected in this step. Then the predictive ability of *S* is evaluated.

$$T(\text{gauge 2} + \text{gauge 5} + \text{gauge 4} + \text{gauge 6} + \text{gauge 3}, \text{gauge 1}) = 1.9657 \text{ bits}$$

Computation of Total Correlation of All in *S* and Each Gauge in *F*   Assume gauge 1 is selected in this step. Then

$$C(\text{gauge 2, gauge 5, gauge 4, gauge 6, gauge 1}) = 5.5462 \text{ bits}$$

Assume gauge 3 is selected in this step. Then

$$C(\text{gauge 2, gauge 5, gauge 4, gauge 6, gauge 3}) = 5.6990 \text{ bits}$$

Computation of Final Score for Each Gauge in *F*
Gauge 1:

H(gauge 2, gauge 5, gauge 4, gauge 6, gauge 1)
   $+ T$(gauge 2 + gauge 5 + gauge 4 + gauge 6 + gauge 1, gauge 3)
   $- C$(gauge 2, gauge 5, gauge 4, gauge 6, gauge 1)

= 2.9647 + 2.1185 − 5.5462

= − 0.4630 bit

Gauge 3:

H(gauge 2, gauge 5, gauge 4, gauge 6, gauge 3)
   $+ T$(gauge 2 + gauge 5 + gauge 4 + gauge 6 + gauge 3, gauge 1)
   $- C$(gauge 2, gauge 5, gauge 4, gauge 6, gauge 3)

= 2.7663 + 1.9657 − 5.6990

= −0.9672 bit

It can be seen that the objective value associated with gauge 1 is larger than that with gauge 3. Therefore, at this step gauge 1 is selected. Since there are only 6 candidate gauges, among which 5 gauges have been ranked, the final rank for these 6 gauges is {gauge 2, gauge 5, gauge 4, gauge 6, gauge 1, gauge 3}.

5.9.4   Surface Water Quality Networks

In water quality management, gathering information on water quality variables and using that information for the planning, design, and operation of water and waste water treatment systems have long been practiced. Recently there has been an increasing concern about the impact of water quality on public health and general life conditions. Therefore, an understanding of how water quality processes evolve in both space and time under natural and artificial conditions is essential. Thus, there is emphasis these days on broader needs for water quality management, including design of monitoring systems and extraction of the most probable information from collected data. These help delineate the general nature and trend in water quality control measures. Sanders et al. (1983) noted that there has been much emphasis on determining how to monitor or how to collect data, but little on why we monitor or how we utilize data and the resulting impact on water quality management. Thus, it is important to better understand how water processes occur in time and space as well as to procure information from available water quality data.

Design of water quality monitoring networks is still a controversial issue, for there are difficulties in the selection of temporal and spatial sampling frequencies, the variables to be monitored, the sampling duration, and the objectives of sampling (Harmancioglu et al., 1999). Water quality data often present problems either in application of current techniques in hydrology or in transferring information between water quality-quantity or some water quality-quality variables. These problems may arise from the quality or quantity of data or from both.

Water quality processes are strongly subject to nonhomogeneities, both artificial and natural. Some water quality variables may be easily monitored, while others require laboratory analyses. Data may be inconsistent due to either errors in laboratory experimental analysis or changes in monitoring or laboratory practices. Nonhomogeneities and inconsistencies in water quality data must be evaluated before modeling or transferring information.

Some water quality variables are regularly monitored, while others are often sampled from time to time for laboratory analysis. In this case samples cover only a short period of observation time or have missing values. Compounding this problem is observation at unequal time intervals. A better understanding of how water quality processes occur under artificial and natural conditions can be realized by their structural analysis (detection of trends, seasonality, and dependence). Structural properties may significantly affect information transfer and may explain what affects relationships between water quality or water quality-quality variables.

Lee and Ellis (1997) compared kriging and the maximum entropy estimator for spatial interpolation and their subsequent use in optimizing monitoring networks. Husain (1989) and Bueso et al. (1999) used entropy theory to illustrate a framework for spatial sampling of a monitoring network. Ozkul et al. (2000) presented a methodology using entropy theory for assessing water quality monitoring networks. The work was a follow-up of earlier work by Harmancioglu and Alpaslan (1992).

Here the focus is on extraction of information contained in the systematic and/or sporadic observations of water quality and quantity variables; feasibility of information transfer from water quantity and water quality variables with long observations to short-length samples; determination of water quality variables that need to be continuously monitored, because of low information transferability by using all the other continuously and/or sporadically observed water quality and quantity variables; and measures of information transfer. The region of frequency of sampling, monitoring effectiveness and monitoring costs, and water quality index are some considerations in design.

5.9.5   Prediction of Water Quality Values at Discontinued Monitoring Stations

One may want to predict water quality changes at some upstream locations (tributaries) after the stations at those locations have been discontinued following a period of data collection sufficient to establish the baseline distribution of the water quality at each station. This may be needed in several cases where budget is either reduced or remains static, meaning a real decrease because of inflation, or transfer of an existing water quality monitoring station from one location to another in response to a more acute need for data at the new location. Furthermore, opportunities can be identified for discontinuing a number of stations when budgetary and other limitations are calling for a rationalization of the network design.

It is assumed that there is an observed change in the distribution of water quality observed at the downstream location, and probability distributions of water quality constituents are known from data prior to their discontinuance. Prior information about distributions of variables X_i, $i = 1, 2, \ldots$, can be considered using the Kullback-Leibler principle of minimum discrimination information (MDI):

$$\min H = \sum p_i \log \frac{p_i}{q_i} \tag{5.36}$$

or

$$\max H = -\sum_{i=1}^{n} p_i \log \frac{p_i}{q_i} \tag{5.37}$$

where q_i is the prior probability of x_i, which can be the probability of x_i known from a previous study or the probability of x_i estimated to be now occurring due to observed changes in the system.

Consider a river basin where water quality data have been collected to define the probability distributions on a number of major tributaries and in the upstream and downstream reaches of the main channel of the river. Consider, as an example, a river with two tributaries, as shown in Fig. 5.5. Assume that water quality at the downstream location a is a function of water quality at the two upstream locations b and c:

$$X_a = f(X_b, X_c) \tag{5.38}$$

This depends on the water quality parameter being monitored and physical conditions, such as flow, distance between stations, nature of pollutants as conservative, and time response of the pollutants as nonconservative. It is implied that there are pollutant inputs between b and c and a. Now the water quality level at the downstream main channel has significantly changed. This also implies that changes in water quality levels are occurring at location b or c or at both. Then one can write

$$\max H = -\sum_{j=1}^{n} \sum_{i=1}^{m} p_{ij} \log \frac{p_{ij}}{q_{ij}/m} \tag{5.39}$$

subject to

$$\sum_{i=1}^{n} p_{ij} = \frac{1}{m} \tag{5.40}$$

$$\frac{\sum_{i=1}^{n} p_{ij} x_{ij}}{\sum_{i=1}^{n} p_{ij}} = \mu_j \qquad j = 1, 2, \ldots, m \tag{5.41}$$

$$\frac{\sum_{i=1}^{n} p_{ij} x_{ij}^2}{\sum_{i=1}^{n} p_{ij}} = \mu_j^2 + \sigma_j^2 \qquad j = 1, 2, \ldots, m \tag{5.42}$$

$$\mu = f(\mu_1, \mu_2, ..., \mu_m) \tag{5.43}$$

$$0 \le p_{ij} \le 1 \qquad \text{for all } i, j \tag{5.44}$$

$$0 \le q_{ij} \le 1 \qquad \text{for all } i, j \tag{5.45}$$

$$\mu_j \ge 0 \qquad \text{for all } j \tag{5.46}$$

$$\sigma_j \ge 0 \qquad \text{for all } j \tag{5.47}$$

where x_{ij} = water quality level i at station j; q_{ij} = prior probability of x_{ij}; μ_j = mean water quality level at station j; μ = observed (changed) level at the downstream location; p_{ij} = probability of x_{ij} to be assigned given the mean water quality level μ; m = number of upstream stations; and n = number of intervals (discrete water quality values) at each station.

This formulation is applied separately for each water quality parameter of interest. For a given observed change in water quality level at a downstream location, probabilities are assigned to each of the possible water quality levels of each of the upstream stations. These new probabilities are then utilized to obtain new, unbiased estimates of the mean values of water quality at the upstream locations. The q_{ij} values can be those probabilities that existed prior to the observed downstream change in water quality, or when changes are known to have occurred upstream, the values of probabilities that can be associated with the new known conditions.

5.9.6 Groundwater Networks

Information measures, including transinformation, information transfer index, and correlation coefficient, have been applied to describe the variability of spatially correlated groundwater data. These measures can be calculated using two types of approaches: discrete and analytical. The discrete approach employs the contingency table, and the analytical approach employs the normal probability density function. Most studies have employed the analytical approach, which presumes knowledge of the probability distributions of random variables under study. The problem of not knowing the probability distributions can, however, be circumvented if a discrete approach is adopted. For the discrete approach, the transinformation and information transfer index can be employed to describe the spatial variability of data which are spatially correlated and which fit the normal distribution. The transinformation model (T model) is a relation between mutual information measures, specifically T, and the distance between wells. Transinformation is useful and comparable with correlation to characterize the spatial variability of data that are correlated with distance. Mogheir and Singh (2002) used entropy to evaluate and assess a groundwater monitoring network by means of marginal entropy contour maps.

To calculate information measures, the joint or conditional probability is needed, and one way is to use a contingency table. A two-dimensional contingency table is given in Table 5.16. To construct a contingency table, let the random variable x have a range of values consisting of v categories (class intervals), while the random variable y is assumed to have u categories (class intervals). The cell density or the joint frequency for (i, j) is denoted by $f_{ij}, i = 1, 2, ..., v; j = 1, 2, ..., u$, where the first subscript refers to the

		Y					
		1	**2**	**3**	**...**	**u**	**Total**
X	1	f_{11}	f_{12}	f_{13}	...	f_{1u}	$f_{1.}$
	2	f_{21}	f_{22}	f_{23}	...	f_{2u}	$f_{2.}$
	3	f_{31}	f_{32}	f_{33}	...	f_{3u}	$f_{3.}$
	.	.	.	.	.	.	.
	.	.	.	.	.	.	.
	.	.	.	.	.	.	.
	v	f_{v1}	f_{v2}	f_{v3}	...	f_{vu}	$f_{v.}$
	Total:	$f_{.1}$	$f_{.2}$	$f_{.3}$	...	$f_{.u}$	f_x or f_y

TABLE 5.16   Two-Dimensional Contingency Table (Frequency)

row and the second subscript to the column. The marginal frequencies are denoted by $f_{i.}$ and $f_{.j}$ for the row and the column values of the variables, respectively. Construction of a two-dimensional contingency table is illustrated using an example.

Questions

1. Take monthly discharge data for several gauging stations along a river. Compute the marginal entropy of monthly discharge at each gauging station, and plot it as function of distance between gauging stations. What do you conclude from this plot? Discuss it. Which station has the maximum entropy?

2. For the discharge data in Question 1, compute the transinformation of monthly discharge among stations. Are all gauging stations needed? What is the redundant information? What can be said about increasing or decreasing the number of gauging stations?

3. Compute the directional transinformation index (DIT) for different station pairs from Question 1.

4. Compute the informational correlation coefficient between different pairs in Question 1 and compare it with Pearson's correlation coefficient. Also compute the amount of nontransferred information.

5. Consider a drainage basin that has a number of rainfall-measuring stations. Obtain annual rainfall values for each gauging station. Using annual rainfall values, determine the marginal entropy at each station; also compute the transinformation among stations. Comment on the adequacy of the rain gauge network.

6. Consider the same basin and the rain gauge network as in Question 5. Now obtain monthly rainfall values and compute the marginal entropy as well as the transinformation. Comment on the adequacy of the rain gauge network. How does the adequacy change with reduced time interval?

7. Compute the directional transinformation index for different gauging station pairs from Question 5.

8. Compute the informational correlation coefficient between different pairs in Question 5 and compare it with Pearson's correlation coefficient. Also compute the amount of nontransferred information.

9. Take water level data for several monitoring wells in a basin. Compute the marginal entropy of the water level at each well. What do you conclude from these entropy values? Which well has the maximum entropy?

10. For the water level data in Question 9, compute the transinformation of the water level among wells. Are all wells needed? What is the redundant information? What can be said about increasing or decreasing the number of wells? Which wells are necessary and which are not?

11. Compute the directional transinformation index for different well pairs from Question 9.

12. Compute the information correlation coefficients between different pairs in Question 9, and compare them with the corresponding Pearson correlation coefficient. Also compute the amount of nontransferred information.

13. Take water quality data for several measuring sites in a basin. Compute the marginal entropy of water quality at each measuring site. What do you conclude from these values? Which site has the maximum entropy?

14. For the water quality data in Question 13, compute the transinformation among measuring sites. Are all measuring sites needed? What is the redundant information? What can be said about increasing or decreasing the number of measuring sites?

15. Compute the directional transinformation index for different measuring site pairs from Question 13.

16. Compute the information correlation coefficient between different site pairs in Question 13, and compare it with Pearson's correlation coefficient. Also compute the amount of nontransferred information.

17. Can entropy be employed for designing a monitoring network? If yes, how? Can entropy be employed for evaluating the adequacy of an existing network? If yes, how?

18. Consider a watershed that has a network of rainfall-measuring stations. Comment on the adequacy of the rain gauge network. Then design a rainfall network for this watershed.

19. Consider a drainage basin that has a network of stream flow measuring stations. Comment on the adequacy of stream gauging network. Then design a stream flow network for the basin.

20. Consider a watershed that has a number of groundwater quality monitoring stations. Comment on the adequacy of the monitoring network. Then design a groundwater quality monitoring network for the watershed.

21. Consider a watershed that has a network of soil moisture measuring stations. Comment on the adequacy of the soil moisture measuring network. Then design a soil moisture network for the watershed.

References

Adamowski, K. (1989). A Monte Carlo comparison of parametric and nonparametric estimation of flood frequencies. *Journal of Hydrology*, vol. 108, pp. 295–309.

Alfonso, J. L. (2010). *Optimisation of monitoring networks for water systems*. PhD dissertation, Delft University of Technology, Delft, The Netherlands, p. 185.

Alfonso, J. L., Lobbrecht, A., and Price, R. (2010). Optimization of water level monitoring network in polder systems using information theory. *Water Resources Research*, vol. 40, W12553, doi: 10.1010.1029/2009WR008953.

Al-Zahrani, M., and Husain, T. (1998). An algorithm for designing a precipitation network in the south-western region of Saudi Arabia. *Journal of Hydrology*, vol. 295, no. 3-4, pp. 205–216.

Bras, R. L., and Rodriguez-Iturbe, I. (1985). *Random Function and Hydrology*. Addison-Wesley, 559 pages.

Bueso, M. C., J. M. Angulo, J. Cruz-Sanjulian, and J. L. Carcia-Arostegui, (1999). Optimal spatial sampling design in a multivariate framework. *Mathematical Geology*, vol. 31, no. 5, pp. 507–525.

Burn, D. H., and Goulter, I. C. (1991). An approach to the rationalization of streamflow data collection networks. *Journal of Hydrology*, vol. 122, pp. 71–91.

Cacoullos, T. (1966). Estimation of a multivariate density. *Annals of the Institute of Statistical Mathematics*, vol. 18, pp. 179–189.

Cressie, N. C. (1990). *Statistics for Spatial Data*, rev. ed. Wiley, New York.

Guiasu, S. (1977). *Information Theory with Applications*. McGraw-Hill, London.

Harmancioglu, N. B. (1981). Measuring the information content of hydrological processes by the entropy concept. *Journal of Civil Engineering*, Faculty of Ege University, Special Issue: Centennial of Ataturk's Birth, Izmir, Turkey, pp. 13–40.

Harmancioglu, N. B., and Alpaslan, N. (1992). Water quality monitoring network design. *Water Resources Bulletin*, vol. 28, no. 1, pp. 179–192.

Harmancioglu, N. B., N. Alpaslan, and V. P. Singh (1992a). Application of the entropy concept in design of water quality monitoring networks. In: *Entropy and Energy Dissipation in Water Resources*, V. P. Singh and M. Fiorentino, eds. Kluwer Academic Publishers, Dordrecht, The Netherlands, pp. 283–302.

Harmancioglu, N. B., Singh, V. P., and Alpaslan, N. (1992c). Design of water quality monitoring networks. In: *Geomechanics and Water Engineering in Environmental Management*, R. N. Chowdhury, ed. A. A. Balkema Publishers, Rotterdam, The Netherlands, Chap. 8, pp. 267–296.

Harmancioglu, N. B., Alpaslan, N., Ozkul, S. D., and Singh V. P. (1997). *Integrated Approach to Environmental Data Management Systems*. NATO ASI Series, vol. 31, Kluwer Academic Publishers, Dordrecht, The Netherlands.

Harmancioglu, N. B., Fistikoglu, O., Ozkul, S. D., Singh, V. P., and Alpaslan, M. N. (1999). *Water Quality Monitoring Network Design*. Kluwer Academic Publishers, Boston.

Harmancioglu, N. B., and V. P. Singh (1990). *Design of Water Quality Networks*. Technical Report WRR14, Water Resources Program, Department of Civil Engineering, Louisiana State University, Baton Rouge.

Harmancioglu, N. B., and Singh V. P. (1991). An information based approach to monitoring and evaluation of water quality data. In: *Advances in Water Resources Technology*, G. Tsakiris, ed. A. A. Balkema, Rotterdam, The Netherlands, pp. 377–386.

Harmancioglu, N. B., and Singh V. P. (1998). Entropy in environmental and water resources. In: *Encyclopedia of Hydrology and Water Resources*, D. R. Herschy, ed. Kluwer Acdemic Publishers, Dordrecht, The Netherlands, pp. 225–241.

Harmancioglu, N. B., and Singh V. P. (2002). Data accuracy and validation. In: *Encyclopedia of Life Support Systems*, A. Sydow, ed. EOLSS Publishers Co., Ltd., Oxford, United Kingdom.

Harmancioglu, N. B., and Yevjevich V. (1985). Transfer of hydrologic information along rivers fed by karstified limestones. *Proceedings of the Ankara-Antalya Symposium on Karst Water Resources*, IAHS Publication 61, pp. 151–161.

Harmancioglu, N. B., and Yevjevich, V. (1987). Transfer of hydrologic information among river points. *Journal of Hydrology*, vol. 91, pp. 103–118.

Husain, T. (1987). Hydrologic network design formulation. *Canadian Water Resources Journal*, vol. 12, no. 1, pp. 44–63.

Husain, T. (1989). Hydrologic uncertainty measure and network design. *Water Resources Bulletin*, vol. 25, no. 3, pp. 527–535.

Kraskov, A. (2003). Hierarchical clustering based on mutual information. ArXiv preprint q-bio. QM/0311039.

Kraskov, A., Stogbauer, H., Andrzejak, R.G. and Grassberger, P. (2005). *Hierarchical clustering based on mutual information*. arXiv.q-bio/0311039v2 [q-bio.QM] 1 Dec. 2003.

Krstanovic, P. F., and Singh V. P. (1988). *Application of Entropy Theory to Multivariate Hydrologic Analysis*. Technical Reports WRR8 and WRR9, Department of Civil Engineering, Louisiana State University, Baton Rouge.

Krstanovic, P. F., and Singh V. P. (1992a). Evaluation of rainfall networks using entropy: 1. Theoretical development. *Water Resources Management*, vol. 6, pp. 279–293.

Krstanovic, P. F., and Singh V. P. (1992b). Evaluation of rainfall networks using entropy: 1. Application. *Water Resources Management*, vol. 6, pp. 295–315.

Lathi, B. P. (1969). *An Introduction to Random Signals and Communication Theory*. International Textbook Company, Scanton, Pa.

Lee, Y., and J. H. Ellis (1997). On the equivalence of Kriging and maximum entropy estimators. *Mathematical Geology*, vol. 29, no. 1, pp. 131–152.

Linfoot, E. H. (1957). An informational measure of correlation. *Information and Control*, vol. 1, pp. 85–89.

McGill, W. J. (1954). Multivariate information transmission. *Psychometrika*, vol. 19, pp. 97–116.

Mogheir, Y., and Singh V. P. (2002). Application of information theory to groundwater quality monitoring networks. *Water Resources Management*, vol. 16, no. 1, pp. 37–49.

Ozkul, S., Fistikoglu, O., Harmancioglu, N. B. and Singh V. P. (1996). Statistical evaluation of monitoring networks in space/time dimensions. *Proceedings, IAHR International Symposium on Stochastic Hydraulics*, K. S. Tickle, I. C. Goulter, C. C. Xu, S. A. Wasimi, and F. Bouchart, eds., pp. 357–365.

Ozkul, S., Harmancioglu, N. B., and Singh V. P. (2000). Entropy-based assessment of water quality monitoring networks. *Journal of Hydrologic Engineering*, ASCE, vol. 5, no. 1, pp. 90–100.

Ozkul, S., Harmancioglu, N. B., and Singh V. P. (2000a). Entropy-based assessment of water quality monitoring networks. *Journal of Hydrologic Engineering*, ASCE, vol. 5, no. 1, pp. 90–100.

Ozkul, S., Harmancioglu, N. B., and Singh V. P. (2000b). Entropy-based assessment of water quality monitoring networks in space/time dimensions. *Journal of Hydrologic Engineering, ASCE*, vol. 5, no. 1, pp. 90–100.

Parzen, E. (1962). On estimation of a probability density function and mode. *Annals of Mathematical Statistics*, vol. 33, pp. 1065–1076.

Rodriguez-Iturbe, I., and J. M. Mejia (1974). The design of rainfall networks in time and space. *Water Resources Research*, vol. 10, no. 4, pp. 713–728.

Sanders, T., Ward, R. C., Loftis, J. C., Steele, T. D., Adrian, D. D., and Yevjevich V. (1983). *Design of Networks for Monitoring Water Quality*. Water Resources Publications, Littleton, Colo.

Scott, D. W. (1992). *Multivariate Density Estimation: Theory, Practice and Visualization*. Wiley, New York.

Singh, V. P. (1998). *Entropy-Based Parameter Estimation in Hydrology*. Kluwer Academic Publishers, Boston.

Uslu, O., and A. Tanriover (1979). Measuring the information content of hydrological process. *Proceedings of the First National Congress on Hydrology*, Istanbul, pp. 437–443.

Watanabe, S. (1960). Information theoretical analysis of multivariate correlation. *IBM Journal of Research and development*, vol. 6, p. 66.

Wertz, W. (1979). *Statistical Density Estimation: A Survey*. Vandenhoeck & Ruprecht, Gottingen.

Yang, Y., and Burn D. H. (1994). An entropy approach to data collection network design. *Journal of Hydrology*, vol. 157, pp. 307–325.

Further Reading

Altiparmak, F., and B. Dengiz (2009). A cross entropy approach to design of reliable networks. *European Journal of Operations Research*, vol. 199, pp. 542–552.

Andricevic, R. (1990). Cost-effective network design for groundwater flow monitoring. *Stochastic Hydrology and Hydraulics*, vol. 4, pp. 27–41.

Bardsley, W. E., and Manly B. F. J. (1985). Note on selecting an optimum rain gauge subset. *Journal of Hydrology*, vol. 76, pp. 197–201.

Berthouex, P. M., and Brown, L. C. (1994). *Statistics for Environmental Engineers*. CRC Press, Boca Raton, Fla.

Bras, R. L., and I. Rodriguez-Iturbe (1976a). Rainfall network design for runoff prediction. *Water Resources Research*, vol. 12, no. 6, pp. 1197–1208.

Bras, R. L., and Rodriguez-Iturbe, I. (1976b). Network design for the estimation of areal mean rainfall. *Water Resources Research*, vol. 12, no. 6, pp. 1185–1195.

Caselton, W. F., and Husain, T. (1980). Hydrologic networks: Information transinformation. *Journal of Water Resources Planning and Management, ASCE*, vol. 106, no. WR2, pp. 503–529.

Chang, Y.-C., Wei, C. and Yeh H. C. (2008). Rainfall network design using kriging and entropy. *Hydrological Processes*, vol. 22, pp. 340–346.

Chapman, T.G. (1986). Entropy as a measure of hydrologic data uncertainty and model performance. *Journal of Hydrology*, vol. 85, pp. 111–126.

Cleveland, T. G., and Yes W. G. (1990). Sampling network design for transport parameter identification. *Journal of Water Resources Planning and Management*, vol. 116, no. 6, pp. 764–783.

Dymond, J. (1982). Rain gauge network reduction. *Journal of Hydrology*, vol. 57, pp. 81–91.

Eddy, A. (1976). Optimal rain gauge densities and accumulation times: A decision-making procedure. *Journal of Applied Meteorology*, vol. 13, pp. 962–971.

Fuentes, M., Chaudhuri, A. and Holland, D. M. (2007). Bayesian entropy for spatial sampling design of environmental data. Environmental and Ecological Statistics, vol. 14, pp. 323–340.

Gelati, G., and Rossi, G. (1986). Optimization of a snow network by multivariate statistical analysis. *Hydrological Sciences Journal*, vol. 33, no. 1/3, pp. 93–108.

Harmancioglu, N. B. (1984). Entropy concept as used in determination of optimum sampling intervals. *Proceedings of Hydrosoft '84*, International Conference on Hydraulic Engineering Software, Portoroz, Yugoslavia, pp. 6-99 to 6-110.

Harmancioglu, N. B., Alpaslan, N., and Singh V. P. (1992a). Design of water quality monitoring networks. In: *Geomechanics and Water Engineering in Environmental Management*, R. N. Chowdhury, ed. A. A. Balkema, Rotterdam, The Netherlands, Chap. 6, pp. 267–296.

Harmancioglu, N. B., Singh, V. P. and Alpaslan, N. (1992b). Versatile uses of the entropy concept in water resources. In: *Entropy and Energy Dissipation in Water Resources*, V. P. Singh and M. Fiorentino, eds. Kluwer Academic Publishers, Dordrecht, The Netherlands, pp. 91–118.

Harmancioglu, N. B., Alpaslan, N., and Singh V. P. (1994). Assessment of the entropy principle as applied to water quality monitoring network design. In: *Stochastic and Statistical Methods in Hydrology and Environmental Engineering*, vol. 3, K. W. Hipel, U. S. Panu, and V. P. Singh, eds., pp. 135–148.

Harmancioglu, N. B., N. Alpaslan, and Singh V. P. (1994). *Assessment of Entropy Principle as Applied to Water Quality Monitoring Network Design. Time Series Analysis in Hydrology and Environmental Engineering*, vol. 3, K. W. Hipel, A. I. McLeopd, U. S. Panu, and V. P. Singh, eds. Kluwer Academic Publishers, Dordrecht, The Netherlands, pp. 135–148.

Harmancioglu, N. B., Alkan, A., Singh, V. P. and Alpaslan N. (1996). Entropy-based approaches to assessment of monitoring networks. *Proceedings, IAHR International Symposium on Stochastic Hydraulics*, K. S. Tickle, I. C. Goulter, C. C. Xu, S. A. Wasimi, and F. Bouchart, pp. 183–190.

Harmancioglu, N. B., Fistikoglu, O. and Singh, V. P. (1998). Modeling of environmental processes. In: *Environmental Data Management*, N. B., Harmancioglu, V. P. Singh, and M. N. Alpaslan, eds. Kluwer Academic Publishers, Dordrecht, The Netherlands, Chap. 9, pp. 213–242.

Harmancioglu, N. B., Singh, V. P., and Alpaslan, N., eds. (1997). *Environmental Data Management*. Kluwer Academic Publishers, Dordrecht, The Netherlands.

Harmancioglu, N. B., M. N. Alpaslan, P. Whitfield, V. P. Singh, P. Literathy, N. Mikhailov, and M. Fiorentino, (1998). *Assessment of Water Quality Monitoring Networks—Design and Redesign*. Final Report to NATO, Brussels, Belgium.

Harmancioglu, N. B., and Singh V. P. (1999). On redesign of water quality networks. In: *Environmental Modeling*, V. P. Singh, I. L. Seo, and J. H. Sonu, eds. Water Resources Publications, Littleton, Colo., pp. 47–60.

Husain, T. (1979). Shannon's information theory in hydrologic design and estimation. Unpublished Ph.D., University of British Columbia, Vancouver, Canada.

Husain, T., and H. U. Khan (1983). Shannon's entropy concept in optimum air monitoring network design. *The Science of the Total Environment*, vol. 30, pp. 181–190.

Husain, T., and Ukayli , M. A. (1983). Meteorological network expansion for Saudi Arabia. *Journal of Research in Atmosphere*, vol. 16, pp. 281–295.

Husain, T., Ukayli, M. A. and Khan, H. U. (1986). Meteorological network expansion using information decay concept. *Journal of Atmospheric and Oceanic Technology, AMA*, vol. 3, no. 1, pp. 27–37.

Jessop, A. (1995). *Informed Assessments: An Introduction to Information, Entropy and Statistics*. Ellis Horwood, New York.

Kapur, J. N., and Kesavan, H. K. (1992). *Entropy Optimisation Principles with Applications*. Academic, San Diego, Calif.

Krstanovic, P. F., and Singh, V. P. (1992). Transfer of information in monthly rainfall series of San Jose, California. In: *Entropy and Energy Dissipation in Water Resources*, V. P. Singh and M. Fiorentino, Kluwer Academic Publishers, Rotterdam, The Netherlands, pp. 155–173.

Lubbe, C. A. (1996). *Information Theory*. Cambridge University Press, Cambridge, United Kingdom.

Markus, M., Knapp, V., and Tasker G. D. (2003). Entropy and generalized least square methods in assessment of the regional value of stream gauges. *Journal of Hydrology*, vol. 283, pp. 107–121.

Martin, W. D. (1996). An Evaluation Procedure for Determining the Adequacy of Alluvial River Sediment Data Sets. Unpublished Ph.D. dissertation, University of Memphis, Memphis, Tenn.

Masoumi, F., and Kerachian, R. (2010). Optimal redesign of groundwater quality monitoring networks: A case study. *Environmental Monitoring Assessment*, vol. 161, pp. 247–257.

Mishra, A. K., and Coulibaly, P. (2009). Developments in hydrologic network design: A review. *Reviews of Geophysics*, vol. 47, RG2001, Paper no. 2007RG000243, pp. 1–25.

Moss, M. E. (1982). Concepts and techniques in hydrological network design. *Operational Hydrology Report* no. 19, World Meteorological Organization, Geneva, Switzerland.

Moss, M. E., and Tasker, G. D. (1991). An intercomparison of hydrological network-design technologies. *Hydrological Sciences Journal*, vol. 36, no. 3/6, pp. 209–221.

Sarlak, N., and Sorman, A. U. (2006). Evaluation and selection of streamflow network stations using entropy methods. *Turkish Journal of Engineering and Environmental Science*, vol. 30, pp. 91–100.

Shamsi, U. M., Quimpo, R. G., and Yoganarasimhan, G. N. (1988). A application of kriging to rainfall network design. *Nordic Hydrology*, vol. 19, pp. 137–152.

Sorman, U., and Balkan, G. (1983). An application of network design procedures for redesigning Kizilimark River basin rain gauge network, Turkey. *Hydrological Sciences Journal*, vol. 28, no. 2/6, pp. 233–246.

Stol, P. T. (1972). The relative efficiency of the density of rain gauge networks. *Journal of Hydrology*, vol. 15, pp. 193–208.

Warrick, A. W., and Myers, D. E. (1987). Optimization of sampling locations for variogram calculations. *Water Resources Research*, vol. 23, no. 3, pp. 496–500.

Yoo, C., Jung, K., and Lee, J. (2008). Evaluation of rain gauge network using entropy theory: Comparison of mixed and continuous distribution function applications. *Journal of Hydrologic Engineering*, vol. 13, no. 4, pp. 226–235.

Rainfall and Evapotranspiration

CHAPTER 6

Precipitation Variability

Long-term hydrometeorological data sets are needed in climate studies. The variability of future climate is likely to be different from that of the past climate, particularly in terms of the rate and magnitude of changes in meteorological, hydrological, and environmental variables or parameters. One of the major consequences of climate change is the change in the hydrologic cycle that will have a significant impact on the amount, timing, duration, frequency, and space-time distribution of rain, evaporation, temperature, snowfall, and runoff, leading to the changes in the availability and distribution of water as well as in the competition for water resources. Changes are also likely in the timing, intensity, duration, and areal extent of water-related disasters, i.e., floods, droughts, debris flow, landslides, and associated changes in water quality. In general, global warming intensifies the global hydrologic cycle, thus increasing the globally averaged precipitation, evaporation, and runoff. The increase in precipitation, however, is not uniformly distributed across the world. Precipitation is expected to increase at higher latitudes and decrease at lower latitudes. One example of this change is the multidecadal decline in precipitation observed in the Sahel; the 1961–1990 average precipitation has decreased by 25 percent compared to earlier decades.

Trends of heavy precipitation events of 1 day (d) and multiple days in the United States and other countries show a tendency toward more days with heavy precipitation totals over the twentieth century. In some areas, such as Japan (Easterling et al., 2000a), there is, however, no increase in the seasonal total even though there is still an increase in the frequency of 1-d heavy precipitation events. Recent studies have reported large increases in precipitation and stream flow across the United States over the second half of the twentieth century, with the largest increases generally being reported in the fall precipitation and low flows.

Large increases have been observed in the fall precipitation and low flow at great many monitoring stations and in watersheds as well as in regional and national averages. Taken together, the preponderance of evidence suggests that trends in precipitation are spatially correlated across a large area of the central United States. The observed quasi-periodic variations in the climate system seem to be associated with episodic decadal to multidecadal periods of anomalous precipitation across the United States.

Since precipitation variability is one of the key variables driving various hydrologic processes, its effect on the global hydrologic and energy cycles is fundamental to understanding the behavior of the earth's climatic system. In his analysis of global data for low-frequency (decadal to multidecadal) variability of precipitation, Tsonis (1996) found no change in the global mean precipitation. However, fluctuations about the mean have significantly increased. The interannual variability of precipitation is the result of the internal dynamics of the atmosphere and the variability of the boundary forcings. This chapter discusses the utility of entropy in analyzing the variability of precipitation.

6.1 Measures of Variability

Generally, variability is construed as the quality of being uneven or lacking uniformity over different space and time scales. Spatial variability is characterized by different values of an observed precipitation or property measured at different locations. Thus, the spatial variability of precipitation in a region can be evaluated by considering individual contributions of each precipitation station to the overall precipitation, or spatially comparing the variability within precipitation records of individual stations. Temporal variability measures the unevenness of a precipitation time series over different time intervals and is often measured by different descriptive statistics, such as the range, mean, standard deviation, and coefficient of variability. Some of the methods commonly used for determining variability are now briefly discussed.

6.1.1 Variance

The most common statistic used to describe variability is the *variance* (and its derivatives, the standard deviation and coefficient of variation), which measures the spread in a data set. An example of a common approach based on variance is the principal-component analysis (PCA). Different variance methods have been widely used for analyzing the variability of rainfall, stream flow, runoff, evaporation, temperature, soil moisture, large-scale climate indices, drought indices, snow water equivalent, sea-level pressure, and sea level for mesoscale variability.

6.1.2 Diversity Indices

Diversity indices have been used to measure the biodiversity of an ecosystem, i.e., to assess the diversity of any population in which different members belong to unique species. Diversity increases as the number of species per sample increases and as the abundances of species within a sample become more even. Commonly used diversity measures include the Simpson, McIntosh, Berger-Parker, and Brillouin indices. These indices can also be employed for analyzing precipitation variability.

Simpson Index

The *Simpson index* (SI) of diversity (Simpson, 1949) characterizes the diversity or concentration of species in a community, and its numerical value increases as the diversity decreases. The index measures the probability that two individuals randomly selected from a sample will belong to the same species (or some other category of species). The proportion p_i of species i relative to the total number of species N is calculated and squared:

$$\text{SI} = 1 - \sum_{i=1}^{N} p_i^2 \tag{6.1}$$

SI varies on a probability scale from 0.0 to 1.0 in ascending order with increasing diversity.

Example 6.1. Compute the Simpson index for the data given in Table 6.1.

Solution. Calculation of the Simpson index is shown in Table 6.2.

$$\text{Simpson index} = 1 - 0.19 = 0.81$$

Year	January	February	March	April	May	June
1940	0.47	0.65	0.32	1.75	4.31	3.62
1941	0.51	1.55	3.61	3.46	1.76	5.42
1942	0.51	5.02	3.21	1.51	3.35	2.84
1943	0.97	0.55	0.64	1.95	2.73	3.65
1944	2.65	4.68	3.61	3.66	0.74	0.78
1945	0.56	2.1	1.05	2.61	2.39	0.06
1946	1.94	1.65	0.98	2.9	1.05	1.49
1947	1.54	3.76	0.82	0.24	1.99	0.54
1948	0.29	0	1.24	0.52	1.2	0.48
1949	2.6	0.94	1.16	0.26	1.28	0.98
1950	0.45	0.42	0.34	0.86	0.75	3.89
1951	0.71	0.19	0.39	1.92	4.56	2.83
1952	0.66	1.21	0.63	4.06	2.47	3.53
1953	1.43	0.91	0.59	1.17	3.58	2.26
1954	2.81	1.85	0.75	2.76	1.94	0.27
1955	6.52	1.44	3.32	1.68	1.45	1.17
1956	3.78	1.6	1.65	0.53	0.99	0.82
1957	4.3	2.68	2.38	1.16	0.05	1
1958	3.06	0.58	1.11	1.24	1.31	0.6
1959	4.68	0.61	0.01	1.25	0.98	0.89
1960	3.92	0.33	0.08	1.27	0.91	1.54
1961	0.24	1.38	0.73	3.53	0.55	3.91
1962	0.53	1.03	0.6	2.84	0.97	1.41
1963	0.53	0.42	1.17	0.61	1.83	0.77

TABLE 6.1   Monthly Rainfall (in.) for January to June from 1940 to 1963

McIntosh Index

The *McIntosh index* MI (McIntosh, 1967) is another measure of dominance, but it is not widely used, probably because its interpretation is not straightforward. Mathematically it is expressed as

$$\text{MI} = \frac{N - \left(\sum_{i=1}^{m} n_i^2 \right)^{1/2}}{N - N^{0.5}} \tag{6.2}$$

where n_i is the number of counts within the ith class interval, m is the number of class intervals, and N is the total number of counts consisting of all class intervals.

Example 6.2.   Compute the McIntosh index, using the data in Table 6.1.

Bin Number	Number of Counts n_i	p_i = Number of Counts/Total Number	p_i^2
1	40	0.266	0.071
2	41	0.273	0.074
3	23	0.153	0.023
4	9	0.060	0.003
5	12	0.080	0.006
6	15	0.100	0.010
7	5	0.033	0.001
8	3	0.020	0.0004
9	1	0.006	0.00004
10	1	0.006	0.00004
Total:	150		Sum = 0.19

TABLE 6.2   Calculation of the Simpson Index

Solution.   Here $N = 150$.

$$\sum n_i^2 = (40)^2 + (41)^2 + (23)^2 + (9)^2 + (12)^2 + (15)^2 + (5)^2 + (3)^2 + (1)^1 + (1)^1 = 4296$$

By substituting in Eq. (4.2), the McIntosh index is found to be

$$\mathrm{MI} = \frac{150 - (4296)^{0.5}}{150 - (150)^{0.5}} = 0.6131$$

Berger-Parker Index

The *Berger-Parker index* (BPI) (Berger and Parker, 1970) is a dominance measure, calculated as the ratio of the highest number of events based on rainfall amounts $n_{\max}$ within a class interval to the total number of the rainfall events N. Mathematically it is denoted as

$$\mathrm{BPI} = 1 - \frac{n_{\max}}{N} \tag{6.3}$$

Example 6.3.   Compute the Berger-Parker index for the data in Table 6.1.

Solution.   Here $N = 150$ and $n_{\max} = 41$. Therefore,

$$\mathrm{BPI} = 1 - \frac{n_{\max}}{N} = 0.7267$$

Brillouin Index

The *Brillouin index* (BI) (Berger and Parker, 1970) assumes a nonrandom sampling and is mathematically expressed as

$$\mathrm{BI} = \frac{\ln(N!) - \sum_{i=1}^{m} \ln n_i}{N} \tag{6.4}$$

where n_i counts the number of rainfall events within the ith individual class interval having a lower and upper limit and N is the total number of counts consisting of all class intervals.

Example 6.4. Compute the Brillouin index for the data in Table 6.1.

Solution. From Table 6.1,

$$\frac{\ln(N!) - \sum \ln n_i}{N} = \frac{\ln(150!) - \ln 40 + \ln 41 + \cdots + \ln 1}{150} = \frac{605.02 - 343.22}{150} = 1.74$$

6.2   Trend and Persistence

6.2.1   Mann-Kendall Test

The Mann-Kendall test is a nonparametric test (Mann, 1945; Kendall, 1975), and it has been widely used to test randomness versus trend. This test is robust to the influence of extremes, performs well with skewed variables due to its rank-based procedure, and has an ability to cope with missing values. The test statistic has zero mean, and its variance is calculated as

$$S = \sum_{k=1}^{n-1} \sum_{j=k+1}^{n} \text{sgn}(x_j - x_k) \tag{6.5}$$

$$\text{sgn}(x) = \begin{cases} 1 & \text{if} & x > 0 \\ 0 & \text{if} & x = 0 \\ -1 & \text{if} & x < 0 \end{cases} \tag{6.6}$$

$$\text{Var}(S) = \frac{1}{18}\left[n(n-1)(2n+5) - \sum_t t(t-1)(2t+5) \right] \tag{6.7}$$

where x_j and x_k are the sequential data values, n is the length of the data record, and t is the extent of any given time. The standard normal variate z is computed as

$$z = \begin{cases} \dfrac{S-1}{\sqrt{\text{Var}(S)}} & \text{if} & S > 0 \\ 0 & \text{if} & S = 0 \\ \dfrac{S+1}{\sqrt{\text{Var}(S)}} & \text{if} & S < 0 \end{cases} \tag{6.8}$$

The null hypothesis H_0 should be accepted if $|z| \le \alpha_0/2$ at the α_0 level of significance in a two-sided test for trend. A positive value of S indicates an upward trend, and a negative value indicates a downward trend.

Example 6.5.   A set of monthly rainfall data given in Table 6.3 is selected on a continuous time scale for trend analysis. The location is grid angle 105 (http://midgewater.twdb.texas.gov/Evaporation/evap.html), and units are inches. The selected data are for the month of January. Compute to see if there is a trend, using the Mann-Kendall test.

Year	Precipitation (in.)	Year	Precipitation (in.)	Year	Precipitation (in.)
1940	0.47	1948	1.54	1956	6.52
1941	0.51	1949	0.29	1957	3.78
1942	0.50	1950	2.60	1958	4.3
1943	0.97	1951	0.45	1959	3.06
1944	2.65	1952	0.71	1960	4.68
1945	1.0	1953	0.66	1961	3.92
1946	0.56	1954	1.43	1962	0.24
1947	1.94	1955	2.81	1963	0.21

Table 6.3 Precipitation for January from 1940 to 1963

Solution. The step-by-step method of calculation is as follows.

Step 1. The trend in the data depends on the value of S, with a high positive value indicating an increasing trend and a low negative value indicating a decreasing trend. Then for the Mann-Kendall statistic (S) Table 6.4 is constructed.

Step 2. Calculate the variance of the score Var(S).

$$\text{VAR}(S) = \frac{1}{18}\left[n(n-1)(2n+5) - \sum_{p=1}^{g} t_p (t_p - 1)(2t_p + 5)\right] = \frac{1}{18}\left[24(24-1)(2 \times 24 + 5) - 0\right] = 1625.33$$

In this example, n is 24 (total number of data points); g is 0 (number of tied group), and t_p is 0 (number of data points in the pth group). Using these values, one finds the second term in the equation is 0 and VAR(S) is equal to 1625.33.

Step 3. The next step is to calculate the normalized test statistic z, which is computed as

$$z = \begin{cases} \dfrac{S-1}{\sqrt{\text{Var}(S)}} & \text{if} \quad S > 0 \\ 0 & \text{if} \quad S = 0 \\ \dfrac{S+1}{\sqrt{\text{Var}(S)}} & \text{if} \quad S < 0 \end{cases} = \begin{cases} \dfrac{78-1}{\sqrt{1625.33}} = 1.91 \end{cases}$$

Since the calculated S value (= 78) is positive, by using the appropriate equation ($S > 0$) the z value is calculated as $78 - 1/\sqrt{1625.33} = 1.91$.

Step 4. Now the probability distribution function $F(z)$ associated with Z is computed. For a normal distribution, the probability density function with mean of 0 and standard deviation of 1 is given as

$$f(z) = \frac{1}{\sqrt{2\pi}}\exp\left(-\frac{z^2}{2}\right) = \frac{1}{\sqrt{2\pi}}\exp\left(-\frac{1.91^2}{2}\right) = 0.0644$$

Step 5. In general the probability level of significance is chosen as 95 percent. The trend is said to be decreasing if z is negative and the computed probability is greater than the level of significance. The trend is said to be increasing if z is positive and the computed probability is greater than the level of significance. If the computed probability is less than the level of significance, there is no trend. In this case the trend is increasing.

Data	$\sum sgn(x_j - x_k)$	Net Value of $\sum sgn(x_j - x_k)$
0.47	0	0
0.51	+1	1
0.50	+1−1	0
0.97	+1+1+1	3
2.65	+1+1+1+1	4
1	+1+1+1+1−1	3
0.56	+1+1+1−1−1−1	0
1.94	+1+1+1+1−1+1+1	5
1.54	+1+1+1+1−1+1+1−1	4
0.29	−1−1−1−1−1−1−1−1−1	−9
2.6	+1+1+1+1−1+1+1+1+1+1	8
0.45	−1−1−1−1−1−1−1−1−1+1−1	−9
0.71	+1+1+1−1−1−1+1−1−1+1−1+1	0
0.66	+1+1+1−1−1−1+1−1−1+1−1+1−1	−1
1.43	+1+1+1+1−1+1+1−1−1+1−1+1+1+1	6
2.81	+1+1+1+1+1+1+1+1+1+1+1+1+1+1+1	15
6.52	+1+1+1+1+1+1+1+1+1+1+1+1+1+1+1+1	16
3.78	+1+1+1+1+1+1+1+1+1+1+1+1+1+1+1+1−1	15
4.3	+1+1+1+1+1+1+1+1+1+1+1+1+1+1+1+1−1+1	16
3.06	+1+1+1+1+1+1+1+1+1+1+1+1+1+1+1−1−1−1−1	13
4.68	+1+1+1+1+1+1+1+1+1+1+1+1+1+1+1+1−1+1+1+1	18
3.92	+1+1+1+1+1+1+1+1+1+1+1+1+1+1+1+1−1+1−1+1−1	15
0.24	−1−1−1−1−1−1−1−1−1−1−1−1−1−1−1−1−1−1−1−1−1−1	−22
0.21	−1−1−1−1−1−1−1−1−1−1−1−1−1−1−1−1−1−1−1−1−1−1−1	−23
		Total S = 78

TABLE 6.4 Values of $sgn(x_j - x_k)$ and Sum of $sgn(x_j - x_k)$

6.2.2 Hurst Exponent

The *Hurst exponent* (HE) (Hurst, 1951), as an indicator of the persistence of a time series, is defined as the relative tendency of a time series to either regress to a long-term mean value or "cluster" in a particular direction. The values of the Hurst exponent range between 0 and 1. A Hurst exponent value HE close to 0.5 indicates a random walk (a brownian time series) where there is no correlation between any element and a future element and there is a 50 percent probability that future values will go either up or down. Series of this type are hard to predict. An HE value between 0 and 0.5 exists for time series with *anti-persistent behavior*. This means that an increase will tend to be followed by a decrease (or a decrease will be followed by an increase). This behavior is sometimes called *mean reversion*, which means that future values will have a tendency to return to a longer-term mean value. The strength of this mean reversion increases as HE approaches 0. An HE value between 0.5 and 1 indicates *persistent behavior*, suggesting that the time series is trending. If there is an increase from time step $t - 1$ to t, there will probably be an increase from t to $t + 1$. The same is true of decreases, where a decrease will tend to follow a decrease. The larger the HE value is, the stronger is the trend. Series of this type are easier to predict than series in the other two categories. The Hurst exponent for a time series can be computed as follows:

1. The time series is divided into d contiguous subseries of length n, where $d \times n = N$, the total length of the time series.

2. For each subseries m, where $m = 1, ..., d$, determine the mean E_m and the standard deviation S_m.

3. Normalize the data x_{im} of each subseries by subtracting the mean from each data point:

$$z_{im} = x_{im} - E_m \qquad i = 1, 2, ..., n$$

4. Using the normalized data, create a cumulative time series by consecutively summing the data points $y_{im} = \sum_{j=1}^{i} z_{jm}$.

5. Using the new cumulative series, find the range by subtracting the minimum value from the maximum value: $R_m = \max\{y_{1,m}, ..., y_{n,m}\} - \min\{y_{1,m}, ..., y_{n,m}\}$.

6. Rescale the range by dividing it by the standard deviation S_m.

7. Calculate the mean of the rescaled range for all subseries of length n:

$$\left(\frac{R}{S}\right)_n = \frac{1}{d} \sum_{m=1}^{d} \frac{R_m}{S_m} \tag{6.9}$$

8. The length of n must be increased to the next-higher value, where $d \times n = N$ and d is an integer value. Steps 1 to 6 are then repeated; these steps should be repeated until $n = N/2$.

9. Finally, the independent variable log n is plotted versus the dependent variable $\log(R/S)_n$, and a straight line is fitted using an ordinary least squares regression. The slope of the resulting line is the estimate of the Hurst exponent HE.

Example 6.6.   Compute the Hurst exponent for the data in Table 6.1.

Solution.   The Hurst exponent is computed in two ways, and both are illustrated here.

Method 1.   It entails the following steps.

1. A time series of full length N (= 24) is divided into a number of shorter time series of length $n = N, N/2, N/4, ...$ The average rescaled range is then calculated for each value of n. For illustrative purposes, here $n = N$ for the whole series is considered.

2. Determine the mean E_m and the standard deviation S_m of series $m = n$. $E_m = 1.908$ and $S_m = 1.728$.
3. Then normalize the data z_{im} by subtracting the mean from each data point, $z_{im} = x_{im} - E_m$, for $i = 1, \ldots, n$.
4. Using the normalized data, create a cumulative time series by consecutively summing the data points $y_{im} = \sum_{j=1}^{i} z_{jm}$. This is shown in Table 6.5.

Actual Data x_i	Normalized Data z_{im}	Cumulative Time Series y_{im}	y_{im} in Increasing Order	Calculate Range R_m
0.47	−1.44	−1.44	−12.35	3.37 − (−12.35)
0.51	−1.40	−2.84	−11.87	= 15.71
0.50	−1.41	−4.25	−11.44	
0.97	−0.94	−5.18	−10.62	
2.65	0.74	−4.44	−9.42	
1.00	−0.91	−5.35	−8.65	
0.56	−1.35	−6.70	−7.96	
1.94	0.03	−6.67	−7.04	
1.54	−0.37	−7.04	−6.83	
0.29	−1.62	−8.65	−6.70	
2.6	0.69	−7.96	−6.67	
0.45	−1.46	−9.42	−5.35	
0.71	−1.20	−10.62	−5.18	
0.66	−1.25	−11.87	−4.96	
1.43	−0.48	−12.35	−4.44	
2.81	0.90	−11.44	−4.25	
6.52	4.61	−6.83	−2.84	
3.78	1.87	−4.96	−2.57	
4.3	2.39	−2.57	−1.44	
3.06	1.15	−1.42	−1.42	
4.68	2.77	1.36	0.00	
3.92	2.01	3.37	1.36	
0.24	−1.67	1.70	1.70	
0.21	−1.70	0.00	3.37	
Mean = 1.908 Standard deviation = 1.728				

TABLE 6.5 Calculation of Range

Number of Subseries d	Length of Each Subseries n	$(R/S)n$	log n	log$(R/S)n$
1	24	9.09	3.18	2.21
2	12	Average (3.17 + 5.00) = 4.08	2.48	1.40
3	8	Average (2.30 + 2.18 + 2.87) = 2.45	2.07	0.89
4	6	Average (1.95 + 1.54 + 2.28 + 2.48) = 2.06	1.79	0.72
6	4	Average (1.49 + 1.61 + 1.28 + 1.43 + 1.41 + 1.71) = 1.48	1.38	0.39

TABLE 6.6 Calculation of R/S

5. Using the new cumulative series, find the range by subtracting the minimum value from the maximum value: $R_m = \max\{y_{1,m}, \dots, y_{n,m}\} - \min\{y_{1,m}, \dots, y_{n,m}\}$. This step is shown in Table 6.5.

6. Rescale the range R/S by dividing the range by the standard deviation: 15.71/1.728 = 9.09.

7. The value of R/S is calculated using the above steps for different subseries (d = 2) of length (n), and their mean is calculated as $(R/S)_n = 1/d \sum_{m=1}^{d} (R_m/S_m)$.

The calculation is shown in Table 6.6. Considering the length of time series is 24, and d = 2, there will be 24/2 = 12 (= n) values in each subseries. The first subseries will consist of the first 12 values of the data sets, and the second subseries will consist of the second half of the data sets consisting of 12 values.

8. Finally, log$(R/S)_n$ is plotted versus log n, and a straight line is fitted using a least squares regression. The slope of the resulting line is the estimate of the Hurst exponent HE, as shown in Fig. 6.1. The value of HE is 1.0, which signifies persistence.

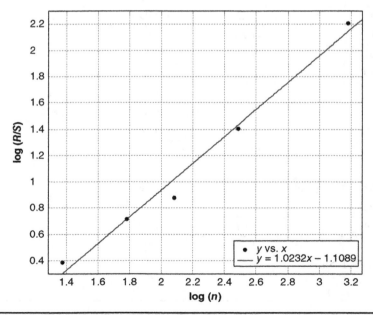

FIGURE 6.1 log (R/S) versus log n [x = log n and y = log(R/S)].

Length of Subseries n and Time Period Considered	R/S	log n	log (R/S)
5 (first five values of time series)	1.74	1.61	0.56
6 (first six values of time series)	1.95	1.79	0.67
7	2.26	1.95	0.81
8	2.30	2.08	0.83
9	2.68	2.20	0.99
10	3.21	2.30	1.17
11	2.63	2.40	0.97
12	3.18	2.48	1.16
13	3.55	2.56	1.27
14	3.93	2.64	1.37
15	3.83	2.71	1.34
16	2.68	2.77	0.98
17	4.10	2.83	1.41
18	5.17	2.89	1.64
19	6.21	2.94	1.83
20	6.87	3.00	1.93
21	7.76	3.04	2.05
22	8.49	3.09	2.14
23	8.79	3.14	2.17
24	9.09	3.18	2.21

TABLE 6.7   Calculation of R/S

Method 2.   Now another way to calculate the Hurst coefficient for the same set of data is illustrated here. By taking the length of data sets from 5 (first five values, 0.47, 0.51, 0.50, 0.97, and 2.65) to 6 (first six values), ... to 24 (first 24 values), R/S is computed following the same procedure as before, as shown in Table 6.7.

Using values from Table 6.7, a plot is constructed of $\log(R/S)_n$ versus $\log n$, and a straight line is fitted by the least square method, as shown in Fig. 6.2. The slope of the resulting line is an estimate of the Hurst exponent HE. From this figure, the value of HE is 1.0, which means persistence. Using both types of subseries, the result appears to be same.

6.3   Entropy Measures

The spatial and temporal variability of precipitation can be quantitatively measured by using entropy. Different precipitation time series (annual, seasonal, and monthly) can be considered individually, because it is useful to understand the uncertainty or variability of each time series and compare these series in terms of their variability. Then the questions concerning (1) the information of a precipitation time series, (2) the uncertainty associated with this information based on different temporal scales (decadal, yearly, seasonal, and monthly), (3) the precipitation variables having maximum variability, and (4) the effect of annual variability on water resources sectors can be further analyzed for decision making.

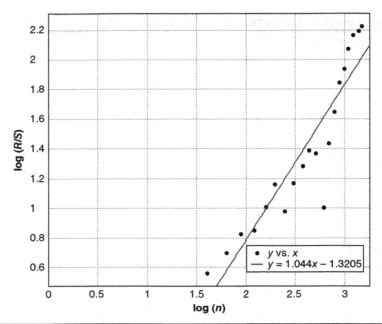

FIGURE 6.2   Plot of log (R/S) versus log n [$x = \log n$ and $y = \log(R/S)$].

6.3.1   Marginal Entropy

Entropy is a measure of uncertainty, dispersion, or diversification. A discrete form of the Shannon entropy $H(x)$ (Shannon, 1948) can be written as

$$H(X) = -\sum_{k=1}^{K} p(x_k) \log p(x_k) \tag{6.10}$$

where k denotes a discrete data interval; K is the number of intervals or outcomes that variable X can have; x_k is an outcome corresponding to interval k; and $p(x_k)$ is the probability of x_k, where the probability $p(x_k)$ can be based on the empirical frequency of the values of variable X. For continuous variables, such as stream flow discharge, a finite number of class intervals K must be chosen. Entropy $H(X)$ is also called the *marginal entropy* of a single variable X. The marginal entropy (ME) $H(X)$ of a single time series gives randomness associated with the entire length of the time series and can be used for any type of data set, e.g., yearly, monthly, seasonal, or daily.

Example 6.7.   Long-term records of monthly total precipitation from a network, based on a relatively uniform grid across Texas, have been collected from data sources for the Texas Water Development Board (TWDB) Freshwater Inflow Studies (http://midgewater.twdb.texas.gov/Evaporation/evap.html). For location, the grid angle of 105 is chosen. The values are in inches, and the monthly time scale is from January 1940 to June 1963. The data are given in Table 6.1. Compute the marginal entropy of the monthly rainfall data.

Bin Number	Bin Size	Bin Location (Center Point)	Number of Counts Bin	p_i = Number of Counts/Total Number	$-(p_i \log_2 p_i)$
1	0–0.652	0.326	40	0.266	0.352
2	0.652–1.304	0.978	41	0.273	0.355
3	1.304–1.956	1.63	23	0.153	0.288
4	1.956–2.608	2.282	9	0.060	0.169
5	2.608–3.260	2.934	12	0.080	0.202
6	3.260–3.912	3.586	15	0.100	0.230
7	3.912–4.564	4.238	5	0.033	0.113
8	4.564–5.216	4.89	3	0.020	0.078
9	5.216–5.868	5.542	1	0.006	0.033
10	5.868–6.520	6.194	1	0.006	0.033
		Total:	150		Sum = 1.854

TABLE 6.8 Empirical Frequency Table

Solution. In discrete form, the Shannon entropy $H(x)$ is given as

$$H(X) = -\sum_{i=1}^{K} p_i \log_2 p_i$$

where i denotes a discrete data interval, and p_i is an outcome corresponding to interval i, which is based on the empirical frequency of the variable X values, as computed in Table 6.8.

The marginal entropy is 1.845 bits.

6.3.2 Intensity Entropy

The *intensity entropy* (IE) is used for analyzing the disorder of the monthly rainfall amount or intensity. Considering the intensity (or amount) of monthly rainfall as a random variable, the IE at a rain gauge can be estimated as follows. Let the total number of rainy days in a 1 yr be denoted as N, the number of rainy days in a month i, for $i = 1, 2, 3, ..., 12$, of the year as n_i. Then the relative frequency for month i will be $p_i = n_i/N$. Probabilities p_i for a particular rain gauge are expressed in discrete form by taking into account all the rainy days at that rain gauge and their occurrence probabilities. Using the Shannon entropy, IE can be calculated as

$$\text{IE} = -\sum_{i=1}^{m} \frac{n_i}{N} \log_2 \frac{n_i}{N} \tag{6.11}$$

where m is the number of months. IE is uniquely estimated for individual months for the respective rain gauges.

6.3.3 Apportionment Entropy

If r_i is the aggregate rainfall (monthly rainfall) during the ith month in a year, the aggregate rainfall during the year (annual rainfall) R can be expressed by the sum of r_i from $i = 1$ to $i = 12$:

$$R = \sum_{i=1}^{12} r_i \tag{6.12}$$

The *apportionment entropy* (AE) can be written as

$$AE = -\sum_{i=1}^{12} \frac{r_i}{R} \log_2 \frac{r_i}{R} = -\sum_{i=1}^{12} p_i \log_2 p_i \tag{6.13}$$

AE measures the temporal variability of monthly rainfall over 1 yr (i.e., monthly-based 1-yr apportionment of annual rainfall). By definition, Eq. (6.13) states that when the annual rainfall amount is quite evenly apportioned to each of the 12 months with the probability of $\frac{1}{12}$, AE takes on its maximum value of $H = \log_2 12$. The minimum value of $AE = 0$ occurs when the apportionment is made to only 1 out of the 12 months with a probability of 1. This indicates that AE takes on a value within a finite range of 0 and $\log_2 12$.

6.3.4 Decadal Apportionment Entropy

Using the *decadal apportionment entropy* (DAE), the randomness of the time series data on a decadal basis can be measured, and this can be applied to time series of any hydrometeorological variable. Considering the annual time series of a station (a_i being the annual rainfall for year i), the aggregate *decadal rainfall* (DR) over 10 yr can be expressed by the sum of r_i from $i = 1$ to $i = 10$:

$$DR = \sum_{i=1}^{10} a_i \tag{6.14}$$

The ratio a_i/DR thus becomes an occurrence probability of the outcome. Letting this ratio be d_i and employing the Shannon discrete entropy, DAE can be written for measuring the temporal variability of decadal annual rainfall over 10 yr (i.e., yearly-based, over-10-yr apportionment of annual rainfall)

$$DAE = \sum_{i=1}^{10} d_i \log d_i = -\sum_{i=1}^{10} \frac{a_i}{DR} \log_2 \frac{a_i}{DR} \tag{6.15}$$

DAE can be applied to months as well. For example, by considering individual month time series (i.e., taking all January months) of all years, it is possible to explore how much the randomness occurred for that month over a decade.

6.3.5 Disorder Index

Here the variability is expressed by the *disorder index*, defined as

Disorder index (DI) = (maximum possible entropy value under evenly apportioned state)

– (actual entropy value obtained for time series)

The maximum possible value of marginal entropy for bin size $= 15$ is $\log_2 15 = 3.9069$; AE and IE $= 3.5850 \log_2 12$; DAE $= 3.3219 \log_2 10$. When the disorder index is calculated from the marginal entropy, it is known as the *marginal disorder index* (MDI); likewise, it is known as the *apportionment disorder index* (ADI) or *intensity disorder index* (IDE) based on the apportionment entropy or intensity entropy. When the decadal analysis is carried out based on decadal apportionment entropy, it is known as the *decadal apportionment disorder index* (DADI).

The higher the disorder index value is, the higher will be the variability. The marginal disorder index is calculated based on marginal entropy for annual, seasonal, and monthly time series. This determines the variability associated with individual series. Spatial and temporal variability can be compared based on the *average disorder index*, defined as

$$ \text{MDI} = \frac{1}{N} \sum_{i=1}^{N} \text{DI}_i \qquad (6.16) $$

where MDI is the mean disorder index and N is defined as the length of entropy time scales in the spatial or temporal domain.

6.4   Comparison of Entropy and Diversity Indices

A diversity index value equal to 1 implies complete evenness (i.e., equivalent amount of rain in each month), and an index equal to 0 implies complete unevenness (i.e., all rain in one month). Dominance measures weight the diversity index toward the dominant month (i.e., the month with the most rainfall). This means that as the amount of rainfall increases in one month, and the amounts in all the other months stay constant, a dominance diversity index decreases faster than does a nondominance measure (i.e., the more uneven rainfall will appear).

Entropy measures use the natural logarithm of monthly proportional rainfall. Both dominance and information measures account for the number of months and the proportion of the total amount of rainfall in each month. Dunbar (1992) used the Simpson index to represent the evenness of rainfall in a study of the correlation between meteorological variation and behavioral variation in baboons. There are two major disadvantages of the Simpson index: (1) its potential maximum is always less than 1, and (2) it is dependent on the number of months with rainfall. Even if rain fell in every month of the year, the range of the Simpson index is not from 0 to 1, but from 0 to a maximum value of $11/12$, or 0.917. Because the Simpson index does not take dry months into account at all, it actually calculates evenness only across months with rain. In addition, the spread of the McIntosh index is dependent on the total amount of rainfall, a problem circumvented by the indices that use proportional rainfall. The Brillouin index has a spread similar to the Shannon entropy, and it also ranges from 0 to 1, but it is not well suited to large amounts of rainfall and has no obvious benefit over the Shannon entropy. It is found that the normalized Shannon entropy is correlated with other normalized diversity indices. The coefficient of correlation between the Shannon entropy and other indices can be as high as 0.99 with the Brillouin index; 0.91 with the Simpson index; 0.94 with the McIntosh index; 0.78 with the Berger-Parker index; and -0.85 with the coefficient of variation.

6.5 Comparison of Entropy and Variance

The variance is a measure of dispersion, and its simplicity remains its major attraction. Historically, the variance has had a primordial role in the analysis of variability. Both entropy and variance reflect concentration, but their metrics for concentration are different. Unlike the variance, which measures concentration only around the mean, entropy measures the diffuseness of a probability density function irrespective of the location of concentration. However, there is no universal relationship between these two measures in terms of the ordering of distributions.

Ebrahimi et al. (1999) found that, under certain conditions, the order of variance and entropy is similar when continuous variables are transformed; but entropy could offer, unlike variance, a much closer characterization of the probability density function, since it uses much more information about the probability distribution than does variance. Another advantage of entropy over variance is that entropy accounts for higher-order moments of a probability distribution function. In terms of mathematical properties, entropy is nonnegative in the discrete case. Also entropy is invariant under a one-to-one transformations of X, but variance is not. For the continuous case, entropy can be made invariant under one-to one transformation, but variance cannot (Dionisio et al., 2006).

It is important to note some properties of the variance (and standard deviation) and entropy as measures of uncertainty. The standard deviation is a convex function that, according to the Jensen inequality $E[\sigma(X)] \geq \sigma[E(X)]^3$, where E is the expectation, X is a random variable, and σ is a convex function, that allows the variance and the standard deviation to be used as risk measures in stock portfolios, since they take into account the effect of diversification.

6.6 Rainfall Disaggregation

Since continuous precipitation data at fine time scales are often needed for water resources planning, design, and management, a number of disaggregation techniques have been developed that generate fine-scale precipitation data from available coarser-scale observed data. Here we focus on three entropy-based disaggregation techniques. The second methodology follows Montesarchio (2011, 2012). This methodology makes use of the additive property of Shannon (1948) entropy to develop a regression equation between the apportionment entropy AE and various time scales. By building these regression equations it becomes possible to obtain daily rainfall probabilities given monthly AE, and thus ultimately the daily rainfall amounts. This is a single-site rainfall disaggregation methodology. In the third case, the influence of the neighboring stations is considered. Hence, regression equations are determined by considering data from two of the nearest rainfall stations.

6.6.1 Random Allocation of Monthly Rainfall

This methodology follows the assumption that rainfall is a random process and can be likened to randomly picking up lottery tickets from a set. If monthly rainfall is to be disaggregated into daily values, then the method involves a random allocation of rainfall to days in a month. To illustrate, a simple example is considered. Assume that a particular month received 60 mm of rainfall. Let it be said that the month has 30 days, and let n be the number of trials for which the experiment of allocating rainfall to days is repeated. Each day can be allotted an amount of $60/n$ mm of rainfall. Now consider these 30 days as 30 lottery tickets, and randomly pick a day. Let us say we picked up the 10th day. Then this 10th day is allotted $60/n$ mm of rainfall, and the procedure is continued for n number of trials. A single day can be picked any number of times. Since rainfall is a random process, maximizing the randomness means placing minimum constraints on the unknown daily rainfall time series and hence subjecting it to minimum bias. The distribution that the final time series follows will depend on the number of trials for which the experiment is repeated. The higher the number of trials, the more uniform the time series becomes. There appear, however, no rigid guidelines to fix a priori the number of trials. It would be a location-dependent factor.

6.6.2 Single-Site Disaggregation

Let R be the total monthly rainfall and $r_1, r_2, ..., r_k$ be rainfall for 1, 2, ..., r days, respectively. The relative frequency (or probability) can be calculated as

$$p_i = \frac{r_i}{R} \tag{6.17}$$

where r_i is the rainfall for the ith day and p_i is the rainfall occurrence probability for the ith day. The apportionment entropy AE for the month of 30 days can be calculated as

$$\text{AE} = -\sum_{i=1}^{30} \frac{r_i}{R} \log_2 \frac{r_i}{R} = -\sum_{i=1}^{30} p_i \log_2 p_i \tag{6.18}$$

The additivity property of entropy can be invoked to develop a framework for monthly to daily disaggregation (Jaynes, 2003).

$$H(p_1, p_2, ..., p_n) = H(w_1, w_2, ..., w_k) + w_1 H\left(\frac{p_1}{w_1}, \frac{p_2}{w_1}, ..., \frac{p_k}{w_1}\right)$$

$$+ w_2 H\left(\frac{p_{k+1}}{w_2}, ..., \frac{p_{k+w}}{w_2}\right) + \cdots \tag{6.19}$$

where w_i represents the probability for each possible alternative considered for disaggregation. The monthly probabilities can thus be broken down as shown in Table 6.9.

For disaggregation from 30 to 15 d, two partitions are involved, and one can write

$$H(p_1, p_2) = -p_1 \log_2 p_1 - p_2 \log_2 p_2 \qquad p_1 + p_2 = 1 \tag{6.20}$$

Likewise, for disaggregation from 15 to 5 d, three partitions are involved for each of the two parts:

$$H(p_1, p_5, p_6) = H(p_1, p_2) + p_2 H\left(\frac{p_5}{p_2}, \frac{p_6}{p_2}\right) \qquad p_5 + p_6 = p_2 \tag{6.21}$$

$$H(p_2, p_3, p_4) = H(p_1, p_2) + p_1 H\left(\frac{p_3}{p_1}, \frac{p_4}{p_1}\right) \qquad p_3 + p_4 = p_1 \tag{6.22}$$

In a similar manner, disaggregation to smaller number of days can be achieved. By building a regression equation linking the entropy values with various time scales ranging from 15 d to 1 d, it is possible to obtain daily rainfall values from a monthly value.

p_{15}	p_{16}	p_{17}	p_{18}	p_{19}	p_{20}	p_{21}	p_{22}	p_{23}	p_{24}	p_{25}	p_{26}	p_{27}	p_{28}	p_{29}	p_{30}	p_{31}	p_{32}	p_{33}	p_{34}	p_{35}	p_{36}	p_{37}	p_{38}
p_7			p_8		p_9		p_{10}					p_{11}			p_{12}			p_{13}			p_{14}		
p_3					p_4							p_5						p_6					
p_1												p_2											

TABLE 6.9 Partitioning of Probabilities

Consider now the case of disaggregation of a 24-h value into hourly values. As shown in Table 6.9, terms p_1 and p_2 are the 12-h rainfall probabilities. Knowing the entropy associated with daily rainfall for the station under consideration, p_1 and p_2 can be calculated using the system of two equations in Eq. (6.20) where $H(p_1, p_2)$ is the 24-h rainfall entropy. Solution of these two equations will give the occurrence frequencies for 12-h rainfall. The 12-h rainfall probabilities can be further broken down into 6-h probabilities. Let p_1 be split up into p_3 and p_4, and let p_2 be split into p_5 and p_6. Then p_3, p_4, p_5, and p_6 are the 6-h occurrence probabilities whose values can be obtained by solving the system of four equations in Eqs. (6.21) and (6.22). Solution of these four equations will give the 6-h rainfall occurrence probabilities. These probabilities can further be broken down to 3-h rainfall probabilities, namely, from p_7 to p_{14}. Finally the 3-h occurrence probabilities can be broken down to 1-h probabilities, namely, p_{15} to p_{38}, as shown in Table 6.9.

6.6.3 Disaggregation Considering the Influence of Neighboring Stations

This case follows the same procedure as described above. Additionally the influence of neighboring rain gauge stations (the nearest two) is also taken into account while deriving the regression equation between entropy and time scales. For a rain gauge station, Liberty County (ID: 415196) in Texas, for the month of January for the year 1995, the methods of disaggregation were applied. The total rainfall for the month was 36 in., which was divided into $36/n$, where n is the number of trials for which the experiment was conducted. This amount was allocated to days in January by randomly choosing the days. Each time a day was picked it was allotted a rainfall amount of $36/n$ mm. If a day was chosen twice, then the rainfall amount for the day became $72/n$ mm, and so on. The experiment was conducted for n values ranging from 5 to 100. The number of trials for which the result closely matched the actual distribution was finally chosen. In the case of January 1995, an n value of 25 gave a better match with the actual daily rainfall distribution. The procedure was repeated for the January months belonging to several years, so that an idea of how the optimal number of trials varied was arrived at.

For the month of January at Liberty, the regression equation linking entropy values with different time scales t is plotted in Fig. 6.3.

$$H = -0.8 \log t + 3.592 \tag{6.23}$$

Considering the data obtained from two of the nearest stations, the regression model became

$$H = -1.07 \log t + 4.634 \tag{6.24}$$

Figure 6.4 shows the regression of Eq. (6.24).

Example 6.8.   Using rainfall data for January 27, 1949, from rain gauge station Abilene (TX410016), Texas, disaggregate the 24-h value into hourly values, using the random allocation method, and compare with the disaggregated values obtained, using the soil conservation service (SCS) method of rainfall disaggregation. Table 6.10 shows the rainfall data.

In the SCS method of rainfall distribution, daily (24-h) rainfall is distributed over hours based on the SCS rainfall curves. For the Texas region within which the station lies, the rainfall distribution follows a type 2 curve; and from its tabulated values, the fraction of cumulative rainfall for each hour can be determined. Table 6.11 shows the percentage of rainfall and cumulative rainfall distributed for each hour over a 24 h period.

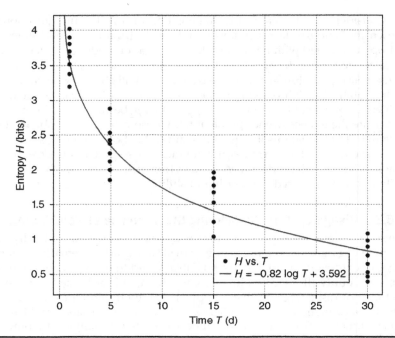

FIGURE 6.3   Regression between entropy and time scales for January for Liberty (x = time in days and y = entropy in bits).

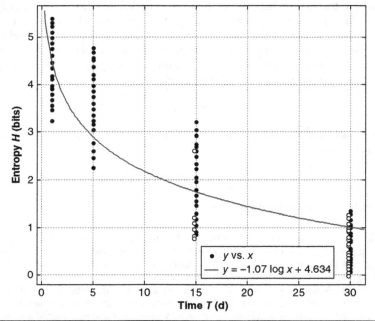

FIGURE 6.4   Regression model between entropy and time scales for January considering the influence of neighboring stations (x = time in days and y = entropy in bits).

Hour	Rainfall (mm)	Hour	Rainfall (mm)	Hour	Rainfall (mm)	Hour	Rainfall (mm)
1	1	7	3	13	0	19	0
2	1	8	2	14	0	20	0
3	2	9	0	15	0	21	0
4	1	10	0	16	0	22	0
5	0	11	0	17	0	23	0
6	1	12	0	18	0	24	0

TABLE 6.10   Rainfall Data for January 27, 1949

Hour	Percent of Hourly Rainfall	Percent of Cumulative Rainfall	Hour	Percent of Hourly Rainfall	Percent of Cumulative Rainfall	Hour	Percent of Hourly Rainfall	Percent of Cumulative Rainfall
1	1.33	1.33	9	2.67	14	17	2	90.66
2	1.33	2.66	10	4	18	18	1.34	92
3	1.34	4	11	6	24	19	1.33	93.33
4	1.33	5.33	12	26	50	20	1.33	94.66
5	1.33	6.66	13	26	76	21	1.34	96
6	1.34	8	14	6	82	22	1.33	97.33
7	1.33	9.33	15	4	86	23	1.33	98.66
8	2	11.33	16	2.66	88.66	24	1.34	100

TABLE 6.11   Rainfall Distribution Based on SCS Type 2 Curve

A sample calculation for disaggregation is now illustrated. For observation number 1, rainfall for hour 1 is 1 mm and the total daily rainfall is 11 mm. According to the SCS curve, the percentage of total rainfall assigned for hour 1 is 1.33. The disaggregated cumulative rainfall for hour 1 is 1.33 × 11/100 = 0.1463 mm. Table 6.12 shows the disaggregated hourly rainfall values using the SCS method.

Figure 6.5 shows the observed and disaggregated rainfall values, and Fig. 6.6a shows the cumulative rainfall for the observed and disaggregated values.

Table 6.13 shows error statistics, including the mean, variance, and skewness, for the SCS method of rainfall disaggregation.

Now the random allocation method is illustrated. To that end, a simple example is considered first. Assume that a particular day received 24 mm of rainfall, and let there be n number of trials for which the experiment is repeated. This means that each hour can be allotted $24/n$ mm of rainfall. Now consider these 24 h as 24 lottery tickets, and randomly pick a day. Let us say we picked the hour 10. Then that hour is allotted $24/n$ mm of rainfall, and the procedure is continued for the n number of trials. A single hour can be picked any number of times. In this method we are assuming rainfall as a random process. The disaggregation will depend on the number of trials for which the experiment is repeated. The higher the number of trials, the more uniform the distribution of disaggregated values will become. There appear, however, no rigid guidelines to fix a priori the number of trials. It would be a location-dependent factor. Now, for the same sample data given in Table 6.10, the randomly picked hours for different trial numbers that will be used are shown in Table 6.14. In the first case

Hour	Actual Rainfall (mm)	Cumulative Disaggregated Rainfall (mm)	Disaggregated Rainfall (mm)	Error (mm)
1	1	0.146	0.146	−0.854
2	1	0.2936	0.146	−0.854
3	2	0.440	0.147	−1.853
4	1	0.586	0.146	−0.854
5	0	0.733	0.146	0.146
6	1	0.880	0.147	−0.853
7	3	1.026	0.146	−2.854
8	2	1.246	0.220	−1.780
9	0	1.540	0.294	0.294
10	0	1.980	0.440	0.440
11	0	2.640	0.660	0.660
12	0	5.500	2.860	2.860
13	0	8.360	2.860	2.860
14	0	9.020	0.660	0.660
15	0	9.460	0.440	0.440
16	0	9.753	0.293	0.293
17	0	9.973	0.220	0.220
18	0	10.120	0.147	0.147
19	0	10.266	0.146	0.146
20	0	10.413	0.146	0.146
21	0	10.560	0.147	0.147
22	0	10.706	0.146	0.146
23	0	10.853	0.146	0.146
24	0	11.000	0.147	0.147

TABLE **6.12** Disaggregation Using the SCS Method

where $n = 5$, each hour, when picked once, will be assigned $11/5 = 2.2$ mm of rainfall. In the second case where $n = 7$, each hour, when picked, will be assigned $11/7 = 1.572$ mm of rainfall. In the third case where $n = 10$, each hour, when picked, will be assigned, $11/10 = 1.1$ mm of rainfall. Table 6.15 shows the disaggregated values for each of the trials.

Figure 6.6*b* shows the observed and disaggregated rainfall values, and Fig. 6.7 shows the cumulative rainfall for observed and disaggregated values.

Table 6.16 shows error statistics, including the mean, variance, and skewness, for the random allocation method of rainfall disaggregation.

Example 6.9. Disaggregate the rainfall given in Example 6.8 using the entropy-based single-site disaggregation.

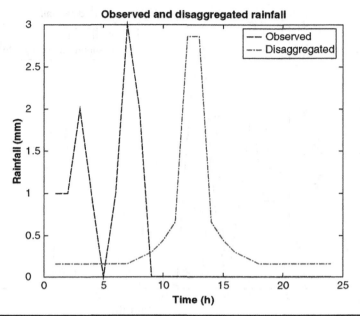

FIGURE **6.5**   Observed and disaggregated rainfall values.

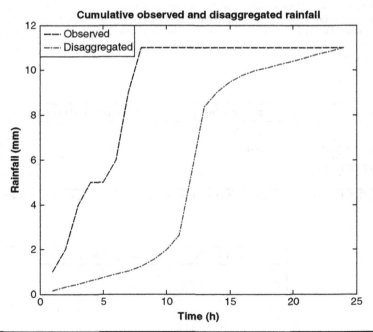

FIGURE **6.6***a*   Cumulative observed and disaggregated rainfall values.

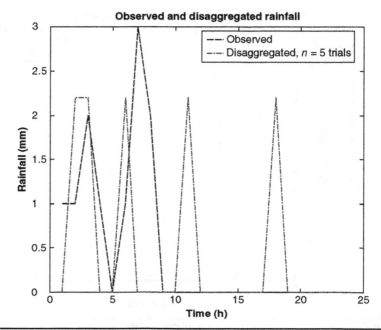

FIGURE **6.6b**   Observed and disaggregated rainfall values.

Solution.   The first step is to calculate the entropy values for different time scales from the available hourly rainfall values. The sample data for all the days in January for the station Abilene can be considered for calculation of the entropy values for different time scales. A small portion of the data is shown below to illustrate the necessary calculations for obtaining the entropy values for different time scales.

	Average (mm)	Standard Deviation (mm)	Skewness
Observed	0.458	0.833	1.867
Disaggregated	0.458	0.756	2.993
Error percentage	0	9.238	60.303

TABLE **6.13**   Comparison of Rainfall Statistics for Observed and Disaggregated Values

Number of Trials	Hours Chosen Randomly
5	2, 3, 6, 11, 18
7	1, 4, 6, 9, 13, 15, 21
10	1, 1, 2, 3, 11, 13, 17, 19, 22, 23

TABLE **6.14**   Randomly Chosen Hours for Different Trials

Observed (mm)	Disaggregation, n = 5 (mm)	Error	Disaggregation, n = 7 (mm)	Error	Disaggregation, n = 10 (mm)	Error
1	0	−1	1.571	0.571	2.2	1.2
1	2.2	1.2	0.000	−1	1.1	0.1
2	2.2	0.2	0.000	−2	1.1	−0.9
1	0	−1	1.571	0.571	0	−1
0	0	0	0.000	0	0	0
1	2.2	1.2	1.571	0.571	0	−1
3	0	−3	0.000	−3	0	−3
2	0	−2	0.000	−2	0	−2
0	0	0	1.571	1.571	0	0
0	0	0	0.000	0	1.1	1.1
0	2.2	2.2	0.000	0	0	0
0	0	0	0.000	0	1.1	1.1
0	0	0	1.571	1.571	0	0
0	0	0	0.000	0	0	0
0	0	0	1.571	1.571	0	0
0	0	0	0.000	0	1.1	1.1
0	0	0	0.000	0	0	0
0	2.2	2.2	0.000	0	1.1	1.1
0	0	0	0.000	0	0	0
0	0	0	1.571	1.571	0	0
0	0	0	0.000	0	1.1	1.1
0	0	0	0.000	0	1.1	1.1
0	0	0	0.000	0	0	0
0	0	0	0.000	0	0	0

TABLE 6.15 Comparison of Observed and Disaggregated Rainfall Values

Calculation of the 24-h entropy for sample rainfall data for the month of January is shown in Table 6.17. A sample calculation for rainfall of 3 mm that fell on day 9 is shown here. The cumulative monthly rainfall is 178 mm; the relative frequency of rainfall $p_i = 3/178 = 0.0168$; entropy, $H = -p_i \log_2 p_i = -0.0168 \times \log_2 0.0168 = 0.099$ bit; and average 24-h entropy = averaged value of all the daily entropy values: 0.184 bit. Table 6.18 shows the sample data used for the calculation of 12-h and lower time scale entropy values.

Now calculation of the 12-h entropy is illustrated. Let the 24-h entropy H be divided into the 12-h entropy value. Let p_1 and p_2 be the 12-h relative rainfall frequencies. The total 24-h rainfall for the sample data is 11 mm. Then p_1 and p_2 can be calculated by summing the rainfall value for 12-h obtained from Table 6.18 and dividing it by the total daily rainfall.

$$p_1 = (1 + 1 + 2 + 1 + 0 + 1 + 3 + 2 + 0 + 0 + 0 + 0)/11 = 1$$
$$p_2 = (0 + 0 + 0 + 0 + 0 + 0 + 0 + 0 + 0 + 0 + 0 + 0)/11 = 0$$

12-h entropy $H_{12} = -p_1 \log_2 p_1 - p_2 \log_2 p_2 = -1 \times \log_2 1 - 0 \times \log_2 0 = 0$ bits

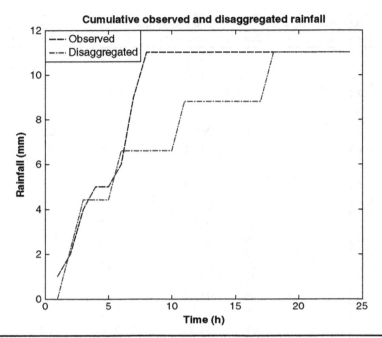

FIGURE 6.7 Cumulative observed and disaggregated rainfall values.

For calculating the 6-h entropy, the 12-h entropy can be further divided into the 6-h entropy. Let $p_3, p_4, p_5,$ and p_6 be the 6-h relative frequencies. Terms $p_3, p_4, p_5,$ and p_6 can be obtained by summing the rainfall values for 6-h and then dividing by the total daily rainfall.

$$p_3 = (1 + 1 + 2 + 1 + 0 + 1)/11 = 0.5454$$
$$p_4 = (3 + 2 + 0 + 0 + 0 + 0)/11 = 0.4545$$
$$p_5 = (0 + 0 + 0 + 0 + 0 + 0)/11 = 0$$
$$p_6 = (0 + 0 + 0 + 0 + 0 + 0)/11 = 0$$

6-h entropy $H_6 = -p_3 \log_2 p_3 - p_4 \log_2 p_4 - p_5 \log_2 p_5 - p_6 \log_2 p_6$
$$= -0.5454 \times \log_2 0.5454 - 0.4545 \times \log_2 0.4545 - 0 - 0 = 0.99403 \text{ bit}$$

For calculating the 3-h entropy, the 6-h entropy can further be divided into the 3-h entropy. Let $p_7,$ $p_8, p_9, p_{10}, p_{11}, p_{12}, p_{13},$ and p_{14} be the relative frequencies for 3-h rainfall. This can be obtained by summing the 3-h rainfall values obtained from Table 6.18 and dividing by daily rainfall.

	Average (mm)	Standard Deviation (mm)	Skewness
Observed	0.458	0.833	1.867
Disaggregated	0.458	0.913	1.534
Error percentage	0	9.568	17.828

TABLE 6.16 Comparison of Rainfall Statistics for Observed and Disaggregated Cases

Day	Rainfall (mm)	Relative Frequency p_i	Entropy H (bits)	Day	Rainfall (mm)	Relative Frequency p_i	Entropy H (bits)
1	0	0	0	18	2	0.011	0.073
9	3	0.017	0.099	22	42	0.236	0.492
10	11	0.062	0.248	23	1	0.006	0.042
11	11	0.062	0.248	24	12	0.067	0.262
12	11	0.062	0.248	25	44	0.247	0.498
13	2	0.011	0.073	26	1	0.006	0.042
14	1	0.006	0.042	27	11	0.062	0.248
15	5	0.028	0.145	29	1	0.006	0.042
16	2	0.011	0.073	30	6	0.034	0.165
17	12	0.067	0.262				

Table 6.17 Data for Illustration of 24-h Entropy Calculation

Hours	Rainfall (mm)	Hours	Rainfall (mm)	Hours	Rainfall (mm)	Hours	Rainfall (mm)
1	1	7	3	13	0	19	0
2	1	8	2	14	0	20	0
3	2	9	0	15	0	21	0
4	1	10	0	16	0	22	0
5	0	11	0	17	0	23	0
6	1	12	0	18	0	24	0

Table 6.18 Sample 24-h Data for Day 27 Used for Illustration of Subdaily Rainfall Disaggregation

$$p_7 = (1 + 1 + 2)/11 = 0.3636 \qquad p_8 = (1 + 0 + 1)/11 = 0.1818$$
$$p_9 = (3 + 2 + 0)/11 = 0.4545 \qquad p_{10} = 0$$
$$p_{11} = 0 \qquad p_{12} = 0$$
$$p_{13} = 0 \qquad p_{14} = 0$$
$$\text{3-h entropy} = H_3 = -p_7 \log_2 p_7 - p_8 \log_2 p_8 - p_9 \log_2 p_9 - p_{10} \log_2 p_{10} - p_{11} \log_2 p_{11} - p_{12} \log_2 p_{12}$$
$$- p_{13} \log_2 p_{13} - p_{14} \log_2 p_{14} = -0.3636 \times \log_2 0.3636 - 0.1818 \times \log_2 0.1818$$
$$- 0.4545 \times \log_2 0.4545 = 1.4949 \text{ bits}$$

For calculating 1-h entropy, 1-h entropy values are calculated in the same way as shown in Table 6.19.

1-h entropy = sum of all values in rightmost column
$$= 0.3145 + 0.3145 + 0.4471 + 0.3145 + 0 + 0.3145 + 0.5112 + 0.4471$$
$$+ 0 + 0 + 0 + 0 + 0 + 0 + 0 + 0 + 0 + 0 + 0 + 0 + 0 + 0 = 2.6635 \text{ bits}$$

Knowing the entropy values at different time scales, a regression equation can be determined between the entropy values and the time scales, as shown in Fig. 6.8.

Having obtained the regression equation for January of the station considered, we can calculate the entropy values corresponding to different time scales by substitution. The regression equation to

Hours	Rainfall (mm)	Relative Frequency p_i	Entropy H (bits)	Hours	Rainfall (mm)	Relative Frequency p_i	Entropy H (bits)
1	1	0.091	0.3145	13	0	0	0
2	1	0.091	0.3145	14	0	0	0
3	2	0.182	0.4471	15	0	0	0
4	1	0.091	0.3145	16	0	0	0
5	0	0	0	17	0	0	0
6	1	0.091	0.3145	18	0	0	0
7	3	0.273	0.5112	19	0	0	0
8	2	0.182	0.4471	20	0	0	0
9	0	0	0	21	0	0	0
10	0	0	0	22	0	0	0
11	0	0	0	23	0	0	0
12	0	0	0	24	0	0	0

TABLE 6.19 The 1-h Entropy Values

be used will be $H = -0.5915 \times \log T + 6.328$, where H is the entropy in bits corresponding to a time scale of T min.

$$H_{24} \text{ h: } -0.5915 \times \log 1440 + 6.328 = 0.122 \text{ bit}$$
$$H_{12} \text{ h: } -0.5915 \times \log 720 + 6.328 = 0.714 \text{ bit}$$
$$H_6 \text{ h: } -0.5915 \times \log 360 + 6.328 = 1.305 \text{ bits}$$
$$H_3 \text{ h: } -0.5915 \times \log 180 + 6.328 = 1.897 \text{ bits}$$
$$H_1 \text{ h: } -0.5915 \times \log 60 + 6.328 = 2.834 \text{ bits}$$

Now we back-calculate the relative frequency values, knowing the entropy values. A sample calculation for 24- to 12-h disaggregation and from 12- to 6-h disaggregation is illustrated below:

24-h to 12-h disaggregation: Let us refer to the calculation of p_1 and p_2.

$$H_{12} \text{ hour entropy} = 0.714 = -p_1 \log_2 p_1 - p_2 \log_2 p_2$$
$$p_1 + p_2 = 1$$

Solving these two equations, we get

$$P_1 = 0.8039 \qquad P_2 = 0.1961$$

The 12-h rainfall can be disaggregated from 24-h rainfall of 11 mm as $11 \times 0.1961 = 2.1571$ mm and $0.8039 \times 11 = 8.8429$ mm.

From 12-h rainfall to 6-h rainfall: Let us refer to solving the 6-h rainfall occurrence probabilities $p_3, p_4, p_5,$ and p_6.

$$H(p_1, p_5, p_6) = H_{12} + p_2 \times H\left(\frac{p_5}{p_2}, \frac{p_6}{p_2}\right)$$

$$H(p_2, p_3, p_4) = H_1 + p_1 \times H\left(\frac{p_3}{p_1}, \frac{p_4}{p_1}\right)$$

$$p_5 + p_6 = p_2$$
$$p_3 + p_4 = p_1$$

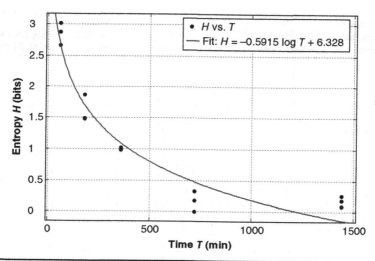

FIGURE 6.8   Regression between entropy and time scales (t = time in minutes and H = entropy in bits).

Known values: $p_1 = 0.8039$, $p_2 = 0.1961$, $H_{12} = 0.714$ bit

$$-0.8039 \times \log_2 0.8039 - P_5 \log_2 P_5 - P_6 \log_2 P_6 = 0.714 + 0.1961 \times H\left(\frac{p_5}{p_2}, \frac{p_6}{p_2}\right)$$

$$p_5 + p_6 = 0.1961$$

Solving the two equations, one gets

$$p_5/p_2 = 0.362$$
$$p_6/p_2 = 0.638$$
$$p_5 = 0.362 \times 0.196 = 0.0709$$
$$p_6 = 0.638 \times 0.196 = 0.125$$

Similarly, solving for p_3 and p_4

$$-0.196 \log_2 0.196 - P_3 \log_2 P_3 - P_4 \log_2 P_4 = 0.714 + 0.1961 \times H\left(\frac{p_3}{p_1}, \frac{p_4}{p_1}\right)$$

$$p_3 + p_4 = 0.8039$$

Solving these equations, we obtain

$$p_3/p_1 = 0.2067$$
$$p_4/p_1 = 0.7933$$
$$p_3 = 0.2067 \times 0.8039 = 0.166$$
$$p_4 = 0.7933 \times 0.8039 = 0.6377$$

Hence the 6-h rainfall distribution can be given as $0.0709 \times 11 = 0.7799$ mm.

$$0.125 \times 111 = 1.3761 \text{ mm}$$
$$0.166 \times 11 = 1.826 \text{ mm}$$
$$0.6377 \times 11 = 7.0147 \text{ mm}$$

Table 6.20 gives the disaggregated values of rainfall.

Figure 6.9 shows observed and disaggregated rainfall values, and Fig. 6.10 shows the cumulative observed and disaggregated rainfall values.

Hour	Actual Rainfall (mm)	Disaggregated Rainfall (mm)	Error	Hour	Actual Rainfall (mm)	Disaggregated Rainfall (mm)	Error
1	1	0.51	−0.49	13	0	0.195	0.195
2	1	0.412	−0.588	14	0	0.24	0.24
3	2	0.74	−1.26	15	0	0.27	0.27
4	1	0.164	−0.836	16	0	0.105	0.105
5	0	0	0	17	0	0	0
6	1	0	−1	18	0	0	0
7	3	3.54	0.54	19	0	0.521	0.521
8	2	3.274	1.274	20	0	0.472	0.472
9	0	0.1	0.1	21	0	0.36	0.36
10	0	0	0	22	0	0	0
11	0	0.1	0.1	23	0	0	0
12	0	0	0	24	0	0	0

TABLE **6.20** Hourly Disaggregated Values

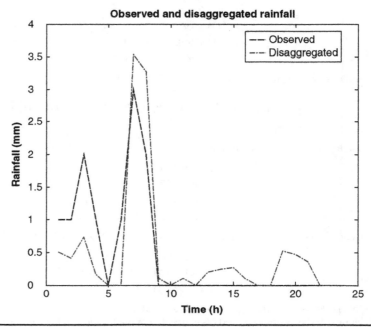

FIGURE **6.9** Observed and disaggregated rainfall.

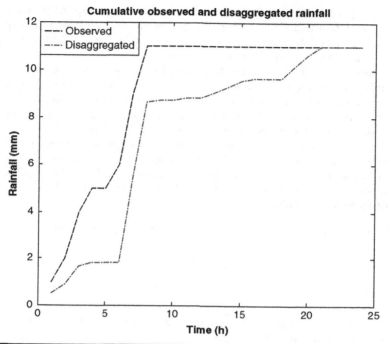

FIGURE 6.10 Cumulative observed and disaggregated rainfall.

	Average (mm)	**Standard Deviation (mm)**	**Skewness**
Observed	0.458	0.833	1.866
Disaggregated	0.458	0.933	2.948
Error percentage	0	12.04	57.957

TABLE 6.21 Comparison of Rainfall Statistics for Observed and Disaggregated Rainfall Values

Table 6.21 shows error statistics, including the mean, variance, and skewness, for the disaggregation method of rainfall distribution.

6.7 Rainfall Complexity

Like the climate itself and its many variables, rainfall is highly complex. Estimation of complexity is important for forecasting the likely change in the rainfall regime. Different methods have been proposed to determine complexity. Examples of such methods include the Lyapunov exponent, correlation dimension, and Kolmogorov complexity measure, to name but a few. In recent years, entropy theory has been applied to quantify complexity. The theory has led to the development of the concepts of approximate entropy (ApEn) (Pincus, 1991) and sample entropy (SampEn) (Pincus, 1995). Because ApEn is dependent on the record length, is uniformly lower than expected for short records, lacks relative consistency (Richman and Moorman, 2000), and has an inherent bias, SampEn is preferred and is discussed here.

Since the method of computing SampEn is discussed in Chap. 1, which involves defining dimension m and filter r, it is not repeated here. When computing SampEn, it may be observed that each type of observation data may exhibit different characteristics at different scales, and valuable information may be obtained by mining characteristics of observations at various scales (Costa et al., 2002). This suggests computing *multiscale entropy* which may overcome the limitations of single-scale entropy measure. Note that greater entropy values do not always correspond to greater complexity.

Multiscale entropy can be computed as follows. Let $Y: \{y(1), y(2), ..., y(N)\}$ be the time series of data of length N. Let τ be the scale factor. Then the time series for various scale factors can be obtained as follows.

1. Apply a course graining process to the time series. Let

$$x_j^\tau = \frac{\sum\limits_{i=(j-1)\tau}^{j\tau} y(i)}{\tau} \qquad i = (j-1)\tau + \tau, ..., j\tau; \quad j = 1, 2, ..., N/\tau \qquad (6.25)$$

   The data length of the time series equals the length of the time series divided by the scale factor τ.

2. Compute the sample entropy for $\{x^\varsigma\}$ for all scale factors.

If data are available for more than one station, then the mean areal rainfall can be computed. The multiscale entropy is computed for rainfall time series of all stations as well as of mean rainfall for all scale factors. The number of dimensions m may be chosen as $m = 2$, and the value of threshold r is 0.2 times the standard deviation (SD). The multiscale entropy values can be plotted versus the scale factor, and complexity can be inferred from the plots. It is likely that the multiscale entropy increases as the scale factor increases, indicating that the rainfall time series exhibits self-similarity and higher complexity. If the entropy measures of average rainfall and of rainfall time series at different stations are more or less the same, then this suggests that averaging does not increase rainfall complexity.

If the rainfall time series at one station has a higher entropy measure than at another station for all scale factors, then that series is said to be more complex than the other series. The multiscale entropy analysis seems more consistent for various values of m and r; that is, if one series has a higher entropy measure than another series, then the results will be consistent for various m and r values. Data length influences the SampEn values. For the same time series, the fewer the data points, the smaller the SampEn values. For a nonstationary system, SampEn gradually decreases with increasing filter r. Parameter r corresponds to prediction precision—a larger r corresponds to a lower prediction precision.

6.8   Entropy Scaling

A number of linear and nonlinear statistical characteristics of rainfall depend on aggregation at increasing time scales, and their dependence then reflects the temporal memory and the Hurst effect. Some of the statistical characteristics are the scale of fluctuation, variance, Hurst effect, and entropy among others. Poveda (2011) investigated some of these dependencies, and their work is followed here.

Consider rainfall intensity $I(t)$ as a continuous stationary stochastic process. Let τ denote the lag time or simply lag, and let $p(\tau)$ denote the autocorrelation function as a function of lag, which gives the coefficient of correlation at a particular value of lag. The scale of fluctuation θ (Poveda, 2011) can be defined for both continuous and discrete processes, as discussed in Chap. 1. The scale of fluctuation can be employed to determine the time interval between observations when they become independent, i.e., when the correlation coefficient becomes zero. This time interval is referred to as twice the integral time scale.

If H is the Shannon entropy of the intensity variable and T is the mutual information or entropy of intensity process sampled at a given time and a lag apart, that is, $I(t)$ and $I(t - \tau)$, then mutual information $T[I(t), I(t - \tau)]$ can be calculated as a function of lag— that is, $T(\tau) = T[I(t), I(t - \tau)]$ defines the sample autocorrelation. If $\tau = 0$, this leads to the Shannon entropy H. Following Poveda (2011), the scale of information is defined as a nonlinear analog of the scale of fluctuation.

If there are two time series, say, rainfall intensity and depth or duration, $\{X(\tau)\}_{\tau=1}^{n}$ and $\{Y(\tau)\}_{\tau=1}^{n}$, sampled at fixed times τ, then their lag-τ mutual information can be defined as

$$T_{X,Y}(\tau) = T[X(\tau), Y(\tau)] \tag{6.26}$$

Using Eq. (6.26), the normalized mutual information can be defined as

$$T_{X,Y}^{*}(\tau) = \frac{T_{X,Y}(\tau)}{\sqrt{S(X)}\sqrt{S(Y)}} \tag{6.27}$$

which varies between 0 and 1.

Now the scale of information can be redefined for the continuous case as

$$\psi = \int_{-\infty}^{\infty} T_{X,Y}^{*}(\tau)d\tau \tag{6.28}$$

and for the discrete case as

$$\psi(k) = \left[H + 2\sum_{\tau=1}^{k} T^{*}(\tau) \right]\tau \qquad k \to \infty \tag{6.29}$$

Poveda (2011) computed entropy H for rainfall in the Andes for aggregation times varying from $D = 15$ min to 64 d and found by curve fitting a power relation

$$H(D) = cD^{b} \tag{6.30}$$

with $b = 0.51$ [0.47 to 0.56] and c a constant. It was found that the scaling regime prevailed up to a time scale (70 to 202 h) and at that time the entropy reached a maximum value and then leveled off, that is, $b = 0$. Since rainfall at short time scales may contain zeros, the amount of zeros may exhibit two different scaling regimes with D separated by a particular value. Nevertheless power law relates both time scales. When entropy is scaled with a time scale, several questions need to be investigated for many basins and climate types. For example, does the scaling exponent depend on the number of bins used for the computation of entropy, and does parameter c have any relation with rainfall regime and climate type? How sensitive are b and c to record length? Does the scaling exponent vary with rainfall type and climate type? When does the exponent reach zero and why? Does this time remain the same for all basins and climates? What is the impact of zeros in rainfall series on the scaling law?

6.9 Case Study 1: Precipitation Variability

Mishra and Singh (2010) investigated the variability of precipitation in Texas. Owing to its location along the warm Gulf of Mexico, and being not so far distant from the Pacific Ocean, Texas has varied physiography, from the forests in the east and the Coastal Plain in the south to the elevated plateaus and basins in the north and west. Because of its expansive and diverse topography, Texas offers continental, marine, and mountain-type climates (Handbook of Texas online, 2008). The variability in climate is a result of the interactions between Texas' unique geographic location and the movements of seasonal air masses, such as arctic fronts, jet streams, subtropical west winds, tropical storms, and a subtropical high-pressure system known as the Bermuda High. The Gulf of Mexico is a dominant geographical feature, moderating temperatures along the Gulf Coast and providing the major source of moisture for the state. Texas is prone to hurricanes that find their way into the Gulf of Mexico during the hurricane season.

The mean annual precipitation varies from a statewide maximum of 59.20 in. at Orange in the lower Sabine River valley of East Texas to a minimum of 7.82 in. at El Paso at the western tip of the state. The mean annual rainfall distribution correlates roughly with longitude and varies little from north to south across Texas. Winter is the driest time of the year in nearly all Texas. Except for east Texas, rainfall typically is least substantial in July and August. December or January is normally the driest month on the High and Low Rolling Plains, as well as on the Edwards Plateau. Mishra and Singh (2011) analyzed long-term records of monthly total precipitation from a network of 43 stations having a common data period from 1925 to 2005, distributed relatively uniformly across Texas.

6.9.1 Variability in Precipitation

The values of marginal disorder index (MDI) show that the spatial variability of annual time series (mean = 0.4524 bit, standard deviation = 0.1887 bit, and range = 0.97 bit) is higher in west Texas than east Texas. High Plains, which belongs to the continental Steppe region, has the highest variability. The other high-variability regions for annual rainfall series are the intersection of Edwards Plateau, Low Rolling Plains, and top right corner of North Central Texas. These two areas lie in the subtropical subhumid zone.

The variability of precipitation is greater in winter with higher standard deviation and range than in other seasons. Winter is the driest time of the year in nearly all Texas, and its variability is higher than in any of the other seasons. Winter precipitation has high variability in South Central Texas, which belongs to both subtropical humid and subtropical subhumid zones. All other parts of Texas have more or less the same variability. Interestingly the region of variability for spring rainfall is different from that for winter rainfall. The spring rainfall variability seems to be higher in the High Plains and Trans-Pecos regions which belong to subtropical arid zones. The variability of summer precipitation is high in a few pockets in the region, i.e., low rolling plains. The variability of fall rainfall seems to occur in High Plains, Trans-Pecos, and Lower South Texas.

It is observed that the month of April dominates the variability for most of the stations with high peaks. The variability of the summer season is less than that of the contributing months, and July seems to be contributing more to the variability of the summer season. The variability of the winter season along with constituent months shows that the no particular months have dominant peaks. Based on the standard deviation, January contributes greater variability than does December for winter precipitation. The variability of the winter season is less than that of the contributing months, and all 3 months have more or less the same contribution to the variability of fall precipitation.

6.9.2   Variability Indices

The values of apportionment disorder index (ADI) show that changes in the distribution of monthly rainfall change within the year. Significant interannual (4- to 5-yr) oscillations occurred around the 1970s, while interdecadal (15- to 17-yr) oscillations were active from 1945 to the 1960s, which is considered as a severe drought period in Texas.

There seems to be an interesting observation in the patterns of ADI, which depicts the variability of monthly rainfall within the year. The observed variability seems to be increasing from east Texas to west Texas. The highest variability seems to be observed for High Plains, Trans-Pecos, and the western part of southern Texas region. These regions belong to the continental steppe, subtropical arid, and subtropical steppe zones.

There is a distinct pattern of increase in the variability from east Texas to west Texas. When the variability is compared with the distribution of precipitation; annual precipitation decreases about 1 in. for each 15-mi displacement from east to west; and the variability increases from east to west, which shows that a decrease in precipitation and an increase in variability go together. The variability is high for Trans-Pecos, which is the driest region in the state. At most locations rainfall for any single month will vary appreciably from the normal. Likewise, the number of days with precipitation usually is significantly abnormal. Moreover, the number of "rain days" follows the general trend of rainfall totals in that seasonal frequencies of rainy days are the lowest when rainfall totals are the lowest.

The values of intensity disorder index (IDI) show that there is a distinct pattern of increase in the variability from east Texas to west Texas. The spatial variability in the number of monthly rainy days within a year seems to be partly different from that of the variability in the amount of monthly rainfall within a year.

The values of DADI for annual time series of precipitation show that the maximum variability is for the decade of 1955 to 1964. The variability of annual precipitation series as well as the September precipitation time series is less than that for all other time series. There is markedly similar deviation for all decades in these two time series (annual and September), whereas for all other time series there is a large variation within the decades.

When the number of rainy days is considered for decadal analysis, the mean decadal intensity disorder index shows that the variability is less for annual time series. It is interesting to note that the variability of precipitation amount and that of the number of rainy days are of contrasting nature in September. When the variabilities for the decades of 1985 to 1994 and 1995 to 2004 are compared, the variability of the decade 1995 to 2004 is greater for almost all time series.

6.9.3   Trend and Persistence of Variability

The values of the Hurst exponent (HE) of ADI time series of precipitation show that out of 43 stations considered, 7 stations have HE below 0.5 and all other stations have the Hurst exponent greater than 0.5. As the HE value greater than 0.5 indicates persistence or positive long-term dependence for most of the time series, this means that there is a long-term persistence in the variability of ADI. The implication is that variability in a year is more likely to be followed by another year of disorder. In other words, the occurrence of disorder has a tendency to appear in clusters over different parts of Texas, and the tendency is rather strong, as indicated by the high values of H for most of the stations, and these ADI time series seem to have similar long-term memory. Because the Hurst exponent captures the long-term persistence in the data series, a higher HE value indicates that the previous variability record will positively affect the future climatic conditions. Thus an extreme event would have higher probability of being followed by other extreme events.

The Mann-Kendall test of MDI time series reveals that mostly there is a no trend throughout the state, but some upward trends can be seen in the Trans-Pecos and South Central parts of Texas, which indicates that variability continues to increase in the future. Both persistence and trend results show that the westernmost part of Texas is undergoing long-term persistence in variability as well as some upward trend.

6.9.4   Implications for Drought

From a historical drought point of view, during the 1930s, the northern and western parts of Texas, along with a large portion of the central United States, experienced a drought compounded by poor agricultural management practices that stripped away large volumes of topsoil, known as the Dust Bowl. The most severe statewide drought occurred during the 1950s when precipitation was about 30 percent below average. This drought persisted for 5 to 7 yr for much of the state. More recently, portions of South, Central, and West Texas have experienced recurrent periods of drought from the 1990s through 2006. When compared with the disorder in rainfall amount within a year, which is represented by ADI, it was observed that the variability was high during the 1950s and around 1995s and these periods are said to be the drought-affected years in Texas. When compared with the disorder of rainy days, it can be seen that there is high variability in the number of rainy days between 1945 and 1952, causing the Texas drought of the 1950s. Similar observations can be made about the drought during the 2000s. This indicates there is a close agreement between the drought in the 1950s and that in the 2000s and the high variability in rainy days in those periods.

There is a significant increase in annual drought severity in two regions, namely, Trans-Pecos and Lower valley. The increasing drought severity is characterized by the following: (1) There is high disorder in monthly precipitation within the year, which can be seen because of higher ADI values in Trans-Pecos region than in other parts of Texas; (2) the variability of monthly rainy days within the year seems to be maximum for the Trans-Pecos region; and (3) the high variability of the spatial apportionment disorder index seems to be very high for the Trans-Pecos region.

6.10   Case Study 2: Rainfall Disaggregation

Rajasekhar and Singh (2013) evaluated the methods of disaggregation for 30 stations spread across Texas for a time period from 1980 to 2000. They employed the mean, variance, skewness, dependence structure of rainfall, and cumulative rainfall of rainfall events to assess the performance of the disaggregation schemes. For the method of allocating rainfall by randomly choosing days of a month, Fig. 6.11 compares observed and disaggregated rainfall events.

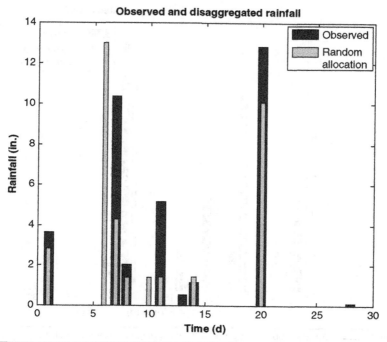

FIGURE **6.11**   Comparison of observed and disaggregated rainfall events.

For the method of disaggregation involving single-site entropy, Fig. 6.12 compares observed and disaggregated rainfall events.

For the method involving entropy disaggregation taking into account the influence of neighboring stations, Fig. 6.13 compares observed and disaggregated rainfall events.

Comparison shows that mean values are perfectly preserved by all methods for all regions. Rainfall correlation structure is better preserved when the influence of neighboring stations is also accounted for. The error in variance and skewness is higher in arid and continental steppe regions. A plausible reason may be that the west experiences an extremely dry season and greater variance in rainfall which might not be well captured by the disaggregation schemes. Monthwise higher errors are observed in March, July, August, October, and November. The period of June through November is the hurricane season in Texas and may be the reason for higher errors in variance. In the continental steppe region, January months show slightly higher errors, possibly influenced by snowfall. The March, April, and May months in humid zones also show higher errors in variance, possibly because of increased frequency of thunderstorms in those regions. No method is found to be consistently superior to others. At most locations consideration of the influence of neighboring stations seems to slightly reduce the error. Overall the three methods are plausible means for disaggregation, since the mean is well preserved and the variance and skewness errors never exceed 21.09 and 26.78 percent, respectively. The number of trials needed in method 3 mostly falls within the range of 5 to 30, with a higher number of trials for eastern parts. On average the number of trials needed is greater for March, August, October, and November, possibly because of more rainy days.

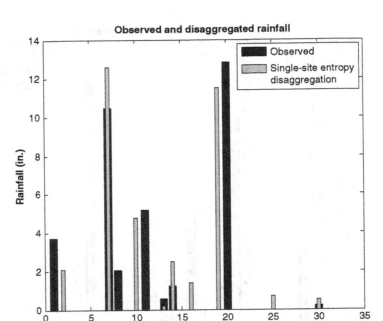

FIGURE 6.12 Comparison of observed and disaggregated rainfall events.

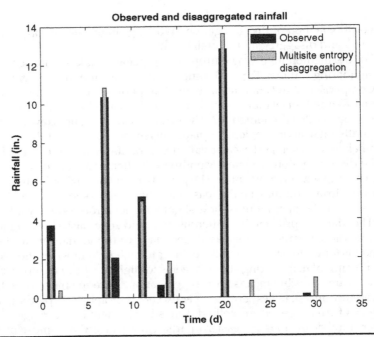

FIGURE 6.13 Comparison of observed and disaggregated rainfall events.

Questions

1. Take long-term monthly precipitation data for several (say, 5) rain gauge stations from a basin. Compute the marginal entropy and variance of monthly rainfall at each of the stations.

2. Compute the diversity indices (Simpson, McIntosh, Berger-Parker, and Brillouin) for each of the stations in Question 1.

3. Compute the Hurst exponent for each of the stations in Question 1.

4. Determine the Mann-Kendall statistic and its variance for each of the stations. Check if there is any trend in the data in Question 1.

5. Compute the intensity entropy for each of the stations in Question 1.

6. Compute the apportionment entropy for each of the stations in Question 1.

7. Compute the decadal apportionment entropy for each of the stations in Question 1.

8. Compute the disorder index for each of the stations in Question 1.

References

Berger, W. H., and Parker, F. L. (1970). Diversity of planktonic Foraminifera in deep sea sediments. *Science*, vol. 168, pp. 1345–1347.

Costa, M., Goldberger, A. L., and Peng, C. K. (2002). Multiscale entropy analysis of biological signals. *Physical Review E*, vol. 71, pp. 021906:1–021906:18.

Dionisio, A., Menezes, R., and Mendes, D. A. (2006). An econophysics approach to analyse uncertainty in financial markets: An application to the Portuguese stock market. *European Physics Journal, B*, vol. 50, pp. 161–164.

Dunbar, R. I. M. (1992). Time: a hidden constraint on the behavioural ecology of baboons. *Behavioral Ecological Sociobiology*, vol. 31, pp. 35–49.

Easterling D. R., Evans, J. L., Groisman, P. Y, Karl, T. R., Kunkel, K. E., and Ambenje, P. (2000a). Observed variability and trends in extreme climate events: A brief review. *Bulletin of the American Meteorological Society*, vol. 81, no. 3, pp. 417–425.

Easterling, D. R., Meehl, G. A., Parmesan, C., Changnon, S. A., Karl, T. R., and Mearns, L. O. (2000b). Climate extremes: Observations, modeling, and impacts. *Science*, vol. 289, 2068. DOI: 10.1126/science.289.5487.2068.

Ebrahimi, N., Maasoumi, E., and Soofi, E. (1999). Ordering univariate distributions by entropy and variance. *Journal of Econometrics*, vol. 90, no. 2, pp. 317–336.

Handbook of Texas Online, http://www.tshaonline.org/handbook/online/articles/WW/yzw1.html (accessed May 14, 2008).

Hurst, H. E. (1951). Long term storage capacity of reservoirs. *Transactions of ASCE*, vol. 116, pp. 776–808.

Jaynes, E. T. (2003). *Probability Theory. The Logic of Science*. Cambridge University Press, Cambridge, United Kingdom.

Kendall, M. G. (1975). *Rank Correlation Methods*. Charles Griffin, London.

Koutsoyiannis, D. (2003). Rainfall disaggregation methods: Theory and applications. Workshop on Statistical and Mathematical Methods for Hydrological Analysis, Rome, May.

Mann, H. B. (1945). Nonparametric tests against trend. *Econometrica*, vol. 13, pp. 245–259.

McIntosh, R. P. (1967). An index of diversity and the relation of certain concepts to diversity. *Ecology*, vol. 48, pp. 392–404.

Montesarchio, V. and Napolitano, F. (2010). A single site rainfall disaggregation model based on entropy. *Proceedings of International workshop on Advances in Statistical Hydrology*, May 23–25, 2010, Italy.

Montesarchio, V., Napolitano, F., Ridolfi, E., and Ubertini, L. (2012). A comparison of two rainfall disaggregation models. *Numerical analysis and Applied Mathematics*, Proceedings, 1479, pp. 1796–1799.

Mishra, A. K., and Singh, V. P. (2010). Changes in extreme precipitation in Texas. *Journal of Geophysical Research*, vol. 115, D14106, doi: 10.1029/2009JD013398.

Pincus, S. M. (1991). Approximate entropy as a measure of system complexity. *Proceedings of National Academy of Science (USA)*, vol. 88, no. 6, pp. 2297–2301.

Pincus, S. M. (1995). Approximate entropy (ApEn) as a complexity measure. *Chaos*, vol. 5, no. 1, pp. 110–117.

Poveda, G. (2011). Mixed memory, (non) Hurst effect, and maximum entropy of rainfall in the tropical Andes. *Advances in Water Resources*, vol. 34, pp. 243–256.

Rajasekhar, D., and Singh, V. P. (2013). *An entropy approach to rainfall disaggregation*. Unpublished Note, 28 p., Department of Biological & Agricultural Engineering, Texas A&M University, College Station, Texas.

Richman, J. S., and J. R. Moorman, (2000). Physiological time-series analysis using approximate entropy and sample entropy. *American Journal of Physiology: Heart and Circulatory Physiology*, vol. 278, pp. 2039–2049.

Salas, H. D., and Poveda, G. (2013). Scaling of entropy and multi-scaling of the time generalized q-entropy in rainfall and streamflows. *Water Resources Research*, vol. 49.

Shannon, C. E. (1948). A mathematical theory of communication. *Bell System Technical Journal*, vol. 27, pp. 379–423 and 623–656.

Simpson, E. H. (1949). Measurement of diversity. *Nature*, vol. 163, p. 688.

Tsonis, A. A. (1996). Widespread increases in low-frequency variability of precipitation over the past century. *Nature*, vol. 382, pp. 700–702.

Further Reading

Bartlein, P. J. (1982). Streamflow anomaly patterns in the U.S.A. and southern Canada—1951–1970. *Journal of Hydrology*, vol. 57, pp. 49–63.

Bo, Z., Islam, S., and Eltahir, E. A. B. (1994). Aggregation-disaggregation properties of a stochastic rainfall model. *Water Resources Research*, vol. 30, no. 12, pp. 3423–3435.

Brocca, L., Morbidelli, R., Melone, F., and T. Moramarco (2007). Soil moisture spatial variability in experimental areas of central Italy. *Journal of Hydrology*, vol. 333, pp. 356–373.

Cayan, D. R. (1996). Interannual climate variability and snowpack in the western United States. *Journal of Climate*, vol. 9, pp. 928–948.

Clark, P. U., Alley, R. B., and Pollard, D. (1999). Northern Hemisphere ice-sheet influences on global climate change. *Science*, vol. 286, pp. 1104–1111.

Cook, E. R., Meko, D. M., Stahle, D. W., and Cleaveland, M. K. (1999). Drought reconstructions for the continental United States. *Journal of Climate*, vol. 12, pp. 1145–1162.

Costa, M., Goldberger, A. L., and Peng, C. K. (2002). Multiscale entropy analysis using complex physiographic time series. *Physical Review Letters*, vol. 89.

Dai, A., Fung, I. Y. and DelGenio, A. D. (1997). Surface observed global land precipitation variations during 1900–88. *Journal of Climate*, vol. 10, no. 11, pp. 2943–2962.

Derksen, C., Misurak, K., LeDrew, E., Piwowar, J., and Goodison, B. (1997). Relationship between snow cover and atmospheric circulation, central North America, winter 1988. *Annals of Glaciology*, vol. 25, pp. 347–352.

Douglas, E. M., Vogel, R. M. and Kroll, C. N. (2000). Trends in floods in the United States: Impact of spatial correlation. *Journal of Hydrology*, vol. 240, pp. 90–105.

Feldstein, S. B. (2002). The recent trend and variance increase of the annular mode, *Journal of Climate*, vol. 15, pp. 88–94.

Grinsted, A, Moore, J. C., and Jevrejeva, S. (2004). Application of the cross wavelet transform and wavelet coherence to geophysical time series. *Nonlinear Processes in Geophysics*, vol. 11, pp. 561–566.

Guetter, A. K., and K. P. Georgakakos (1993). River outflow of the conterminous United States, 1939–1988. *Bulletin of American Meteorological Society*, vol. 74, pp. 1873–1891.

Haylock, M. R., Jones, P. D., Allan, R. J., and Ansell, T. J. (2007). Decadal changes in 1870–2004 Northern Hemisphere winter sea level pressure variability and its relationship with surface temperature. *Journal of Geophysical Research*, vol. 112, D11103, doi:10.1029/2006JD007291.

Huh, S., Dickey, D. A., Meador, M. R., and Ruhl, K. E. (2005). Temporal analysis of the frequency and duration of low and high streamflow: Years of record needed to characterize streamflow variability. *Journal of Hydrology*, vol. 310, pp. 78–94.

Hulme, M., Doherty, R., Ngara, T., New, M., and Lister, D. (2001). African climate change: 1900–2100. *Climate Research*, vol. 17, pp. 145–168.

Hulme, M., Osborn, T. J., and Johns, T. C. (1998). Precipitation sensitivity to global warming: Comparison of observations with HadCM2 simulations. *Geophysical Research Letters*, vol. 25, no. 17, pp. 3379–3382.

Kalayci, S., and Kahya, E. (2006). Assessment of streamflow variability modes in Turkey: 1964–1994. *Journal of Hydrology*, vol. 324, no. 1–4, pp. 163–177.

Karl, T. R., and Knight, R. W. (1998). Secular trends of precipitation amount, frequency, and intensity in the United States. *Bulletin of American Meteorological Society*, vol. 79, no. 2, pp. 231–241.

Kawachi, T., Maruyama, T., and Singh, V. P. (2001). Rainfall entropy for delineation of water resources zones in Japan. *Journal of Hydrology*, vol. 246 , pp. 36–44.

Koutsoyiannis, D., and Onof, C. (2001). Rainfall disaggregation using adjusting procedures on a Poisson cluster model. *Journal of Hydrology*, vol. 246, pp. 109–122.

Kutzbach, J. E. (1967). Empirical eigenvectors of sea-level pressure, surface temperature and precipitation complexes over North America. *Journal of Applied Meteorology*, vol. 6, pp. 791–802.

Larkin, T. J., and Bomar, G. W. (1983). Climatic atlas of Texas. Texas Water Development Board Limited Publication 192.

Lenters, J. D., Kratz, T. K., and Bowser, C. J. (2005). Effects of climate variability on lake evaporation: Results from a long-term energy budget study of Sparkling Lake, northern Wisconsin (USA). *Journal of Hydrology*, vol. 308, pp. 168–195.

Lettenmaier, D. P., Wood, E. F., and Wallis, J. R. (1994). Hydro-climatological trends in the continental United States, 1948–1988. *Journal of Climate*, vol. 7, pp. 586–607.

Lins, H. F. (1997), Regional streamflow regimes and hydroclimatology of the United States. *Water Resources Research*, vol. 33, pp. 1655–1667.

Lins, H. F. and Slack, J. R. (1999). Streamflow trends in the United States. *Geophysical Research Letters*, vol. 26, no. 2, pp. 227–230.

Maasoumi, E. (1993). A compendium to information theory in economics and econometrics. *Econometric Reviews*, vol. 12, no. 2, pp. 137–181.

Maasoumi, E., and Racine, J. (2002). Entropy and predictability of stock market returns. *Journal of Econometrics*, vol. 107, pp. 291–312.

Mann, M., and Park, J. (1999). Oscillatory spatiotemporal signal detection in climate. *Studies, Advanced Geophysics*, vol. 41, pp. 1–131.

Mann, M. E., and Park, J. (1994). Global modes of surface temperature variability on interannual to century time scales. *Journal of Geophysical Research*, vol. 99 (D12), pp. 25819–25833.

Maruyama, T., and Kawachi, T. (1998). Evaluation of rainfall characteristics using entropy. *Journal of Rainwater Catchment Systems*, vol. 4, no. 1, pp. 7–10.

Maruyama, T., Kawachi, T., and Singh, V. P. (2005). Entropy-based assessment and clustering of potential water resources availability. *Journal of Hydrology*, vol. 309, pp. 104–113.

Mauget, S. A. (2003). Intra- to multidecadal climate variability over the Continental United States: 1932–99. *Journal of Climate*, vol. 16, no. 13, pp. 2215–2231.

Maurer, E. P., Lettenmaier, D. P., and Mantua, N. J. (2004). Variability and potential sources of predictability of North American runoff. *Water Resources Research*, vol. 40, W09306, doi:10.1029/2003WR002789.

McCauley, J. (2003). Thermodynamic analogies in economics and finance: Instability of markets. *Physica A*, vol. 329, pp. 199–212.

Michaels, P. J., Balling, Jr., R. C., Vose, R. S., and Knappenberger, P. C. (1998). Analysis of trends in the variability of daily and monthly historical temperature measurements. *Climate Research*, vol. 10, pp. 27–33.

Milly, P. C. D., Wetherald, R. T., Dunne, K. A., and Delworth, T. L. (2002). Increasing risk of great floods in a changing climate. *Nature*, vol. 415, pp. 514–517.

Mishra, A. K, Özger, M and Singh, V. P. (2009). An entropy-based investigation into the variability of precipitation. *Journal of Hydrology*, vol. 370, pp. 139–154.

Piechota, T. C., Dracup, J. A., and Fovell, R. G. (1997). Western US streamflow and atmospheric circulation patterns during El Nino-Southern Oscillation, *Journal of Hydrology*, vol. 201, pp. 249–271.

Pincus, S. M., and Viscarello, R. R. (1992). Approximate entropy: A regulatory measure for fetal heart rate analysis. *Obstetrics and Gynecology*, vol. 79, pp. 249–255.

Raible, C. C., Luksch, U., Fraedrich, K., and Voss, R. (2001). North Atlantic decadal regimes in a coupled GCM simulation. *Climate Dynamics*, vol. 18, pp. 321–330.

Reale, O., and Dirmeyer, P. (2002). Modeling the effect of land surface evaporation variability on precipitation variability. Part I: General response. *Journal of Hydrometeorology*, vol. 3, pp. 433–450.

Rodriguez-Iturbe, I., and Eagleson, P. (1987). Mathematical models of rainstorm events in space and time. *Water Resources Research*, vol. 23, no. 1, pp. 181–190.

Sakalauskienè, G. (2003). The Hurst phenomenon in hydrology. Environmental Research, *Engineering and Management*, vol. 3, no. 25, pp. 16–20.

Singh, V. P. (1997). The use of entropy in hydrology and water resources. *Hydrological Processes*, vol. 11, pp. 587–626.

Small, D., Islam, S., and Vogel, R. (2006). Trends in precipitation and streamflow in the eastern U.S.: Paradox or perception? *Geophysical Research Letters*, vol. 33, L03403, doi:10.1029/2005GL024995.

Soofi, E. (1997). Information theoretic regression methods. In: *Advances in Econometrics—Applying Maximum Entropy to Econometric Problems,* vol. 12 edited by T. Fomby, and R. Carter Hill Jai Press Inc., London.

Thompson, K. R., and Demirov, E. (2006). Skewness of sea level variability of the world's oceans. *Journal of Geophysical Research,* vol. 111, C05005, doi:10.1029/2004JC002839.

Torrence, C., and Compo, G. P. (1998). A practical guide to wavelet analysis. *Bulletin of American Meteorological Society,* vol. 79, pp. 61–78.

Walter, K., and Graf, H. -F. (2002). On the changing nature of the regional connection between the North Atlantic oscillation and sea surface temperature. *Journal of Geophysical Research,* vol. 107(D17), 4338, doi:10.1029/2001JD000850.

Wang, X. L., and Zwiers, F. W. (1999). Interannual variability of precipitation in an ensemble of AMIP climate simulations conducted with the CCC GCM2. *Journal of Climate,* vol. 12, no. 5, pp. 1322–1335.

Water for Texas (2007). Texas Water Development Board Report, Document no: GP-8-1.

Wittrock, V., and Ripley, E. A. (1999). The predictability of autumn soil moisture levels on the Canadian prairies. *International Journal of Climatology,* vol. 19, pp. 271–289.

Zhang, Z., and Mann, M. E. (2005). Coupled patterns of spatiotemporal variability in Northern Hemisphere sea level pressure and conterminous U.S. drought. *Journal of Geophysical Research,* vol. 110, D03108, doi:10.1029/2004JD004896.

CHAPTER 7

Rainfall Frequency Distributions

E
xtreme rainfall values, such as annual rainfall maxima, are of interest in modeling floods and quantifying the effects of climate change. Their frequency analysis is used for constructing intensity-duration-frequency (IDF) curves which are needed for a range of hydrologic designs, including drainage systems, culverts, roadways, parking lots, runways, and urban development. Extreme rainfall exhibits different properties for different durations in different regions. Analysis of rainfall data from many regions of the world shows that the form of the frequency distribution of annual rainfall maxima changes with time scale (or duration), climate type, and distance from the sea. Therefore, determining rainfall characteristics is important for choosing a suitable rainfall distribution and consequently estimating rainfall quantiles. From the fitted distribution, statistical properties of extreme rainfall values can be investigated and extrapolated beyond the available data for engineering purposes. The objective of this chapter is to discuss the use of entropy theory for deriving frequency distributions for annual rainfall maxima that hopefully apply across different time scales, climate types, and distances from the sea, through a case study application of the theory (Hao and Singh, 2013).

7.1 Factors Affecting Rainfall Frequency Distributions

The shape of the annual maximum rainfall frequency distribution seems to depend on the duration, climate, and distance from the sea. Consider, e.g., the state of Texas (longitude: 93°31'W to 106°38'W, latitude: 25°50'N to 36°30'N), as shown in Fig. 7.1. The climate of Texas is dominated by the passage of two competing influences due to frontal systems from the northwest and the moist air moving inland from the Gulf of Mexico. Texas has three main types of climate: continental, mountain, and modified marine with no clearly distinguishable boundaries. The modified marine zone is further classified into four "subtropical" zones, as shown in Fig. 7.1. The mountain climate is dominant in several mountains of the Trans-Pecos region. Different zones of continental and modified marine climate are abbreviated as continental steppe (CS), subtropical arid (SA), subtropical humid (SH), subtropical subhumid (SSH), and subtropical steppe (SST) zone. In addition, the U.S. National Weather Service has divided Texas into 10 climate divisions (including Upper Coast, East Texas, High Plain, Trans-Pecos, and so on).

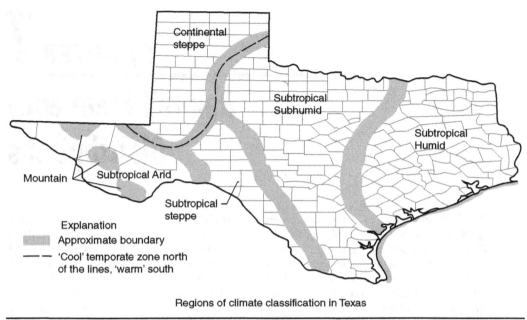

Regions of climate classification in Texas

FIGURE 7.1   Regions of climate zones in Texas (Larkin and Bomar, 1983).

7.1.1   Change in Distribution Form with Time Scale

Histograms of annual rainfall maxima of different durations show that frequency distributions for short durations are more skewed with sharp peaks but tend to be less skewed as the duration increases, as shown in Fig. 7.2. For example, annual rainfall maxima for a station in subhumid tropical climate in Texas have a skewness value of 2.7 for 15-min data but 1.1 for 30-d data (not shown). The box plot of skewness values for data sets of different durations, given in Fig. 7.3, shows that the 75 percentile of skewness of the 15-min duration is around 3.2 while that for the 30-d duration is 1.2. This is so partly because for short durations such as 15-min, a large amount of rainfall may occur within a short time in certain cases exhibiting large skewness, whereas for long durations, such as 30-d, the data are averaged and thus they exhibit less skewness.

7.1.2   Change in Distribution Form with Climatic Zone

The subtropical humid zone lies in the eastern part of Texas which is mostly noted for warm summers. This zone includes most parts of Upper Coast and East Texas divisions. Four rainfall generating mechanisms lead to varying patterns from year to year. In the Upper Coast area, typical thunderstorms are expected slightly inland in May while the belt of maximum activity is along the coast by July; in September tropical disturbances can cause very heavy rains for some years, while in December the frontal activity affects the region. The East Texas division is characterized by a fairly uniform seasonal rainfall with slight maxima occurring in May and December, and there is little variation in the weather in the summer season, because the influence of the Gulf of Mexico is dominant. The most widespread and lengthy precipitation periods in East

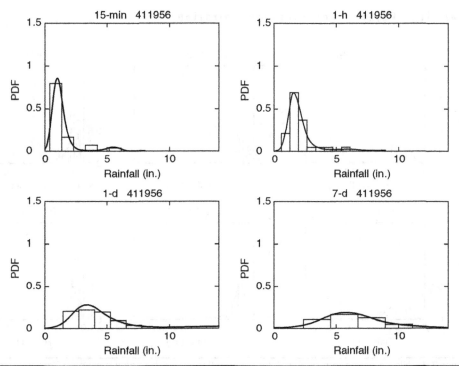

FIGURE 7.2 Histograms and probability density functions of rainfall data of different durations for station 411956 in the subtropical humid (SH) climate (after Hao and Singh, 2013).

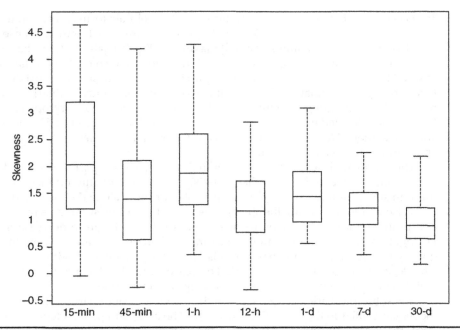

FIGURE 7.3 Skewness of rainfall of different durations (40 data sets for each duration) (after Hao and Singh, 2013).

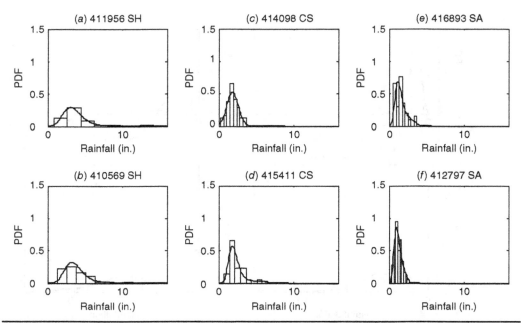

FIGURE 7.4 Histograms and probability density functions of 12-h rainfall maximum from different climate regions.

Texas occur during spring and autumn when the cold air forms a barrier, forcing the overriding moist Gulf air to be deflected upward where it cools and condenses.

Histograms of 12-h annual rainfall maxima in Fig. 7.4*a* and *b* show that frequency distributions are smooth. Because this region is along the coast, the rainfall pattern is affected by the Gulf of Mexico. Since the proximity to the coast is the most determining factor for regional climate differences, the reason for this frequency distribution pattern may be the moderating moisture from the Gulf of Mexico. The subtropical subhumid (SSH) zone is located in the central part of Texas which is characterized by hot summers and dry winters. No clear pattern is discernible from the frequency distribution of several stations in this climate zone.

The continental steppe (CS) zone lies in the northwestern part of Texas and includes the regions similar to the High Plains division. The rainfall amount increases steadily through spring and reaches a maximum in May or June, while the thunderstorm activity is also on the rise during the spring season. In this region, summer is the wet season, and thunderstorms are numerous in June and July but begin to decrease in August. Histograms for 12-h annual rainfall maxima, seen in Fig. 7.4*c* and *d*, show that the frequency distributions in this part are relatively sharp, compared with those in the SH climate zone (Hao and Singh, 2013). The reason may be that the maximum rainfall mainly comes from thunderstorms during the summer season.

The subtropical arid (SA) zone lies in the extreme western part of Texas and includes the region similar to the Trans-Pecos division. The basin and plateau region of the Trans-Pecos features a subtropical arid climate, which is marked by summertime rainfall anomalies of the mountain relief. Rainfall reaches its maximum in July and in summer, where the rain comes mainly from thunderstorms, often affected by local topography. In the Trans-Pecos region, the biggest percentage of rainfall occurring in this area is due to convective showers and thundershower activity, while the thundershower activity is the primary contributor of rainfall during late summer and early autumn months. The histograms for the 12-h annual rainfall maxima, seen in Fig. 7.4*e* and *f*, show that the frequency distributions are relatively sharp compared with those in the SH climate zone. The reason for the variation of rainfall may be that the heavy rainfall in the SA zone is mainly produced by convective showers and the thundershower activity.

From the mid-Rio Grande Valley to the Pecos Valley, the broad swath of Texas has a subtropical steppe (SST) climate and is typified by semiarid to arid conditions. No clear pattern of frequency distributions in this zone is found from the data of several stations. In general, frequency distributions for regions in extreme northern and western parts (or the CS and SA climate zones) are sharp; however, those for the regions in the southeast near the Gulf of Mexico (or the SH climate zone) are rather smooth.

7.1.3   Influence of Distance from the Sea

The Gulf of Mexico is particularly important for the climate of Texas, as it provides the source of moisture and modulates the average seasonal and diurnal weather cycles, particularly in the coastal regions. In general, the average annual rainfall decreases with increasing distance from the Gulf of Mexico. The histograms of 12-h maximum rainfall in Fig. 7.5 show that the frequency distributions in group II (more than 250 mi away from the Gulf) are not as smooth as those in group I (within 60 mi of the Gulf) (Hao and Singh, 2013). The smoothness of frequency distributions in group I is partly due to the closeness to the Gulf of Mexico. The effect of the Gulf of Mexico is reduced with the distance away from it, and the topography may play an important role.

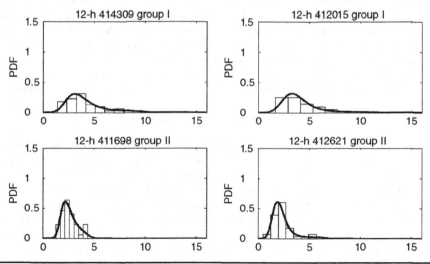

FIGURE 7.5   Histograms and probability density functions of 12-h rainfall maximum for different distances from the Gulf of Mexico (station 414309, 60 mi; station 412015, 20 mi; station 411698, 480 mi; and station 412621, 450 mi).

7.2   Derivation of Annual Maximum Rainfall Distributions

It is assumed that the annual maximum rainfall for a given duration is a continuous random variable $X \in [a, b]$ with a probability density function (PDF), of $f(x)$. The objective is to derive the PDF $f(x)$ using entropy theory. For $f(x)$, the Shannon entropy H can be defined as (Shannon, 1948; Shannon and Weaver, 1949)

$$H = -\int_a^b f(x)\ln f(x)\,dx \qquad (7.1)$$

where x is a value of random variable X with lower limit a and upper limit b. To derive the least-biased PDF using the principle of maximum entropy (POME) (Jaynes, 1957), the constraints can be expressed in general form as

$$\int_a^b g_r(x)f(x)\,dx = \overline{g}_r, \qquad r = 0, 1, 2, \dots, m \qquad (7.2)$$

where function $g_r(x)$ in Eq. (7.2) is the known function with $g_0(x) = 1$; $\overline{g}_r$ is the rth expected value obtained from observations with $g_0 = 1$; and m is the number of constraints.

Following the discussion in Chap. 1, the maximum entropy-based probability density function can then be obtained by maximizing the entropy in Eq. (7.1), subject to Eq. (7.2), using the method of Lagrange multipliers (Kesavan and Kapur, 1992):

$$f(x) = \exp[-\lambda_0 - \lambda_1 g_1(x) - \lambda_2 g_2(x) - \dots - \lambda_m g_m(x)] \qquad (7.3)$$

where $\lambda_r (r = 0, 1, \dots, m)$ are the Lagrange multipliers.

7.2.1   Maximum Entropy-Based Distribution with Moments as Constraints

For simplicity, constraints are defined in terms of moments, and no more than four moments are used in practice. It has been shown by Matz (1978) that the first four moments would suffice in most cases for deriving $f(x)$. Therefore, the first four moments can be defined by expressing function $g(x)$ in Eq. (7.2) as $g_i(x) = x_i$ ($i = 1, 2, 3,$ and 4), and these moments constitute constraints. Then the maximum entropy-based probability density function (denoted ENT4), defined on the interval $[a, b]$, can be expressed as

$$f(x) = \exp(-\lambda_0 - \lambda_1 x - \lambda_2 x^2 - \lambda_3 x^3 - \lambda_4 x^4) \qquad (7.4)$$

The lower bound of the interval a can be set to be zero, while the upper bound b can be set to be 20 times the observed maximum value. Since higher moments are involved in Eq. (7.4), a relatively large data set would be needed for the accuracy of moment estimation.

With the first four moments as constraints, the skewness, kurtosis, and multiple modes can be included in the resulting maximum entropy-based distribution (Zellner and Highfield, 1988). Each maximum of the polynomial inside the exponential corresponds to one mode, and thus multiple modes may exist in the maximum entropy-based distribution (Smith, 1993). This distribution has been applied for fitting bimodal distributions (Eisenberger, 1964). Matz (1978) examined this distribution, and application of this distribution shows that it fits the observed frequencies well. Comparing this distribution with the Pearson distribution, Zellner and Highfield (1988) showed that it provides a better fit, especially at the tails. Smith (1993) used the maximum entropy-based distribution with moments as constraints for decision analysis to construct the distribution of value lottery and showed the distribution with the first four moments as constraints performed well.

The entropy-based distribution with the first three moments as constraints can also be considered as a candidate for modeling extreme rainfall values. From Eq. (7.3), this distribution can be expressed as

$$f(x) = \exp(-\lambda_0 - \lambda_1 x - \lambda_2 x^2 - \lambda_3 x^3)$$ (7.5)

There are three parameters associated with this distribution, which is denoted as ENT3.

7.2.2 Estimation of Parameters

The Lagrange multipliers of Eq. (7.3) have to be determined using Eq. (7.2) where $E[g_r]$ ($r = 1,\ldots, 4$) is the expectation of the first four noncentral moments. Generally, the analytical solution does not exist, and numerical estimation of the Lagrange multipliers is the only resort. Two methods are presented here.

Method 1

Substitution of Eq. (7.3) in Eq. (7.2) with $r = 0$ leads to

$$\int_a^b \exp\left[-\lambda_0 - \sum_{r=1}^m g_r(x)\right] dx = 1$$ (7.6a)

Equation (7.6a) leads to

$$\lambda_0 = \int_a^b \exp\left[-\sum_{r=1}^m g_r(x)\right] dx$$ (7.6b)

Adding $\sum_{r=1}^4 \lambda_r \overline{g_r}$ to both sides, one obtains

$$\Gamma = \lambda_0 + \sum_{r=1}^4 \lambda_r \overline{g_r} = \ln \int_a^b \exp\left[-\sum_{r=1}^4 \lambda_r g_r(x)\right] dx + \sum_{r=1}^4 \lambda_r \overline{g_r}$$ (7.6c)

One can maximize the function Γ (Mead and Papanicolaou, 1984; Wu, 2003). For determining the Lagrange multipliers, the maximization can be achieved by employing Newton's method. Starting from some initial value $\lambda_{(0)}$, one can solve for the Lagrange parameters by updating $\lambda_{(1)}$ through the equation

$$\lambda_{(1)} = \lambda_{(0)} - H^{-1} \frac{\partial \Gamma}{\partial \lambda_i} \qquad i = 1, 2, 3, 4$$ (7.7a)

where Γ is the gradient expressed as

$$\frac{\partial \Gamma}{\partial \lambda_i} = \overline{g_i} - \int_a^b \exp\left[-\sum_{r=0}^4 \lambda_r g_r(x)\right] g_i(x) dx \qquad i = 1, 2, \ldots, 4$$ (7.7b)

and H is the hessian matrix whose elements are expressed as

$$H_{i,j} = \int_a^b \exp\left(-\sum_{r=0}^4 \lambda_r g_r(x)\right) g_i(x) g_j(x) dx$$

$$-\int_a^b \exp\left(-\sum_{r=0}^4 \lambda_r g_r(x)\right) g_i(x) dx \bullet \int_a^b \exp\left(-\sum_{r=0}^4 \lambda_r g_r(x)\right) g_i(x) dx, i, j = 1, 2, \ldots, 4 \qquad (7.7c)$$

Method 2

This method leads to a system of equations that can be solved numerically. To that end, substitution of Eq. (7.3) in Eq. (7.2) with $r = 0$ yields

$$\int_a^b \exp[-\lambda_0 - \lambda_1 g_1(x) - \lambda_2 g_2(x) - \cdots - \lambda_m g_m(x)] dx = 1 \qquad (7.8a)$$

Taking $\exp[-\lambda_0]$ on the left side and then taking the logarithms, Eq. (7.8a) can be cast as

$$\lambda_0 = \ln \int_a^b \exp[-\lambda_1 g_1(x) - \lambda_2 g_2(x) - \cdots - \lambda_m g_m(x)] dx \qquad (7.8b)$$

Differentiating Eq. (7.8b) with $\lambda_j, j = 1, 2, \ldots, m$ and equating each derivative to Eq. (7.2), one finds

$$\frac{\partial \lambda_0}{\partial \lambda_j} = -\overline{g_j(x)} \qquad j = 1, 2, \ldots, m \qquad (7.9a)$$

Now recall that λ_0 is a function of other Lagrange multipliers and can be expressed as

$$\lambda_0 = \lambda_0(\lambda_1, \lambda_2, \ldots, \lambda_m) \qquad (7.9b)$$

Equation (7.9b) can be explicitly obtained from the solution of Eq. (7.8b). Then differentiating Eq. (7.9b) with $\lambda_j, j = 1, 2, \ldots, m$, and equating each derivative to Eq. (7.9a), one finds a system of equations involving the Lagrange multipliers and constraints which can then be solved for the unknown multipliers.

7.3 Three Generalized Distributions

Many frequency distributions have been used for rainfall frequency analysis. Three generalized distributions, including the generalized extreme value (GEV) distribution, generalized gamma (GGM) distribution, and generalized beta (GBT) distribution, are selected for comparison with the entropy-based distribution. These distributions have been applied extensively in hydrology. By entropy maximizing, Papalexiou and Koutsoyiannis (2012) derived three-parameter generalized gamma and four-parameter generalized beta of the second kind (GB2) which are flexible distributions for analyzing daily rainfall. They tested these distributions with 11,519 daily rainfall records from across the world and found that the GB2 distribution was able to fit all records. Two special cases of the GB2 distribution, the GGM and Burr type XII distributions, fitted nearly 98 and 88 percent of the records. In another study, Papalexiou and Koutsoyiannis (2013) analyzed three probability distributions, based on extreme value theory, for annual maxima of daily rainfall: (1) type I or Gumbel, (2) type II or Frechet, and (3) type III or reverse Weibull. These three extreme value distributions are special cases of the generalized extreme value (GEV) distribution. Analyzing the annual maxima of daily rainfall from 15,137 stations from around the world with record length varying from 40 to 163 yr, they found that the GEV shape parameter varied from one location to another and was precipitously influenced by the record length. Now a short discussion of the three extreme value distributions is given below.

7.3.1 Generalized Extreme Value Distribution

The GEV distribution has a probability density function defined as

$$
f(x;\mu,\sigma,k) = \begin{cases} \dfrac{1}{\sigma}\left[1+k\left(\dfrac{x-u}{\sigma}\right)\right]^{-1/k-1} \exp\left\{-\left[1+k\left(\dfrac{x-u}{\sigma}\right)\right]^{-1/k}\right\} & k \neq 0 \\[3mm] \dfrac{1}{\sigma}\exp\left[-\dfrac{x-u}{\sigma}-\exp\left(-\dfrac{x-u}{\sigma}\right)\right] & k = 0 \end{cases}
\tag{7.10}
$$

where u, σ, and k are the location, scale, and shape parameters, respectively. One problem with fitting the GEV distribution to the annual rainfall maxima is that the resulting quantile estimators have large variance and may be seriously biased. In that case, regional frequency analysis can be applied to increase the accuracy of the estimates by assuming that certain parameters are constant over the region (Overeem et al., 2009). Many methods have been proposed for parameter estimation, such as the maximum likelihood method (Smith, 1985), method of moment (Madsen et al., 1997), L-moment method (Hosking, 1990), partial probability weighted moment method (Wang, 1990), and generalized maximum likelihood method (Martins and Stedinger, 2000). This distribution can also be obtained by employing the principle of maximum entropy subject to the constraints as follows (Singh, 1998):

$$
E\left[\ln\left[1+k\left(\frac{x-u}{\sigma}\right)\right]\right] = \overline{\ln\left[1+k\left(\frac{x-u}{\sigma}\right)\right]}
\tag{7.11}
$$

$$
E\left[1+k\left(\frac{x-u}{\sigma}\right)\right]^{-1/k} = \overline{\left[1+k\left(\frac{x-u}{\sigma}\right)\right]^{-1/k}}
\tag{7.12}
$$

Derived from the extreme value theory, GEV encompasses as special cases three types of distributions, namely, the Gumbel, Frechet, and Weibull. This distribution has been used for extreme rainfall frequency analysis in different areas of the world. Schaefer (1990) used the GEV distribution for frequency analysis of rainfall annual maxima of durations of 2, 6, and 24 h for the state of Washington. Huff and Angel (1992) selected the GEV distribution to model the distribution of annual rainfall maxima for durations from 5 min to 10 d, in the midwestern United States. Parrett (1997) also used the GEV distribution to construct dimensionless frequency curves of annual rainfall maxima of durations of 2, 6, and 24 h within each region in Montana. Using the

L-moment ratio diagram, Asquith (1998) determined that the GEV distribution was an appropriate distribution for modeling the annual maxima for durations from 1 to 7 d. Alila (1999) showed that the annual rainfall extremes for durations from 5 min to 24 h in Canada were better described by the GEV distribution than other distributions, such as the generalized logistic (GLO), Pearson type 3 (P3), and EV1 distributions.

7.3.2   Generalized Gamma Distribution

The generalized gamma (GGM) distribution, discussed by Stacy (1962), is defined as

$$f(x) = \frac{px^{d-1}}{a^d \Gamma(d/p)} \exp\left[-\left(\frac{x}{a}\right)^p\right] \qquad x \geq 0 \qquad (7.13)$$

where a, d, and p are positive parameters. The generalized gamma distribution is flexible in that it includes many distributions as special cases, such as the gamma, chi-square, Weibull, and exponential distributions (Stacy, 1962; Kleiber and Kotz, 2003). Papalexiou and Koutsoyiannis (2012) derived this distribution using the entropy maximizing method and applied it to investigate the probability distribution of rainfall. However, the difficulty of parameter estimation of the generalized gamma distribution is still an unresolved issue. The moment method is difficult to use, since different sets of parameters result in the same probability density function, while the maximum likelihood method is not stable because of the parameter confounding (Kleiber and Kotz, 2003; Gomès et al., 2008; Chen and Lio, 2009). This distribution can also be obtained with the principle of maximum entropy with the constraints as follows with certain reparameterization (Koutsoyiannis, 2010):

$$E[\ln x] = \overline{\ln x} \qquad E[x^q] = \overline{x^q} \qquad (7.14)$$

7.3.3   Generalized Beta Distribution

The generalized beta (GBT) distribution of the second kind is defined as

$$f(x) = \frac{\gamma_3}{\gamma_4^{\gamma_1 \gamma_3} B(\gamma_1, \gamma_2)} x^{\gamma_1 \gamma_3 - 1} \left[1 + \left(\frac{x}{\gamma_4}\right)^{\gamma_3}\right]^{-(\gamma_1 + \gamma_2)} \qquad (7.15)$$

where $B(\ldots)$ is the beta function; parameters γ_1, γ_2, γ_3 are the shape parameters; γ_4 is the scale parameter; and all the parameters are positive. Mielke and Johnson (1974) applied this distribution in hydrology and meteorology, and Koutsoyiannis (2010) derived this distribution with the entropy maximizing method. This distribution is also quite flexible and includes other distributions as special or limiting cases, such as the Pearson VI distribution and generalized gamma distribution (Kleiber and Kotz, 2003). It has been applied to different areas, such as the income distribution (McDonald, 1984). The distribution can be obtained using the principle of maximum entropy with the constraint reparameterization as

$$E[\ln x] = \overline{\ln x} \qquad (7.16)$$

$$E\left[\frac{\ln(1 + px^q)}{p}\right] = \frac{\overline{\ln(1 + px^9)}}{p} \qquad (7.17)$$

where p and q are parameters.

7.4 Case Study 1: Evaluation of Entropy-Based Distributions and Comparison Using Observed Rainfall Data

Hao and Singh (2013) have evaluated the performance of the entropy-based distribution in modeling extreme rainfall quantiles using the root mean square and relative root mean square error and have compared it with the generalized extreme value for both observed data and Monte Carlo simulated data. Their discussion is followed here.

7.4.1 Root Mean Square Error

The root mean square error (RMSE) and relative root mean square error (RRMSE) are defined as

$$\text{RMSE} = \sqrt{\frac{1}{n}\sum_{i=1}^{n}(x_i - o_i)^2} \qquad \text{RRMSE} = \sqrt{\frac{1}{n}\sum_{i=1}^{n}\left(\frac{x_i - o_i}{o_i}\right)^2} \qquad (7.18)$$

where n is length of observed data; the x_i are the quantiles estimated from the entropy-based distribution; and the o_i are the observed quantiles obtained from the fitted distribution corresponding to the empirical nonexceedance probabilities estimated by the plotting position formula, such as the Gringorten plotting position formula (Gringorten, 1963)

$$P = \frac{i - 0.44}{n + 0.12} \qquad (7.19)$$

where i is the rank of the observed values and n is the length of the observed data.

7.4.2 Modeling of Extreme Values

Two typical annual rainfall maxima data sets of 1-h duration from two sample stations (417174, 419532) were used to compare the performance in modeling the extreme values. One of the differences between the two data sets is that several extreme quantiles (defined as the large quantiles deviating greatly from other data) exist for the data of station 419532, while the data for station 417174 are relatively smooth. Empirical histograms and fitted PDFs of the four distributions for the two stations are shown in Figs. 7.6 and 7.7. For the data of station 417174, the PDF curves of the four distributions are close to one another for the most part. For the data of station 419532, the PDF curves of the ENT distribution differ from other distributions, especially at the right tail where there is a large quantile of 6.20 in. Interestingly, the PDF curves of GEV, GGM, and GBT all decrease gradually to zero at the right tail, while the PDF curve of ENT distribution decreases at a slower rate at the tail.

The Q-Q (quantile-quantile) plots for observed and estimated quantiles were constructed for the two data sets for further comparison, as shown in Figs. 7.8 and 7.9. For station 417174, quantiles from all four distributions fit the observed values relatively well for the whole range of data, as shown in Fig. 7.8. For station 419532, quantiles estimated from the four distributions fit the observations well for most of the data. However, only the ENT distribution performs well for extreme quantiles, while the other three distributions perform poorly. The RMSE and RRMSE values for the four

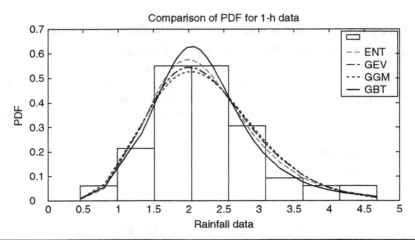

FIGURE 7.6   Comparison of theoretical density and histogram of 1-h rainfall data from station 417174 in subtropical humid (SH) climate and 419532 in subtropical subhumid (SSH) climate.

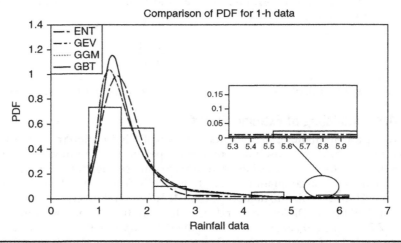

FIGURE 7.7   Comparison of theoretical density and histogram of 1-h rainfall data from station 419532 in subtropical subhumid (SSH) climate.

distributions are shown in Table 7.1. For these two data sets, RMSE and RRMSE of the ENT distribution are the smallest among the four distributions. For example, the RMSE value for the ENT distribution is 0.12 in., which is smaller than 0.25 for other distributions. Results from more data sets based on the Q-Q plots, RMSE, and RRMSE also show that generally the ENT distribution can model the extreme values better than the other distribution and has fewer RMSE and RRMSE values.

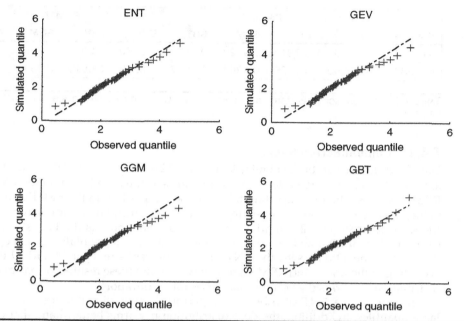

FIGURE 7.8 Comparison of Q-Q plots of 1-h rainfall data from station 417174 in subtropical humid climate.

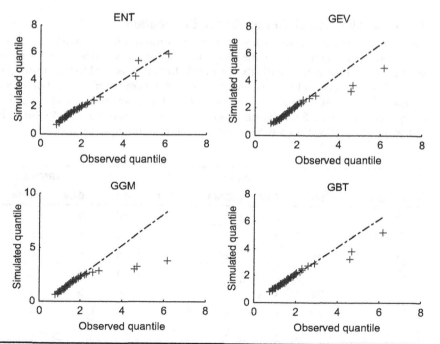

FIGURE 7.9 Comparison of Q-Q plots of 1-h rainfall data from station 419532 in subtropical subhumid climate.

Stations	RMSE				RRMSE			
	ENT	GEV	GGM	GBT	ENT	GEV	GGM	GBT
417174	0.1209	0.2819	0.4423	0.2551	0.0560	0.2070	0.3967	0.1664
419532	0.0959	0.1135	0.1244	0.1017	0.0148	0.0519	0.0640	0.0861

TABLE 7.1   Comparison of Performance of ENT, GEV, GGM, and GBT of Two Stations for 1-h Rainfall

7.4.3   General Performance

Data for different durations (15 min, 45 min, 1 h, 12 h, 1 d, 7 d, and 30 d) from 40 stations are used to compare the performance for quantile estimation based on the RMSE and RRMSE values. The four distributions are ranked according to the RMSE and RRMSE values as shown in Table 7.2, where the first ranked distribution is the one with the least RMSE or RRMSE. For all durations, the ENT distribution has the largest number of cases in which it is ranked first. For example, for the annual rainfall maxima of the 1-h duration of the 40 data sets, the ENT distribution ranks first for 34 and 29 data sets according to the RMSE and RRMSE, respectively. From these results, the entropy-based distribution performs the best for most cases for all durations.

In general, the ENT distribution outperforms other distributions for estimating large quantiles. In addition, the ENT distribution performs better than other distributions for the quantile estimation for different durations. Thus, the ENT distribution can be considered as a good candidate for rainfall analysis.

7.4.4   Application of Entropy-Based Distribution

The entropy-based distribution was used to fit the rainfall data, as shown in Figs. 7.10, 7.11, and 7.12 together with the empirical histograms. These figures show that the entropy-based distribution fits the empirical histograms well for the rainfall data of different durations, climate zones, and different distances from the Gulf of Mexico and can be used for rainfall frequency analysis. The GEV distribution is also applied here for comparison with the ENT distribution. For each duration (15 min, 45 min, 1 h, 12 h, 1 d, 7 d, and 30 d), 10 stations were used in each climate zone (except that for SA climate

Duration	RMSE				RRMSE			
	ENT	GEV	GGM	GB T	ENT	GEV	GGM	GBT
15 min	30	2	0	8	27	5	1	7
45 min	31	1	2	6	24	6	1	9
1 h	34	1	2	3	28	5	4	3
12 h	31	1	2	6	24	5	2	9
1 d	32	4	2	2	30	3	5	2
7 d	29	6	4	1	23	9	5	3
30 d	31	6	1	2	21	4	5	10

TABLE 7.2   Number of the First Ranking of Performance for Different Durations

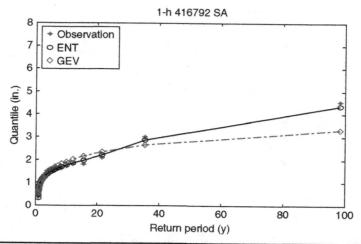

Figure 7.10 Quantiles and return periods of 1-d rainfall data for different distributions of station 416792 in SA climate zone (after Hao and Singh, 2013).

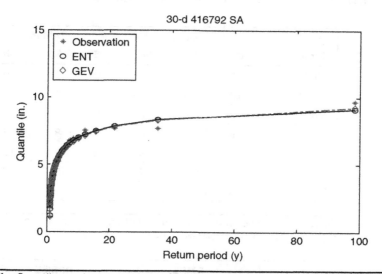

Figure 7.11 Quantiles and return periods of 1-h rainfall data for different distributions of station 416792 in SA climate zone (after Hao and Singh, 2013).

zone, 6 stations were used for 15-min and 45-min duration due to limited data). The number of the cases in which ENT performs better than GEV in different climate zones is shown in Table 7.3. The ENT distribution is also compared with the GEV distribution for different distances from the sea (group I and group II) with 10 stations in each group. For the lack of sufficient stations with a relatively long record of 15-min data in group I, only the hourly (1-h and 12-h) and daily data (1-d, 7-d, and 30-d) are used. The number of the cases in which ENT performs better than GEV in different distances from the sea is shown in Table 7.4.

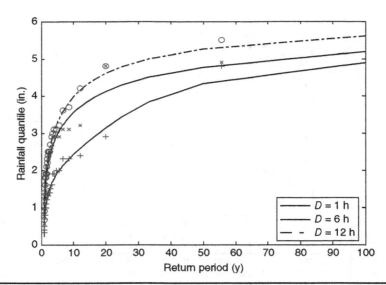

FIGURE 7.12   IDF curves for different durations for station 418583.

Duration	RMSE			RRMSE		
	CS	SA	SH	CS	SA	SH
15 min	9	4	10	10	3	10
45 min	10	6	9	8	4	8
1 h	9	9	10	10	7	7
12 h	9	10	8	7	8	8
1 d	8	6	9	8	5	10
7d	8	8	7	8	5	7
30 d	8	8	7	8	5	5

TABLE 7.3   Number of the First Ranking of ENT Performance for Different Climate Zones (10 Stations in Each Zone)

Duration	RMSE		RRMSE	
	Group I	Group II	Group I	Group II
1 h	10	8	7	8
12 h	6	9	5	10
1 d	9	9	9	9
7d	10	9	9	7
30 d	10	9	7	9

TABLE 7.4   Number of the First Ranking of Performance for Different Distances from the Sea (10 Stations in Each Group)

First, the performance of ENT and GEV is assessed for different durations. Taking the CS climate zone as an example, for all the durations, the ENT distribution performs better for at least 8 out of 10 data sets based on RMSE and for at least 7 out of 10 data sets based on RRMSE.

Second, the performance of ENT and GEV distributions was also assessed for different climate zones. Taking 1-h data as an example, out of 10 data sets the ENT distribution has less RMSE for 9, 9, and 10 data sets for the CS, SA, and SH climate zones, respectively. The ENT distribution has less RRMSE for at least 7 out of 10 data sets for different climate zones. Generally the difference between the RRMSE values of ENT and GEV becomes smaller for 30-d duration than that for 1-h duration.

Third, the performance of ENT and GEV distributions is assessed for stations with different distances from the sea. Taking 1-h data as an example, the ENT distribution has less RMSE for 10 and 8 cases and less RRMSE for 7 and 8 cases for groups I and II, respectively. The RRMSE values for hourly data (1-h and 12-h) and daily data (1-d and 30-d) for different distances from the sea generally show that RRMSE is smaller for 30-d duration than that for 1-h duration.

7.5   Frequency Analysis and IDF Curves

The ENT and GEV distributions are used for frequency analysis, and the return period is computed. Comparison of results from the ENT and GEV distributions for 1-h rainfall data for station 416792 is shown in Fig. 7.10. The ENT distribution fits the observations well for the whole range of data, while the GEV distribution does not fit as well for long return periods (longer than 50 yr). Comparison of the ENT and GEV distributions for the 30-d rainfall data from station 416792 is shown in Fig. 7.11. It can be seen that both ENT and GEV distributions fit the empirical data well. Data from other stations of different climate zones and different distances from the sea for different durations also show similar results. This further implies that for the rainfall analysis of long durations (say, 30-d) data, the GEV distribution can also model the data relatively well and may be preferred, since it has fewer parameters than the ENT distribution.

The above results show that the ENT distribution is applicable for rainfall analysis of data sets of different durations in different climate zones or different distances from the sea, although the rainfall distribution pattern varies. For long durations, the difference between performances of ENT and GEV distributions is relatively small, and thus the GEV distribution can also be used for rainfall analysis from a parsimonious point of view.

The annual maximum rainfall distribution can be employed for the construction of intensity-duration-frequency (IDF) curves (Singh, 1992). An *IDF curve* is defined as a relationship of rainfall intensity occurring over a certain duration d with different return periods. The hourly annual rainfall data for station 418583 are used to construct the IDF curves, as shown in Fig. 7.12. The empirical return period (T_E) is obtained from the Gringorten plotting position formula as $T_E = 1/(1-P)$, where P is the nonexceedance probability. The empirical return periods are also plotted on the IDF curves. Note that the accuracy of the empirical return period for the highest-ranked peak flows is limited (Stedinger, 1993; Beckers and Alila, 2004). Generally the return period from the IDF curves fits the empirical return period well. For example, for the return period of 20 yr, the theoretical rainfall quantile from the ENT distribution is 4.6 in. while the observed quantile is 4.8 in.

7.6   Case Study 2: Evaluation of Entropy-Based Distributions and Comparison Using Synthetic Data

Hao and Singh (2013) carried out Monte Carlo experiments to compare the quantiles estimated from the GEV, ENT4, and ENT3 distributions. Three Monte Carlo simulations were conducted and random numbers were generated from the known GEV, gamma, and lognormal distributions. The parent distributions used for random number generation are shown in Fig. 7.13.

For the first simulation (S_1), the quantiles corresponding to different return periods (5, 10, 25, 50, 100, 200 yr) were first assessed with the random number generated from the GEV distribution. For the second and third simulations, the quantiles corresponding to relatively long return period ($T = 100$ and 200 yr) for the three distributions were assessed with the synthetic data from the gamma distribution (S_2) and lognormal distribution (S_3) with different skewness (sk) values.

First, synthetic data were generated from the GEV distribution with different data lengths ($n = 40, 70, 100$). The GEV, ENT4, and ENT3 distributions were then fitted to these data from the GEV distribution, and the quantiles corresponding to different return periods were obtained. A total of 1000 data sets with different sample sizes were generated. The median and the relative RMSE for simulation S_1 are shown in Tables 7.5 and 7.6. From the median values, it can be seen that for a short return period ($T < 100$), the median values from the ENT4 and GEV distributions are close to each other for each sample size. For example, for sample size $n = 100$, the median values from GEV and ENT4 are 3.78 and 3.78, while the observed value is 3.82. The ENT3 distribution also estimates the quantiles well for relatively short periods ($T < 100$), while it does not model the quantiles well corresponding to relatively long return periods. From the RMSE values, it can be seen that the relative error for each distribution is within 50 percent for most cases. Generally ENT4 models quantiles well, especially when the

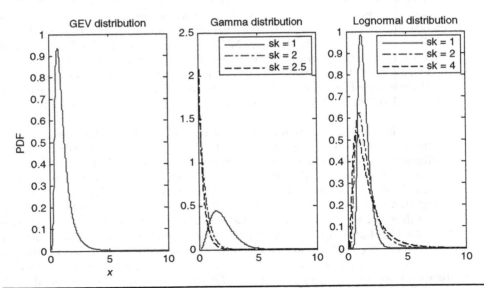

FIGURE 7.13   Three parent distributions used for Monte Carlo simulation.

Sample Size	Return Period	5	10	25	50	100	200
	Observation	**1.50**	**1.94**	**3.16**	**3.82**	**4.57**	**7.73**
$N = 40$	GEV	1.47	1.89	3.07	3.72	4.48	5.61
	ENT	1.47	1.91	2.97	3.20	3.35	3.50
	ENT3	1.65	1.97	2.54	2.74	2.93	3.16
$N = 70$	GEV	1.49	1.91	3.11	3.75	4.49	5.58
	ENT	1.47	1.87	3.18	3.57	3.81	4.01
	ENT3	1.68	2.02	2.62	2.84	3.04	3.29
$N = 100$	GEV	1.49	1.93	3.14	3.78	4.52	5.67
	ENT	1.48	1.88	3.20	3.78	4.00	4.22
	ENT3	1.70	2.05	2.68	2.90	3.11	3.36

TABLE 7.5 Median of Observed and Generated Quantiles from GEV for Different Data Length

Sample Size	Return Period	5	10	25	50	100	200
$N = 40$	GEV	0.10	0.13	0.28	0.38	0.52	0.77
	ENT	0.11	0.16	0.39	0.36	0.36	0.42
	ENT3	0.18	0.18	0.26	0.32	0.39	0.47
$N = 70$	GEV	0.08	0.10	0.19	0.25	0.31	0.41
	ENT	0.08	0.11	0.29	0.48	0.42	0.41
	ENT3	0.18	0.17	0.24	0.34	0.42	0.48
$N = 100$	GEV	0.06	0.08	0.15	0.19	0.24	0.31
	ENT	0.07	0.09	0.23	0.37	0.34	0.34
	ENT3	0.17	0.15	0.20	0.26	0.33	0.41

TABLE 7.6 RMSE of Observed and Generated Quantiles from GEV for Different Data Lengths

sample size was relatively large, while the ENT3 distribution does not perform as well as the ENT4 distribution for the quantiles corresponding to relatively long return periods.

Second, random numbers were generated from the gamma distribution with different skewness values. It was shown earlier that the skewness of the annual rainfall maximum decreases with the increase of the duration. Clearly, for minute and hourly data, the skewness is relatively high (>2) for many cases. The evaluation of the three distributions in modeling data with high skewness is important. To generate synthetic data for different skewness values, the gamma distribution is selected as the parent distribution for random number generation for the second simulation. Random values with different sample sizes (n = 40, 70, and 100) with different skewness values of 1, 2, and 2.5

Sample Size	Skewness	k = 1		k = 2		k = 2.5	
	Quantile	x_{100}	x_{200}	x_{100}	x_{200}	x_{100}	x_{200}
	Observation	**5.02**	**5.49**	**2.30**	**2.65**	**1.86**	**2.18**
N = 40	GEV	5.06	5.60	3.88	5.38	7.25	13.09
	ENT	4.60	4.82	2.00	2.14	1.53	1.64
	ENT3	4.36	4.62	2.02	2.20	1.54	1.68
N = 70	GEV	5.14	5.66	3.95	5.53	7.56	14.14
	ENT	4.80	5.05	2.11	2.28	1.70	1.85
	ENT3	4.46	4.73	2.08	2.30	1.70	1.86
N = 100	GEV	5.13	5.69	4.07	5.76	7.72	14.35
	ENT	4.88	5.16	2.19	2.39	1.75	1.90
	ENT3	4.43	4.71	2.15	2.40	1.75	1.91

TABLE 7.7 Median of ENT4, GEV, and ENT3 with Random Values Generated from the Gamma Distribution

are generated and used for comparison. The median and relative RMSE values of the quantiles x_{100} and x_{200} corresponding to nonexceedance probability values 0.99 and 0.995 are shown in Tables 7.7 and 7.8. Take the skewness $k = 1$, for example. For $n = 40$, 70, and 100, the median values from the ENT4 distribution are not as close to the observed values as for the GEV distribution. However, the RMSE values for ENT are not larger than for the GEV distribution for all n values. For example, for $n = 100$, the RMSE values for the ENT4 and ENT3 distributions are 0.12, and 0.14, while those for the GEV distribution are 0.12 and 0.15. Comparison of ENT3 and ENT4 shows that the median value from the ENT4 distribution is closer to the observed value than that from the ENT3 distribution. Generally, the three distributions are comparable in terms of the RMSE values, since in most cases the relative RMSE values are lower than 20 percent.

Sample Size	Skewness	k = 1		k = 2		k = 2.5	
	Quantile	x_{100}	x_{200}	x_{100}	x_{200}	x_{100}	x_{200}
N = 40	GEV	0.22	0.29	1.97	3.72	20.01	67.87
	ENT	0.18	0.19	0.28	0.29	0.32	0.34
	ENT3	0.16	0.18	0.25	0.28	0.32	0.34
N = 70	GEV	0.15	0.19	1.28	2.09	8.44	20.29
	ENT	0.14	0.16	0.23	0.25	0.26	0.28
	ENT3	0.14	0.16	0.21	0.23	0.25	0.28
N = 100	GEV	0.12	0.15	1.08	1.72	7.11	16.87
	ENT	0.12	0.14	0.19	0.22	0.23	0.25
	ENT3	0.13	0.15	0.17	0.20	0.22	0.25

TABLE 7.8 Relative RMSE of ENT4, GEV, and ENT3 with Random Values Generated from the Gamma Distribution

The performance of ENT4 improves with the increase of the sample size based on the median values or the relative RMSE values. For skewness $k = 2$, the median quantiles from the GEV distribution are overestimated compared with the observed quantiles. The median values from ENT4 and ENT3 are closer to the observed values, demonstrating that in this case the ENT4 and ENT3 distributions perform better than does the GEV distribution. Further comparison between ENT4 and ENT3 distributions shows that median values from ENT4 and ENT3 are not close to each other for sample size $n = 40$. However, median values from these three distributions are close to one another for sample size $n = 70$ and 100. The RMSE values also show that the GEV distribution does not perform as well as the other three distributions, since it has the largest relative RMSE for all sample sizes. The relative RMSE values for the ENT4 and ENT3 distributions for all sample sizes are comparable, while those from ENT3 are slightly smaller than from ENT4. The median and RMSE for the case $k = 2.5$ show similar results as for the case $k = 2$.

For simulation S_3 with random numbers generated from the lognormal distribution, the medians of the simulated quantile and RMSE are shown in Tables 7.9 and 7.10. For the case $k = 1$, the median of quantiles from the ENT4 distribution is not as close to the observed values as from the GEV distribution. However, for relatively large sample sizes, the median values from the ENT4 distribution are close to those from the GEV distribution. For example, for $n = 100$, the medians from the GEV and ENT4 distributions are 2.79 and 2.72 with the observed value of 2.80. The median value from ENT3 is somewhat underestimated. The RMSE values of these three distributions are relatively close to one another with the relative RMSE being lower than 20 percent. For relative skewness values $k = 2$ and 4, the median values from the GEV distribution are again overestimated, while those from the ENT3 distribution are underestimated. The median values from the ENT distribution are close to the observed values, better than from the

	Skewness	$k = 1$		$k = 2$		$k = 4$	
	Quantile	x_{100}	x_{200}	x_{100}	x_{200}	x_{100}	x_{200}
Sample Size	Observation	2.80	3.03	4.87	5.59	9.36	11.52
$N = 40$	GEV	2.74	2.95	5.07	5.99	11.44	17.51
	ENT	2.61	2.71	4.34	4.55	7.67	8.05
	LP3	2.77	2.99	4.78	5.47	9.16	11.16
	ENT3	2.44	2.55	3.81	4.08	7.23	8.01
$N = 70$	GEV	2.76	2.97	5.16	6.07	11.83	16.22
	ENT	2.68	2.81	4.65	4.93	8.82	9.41
	LP3	2.77	2.99	4.85	5.56	9.27	11.54
	ENT3	2.46	2.57	3.90	4.17	7.63	8.60
$N = 100$	GEV	2.79	3.01	5.20	6.15	11.55	17.75
	ENT	2.72	2.86	4.78	5.08	9.33	10.12
	LP3	2.80	3.02	4.87	5.60	9.20	11.23
	ENT3	2.47	2.59	3.95	4.22	7.82	8.86

TABLE 7.9 Median of ENT4, GEV, and ENT3 with Random Values Generated from Lognormal Distribution

Sample Size	Skewness Quantile	k = 1		k = 2		k = 4	
		x_{100}	x_{200}	x_{100}	x_{200}	x_{100}	x_{200}
N = 40	GEV	0.16	0.20	0.35	0.47	0.84	1.32
	ENT	0.15	0.17	0.26	0.28	0.47	0.45
	LP3	0.13	0.15	0.23	0.28	0.37	0.46
	ENT3	0.14	0.17	0.26	0.30	0.45	0.47
N = 70	GEV	0.12	0.15	0.25	0.33	0.59	0.86
	ENT	0.13	0.15	0.29	0.28	0.45	0.41
	LP3	0.10	0.12	0.18	0.22	0.27	0.32
	ENT3	0.13	0.16	0.23	0.27	0.32	0.39
N = 100	GEV	0.09	0.11	0.20	0.27	0.49	0.71
	ENT	0.11	0.13	0.23	0.24	0.44	0.41
	LP3	0.08	0.10	0.14	0.17	0.23	0.27
	ENT3	0.12	0.15	0.21	0.26	0.27	0.35

TABLE 7.10 Relative RMSE of ENT4, GEV, and ENT3 with Random Values Generated from Lognormal Distribution

GEV and ENT3 distributions. For example, for $n = 100$ and $k = 2$, the median values from GEV, ENT4, and ENT3 are 5.20, 4.78, and 3.95, respectively, whereas the observed value is 4.87. The RMSE values are relatively close for these three distributions for $k = 2$, while for $k = 4$ the RMSE values for the GEV distribution are relatively large.

The Monte Carlo simulation of S_1 shows that the ENT distribution could be comparable to GEV. The Monte Carlo simulation of S_2 and S_3 shows that the performance of the ENT distribution is comparable with GEV for low skewness, especially when the sample sizes are relative large ($n = 100$). When the skewness is relatively high (>2), the GEV distribution does not perform well for estimating quantiles corresponding to relatively long return periods, and the ENT4 distribution is preferable.

To summarize, the Monte Carlo simulation shows that generally the ENT4 distribution is comparable to the GEV distribution in modeling extreme rainfall values. Since the GEV distribution has been extensively applied for modeling extreme values, the results from simulation show that the ENT distribution would also be a candidate for modeling extreme values. The Monte Carlo simulation S_2 showed that the performance of the ENT distribution was comparable with the GEV for low skewness, especially when the sample sizes were relatively large ($n = 70$). When the skewness was relatively high (>2), the ENT4 distribution performed relatively better than the GEV distribution for estimating quantiles corresponding to relatively long return periods, especially when the sample size was large. Botero and Francés (2010) also found that the GEV distribution led to large errors for quantile estimation corresponding to long return periods for high skewness. Synthetic data from other distributions (e.g., gamma distribution) were also used for comparison, and generally similar results were obtained. Thus it can be concluded from the Monte Carlo simulation that generally the ENT4 distribution provided an alternative to the commonly used GEV distribution and should be preferable for observations with higher skewness. The ENT3 distribution was not suitable for modeling extreme values.

Questions

1. Obtain annual maximum rainfall data corresponding to different durations for a number of rain gauge stations from a coastal state. These stations should be at varying distances from the sea as well as from different climatic zones. These data could also be from different countries. The durations can be 15 min, 30 min, 1 h, 2 h, 4 h, 6 h, 12 h, and 24 h. Note there will only one value for a given duration for a year. Compute the mean, coefficient of variation, standard deviation, coefficient of skewness, and kurtosis for each duration and each station. Tabulate these values and see if there is any pattern.

2. Plot a frequency histogram of annual maximum rainfall for each duration for each station in Question 1, and discuss if these histograms vary with duration, climate type, and distance from the sea. What do these histograms look like? Do they resemble any known distribution?

3. Assume that, for the purposes of computing the probability distribution of annual rainfall maxima for a given duration, it will suffice to know the mean. Then compute the probability distribution and fit it to the histogram for each duration and for each station in Question 1. How good is the fit?

4. Now assume that it will be better to specify the mean and the mean of log values. Then compute the probability distribution and fit it to the histogram for each duration and for each station in Question 1. How good is the fit?

5. Now assume that it will be better to specify the mean and the mean of values raised to some power. Then compute the probability distribution and fit it to the histogram for each duration and for each station in Question 1. How good is the fit?

6. Now assume that it will be better to specify the first three moments. Then compute the probability distribution and fit it to the histogram for each duration and for each station in Question 1. How good is the fit?

7. Now assume that it will be better to specify the first four moments. Compute the probability distribution and fit it to the histogram for each duration and for each station in Question 1. How good is the fit?

8. Use two probability distributions that can represent the histograms of annual rainfall maxima in Question 1. Fit these distributions and discuss how well they fit.

References

Alila, Y. (1999). A hierarchical approach for the regionalization of precipitation annual maxima in Canada. *Journal of Geophysical Research*, vol. 104(D24), no. 31, pp. 645–651, 657.

Asquith, W. (1998). Depth-duration frequency of precipitation for Texas. *USGS Water Resources Investigations Report 98-4044*.

Beckers, J., and Alila Y. (2004). A model of rapid preferential hillslope runoff contributions to peak flow generation in a temperate rain forest watershed. *Water Resources Research*, vol. 40, no. 3, W03501.

Botero, B., and Francés F. (2010). Estimation of high return period flood quantiles using additional non-systematic information with upper bounded statistical models. *Hydrology and Earth System Sciences*, vol. 14, no. 12, pp. 2617–2628.

Eisenberger, I. (1964). Genesis of bimodal distributions. *Technometrics*, vol. 6, no. 4, pp. 357–363.

Gringorten, I. I. (1963). A plotting rule for extreme probability paper, *Journal of Geophysical Research*, vol. 68, p. 813.

Hao, Z., and Singh V. P. (2013). Entropy-based method for extreme rainfall analysis in Texas. *Journal of Geophysical Research*, vol. 118, pp. 263–273, doi: 10.1029/2011JD017394.

Huff, F., and Angel J. (1992). Rainfall frequency atlas of the midwest. Midwestern Climate Center, *Bulletin 71*, MCC Research Report 92–03, Champaign, Ill.

Jaynes, E. (1957). Information theory and statistical mechanics. *Physical Review*, vol. 106, no. 4, pp. 620–630.

Kesavan, H., and Kapur J. (1992). *Entropy Optimization Principles with Applications*, Academic Press, New York.

Larkin, T. and Bomar G. (1983). *Climatic Atlas of Texas*. Texas Department of Water Resources, Austin.

Matz, A. (1978). Maximum likelihood parameter estimation for the quartic exponential distribution. *Technometrics*, vol. 20, no. 4, pp. 475–484.

Madsen, H., Rasmussen, P., and Rosbjerg D. (1997). Comparison of annual maximum series and partial duration series methods for modeling extreme hydrologic events: 1. At-site modeling, *Water Resources Research*, vol. 33, no. 4, pp. 747–757.

Madsen, H., Pearson, C. and Rosbjerg D. (1997). Comparison of annual maximum series and partial duration series methods for modeling extreme hydrologic events 2. Regional modeling. *Water Resources Research*, vol. 33, no. 4, pp. 759–769.

Martins, E., and Stedinger J. (2000). Generalized maximum-likelihood generalized extreme-value quantile estimators for hydrologic data. *Water Resources Research*, vol. 36, no. 3, pp. 737–744.

McDonald, J. (1984). Some generalized functions for the size distribution of income. *Econometrica, Journal of the Econometric Society*, pp. 647–663.

Mead, L., and N. Papanicolaou (1984). Maximum entropy in the problem of moments. *Journal of Mathematical Physics*, vol. 25, no. 8, pp. 2404–2417.

Mielke, P. and Johnson E. (1974). Some generalized beta distributions of the second kind having desirable application features in hydrology and meteorology. *Water Resources Research*, vol. 10, no. 2, pp. 223–226.

Overeem, A., Buishand, T., and Holleman I. (2009). Extreme rainfall analysis and estimation of depth-duration-frequency curves using weather radar. *Water Resources Research*, vol. 45, no. 10, W10424, doi:10410.11029/12009WR007869.

Papalexiou, S. M., and Koutsoyiannis D. (2012). Entropy based derivation of probability distributions: A case study to daily rainfall. *Advances in Water Resources*, vol. 45, pp. 51–57.

Papalexiou, S. M., and Koutsoyiannis D. (2013). Battle of extreme value distributions: A global survey on extreme daily rainfall. *Water Resources Research*, vol. 49, pp. 187–201, doi: 10.1029/2012WR012557.

Parrett, C. (1997). *Regional Analysis of Annual Precipitation Maxima in Montana*. USGS Water-Resources Investigations Report, 51.

Schaefer, M. (1990). Regional analyses of precipitation annual maxima in Washington State. *Water Resources Research*, vol. 26, no. 1, pp. 119–131.

Shannon, C. and Weaver W. (1949). *The Mathematical Theory of Communication*, Univ. of Illinois Press, Urbana.

Shannon, C. E. (1948). A mathematical theory of communications. *Bell System Technical Journal*, vol. 27, no. 7, pp. 379–423.

Singh, V. P. (1992). *Elementary Hydrology*, Prentice Hall, Upper Saddle River, N.J.

Singh, V. P. (1998). *Entropy-based parameter estimation in hydrology*. Kluwer Academic Publishers (Now Springer), Dordrecht, The Netherlands.

Smith, J. (1993). Moment methods for decision analysis. *Management Science*, vol. 39, no. 3, pp. 340–358.

Smith, R. (1985). Maximum likelihood estimation in a class of nonregular cases. *Biometrika*, vol. 72, no. 1, p. 67.

Stacy, E. (1962). A generalization of the gamma distribution. *Annals of Mathematical Statistics*, vol. 33, no. 3, pp. 1187–1192.

Stedinger, J. R., Ed (1993). Frequency analysis of extreme events. In: *Handbook of Hydrology*, edited by D. R. Maidment, McGraw-Hill, New York, Chap. 18.

Wang, Q. (1990). Estimation of the GEV distribution from censored samples by method of partial probability weighted moments. *Journal of Hydrology*, vol. 120, no. 1-4, pp. 103–114.

Wu, X. (2003). Calculation of maximum entropy densities with application to income distribution. *Journal of Econometrics*, vol. 115, no. 2, pp. 347–354.

Zellner, A., and Highfield R. (1988). Calculation of maximum entropy distributions and approximation of marginal posterior distributions. *Journal of Econometrics*, vol. 37, no. 2, pp. 195–209.

Further Reading

Carr, J. (1967). *The Climate and Physiography of Texas*. Texas Water Development Board, Austin.

Center, N. F. I. (1987). *The Climate of Texas Counties*. University of Texas, Austin and Texas A & M University, College Station.

Chen, D., and Lio Y. (2009). A note on the maximum likelihood estimation for the generalized gamma distribution parameters under progressive type censoring. *International Journal of Intelligent Technologies*, vol. 2, no. 2, pp. 145–152.

Gomès, O., Combes, C. and Dussauchoy A. (2008). Parameter estimation of the generalized gamma distribution. *Mathematics and Computers in Simulation*, vol. 79, no. 4, pp. 955–963.

Hosking, J. (1990). L-moments: Analysis and estimation of distributions using linear combinations of order statistics. *Journal of the Royal Statistical Society, Series B (Methodological)*, vol. 52, no. 1, pp. 105–124.

Kleiber, C., and Kotz S. (2003). *Statistical Size Distributions in Economics and Actuarial Sciences*, Wiley-Interscience, New York.

Li, C., Singh, V. P., and Mishra A. K. (2012). Simulation of the entire range of daily precipitation using a hybrid probability distribution. *Water Resources Research*, vol. 48, W03521, doi: 10.1029/2011WR0011446.

Li, C., Singh, V. P., and Mishra A. K. (2013). A bivariate mixed distribution with a heavy-tailed component and its application to single-site daily rainfall distribution. *Water Resources Research*, vol. 49, pp. 767–789, doi: 10.1002/wrcr.20063.

Mead, L., and Papanicolaou N. (1984). Maximum entropy in the problem of moments. *Journal of Mathematical Physics*, vol. 25, no. 8, pp. 2404–2417.

Narasimhan, B., Srinivasan, R., Quiring, S., and Nielsen-Gammon J. (2008). *Digital Climatic Atlas of Texas*. Texas A&M University, Texas Water Development Board Contract, Report 2005-483-5591.

North, G., Schmandt, J., and Clarkson J. (1995). *The Impact of Global Warming on Texas*. University of Texas Press, Austin.

Pilon, P., Adamowski, K., and Alila Y. (1991). Regional analysis of annual maxima precipitation using *L*-moments. *Atmospheric Research*, vol. 27, no. 1-3, pp. 81–92.

Svenson, C., and Jones D. A. (2010). Review of rainfall frequency estimation methods. *Journal of Flood Risk Management*, vol. 3, pp. 296–313.

CHAPTER 8

Evaluation of Probabilistic Precipitation Forecasting

In forecasting precipitation, temperature, soil moisture, water level, river flow, pollution levels, traffic jam, and so on, two issues are of main concern: (1) forecast quality and (2) whether a forecast value is above or below some critical value. The purpose of a forecast is to provide information for practical use. The quality of a forecast or the quality of information provided by the forecast as well as its forecasting scheme is assessed by comparing forecasts with observations, and this can be done in many ways. Uncertainties in initial conditions and errors in methods of forecasting result in uncertainties in forecasts, whether deterministic or probabilistic. The amount of information contained in a probabilistic forecast is greater than that in a single deterministic forecast. Indeed, it may be possible to determine the economic benefit of this greater information to the user in decision making. Since information is the underpinning of a forecasting scheme and the forecast it provides and the verification thereof, it is logical to employ information theory. Because entropy and information are intertwined, entropy theory can provide a way to measure the quality of forecast and that of the forecasting scheme. Furthermore, it can also be employed to evaluate the quality of forecasts of existing schemes. Therefore, this chapter discusses the use of entropy for evaluating precipitation forecasting.

8.1 Notation

Consider a time series of observed precipitation $\{x_{0,i}; i = 1, 2, ..., n\}$, where i indicates the time (or occasion) of observed precipitation, say, hour or day or month or year; subscript 0 denotes the observed precipitation, and n denotes the length of time of observations. In this case, continuous time is being discretized into discrete values using a specified time interval. Let the precipitation series simulated by a single model be denoted as $\{x_{m,i}; i = 1, 2, ..., n\}$, where subscript m refers to model-simulated. Of common interest in precipitation forecasting is whether a forecast value is above or below a given threshold value c (for example, $c = 50$ mm/d for rainfall intensity) and whether a warning should be issued. Thus, two forecast categories are $x_{m,i} \leq c$ and $x_{m,i} > c$, and the corresponding probabilistic forecasts are $p_i = P(x_{m,i} \geq c)$ and $w_i = P(x_{m,i} \leq c)$ or $1 - p_i$. These forecasts must be compared with the observed occurrences $q_i = 1$ if $x_{o,i} \geq c$ (i.e., the event has occurred) and $q_i = 0$ if $x_{o,i} < c$ (i.e., the event has not occurred). In this case the rainfall occurrence is being considered as a binary event, and q_i is not a probability value.

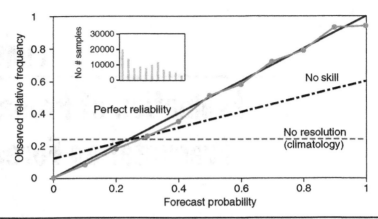

Figure 8.1   Reliability diagram.

A probabilistic forecast yields a value of probability of occurrence of an event between 0 and 1, or 0 and 100 percent. In general, a set of probabilistic forecasts is verified by comparing with observations of whether those events occurred. The agreement between forecast probabilities and mean observed frequencies defines the reliability. Thus, the reliability curve or diagram is the plot between observed frequency and forecast probability, in which the range of forecast probabilities is divided into a number of bins, such as 0 to 0.05, 0.05 to 0.10, 0.10 to 0.15, etc. The sample size in each bin is often included as a histogram or values besides the data points. The reliability curve is shown in Fig. 8.1. The reliability is measured by the closeness of the plotted curve to the diagonal, with the deviation indicating the conditional bias. For example, if the curve is below the line, then this indicates overforecasting, i.e., probabilities are too high. If the curve is above the line, then probabilities are too low, indicating underforecasting. The reliability diagram is conditioned on forecasts. That is, given that the occurrence of an event was predicted, what was the outcome? The diagram can yield information on the real meaning of the forecast.

Sharpness is usually defined by the tendency to forecast probabilities near 0 or 1 versus values clustered around the mean. Another term used in forecasting is *resolution*, which is defined by the ability of the forecast to resolve the set of sample events into subsets with characteristically different outcomes. If the curve in the reliability diagram is flatter, then it has less resolution. Resolution can also be measured by the relative operating characteristic (ROC), which measures the ability of the forecast to discriminate between two alternative outcomes. ROC is a plot between the probability of detection (POD), or hit rate, and false-alarm rate or probability of false detection (POFD), as shown in Fig. 8.2. Using a set of increasing probability thresholds (for example, 0.05, 0.1, 0.15, 0.2, etc.), ROC can be employed to make the yes/no decision. The area under the ROC curve is frequently used as the score, which ranges from 0 to 1, with 0.5 indicating no skill and 1 indicating a perfect score. It is not sensitive to bias in the forecast and does not say anything about reliability. A biased forecast may still have good resolution and yield a good ROC. This means that the forecast can be improved through calibration. Thus, ROC can be considered as a measure of potential usefulness. It is conditioned on observations. That is, given that an event occurred, what was the forecast? It can be a

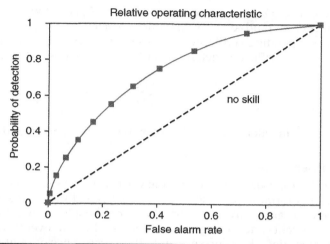

FIGURE 8.2   Relative operating characteristic.

good companion to the reliability diagram which is conditioned on the forecasts. Mason (1982) and Jolliffe and Stephenson (2003) have given details on ROC.

Thus, there are four main items: event (or random variable value), denoted by x; number of binary outcomes or categories of occurrence, denoted by r (occurrence defined as 1, $r = 1$, and no occurrence defined as 0, $r = 2$); a threshold, denoted by c; and the forecast probability of occurrence defined as $p = P(x \geq c)$ or forecast probability of no occurrence as $1 - p = P(x \leq c)$. The observed outcome or the occurrence of an event is defined as q: $q = 1$ if the event occurs and $q = 0$ if it does not.

8.2   Methods for Evaluation of Probabilistic Forecasting

To assess the quality of probabilistic forecasts, some current methods in use include the Brier score (Brier, 1950), ranked probability score (Epstein, 1969; Murphy, 1971), logarithmic scoring rule (Roulston and Smith, 2002), relative operating characteristics (Swets, 1973; Mason, 1982), ranked histograms (Anderson, 1996; Hamil and Colucci, 1996; Talagrand et al., 1997), generalization of ranked histograms to higher dimensions (Smith, 2000), and information theory (Roulton and Smith, 2002). The magnitude of probability forecast errors is measured by the mean square probability error, which can be partitioned into three terms (Murphy, 1973): reliability, resolution, and uncertainty. The magnitude ranges from 0 to 1. This aspect will be discussed as and when appropriate. Further, the forecast quality or the prediction capability of a forecast system is often assessed by skill scores. Some of these methods are now discussed.

To define a probabilistic forecast of a variable of interest, say, rainfall or temperature, let X denote the observed value of that variable and x denote any possible value in the range of X. The probabilistic forecasts are in the form of a probability density function $p(x)$ which expresses the uncertainty over what the possible value of X will be. The notation $p(X)$ or $q(X)$ denotes the value of the function at the particular observation X. A *score* is a comparison between X and each of the probabilistic forecasts. Note that $p(x)$

and x are unlike quantities. A score is a function $S(p(x), X)$ which may depend on the whole function $p(x)$ and acts as an operator on it. If there are n forecasts $\{p_i(x), i = 1, 2, ..., n\}$ and the corresponding verifications or observations $\{x_i, i = 1, 2, ..., n\}$, then an empirical skill score (SS) is defined to determine the value of the forecast system as

$$SS = \frac{1}{n} \sum_{i=1}^{n} S(p_i(x), x_i) \tag{8.1}$$

Scores are cost functions; i.e., smaller numerical values correspond to better forecasts.

8.2.1   Brier Score

Brier (1950) proposed a formula for the verification of precipitation forecasts for 6-h periods. This formula provides a score, now called the *Brier score*, for judging the goodness of weather forecasts, and it has received a lot of attention in the literature. Consider n events and r categories or classes where an event can occur in only one of these categories: (1) occurrence and (2) no occurrence of rainfall. In tossing a coin there are two categories of occurrence: (1) head and (2) tail. The throw of a die can have six categories of occurrence, designated as 1, 2, 3, 4, 5, and 6. These r categories of classes are mutually exclusive. Now consider an event or occasion i. The forecast probabilities that event i will occur in categories 1, 2, 3, ..., r are denoted as $p_{i1}, p_{i2}, ..., p_{ir}$. Then for the ith event (or occasion) the total probability for the r categories can be written as

$$\sum_{j=1}^{r} p_{ij} = 1 \qquad i = 1, 2, ..., n \tag{8.2}$$

where p_{ij} denotes the probability of the ith event occurring in the jth category.

Let q_{ij} denote the corresponding observation (or outcome) that takes on the value of 1 if event i occurred in category j or the value 0 if the event did not occur. The Brier score, designated as BS, for a single forecast i is defined as

$$BS_i = 2(p_{ij} - q_{ij})^2 \tag{8.3a}$$

The quantity 2 in Eq. (8.3a) arises because the scores associated with the occurrence and non-occurrence of the events are equal. Equation (8.3a) expresses the square of the distance the between the observation and forecast probability. For BS to be practically useful, it should be computed for a series of forecasts and observations and then averaged over the Brier scores for individual forecasts as

$$BS = \frac{1}{n} \sum_{i=1}^{n} BS_i = \frac{2}{n} \sum_{i=1}^{n} \sum_{j=1}^{r} (p_{ij} - q_{ij})^2 \tag{8.3b}$$

The *Brier score* is the mean squared difference between probabilities of issued forecasts and observed binary outcomes (e.g., rain or no rain). It is one of the oldest scores for evaluating the skill of probabilistic forecasts, and it constitutes the basis for several other scores, such as ranked probability scores (Epstein, 1969). It is a negatively oriented score yielding smaller values for better forecasts.

The Brier score is also defined (Wilks, 2006) as

$$BS = \frac{1}{n}\sum_{i=1}^{n}\sum_{j=1}^{r}(p_{ij} - q_{ij})^2 \qquad (8.4a)$$

Equation (8.4a) defines the *half Brier score,* and this definition of the Brier score is almost always used these days (Ahrens and Walser, 2008). Here, for the jth category, p_i defines the forecast probability of ith event, and q_i denotes the observed occurrence of the ith event. Here q_i can also be interpreted as the ex post probabilities. The Brier score as given by Eq. (8.4a) is the expectation of the square of differences between forecasts and observations.

It is clear from Eq. (8.3b) that BS = 0 if $p_{ij} = q_{ij}$. This means that the forecast is perfect. In other words, the event is forecasted with probability 1 or with 100 percent confidence. On the other hand, the worst possible forecast occurs when the event is forecast with probability 1 and the event actually did not occur. In this case, BS = 2.0. Thus, $0 \leq BS \leq 2.0$. Since this forecast system does not employ any probabilities between 0 and 1, it does not consider any uncertain cases and hence can be considered as a deterministic binary forecast system.

Sometimes the Brier score is also expressed as

$$BS = \frac{1}{n}\sum_{i=1}^{n}\sum_{j=1}^{r}(p_i - \delta_{ij})^2 \qquad (8.4b)$$

where $\delta_{ij} = 0$ when $i \neq j$ and $\delta_{ij} = 1$ when $i = j$; j is the actual outcome. Thus, Eq. (8.4b) can be written as

$$BS = \frac{1}{n}\sum_{i=1}^{n}\sum_{j=1}^{r}(p_{ij}^2 + \delta_{ij}^2 - 2p_{ij}\delta_{ij})$$

$$= \frac{1}{n}\left[\sum_{i=1}^{n}(p_i^2 + 1 - 2p_j)\right] \qquad (8.4c)$$

If $f_1, f_2, \ldots, f_r$ are the relative frequencies that the event occurred in categories $1, 2, \ldots, r$, then forecasting the same thing on every occasion leads to the minimum Briar score, i.e.,

$$p_{ij} = f_j \qquad i = 1, 2, \ldots, n \qquad (8.5a)$$

For constant values of $p_{1j} = p_{2j} = \cdots = p_{nj}$, the score will be minimized and the mean value of the Brier score from Eq. (8.4a) will be

$$BS_m = 1 - \sum_{j=1}^{r} f_j^2 \qquad (8.5b)$$

BS has a number of interesting properties, but it would be interesting to consider illustrative examples first.

Example 8.1. Let there be five occasions on which actual forecasts ($n = 5$) of rain or no rain ($r = 2$) are made as shown in Table 8.1. For each forecast, a probability p_{ij} is computed. In the category of rain, following the definition of q_{ij}, $q_{ij} = 1$ if rain occurs and $q_{ij} = 0$ if it does not. Likewise, in the category of no event, if the event is no rain, then $q_{ij} = 1$; and $q_{ij} = 0$ if there is rain. Compute the Brier score.

	Rain		No Rain	
Occasion *i*	Forecast Probability p_{ij}	Observed Outcome q_{ij}	Forecast Probability p_{ij}	Observed Outcome q_{ij}
1	0.2	0.0	0.8	1.0
2	0.5	1.0	0.5	0.0
3	0.8	1.0	0.2	0.0
4	0.1	0.0	0.9	1.0
5	0.0	0.0	1.0	1.0

TABLE 8.1   Forecasts in Terms of Probabilities

Solution.   Here there are two categories $r = 2$: 1 and 2. From Eq. (8.4a), BS is computed as

$$BS = \frac{1}{5}\left[\sum_{i=1}^{5}(p_{i1} - q_{i1})^2 + \sum_{i=1}^{5}(p_{i2} - q_{i2})^2\right]$$

$$= \frac{1}{5}[(p_{11} - q_{11})^2 + (p_{21} - q_{21})^2 + (p_{31} - q_{31})^2 + (p_{41} - q_{41})^2 + (p_{51} - q_{51})^2$$

$$+ (p_{12} - q_{12})^2 + (p_{22} - q_{22})^2 + (p_{32} - q_{32})^2 + (p_{42} - q_{42})^2 + (p_{52} - q_{52})^2]$$

$$= \frac{1}{5}[(0.2 - 0)^2 + (0.5 - 1.0)^2 + (0.8 - 1.0)^2 + (0.1 - 0.0)^2 + (0.0 - 0.0)^2$$

$$+ (0.8 - 1.0)^2 + (0.5 - 0.0)^2 + (0.2 - 0.0)^2 + (0.9 - 1.0)^2 + (1.0 - 1.0)^2]$$

$$= \frac{1}{5}(0.2^2 + 0.5^2 + 0.2^2 + 0.1^2 + 0.0^2 + 0.2^2 + 0.5^2 + 0.2^2 + 0.1^2 + 0.0^2)$$

$$= \frac{1}{5}(0.04 + 0.25 + 0.04 + 0.01 + 0.0 + 0.04 + 0.25 + 0.04 + 0.01 + 0.0) = 0.68$$

Example 8.2.   Consider the data in Example 8.1. In actual practice, the relative frequencies f_1 and f_2 are not known, so one can use the best estimate of these based on climatological records. Out of 5 occasions it rained on 2 occasions so $2/5 = 0.4$ can be used as the forecast probability of rain and $1 - 0.4 = 0.6$ as probability of no rain. Likewise, use the same forecast each time with a forecast probability of 0.3. Then compute the mean Brier score.

Solution.   In this case there are two categories: rain and no rain; hence $r = 2$. The use of 0.4 probability of rain on every occasion yields

$$BS_m = 1 - (0.4)^2 - (0.6)^2 = 0.48$$

If the forecast probability of 0.3 is used, then

$$BS_m = 1 - (0.3)^2 - (0.7)^2 = 0.42$$

The difference between the two values 0.48 and 0.42 suggests that the Brier score can be reduced by making better forecasts.

Example 8.3.   Table 8.2 shows forecasts in terms of probabilities. Compute the Brier score and the mean Brier score.

Solution. $BS = \frac{1}{5}(0.5^2 + 0.5^2 + 0.5^2 + 0.5^2 + 0.5^2 + 0.5^2 + 0.5^2 + 0.5^2 + 0.5^2 + 0.5^2)$

$$= \frac{1}{5}(5 \times 0.5^2 + 5 \times 0.5^2) = 0.25 + 0.25 = 0.5$$

	Rain		No Rain	
Occasion i	Forecast Probability p_{ij}	Observed Outcome q_{ij}	Forecast Probability p_{ij}	Observed Outcome q_{ij}
1	0.5	0.0	0.5	1.0
2	0.5	1.0	0.5	0.0
3	0.5	1.0	0.5	0.0
4	0.5	1.0	0.5	0.0
5	0.5	1.0	0.5	0.0

TABLE 8.2   Forecasts in Terms of Probabilities

In this case the forecast probability for rain is 0.5 on each of the five occasions, and so is the no-rain forecast probability. The mean Brier score is

$$BS_m = 1 - (0.5)^2 - (0.5)^2 = 1 - 0.25 - 0.25 = 0.50$$

Example 8.4.   In Table 8.2, take the forecast probability as 0.3 in place of 0.5 for each occasion. Then compute the Brier score and the mean Brier score. Consider another case where the forecast probability is 0.4 and when it is 0.8. Then compute the Brier scores and the mean Brier scores.

Solution.   In the first case the forecast probability is 0.3 for each occasion. The observed probabilities are the same as in Table 8.2. The Brier score is

$$BS = \frac{1}{5}(0.3^2 + 0.7^2 + 0.7^2 + 0.7^2 + 0.7^2 + 0.3^2 + 0.7^2 + 0.7^2 + 0.7^2 + 0.7^2)$$

$$= \frac{1}{5}(0.09 + 8 \times 0.7^2 + 0.09) = \frac{1}{5}(0.18 + 3.92) = \frac{4.10}{5} = 0.82$$

The mean Brier score is

$$BS_m = 1 - (0.3)^2 - (0.7)^2 = 1 - 0.09 - 0.49 = 0.42$$

When the forecast probability is 0.4 for each occasion, the Brier score is

$$BS = \frac{1}{5}(0.4^2 + 0.6^2 + 0.6^2 + 0.6^2 + 0.6^2 + 0.4^2 + 0.6^2 + 0.6^2 + 0.6^2 + 0.6^2)$$

$$= \frac{1}{5}(0.16 + 8 \times 0.6^2 + 0.16) = \frac{1}{5}(0.32 + 2.88) = \frac{3.20}{5} = 0.64$$

The mean Brier score is

$$BS_m = 1 - (0.4)^2 - (0.6)^2 = 1 - 0.16 - 0.36 = 0.48$$

When the forecast probability is 0.8 for each occasion, the Brier score is

$$BS = \frac{1}{5}(0.8^2 + 0.2^2 + 0.2^2 + 0.2^2 + 0.2^2 + 0.8^2 + 0.2^2 + 0.2^2 + 0.2^2 + 0.2^2)$$

$$= \frac{1}{5}(2 \times 0.64 + 8 \times 0.04) = \frac{1}{5}(1.28 + 0.32) = \frac{1.60}{5} = 0.32$$

The mean Brier score is

$$BS_m = 1 - (0.8)^2 - (0.2)^2 = 1 - 0.64 - 0.04 = 0.32$$

Brier (1950) emphasized that two conditions must be satisfied for the score to be useful: there must be (1) a strong correlation between forecast and observed probabilities and (2) an increased frequency or proportion of forecasts in various categories, especially in extreme cases. Then the score would be small. Consider a reference forecast system denoted as B_{ref}. Then the Brier score BS can be compared with the reference forecast system by defining the *Brier skill score* (BSS) as

$$BSS = \frac{BS - B_{ref}}{0 - B_{ref}} = 1 - \frac{BS}{B_{ref}} \tag{8.6a}$$

BSS is 1 if BS = 0 (i.e., perfect forecast) and is 0 or negative if BS = B_{ref} or BS < B_{ref} (i.e., poor or poorer than the reference system). Unlike BS, BSS is positively oriented; i.e., greater values point to better forecasts. The relative skill score of the probability forecast over that of climatology in terms of predicting whether an event occurred ranges from $-\infty$ to 1, with 0 indicating no skill when compared to the reference forecast and 1 indicating a perfect score. The reference system is often taken as the climatological forecast system that usually has low skill. In this case, $p_{ij} = s$ for all j, where $s = p(q = 1)$ is the base rate probability for the occurrence of the event. For a large sample, the Brier score for such reference forecasts can be expressed asymptotically as

$$BS = B_{ref} = s(1 - s) \tag{8.6b}$$

Climatological forecasts have perfect reliability since the expectation of s is s, but have no resolution, since variance of s is zero.

The Brier score can also be applied to cases where an event may have multiple categories of outcomes and not just binary outcomes, although it is not recommended if the occurrence of an event has more than two categories (Daan, 1985; Stanski et al., 1989) for two reasons. First, it strongly depends on the category boundaries. Second, it does not consider the distance between forecast categories and observed categories. For an event, let r be mutually exclusive (not necessarily ordered) outcomes of which only one is always observed. For such an event, a probabilistic forecast comprises an r vector of probabilities p_j, $j = 1, 2, ..., r$, such that $\sum_{j=1}^{r} p_j = 1$. The Brier score for probabilistic forecasts of r categories can be written as

$$BS = E\left[\frac{1}{r}\sum_{j=1}^{r}(p_j - q_j)^2\right] \tag{8.7a}$$

where $q_j = 1$ if the observed outcome is that event E_k occurred and $q_j = 0$ otherwise. The observation can also be defined as an r vector containing $r - 1$ zeros and only a single 1. In the vector notation, the Brier score can be written as

$$BS = E\left[\frac{\|p - q\|^2}{r}\right] \tag{8.7b}$$

where the symbol $\|\ \|$ denotes the euclidean norm. Equation (8.7a) shows that the Brier score for r categories is the arithmetic mean of the binary Brier scores for each outcome E_k. Zhu et al. (1996) and Toth et al. (1998) have discussed the use of the multiple-category Brier score.

8.2.2   Ranked Probability Score

Epstein (1969) presented a scoring rule that accounts for several alternative weather states as an extension of the method discussed by Murphy (1966). Jolliffe and Stephenson (2003) employed the *ranked probability score* (RPS) for assessing the forecast quality if more than one threshold is to be used. Let there be multiple thresholds $c_j : j = 1, 2, ..., r$. For these thresholds the cumulative forecast probabilities P_{ij} for $x_{i,m}$ can be expressed as

$$P_{ij} = P(x_{i,m} \le c_j) = \sum_{k=1}^{j} p_{ik} \tag{8.8a}$$

where p_{ik} is the probability of forecast of the ith event for the kth threshold in the jth category. Likewise, the ex post probabilities can be expressed as

$$Q_{ij} = P(x_{i,0} \le c_j) = \sum_{k=1}^{j} q_{ik} \tag{8.8b}$$

where q_{ik} is the occurrence of the ith event in the jth category. The ranked probability score (RPS) compares the forecast cumulative probability distribution P_{ij} with the observed cumulative probability distribution Q_{ij} and represents the sum of the squares of differences between forecast and observed cumulative probability mass functions. It can be expressed as

$$\text{RPS}(P,Q) = \frac{1}{n} \sum_{i=1}^{n} \sum_{j=1}^{r} (P_{ij} - Q_{ij})^2 \tag{8.9a}$$

or

$$\text{RPS}(P,Q) = \frac{1}{n} \sum_{i=1}^{n} \sum_{j=1}^{r} \left[\left(\sum_{k=1}^{j} p_{ik} \right) - \left(\sum_{k=1}^{j} q_{ik} \right) \right]^2 \tag{8.9b}$$

Equation (8.10a) shows that RPS is a measure of the distance applied to the time series of cumulative probabilities P and Q. If $r = 1$, a special case, then Eq. (8.10b) reduces to the Brier score given by Eq. (8.3b). Note that the probability score (PS) is also concerned with the sum of squares of differences between forecast and observed probability mass functions:

$$\text{PS} = \sum_{j=1}^{r} [p(x_j) - q(x_j)]^2 \tag{8.10a}$$

where x_j denotes the jth class of the variable of concern X.

For two forecast categories, RPS is the same as the Brier score. Its continuous version can be expressed as

$$\text{PS} = \int_{-\infty}^{\infty} [P(x) - q(x)]^2 \, dx \tag{8.10b}$$

The *ranked probability skill score* (RPSS) can be expressed as

$$\text{RPSS} = \frac{\text{RPS} - \text{RPS}_{\text{ref}}}{0 - \text{RPS}_{\text{ref}}} = 1 - \frac{\text{RPS}}{\text{RPS}_{\text{ref}}} \tag{8.10c}$$

RPS penalizes forecasts more severely when their probabilities are farther from the actual outcome. RPSS ranges from $-\infty$ to 1, with 0 indicating no skill when compared to the reference forecast and 1 indicating a perfect score. It measures the improvement of a multicategory probabilistic forecast relative to a reference forecast which usually is the long-term or sample climatology.

8.2.3   Logarithmic Scoring Rule

Roulston and Smith (2002) discussed the *logarithmic scoring rule* (LSR) in the context of information theory. The underlying stipulation is that the best forecast would correspond to the greatest level of data compression. The objective is to derive a forecast skill score that evaluates how close the forecast PDF is to the true PDF which is considered as the PDF of consistent initial conditions.

Let there be two PDFs defined by vectors P: $\{p_i, i = 1, 2, ..., n\}$ and Q: $\{q_i, i = 1, 2, ..., n\}$. The ith component of these vectors corresponds to the ith event occurring. This means that one can write $\sum_{i=1}^{n} p_i = \sum_{i=1}^{n} q_i = 1$. The distance between P and Q can be expressed by the relative entropy $D(P \mid Q)$ as

$$D(P \mid Q) = \sum_{i=1}^{n} p_i \log \frac{p_i}{q_i} = \sum_{i=1}^{n} (p_i \log_2 p_i - p_i \log_2 q_i) \tag{8.11}$$

Consider that every event is classified into one and only one possible outcome. The forecast model yields a forecast of the outcome of any event, i.e., the probability p_i of the ith outcome ($i = 1, 2, 3, ..., n$). Roulston and Smith (2002) discussed a data compression scheme. Following them, the optimal data compression scheme is based on information theory (Shannon, 1948). Accordingly, B_i bits are assigned to outcome i where the probability of the ith outcome is specified by p_i, $i = 1, 2, ..., n$, as

$$B_i = -\log_2 p_i \tag{8.12}$$

If $p_i = 0$, $B_i \to -\infty$, because the encoding scheme cannot encode an outcome that is deemed impossible a priori. This suggests that zero forecast probabilities should be replaced by small probabilities on the basis of uncertainties in forecast probabilities or PDF.

Now, the ignorance (IGN) score can be formulated as

$$IGN = -\log_2 p_i \tag{8.13}$$

This value of ignorance is for one forecast and realization and can be used as a scoring rule (Murphy, 1997). IGN is based on information theory and has a simple interpretation. Consider that two persons A and B have probabilistic forecasts f_i ($i = 1, 2, ..., n$). Assume that person A has come to know the actual outcome and wants to transmit a message about the outcome. Both A and B have agreed to use an optimal data encoding scheme defined by $\log f_i$. The number of bits that A must send B is given by Eq. (8.13): $IGN = -\log_2 f_j$ if the actual outcome is j. This can be construed as the information deficit, or ignorance, of person B having the probabilistic forecast before A transmitted the message about the actual outcome.

The IGN score should be averaged over the entire set of forecast realizations as

$$\overline{IGN} = -\frac{1}{n} \sum_{i=1}^{n} \log_2 p(i, j(i)) \tag{8.14}$$

where $p(i, j(i))$ is the probability of outcome j according to the probabilistic forecast i and $j(i)$ is the corresponding actual outcome. This defines the number of bits that A must transmit to B about the true outcome averaged over all verification forecasts. If the true PDF $G: \{g_i, i = 1, 2, ..., n\}$, which is unknown, is defined by g_i, then the average value of IGN of forecasts p_i is given by

$$E[\text{IGN}] = -\sum_{i=1}^{n} g_i \log_2 p_i \tag{8.15}$$

Equation (8.15) can be written as

$$E[\text{IGN}] = -\sum_{i=1}^{n} g_i \log_2 \left(\frac{p_i}{g_i} g_i \right) = -\sum_{i=1}^{n} g_i \log_2 g_i - \sum_{i=1}^{n} g_i \log_2 \left(\frac{g_i}{p_i} \right)$$
$$= H(G) - D(G|P) \tag{8.16}$$

where $H(G)$ is the Shannon entropy of the true PDF, which cannot be determined in reality since G is unknown, and $D(G|P)$ is the cross entropy of G and P:

$$D(G|P) = H(G) - E[\text{IGN}] \tag{8.17}$$

where $P: \{p_i, i = 1, 2, ..., n\}$. Equation (8.17) is the relative entropy of the true and forecast PDFs. Therefore, the model $D(G|P) = 0$ cannot be attained. If $g_i = p_i$, then H defines the maximum number of bits required to represent the data.

Consider a best estimate of g_i as f_i and the forecast intended for issuance of p_i. Then the predicted ignorance of the forecast can be expressed as

$$\text{Predicted } E[\text{IGN}] = -\sum_{i=1}^{n} f_i \log_2 p_i \tag{8.18}$$

Equation (8.18) has a minimum where $p_i = f_i$:

$$\text{Predicted } E[\text{IGN}] = -\sum_{i=1}^{n} f_i \log_2 f_i = H(F) \tag{8.19}$$

This shows that the IGN score cannot be reduced by issuing a PDF different from the best PDF. Equation (8.19) shows that ignorance would be the same as the entropy of forecast, if averaged over many forecasts. It has been suggested as a predictor of forecast skill (Stephenson and Doblas-Reyes, 2000).

Now the question concerns how the ignorance score is related to the forecast quality. IGN does not make any assumption about the shape of the PDF. Assume that the forecast variable X is continuous and that both the true (tr) PDF g and the forecast (m) PDF p are normal distributions defined as

$$g(x) = \frac{1}{\sigma_{tr} \sqrt{2\pi}} \exp\left[-\frac{(x - \overline{x_{tr}})^2}{2\sigma_{tr}^2} \right] \tag{8.20a}$$

and

$$p(x) = \frac{1}{\sigma_{m} \sqrt{2\pi}} \exp\left[-\frac{(x - \overline{x_{m}})^2}{2\sigma_{m}^2} \right] \tag{8.20b}$$

The expected ignorance can be expressed as

$$E[\text{IGN}] = -\frac{1}{\ln 2} \int_{-\infty}^{\infty} g(x) \ln p(x) dx \qquad (8.21)$$

Substitution of Eqs. (8.20a) and (8.20b) in Eq. (8.21) leads to

$$E[\text{IGN}] = \frac{1}{2\ln 2}\left[\ln(2\pi) + \ln \sigma_m^2 + \frac{\sigma_{tr}^2 + (\overline{x}_{tr} - \overline{x}_m)^2}{2\sigma_m^2} \right] \qquad (8.22)$$

Equation (8.22) is affected by the mean and variance of true and forecast PDFs. An increase in ignorance is caused by the bias in forecast $\overline{x}_{tr} - \overline{x}_m$. The uncertainty in the true PDF σ_{tr}^2 results in higher $E[\text{IGN}]$. The influence of σ_m on $E[\text{IGN}]$ is not monotonic. However, $E[\text{IGN}]$ achieves a unique minimum when $\sigma_m^2 = \sigma_{tr}^2 + (\overline{x}_{tr} - \overline{x}_m)^2$. If the forecast is unbiased, $\overline{x}_m = \overline{x}_{tr}$, then the variance of the forecast is the same as the variance of the perfect forecast PDF. On the other hand, for the biased forecast, the ignorance is minimized if the forecast variance is increased to partially make up for the bias.

Ignorance can be construed as a measure of reliability and is minimized if the true PDF is close to the forecast PDF. The accuracy (or skill) of forecast by comparison with other forecasts can be obtained by computing the difference in the ignorance of different forecasts. The reason is that ignorance is information which is additive. Each bit of ignorance denotes a factor-of-2 increase in uncertainty.

8.2.4 Relation between Ignorance Score and Brier Score

A forecast scheme having a lower Brier score than another forecast scheme may not necessarily possess a lower value of expected ignorance. In the two-outcome case, ignorance turns out to be a double-valued function of the Brier score, because $E[\text{BS}]$ is symmetric in p but ignorance is asymmetric in p. If the true probability distribution is g_i, then the expected value of the Brier score $E[\text{BS}]$ can be written as given by Eq. (8.4c):

$$E[\text{BS}] = \sum_{j=1}^{n} g_j \frac{1}{n} \left(\sum_{i=1}^{n} p_i^2 - 2p_j + 1 \right) = \frac{1}{n}\left(\sum_{i=1}^{n} p_i^2 - 2\sum_{i=1}^{n} g_i p_i + 1 \right) \qquad (8.23)$$

Now let there be only two possible outcomes and the true probability distribution be given by $(g, 1-g)$. Since the model probability distribution is given by $(p, 1-p)$, it is possible to write $p = g + \Delta$, where Δ is the deviation. Then Eq. (8.23) can be recast as

$$E[\text{BS}] = \frac{1}{2}[(g + \Delta)^2 + (1 - g - \Delta)^2 - 2g(g + \Delta) - 2(1 - g)(1 - g - \Delta) + 1] \qquad (8.24)$$

Equation (8.24) can be simplified as

$$E[\text{BS}] = \Delta^2 - g^2 + g \qquad (8.25)$$

The deviation Δ can be written from Eq. (8.25) as

$$\Delta = \pm\sqrt{E[\text{BS}] + g^2 - g} \qquad (8.26)$$

Equation (8.26) shows that two models can have the same Brier score, with the given Δ values.

The expected ignorance of each of these two models can be expressed as

$$E[IGN] = -g\log_2(g + \Delta) - (1 - g)\log_2(1 - g - \Delta) \tag{8.27}$$

The difference in the IGN values of these models can be expressed as

$$E[IGN]_+ - E[IGN]_- = g\log_2\left(\frac{g - |\Delta|}{g + |\Delta|}\right) - (1 - g)\log_2\left(\frac{1 - g - |\Delta|}{1 - g + |\Delta|}\right) \tag{8.28}$$

Equation (8.28) shows that the two values of Δ for a single value of the expected value of the expected Brier score yield two different values of ignorance. If $\Delta = 0$ ($g = p$), $E[IGN]_+ = E[IGN]_-$, indicating that a perfect model will have both a lower BS and a lower $E[IGN]$ than any other imperfect model. For $g = 0.5$, the two branches are coincident; thus, in this case a lower expected BS means a lower $E[IGN]$. For $g \neq 0.5$, the branches are not the same; they correspond to underestimating p and overestimating p, respectively. The implication here is that in comparing two models, one model may be better if judged by the Brier score and the other model may be better if judged by the ignorance score.

8.2.5   Ignorance Score and Continuous Variables

Let there be an ensemble of N forecasts of variable X: x_i ($i = 1, 2, ..., N$). Let the forecast PDF be approximately a uniform distribution with X_{min} and X_{max} as lower and upper limits, respectively. The forecast PDF between x_i and x_{i+1} can be written as

$$p_i = \frac{1}{(N + 1)\Delta x_i} \tag{8.29}$$

in which

$$\Delta x_i = x_{i+1} - x_i \qquad 0 \leq i \leq N$$
$$\Delta x_0 = x_{i+1} - x_{min} \qquad \Delta x_N = x_{max} - x_N \tag{8.30}$$

where $[x_{min}, x_{max}]$ is the a priori interval in which X is expected to be. If the true forecast lies in the jth interval, then the ignorance can be expressed as

$$IGN = \log_2(N + 1) + \log_2 \Delta x_j \tag{8.31}$$

denoting the number of bits required to specify the true forecast. The first term in Eq. (8.31) represents the number of bits required to specify the two ensemble members between which the true forecast lies. The second term represents the number of bits required to specify where the true forecast lies in the interval.

The expected ignorance is the expected value of the minimum number of bits required, which depends on the choice of x_{min} and x_{max}. When ignorance is calculated from rank histograms, if the same resolution of forecasts is not used, categorical forecasts should be used. The categories of a continuous forecast variable can be derived climatologically where each category is equally probable. If there are r categories, the ignorance of a forecast will be $\log_2 r$. The *fractional ignorance* can be defined as

$$IGN = \frac{\log_2 p_i}{\log_2 r} \tag{8.32}$$

where p_i is the probability assigned to the actual outcome. Three cases can be distinguished from Eq. (8.31). First, if IGN < 1, then the forecast contains more information than climatology. Second, if IGN = 1, then the forecast is no better than the forecast based on climatology. If IGN > 1, the forecast has less information than the climatology. In other words, a forecast is more precise or confident but less accurate than a climatological forecast.

The skill score values are compared with score values corresponding to reference forecasts, often chosen to be observed climatology $\bar{q}$, frequently expressed by the cumulative probabilities $\bar{Q}_j$ in case of RPS. Let $S(Q, Q)$ be the score value of a perfect forecast $P = Q$ (that is, 0 for BS and RPS). The skill score (SS) can be defined as

$$SS = \frac{S(P, Q) - S(\bar{Q} - Q)}{S(Q, Q) - S(\bar{Q} - Q)} \qquad (8.33a)$$

for scores with cumulative probabilities. An equivalent score can be defined using lowercase letters for probability masses:

$$SS = \frac{S(p, q) - S(\bar{q} - q)}{S(q, q) - S(\bar{q} - q)} \qquad (8.33b)$$

If the forecast value is equal to the reference or observed forecast, SS = 0.

The skill score, defined as above, permits easy comparison of forecast results for different cases, such as season, region, and reference data sets. In this manner, BS and RPS can be compared with scores for climatological forecast by score normalization with a measure of uncertainty inherent in the events being forecast $[\bar{Q}, (1 - \bar{Q})]$ for the Brier skill score with $\bar{Q}$, the observed climatology.

Since evaluation of a forecast must deal with uncertainty, it is logical to employ skill scores based on the information content, or surprise I, in the realization of an event in the jth category, $c_{j-1} \leq x_{jt} \leq c_j$, which can be defined as

$$I(X = x_{jt}) = \log \frac{1}{p(x_{jt})} = -\log p(x_{jt}) \qquad (8.34)$$

Hence, the process can be either forecasting or observation; thus, $P(x_{jt})$ is p_{jt} or q_{jt}, respectively. For example, for $P(x_{jt}) = 1$, the event is certain, and $P(x_{jt}) = 0$ for $i \neq t$. The future outcome will not be surprising, and its occurrence gives no information. Furthermore, no information loss occurs in any realization, and no negative surprise occurs. Hence, positive information is derived by the forecast.

For one forecast and observation pair, the ignorance is expressed as

$$\text{IGN}_t = -\log_2 p_{j(t)t} \qquad (8.35a)$$

where $p_{j(t)t}$ is the probability mass of the forecast in the observed category $j(t)$ at date t. Ignorance IGN_t is a statistic—a scoring rule—defined for single dates (Murphy, 1997). To be representative, IGN_t must be averaged over a longer data set of forecast observation pairs.

$$\text{IGN} = -\frac{1}{n} \sum_{t=1}^{n} \log_2 p_{j(t)t} \qquad (8.35b)$$

If the forecast is perfect, $p_{j(t)t} = 1$ and IGN = 0. If the forecast is always $p_{j(t)t} = \varepsilon$, ignorance is minimized at $-\log \varepsilon$.

8.3   Decomposition of Brier Score

8.3.1   Murphy's Decomposition

Murphy (1986) decomposed the Briar score into two components: the variance of forecast error and the square of the mean error. This decomposition is insightful and is discussed here. Let there be a sample of n forecasts and the corresponding observations. Let p_i denote the forecast probability of precipitation on the ith occasion, and let q_i denote the observation on the ith occasion. As before, $q_i = 1$ if precipitation occurs and $q_i = 0$ if precipitation does not occur. Let there be a two-event scenario—occurrence or no occurrence—for which the Briar score is given by Eq. (8.4a).

Because only one of two events can happen on any occasion, it is convenient to divide the sample of size n into two subsamples: n_0 denotes the number of occasions when $q_i = 0$, and n_1 denotes the number of occasions when $q_i = 1$. Clearly, $n = n_0 + n_1$. Accordingly, let p_{1i} denote the forecast probability of precipitation on the ith of n_1 occasions, and p_{0j} is the forecast probability of no precipitation on the jth of n_0 occasions. Then BS can be expressed as

$$BS = \frac{2}{n}\left[\sum_{i=1}^{n_1}(p_{1i}-1)^2 + \sum_{j=1}^{n_0}(p_{0j}-0)^2\right] \tag{8.36a}$$

Equation (8.36a) is based on two conditional distributions: the distribution of forecast probability given the occurrence of precipitation and the distribution of forecast probability given the occurrence of no precipitation. Equation (8.36a) can be written as

$$BS = \frac{2}{n}\left(\sum_{i=1}^{n_1}p_{1i}^2 + \sum_{j=1}^{n_0}p_{0j}^2 - 2\sum_{i=1}^{n_1}p_{1i} + n_1\right) \tag{8.36b}$$

The third term on the right side of Eq. (8.36b) can be expressed in terms of the mean forecast probability as

$$\overline{p_1} = \frac{1}{n_1}\sum_{i=1}^{n_1}p_{1i} \tag{8.37}$$

Substituting Eq. (8.37) in Eq. (8.36b), one can write

$$BS = \frac{2}{n}\left(\sum_{i=1}^{n_1}p_{1i}^2 + \sum_{j=1}^{n_0}p_{0j}^2 - 2n_1\overline{p_1} + n_1\right) \tag{8.38}$$

Also

$$\overline{p_0} = \frac{1}{n_0}\sum_{j=1}^{n_0}p_{0j} \tag{8.39}$$

Now, by adding and subtracting $(2/n)(n_1\overline{p_1}^2 + n_0\overline{p_0}^2)$, Eq. (8.38) can be written as

$$BS = \frac{2}{n}\left[\sum_{i=1}^{n_1}(p_{1i}^2 - n_1\overline{p_1}^2) + \sum_{j=1}^{n_0}(p_{0j}^2 - n_0\overline{p_0}^2)\right] + \frac{2}{n}\left(n_1\overline{p_1}^2 - 2n_1\overline{p_1} + n_1 + n_0\overline{p_0}^2\right) \tag{8.40}$$

Recalling the definition of variance (Var), one can write

$$\text{Var}(p_1) = \frac{1}{n_1}\left(\sum_{i=1}^{n_1} p_{1i}^{\,2} - n_1 \overline{p_1}^{\,2}\right) \tag{8.41a}$$

$$\text{Var}(p_0) = \frac{1}{n_0}\left(\sum_{j=1}^{n_0} p_{0j}^{\,2} - n_0 \overline{p_0}^{\,2}\right) \tag{8.41b}$$

Therefore, Eq. (8.40) can be cast as

$$\text{BS} = \frac{2}{n}[n_1 \text{Var}(p_1) + n_0 \text{Var}(p_0)] + \frac{2}{n}\left[n_1(\overline{p_1} - 1)^2 + n_0(\overline{p_0} - 0)^2\right] \tag{8.42a}$$

Let $\overline{q_1} = n_1 / n$ and $\overline{q_0} = n_0 / n$. Note that $\overline{q_1} + \overline{q_0} = 1$. Then Eq. (8.42a) can be written as

$$\text{BS} = 2[\overline{q_1}\text{Var}(p_1) + \overline{q_0}\text{Var}(p_0)] + 2\left[\overline{q_1}(\overline{p_1} - 1)^2 + \overline{q_0}(\overline{p_0} - 0)^2\right] \tag{8.42b}$$

The first term on the right side of Eq. (8.42b) is 2 times the weighted average of the variances of the two distributions where the weights are given by the sample frequencies of precipitation and no precipitation, respectively. This term varies from 0 when $p_{1i} = \overline{p_1}$ for all i and $p_{0i} = \overline{p_0}$ for all i to 0.5 when $p_{1i} = \overline{p_1}$ on $n_1/2$ occasions and $p_{0i} = 0$ on $n_1/2$ occasions, $p_{0j} = 1$ on $n_0/2$ occasions and $p_{0j} = 0$ on $n_0/2$ occasions. This shows that a decrease in the variance of forecasts in either subsample will result in a decrease in the Briar score.

The second term is a mean error term and is 2 times the weighted average of the squared differences between the average subsample forecast probability and the corresponding subsample observation. This term ranges from 0 when $p_{1i} = 0$ for all i and $p_{0j} = 1$ for all j to 2 when $p_{1i} = 0$ for all i and p_{0j} for all j. It shows that a reduction in mean error will result in a reduction in the Briar score.

If $\text{Var}(p) = \overline{q_1}\text{Var}(p_1) + \overline{q_0}\text{Var}(p_0)$ and $[E(p)]^2 = \overline{q_1}(\overline{p_1} - 1)^2 + \overline{q_0}(\overline{p_0} - 1)^2$, then Eq. (8.42b) can be written as

$$\text{BS} = 2\,\text{Var}(p) + 2[E(p)]^2 \tag{8.43a}$$

Equation (8.43a) represents a special case of the mean square error, which is the sum of the variance of errors and the square of the mean error. Note that Eq. (8.42b) can also be written as

$$\text{BS} = \frac{2}{n}\sum_{i=1}^{n} p_i^2 + 2\overline{q_1}(1 - 2\overline{p_1}) \tag{8.43b}$$

In Eq. (8.43b), the first term is 2 times the mean square of all forecast probabilities $\{p_i, i = 1, 2,..., n\}$. It is always positive and reaches a minimum when $p_i = \overline{p}$ for all i, where $\overline{p} = \frac{1}{n}\sum_{i=1}^{n} p_i$. The second term is a measure of the mean error for the subsample of forecasts for which $q_i = 1$. This is negative when $\overline{p_1} > 0.5$ and positive when $\overline{p_1} < 0.5$.

8.3.2 Decomposition of Brier Score

Toth et al. (2003) decomposed the Brier score into the sum of reliability, resolution, and the underlying uncertainty of observations (Murphy, 1973). To do the decomposition, it is assumed that the forecast probabilities can take on any continuous value of p in the range of 0 to 1 and that p has probability density function (PDF) $f(p)$ such that

$$\int_0^1 f(p)\,dp = 1 \tag{8.44}$$

Let the climatological base rate of the event be denoted as $b = f(X = 1)$, which can be expressed as

$$b = f(X = 1) = \int_0^1 f(X = 1|p) f(p)\,dp = \int_0^1 f_c(p) f(p)\,dp \tag{8.45a}$$

where $f(X = 1|p) = f_c(p)$ is the conditional distribution. Equation (8.45a) shows that the base rate can be shown as the expectation of $f_c(p)$ over all values of p: $b = E_p[f_c(p)]$. This equation can also be expressed in terms of expectations over X and p as

$$b = E[X] = \int_0^1 E_X[X|p] f(p)\,dp = E_p[E_X(X|p)] \tag{8.45b}$$

Equations (8.44) through (8.45b) show that functions $f_c(p)$ (calibration function) and $f(p)$ (refinement function) entirely determine the performance of the probability forecast system.

Conditioning on the forecast probabilities, the Brier score can now be cast as

$$BS = E[(f_i - X)^2] = E_p[E_X[(p - X)^2|p]] \tag{8.46a}$$

Recalling the definition of $f_c(p) = p(X = 1|p)$, the expectation over X can be expressed as

$$E_X[(p - X)^2|p] = (p - 0)^2[1 - f_c(p)] + (p - 1)^2 f_c(p) = [p - f_c(p)]^2 + f_c(p)[1 - f_c(p)] \tag{8.46b}$$

Now taking the expectation of Eq. (8.46b) over all possible p values leads to the decomposition of the Brier score:

$$BS = E_p[(p - f_c(p))^2] - E_p[(f_c(p) - b)^2] + b(1 - b) \tag{8.47}$$

The right side of Eq. (8.47) is the sum of three terms. The first term, $E_q[(p - f_c(p))^2]$, is an overall measure of reliability and equals the mean square deviation of the reliability curve from the diagonal. If $f_c(p) = p$, that is, the system is perfectly reliable, then this term is zero. The second term, $E_q[(f_c(p) - b)^2]$, is an overall measure of resolution and equals the variance of p, $Var_p[f_c(p)]$. It measures how the different forecast events are classified or resolved by a forecast system, hence its name. Resolution depends entirely on the conditional probabilities $f_c(p)$ and is independent of the actual marginal probability distribution of forecast probability values and is hence of reliability. A larger value of resolution implies that the forecast system a priori can better determine situations that result in the occurrence or non-occurrence of the event in the future. A forecast system with good resolution has $f_c(p)$ that differs from the climatological base rate b. The third term, $b(1 - b)$, is referred to as *uncertainty* and equals the variance of observations $Var(X)$. This term does not depend on the forecast system and cannot be improved by it.

If the climatological probability is close to 0.5, then a large uncertainty in Eq. (8.47) occurs. On the other hand, if the climatological probability is 0 or 1, then the uncertainty term will be small.

It may be instructive to compare the three terms with another and to construct relative measures. Considering the uncertainty terms as reference, the relative measures of reliability and resolution can be defined as

$$B_{rel} = \frac{E_p[(p - f_c(p))^2]}{b(1-b)}$$ (8.48a)

$$B_{res} = 1 - \frac{E_p[(f_c(p) - b)^2]}{b(1-b)}$$ (8.48b)

Equations (8.48a) and (8.48b) are 0 for a perfect forecasting system and are hence negatively oriented. Taking the climatological system as a reference, they are connected to the Brier skill score BSS as

$$BSS_c = 1 - B_{rel} - B_{res}$$ (8.49)

8.3.3   General Decomposition

Stephenson et al. (2008) have decomposed the Brier score into different components using Eqs. (8.36a), (8.36b), (8.38), (8.40), (8.42a), (8.42b), (8.43a), and (8.43b). Consider a sample of binary observations q and probability forecasts p. Let the probability unit interval [0,1] be partitioned into m mutually exclusive subintervals or bins labeled by index j, $j = 1, 2, ..., m$. Let there be n_j probability forecasts in the jth bin denoted as p_{jk}, where $k = 1, 2, ..., n_j$. The total number of forecasts n in all the bins can be expressed as

$$n = \sum_{j=1}^{m} n_j$$ (8.50)

Let the average probability of all the probabilistic forecasts in the jth bin be defined as

$$\bar{p}_j = \frac{1}{n_j} \sum_{k=1}^{n_j} p_{jk}$$ (8.51)

Let q_{jk} denote the binary outcome (0 or 1) of the observed event corresponding to the kth probability in the jth bin, i.e., the one for which the probability of forecast is p_{jk}. Then the forecast error for the kth forecast in the jth bin or the Brier score can be expressed as $p_{jk} - q_{jk}$. Then the mean square error or Brier score can be written as

$$BS = \frac{1}{n} \sum_{j=1}^{m} \sum_{k=1}^{n_j} (p_{jk} - q_{jk})^2$$ (8.52)

Equation (8.52) can be expressed as a nested mean of within-bin averages as

$$BS = \frac{1}{n} \sum_{j=1}^{m} n_j \left[\frac{1}{n_j} \sum_{k=1}^{n_j} (p_{jk} - q_{jk})^2 \right]$$ (8.53)

In Eq. (8.53), the quantity inside the brackets is the mean of the squared forecast error, and it can be simplified further. Let $x_{jk} = p_{jk} - q_{jk}$. Then the term within brackets is the expectation of the square of x_{jk} within the bin, since summation is from $k = 1$ to n_j. Now, recalling the definition of variance (variance of x = mean of the square of x minus the square of the mean of x), the bracketed quantity can be written as the square of the mean forecast error plus the within-bin variance of the forecast error, that is, $[E(x)]^2 + \mathrm{Var}(x) = E(x^2)$. Therefore, Eq. (8.53) can be written as

$$\mathrm{BS} = \frac{1}{n} \sum_{j=1}^{m} n_j \left\{ (\overline{p}_j - \overline{q}_j)^2 + \frac{1}{n_j} \sum_{k=1}^{n_j} \left[(p_{jk} - q_{jk}) - (\overline{p}_j - \overline{q}_j) \right]^2 \right\} \qquad (8.54)$$

where

$$\overline{q}_j = \frac{1}{n_j} \sum_{k=1}^{n_j} q_{jk} \qquad (8.55)$$

denotes the relative frequency with which the observed event occurred at times when the forecast probabilities were in the range $[p_j, p_{j+1}]$. The second square on the right side of Eq. (8.54) can be expanded as two sums of square terms (variance) and a cross-product (covariance) term. Thus, Eq. (8.54) becomes

$$\mathrm{BS} = \frac{1}{n} \sum_{j=1}^{m} n_j \left[(\overline{p}_j - \overline{q}_j)^2 + \frac{1}{n_j} \sum_{k=1}^{n_j} (q_{jk} - \overline{q}_j)^2 + \frac{1}{n_j} \sum_{k=1}^{n_j} (p_{jk} - \overline{p}_j)^2 - \frac{2}{n_j} \sum_{k=1}^{n_j} (q_{jk} - \overline{q}_j)(p_{jk} - \overline{p}_j) \right] \qquad (8.56)$$

Now consider each term in the Brier score expressed by Eq. (8.56). The first term $\frac{1}{n} \sum_{j=1}^{m} n_j (\overline{p}_j - \overline{q}_j)^2$ is known as the *reliability of forecasts* (REL) and measures the bias (unconditional and conditional) in the probability estimates of forecasts. Murphy (1986) noted that good calibration of forecasts can reduce bias and increase reliability.

The second term $\frac{1}{n} \sum_{j=1}^{m} \sum_{k=1}^{n_j} (q_{jk} - \overline{q}_j)^2$ can be written as

$$\frac{1}{n} \sum_{j=1}^{m} \sum_{k=1}^{n_j} n_j (q_{jk} - \overline{q}_j)^2 = \frac{1}{n} \sum_{j=1}^{m} \sum_{k=1}^{n_j} n_j (q_{jk} - \overline{q})^2 - \frac{1}{n} \sum_{j=1}^{m} n_j (\overline{q}_j - \overline{q})^2 \qquad (8.57)$$

where

$$\overline{q} = \frac{1}{n} \sum_{j=1}^{m} n_j \overline{q}_j \qquad (8.58)$$

denotes the mean probability or climatological base rate for the event to occur. The right side of Eq. (8.57) contains two parts. The first part denotes the total variance of observed outcomes, and the second part denotes the variance of the within-bin means of observed outcomes. The first part can be expressed as $\overline{q}(1 - \overline{q})$ by expanding the square and recalling that $q_{jk}^2 = q_{jk}$ for binary variables. Therefore, the second term in Eq. (8.56) can now be written as

$$\frac{1}{n} \sum_{j=1}^{m} \sum_{k=1}^{n_j} (q_{jk} - \overline{q}_j)^2 = \overline{q}(1 - \overline{q}) - \frac{1}{n} \sum_{j=1}^{m} n_j (\overline{q}_j - \overline{q})^2 \qquad (8.59)$$

In Eq. (8.59) the first term $\overline{q}(1 - \overline{q})$ denotes the observational uncertainty (UNC); i.e., it measures the uncertainty inherent in the process that is forecast. In Eq. (8.59) the second term, $\frac{1}{n} \sum_{j=1}^{m} n_j (\overline{q}_j - \overline{q})^2$, denotes the forecast resolution (RES) which measures the amount of uncertainty explained by the forecast. This shows that the first and second terms in Eq. (8.56) yield the traditional REL – RES + UNC components of the Brier score. A perfect forecast means that its resolution equals uncertainty and it has a perfect reliability.

The third term, $\dfrac{1}{n}\sum_{k=1}^{n_j}\dfrac{1}{n}\sum_{j=1}^{m}\sum_{k=1}^{n_j}(p_{jk}-\overline{p}_j)^2$, denotes the pooled average of within-bin variances (WBV) of forecasts. Likewise, the fourth term, $\dfrac{2}{n}\sum_{j=1}^{m}\sum_{k=1}^{n_j}(q_{jk}-\overline{q}_j)(p_{jk}-\overline{p}_j)$, represents the pooled average of the within-bin covariances (WBC) between forecasts and observations. If only one value is forecast for each bin ($p_{jk}=\overline{p}_j$ for all k), then both WBV and WBC will vanish. This means that no within-bin variation among forecasts occurs. If the variation in forecast probabilities within any of the bins does occur, then the Brier score is

$$\mathrm{BS}=\frac{1}{n}\sum_{j=1}^{m}n_j\left[(\overline{p}_j-\overline{q}_j)^2-\frac{1}{n}\sum_{k=1}^{n_j}n_j(\overline{q}_j-\overline{q})^2+\overline{q}(1-\overline{q})+\frac{1}{n}\sum_{j=1}^{m}\sum_{k=1}^{n_j}(p_{jk}-\overline{p}_j)^2\right]$$

$$-\frac{2}{n}\sum_{j=1}^{m}\sum_{k=1}^{n_j}(q_{jk}-\overline{q}_j)(p_{jk}-\overline{p}_j)$$

(8.60)

Equation (8.60) expresses the Brier score BS as the sum of five components:

$$\mathrm{BS} = \mathrm{REL} - \mathrm{RES} + \mathrm{UNC} + \mathrm{WBV} - \mathrm{WBC}$$

(8.61)

If the bin size is increased, the resolution decreases. The two within-bin components, WBV and WBC, help make up for the increased resolution. Thus, the resolution term can be generalized by including the two within-bin terms as

$$\frac{1}{n}\sum_{j=1}^{m}n_j\left[(\overline{q}_j-\overline{q}_j)^2-\frac{1}{n}\sum_{j=1}^{m}n_j\frac{1}{n_j}\sum_{k=1}^{n_j}(pjk-\overline{p}_k)-\frac{2}{n_j}\sum_{k=1}^{n_j}(q_{jk}-\overline{q}_j)(p_{jk}-\overline{p}_j)\right]$$

(8.62)

That is,

$$\mathrm{GRES} = \mathrm{RES} - \mathrm{WBV} + \mathrm{WBC}$$

(8.63)

Thus, the Brier score can be expressed as

$$\mathrm{BS} = \mathrm{REL} - \mathrm{GRES} + \mathrm{UNC}$$

(8.64)

It is preferable to include WBV and WBC in RES and use GRES. Terms WBV and WBC can be forced to vanish by mapping forecast probabilities to a smaller number of bin location values. The generalized resolution component is less sensitive to bin width. The Brier score and uncertainty terms are independent of bin width. If bin width decreases, GRES must increase to compensate for the increase in reliability. This shows that the difference between GRES and RES is the amount of increase in the Brier score due to the binning of forecasts—a measure of the loss of skill due to the binning of probability forecasts. Note that the multiple-category Brier score can also be decomposed into reliability-resolution-uncertainty components. This can be achieved by averaging the components of the binary Brier scores for each individual category.

8.4   Kullback-Leibler Divergence

The relative entropy or the Kullback-Leibler divergence was employed by Weijs et al. (2010a, 2010b) and Weijs and van de Giesen (2011) as a measure of forecast verification. As the name suggests, it measures the divergence of observations from forecasts, called the *divergence score* (DS), and can be used as a forecast skill score. If observations are error-free or perfect, then the divergence will equal the ignorance score or logarithmic score. Analogous to the decomposition of the Brier score (Murphy, 1973), Weijs et al. (2010a) presented a decomposition of divergence into uncertainty, resolution, and reliability.

Consider the occurrence of rainfall as a binary event: (no occurrence, occurrence). The probability mass function (PMF) can be denoted as a 2D vector $Q = [1 - q, q]^T$, where T is the transpose. If the event has already occurred or for observations, q can be represented as $q \, \varepsilon \, [0, 1]^T$; this means $Q = [0,1]^T$ if the rainfall event occurred and $Q = [1,0]^T$ if it did not. Now, a probabilistic forecast of the rainfall occurrence is made in terms of the probability of occurrence p, with the forecast PMF written as $P = [1 - p, p]^T$, where $p \varepsilon [0,1]$. If rainfall occurs with a 30% probability, then $P = [0.7, 0.3]^T$.

For a single ith forecast, DS can be defined as the divergence of the forecast distribution from the observation distribution as

$$\text{DS}_i = \sum_{j=1}^{r} q_{ij} \log_2 \left(\frac{q_{ij}}{p_{ij}} \right) \tag{8.65}$$

where $r = 2$ defines the number of binary possibility of events, that is, $j = 1, 2$. For a series of forecasts and observations, DS can be formulated as the average divergence as

$$\text{DS} = \frac{1}{n} \sum_{i=1}^{n} \text{DS}_i = \frac{1}{n} \sum_{i=1}^{n} \sum_{j=1}^{r} q_{ij} \log \left(\frac{q_{ij}}{p_{ij}} \right) \tag{8.66}$$

The divergence score is not symmetric and is not a true distance. It can also be interpreted as a measure of missing information when distribution P is assumed but the true distribution is Q. Equation (8.66) indicates the gain in information as the prior forecast distribution P is replaced by the distribution of observations Q. If there were no gain in the information, this would mean that the forecast was perfect, i.e., the forecast distribution had all the information that was in the observation distribution. On the other hand, if the information gain from the forecast to the observation was the same as the climatological uncertainty, then the forecast was useless because it had no more information than the climate. Put another way, DS can be understood as the uncertainty remaining in the true outcome after receiving the forecast, as shown in Fig. 8.1.

Recall the decomposition of the Briar score into reliability (REL), resolution (RES), and uncertainty (UNC):

$$\text{BS} = \text{REL}_{\text{BS}} - \text{RES}_{\text{BS}} + \text{UNC}_{\text{BS}} \tag{8.67}$$

That is,

$$\text{BS} = \frac{1}{2n} \sum_{k=1}^{K} n_k (p_k - \overline{q_k})^2 - \frac{1}{n} \sum_{k=1}^{K} n_k (\overline{q_k} - \overline{q})^2 + \overline{q}(1 - \overline{q}) \tag{8.68}$$

where n is the total number of forecasts issued, K is the number of unique forecasts issued (or number of bins), $\overline{q} = \frac{1}{n} \sum_{i=1}^{n} q_i$ is the average observed climatological base rate with which the event occurs, n_k is the number of forecasts having the same probability category, and $\overline{q}_k$ is the average observed frequency, given forecasts of probability p_k. The first term in Eq. (8.68) is the reliability expressed by the summation of differences or distances between probabilities of binary events, and the second term is the resolution, which also has the same connotation. The last term expresses the uncertainty in the climate distribution. These three terms can now be formulated in terms of the relative entropy.

The relative entropy is the divergence of the observed probability distribution from the forecast probability distribution, both stratified (conditioned) on all issued forecast probabilities; i.e., reliability measures the conditional bias in the forecast probability. If both probability distributions were the same, i.e., the issued forecast probabilities were the same as the observed frequencies, then the reliability value would be 0. This case is referred to as a *perfectly calibrated* forecast. The reliability can now be expressed as the expected divergence:

$$\text{REL}_{\text{DS}} = \frac{1}{n}\sum_{k=1}^{K} n_k D_{KL}(\overline{q}_k | p_k) \tag{8.69}$$

where n is the total number of forecasts, K is the number of unique forecasts issued, $\overline{q}_k$ is the observed frequency distribution (PMF) for forecasts in group k, and p_k is the forecast PMF for group k.

Resolution is a measure of the reduction in climatic uncertainty or the amount of information in the forecast. In other words, it measures the amount of uncertainty in the observation that is explained by the forecast. From a divergence perspective, it is the expected divergence of conditional frequencies from the marginal frequency of occurrence. If the climatological probability is always forecasted or forecasts are entirely random, the minimum resolution is zero. Ideally, the resolution equals the uncertainty; in this case all uncertainty is explained. This is possible only for a deterministic forecast which is ether always right or always wrong. If it is wrong, it needs to be recalibrated:

$$\text{RES}_{\text{DS}} = \frac{1}{n}\sum_{k=1}^{K} n_k D_{KL}(\overline{q}_k | \overline{q}) \tag{8.70}$$

Equation (8.70) shows that resolution is the expectation over all forecast probabilities of the divergence from the conditional probability of occurrence to the marginal probability of occurrence. This can also be characterized as the mutual information I between forecasts and observations:

$$\text{RES}_{\text{DS}} = \underset{k}{E}\{D_{KL}(\overline{q}_k | \overline{q})\} = \underset{k}{E}\{D_{KL}[(\overline{q}|p_k)|\overline{q}]\} = I(P;Q) \tag{8.71}$$

Uncertainty is a measure of the initial uncertainty about the event, which can be measured by the entropy of the climatological distribution $H(\overline{q})$. Since this is observational uncertainty, it does not depend on the forecast and is a function of the climatological base rate only. From entropy theory, it is known that uncertainty is maximum for the probability of occurrence being 0.5, and 0 for the probability being 0 or 1:

$$\text{UNC}_{\text{DS}} = H(\overline{q}) = -\sum_{j=1}^{r} \{[\overline{q}]_j \log [\overline{q}]_j\} \tag{8.72}$$

For a single forecast-observation pair, the uncertainty and resolution vanish and the total score equals the reliability. For a large number of forecasts, uncertainty approaches climatic uncertainty and reliability tends to 0, provided the forecast is well calibrated.

Now it is shown that DS can be partitioned into three components: reliability, resolution, and uncertainty. To that end, first reliability and resolution can be combined:

$$\text{REL} - \text{RES} = \sum_{k=1}^{K} n_k \{ D_{\text{KL}}(\overline{q}_k | p_k) - D_{\text{KL}}(\overline{q}_k | \overline{q}) \}$$

$$= \sum_{k=1}^{K} n_k \sum_{j=1}^{r} [\overline{q}_k]_j \left\{ \log \frac{[\overline{q}_k]_j}{[p_k]_j} - \log \frac{[\overline{q}_k]_j}{[\overline{q}]_j} \right\}$$

$$= \sum_{k=1}^{K} n_k \sum_{j=1}^{r} [\overline{q}_k]_j \left\{ \log \frac{[\overline{q}_k]_j}{[p_k]_j} \right\} \tag{8.73}$$

Note that K is the number of different p_i, and by taking one bin for each p_i, both bins and outcomes can be labeled by k. Let the outcomes in a bin be labeled as $[q_i]_{km_k}$, with $m_k = 1, 2, \ldots, n_k$. Now Eq. (8.73) can be written as

$$\text{DS} = \frac{1}{n} \sum_{j=1}^{r} \sum_{k=1}^{K} \sum_{m_k=1}^{n_k} [q_{k,m_k}]_j \left\{ \log \frac{[q_{k,m_k}]_j}{[p_k]_j} \right\} \tag{8.74}$$

Equation (8.74) can be expressed as

$$\text{DS} = \frac{1}{n} \sum_{j=1}^{r} \sum_{k=1}^{K} \left(\sum_{m_k=1}^{n_k} [q_{k,m_k}]_j \left\{ \log \frac{[q_{k,m_k}]_j}{[p_i]_j} - \log \frac{[q_{k,m_k}]_j}{[\overline{q}]_j} + \log \frac{[q_{k,m_k}]_j}{[\overline{q}]_j} \right\} \right)$$

$$= \frac{1}{n} \sum_{j=1}^{r} \sum_{k=1}^{K} \left(\sum_{m_k=1}^{n_k} [q_{k,m_k}]_j \left\{ \log \frac{[\overline{q}]_j}{[p_i]_j} + \log \frac{[q_{k,m_k}]_j}{[\overline{q}]_j} \right\} \right) \tag{8.75}$$

$$= \frac{1}{n} \sum_{j=1}^{r} \sum_{k=1}^{K} \left(n_k [\overline{q}_k]_j \left\{ \log \frac{[\overline{q}]_j}{[p_k]_j} \right\} + \frac{1}{n} \sum_{j=1}^{r} \sum_{k=1}^{K} \left(\sum_{m_k} [q_{k,m_k}]_j \left\{ \log \frac{[q_{k,m_k}]_j}{[\overline{q}]_j} \right\} \right) \right)$$

The first term on the right side is equal to REL $-$ RES. Therefore,

$$\text{DS} - (\text{REL} - \text{RES}) = \frac{1}{n} \sum_{j=1}^{r} \sum_{k=1}^{K} \left(\sum_{m_k=1}^{n_k} [q_{k,m_k}]_j \left\{ \log \frac{[q_{k,m_k}]_j}{[\overline{q}]_j} \right\} \right)$$

$$= \frac{1}{n} \sum_{i=1}^{n} \left(\sum_{j=1}^{r} [q_i]_j \left\{ \log \frac{[q_i]_j}{[\overline{q}]_j} \right\} \right) = \frac{1}{n} \sum_{i=1}^{n} D_{\text{KL}}(q_i | \overline{q}) \tag{8.76}$$

Note that $\lim_{x \to 0} x \log x = 0$, and for $r = 2$, $Q_i \varepsilon \{ (1,0)^T, (0,1)^T \}$, it is found that

$$\sum_{j=1}^{r} [q_i]_j \left\{ \log \frac{[q_i]_j}{[\overline{q}]_j} \right\} = \sum_{j=1}^{r} \{ [q_i]_j \log [q_i]_j - [q_i]_j \log [\overline{q}]_j \}$$

$$\tag{8.77}$$

$$= \sum_{j=1}^{r} [q_i]_j \log [\overline{q}]_j$$

Therefore,

$$\sum_{i=1}^{n}\left(\sum_{j=1}^{r}[q_i]_j\left\{\log\frac{[q_i]_j}{[\bar{q}]_j}\right\}\right) = n\sum_{j=1}^{r}[\bar{q}]_j\log[\bar{q}]_j \qquad (8.78)$$

Hence,

$$\text{DS} - \text{REL} + \text{RES} = -\sum_{j=1}^{r}[\bar{q}]_j\log[\bar{q}]_j = H(\bar{q}) = \text{UNC} \qquad (8.79)$$

Equation (8.79) is valid for the forecast of events with an arbitrary number of categories.

8.4.1 Relation to Brier Score

Since the Brier score is valid for only binary events, the discussion will be restricted to such events. Benedetti (2009) has noted the Brier score to be a second-order approximation of the divergence score. Both scores achieve their minimum values only for perfect forecasts. The Brier score is symmetric, but the divergence score is not, except for the case when the true forecast probability equals 0.5. Roulston and Smith (2002) found the divergence score to be a double-valued function of the Brier score. When two forecast systems are compared, the forecast system with a higher Brier score may have a lower divergence score.

Reliability can reach infinity in the divergence score; this occurs when wrong deterministic forecasts are issued. This suggests that REL in DS is more sensitive to events with near deterministic wrong forecasts. In the Brier score REL varies between 0 and 1.

Resolution is a mean of divergences in DS but is the variance of conditional probabilities in BS or a mean of squared deviations from the climate probability. It can be interpreted as the amount of uncertainty explained. The range of values that RES can take on in both BS and DS is between 0 and uncertainty. RES in BS is the second-order approximation of that in DS and satisfies the condition that its minimum value be 0 and its maximum value equal the inherent uncertainty of the forecast event.

The uncertainty in BS is the second-order approximation of that in DS (entropy) and attains its maximum of 0.5. Both the zero uncertainty (probability of one of the two events being 1) and the maximum uncertainty (equally likely events with probability being 0.5) are located at the same place. In DS the maximum uncertainty is 1 bit.

8.4.2 Normalization to a Skill Score

The Brier score depends on the climatological probability but not on the forecast quality. Taking the climatological forecast as a reference, BS is normalized to Brier skill score (BSS). Another reference is a perfect forecast. The divergence score can be normalized to the *divergence skill score* (DSS) as

$$\text{DSS} = \frac{\text{DS} - \text{DS}_{\text{ref}}}{\text{DS}_{\text{perf}} - \text{DS}_{\text{ref}}} = 1 - \frac{\text{DS}}{\text{DS}_{\text{ref}}} \qquad (8.80)$$

For a perfect forecast the score is 0. For the climatological forecast (DS_{ref}), both RES and REL are 0 (perfect reliability, no resolution). Therefore, DSS becomes

$$\text{DSS} = 1 - \frac{\text{UNC} - \text{RES} + \text{REL}}{\text{UNC}} = \frac{\text{RES} - \text{REL}}{\text{UNC}} \qquad (8.81)$$

Equation (8.81) yields DSS as a positively oriented skill score having 0 for a forecast based on climatological probability and 1 for a perfect forecast. If the forecast is completely random having a marginal distribution as climatology, the score will be 0.

8.4.3 Relation to Ranked Mutual Information Skill Score

The ranked mutual information skill score (RMIS), proposed by Ahrens and Walser (2008), is for multicategory forecasts. For forecasts of binary events $RMIS_b$ can be expressed as the mutual information between forecasts and observations divided by the entropy of the observations:

$$RMIS_b = \frac{I(P,Q)}{H(\overline{Q})} \tag{8.82}$$

Comparing Eq. (8.72) with Eqs. (8.81), (8.82), and (8.81), it is seen that

$$RMIS_b = \frac{RES_{DSS}}{UNC_{DSS}} \tag{8.83}$$

If reliability is perfect (zero), $RMIS_b$ will be the same as DSS. If the forecast is not well calibrated, $RMIS_b$ measures the amount of information after calibration and ignores the reliability component. On the other hand, DSS measures the information based on the forecasts, and its individual components indicate the information that can potentially be extracted, as measured by $RMIS_b$.

8.4.4 Relation to Ignorance Score

The ignorance score rule, developed by Roulston and Smith (2002), is a measure of the ignorance or information deficit of a forecaster by comparison with the person knowing the true outcome of the event (i). The ignorance score can be characterized as a reinterpretation of the logarithmic score proposed by Good (1952). To establish the relation between IGN and DS, the relative entropy measure is expanded as

$$D_{KL}(Q_i|P_i) = \sum_{j=1}^{r} q_j \log\left(\frac{q_j}{p_j}\right) = q_{j\neq i} \log\left(\frac{q_{j\neq i}}{p_{j\neq i}}\right) + q_{j=i} \log\left(\frac{q_{j=i}}{p_{j=i}}\right) \tag{8.84}$$

Noting that $q_{j\neq i} = 0$ and $q_{j=i} = 1$, Eq. (8.84) reduces to

$$D_{KL}(Q_i|P_i) = 0 \log\left(\frac{0}{p_{j\neq i}}\right) + 1 \log\left(\frac{1}{p_{j=i}}\right) = -\log p_j = IGN \tag{8.85}$$

Equation (8.85) shows that the divergence from certain observation PMF(Q) to the forecast PMF(P) is the same as IGN, implying that DS is actually equal to the IGN. In a way, IGN not only is a scoring rule related to relative entropy but also is a relative entropy.

Expressing the divergence between the unknown "true" distribution Q and the forecast distribution P,

$$D_{KL}(Q|P) = E[IGN] - H(Q) \tag{8.86}$$

Equation (8.86) shows a relation of IGN to divergence (Roulston and Smith, 2002). The divergence used in DS is computed from the PMF after the observation (Q) in place of P, making the second term vanish and IGN equal to the divergence. This discussion also suggests that the IGN score can also be decomposed into three components: reliability, resolution, and uncertainty.

Example 8.6. Suppose we have a forecaster who forecasts every day the probability of rain the next day. Table 8.4 gives the forecast probabilities and observations for 21 events. Assess the quality of forecasts by comparing with observations. Compute the divergence score and its three components: reliability, resolution, and uncertainty. (This example was furnished by Dr. S. Weijs from Delft University of Technology, Delft, The Netherlands, 2010.)

Solution. First, all data are converted to probability mass functions (PMFs), which are discrete probability distributions. For example, on day 5 the forecast probability (see column of Table 8.4) is 0.7 (that is, $p_5 = 0.7$). Then the probability of no rain is $1 - 0.7 = 0.3$; only two possibilities are being considered here. Thus, the probability vector or distribution (PMF) for day 5 is $P = \{0.3, 0.7\}$. The corresponding observation for that day was that it rained, that is, $q_5 = 1$ (see column 5 of Table 8.4).

1	2	3	4	5	6	7
Day i	Forecast p_i	p_i^{cal}	p_i^{clim}	Observed q_i	DS$_i$	DSS$_i$
1	0.1	0	0.571	0	0.152	0.85
2	0.1	0	0.571	0	0.152	0.85
3	0.1	0	0.571	0	0.152	0.85
4	0.2	0.5	0.571	1	2.32	−1.36
5	0.7	1	0.571	1	0.515	0.48
6	0.9	0.8	0.571	1	0.152	0.85
7	0.9	0.8	0.571	1	0.152	0.85
8	0.8	1	0.571	1	0.322	0.67
9	0.9	0.8	0.571	1	0.152	0.85
10	0.2	0.5	0.571	0	0.322	0.67
11	0.2	0.5	0.571	1	2.322	−1.36
12	0.4	1	0.571	1	1.322	−0.34
13	0.3	0.5	0.571	1	1.737	−0.76
14	0.3	0.5	0.571	0	0.515	0.48
15	0.2	0.5	0.571	0	0.322	0.67
16	0.8	1	0.571	1	0.322	0.67
17	0.6	0.5	0.571	0	1.322	−0.34
18	0.5	0	0.571	0	1	−0.01
19	0.9	0.8	0.571	0	3.322	−2.37
20	0.6	0.5	0.571	1	0.737	0.25
21	0.9	0.8	0.571	1	0.152	0.85
Average	0.5048	0.5714	0.5714	0.5714	0.8316	0.1559

Table 8.4 Forecasts A and B and Corresponding Observations

This can be interpreted as being after observation since it is certain with probability 1 that it rained on that day. Thus, the observation can be cast as a vector of probabilities or probability distribution (PMF): $Q = \{0, 1\}$. This PMF contains 0 probability for no rain and probability of 1 for rain.

Now the divergence score is computed. To illustrate, consider day 5 and day 10. This is the Kullback-Leibler divergence or relative entropy or cross entropy from the observation to the forecast. Using Eq. (8.65), for day 5

$$DS_5 = \sum_{j=1}^{2} q_j \log_2\left(\frac{q_j}{p_j}\right) = 0 \times \log\left(\frac{0}{0.3}\right) + 1 \times \log\left(\frac{1}{0.7}\right) = -\log 0.7 = 0.5146 \text{ bit}$$

and for day 10

$$DS_{10} = \sum_{j=1}^{2} q_j \log_2\left(\frac{q_j}{p_j}\right) = 1 \times \log\left(\frac{1}{0.8}\right) + 1 \times \log\left(\frac{0}{0.2}\right) = -\log 0.8 = 0.3219 \text{ bit}$$

In this manner column 6 of Table 8.4 can be calculated for each day or event.

The divergence score for the complete set of 21 forecast-observation pairs given in Table 8.4 is the average of all DS_i scores corresponding to individual pairs. The sum of DS_i scores in column 6 is computed, and when divided by 21, the average DS score is 0.8316 as given in the last row and column 6. Physically, this DS value measures the average uncertainty remaining in the observations after the issuance of forecast.

Now each component of DS is computed. The purpose is to analyze the degree to which the information that the forecasts contain matches the information they convey. To that end, the first step is to count the number of unique probabilities and the corresponding number of times the event occurred. From Table 8.4, it is seen that $p = 0.1$ was forecasted 3 times and the event occurred 0 times; $p = 0.2$ was forecasted 4 times and the event occurred 2 times; $p = 0.3$ was forecasted 2 times and the event occurred 1 time; $p = 0.4$ was forecasted 1 time and the event occurred 1 time; $p = 0.5$ was forecasted 1 time and the event occurred 1 time; $p = 0.6$ was forecasted 2 times and the event occurred 1 time; $p = 0.7$ was forecasted 1 time and the event occurred 1 time; $p = 0.8$ was forecasted 2 times and the event occurred 2 times; and $p = 0.9$ was forecasted 5 times and the event occurred 4 times.

The next step is to compute the conditional distribution. For each unique forecast probability p_k, the probability that is forecasted is compared with the conditional probability of observing rain the next day q_k. To illustrate, consider the forecast probability of 0.9 which was forecast 5 times and the rain actually occurred only 4 times. This yields the conditional probability of $4/5 = 0.8$. In this manner, the conditional distribution was computed as shown in column 4 of Table 8.5.

Now the uncertainty component is computed. This quantifies the uncertainty inherent in the climate and is independent of the forecasts. The meaning of climate here is the average PMF of observations. For the data in Table 8.4, it rained 12 times out of 21 occasions (days). In other words, it rained on $12/21 \times 100 = 57$ percent of the days. Thus, average PMF is $\bar{q} = \frac{1}{n}\sum_{i=1}^{n} q_i = 0.5714$. Therefore, $\bar{Q} = (0.4286, 0.5714)$. The entropy of this distribution gives the uncertainty component. Therefore,

$$H(\bar{Q}) = -\sum_{j=1}^{r} \bar{q}_j \log \bar{q}_j = 0.4286 \log 0.4286 + 0.5714 \log 0.5714 = 0.9852 \text{ bit}$$

This is the uncertainty about the true outcome, given the climate but not the forecast.

It can be shown that this uncertainty is equal to the divergence score of a forecaster who always forecasts the climate probability as shown in column 4, $p_i^{\text{clim}} = \bar{p}$, of Table 8.4:

$$H(\bar{Q}) = \frac{1}{n}\sum_{i=1}^{n} D_{KL}(Q_i \mid \bar{Q}) = \frac{1}{n}\sum_{i=1}^{n}\sum_{j=1}^{r} q_{ij} \log \frac{q_{ij}}{\bar{q}_j} = -\sum_{j=1}^{r} \bar{q}_j \log \bar{q}_j$$

$$= -\frac{9}{21} \times 1 \times \log\left(\frac{1}{0.4286}\right) - \frac{12}{21} \times 1 \times \log\left(\frac{1}{0.5714}\right) = 0.9852$$

which is the same as before.

Now the resolution component is computed. The resolution is a measure of the conditioning power of the forecasts. Each time a user hears a forecast, his or her probability estimate becomes conditional on that forecast. If the user believes the forecast, the conditional estimates are equal to the forecast probabilities. For computing resolution, only the conditioning power of the forecasts for the observations needs to be considered. To illustrate, consider row 9 of Table 8.5. The forecast PMF is $(0.1, 0.9)$ and the average observed PMF is $(0.2, 0.8)$, because it rained 4 out of 5 times following that forecast. The divergence from this conditional distribution to the marginal (climatic) distribution of observations is computed as

$$D_{KL}(\overline{Q_k}|\overline{Q}) = D_{KL}[(0.2, 0.8)]\|(0.4286, 0.5714)$$

$$= 0.2 \times \log\left(\frac{0.2}{0.4286}\right) + 0.8 \times \log\left(\frac{0.8}{0.5714}\right) = 0.1685$$

Likewise, the divergence is computed for forecast-observation pairs. The resolution is the average of these divergence values. The contribution to the resolution of all forecasts $p_i = (0.1, 0.9)$ is therefore

$$\frac{n_k}{n} D_{KL}(\overline{Q_k}|\overline{Q}) = \frac{5}{21} \times 0.1684 = 0.0401$$

In a similar way, the contribution to the resolution is computed for every group $k\varepsilon[1,9]$, and then the contributions are summed, as shown in Table 8.5. This results in RES = 0.4324. This can also be obtained by dividing the number in the lower right of Table 8.5 by n: $9.0802/21 = 0.4324$.

Now the reliability is considered. This component compares each unique forecast probability p_k with the conditional probability of observation that follows the forecast q_k. It measures the extra uncertainty introduced for the user, because the forecasts are not perfectly reliable. Note that a perfect reliability is a reliability of 0. If the forecast probability is not 0 or 1, which is usually the case, then there is some uncertainty for the user. The remaining uncertainty after a forecast is thus at least the entropy of the forecast distribution $H(p_i)$. This however is only the remaining uncertainty if the forecast is perfectly reliable. If the probability a forecaster issues does not match the probability of the event, then the uncertainty that the user could hypothetically have if the forecast were perfectly calibrated was $H(\overline{q_k})$, and an extra uncertainty for the user is introduced by the unreliable forecast

1	2	3	4	5	6	7	8
K	n_k	p_k	$\overline{q}_k$	$D_{KL}(\overline{q_k}|p_k)$	$n_k D_{KL}(\overline{q_k}|p_k)$	$D_{KL}(\overline{q_k}|\overline{q})$	$n_k D_{KL}(\overline{q_j}|\overline{q})$
1	3	0.1	0	0.152	0.4560	1.2224	3.6672
2	4	0.2	0.5	0.322	1.2877	0.0149	0.0595
3	2	0.3	0.5	0.126	0.2515	0.0149	0.0297
4	1	0.4	1	1.322	1.3219	0.8074	0.8074
5	1	0.5	0	1.000	1.0000	1.2224	1.2224
6	2	0.6	0.5	0.029	0.0589	0.0149	0.0297
7	1	0.7	1	0.515	0.5146	0.8074	0.8074
8	2	0.8	1	0.322	0.6439	0.8074	1.6147
9	5	0.9	0.8	0.064	0.3203	0.1684	0.8422
Sum	$n = 21$				5.8548		9.0802
$\frac{1}{n}\sum_{k=1}^{K} n_k^*$		$p = 0.5048$	$\overline{q} = 0.5714$	REL = 0.2788		RES = 0.4324	

Table 8.5   Statistics of Unique Forecast Probabilities

equal to $D_{KL}(\overline{q_k}|p_k)$. It is then logical that $H(p_i)$ cannot be a measure of forecast quality. A forecaster could pretend that he is certain and only gives 10 or 90 percent chance forecasts. This, however, does not mean that he truly reduces uncertainty, and this can be determined only after comparing his forecasts with observations. Therefore, $H(p_i)$ can be seen as the uncertainty a forecaster thinks he has, while $DS_i = D_{KL}(q_i|p_i)$ is the uncertainty the forecaster turned out to have relative to the true outcome.

To illustrate, consider the row for $k = 9$ in Table 8.5. The forecaster thinks he has uncertainty of $H(p_k) = 0.4690$, the uncertainty he could have if perfectly calibrated is $H(\overline{q_k}) = 0.7219$, and the extra uncertainty his miscalculation introduces is $D_{KL}(\overline{q_k}|p_k) = 0.0641$. On average his remaining uncertainty, given a forecast $p_k = 0.9$, is $H(\overline{q_k}) + D_{KL}(\overline{q_k}|p_k) = 0.7219 + 0.0641 = 0.7860$. This is also known as the cross entropy between $\overline{q_k}$ and p_k and can be computed directly as

$$H(\overline{q_k}, p_k) = -\sum_{j=1}^{r} \overline{q_{kj}} \log p_{kj} = 0.2 \log 0.1 - 0.8 \log 0.9 = 0.786$$

This is exactly equal to the average divergence score over this particular forecast, $p_k = (0.1, 0.9)$. In case it rains, the divergence score becomes

$$D_{KL}(0,1)(0.1, 0.9) = 0 \log \frac{0}{0.1} + 1 \log \frac{1}{0.9} = -\log 0.9 = 0.1520$$

and when it is dry, the divergence score is

$$D_{KL}(1,0)(0.1, 0.9) = 1 \log \frac{1}{0.1} + 0 \log \frac{0}{0.9} = -\log 0.1 = 3.3219$$

Because it rained 4 out of 5 times after these forecasts, the average divergence score is therefore

$$\frac{4}{5}(0.1520) + \frac{1}{5}(3.3219) = 0.7860$$

This example is for a short time series in which reliability is overestimated because the conditional distributions $\overline{q_k}$ are based on too few data realistically to compare them with forecast probabilities. This, however, is compensated for by the resolution which is also overestimated. The results for a more realistic 3-yr data set of daily rainfall forecasts are shown in Table 8.6. It is seen from the column for $\overline{q_k}$ that the error in probability is less than 0.1 for all unique forecasts. The uncertainty term is

1	2	3	4	5	6	7	8
K	n_K	p_k	$\overline{q_k}$	$D_{KL}(\overline{q_k}|f_k)$	$n_k D_{KL}(\overline{q_k}|p_k)$	$D_{KL}(\overline{q_k}|\overline{q})$	$n_k D_{KL}(\overline{q_k}|\overline{q})$
1	299	0.1	0.04	0.0363	10.8594	0.6720	200.9343
2	107	0.2	0.27	0.0210	2.2522	0.1163	12.4448
3	92	0.3	0.32	0.0008	0.0725	0.0686	6.3130
4	82	0.4	0.46	0.0119	0.9767	0.000	0.0027
5	60	0.5	0.42	0.0201	1.2079	0.0073	0.4396
6	53	0.6	0.68	0.0194	1.0279	0.1324	7.0168
7	66	0.7	0.74	0.0064	0.4201	0.2267	14.9591
8	92	0.8	0.76	0.0066	0.6070	0.2597	23.8894
9	233	0.9	0.94	0.0115	2.6778	0.7422	172.9364
Sum	$n = 1084$				20.1014		438.9360
$\frac{1}{n}\sum_{k=1}^{K} n_k^*$		$p = $ 0.4640	$\overline{q} = $ 0.4668	REL = 0.0185		RES = 0.4049	

TABLE 8.6 Information Measures for All Unique Forecast Probabilities

$H(0.5332, 0.4668) = 0.9968$. The divergence score for the whole series is 0.6104. The skill [(RES − REL)/ UNC] is 0.3876 and the potential skill (for hypothetical perfect calibration without loss of resolution: RES/UNC) is 0.4062.

Questions

1. Compute the Briar score and the mean Briar score for the forecasts in Table 8.7.
2. Compute the Briar score and the mean Briar score for the forecasts in Table 8.8.
3. Ranked probability forecast problem.
4. IGN
5. Decomposition of Briar score.
6. Consider Example 8.6. What would be the score DS for day 19, if on that day the forecast had been 100 percent chance of rain instead of 90 percent chance of rain? How would you interpret this?
7. Mutual information between X and Y is defined as the expected Kullback-Leibler divergence from the marginal distribution of X to all conditional distributions $X|Y$, written as

$$I(X, Y) = \sum_{x,y} p(x, y) \log \frac{p(x|y)}{p(x)}$$

	Rain		No Rain	
Occasion *i*	Forecast Probability p_{ij}	Observed Outcome q_{ij}	Forecast Probability p_{ij}	Observed Outcome q_{ij}
1	0.3	1.0	0.7	0.0
2	0.7	1.0	0.3	0.0
3	0.9	0.0	0.1	1.0
4	0.2	1.0	0.8	0.0
5	0.1	0.0	0.9	1.0

TABLE 8.7 Forecasts in Terms of Probabilities

	Rain		No Rain	
Occasion *i*	Forecast Probability p_{ij}	Observed Outcome q_{ij}	Forecast Probability p_{ij}	Observed Outcome q_{ij}
1	0.2	0.0	0.8	1.0
2	0.2	1.0	0.8	0.0
3	0.2	1.0	0.8	0.0
4	0.2	0.0	0.8	1.0
5	0.2	0.0	0.8	1.0

TABLE 8.8 Forecasts in Terms of Probabilities

Which of the three terms of the decomposition could be written as mutual information and between which variables is this mutual information to be measured?

8. Suppose the forecaster cannot be held accountable for the forecasts on Sundays (days 7, 14, and 21) and these forecasts and corresponding observations are excluded from analysis. What would be the new values (calculated from the remaining 18 forecasts and observations) for the total divergence score, the resolution, the uncertainty, and the reliability?

9. When the decomposition is applied to only one forecast, what can be said about the values of DS, REL, RES, and UNC? Does the decomposition still hold?

10. What is the maximum value the uncertainty could have? How would this change if there were four forecast categories (e.g., no rain, little rain, moderate rain, heavy rain)?

References

Ahrens, B., and Walser A. (2008). Information-based skill scores for probabilistic forecast. *Monthly Weather Review*, vol. 136, pp. 352–363.

Anderson, J. L. (1996). A method for producing and evaluating probabilistic forecasts from ensemble model integration. *Journal of Climate*, vol. 9, pp. 1518–1530.

Benedetti, R. (2009). Scoring rules for forecast verification. *Monthly Weather Review*, doi: http://dx.doi.org/10.1175%2F2009MWR2945.1, preprint.

Brier, G. W. (1950). Verification of forecasts expressed in terms of probability. *Monthly Weather Review*, vol. 78, no. 1, pp. 1–3.

Daan, H. (1985). Sensitivity of verification scores to the classification of the predictant. *Monthly Weather Review*, vol. 113, pp. 1384–1392.

Epstein, E. (1969). A scoring system for probability forecasts of ranked categories. *Journal of Applied Meteorology*, vol. 8, pp. 985–987.

Good, I. J. (1952). Rational decisions. *Journal of the Royal Statistical Society*, Series B (Methodological), vol. 14, no. 1, pp. 107–114.

Hamil, T. M., and Colucci S. J. (1996). Random and systematic error in NMC's short range Eta ensembles. *Proceedings, 13th Conference on Probability and Statistics in the Atmospheric Sciences*, San Francisco, American Meteorological Society, pp. 51–56.

Jolliffe, I., and Stephenson, D. B. eds. (2003). *Forecast Verification: A Practitioner's Guide in Atmospheric Sciences*, Wiley, New York.

Mason, I. B. (1982). A model for assessment of weather forecasts. *Australian Meteorological Magazine*, vol. 30, pp. 291–303.

Murphy, A. H. (1966). A note on the utility of probabilistic predictions and the probability score in the cost-loss ratio decision situation. *Journal of Applied Meteorology*, vol. 5, pp. 534–537.

Murphy, A. H. (1971). A note on the ranked probability score. *Journal of Applied Meteorology*, vol. 10, pp. 155–156.

Murphy, A. H. (1973). A new vector partition of the probability score. *Journal of Applied Meteorology*, vol. 12, pp. 595–600.

Murphy, A. H. (1986). A new decomposition of the Brier score: Formulation and interpretation. *Monthly Weather Review*, vol. 114, pp. 2671–2673.

Murphy, A. H. (1997). Forecast verification. In: *Economic Value of Weather and Climate Forecasts*, edited by A. H. Murphy and R. W. Katz. Cambridge University Press, pp. 19–70.

Richardson, D. S. (2000). Sill and relative economic value of the ECMWF ensemble prediction system. *Quarterly Journal of Royal Meteorological Society*, vol. 126, pp. 649–667.

Richardson, D. S. (2001). Measures of skill and value of ensemble prediction systems, their interrelationship and the effect of ensemble size. *Quarterly Journal of Royal Meteorological Society*, vol. 127, pp. 2473–2489.

Roulston, M. S., and Smith L. A. (2002). Evaluating probabilistic forecasts using information theory. *Monthly Weather review*, vol. 130, no. 6, pp. 1653–1660.

Shannon, C. E. (1948). A mathematical theory of communication. *Bell System Technical Journal*, vol. 27, pp. 379–423.

Smith, L. A. (2000). Disentangling uncertainty and error: On the predictability of nonlinear systems. *Nonlinear Dynamics and Statistics*, edited by A. I. Mees, Birkhauser, MA pp. 31–64.

Stanski, H., Wilson, L., and Burrows W. (1989). Survey of common verification methods in meteorology. Technical Report WMO/TD 358, WMO Weather Watch.

Stephenson, D. B., Coelho, C. A. S. and Jolliffe, I. T. (2008). Two extra components in the Brier score decomposition. *Weather and Forecasts*, vol. 23, pp. 752–757.

Stephenson, D. B., and Doblas-Reyes F. J. (2000). Statistical methods for interpreting Monte Carlo ensemble forecasts. *Tellus*, vol. 52A, pp. 300–322.

Swets, J. A. (1973). The relative operating characteristic in psychology. *Science*, vol. 182, pp. 990–999.

Talagrand, O., Vautard, R. and Strauss B. (1997). Evaluation of probabilistic prediction systems. *Proceedings, ECMFW Workshop on Predictability*, U. K., ECMWF, pp. 1–25.

Toth, Z., Talagrand, O. Candille, G. and Zhu Y. (2003). Probability and ensemble forecast. Chapter 7 in: *Forecast Verification: A Practitioner's Guide in Atmospheric Science*, edited by I. T. Jolliffe and D. B. Stephenson. Wiley, Chap. 7, pp. 137–163.

Toth, Z., Zhu, Y., Marchok, T., Tracton, S. and Kalnay E. (1998). *Verification of the NCEP global ensemble forecasts*. Preprint. 12th Conference on Numerical Weather Prediction, Phoenix, Arizona, American Meteorological Society, pp. 286–289.

Weijs, S. V., van Nooijen, R. and van de Giesen N. (2010a). Kullback-Leibler divergence as a forecast skill score with classic reliability-resolution-uncertainty decomposition. *Monthly Weather Review*, vol. 138, pp. 3387–3399, doi: 10.1175/2010MWR3229.1.

Weijs, S. V., Schoups, G. and van de Giesen N. (2010b). Why hydrological predictions should be evaluated using information theory? *Hydrology and Earth System Science*, vol. 14, No. 12, pp. 2545–2558.

Weijs, S. V., and van de Giesen N. (2011). Accounting for observational uncertainty in forecast verification: Information-theoretic view on forecast, observations, and truth. *Monthly Weather Review*, vol. 139, pp. 2156–2162, doi: 10.1175/2011MWR3573.1.

Wilks, D. S. (2006). *Statistical Methods in the Atmospheric Sciences*. Academic Press, New York.

Zhu, Y., Iyengar, G., Toth, Z., Tracton, M.S. and Marchok T. (1996). *Objective evaluation of the NCEP global ensemble forecasting systems*. Preprint, 15th Conference on Weather Analysis and Forecasting, Norfolk, Virginia, American Meteorological Society, J79–J82.

Further Reading

Atger, F. (2003). Spatial and interannual variability of the reliability of ensemble-based probabilistic forecasts: Consequences for calibration. *Monthly Weather Review*, vol. 131, pp. 1509–1523.

Benedetti, R. (2010). Scoring rules for forecast verification. *Monthly Weather Review*, vol. 138, pp. 203–211.

Briggs, W., Pocernich, M. and Ruppert D. (2005). Incorporating misclassification error in skill assessment. *Monthly Weather Review*, vol. 133, pp. 3382–3392.

Brocker, J. (2009). Reliability, sufficiency, and the decomposition of proper scores. *Quarterly of Royal Meteorological Society*, vol. 135, no. 643, pp. 1512–1519.

Brocker, J., and Smith L. (2007). Scoring probabilistic forecasts: The importance of being proper. *Weather Forecasting*, vol. 22, pp. 382–388.

DelSole, T. (2004). Predictability and information theory. Part 1: Measures of predictability. *Journal of the Atmospheric Sciences*, vol. 61, pp. 2425–2440.

Kleeman, R. (2002). Measuring dynamical prediction utility using relative entropy. *Journal of the Atmospheric Sciences*, vol. 59, pp. 2057–2072.

Kelly, J. (1956). A new interpretation of information rate. *IEEE Transactions: Information Theory*, vol. 2, No. 3, pp. 185–189.

Laio, F., and Tamea S. (2007). Verification tools for probabilistic forecasts of continuous hydrological variables. *Hydrology and Earth System Science*, vol. 11, pp. 1267–1277.

Leung, L., and North G. (1990). Information theory and climate prediction. *Journal of Climate*, vol. 3, pp. 5–14.

Mason, S. J. (2004). On using "climatology" as a reference strategy in the Brier and ranked probability skill scores. *Monthly Weather Review*, vol. 132, pp. 1891–1895.

Murphy, A. H. and Winkler R. L. (1987). A general framework for forecast verification. *Monthly Weather Review*, vol. 115, pp. 1330–1338.

Peirolo, R. (2010). Information gain as a score for probabilistic forecasts. *Meteorological Applications*, vol. 18, pp. 9–17, doi: 10.1002/met.188.

Sanders, F. (1963). On subjective probability forecasting. *Journal of Applied Meteorology*, vol. 2, no. 2, pp. 191–201.

Selten, B. (1998). Axiomatic characterization of the quadratic scoring rule. *Experimental Economics*, vol. 1, pp. 43–62.

Stephenson, D. B., Coelho, C. A. S. Doblas-Reyes, F. J. and Balmaseda M. (2005). Forecast assimilation: A unified framework of the combination of multi-model weather and climate predictions. *Tellus*, vol. 57A, no. 3, pp. 253–264.

Weigel, A. P., Liniger, M. A. and Appenzeller C. (2007). Generalization of the discrete Brier and ranked probability skill scores for weighted multimodel ensemble forecasts. *Monthly Weather Review*, vol. 135, pp. 2778–2785.

CHAPTER 9

Assessment of Potential Water Resources Availability

Rainfall constitutes the primary input to the hydrologic cycle, and it is thus vital for the assessment of the potential availability of water resources in an area or country. Although rainfall is cyclic, its distribution in both space and time is highly erratic, leading to unevenly distributed water resources. The uncertainty (or disorder) in the occurrence of rainfall, especially intensity, amount and duration, in time and space is one of the primary constraints to the development and use of water resources. When one is creating a basinwide, regional, or nationwide strategy for water resources development as well as for meeting current and future water demands, the uncertainty in rainfall occurrence over a given area (e.g., basin, region, or country) can be the determining factor in making a decision on the priorities for areawide development or demarcating the boundaries to establish the feasibility and necessity of development.

The development and management of water resources require not only the aggregate rainfall data but also their variability. The stability of the water supply increases with decreasing spatial variability of rainfall. On the other hand, the less temporally variable the rainfall, the more dependable the water supply. In evaluating the availability of water resources (i.e., water resources potential or water supply potential) in a watershed or investigating the relative availability of local or regional water resources, the temporal variability of rainfall becomes a major concern.

As discussed in Chap. 1, the uncertainty or disorder of a rainfall variable can be calculated using entropy (Shannon, 1948), provided the probability distribution function of the rainfall variable under consideration is given. If observed rainfall data are available, then entropy theory can be applied to quantify the disorder in rainfall. The variance, a well-known conventional technique, can also quantify the disorder about the mean of a frequency distribution. However, entropy is more powerful and general than variance. As is well known (Kagan et al., 1973), when the entropy of a random variable is maximized under the constraint that the variance and the mean are known for a set of values of the variable, the probability density function for this set is reduced to the normal distribution. This implicitly means that the use of variance is limited to the case in which the distribution is normal, whereas the entropy-based methodology is applicable to any distribution, whether the distribution is known or unknown a priori.

This chapter discusses an areawide evaluation of the degree of variability of annual rainfall in an area (i.e., basin, region, nation, or even the world). The evaluation would entail quantification of the degree of uncertainty of rainfall occurrence or temporal rainfall apportionment, delineation of areawide entropy distribution of rainfall, contrasting the entropy map with area's well-known climatic division map, and construction of a water availability map by linking entropy with rainfall.

9.1   Entropy as a Measure of Temporal Rainfall Apportionment

Entropy can be applied to determine the uncertainty of rainfall intensity, rainfall amount, and rainfall occurrence (i.e., temporal apportionment). From the standpoint of evaluating the availability of water resources, the uncertainty associated with temporal distribution of rainfall or apportionment is useful. It is premised that null entropy occurs when the rainfall intensity is uniform in time. If rainfall data are available, then probabilities can be defined as frequencies of occurrence of discrete rainfall amounts spread over a given period. Here the focus is on the uncertainty of the temporal rainfall distribution or apportionment and to measure its degree by using entropy expressed in terms of the probability density of rainfall randomly apportioned over fragmented times. This is discussed in what follows.

First, consider a historical rainfall series in a year. Let r_i be the aggregate rainfall (daily rainfall) during the ith day in the year. For example, daily rainfall values on January 1 and on December 31 for the same year can be expressed as r_1 and r_{365}, respectively. The aggregate rainfall during the year (annual rainfall) R can then be expressed by the summation of r_i from $i = 1$ to $i = 365$ as

$$R = \sum_{i=1}^{365} r_i \tag{9.1}$$

where the value of r_i may be zero for some days and is finite for other days.

Second, the entropy of the rainfall distribution within the year can be determined in one of two ways. If the rainfall data are not available for several years, then a random experiment is conducted to interpret the throughout-the-year rainfall variability or disaggregate annual rainfall into daily values. The experiment consists of a number of trials, where each trial is considered in a probabilistic sense. The experiment can be conducted as follows. If the annual rainfall R, is say, 1000 mm, then the experiment may consist of 1000 trials. The value of R is rounded off to the next nearest integer. Each trial contains 365 days (d), January 1 to December 31. From each trial 1 d is randomly selected. That means that any day of the year has the same probability of being selected. The selected day is assigned a value of rainfall of 1 mm. Suppose January 3 happens to be selected during this trial; then it will be assigned a rainfall value of 1 mm. Similarly, the next trial is performed and a day is selected randomly. The selected day will again be assigned a rainfall value of 1 mm. In this manner, 1000 trials are made and the total amount of rainfall associated with the selected days will be 1000 mm. If a particular day happens to be randomly selected 10 times, then it will be assigned a rainfall value of 10 mm (which is the sum of 10 selections). The days not selected will be assigned a zero value. When the rainfall values thus generated are plotted versus days, the result will be the rainfall series for the year.

If rainfall data are available for several years, then one can also use the data in place of the experimentally generated rainfall series. A rainfall series of $r_1, r_2, ..., r_n$ can thus be regarded as accumulated occurrence frequencies of unit rains for the first, second, ..., nth days, respectively, and r_i divided by the sample value of R defines the *relative frequency p_i*:

$$p_i = \frac{r_i}{R} \tag{9.2}$$

This relative frequency, p_i, is regarded as an occurrence probability for the rainfall amount on the ith day, and therefore its distribution represents the probabilistic characteristic of the over-a-year temporal apportionment of annual rainfall, i.e., of the uncertainty of rainfall occurrence.

Substitution of Eq. (9.2) into the entropy definition yields the value of entropy (bit):

$$H = -\sum_{i=1}^{n} \frac{r_i}{R} \log_2 \frac{r_i}{R} \tag{9.3}$$

Equation (9.3) implies that the value of H (in bits) is independent of the sequential order of r_i in the series. It is also seen that H takes on a zero value when R falls only on one day of the year, and a maximum value ($\log_2 n$) when R/n falls equally every day throughout the year (i.e., the closer the entropy H approaches its maximum value, the more uniform the rainfall apportionment is (i.e., the less temporally variable the rainfall is). Thus H can be regarded as a measure of rainfall variability in a scalar sense. Note that the amount-based measure defined here is distinct from the variability measure as a vector component of the timing-based seasonality measures (Magilligan and Graber, 1996; Burn, 1997) which consider the time of occurrence of hydrologic events, such as floods, rainfall, and droughts. When yearly rainfall series for m years is available at the same rain gauge, a better estimate of the annual entropy can be obtained by averaging the entropy values (bit) as

$$\overline{H} = \frac{1}{m} \sum_{j=1}^{m} H_j \tag{9.4}$$

where $\overline{H}$ is the average entropy (in bits) of the m-year record and H_j is the entropy of the jth year rainfall.

9.1.1 Rainfall Data

All rain gauges for the area under consideration are selected. Daily rainfall data for these gauges are obtained. If some yearly rainfall series measured at these gauges have missing values due to the failure of recording and/or some gauges have extremely short duration of data acquisition, then these series may be omitted from the analysis. For a reliable estimation of H and $\overline{H}$ at any rain gauge, rainfall series and rain gauges should be screened, based on two criteria: (1) Any yearly rainfall series to be selected must have a complete set of daily rainfall data, thus having 365 and 366 consecutive data values for the common and leap years, respectively. (2) Any rain gauge to be selected must have at least 8 yr or preferably more than 8 yr of rainfall observations. (This may vary from one basin or country to another.) The gauges must satisfy the above criterion: (1) criterion (2) requiring $m \geq 8$ in Eq. (9.4) is based on the acceptance, such as in Japan, that the meteorological data consecutively observed over 8 years or more can be used for the description of quasi-averaged yearly meteorology.

9.1.2   Calculation of Rainfall Entropy and Isoentropy Lines

The rainfall series with a 1-d resolution can be considered to describe the throughout-the-year rainfall characteristics. The sequence of observed daily rainfall values in a year is described by a probability distribution of rainfall occurrence, and an annual Shannon entropy (Shannon, 1948) value is obtained at each rain gauge station. The analysis is performed for all available yearly rainfall sequences. Then an average of the entropies obtained over the years of interest is considered as the averaged annual entropy at the rain gauge. Calculations show that the use of different averaging periods for different rain gauges has an insignificant effect on the results, since the focus is on the areawide or nationwide gross aspects of hydrologic or water resources issues. The averaged annual entropy values thus obtained for the rain gauges scattered throughout the area are employed to construct an isoentropy map that delineates rainfall variability. Considering the occurrence of these daily rainfall values within a year as random outcomes, entropy as an objective measure of rainfall variability is computed, and isoentropy lines as well as isohyetal lines are then constructed.

9.1.3   Hydrological Zoning

A map, depicting the areawide distribution of averaged annual entropy $\bar{H}$ can be constructed. This map may consist of scattered circles, each of which is drawn at the place of the rain gauge, and different circles, can define the ranks of entropy. One can also compute and summarize basic statistics [e.g., mean, minimum, maximum, standard deviation (SD), and coefficient of variation (CV)] for $\bar{H}$, R, and m selected for the calculation of entropy.

Example 9.1.   Consider the Brazos River basin that has a network of rain gauges. Compute yearly rainfall amounts for each gauge for the entire period of record. Also compute the average yearly rainfall for the entire record. Using daily rainfall values, compute the entropy of each rain gauge for each year. Compute the yearly entropy in two ways. First, compute the probability of rainfall on the ith day by taking the ratio of the daily rainfall to the yearly rainfall. In this manner one will have the rainfall probability distribution of daily rainfall apportionment. Using these probability values, compute the yearly entropy. Then compute the average entropy for the period of record, using these yearly entropy values. Also compute the standard deviation and coefficient of variation of the yearly entropy. Compute the mean, standard deviation, and coefficient of variation of daily rainfall, considering only those days when it rained. In this manner one will have these statistics for each year, and then one can average them for the whole record. Plot yearly rainfall as well as yearly entropy versus the year. Second, conduct a random experiment for rainfall apportionment, and then do the same calculations as for the first case. Now compare the two cases and comment on the results. Place entropy values obtained in two ways on the Brazos River basin map. What do these two maps show?

Solution.   Here 27 stations, having records from 1970 to 2006, are selected. These stations, along with their average annual rainfall amounts, are given in Table 9.1.

By using the daily rainfall values, the entropy of each rain gauge is computed for each year. First, the probability of rainfall on the ith day is computed by taking the ratio of daily rainfall to the yearly rainfall. This leads to the rainfall probability distribution of daily rainfall apportionment. Using these probability values, yearly entropy is computed. Then the average entropy for the period of record using these yearly entropy values is computed, as shown in Table 9.2.

Now the standard deviation and coefficient of variation of yearly entropy are computed as shown in Table 9.3. Also the minimum and maximum entropy values are shown in Tables 9.4 and 9.5.

The mean, standard deviation, and coefficient of variation of daily rainfall are computed, considering only those days when it rained. In this manner these statistics are obtained for each year, and then they are averaged for the whole record, as shown in Tables 9.6 to 9.8.

Node	Name of Station	Total Rainfall Amount (in.)	Average Yearly Rainfall (in.)	Length of Record (yr)
1	Albany	795.47	27.43	29
2	Cameron	1335.50	38.16	37
3	College Station	1646.50	44.50	34
4	Columbus	1645.70	47.02	37
5	Conroe	1768.70	52.02	36
6	Eastland	888.54	29.62	32
7	Freeport	1780.80	53.96	35
8	Hillsboro	1336.00	38.17	37
9	Houston Height	1711.90	61.14	30
10	Jayton	825.23	23.58	37
11	Marlin	1217.20	38.04	34
12	Matador	810.68	23.84	36
13	McGregor	1216.90	36.88	35
14	Morgan Mill	1013.30	31.67	34
15	Olney	947.31	29.60	34
16	Rosebud	998.44	39.94	27
17	Seymour	993.61	29.22	36
18	Thompson	1617.80	49.02	35
19	Tomball	1699.70	51.51	35
20	Waco	1186.60	33.90	37
21	Weatherford	1263.00	36.09	37
22	Snyder	823.67	22.80	37
23	Crosbyton	829.34	22.42	37
24	Plainview	757.37	20.96	37
25	Muleshoe	648.59	17.53	37
26	Ft Sumner	570.19	15.41	37
27	Bell Ranch	516.99	13.97	37

TABLE 9.1   Rainfall for Rain Gauge Stations in Brazos River Basin

Stations	Average Entropy (bits)						
1–7	4.946	5.314	4.968	5.561	5.474	5.066	5.462
8–14	5.297	5.513	4.696	5.285	4.973	5.296	4.993
15–21	4.917	5.038	5.037	5.523	5.434	5.195	5.205
22–27	4.957	4.817	4.893	4.516	4.625	4.987	—

TABLE 9.2   Average Yearly Entropy of Each Station

Stations	Standard Deviation (bits)						
1–7	0.424	0.224	0.315	0.178	0.297	0.270	0.280
8–14	0.225	0.237	0.353	0.277	0.257	0.227	0.345
15–21	0.451	0.474	0.318	0.220	0.254	0.287	0.201
22–27	0.429	0.453	0.448	0.505	0.497	0.389	—

TABLE **9.3**   Standard Deviation of Yearly Entropy of Each Station

Stations	Minimum Yearly Entropy (bits)						
1–7	3.165	4.926	4.176	5.024	4.647	4.346	4.612
8–14	4.885	4.814	3.900	4.568	4.331	4.800	3.750
15–21	3.196	3.065	4.240	5.096	4.823	4.376	4.723
22–27	3.548	3.216	3.357	3.193	3.202	3.697	—

TABLE **9.4**   Minimum Yearly Entropy of Each Station

Stations	Maximum Yearly Entropy (bits)						
1–7	5.686	5.788	5.624	5.865	6.024	5.493	5.949
8–14	5.802	6.007	5.287	5.823	5.479	5.655	5.465
15–21	5.471	5.663	5.775	5.917	5.959	5.618	5.663
22–27	5.710	5.416	5.608	5.219	5.338	5.746	—

TABLE **9.5**   Maximum Yearly Entropy of Each Station

Stations	Mean Daily Rainfall (in.)						
1–7	0.407	0.459	0.464	0.416	0.579	0.465	0.526
8–14	0.477	0.708	0.576	0.460	0.361	0.484	0.546
15–21	0.460	0.677	0.448	0.490	0.609	0.419	0.491
22–27	0.512	0.631	0.583	0.548	0.587	0.576	—

TABLE **9.6**   Mean Daily Rainfall of Each Station

Stations	Standard Deviation (in.)						
1–7	0.549	0.651	0.723	0.670	0.853	0.599	0.887
8–14	0.659	0.976	0.608	0.641	0.525	0.629	0.672
15–21	0.659	0.747	0.612	0.731	0.849	0.609	0.670
22–27	0.694	0.723	0.594	0.607	0.718	0.659	—

TABLE **9.7**   Standard Deviation of Daily Rainfall of Each Station

Stations	Coefficient of Variation						
1–7	0.743	0.704	0.643	0.621	0.680	0.776	0.593
8–14	0.724	0.725	0.948	0.717	0.689	0.770	0.812
15–21	0.699	0.906	0.732	0.671	0.718	0.674	0.733
22–27	0.738	0.873	0.982	0.902	0.817	0.874	—

TABLE 9.8　Coefficient of Variation of Rainfall of Each Station

Now, a random experiment is conducted for rainfall apportionment, and then calculations are done as for the first case. The average entropy of each station is found as shown in Table 9.9.

The standard deviation and coefficient of variation of yearly entropy are computed as shown in Table 9.10.

The minimum and maximum values of yearly entropy are computed, as shown in Tables 9.11 and 9.12, respectively.

Now the mean, standard deviation, and coefficient of variation of daily rainfall are computed, considering only those days when it rained. In this manner, these statistics for each year are obtained, and then they are averaged for the whole record. These statistics are given in Tables 9.13 to 9.15.

Now, comparing the two cases, it is seen that the entropy calculated with the random experiment is larger than that for the first case using actual values. The reason is that in the second case the

Stations	Mean Entropy (bits)						
1–7	7.337	7.791	7.485	7.882	7.929	7.556	7.967
8–14	7.728	8.021	7.369	7.744	7.387	7.711	7.615
15–21	7.554	7.747	7.544	7.905	7.928	7.654	7.668
22–27	7.408	7.497	7.527	7.176	7.312	7.542	—

TABLE 9.9　Average Entropy of Each Station

Stations	Standard Deviation (bits)						
1–7	0.213	0.142	0.165	0.162	0.172	0.227	0.140
8–14	0.173	0.132	0.282	0.156	0.238	0.157	0.220
15–21	0.281	0.232	0.186	0.168	0.158	0.216	0.194
22–27	0.265	0.271	0.192	0.248	0.223	0.214	—

TABLE 9.10　Standard Deviation of Entropy of Each Station

Stations	Minimum Yearly Entropy (bits)						
1–7	6.826	7.503	7.198	7.346	7.430	7.153	7.677
8–14	7.309	7.685	6.654	7.449	6.824	7.371	7.103
15–21	6.602	6.988	7.072	7.420	7.497	7.169	6.979
22–27	7.166	6.958	6.596	7.091	6.857	6.618	—

TABLE 9.11　Minimum Yearly Entropy of Each Station

Stations	Maximum Yearly Entropy (bits)						
1–7	7.846	8.075	7.995	8.161	8.214	7.998	8.282
8–14	8.052	8.212	7.835	8.026	7.819	8.012	7.932
15–21	7.966	8.111	7.926	8.174	8.187	8.062	7.946
22–27	7.986	8.019	8.078	7.498	7.894	7.887	—

TABLE 9.12   Maximum Yearly Entropy of Each Station

Stations	Mean Rainfall (in.)						
1–7	0.150	0.162	0.171	0.177	0.186	0.147	0.190
8–14	0.159	0.203	0.137	0.160	0.136	0.157	0.149
15–21	0.147	0.163	0.145	0.179	0.185	0.155	0.156
22–27	0.198	0.197	0.203	0.166	0.198	0.190	—

TABLE 9.13   Mean of Daily Rainfall

Stations	Standard Deviation (in.)						
1–7	0.064	0.085	0.088	0.095	0.104	0.073	0.107
8–14	0.082	0.113	0.065	0.085	0.063	0.081	0.075
15–21	0.073	0.088	0.070	0.098	0.103	0.080	0.082
22–27	0.086	0.079	0.078	0.065	0.081	0.076	—

TABLE 9.14   Standard Deviation of Daily Rainfall

Stations	Coefficient of Variation						
1–7	2.343	1.918	1.939	1.855	1.787	2.001	1.771
8–14	1.931	1.794	2.101	1.882	2.159	1.936	1.991
15–21	2.011	1.850	2.062	1.835	1.798	1.940	1.908
22–27	2.319	2.498	2.608	2.554	2.439	2.501	—

TABLE 9.15   Coefficient of Variation of Daily Rainfall

experiment is random, wherein the frequency of occurrence of rainfall tends to be uniform. This also explains the case of average entropy and yearly entropy. Due to random sampling, the standard deviation is less for the second case.

For rainfall in the second case, the mean is less and the standard deviation is also less than that in the first case. This is so because the random sampling will spread the distribution of the amount of rainfall uniformly during the record and thus will decrease the mean. For the same reason, the standard deviation is less, which means the data are "smooth" or uniform during the record. In other words, the random sampling of rainfall during the record can decrease the mean and standard deviation of rainfall, decrease the standard deviation of entropy, and increase the entropy (including average entropy and yearly entropy).

Now entropy values obtained in the two ways are placed on the Brazos River basin map, as shown in Figs. 9.1 and 9.2.

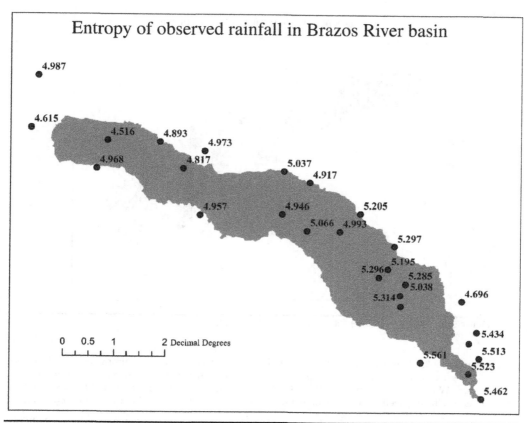

FIGURE 9.1 Entropy of rainfall from observations for the Brazos River basin.

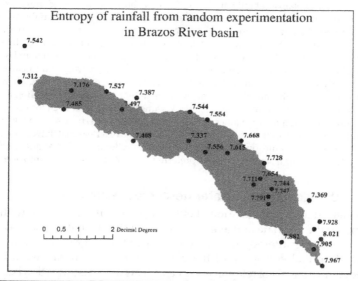

FIGURE 9.2 Entropy of rainfall from random experimentation for the Brazos River basin.

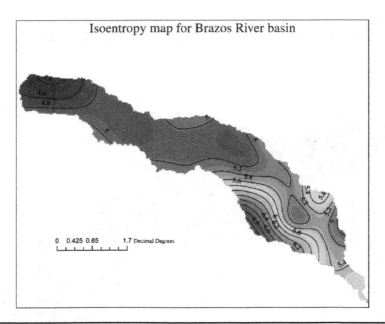

FIGURE **9.3**   Isoentropy lines for the Brazos River basin.

Example 9.2.   Using the results from Example 9.1, construct a map of isoentropy lines for the Brazos River basin, and comment on the information obtained from the map. Also construct an isohyetal map. What does this map reflect? Construct a map of the coefficient of variation. What does this map reflect? Taken together, what do these three maps convey? Which map should be preferred and why?

Solution.   The isoentropy map is constructed as shown in Fig. 9.3, in which entropy indicates significant zones over the whole Brazos River basin and can delineate a plausible climatic map that qualitatively explains rainfall variability. The distribution of averaged annual rainfall, delineated by isohyetal lines, is given in Fig. 9.4. This map indicates the spatial variability of mean rainfall. The distribution of averaged annual rainfall can be related to the zones delineated by isoentropy lines. The potential availability of water resources can be assessed or categorized by isoentropy lines. The spatial distribution of the coefficient of variation of daily rainfall is shown in Fig. 9.5. The CV of daily rainfall at each rain gauge is computed in the same way as entropy. Eliminating the zero values, the mean daily value of rainfall is obtained for each rain gauge for each year. Then the CV is computed for each year. By summing the CV values obtained for each year and dividing the sum by the number of years of record, the annually averaged CV value is obtained for each rain gauge. As compared with the entropy-based map, the values of CV show a similar pattern, as seen in Fig. 9.5, and one can make a clear classification of the region. As a result, it does have a clear match with the climate division map. Taken together, the entropy can delineate a plausible climatic map that qualitatively explains rainfall characteristics. The isoentropy map is preferred, since it shows significant zones over the river basin.

9.1.4   Categorization of Water Resources Availability

Of further interest is the relation of entropy to rainfall. A diagram with averaged annual entropy on the ordinate and averaged annual rainfall on the abscissa can be constructed which will be a scattered diagram. Glancing at the diagram, one might observe that the averaged annual entropy and the averaged annual rainfall would be less mutually related, with a small correlation coefficient. This demonstrates that besides the aggregate

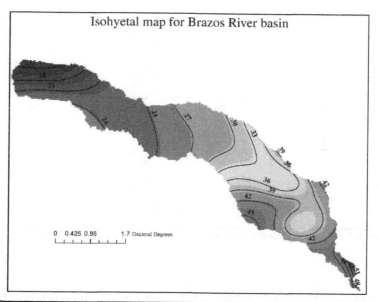

FIGURE 9.4   Isohyets for the Brazos River basin.

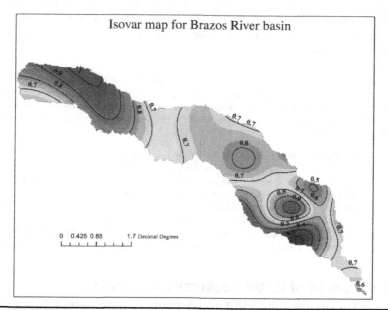

FIGURE 9.5   Coefficient of variation for the Brazos River basin.

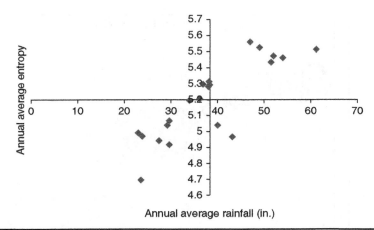

FIGURE **9.6** Relation between average yearly entropy and yearly rainfall.

of rainfall, its temporal apportionment can be a significant aspect of rainfall data. When coupled, entropy and rainfall on a yearly basis can become a measure of the throughout-the-year potential availability of water resources. To explain it qualitatively, the whole plotted area can be divided into, say, four parts, each delineated with two intersecting dividing lines that pass through the means of the respective two variables. Then, in terms of water resources availability, the respective quadrants, A, B, C and D, can be comparatively categorized. This is explained using the Brazos River basin as an example.

Example 9.3. Using the values obtained in Example 9.1, plot the average annual entropy versus the average the annual rainfall or precipitation on rectangular paper. Then divide the plot into four parts based on the lines issuing from the mean annual rainfall and the mean entropy. What do you conclude from each part? Should the basin be divided into more than four parts? If so, what should those parts be?

Solution. The first quadrant shows that the average rainfall varying between 45 and 60 in. has high entropy with not very much variability. This suggests that the available water resources are reliable. The second quadrant shows that the annual rainfall is between 34 and 40 in. and its entropy is low with quite low variability. This suggest that water needs to be managed properly. The third quadrant shows that yearly rainfall varies between 20 and 35 in. but mostly between 20 and 30 in. and its variability is high. Its entropy has a high range. This suggests that water needs to be stored for future use. The fourth quadrant has rainfall varying between 40 and 45 in. Its entropy is also not highly variable. This suggests that water availability is reliable but needs proper management. See Fig. 9.6.

9.2 Assessment of Water Resources Availability

Let the total amount of rainfall over 1 month or simply monthly rainfall (or intensity) be a random variable. Then its probability distribution can be derived from data. The entropy is then obtained, and it is referred to as the *intensity entropy* (IE). The ratio of monthly rainfall to the sum of monthly rainfall values over a year (i.e., to annual rainfall) can also be considered as a random variable. These relative rainfall intensities over a year reflect the probabilistic characteristics of rainfall occurrence in the year. Since these ratios comprise rainfall apportionment rates for all months in the year, the entropy so calculated is called the *apportionment entropy* (AE).

The potential water resources availability (PWRA) in an area can be assessed in terms of disorder in intensity and over-a-year apportionment of monthly rainfall. The disorder can be measured by the above two entropies, intensity entropy and apportionment entropy. These entropies can be standardized, and pairs of standardized IE and AE for different locations of rain gauges can be plotted. Then simple clustering and K-means clustering can be considered for delineating PWRA distributed over an area of interest and for classifying regional attributes of PWRA.

9.2.1 Selection of Rain Gauges

A country or region is considered as an area where the spatial distribution of PWRA needs to be evaluated in terms of rainfall. To that end, all the monthly (or daily) rainfall data available at rain gauge sites in the entire area of interest are collected. A critical period of observations (number of years) is then defined. Rain gauges having less than the designated critical number of observation years are screened out. The remaining rain gauges have equal to or more than the designated critical number of years of observation and are considered as qualified, although these may not have the same number of observation years. If the available data are daily rainfall, these must be employed to obtain monthly rainfall values. From these monthly values, the values of IE and AE are calculated for each rain gauge.

9.2.2 Intensity Entropy

IE can be evaluated as follows:

1. Gather all the complete-year sets of monthly rainfall data available at a rain gauge. Then the $12 \times m$ ($= N$) monthly rainfall data included in the m-year sets are lumped under one data set without any sequence attribute.
2. Split up the whole range of monthly rainfall into n classes at an equal interval.
3. Count the frequency m_i for each class i to make a frequency distribution table $\{m_i\}$.
4. Calculate the relative frequency $f_i = m_i/N, i = 1, 2, \ldots, n$, for each class I to obtain a probability density function in discrete form (i.e., probability mass function) for the whole range of monthly rainfall.
5. Calculate IE in terms of these relative frequencies as

$$\text{IE} = -\sum_{i=1}^{n} \frac{m_i}{N} \log_2 \frac{m_i}{N} = -\sum_{i=1}^{n} f_i \log_2 f_i \qquad (9.5)$$

where n is the number of classes, N is the number of values in all classes, and f_i is the relative frequency for class i. Here the unit of IE is the *bit* with the logarithmic base of 2.

IE, defined over a semi-infinite range of $0 \leq \text{IE} < \infty$, is a measure to decipher the disorder of monthly rainfall intensity. Less disordered intensity is measured by smaller IE, pointing to a more skewed distribution of the occurrence frequency of monthly rainfall. On the contrary, more highly disordered intensities result in larger IE, extending over a wider range of monthly rainfall. However, note that the probability density function of rainfall intensity is always defined over a positive abscissa, including its zero origin, due to the nonnegativity of rainfall. It then follows that an increase in IE results in an increase in the expected value of monthly rainfall, flattening the graph of the function. This suggests that IE is positively correlated with the expected amount of monthly rainfall and therefore can be an alternative to the aggregate amount of rainfall.

Example 9.4. Consider monthly rainfall for each station in the Brazos River basin, and determine the probability distribution for some selected stations. Indicate the mean rainfall value for the distribution. Then comment on how the expected monthly rainfall varies with the distribution and consequently with the corresponding entropy.

Solution. Let us consider the first station, Albany in Brazos River basin, for explaining the calculations involved in determining IE. Number of years $m = 37$. Considering $n = 5$ equal classes or bins, frequencies can be computed as shown in Table 9.16a.

The frequency distribution looks like an exponential distribution, as shown in Fig. 9.7.

$$\text{Mean rainfall value for distribution} = \frac{\sum m_i x}{N} = 2.653 \text{ in.}$$

Now, consider the second station, Cameron. The frequencies for this station with $n = 5$ bins can be computed as shown in Table 9.16b.

$$\text{Mean rainfall value for the distribution} = \frac{\sum m_i x}{N} = 1.703 \text{ in.}$$

Example 9.5. Compute the monthly entropy, using probabilities of monthly rainfall from Example 9.4. What does it reflect?

Bins	x (in.)	Frequency m_i	$m_i \times x$	$f_i = m_i/N$	$IE = -f_i \log_2 f_i$
0–2.696	1.348	295	397.66	0.664	0.392
2.696–5.392	4.044	98	396.312	0.221	0.481
5.392–8.088	6.74	38	256.12	0.086	0.304
8.088–10.784	9.436	11	103.796	0.025	0.132
10.784–13.48	12.132	2	24.264	0.005	0.035

TABLE 9.16a Frequency Distribution Table for Albany

Bins	x (in.)	Frequency m_i	$m_i \times x$	$f_i = m_i/N$	$IE = -f_i \log_2 f_i$
0–2.324	1.162	388	450.856	0.874	0.167
2.324–5.448	3.886	31	120.466	0.070	0.268
5.448–7.372	6.61	18	115.38	0.041	0.187
7.372–10.896	9.334	5	45.67	0.011	0.073
10.896–12.82	12.058	2	23.716	0.005	0.035

TABLE 9.16b Frequency Distribution Table for Cameron

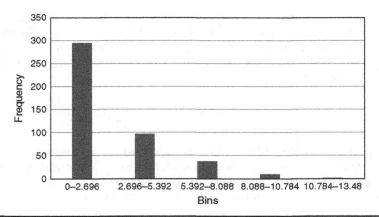

Figure 9.7   Frequency histogram for Example 9.4.

Solution.   It can be seen from the frequency values in both tables that a smaller IE value corresponds to a more skewed distribution, and a higher IE value, as seen in the first case, occurs for a less skewed distribution. It can also be seen from the values of mean monthly rainfall in either case that a positive correlation exists between IE and rainfall amount. Hence, IE can be used as an alternative to the aggregate rainfall at that rain gauge location. Using Eq. (9.5),

$$IE = -\sum_{i=1}^{n} \frac{m_i}{N} \log_2 \frac{m_i}{N} = -\sum_{i=1}^{n} f_i \log_2 f_i$$

$$= 1.3438 \text{ bits}$$

$$IE = -\sum_{i=1}^{n} \frac{m_i}{N} \log_2 \frac{m_i}{N} = -\sum_{i=1}^{n} f_i \log_2 f_i$$

$$= 0.733 \text{ bit}$$

The entropy values corresponding to each class interval can be obtained from the last columns of Tables 9.16a and 9.16b, respectively, for the Albany and Cameron stations. Using Eq. (9.5), IE is 1.343 bits at Albany and 0.733 bit at Cameron. IE is an indicator of the rainfall intensity at a particular location. It has been shown in the previous example that a positive correlation exists between IE and the expected rainfall amount.

9.2.3   Apportionment Entropy

Using the Shannon entropy, Kawachi et al. (2001) defined the apportionment entropy for daily rainfall. In a similar manner, AE can be defined for monthly rainfall. Let r_i be the aggregate rainfall (monthly rainfall) during the ith month in a year. The aggregate rainfall during the year (annual rainfall) R can be expressed by the sum of r_i from $i = 1$ to $i = 12$ as

$$R = \sum_{i=1}^{12} r_i \tag{9.6}$$

Now consider an independent trial as follows. A unit of rainfall is taken out from rainfall values that form the annual rainfall. Then the month in which that unit of rainfall occurred is considered. The outcome events of this trial are integers from 1 to 12. The ratio r_i/R thus becomes an occurrence probability of the outcome. By letting this ratio be p_i and employing the Shannon entropy, AE that measures the temporal variability of monthly rainfall over a year (i.e., monthly-based over-a-year apportionment of annual rainfall) can be written as

$$\text{AE} = -\sum_{i=1}^{12} p_i \log_2 p_i = -\sum_{i=1}^{12} \frac{r_i}{R} \log_2 \frac{r_i}{R} \tag{9.7}$$

The previously defined IE is uniquely estimated for the respective rain gauges, while AE defined for a year is estimated for each of the years considered for the respective rain gauges. Therefore, the AE values obtained for a certain rain gauge are simply averaged to have a representative AE for that rain gauge. Hereinafter, the notation AE is used in the sense of such an average.

AE measures the over-a-year disorder (or temporal evenness) of monthly rainfall events over time. By definition, Eq. (9.7) states that when the annual rainfall amount is quite evenly apportioned to each of the 12 months with the probability of $\frac{1}{12}$, AE takes on its maximum value of $H = \log_2 12$. The minimum AE, $H = 0$, occurs when the apportionment is made to only 1 out of the 12 months with a probability of 1. This indicates that AE takes on a value within a finite range of 0 to $\log_2 12$. Higher AE indicates higher in-time PWRA with less monthly or seasonal rainfall variability.

Example 9.6.   Compute the apportionment entropy, using probabilities of monthly rainfall in the Brazos River basin. The method of calculation is the same as for daily values. What do these values reflect?

Solution.   Let us consider station Albany for explaining the steps involved in the calculation of AE. The annual rainfall values at the station location for the years 1970–2006 are given in Table 9.17a. The AE values for each year are computed as shown in Table 9.17b.
Representative average AE of Albany is 2.3 bits.

1970–1978	16.54	25.28	28.87	26.75	35.57	19.07	27.55	18.75	17.19
1979–1987	23.48	28.6	27.13	26.49	21.43	25.27	30.67	38.11	27.96
1988–1996	15.4	27.65	44	43.78	34.34	24.91	26.73	31.68	26.96
1997–2005	33.18	17.96	23.99	17.07	25.1	38.25	19.06	42.12	22.32
2006	20.58	—	—	—	—	—	—	—	—

TABLE 9.17a   Annual Rainfall Amounts at Albany (in.)

1970–1978	2.175	2.349	2.471	2.387	2.502	2.213	2.405	2.204	2.198
1979–1987	2.302	2.351	2.342	2.340	2.240	2.202	2.361	2.405	2.343
1988–1996	2.163	2.340	2.495	2.49	2.40	2.331	2.340	2.363	2.341
1997–2005	2.392	2.20	2.304	2.190	2.348	2.409	2.213	2.489	2.294
2006	2.223	—	—	—	—	—	—	—	—

TABLE 9.17b   AE Values for Each Year at Albany (bits)

9.2.4   Standardization

Based on Eqs. (9.5) and (9.7), a set of IE and AE values can be calculated for every rain gauge. These constitute two different sets. To pair these sets on equal terms, they are standardized so that each set may have the mean value of 0 and the standard deviation of 1. This is implemented by standardizing the component values as

$$X^* = \frac{X - m_X}{\sigma_X} \tag{9.8}$$

where X^* is the standardized value (IE* or AE*), X is the estimated value in bits (IE or AE), and m_x and σ_x are the mean and the standard deviation for the set X (IE or AE), respectively. IE* and AE* including all the estimated IE and AE values as their components, respectively, can be obtained. By this standardization, a scatter diagram can be drawn to illustrate the location-dependency of the relation between IE and AE. Then PWRA can be assessed in a relative sense within the area of interest. Hereafter, the notations IE and AE are used for standardized entropies, dropping their primes.

Example 9.7.   Compute the values of IE and AE for the Brazos River basin, and then compute their standardized values. Put the values of IE and AE on the Brazos River basin map.

Solution.   The standardized values of IE and AE are computed as given in Table 9.18. Values of AE and IE are shown on the Brazos River basin map in Figs. 9.8 and 9.9.

9.2.5   Simple Clustering

Different pairs of IE and AE values for different locations of rain gauges are plotted as a scatter diagram. The coordinates of a point on this diagram characterize PWRA at the location of the corresponding rain gauge. Note that the axes of the diagram using the standardized IE and AE values become the lines of the mean values of IE and AE, respectively. Clustering or grouping of the points may lead to a significant classification and delineation of water resources zones. The simplest way of clustering is to use line(s) passing through the origin of the coordinate axes of the diagram. The more the number of lines is increased, the more the clustering is refined.

Example 9.8.   Cluster the values of IEs and AEs based on simple clusters: 2, 4, and 8 using the lines going through the origin. Show these clusters of points on the Brazos River basin. What do you conclude from these cluster diagrams as regards the potential water resources availability?

Solution.   Clusters of IEs and AEs are shown in Figs. 9.10 to 9.12.
For the tetramerous classification, the first quadrant where both values of IE and AE are higher than their mean values can be categorized as a rich-in-water zone and with more PWRA; the third quadrant, where both values of IE and AE are less than their mean values, can be categorized as a poor-in-water zone and with less PWRA. The second and fourth quadrants can be categorized as moderate in PWRA.
A pair of AE and IE can appraise PWRA at a certain location in a relative sense within the whole area of interest. The PWRA classification is useful for developing strategies for water resources planning, development, or management. For the Brazos River basin, those stations whose data fall in the first quadrant can have higher PWRA while those stations whose data fall in the third quadrant can have lower PWRA. For all 21 stations, 5 stations lie in the area that has lower PWRA, while 12 stations lie in the area that has greater PWRA. The other 4 stations are moderate in PWRA. This would be useful in the local water resources management.

Stations	IE (bits)	AE (bits)	Standardized IE	Standardized AE
Cameron	0.713	1.351	−1.629	−1.823
College Station	1.369	2.204	0.312	0.643
Columbus	1.439	2.088	0.520	0.306
Conroe	1.556	2.241	0.877	0.750
Eastland	1.608	2.244	1.020	0.757
Freeport	1.126	2.079	−0.406	0.281
Hillsboro	1.643	2.162	1.123	0.522
Houston Height	1.379	2.212	0.343	0.666
Jayton	1.671	1.962	1.206	−0.058
Marlin	0.598	1.246	−1.970	−2.127
Matador	1.373	2.200	0.323	0.631
McGregor	1.347	2.221	0.246	0.691
Morgan Mill	1.000	2.051	−0.781	0.200
Olney	1.201	2.088	−0.186	0.307
Rosebud	0.767	1.372	−1.471	−1.762
Seymour	1.230	1.586	−0.099	−1.144
Thompson	0.666	1.446	−1.771	−1.550
Tomball	1.596	2.242	0.9834	0.752
Waco	1.625	2.261	1.078	0.808
Weatherford	1.281	2.173	0.053	0.552
Albany	1.343	2.190	0.236	0.601

TABLE 9.18   IE and AE Values for Rain Gauge Stations in Brazos River Basin

9.2.6   K-Means Clustering

Another way of clustering is the K-means clustering which is suitable where there are a host of points to be clustered. Hartigan (1975) describes the K-means algorithm in detail, and Hartigan and Wong (1979) show an efficient K-means algorithm. The aim of the K-means algorithm is to search for a partition with small values of the within-cluster sum of squared euclidean distances between individual points and the cluster center by moving points from one cluster to the other. The cluster centers are the averages of points contained within them.

First, initial cluster centers are required for this method. The initial cluster partition can be obtained, based on the sum of standardized IE and AE at each point in the scatter diagram. Suppose the sum at point I is denoted by SUM(I), having all points over the minimum value MIN and the maximum value MAX. To obtain K initial clusters, point I is set into the Jth cluster, where J is the integer part of $[K(\text{SUM}(I) - \text{MIN})/(\text{MAX} - \text{MIN})] + 1$. However, only the point that takes on MAX belongs to the Kth cluster. The initial cluster centers are obtained from the averages of points contained within these initial clusters.

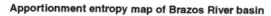

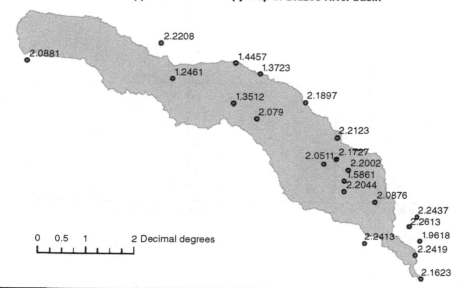

FIGURE 9.8   AE for the Brazos River basin.

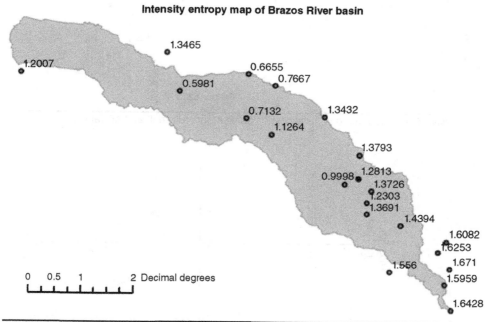

FIGURE 9.9   IE for the Brazos River basin.

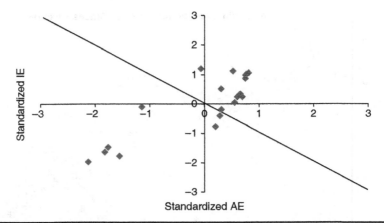

FIGURE **9.10** Binary classification using standardized IE and AE values.

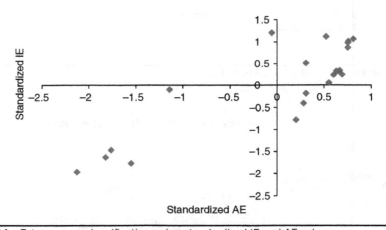

FIGURE **9.11** Tetramerous classification using standardized IE and AE values.

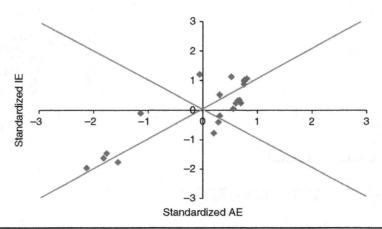

FIGURE **9.12** Octonary classification using standardized IE and AE values.

The next step is to search for a K partition with locally optimal within-cluster sum of squares by moving points from one cluster to another. To that end, the euclidean distance, $D(I, L)$, between point I and the center of the cluster L is calculated. Equation (9.9) is used to judge whether point I should move to the other cluster L from the present cluster M:

$$\frac{N(L)D(I,L)^2}{N(L)+1} - \frac{N(M)D(I,M)^2}{N(M)-1} \tag{9.9}$$

where $N(L)$ and $N(M)$ are the numbers of points in cluster L and cluster M, respectively. This value shows an amount of increase in the total sum of squares, when point I moves from cluster M to cluster L. If the minimum of this quantity over all clusters is negative, the point moves to the minimal cluster and the cluster centers that are related to the movement are adjusted. The procedure ends when no such movement reduces the sum of squares.

For example, consider the case in which all the points are $I_1 = (-2, -1)$, $I_2 = (-1, -2)$, $I_3 = (0, 0)$, $I_4 = (1, 2)$, and $I_5 = (2, 1)$ and they classified into two clusters $(K = 2)$. Then the components of $SUM(I)$ are calculated as $SUM(I_1) = -3$, $SUM(I_2) = -3$, $SUM(I_3) = 0$, $SUM(I_4) = 3$, and $SUM(I_5) = 3$ with $MIN = -3$ and $MAX = 3$. Since the number of initial clusters can be obtained by the integer part of $K[SUM(I) - MIN]/(MAX - MIN) + 1$, each point can be classified as follows:

I_1: $2 \times [-3 - (-3)]/[3 - (-3)] + 1 = 1$ cluster 1

I_2: $2 \times [-3 - (-3)]/[3 - (-3)] + 1 = 1$ cluster 1

I_3: $2 \times [0 - (-3)]/[3 - (-3)] + 1 = 2$ cluster 2

I_4: $2 \times [3 - (-3)]/[3 - (-3)] + 1 = 3$ cluster 2 [= 3 −1, because of $SUM(I_4) = MAX$]

I_5: $2 \times [3 - (-3)]/[3 - (-3)] + 1 = 3$ cluster 2 [= 3 −1, because of $SUM(I_5) = MAX$]

Using Eq. (9.9), the movement of the present cluster to another cluster is judged. For instance, point I_3 remains in cluster 2, taking the quantity calculated from Eq. (9.9) as

$$\frac{N(L)D(I,L)^2}{N(L)+1} - \frac{N(M)D(I,M)^2}{N(M)-1}$$

$$= \frac{2[0-(-1.5)]^2 + [0-(-1.5)]^2}{2+1} - \frac{3[(0-1)^2 + (0-1)^2]}{3-1}$$

$$= 0$$

If this quantity is negative, point I_3 should be moved from cluster 2 to cluster 1. In the case of more than two clusters $(K > 2)$, the minimum of this quantity over all clusters is examined for this judgment. Using this procedure, the points on the scatter diagram are grouped into the designated number of clusters.

9.3   Case Study: Japan

The entire procedure discussed above is now illustrated by using the nation of Japan as a case study in which about 1300 rain gauges cover the entire country. These rain gauges form a central part of the meteorological observation network, called AMeDAS (Automated

Meteorological Data Acquisition System), under the Japan Meteorological Agency. Daily rainfall data effective from 1976 are available (Japan Meteorological Agency, 1996–1998). Here the work of Maruyama and Kawchi (1998), Kawachi et al. (2001), and Maruyama et al. (2005) is followed.

Step 1: Screening of rain gauges: Rain gauges were screened using the criteria discussed earlier. It was found that for the period from 1976 to 1997, only 1107 rain gauges were rendered usable. The remainder of the gauges had to be deleted, because these gauges either have shorter rainfall records or have missing records. The selected 1107 rain gauges provide complete daily rainfall series for 8 yr or more (Japan Meteorological Agency, 1996–1998). In this manner, a total of 16,812 yr of rainfall data was available which yielded an average of 15.19 yr per gauge.

Step 2: Computation of basic statistics: Basic statistics (e.g., mean, minimum, maximum, standard deviation, and coefficient of variation) were computed for H, R, and m selected for the calculation of entropy. For Japan, these statistics are given in Table 9.19. The maximum value of entropy was 7.155 bits, and the minimum value was 5.118 bits; they occurred at Irihirose of Niigata prefecture and Akan of Hokkaido prefecture, respectively, as noted in Fig. 9.13. As shown in the figure, entropy indicates significant zones over the whole country of Japan. This is more easily seen from the isoentropy map shown in Fig. 9.14a that delineates the entropy distribution with the use of isoentropy lines. The distribution of the averaged annual rainfall, delineated by isohyetal lines, is given in Fig. 9.14b.

Now the coefficient of variation values are computed for constructing the CV map, as shown in Fig. 9.15. The average value of CV in Japan is 1.39 and exceeds the value of 1.0. The averaged coefficient of variation of daily rainfall exhibits a significant spatial variability over the country. The values of CV show a complicated distribution compared with the entropy-based map. The CV contours break too often and exhibit many separated islands. Thus, the CV map does not lend itself to a clear classification of the region.

Step 3: Climate division: For comparison purposes the midscale climatic map developed by Suzuki (1962) showing the climate division map was obtained, as shown in Fig. 9.16. This was drawn based on incidences of daily rainfall during the winter season, based on the fact that climatic characteristics in Japan can be explained satisfactorily by the spatial distribution of daily rainfall, especially during the northwest monsoon season. The lines thus depicted represent the rainfall boundaries that divide the whole country into three climatic regions I, II, and III. Two extremities I and III are the regions with and without perceptible winter rainfall, respectively; these are referred to as the *reverse* and *obverse* of Japan, respectively.

Step 4: Comparison with climate division map: It was observed that the CV map did not have a clear match with the climate division map. Figure 9.14 also shows that the average of the entropies of over 8 yr at each observation point is regionally and clearly classified in Japan. Of particular interest is the agreement between Fig. 9.14a and Japan's well-known map of climatic division shown in Fig. 9.15.

In the transitional region II, rain or snow may or may not fall, depending on the atmospheric pressure gradient and wind direction. This map is comparable to the isoentropy map shown in Fig. 9.14a inasmuch as the former is also the result of considering the temporal variability

	Mean	Minimum	Maximum	SD	CV
Entropy $\overline{H}$ (bits)	6.18	5.12	7.16	0.39	0.06
Annual rainfall R (mm/yr)	1696.6	652.3	4390.6	573.6	0.34
Number of yearly rainfall series m	15.19	8	22	3.44	0.23

SD = standard deviation, and CV = coefficient of variation.

TABLE 9.19 Basic Statistics for $\overline{H}$, R, and m

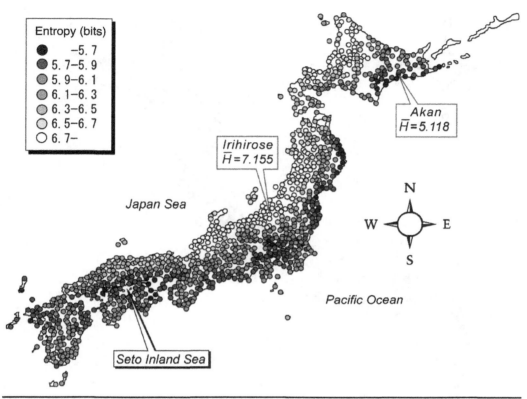

FIGURE 9.13   Entropies of rain gauge stations.

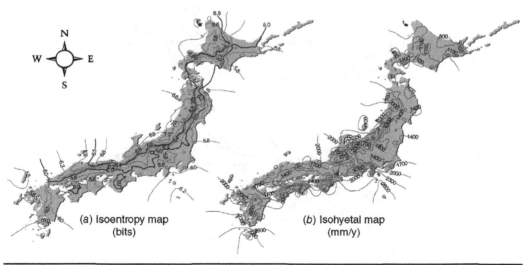

FIGURE 9.14   (a) Isoentropy map and (b) Isohyetal map.

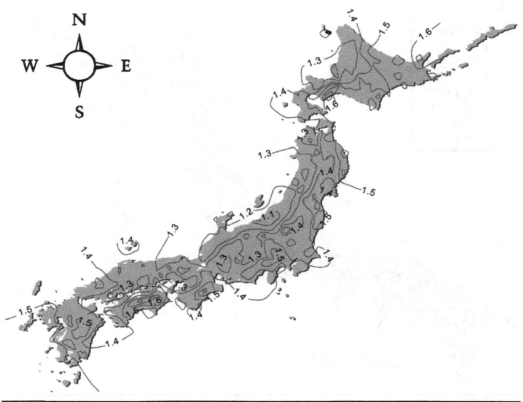

FIGURE 9.15 Contours of coefficient of variation.

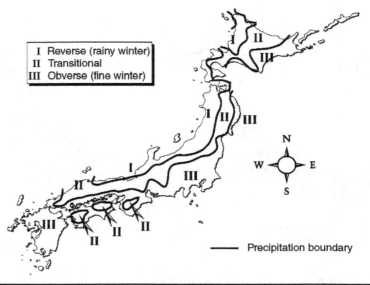

FIGURE 9.16 Midscale climatic division (after Suzuki, 1962).

of rainfall. This comparison implies that with a sufficiently high degree of accuracy the boundary between regions I and II is identified as $\overline{H} = 6.5$ bits, and the boundary between II and III, inclusive of the isolated closed ones, is identified as $\overline{H} = 6.0$ bits. Such an identification is climatologically meaningful not only for quantitatively measuring the rainfall boundaries but also for demonstrating the validity of the isoentropy map obtained. In addition, it justifies the inference by Suzuki (1962) that the entire boundary shown in Fig. 9.16, obtained by using a yearly rainfall series for the year 1955, would be indistinguishably moved even if the series for another year or several years were taken into account.

Step 5: Relation between entropy and rainfall: Depicting the averaged annual entropy on the ordinate and the averaged annual rainfall on the abscissa, Fig. 9.17 was constructed, which is a scattered diagram, plotting values for all the rain gauges selected. Glancing at the diagram, one observes that the average annual entropy and the average annual rainfall are less mutually related, with a small correlation coefficient of $r = 0.19$. This demonstrates that besides the aggregate of rainfall, its temporal apportionment can be a significant aspect of rainfall data. When coupled, entropy and rainfall on a yearly basis become a measure of the throughout-the-year potential availability of water resources. To explain it qualitatively, the whole plotted area of Fig. 9.20 is divided into four parts, each delineated with two intersecting dividing lines that pass through the means of the respective two variables. Then in terms of water resources availability, the respective quadrants A, B, C and D can be comparatively categorized as follows.

Category A: Water resources are abundantly and perennially available. The inhabitants in the areas in this category can use the water resources willfully, being limited only by the permissible water withdrawal from the natural water cycle.

Category B: The water resources availability is identified as moderate, since the rainwater is poor but perennial. For spells of high water demand beyond available capacity, water resources development and management are needed to temporarily increase the availability.

Category C: The water resources availability is so low that permissible water use may be extremely constrained. To meet perennial or intensive water demands, it is necessary to intervene in the water cycle and appropriately increase disposable water resources by the construction of water storage facilities.

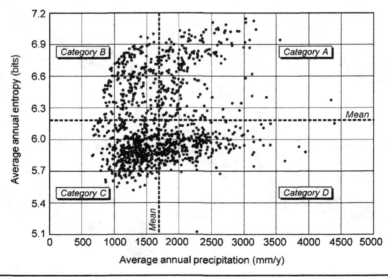

Figure 9.17   Relation between average annual entropy and average annual precipitation showing categories of water resources availability.

Category D: The water resources availability is considered as moderate, in a different sense from that of category B. This category is characterized by concentrated heavy rainfall, so that water control is needed for flood mitigation. Excess water, withheld from wastefully running off and stored in reservoirs, can be used effectively to reduce the temporal variation of water resources and consequently augment the water supply for water users.

Step 6: Construction of entropy-based climate division map: The aforementioned categories are illustrated separately in Fig. 9.18 in the form of a location map of the rain gauges included in the respective categories. From this map, it can be readily seen that rain gauges in the same category are clustered, and therefore the spatial distribution of the water resources availability is zoned with discernible boundaries. Figure 9.18 serves as an improved midscale climatic division map, compounded by rainfall and its temporal apportionment. Although this map is a result of simple clustering based on the mean values of the variables, it is, at the same time, worthy of becoming a water availability map, explaining long-established water use practices well. For example, the strip identified as category A includes the nation's most prominent rice-growing districts, such as Niigata, Yamagata, and Akita prefectures, with heavy snowfall that increases precipitation in winter and consequently decreases the degree of its throughout-the-year variability. This less time-varying rainfall provides one of the natural environments necessary for good rice production, with melted snow and off-winter rainfall supplying sufficient water for soil paddling and rice planting in spring and for rice growing, respectively.

On the other hand, the area belonging to category C has long been suffering from serious water shortages because of the nonuniform and scarce rainfall. To escape from such a predicament and to control the ravages of droughts, a huge number of small-scale irrigation

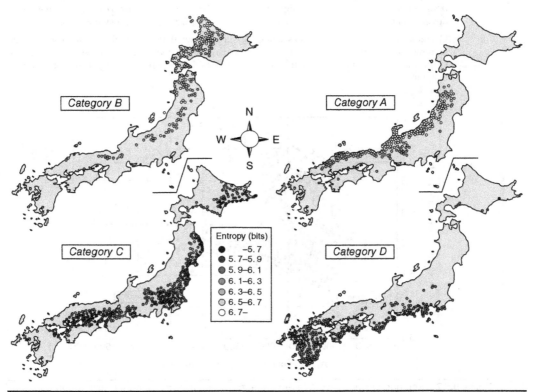

FIGURE 9.18 Entropy-based classification.

tanks have been constructed as a self-reliant storage system all over the area. There exist 110,835 tanks, about one-half of the total 246,158 tanks in Japan, especially on the fringes of the Seto Inland Sea which embraces the prefectures of Osaka, Hyogo, Okayama, Hiroshima, and Kagawa. Even now much effort is still being made to eradicate the dearth of water.

Thus, from rainfall entropy analysis, one concludes that for yearly rainfall, entropy delineates a plausible climatic map that qualitatively explains rainfall characteristics. An entropy-based climatic map (isoentropy map), drawn for the whole country of Japan, is valuable in comparison with the known classical climatic division map, and for coupling with the isohyetal map. Coupling of entropy with annual rainfall enables a relative assessment or categorization of the potential availability of water resources at local, areal or regional levels, considering both the aggregate and the temporal variability of rainfall. See Figs. 9.19 to 9.22.

9.4 Case Study: Worldwide Mapping of PWRA

Maruyama et al. (2005) analyzed monthly rainfall data recorded at 11,260 rain gauges in the world and comparatively mapped different classifications of PWRA. An example application of their methodology to the worldwide PWRA assessment is discussed in this section. In this application, entropy is expressed in the context of the probability of

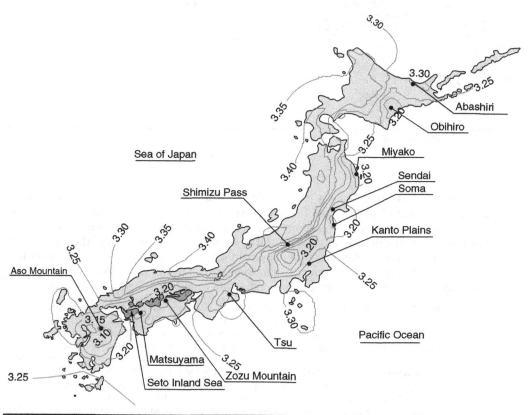

FIGURE 9.19 Isoentropy map of AE.

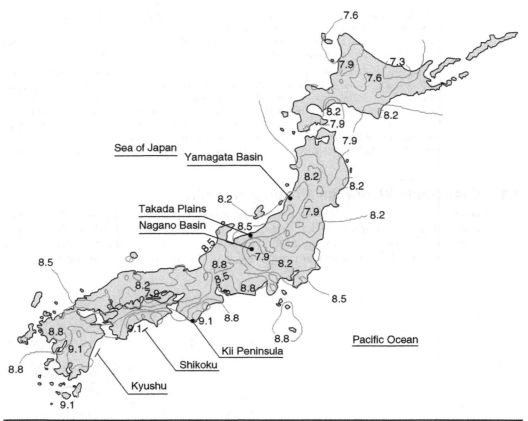

Figure 9.20   Isoentropy map of IE.

occurrence of, say, the intensity of rainfall in the framework of IE. Focusing on the temporal variability of daily rainfall, Maruyama and Kawachi (1998) and Kawachi et al. (2001) employed entropy to measure the degree of uncertainty of rainfall occurrence in time or the over-the-year temporal rainfall apportionment based on the probability density function of rainfall randomly apportioned over fragmented times. In general, it is easier to obtain monthly rainfall data than daily rainfall data. Therefore, AE on a monthly basis, presented here, can be a viable alternative to that on a daily basis.

Step 1: Data set: The data set that includes monthly rainfall data observed at 20,614 rain gauges over the world is available in GHCN Version 2 (Vose et al., 1992). The GHCN is a comprehensive global surface baseline climate data set designed for monitoring and detecting climate change. The data are comprised of surface station observations of temperature, precipitation, and pressure, and all GHCN data are on a monthly basis. (They are downloadable from National Climatic Data Center—NOAA (National Oceanic and Atmospheric Administration) at http://www.ncdc.noaa.gov/oa/ncdc.html. GHCN is produced jointly by the National Climatic Data Center, Arizona State University, and the Carbon Dioxide Information Analysis Center at the Oak Ridge National Laboratory. The GHCN Version 2 precipitation data used here were made available in 1999. The precipitation

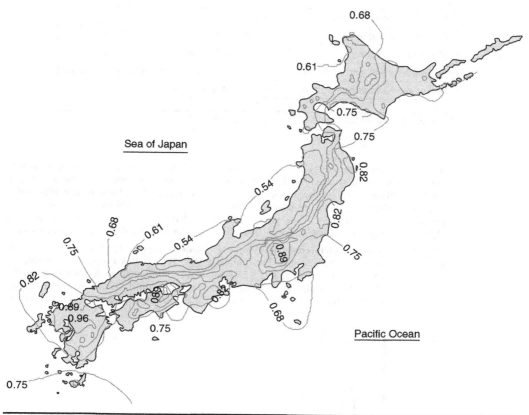

FIGURE 9.21 Iso-CV map.

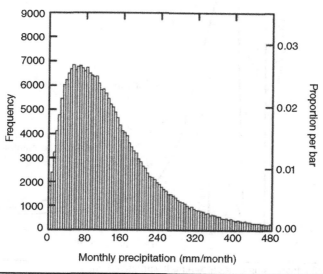

FIGURE 9.22 Frequency distribution of monthly precipitation in Japan.

Total number of rain gauges	11,260
Average observation years	67.55 yr (min. 30 yr, max. 299 yr)
Mean annual rainfall	1003.58 mm/yr (SD = 759.71 mm/yr)
Maximum mean annual rainfall	10,738.52 mm/yr (at Cherrapunji, India)
Minimum mean annual rainfall	0.63 mm/yr (at Dakhla, Egypt)

TABLE 9.20   Basic Statistics of Rain Gauges Selected

data set has data from 20,614 stations over the world. The earliest station data are from 1697 and the most recent are from 2002, which was created from 15 source data sets. One station has 299 yr of data, and another has less than 1 yr of data. Therefore, for reliable evaluation of IE and AE, the rain gauges are selected such that they have longer than 30-yr rainfall series containing a complete set of over-a-year monthly rainfall data. Thus, out of 20,614 rain gauges, 11,260 rain gauges are selected. Table 9.20 shows basic statistics for the rain gauges selected.

Step 2: Computation of intensity and apportionment entropies: At every rain gauge, the values of IE and AE are calculated. For computation of IE, monthly frequency distributions were first determined, as shown in Fig. 9.23. Note that IE was obtained from splitting up the whole range of monthly rainfall at an equal interval of 10 mm, and AE was averaged over all the available yearly rainfall series. The mean values of IE and AE over the worldwide area were

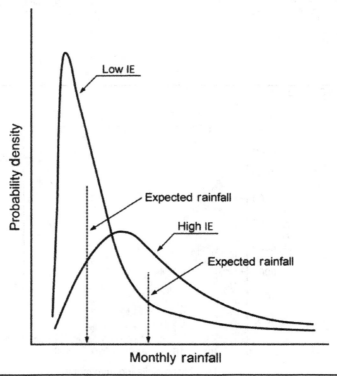

FIGURE 9.23   Increasing expected rainfall with increasing IE.

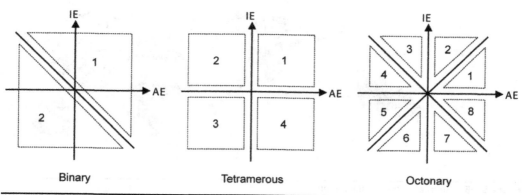

FIGURE 9.24 Rain gauge clustering.

3.845 bits (SD = 0.952) and 2.831 bits (SD = 0.485), respectively, where SD = standard deviation. Correlations of IE and AE with annual rainfall were found to be $r = 0.830$ and $r = 0.228$, respectively. As previously shown, IE is indeed highly correlated with the aggregate rainfall, such as annual rainfall. The correlation between IE and AE is 0.508. Since the probability density function of rainfall intensity is always defined over a positive abscissa, including its zero origin, due to the nonnegativity of rainfall, it then follows that an increase in IE results in an increase in the expected value of monthly rainfall, flattening the graph of the function (see Fig. 9.23). This suggests that IE is positively correlated with the expected amount of monthly rainfall and can therefore be an alternative to the aggregate amount of rainfall, as shown in Fig. 9.23.

Step 3: Clustering of rain gauges and mapping: The calculated values of IE and AE were standardized using Eq. (9.7) to draw a scatter diagram. For comparison, clustering of the rain gauges for classifying PWRA was carried out with both simple clustering and K-means clustering, as shown in Fig. 9.24; and two, four, and eight clusters were considered for each clustering. The results of clustering are shown in the diagram and also on a geographical map of the world so that the cluster in which each rain gauge was categorized can be discerned. A few clustering partitions are shown, including three cases of two, four, and eight clusters. The second clustering categorizes the points against the mean values of IE and AE or against IE = 0 and AE = 0 into four clusters. Thus, the third quadrant, in which both values of IE and AE are less than their mean values in the area, is, in an average sense, categorized as a poor-in-water zone with less amount and high variability of rainfall. The first quadrant is characterized as a zone where abundant and perennial rainfall is available. The second quadrant is a zone where rainwater is relatively rich in amount but concentrated in time. The fourth quadrant is a zone with short but perennial rainfall. Similar explanations can be given to the other clustering partitions.

The results of binary, tetramerous, and octonary classifications of PWRA with the simple clustering method are illustrated in Figs. 9.25 through 9.27, and those with the K-means clustering are shown in Figs. 9.28 through 9.30.

(a) Binary classification (Figs. 9.25 and 9.28): As shown in Fig. 9.25, the K-means clustering method divides the whole area into two clusters with a straight line having a negative slope. Thus, two different methods of clustering yield almost the same classification of PWRA. The lines partitioning the worldwide map of PWRA into two zones have affinities with those climatically partitioning the world into five zones (hyperarid, arid, semiarid, subhumid, and humid zones) using the aridity index proposed by UNESCO (1984), which is expressed in terms of annual rainfall P and annual evapotranspiration E. Comparing Fig. 9.25 with the map of the worldwide distribution of arid regions by UNESCO, the zone in which the rain gauges colored green are included can be identified as the humid zone with $P/E \geq 0.75$. Meanwhile, comparison of Fig. 9.28 and the UNESCO map implies that the K-means clustering method

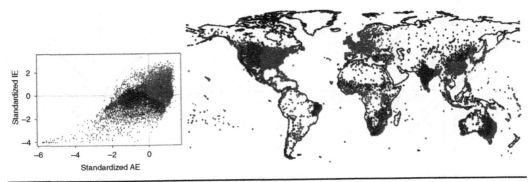

FIGURE 9.25   Simple binary classification.

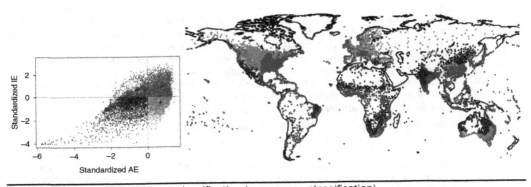

FIGURE 9.26   Simple tetramerous classification (zero-means classification).

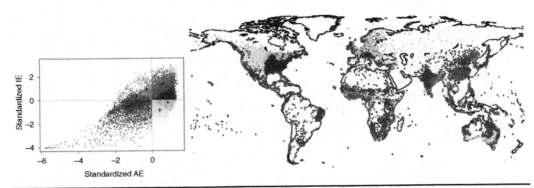

FIGURE 9.27   Simple octonary classification.

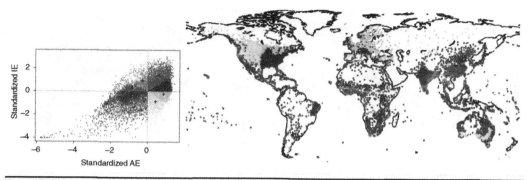

FIGURE 9.28   *K*-means binary classification.

delineates fairly well the boundaries of combined subhumid and humid zones, which are discriminated from the first three climatic zones by $P/E \geq 0.50$.

(b) Tetramerous classification (Figs. 9.26 and 9.29): The simple clustering method, shown in Fig. 9.26, can be referred to as *zero-mean clustering*. It classifies the points into four clusters that are just assigned to the quadrants of the scatter diagram. The regions falling in the first and third quadrants are rich and poor in the sense of PWRA, respectively. The remaining regions clustered in the second and fourth quadrants cannot definitely be identified to be rich or poor, since either IE or AE is negative, or rather can be considered as moderate in water resources availability. In such regions, therefore, storage reservoirs are needed to regulate the undesirability of river or stream flow that is reflected by negative IE or AE. The regions with negative AE, colored green, suffer from concentrated heavy rainfall, so that reservoirs are needed to control floods and the excess water stored can effectively be managed for later-day water use, reducing the temporal variation of the natural water supply. For example, Bangladesh, East India, and the belt situated at around latitude 10°N in West Africa are discerned as such areas. Meanwhile the regions with negative IE, have poor but perennial rainfall and therefore need reservoirs to augment stream flows for meeting spells of high water demand beyond the natural water supply. Such regions can be found, e.g., in Europe and North America.

As the scatter diagram in Fig. 9.29 shows, the *K*-means clustering method characteristically refines the third quadrant, making the extremely low-in-PWRA regions stand out in strong relief. In addition, it indicates two different clusters in roughly the same range of IE. Thus, clusters can be ranked according to their PWRAs as: colored-blue, -orange, -green and -red clusters.

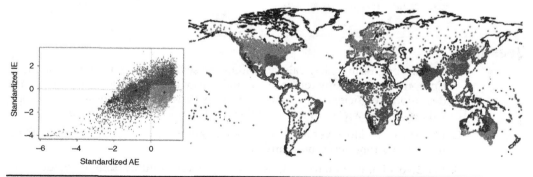

FIGURE 9.29   *K*-means tetramerous classification.

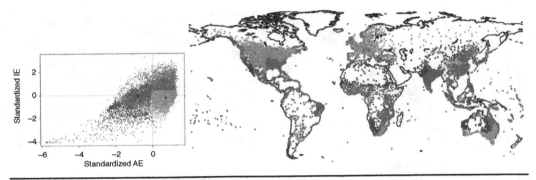

FIGURE 9.30   *K*-means Octonary classification.

This means that this clustering lends itself to worldwide ranking of the priority for water resources development on a regional scale.

(c) *Octonary classification* (**Figs. 9.35 and 9.30**): The simple clustering method (Fig. 9.23), in a straightforward manner, refines the zero-means classification, dividing the respective quadrants into two. The *K*-means clustering (Fig. 9.30) method provides a separate classification with eight clusters of patchy shape. In this case, bold ranking of the clusters or their couples according to the potential water availability can also be made as colored pink, turquoise/brown, cerise/blue, green/orange, and red clusters.

The two different methods of clustering analyses demonstrate possible ways of classifying regional attributes of PWRA. The methodology is applied to develop classification maps of PWRAs distributed over the world. In lieu of conventional statistics, such as the mean, standard deviation, and variance, the methodology described here focuses on the information content (or entropy) generated by variable monthly rainfall for the assessment of PWRA. Random characteristics of intensity and over-a-year apportionment of monthly rainfall are separately and coherently measured in terms of standardized AE and IE. Thus, a pair of AE and IE, or more practically the coordinates of it in the (AE, IE) space, appraise PWRA at a certain location in a relative sense within the whole area of interest. The PWRA classification and its mapping are useful for developing a worldwide, nationwide, or regional strategy of water resources development or management. Of particular significance is the classification by the *K*-means clustering method, because it aids the ranking of regions according to the necessity or feasibility of water resources development. In addition, the worldwide PWRA maps drawn in this section are of value as climate classification maps, although a variety of such maps, based on the total or average precipitation and evapotranspiration (or mean temperature), are currently available.

Questions

1. Obtain long-term yearly rainfall values for several rain gauges from a basin. Compute the entropy of each station. Also compute the mean, standard deviation, coefficient of variation, and coefficient of skewness. Do entropy values have any relation to any of these distribution characteristics?

2. Plot entropy values versus year for each gauging station in Question 1. Does the entropy time series have any pattern?

3. For two of the stations in Question 1, obtain monthly rainfall values and compute the apportionment entropy. What does this entropy tell you?

4. For the two stations in Question 3, obtain monthly rainfall values by random experimentation and then compute the apportionment entropy. How does this entropy compare with that in Question 3?

5. Compute the intensity entropy for the stations in Question 3. What does this entropy tell?

6. Obtain apportionment entropies for rain gauges in Question 1.

7. Obtain intensity entropies for rain gauges in Question 1.

8. Cluster the values of AE and IE obtained in Question 6 and 7 based on simple clusters: two, four, and eight. What can be concluded from these graphs?

9. How well do the clusters compare with meteorological divisions of the basin under consideration?

10. Using *K*-means clustering, cluster the rain gauges in Question 1.

References

Burn, D. H. (1997). Catchment similarity for regional flood frequency analysis using seasonality measures. *Journal of Hydrology*, vol. 202, pp. 212–230.

Hartigan, J. A. (1975). *Clustering Algorithms. Wiley*, New York, pp. 84–112.

Hartigan, J. A., and Wong M. A. (1979). A K-means clustering algorithm. *Applied Statistics*, vol. 28, pp. 100–108.

Japan Meteorological Agency (1996-1998). AMeDAS Data of Japan (CD-ROM), Japan Meteorological Business Support Center.

Kagan, A. M., Linnik, Yu V. and Rao C. R. (1973). *Characterization Problems in Mathematical Statistics*. Wiley, New York, pp. 408–410.

Kawachi, T., Maruyama, T. and Singh V. P. (2001). Rainfall entropy for delineation of water resources zones in Japan. *Journal of Hydrology*, vol. 246, pp. 36–44.

Magilligan, F. J., and Graber B. E. (1996). Hydroclimatological and geomorphic controls on the timing and spatial variability of floods in New England, USA. *Journal of Hydrology*, vol. 178, pp. 159–180.

Maruyama, T., and Kawachi T. (1998). Evaluation of rainfall characteristics using entropy. *Journal of Rainwater Catchment Systems*, vol. 4, no. 1, pp. 7–10.

Maruyama, T., Kawachi, T. and Singh, V. P. (2005). Entropy-based assessment and clustering of potential water resources availability. *Journal of Hydrology*, vol. 309, pp. 104–113.

Shannon, C. E. (1948). A mathematical theory of communication. *The Bell System Technical Journal*, vol. 26, pp. 379–423, 623–656.

Suzuki, H. (1962). Classification of Japanese climates. *Geographical Review of Japan*, vol. 35, pp. 205–211 (in Japanese).

UNESCO (1984). Climate, drought and desertification. *Nature and Resources*, vol. 20, no. 1, pp. 2–8.

Vose, R. S., Richard L. Schmoyer, Peter M. Steurer, Thomas C. Peterson, Richard Heim, Thomas R. Karl, and J. Eischeid (1992). The Global Historical Climatology Network: Long-term monthly temperature, precipitation, sea level pressure, and station pressure data. ORNL/CDIAC-53, NDP-041. Carbon Dioxide Information Analysis Center, Oak Ridge National Laboratory, Oak Ridge, Tenn. Cf. http://cdiac.esd.ornl.gov/ghcn/ghcn.html.

Further Reading

Brunsell, N. A. (2010). A multiscale information theory approach to assess spatial-temporal variability of daily precipitation. *Journal of Hydrology*, vol. 385, pp. 165–172.

Caselton, W. F., and Husain, T. (1980). Hydrological networks: Information transmission. *Journal of Water Resources Planning and Management, ASCE*, vol. 106, no. WR 2, pp. 503–520.

Harmancioglu, N. B., Singh, V. P. and Alpaslan, N. (1992).Versatile uses of the entropy concept in water resources. *Entropy and Energy Dissipation in Water Resources*, edited by V. P. Singh and M. Fiorentino, Kluwer Academic Publishers, pp. 91–117.

Harmancioglu, N. B., and V. P. Singh (1998). Entropy in environmental and water resources. In: *Encyclopedia of Hydrology and Water Resources*, edited by Herschy, R. W. and R. W. Fairbridge. Kluwer Academic Publishers, pp. 225–241.

Husain, T. (1989). Hydrologic uncertainty measure and network design. *Water Resources Bulletin*, vol. 25, no. 3, pp. 527–534.

Krstanovic, P. F., and V. P. Singh (1992a). Transfer of information in monthly rainfall series of San Jose. In: *Entropy and Energy Dissipation in Water Resources*, edited by V. P. Singh and M. Fiorentino. Kluwer Academic Publishers, pp. 155–173.

Krstanovic, P. F., and V. P. Singh (1992b). Evaluation of rainfall networks using entropy I. *Water Resources Management*, vol. 6, pp.279–293.

Krstanovic, P. F., and V. P. Singh (1992c). Evaluation of rainfall networks using entropy II. *Water Resources Management*, vol. 6, pp. 295–314.

Molini, A., La Barbera, P. and Lanza L. G. (2006). Correlation patterns and information flows in rainfall fields. *Journal of Hydrology*, vol. 322, pp. 89–104.

Singh, V. P. (1997). The use of entropy in hydrology and water resources. *Hydrological Processes*, vol. 11, pp. 587–626.

Singh, V. P. (2013). *Entropy Theory and Its Application in Environmental and Water Engineering*. Wiley, New York.

Yang, Y., and Burn D. H. (1994). An entropy approach to data collection network design. *Journal of Hydrology*, vol. 157, pp. 307–324.

CHAPTER 10

Evaporation

Evaporation is a process of conversion of liquid water to water vapor. It couples the dynamics of water and energy cycles over the earth's surface. Evaporation from nonvegetated surfaces (i.e., land without vegetation) and evaporation from vegetated surfaces are different in their underlying physics. The process of evaporation entails four elements (Wang et al., 2004): (1) supply of energy, (2) supply of water, (3) tendency of liquid water molecules to escape, and (4) turbulent transport. Evaporation from vegetated surfaces entails evaporation as well as transpiration and is referred to as evapotranspiration. For transpiration water is extracted from soil by plant roots and is transported through the plant body, and the process of phase change of liquid water occurs at leaf scale. Thus, transpiration involves physical, chemical, and biological processes. In hydrology the focus has been on turbulent transport at the canopy scale, and processes involving the transport of water through the plant body have not received much attention. The process of transpiration is complicated by the complex and diverse features of plant physiology and is strongly affected by stomatal conductance, depending on the sizes and shapes of stomatal apertures. This chapter discusses extremal and minimum entropy production principles for describing the processes of evaporation and transpiration. The discussion is this chapter draws heavily from the works of Wang et al. (2004, 2007) and Wang and Bras (2011).

10.1 Extremum Principle for Evaporation from Land Surfaces

Wang et al. (2004) developed an extremal principle for evaporation modeling, using the argument that under thermodynamic equilibrium, the thermal and hydrologic states of the land surface resulting from the interaction between land and atmospheric processes tend to maximize evaporation. Turning to the four elements of evaporation, the source of energy that is needed for evaporation to occur from land surfaces is solar radiation and can be defined by radiative flux. The supply of water can be both natural in the form of precipitation and artificial, such as irrigation, and determines soil wetness and is characterized by soil moisture. *Fugacity* refers to the tendency of water molecules to escape. It is a function of soil water temperature and water potential, is expressed by the saturated vapor pressure at the liquid-vapor interface, and is determined by the surface soil temperature. The water potential is expressed in terms of volumetric soil water content. The turbulent transport of water vapor and heat is determined by the wind speed and thermal instability of the surface layer. This is defined by turbulent sensible heat flux into atmosphere.

The rate of evaporation for a given radiative energy flux depends on the combination of surface soil moisture, surface soil temperature, and sensible heat flux, and the dynamic feedback among them at the land surface. There can be many combinations of ground and sensible heat fluxes, evaporation rate, and net radiation that can satisfy the energy balance. Wang et al. (2004) hypothesized that the preferred combination is the one that maximizes evaporation.

10.1.1 Evaporation (Latent Heat Flux)

Let the rate of evaporation be denoted by E, surface soil moisture by w, surface soil temperature by T, sensible heat flux into the atmosphere by H, and radiative energy at the land surface by R. The rate of evaporation E can be expressed by a general function as

$$E = E(w, T, H; R) \tag{10.1}$$

The objective of modeling is to determine the functional form E. Note that Eq. (10.1) does not include the water vapor deficit.

10.1.2 Ground Heat Flux

Wang and Bras (1999) have shown that the ground heat flux G can be expressed as

$$G = G(T, w; R) \tag{10.2}$$

The dependence of G is on the time history of T, w, and R.

10.1.3 Sensible Heat Flux

The sensible heat flux can be considered as a state independent variable and employed to represent the transport mechanism. It permits partitioning of radiative energy input flux without explicitly enumerating a mathematical equation for H.

10.2 Extremum Principle

The surface energy balance can be expressed as

$$R_n = G + H + E \tag{10.3}$$

The net radiation R_n at the surface is comprised of both shortwave radiation (incoming R_i and reflected R_r) and longwave radiation (incoming R_l and emitted R_e) components. Except for R_e all other components of net radiation are affected by the surface temperature and soil moisture. Thus, R_n can be expressed as

$$R_n = R_i - R_r + R_l - R_e \tag{10.4}$$

Note that for determination of E, variables T, H, and w can be considered independent. Equation (10.1) can be maximized, subject to the energy balance given by Eq. (10.3), and the maximization problem can be formulated as

$$E = \max[E(w, T, H; R) \mid E + H + G = R_n] \tag{10.5}$$

For all combinations of independent variables w, T, and H. Maximization of E using Eq. (10.5) can be achieved based on the specification of radiative energy R. Wang et al. (2004) investigated three cases: (1) $R = R_n - G$. This case represents the turbulent energy budget, described by the Bowen ratio. (2) $R = R_n$. This case corresponds to the partitioning of the net radiation into latent, sensible, and ground heat fluxes. (3) $R = R_t$. This represents the budget of all surface energy fluxes. These three cases express land-atmosphere interaction.

Evaporation E can be maximized, as given by Eq. (10.5), using the method of Lagrange multipliers. To that end, the lagrangian function L can be written as

$$L(w, T, H, \lambda; R) = E(w, T, H; R) + \lambda[R_n - E(w, T, H; R) - H - G(w, T; R)] \quad (10.6)$$

in which λ is the Lagrange multiplier.

10.2.1   Case 1: Given $R = R_n - G$

Differentiating the lagrangian function partially with respect to the independent variables, including the Lagrange multiplier, and equating each derivative to zero yield

$$\frac{\partial L}{\partial T} = \frac{\partial E}{\partial T} - \lambda \frac{\partial E}{\partial T} = 0 \quad (10.7)$$

$$\frac{\partial L}{\partial w} = \frac{\partial E}{\partial w} - \lambda \frac{\partial E}{\partial w} = 0 \quad (10.8)$$

$$\frac{\partial L}{\partial H} = \frac{\partial E}{\partial H} - \lambda \left(\frac{\partial E}{\partial H} + 1 \right) = 0 \quad (10.9)$$

$$\frac{\partial L}{\partial \lambda} = (R_n - G) - E - H = 0 \quad (10.10)$$

It may be emphasized again that $R = R_n$ is given, and in each equation the variables not considered are nondifferentiating variables of independent variables. For example, in Eq. (10.7) the nondifferentiating variables are w and H. Equations (10.7) to (10.10) yield

$$\frac{\partial E}{\partial T} = 0 \quad (10.11)$$

$$\frac{\partial E}{\partial w} = 0 \quad (10.12)$$

$$\lambda = 1 + \left(\frac{\partial E}{\partial H} \right)^{-1} \quad (10.13)$$

$$R_n - G = E + H \quad (10.14)$$

Equations (10.11) to (10.14) are solved for the optimum combination of T, w, and H for given $R_n - G$ that maximizes E. This case reflects on the balance between turbulent flexes.

10.2.2   Case 2: Given Net Radiation R_n

Differentiating Eq. (10.6) partially with respect to w, T, H, and λ and equating the corresponding derivatives to zero yield

$$\frac{\partial L}{\partial T} = \frac{\partial E}{\partial T} - \lambda\left(\frac{\partial E}{\partial T} + \frac{\partial G}{\partial T}\right) = 0 \tag{10.15}$$

$$\frac{\partial L}{\partial w} = \frac{\partial E}{\partial w} - \lambda\left(\frac{\partial E}{\partial w} + \frac{\partial G}{\partial w}\right) = 0 \tag{10.16}$$

$$\frac{\partial L}{\partial H} = \frac{\partial E}{\partial H} - \lambda\left(\frac{\partial E}{\partial H} + 1\right) = 0 \tag{10.17}$$

$$\frac{\partial L}{\partial \lambda} = R_n - E - H - G = 0 \tag{10.18}$$

Equations (10.15) to (10.18) yield the governing equations for E:

$$\frac{\partial E}{\partial T} = \frac{\partial E}{\partial H}\frac{\partial G}{\partial T} \tag{10.19}$$

$$\frac{\partial E}{\partial w} = \frac{\partial E}{\partial H}\frac{\partial G}{\partial w} \tag{10.20}$$

$$\lambda = \left[1 + \left(\frac{\partial E}{\partial H}\right)^{-1}\right]^{-1} \tag{10.21}$$

$$R_n = E + H + G \tag{10.22}$$

Equations (10.19) and (10.20) show that for maximizing E, the three components of the surface energy balance for the net radiative energy compete. Equation (10.19) shows that greater evaporation due to greater surface temperature (or greater fugacity) is proportional to greater evaporation due to greater turbulent transport. This case reflects on the partitioning of net radiation into turbulent and conductive heat fluxes. It is important to note two points. First, through ground heat flux it highlights the role of land surface in the energy balance. Second, the role of heat flux is comparable to that of turbulent flux in the exchange of water and heat at the land surface.

10.2.3   Case 3: Given Incoming Shortwave Radiation $R = R_i$

Note that other components of solar radiation R_0, R_l, and R_e depend on the surface temperature T and soil moisture w. Following the above procedure, the governing equations for E for R_i can be expressed as

$$\frac{\partial E}{\partial T} = \frac{\partial E}{\partial H}\left[\frac{\partial G}{\partial T} + \frac{\partial R_0}{\partial T} + \frac{\partial(R_0 - R_e)}{\partial T}\right] \tag{10.23}$$

$$\frac{\partial E}{\partial w} = \frac{\partial E}{\partial H}\left[\frac{\partial G}{\partial w} + \frac{\partial R_0}{\partial w} + \frac{\partial(R_0 - R_e)}{\partial w}\right] \tag{10.24}$$

$$\lambda = \left[1 + \left(\frac{\partial E}{\partial H}\right)^{-1}\right]^{-1} \tag{10.25}$$

$$R_i = E + H + G + R_l - R_0 - R_e \tag{10.26}$$

Equations (10.23) to (10.25) describe the portioning of solar radiation into turbulent and nonturbulent energy fluxes for evaporation maximizing. This case connects all fluxes—conductive, turbulent, and radiative—in the energy balance to the surface temperature and soil moisture.

10.2.4   Computation of Derivatives

The conditional derivatives of a flux variable can be determined from its total derivative. To illustrate, consider E. Using Eq. (10.1), the total derivative of E can be expressed as

$$E^i - E^0 = \frac{\partial E}{\partial T}(T^i - T^0) + \frac{\partial E}{\partial w}(w^i - w^0) + \frac{\partial E}{\partial H}(H^i - H^0) \qquad i = 1, 2, \ldots, m \qquad (10.27)$$

where superscript 0 denotes the reference point and i denotes the ith point of a sequence, and m denotes the number of points. Wang et al. (2004) discussed the procedure for computing the derivatives. Their procedure involves the following steps: (1) Time series measurements of fluxes are arranged in ascending or descending order of R. (2) The resulting sequences are divided into subsequences corresponding to a narrow range of a certain R. (3) The conditional derivatives are the regression coefficients as shown by Eq. (10.27).

10.2.5   Application

Priestley and Taylor (1972) presented an expression for potential evaporation E_p (that is, under limitless water supply) as

$$E_p = \alpha \frac{R_n - G}{1 + \gamma/\Delta} \qquad (10.28)$$

where Δ is the saturated vapor pressure gradient with respect to temperature, γ is the psychometric constant, and α is an empirical parameter which is normally assumed to be 1.26. However, field observations show that α varies but within a narrow range.

Using Eq. (10.18), E_p of Eq. (10.28) can be expressed as a function of H and T as

$$E_p = \frac{\alpha H}{1 + \gamma/\Delta - \alpha} \qquad (10.29)$$

Equation (10.28) holds if E is maximized. If $1 + \gamma/\Delta$ is less than α, then H can be negative. Introducing Eq. (10.28) in Eq. (10.19), Wang et al. (2004) obtained α as

$$\alpha = 1 + \frac{\gamma}{\Delta} + H \frac{d}{dT}\left(\frac{\gamma}{\Delta}\right)\left(\frac{dG}{dT}\right)^{-1} \qquad (10.30)$$

Example 10.1.   Data from the Hydrology-Atmosphere Pilot Experiment in the Sahel experiment are used for this example. The Southern Supersite Tiger Bush site selected is located in a bare soil region where turbulent heat fluxes are measured. An interpolation procedure is emplyed to create a time series of sequences to match the temporal resolution of temperature. Take the temperature data, e.g., and verify Eq. (10.11) using the observed data.

Solution.   First, the data used are shown in Fig. 10.1. Second, all measured time series variables are rearranged in ascending order of the given parameter $R_n - G$. Third, the resulting sequences are divided into a number of subsequences with equal number of m data points. In this example, m is equal to 3. For each subsequence, the method for calculating the conditional derivatives is used. For

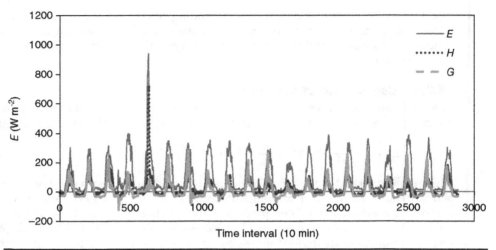

Figure **10.1a**   Time series of variables *E*, *H*, and *G*.

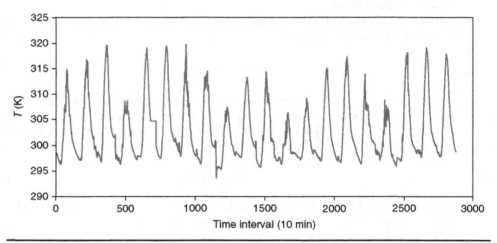

Figure **10.1b**   Time series of variable *T*.

each subsequence, $\partial E/\partial T$ is calculated. The histogram of the conditional derivative in Eq. (10.11) is illustrated in Fig. 10.2. The mean value of $\partial E/\partial T$ is 0.31, and the standard deviation is 0.16.

Example 10.2.   Data from the Southern Great Plains 1997 field experiment is used. The length of the data is from July 1 to July 14, 1997. For soil moisture, the daily data are used. Verify Eq. (10.20), using the soil moisture data.

Solution.   First, a linear method is employed to match the temporal resolution of the temperature and surface heat flux data. Second, the resulting sequences are divided into a number of subsequences with equal number of *m* data points. In this example, *m* is also equal to 4. For each subsequence, the

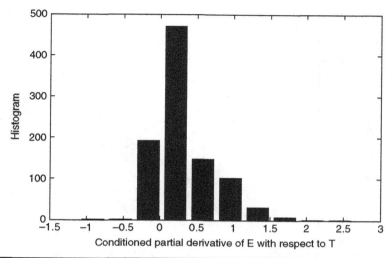

FIGURE 10.2   Histogram of $\partial E / \partial T$ using the observed data from Southern Supersite Tiger Bush site.

conditional derivatives $\partial E / \partial w$, $\partial E / \partial H$, and $\partial G / \partial w$ are calculated for each subsequence using the following equations:

$$E^i - E^0 = \frac{\partial E}{\partial T}(T^i - T^0) + \frac{\partial E}{\partial w}(w^i - w^0) + \frac{\partial E}{\partial H}(H^i - H^0) \qquad i = 1, 2, \dots, m$$

$$G^i - G^0 = \frac{\partial G}{\partial T}(T^i - T^0) + \frac{\partial G}{\partial w}(w^i - w^0) \qquad i = 1, 2, \dots, m$$

Then $\partial E / \partial w$ is plotted in Fig. 10.3, which shows that Eq. (10.20) is valid.

Example 10.3.   Data from the Hydrology-Atmosphere Pilot Experiment in the Sahel experiment are used for this example. The data include latent heat flux E, sensible heat flux H, net radiation R_n, ground heat flux G, and incoming and reflected shortwave radiation R_i and R_r, respectively. Test Eq. (10.23).

Solution.   From Eq. (10.4), $R_n = R_i - R_r + R_l - R_e$, longwave radiation $(R_e - R_l)$ can be obtained. Here $R_e - R_l$ is used for calculation. The conditional derivatives $\partial E / \partial T$ and $\partial E / \partial H$ are calculated by

$$E^i - E^0 = \frac{\partial E}{\partial T}(T^i - T^0) + \frac{\partial E}{\partial w}(w^i - w^0) + \frac{\partial E}{\partial H}(H^i - H^0) \qquad i = 1, 2, \dots, m$$

and $\partial G / \partial T$ and $\partial G / \partial T$ are calculated by

$$G^i - G^0 = \frac{\partial G}{\partial T}(T^i - T^0) + \frac{\partial G}{\partial w}(w^i - w^0) \qquad i = 1, 2, \dots, m$$

where m equals 4. Then $\partial R_0 / \partial T$ and $\partial R_0 / \partial T$, $\partial (R_0 - R_e) / \partial T$, and $\partial (R_0 - R_e) / \partial w$ can be computed in the same way. The values of

$$\frac{\partial E}{\partial T} \qquad \text{and} \qquad \frac{\partial E}{\partial H}\left[\frac{\partial G}{\partial T} + \frac{\partial R_0}{\partial T} + \frac{\partial (R_0 - R_e)}{\partial T}\right]$$

are plotted in Fig. 10.4.

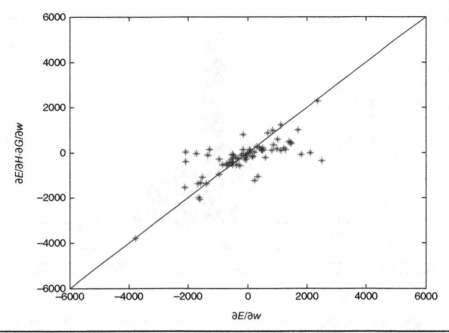

FIGURE 10.3   Conditional derivatives using the observed data from Southern Great Plains 1997 field experiment.

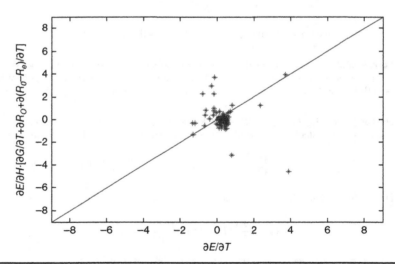

FIGURE 10.4   Testing Eq. (10.23) using the observed data from Southern Great Plains 1997 field experiment.

Example 10.4. Given available energy $R_n - G$ ($R_n - G \approx -57$), calculate conditional derivatives using the hydrometeorological data given in Table 10.1.

Solution. First $E^i - E^0$ is calculated. The first time period is set to be 0. Then ΔE, ΔT, Δw, and ΔH are computed as given in Table 10.2.

A linear regression procedure is used to obtain the three derivative terms with the calculated results

$$\frac{\partial E}{\partial T} = 0.0000 \qquad \frac{\partial E}{\partial T} = 1.3680 \qquad \frac{\partial E}{\partial H} = -0.0001$$

Example 10.5. Let $T = 20°C$, $H = 7$ W m^{-2}, $dG/dT = 1$, and atmospheric pressure $P = 96$ kPa. Calculate α.

Solution. First, the psychrometric constant γ is calculated by

$$\gamma = \frac{C_p \cdot P}{\varepsilon \cdot \lambda}$$

where ε is the ratio of molecular weight vapor of water to dry air (equal to 0.622); C_p is equal to 0.001013 MJ kg^{-1} °C^{-1}; and $\lambda = 6.11$. For this example,

$$\gamma = \frac{C_p \cdot P}{\varepsilon \cdot \lambda} = 0.001013 \times 96/0.622/6.11 = 0.026$$

When $T \geq 0°C$,

$$\Delta = e_b \times \frac{4249.9}{(241.9 + T)^2}$$

in which

$$e_b = 6.11 \times 10^{\frac{7.63T}{(241.9+T)^2}} \qquad (T \geq 0°C)$$

$$\Delta = e_b \times \frac{4249.9}{(241.9+T)^2} = 6.11 \times 10^{7.63 \times 25/266.9^2} \times 4249.9/266.9^2 = 0.367$$

Now, d/dT (γ/Δ) can be calculated and is found to be 0.014. Therefore,

$$\alpha = 1 + \frac{\gamma}{\Delta} + H\frac{d}{dT}\left(\frac{\gamma}{\Delta}\right)\left(\frac{dG}{dT}\right)^{-1} = 1 + \frac{0.026}{0.367} + 7 \times 0.014 \times 1 = 1.2$$

E(Wm^{-2})	T(K)	w(kg/kg)	H(Wm^{-2})	$R_n - G$(Wm^{-2})
13.37	297.38	0.05	−11.850	−57.28
13.56	297.47	0.05	−12.003	−57.15
15.79	298.04	0.0505	−8.339	−57.13
13.76	297.55	0.05	−12.157	−57.03

TABLE 10.1 Hydrometeorologic Data

ΔE	ΔT	Δw	ΔH
0.1933	0.0900	0.0000	−0.1533
2.2267	0.5700	0.0005	3.6643
−2.0333	−0.4900	−0.0005	−3.8177

TABLE 10.2 Differences of E, T, w, and H

10.3 Transpiration

When liquid water in the plant leaf tissues tends to be in equilibrium with water vapor in stomatal cavities, transpiration results. Based on the thermodynamics of liquid-vapor equilibrium, Wang et al. (2007) hypothesized a principle of maximum transpiration similar to that of evaporation. The process of transpiration entails (1) the supply of energy due to solar radiation; (2) the supply of water characterized by leaf water content or potential ψ, fugacity characterized by leaf temperature T_l, transport mechanism characterized by turbulence in plant boundary layer (PBL), and pathway of water vapor into the atmosphere characterized by stomatal aperture conductance g. Considering these four elements, transpiration T_E can be formulated as (Wang et al., 2007)

$$T_E = T_E(T_l, \psi, g, H, R) \tag{10.31}$$

Wang et al. (2007) stated the hypothesis as follows. The dynamics of water and heat at the leaf-atmosphere interface correspond to a state of leaf temperature, leaf water potential, and sensible heat flux into the atmosphere that maximizes the latent heat flux expenditure for transpiration under a given energy input, meteorological conditions, and physical and physiological environment.

To obtain the combination of controlling factors that leads to the maximum transpiration, the method of Lagrange multipliers is employed:

$$T_E = \max[T_E(T_l, \psi, g, H; R) \,|\, T_E + H = R_n] \tag{10.32}$$

for all combinations of T_l, H, and ψ. The Lagrange function can therefore be written as

$$L(T_E, T_l, \psi, g, H; R) = T_E(T_l, \psi, g, H; R) + \lambda[R_n - T_E(T_l, \psi, g, H; R) - H] \tag{10.33}$$

where R_n is the net radiation at the canopy surface; g is assumed to be a species-dependent function of ψ and is hence not an independent variable; and $T_E + H = R_n$ is the energy balance equation and is used as a constraint. Among all possible combinations of T_l, ψ and H that satisfy the energy budget, the combination that maximizes T_E corresponds to the unique partition between T_E and H.

Denoting CO_2 flux by A, Wang et al. (2007) assumed stomatal conductance as

$$g = g(A) \tag{10.34}$$

Then Eq. (10.31) becomes

$$T_E = T_E(T_l, A, H; R) \tag{10.35}$$

The governing equation for T_E can now be expressed as

$$T_E = \max[T_E(T_l, A, H; R) \,|\, T_E + H = R_n] \tag{10.36}$$

for all T_l, A, and H.

For maximizing T_E by Eq. (10.36), the lagrangian function L can be defined as

$$L(T_E, T_l, A) = T_E(T_l, A, H, R) + \lambda[R_n - T_E(T_l, A, H) - H]$$ (10.37)

in which λ is the Lagrange multiplier. Differentiating Eq. (10.35) partially with respect to T_l, A, H and λ and equating each derivative to zero, keeping R_n constant, one obtains

$$\frac{\partial L}{\partial T_l} = \frac{\partial T_E}{\partial T_l} - \lambda \frac{\partial T_E}{\partial T_l} = 0$$ (10.38)

$$\frac{\partial L}{\partial A} = \frac{\partial T_E}{\partial A} - \lambda \frac{\partial T_E}{\partial A} = 0$$ (10.39)

$$\frac{\partial L}{\partial H} = \frac{\partial T_E}{\partial H} - \lambda \left(\frac{\partial T_E}{\partial H} + 1 \right) = 0$$ (10.40)

$$\frac{\partial L}{\partial \lambda} = R_n - T_E - H = 0$$ (10.41)

Equations (10.36) to (10.39), respectively, lead to

$$\frac{\partial T_E}{\partial T_l} = 0$$ (10.42)

$$\frac{\partial T_E}{\partial A} = 0$$ (10.43)

$$\lambda = \left[1 + \left(\frac{\partial T_E}{\partial H} \right)^{-1} \right]^{-1}$$ (10.44)

$$T_E = R_n - H$$ (10.45)

The conditional derivatives can be determined from field observations as before. If the maximal hypothesis holds, then conditional derivatives given by Eqs. (10.42) to (10.45) vanish for all values of R_n.

Example 10.6. Data from a temperate forest area in Harvard Forest of Massachusetts during 2000 are selected. Hourly resolution measurements of surface energy fluxes, CO_2 flux, and air temperature during the period of May 24–June 23, 2000, at Harvard Forest, are used. The air temperature is used as a surrogate for the missing leaf temperature. The average value of air temperature measured at 28 and 22.6 m should be fairly close to the leaf temperature at the 24-m canopy top. Evaluate Eqs. (10.42) and (10.43), using observed values.

Solution. The derivatives of temperature are computed and plotted, as shown in Fig. 10.5. Clearly, the derivatives computed with Eqs. (10.42) and (10.43) are in good agreement with observations, supporting the maximum hypothesis.

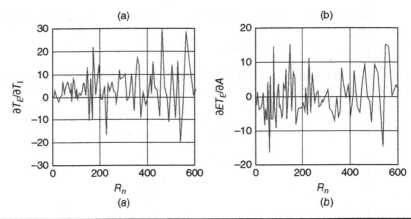

FIGURE **10.5** Testing Eqs. (10.42) and (10.43) using the observed data from Harvard Forest.

Questions

1. Obtain data on the surface heat flux, temperature, and soil moisture for a basin. Given $R = R_n - G$, compute the value of $\partial E/\partial T$ and verify Eqs. (10.11) and (10.22).

2. Using the data from Question 1, given $R = R_n$, compute the values of $\partial E/\partial T$, $\partial E/\partial H$, $\partial G/\partial T$, and so on. Verify Eqs. (10.19) and (10.20).

3. Given incoming shortwave radiation $R = R_t$, verify Eqs. (10.23) and (10.24).

4. According to the equations mentioned above, deduce Eq. (10.30).

5. Let $T = 15°C$, $H = 10$ Wm^{-2}, $dG/dT = 2$, and atmospheric pressure $P = 96$ kPa. Calculate α.

References

Priestley, C. H. B., and Taylor, R. J. (1972). On the assessment of surface heat flux and evaporation using large scale parameters. *Monthly Weather Review,* vol. 100, pp. 81–92.

Wang, J., and Bras, R. L. (1999). Ground heat flux estimated from surface soil moisture. *Journal of Hydrology,* vol. 216, pp. 214–226.

Wang, J., and Bras, R. L. (2011). A model of evapotranspiration based on the theory of maximum entropy production. *Water Resources Research,* vol. 47, W03521, doi: 10.1029/2010WR009392.

Wang, J., Bras, E. L., Lerdau, M., and Salvucci, G. D. (2007). A maximum hypothesis of transpiration. *Journal of Geophysical Research,* vol. 112, G03010, doi: 10.1029/2006JG000255.

Wang, J., Salvucci, G. D., and Bras, R. L. (2004). A extremum principle of evaporation. *Water Resources Research,* vol. 40, W09303, doi: 10.1029/2004WR003087.

Further Reading

Brunsell, N. A., and Anderson, M. C. (2011). Characterizing the multi-scale spatial structure of remotely sensed evapotranspiration with information theory. *Biogeosciences,* vol. 8, pp. 2269–2280.

Pettijohn, J. C., and Salvucci, G. D. (2009). A new two-dimensional physical basis for the complementary relation between terrestrial and pan evaporation. *Journal of Hydrometeorology*, vol. 10, pp. 565–574, doi: 10.1175/2008JHM1026.1.

Wang, J., and Bras, R. L. (2001). Effect of temperature on surface energy balance. *Water Resources Research*, vol. 37, no. 12, pp. 3383–3386.

Wang, J., and Bras, R. L. (2009). A model of surface heat flux based on the theory of maximum entropy production. *Water Resources Research*, vol. 45, W11422, doi: 10.1029/2009WR007900.

Yang, D., Sun, F., Liu, Z., Cong, Z., and Lei, Z. (2006). Interpreting the complementary relationship in non-humid environments based on the Budyko and Penman hypotheses. *Geophysical Letters*, vol. 33, L18402, doi: 10.1029/2006GL027657.

PART III

Subsurface Flow

CHAPTER 11

Infiltration

Infiltration is a key component of the hydrologic cycle. It partitions rainfall into surface runoff and that entering the soil. It is fundamental to determining the runoff hydrograph, soil moisture and groundwater recharge, irrigation efficiency, life span of pavements, and leaching of nutrients. Because of the fundamental role that infiltration plays in hydrology, irrigation engineering, and soil science, it has been a subject of much research for over a century. As a result, a large number of infiltration equations have been developed, some of which are now commonly applied in hydrologic modeling and have been included in popular watershed hydrology models (Singh, 1995; Singh and Frevert, 2002a, 2002b, 2006; Singh and Woolhiser, 2002). Some of these equations are found in Horton (1938), Green-Ampt (1911), Kostiakov (1932), Philip (1957), Overton (1964), Holtan (1961), and Singh and Yu (1990). Although these equations are simple and easy to use, one of the main difficulties is that their parameters have large spatial variability, may vary from one experiment to another at the same site, and need to be calibrated using extensive field or experimental measurements. These equations represent the capacity rate of infiltration at a point. Soil characteristics vary significantly from one place to another, and antecedent soil moisture, which defines the initial infiltration, also significantly varies spatially. The infiltration parameters determined using point measurements are point values, or at best reflect local average values, not the areal or watershed average although frequently they are taken as such. Although large spatial variability in infiltration is recognized, little effort has been made to account for its probabilistic characteristics, except for a few watershed models, such as the BASINS (formerly Stanford Watershed Model) (Crawford and Linsley, 1966; Donigian and Imhoff, 2006). In a series of papers Singh (2010a, 2010b; 2011a, 2011b) derived popular infiltration equations and then developed a general framework for deriving these equations by employing both Shannon entropy theory and Tsallis entropy theory.

The objective of this chapter is to discuss the derivation of the well-known infiltration equations of Horton, Kostiakov, Philip, Green-Ampt, Overton, Holtan, and Singh and Yu using entropy theory, the probability distributions of these equations, and their parameters in terms of constraints (given information on infiltration). The theory leads to a probabilistic basis of infiltration equations and hence an estimate of uncertainty associated with each equation.

11.1 Preliminaries

Consider a soil element on the surface of which there is plenty of water. The water infiltrates the soil. Let the rate of infiltration as a function of time t be defined as $I(t)$. It is assumed that the soil is dry and water is applied to the dry soil. At the beginning, infiltration

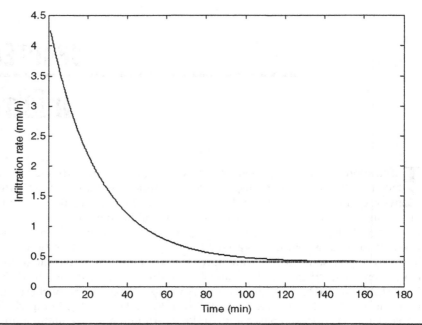

will be high; as time progresses, the rate of infiltration declines and may reach a steady or constant rate denoted by I_c or may approach zero, as shown in Fig. 11.1.

The cumulative infiltration at any time t, denoted by J, is the amount of water that has infiltrated the soil up to that time. Clearly, $J(t)$ depends on $I(t)$. Then for a soil element the continuity equation can be expressed as

$$J(t) = \int_0^t [I(t) - I_c]dt \qquad \frac{dJ}{dt} = I(t) - I_c \tag{11.1}$$

In hydrologic parlance, $I(t)$ is analogous to flux and $J(t)$ is analogous to the concentration of soil moisture. If W is the amount of pore space available for infiltration of water in the soil element at any time, S equals porosity. In other words, $W + J = S$, where J is the cumulative infiltration.

It is assumed that the spatially averaged rate of infiltration $I(t)$ is a random variable with a probability density function defined as $f(I)$. To formulate the entropy theory for infiltration rate I, four parts are involved: (1) Shannon entropy, (2) principle of maximum entropy (POME), (3) specification of information on infiltration in terms of constraints, and (4) maximization of entropy in accordance with POME.

11.1.1   Shannon Entropy

The Shannon entropy for the infiltration rate I, denoted by $H(I)$, can be expressed as

$$H(I) = -\int_a^b f(I) \ln f(I) dI \tag{11.2}$$

where a and b are the limits of integration for I. The infiltration rate $H(I)$ describes the expected value of $-\ln f(I)$. Considering $-\ln f(I)$ as a measure of uncertainty, Eq. (11.2) defines the average uncertainty associated with $f(I)$ and in turn with I. The more uncertain I is, the more information will be needed to characterize it. In other words, information reduces uncertainty.

11.1.2 Principle of Maximum Entropy

The principle of maximum entropy formulated by Jaynes (1957a, b) says that the least-biased probability distribution of I, denoted by $f(I)$, will be the one that will maximize $H(I)$ given by Eq. (11.2), subject to the given information on I expressed as constraints. In other words, if no information other than the given constraints is available, then the probability distribution should be selected such that it is least biased toward what is not known. Such a probability distribution is yielded by the maximization of the Shannon entropy.

11.1.3 Constraints

Information on $I(t)$ can be obtained using the knowledge of soil physics and experimental observations. For a given soil, one frequently measures infiltration and then characterizes the soil infiltration characteristics and more particularly the time rate of infiltration or the infiltration curve under the condition that water is not a limiting factor for the soil. If infiltration rate observations are available, then information on the infiltration capacity rate is expressed in terms of constraints as

$$\int_a^b f(I)dI = 1 \tag{11.3}$$

$$\int_a^b g_r(I) f(I)dI = \overline{g_r(I)} \qquad r = 1, 2, \ldots, n \tag{11.4}$$

where $g_r(I)$, for $r = 1, 2, \ldots, n$, represents some functions of I; n denotes the number of constraints; and $\overline{g_r(I)}$ is the expectation of $g_r(I)$. For example, if $r = 1$ and $g_1(I) = I$, then Eq. (11.4) corresponds to the mean infiltration rate; likewise, for $r = 2$ and $g_2(I) = (I - \bar{I})^2$, where $\bar{I}$ is the mean infiltration rate, it would denote the variance of I. In a similar manner, higher-order moments of $f(I)$ can be defined for higher values of r and appropriate values of $g_r(I)$. For most infiltration equations, however, no more than two constraints are needed.

11.1.4 Maximization of Shannon Entropy

To obtain the least-biased $f(I)$, the entropy given by Eq. (11.2) is maximized, subject to Eqs. (11.3) and (11.4); and one simple way to achieve maximization is through the use of the method of Lagrange multipliers. Maximization leads to the probability distribution of I in terms of the given constraints as

$$f(I) = \exp\left[-\lambda_0 - \sum_{r=1}^n \lambda_r g_r(I)\right] \tag{11.5}$$

where λ_r, $r = 1, 2, \ldots, n$, are Lagrange multipliers which can be determined with the use of Eqs. (11.3) and (11.4). Equations (11.2) to (11.5) constitute the building blocks of the entropy theory of infiltration, which is now applied to derive seven infiltration equations as examples.

11.1.5 Hypothesis on Cumulative Distribution of Infiltration

It is hypothesized that the cumulative probability distribution of excess infiltration $F(I)$ can be defined as the ratio of soil moisture potential W to the maximum soil moisture retention S. Thus, $F(I)$ can also be defined as 1 minus the ratio of the cumulative infiltration to maximum potential cumulative infiltration or maximum soil moisture retention S. Therefore,

$$F(I) = 1 - \frac{I}{S} \tag{11.6a}$$

or

$$F(I) = \frac{W}{S} \tag{11.6b}$$

Differentiation of Eq. (11.6a) yields

$$dF(I)dI = -\frac{dI}{S} \qquad dF(I) = f(I) = -\frac{1}{S}\frac{dI}{dI} \tag{11.7}$$

where $f(I)$ is the probability density function of $I(t)$ which is determined using the entropy theory.

11.2 Derivation of Infiltration Equations

Application of the entropy theory is illustrated with the derivation of infiltration equations discussed by Singh (2010a, b).

11.2.1 Horton Equation

The Horton equation is derived in three different ways, discussed here.

POME-Based Method

Let the beginning infiltration rate be defined as I_0. As time progresses, the rate of infiltration declines and reaches a steady or constant rate denoted by I_c. The objective is to derive the rate of infiltration as a function of time.

The simplest constraint that $f(I)$ must satisfy is

$$\int_{I_c}^{I_0} f(I)dI = 1 \tag{11.8}$$

Applying POME and using the method of Lagrange multipliers (Singh, 1998), one obtains

$$f(I) = \exp(-\lambda_0) \tag{11.9}$$

where λ_0 is the Lagrange multiplier.

Inserting Eq. (11.9) in Eq. (11.8), one gets

$$\int_{I_c}^{I_0} \exp(-\lambda_0)dI = 1 \Rightarrow f(I) = \frac{1}{I_0 - I_c} \tag{11.10}$$

Equation (11.10) states that the probability density function of I is constant or is a uniform distribution. From Eqs. (11.9) and (11.10), one obtains the Lagrange multiplier λ_0 as

$$\lambda_0 = \ln(I_0 - I_c) \tag{11.11}$$

Equation (11.11) expresses the Lagrange multiplier λ_0 in terms of the initial infiltration capacity rate I_0 and the final infiltration capacity rate I_c.

The cumulative distribution function of I obtained by integration of Eq. (11.10) would be linear, expressed as

$$F(I) = \int_{I_c}^{I} f(I)\,dI = \int_{I_c}^{I} \frac{1}{I_0 - I_c}\,dI = \frac{I - I_c}{I_0 - I_c} \tag{11.12}$$

Combining Eqs. (11.7) and (11.9), one obtains

$$\frac{1}{I_0 - I_c}\,dI = -\frac{1}{S}\,dJ \tag{11.13}$$

Woodbury (2012) commented that Eq. (11.7) is not compatible with Eq. (11.10) derived using entropy theory. However, if there is a linear relation between J and I, then the compatibility will be maintained. Integrating Eq. (11.13), one obtains

$$\frac{1}{I_0 - I_c}(I - I_0) = -\frac{J}{S} \quad \text{or} \quad I = I_0 - \frac{(I_0 - I_c)J}{S} \tag{11.14}$$

Recalling the continuity equation, Eq. (11.1), Eq. (11.14) can be recast as

$$\frac{dJ}{dt} + I_c = I_0 - \frac{I_0 - I_c}{S}J \tag{11.15}$$

Solution of Eq. (11.15) is

$$J = S\left[1 - \exp\left(-\frac{I_0 - I_c}{S}t\right)\right] \tag{11.16}$$

Differentiating Eq. (11.16) with respect to t yields

$$\frac{dJ}{dt} = (I_0 - I_c)\exp\left(-\frac{I_0 - I_c}{S}t\right) \tag{11.17}$$

Recalling the continuity Eq. (11.1), Eq. (11.17) can be written as

$$I(t) - I_c = (I_0 - I_c)\exp\left(-\frac{I_0 - I_c}{S}t\right) \tag{11.18}$$

Equation (11.18) can be written as

$$I(t) = I_c + (I_0 - I_c)\exp\left(\frac{-t}{k}\right) \tag{11.19}$$

where

$$k = \frac{S}{I_0 - I_c} \tag{11.20}$$

Equation (11.19) is the Horton equation. Using Eq. (11.2), the entropy value of the Horton equation can be expressed as

$$H(I) = -\int_a^b f(I)\ln f(I)dI = -\int_{I_c}^{I_0} \frac{1}{I_0 - I_c}\ln\frac{1}{I_0 - I_c}dI = \ln(I_0 - I_c) = \lambda_0 \qquad (11.21)$$

In Eq. (11.20), parameter k is expressed as the ratio of the maximum soil moisture retention to the initial infiltration rate minus the steady-state infiltration rate. It has the dimension of time and indicates the time it will take for the infiltrated water to fill the maximum moisture retention space if the capacity rate of infiltration was the initial infiltration rate or the maximum infiltration rate minus the steady rate or the initial excess infiltration rate. Equation (11.9) shows that the probability density function of the infiltration rate associated with the Horton equation requires no constraint other than the total probability theorem, which is not a constraint in a true sense, for all probability distributions must satisfy it.

Equation (11.21) states that for a given soil the uncertainty of the Horton equation is maximum when the spread between the initial infiltration capacity rate and the steady infiltration capacity rate is maximum, which would be the case when the soil is dry, and the uncertainty reduces as soil becomes wetter. This means that when sampling infiltration, greater care should be exercised in the beginning of infiltration and less toward the end. This also means that infiltration observations should be more closely spaced temporally in the beginning, but the time interval between observations can be increased with the progress of infiltration.

Maximum Extended Entropy
Woodbury (2012) suggested the maximum extended entropy approach for the derivation of the Horton equation which is outlined here. Note from Eq. (11.1) that

$$\frac{J(t)}{S} = \frac{1}{S}\int_0^\infty [I(t) - I_c]dt \to 1 \qquad (11.22)$$

as time t gets large. Let that large value of t be denoted by t_L. Then Eq. (11.22) can be expressed as

$$\frac{1}{S}\int_0^{t_L} [I(t) - I_c]dt = 1 \qquad (11.23)$$

Defining $[I(t) - I_c]/S$ as a density function, $g(t)$, from the entropy viewpoint, Eq. (11.23) is cast as

$$\int_0^{t_L} g(t) = 1 \qquad (11.24)$$

Equation (11.24) defines a normalization. Note that $g(t)$ is not necessarily a probability density function. Here time is considered as the independent variable, unlike the infiltration capacity rate considered in the earlier derivation. If nothing is known about $I(t)$, then the application of POME yields, similar to Eq. (11.9),

$$g(t) = \exp(-\lambda_0) \qquad (11.25)$$

Substitution of Eq. (11.25) in Eq. (11.24) yields

$$g(t) = \frac{1}{t_L} \tag{11.26}$$

Then from the definition of $g(t)$,

$$I(t) - I_c = Sg(t) = \frac{S}{t_L} \tag{11.27}$$

Note that $S/t_L = I_0 - I_c$ for a uniform distribution. The mean value of $g(t)$ from Eq. (11.24) is $t_L/2$, which then leads to the mean value of $I(t) - I_c$ in terms of $St_L/2$. This results in the classical ϕ index method for infiltration.

Now let there be a mean travel time constraint defined as

$$\int_0^{t_L} tg(t)dt = \frac{S}{I_0 - I_c} \tag{11.28}$$

Then entropy maximizing for $g(t)$ produces

$$g(t) = \frac{I(t) - I_c}{S} = \frac{I_0 - I_c}{S}\exp\left(-\frac{I_0 - I_c}{S}t\right) \tag{11.29}$$

which leads to the same equation as Eq. (11.18). In this $I(t) - I_c$ represents the case that satisfies the constraints and has the largest spreadout that can be realized in the greatest number of ways. This approach does not have a stochastic interpretation.

Relative Entropy-Based Approach

The second approach (Woodbury, 2012) employs the principle of minimum relative entropy (POMRE) or cross entropy (POMCE). Let the PDF of t, denoted by $f(t)$, be continuous and uniform. Time is considered here as the independent variable. The cross-entropy D can be defined as

$$D(f, q) = \int_0^{t_L} f(t)\ln\left[\frac{f(t)}{q(t)}\right]dt \tag{11.30}$$

where $q(t)$ is the prior PDF of t. The objective is to minimize $D(f, q)$, subject to a given prior PDF and a set of constraints. In the absence of any information on the time history of $I(t)$, a uniform prior PDF for variable t can be assumed: $q(t) = 1/t_L$ for infiltration over the range 0 to ∞ (in actuality the range is from I_c to I_0). Then, using POMCE, the PDF $f(t)$ of t is

$$f(t) = q(t)\exp(-\lambda_0 - \lambda_1 t) = \frac{1}{t_L}\exp(-\lambda_0 - \lambda_1 t) \tag{11.31}$$

The two Lagrange multipliers λ_0 and λ_1 are determined from the given constraints on time as follows. First, integrating Eq. (11.31) gives

$$\int_0^\infty \frac{1}{t_L}\exp(-\lambda_0 - \lambda_1 t)dt = 1 \tag{11.32}$$

Equation (11.32) yields

$$\exp(-\lambda_0) = \lambda_1 t_L \tag{11.33}$$

With the use of Eq. (11.33) in Eq. (11.31), we obtain

$$f(t) = \lambda_1 t_L \exp(-\lambda_1 t) \tag{11.34}$$

If the mean value in the time constraint on $f(t)$ is specified as $S/(I_0 - I_c)$, then POMCE yields

$$f(t) = \frac{I_0 - I_c}{S} \exp\left(-\frac{I_0 - I_c}{S}t\right) \tag{11.35}$$

Equation (11.35) is identical to Eq. (11.29) if

$$\frac{I_0 - I_c}{S} = f(t) = g(t) \tag{11.36}$$

Another way is to use a POMRE approach similar to that of Woodbury and Ulrych (1996). Let a bivariate PDF be $f(I, t) = f(I)f(t)$ in $I(t)$ and t. In this the independent variables are $I(t)$ and t. Let a uniform prior PDF of $I(t)$ be defined as $p(I) = 1/(I_0 - I_c)$ for infiltration over the range from I_0 to I_c. Then the bivariate PDF will take on the form

$$f(I, t) = \frac{1}{I_0 - I_c} \exp(-\lambda_0 - \lambda_1 t) \tag{11.37}$$

The two Lagrange multipliers can be determined as follows. Integrate Eq. (11.37) as

$$\int_{I_c}^{I_0} \int_0^\infty \frac{1}{I_0 - I_c} \exp(-\lambda_0 - \lambda_1 t) \, dI \, dt = 1 \tag{11.38}$$

Equation (11.38) yields $\exp(-\lambda_0) = \lambda_1$. Therefore, Eq. (11.37) becomes

$$f(I, t) = \frac{\lambda_1}{I_0 - I_c} \exp(-\lambda_1 t) \tag{11.39}$$

If the mean value on $f(I, t)$ is defined as $S/(I_0 - I_c)$, then POMRE yields

$$f(I, t) = \frac{1}{S} \exp\left(-\frac{I_0 - I_c}{S}t\right) \tag{11.40}$$

Since Eq. (11.40) is a bivariate PDF, we need to marginalize out either time or infiltration. The marginal mean in $I(t)$ is $(I_0 - I_c)/2$, which is expected in the absence of any supply of new information. Marginalizing out $I(t)$ from $f(I, t)$, we obtain

$$f(t) = \int_{I_c}^{I_0} \frac{1}{S} \exp\left(-\frac{I_0 - I_c}{S}t\right) dt \tag{11.41}$$

Solution of Eq. (11.41) is

$$f(I, t) = \frac{I_0 - I_c}{S} \exp\left(-\frac{I_0 - I_c}{S} t\right) \tag{11.42}$$

which is identical to the univariate case if $q(t) = [I(t) - I_c]/S$. Note that uncertainty in $I(t)$ is removed by marginalization, and only uncertainty in time remains.

Example 11.1. Data on infiltration in field soils have been reported by Rawls et al. (1976) in a report published by the Agriculture Research Service of the U.S. Department of Agriculture. One data set on infiltration in Robertsdale loamy sand in the Georgia Coastal Plain is obtained for illustrative purposes. This data set is given in Table 11.1. For this data set the cover was 70 percent weed, and bare surface is 30 percent. The rainfall intensity was 12.19 cm/h, the initial soil moisture for the 0- to 30.48-cm depth was 4.55 cm and for 30.48- to 91.44-cm depth it was 14.00 cm. The final soil moisture was 6.45 cm for the 0- to 30.48-cm depth and was 16.26 cm for the 30.48- to 91.44-cm depth. The Horton equation has three parameters: I_c, I_0, and k. Compute the Horton equation parameters, using the entropy theory and the least-squares method. Compute the infiltration rate using parameters estimated in these two ways, and compare these rates with observed rates.

Solution. From the given data, the maximum soil moisture retention is 6.45 + 16.26 cm (−4.55 + 14.00 cm) = 4.16 cm, which is less than the accumulated infiltration amount given in the data. This is true because once the soil was saturated, the infiltration rate was gravity drainage and was not retained by the soil. The steady infiltration rate was attained at a time approximately equal to 50 min with a value of 3.10 cm/h, and the initial infiltration rate was 14.00 cm/h (obtained by extrapolation) which is higher than 12.21 cm/h given at a time equal to 4 min. Then using Eq. (11.20), the value of k is obtained as 0.43 h. The three Horton equation parameters I_c, I_0, and k are also obtained by the least-squares method by minimizing the sum of squares of deviations between computed infiltration rates and observed infiltration rates. These parameter values are found to be I_c = 3.02 cm/h, I_0 = 16.21 cm/h, and k = 0.13 h. Figure 11.2 shows observed and computed infiltration rates.

Time from Start of Rain (min)	Infiltration Rate (cm/h)	Accumulated Infiltration (cm)	Time from Start of Rain (min)	Infiltration Rate (cm/h)	Accumulated Infiltration (cm)
4	12.21	0.81	65	3.36	5.10
5	8.55	0.96	70	3.18	5.34
10	5.81	1.60	75	3.31	5.62
15	4.88	2.03	80	3.16	5.88
20	4.78	2.44	85	3.11	6.14
25	4.27	2.82	90	2.89	6.38
30	3.86	3.16	95	2.51	6.58
35	3.51	3.48	100	2.60	6.82
40	3.21	3.75	105	2.68	7.04
45	3.23	4.03	110	2.31	7.21
50	3.10	4.28	115	2.44	7.43
55	3.18	4.56	120	2.42	7.61
60	3.21	4.82			

TABLE 11.1   Observations for Robertsdale Loamy Sand (ID = 09091D) (after Rawls et al., 1976)

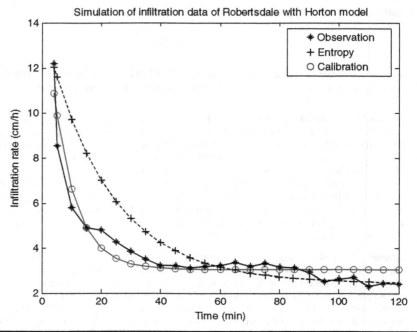

FIGURE 11.2 Comparison of the infiltration capacity rate computed using the Horton equation with parameters determined using the entropy theory and the least square method (calibration) with observed infiltration capacity rate values.

11.2.2 Kostiakov Equation

Let the constraints be defined as Eq. (11.4) and

$$\int_0^{I_0} \ln I f(I) \, dI = \overline{\ln I} \tag{11.43}$$

Using POME and the method of Lagrange multipliers, $f(I)$ is obtained as

$$f(I) = \exp(-\lambda_0 - \lambda_1 \ln I) = I^{-\lambda_1} \exp(-\lambda_0) \tag{11.44}$$

Substituting Eq. (11.44) in Eq. (11.3), one gets

$$\int_{I_c}^{\infty} I^{-\lambda_1} \exp(-\lambda_0) \, dI = 1 \Rightarrow \exp(-\lambda_0) = \frac{-\lambda_1 + 1}{I_c^{-\lambda_1 + 1}} \tag{11.45}$$

Inserting Eq. (11.45) in Eq. (11.44), one obtains

$$f(I) = -\frac{-\lambda_1 + 1}{I_c^{-\lambda_1 + 1}} I^{-\lambda_1} \tag{11.46}$$

For the Kostiakov equation, it turns out that $\lambda_1 = 2$. Then combining Eq. (11.45) with Eq. (11.7), the result with limits on I from I to ∞ and on J from J to 0 is

$$\frac{I_c S}{I} = J \tag{11.47}$$

Recalling that $I = dJ/dt$, Eq. (11.47) can be expressed as

$$\frac{dJ}{dt} = \frac{I_c S}{J} \Rightarrow J = (2 I_c S)^{0.5} t^{0.5} \tag{11.48}$$

Integration of Eq. (11.48) yields

$$J = (2 I_c S)^{0.5} t^{0.5} \tag{11.49}$$

Differentiating Eq. (11.49), one obtains the rate of infiltration

$$I = \frac{1}{2}(2 I_c S)^{0.5} t^{-0.5} \tag{11.50}$$

Equation (11.49) can be written as

$$J = a t^{0.5} \tag{11.51}$$

and Eq. (11.51) as

$$I(t) = 0.5 a t^{-0.5} \tag{11.52}$$

which is the Kostiakov equation. The Kostiakov equation turns out to be twice the product of the steady infiltration and the soil moisture retention capacity. Here

$$a = 2 I_c S \tag{11.53}$$

is parameter, and it has a physical meaning.

The entropy of the Kostiakov equation follows:

$$f(I) = I_c I^{-2} \tag{11.54}$$

$$H(I) = -\int_{I_c}^{\infty} I_c I^{-2} \ln I_c I^{-2}\, dI = -I_c \int_{I_c}^{\infty} I^{-2} \ln I_c\, dI + 2 I_c \int_{I_c}^{\infty} I^{-2} \ln I\, dI = 2 + \ln I_c \tag{11.55}$$

Example 11.2. The Kostiakov equation has two parameters: the exponent and parameter a. From observations given in Example 11.1, compute the Kostiakov equation parameters by using entropy theory and the least-squares method. Compute the infiltration rate using parameters estimated in these two ways, and compare these rates with observed rates.

Solution. For the Kostiakov equation the value of S is 4.17 cm. Then using Eq. (11.52), the value of a is obtained as 6.07 and the exponent is 0.5. For a general value of the exponent, the Kostiakov equation has two parameters, a and the exponent, and they are also obtained by the least-squares method by minimizing the sum of squares of deviations between computed infiltration rates and observed infiltration rates. These parameter values are found to be $a = 3.03$ and exponent $= -0.46$. Figure 11.3 shows observed and computed infiltration rates.

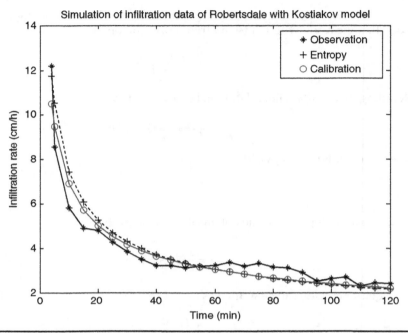

FIGURE 11.3 Comparison of infiltration rates computed using the Kostiakov equation with parameters determined using entropy theory and the least-squares method (calibration) with observed infiltration rate values.

11.2.3 Philip Two-Term Equation

Let the infiltration rate be defined as $I = i - a$, where a is some constant value. Let the constraints be defined by Eq. (11.4) with limits of a to ∞, and

$$\int_{a}^{\infty} \ln If(I)dI = \overline{\ln I} \tag{11.56}$$

Using POME and the method of Lagrange multipliers, $f(I)$ is obtained as

$$f(I) = \exp(-\lambda_0 - \lambda_1 \ln I) = I^{-\lambda_1} \exp(-\lambda_0) \tag{11.57}$$

Substituting Eq. (11.57) in Eq. (11.3), one gets

$$\int_{a}^{\infty} I^{-\lambda_1} \exp(-\lambda_0)dI = 1 \Rightarrow \exp(-\lambda_0) = \frac{-\lambda_1 + 1}{a^{-\lambda_1 + 1}} \tag{11.58}$$

Inserting Eq. (11.58) in Eq. (11.57) and taking $\lambda_1 = 2$, one obtains

$$f(I) = \frac{a}{I^2} \tag{11.59}$$

which is the probability density function of the Philip two-term equation. Then combining Eq. (11.59) with Eq. (11.7), the result with limits on I from I to ∞ and on J from J to 0 is

$$\frac{aS}{J} = J \tag{11.60}$$

Recalling that $I = dJ/dt$, Eq. (11.60) can be expressed as

$$\frac{dJ}{dt} = \frac{aS}{J} \Rightarrow J = (2aS)^{0.5} t^{0.5} \tag{11.61}$$

Differentiating Eq. (11.61), one obtains the rate of infiltration

$$I = \frac{1}{2}(2aS)^{0.5} t^{-0.5} \tag{11.62}$$

Equation (11.62) can be written as

$$i(t) = a + 0.5(2aS)^{0.5} t^{-0.5} = a + bt^{-0.5} \tag{11.63}$$

where

$$b = 0.5(2aS)^{0.5} \tag{11.64}$$

Equation (11.63) is the Philip two-term equation. Interestingly, both Philip equation parameters have physical meaning.

Entropy of the Philip two-term equation can be expressed as

$$H(I) = -\int_{a}^{\infty} I_c I^{-2} \ln I_c I^{-2} \, dI = \ln a + 2 \tag{11.65}$$

Example 11.3. The Philip equation has two parameters, a and b, as shown in Eq. (11.63). From observations given in Example 11.1, compute the Philip two-term equation parameters using the entropy theory and the least-squares method. Compute the infiltration rate, using parameters estimated in these two ways, and compare these rates with observed rates.

Solution. For the Philip equation the value of S is 4.17 cm. Using the entropy method, the value of a is 1.21 and the value of b is 2.15; by the least-squares method and by minimizing the sum of squares of deviations between computed infiltration rates and observed infiltration rates, a is 0.53 and b is 2.55. Figure 11.4 shows the observed and computed infiltration rates.

11.2.4   Green-Ampt Equation

Let the constraints be defined by Eq. (11.4) with limits of b to c where b would tend to ∞, and c to 0, and

$$\int_{c}^{\infty} \ln(I - I_c) f(I) \, dI = \overline{\ln(I - I_c)} \tag{11.66}$$

Using POME and the method of Lagrange multipliers, $f(I)$ is obtained as

$$f(I) = \exp[-\lambda_0 - \lambda_1 \ln(I - I_c)] = (I - I_c)^{-\lambda_1} \exp(-\lambda_0) \tag{11.67}$$

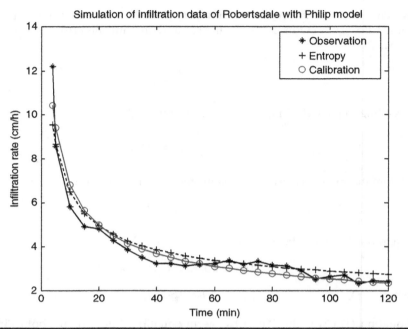

Figure 11.4 Comparison of infiltration rates computed using the Philip two-term equation with parameters determined using entropy theory and by the least-squares method (calibration) with observed infiltration rate values.

Substituting Eq. (11.67) in Eq. (11.3), one gets

$$\int_{c}^{b}(I-I_c)^{-\lambda_1}\exp(-\lambda_0)\,dI = 1 \Rightarrow \exp(\lambda_0) = \frac{(b-I_c)^{-\lambda_1+1}}{-\lambda_1+1} - \frac{(c-I_c)^{-\lambda_1+1}}{-\lambda_1+1} \tag{11.68}$$

Inserting Eq. (11.68) in Eq. (11.67), one obtains

$$f(I) = \frac{(-\lambda_1+1)(I-I_c)^{-\lambda_1}}{(b-I_c)^{-\lambda_1+1}-(a-I_c)^{-\lambda_1+1}} \tag{11.69}$$

Taking $\lambda_1 = 2$, Eq. (11.69) reduces to

$$f(I) = \frac{I_c}{(I-I_c)^2} \tag{11.70}$$

Equation (11.70) is the probability density function of the G-A equation. Note, however, that Eq. (11.70) is valid for $2I_c \leq I < \infty$, not for the entire quadrant in the euclidean space.

Combining Eq. (11.70) with Eq. (11.7), the result is

$$\frac{I_c DI}{(I - I_c)^2} = -\frac{1}{S}\frac{DJ}{DI} \tag{11.71}$$

Integrating with limits for I from 0 to I and for J from 0 to J,

$$\frac{I_c}{I - I_c} = -\frac{J}{S} \tag{11.72}$$

Recalling that $I = dJ/dt$, Eq. (11.72) can be expressed as

$$\frac{dJ}{dt} = \frac{SI_c}{J} + I_c \tag{11.73}$$

Solution of Eq. (11.73), with the condition that $t = 0, J = 0$, is

$$t = \frac{1}{I_c}\left[J - S\log\left(1 + \frac{J}{S}\right)\right] \tag{11.74}$$

Let

$$S = \frac{a}{I_c} \tag{11.75}$$

Then Eq. (11.74) can be expressed as

$$t = \frac{1}{I_c}\left[J - \frac{a}{I_c}\log\left(1 + \frac{J}{a/I_c}\right)\right] \tag{11.76}$$

Equation (11.76) is the Green-Ampt (G-A) equation in which parameter I_c can be interpreted as equal to the saturated hydraulic conductivity and parameter S equal to the product of the capillary suction at the wetting front and the initial moisture deficit.

The entropy of the Green-Ampt equation can be expressed as

$$H(I) = -\int_{2I_c}^{\infty} \frac{I_c}{(I - I_c)^2} \ln\frac{I_c}{(I - I_c)^2}\, dI = -\int_{2I_c}^{\infty} \frac{I_c}{(I - I_c)^2}[\ln I_c - 2\ln(I - I_c)]\, dI$$

$$H(I) = 2 + I_c \ln I_c - \ln I_c + \frac{2}{I_c} \tag{11.77}$$

Example 11.4. The G-A equation has two parameters, a and I_c, as shown in Eq. (11.76). From observations given in Example 11.1, compute the Green-Ampt two-term equation parameters, using entropy theory and the least-squares method. Compute the infiltration rate, using parameters estimated in these two ways, and compare these rates with observed rates.

Solution. For the G-A equation the value of S is 4.17 cm. Using the entropy method, the value of I_c is 2.42 cm/h and the value of a is 18.42; by the least-squares method and by minimizing the sum of squares of deviations between computed infiltration rates and observed infiltration rates, I_c is 1.49 and a is 7.72. Figure 11.5 shows the observed and computed infiltration rates.

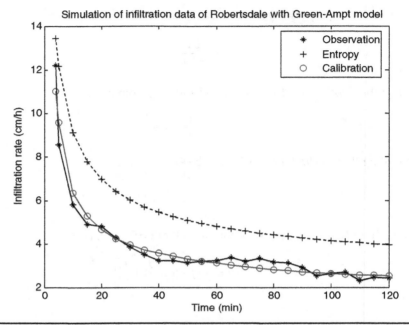

FIGURE 11.5 Comparison of infiltration rates computed using the Green-Ampt equation with parameters determined using the entropy theory and the least-squares method (calibration) with observed infiltration rate values.

11.2.5   Overton Equation

Let the constraints be defined by Eq. (11.3) and

$$\int \ln(I - I_c) f(I) dI = \overline{\ln(I - I_c)} \tag{11.78}$$

Using POME and the method of Lagrange multipliers, $f(I)$ is obtained as

$$f(I) = \exp[-\lambda_0 - \lambda_1 \ln(I - I_c)] = (I - I_c)^{-\lambda_1} \exp(-\lambda_0) \tag{11.79}$$

Substituting Eq. (11.79) in Eq. (11.2), one gets

$$\int_{I_c}^{I_0} (I - I_c)^{-\lambda_1} \exp(-\lambda_0) dI = 1 \Rightarrow \exp \lambda_0 = \frac{(I_0 - I_c)^{-\lambda_1 + 1}}{-\lambda_1 + 1} \tag{11.80}$$

Inserting Eq. (11.80) in Eq. (11.79) gives

$$f(I) = \frac{(-\lambda_1 + 1)(I - I_c)^{-\lambda_1}}{(I_0 - I_c)^{-\lambda_1 + 1}} \tag{11.81}$$

Taking $\lambda_1 = 0.5$, Eq. (11.81) becomes

$$f(I) = \frac{0.5(I - I_c)^{-0.5}}{(I_0 - I_c)^{0.5}} \tag{11.82}$$

Equation (11.82) is the probability density function of the Overton equation.
   Substituting Eq. (11.82) in Eq. (11.7), one obtains

$$-\frac{1}{S} dJ = \frac{0.5(I - I_c)^{-0.5}}{(I_0 - I_c)^{0.5}} dI \tag{11.83}$$

Integration of Eq. (11.83) yields

$$I = \frac{I_0 - I_c}{S^2} J^2 + I_c \tag{11.84}$$

Recalling Eq. (11.1), Eq. (11.85) with limits on t from t to t_c and on J from J to J_c (constant) gives

$$J = J_c - S \sqrt{\frac{I_c}{(I_0 - I_c)}} \tan\left[\frac{\sqrt{I_c(I_0 - I_c)}}{S}(t_c - t)\right] \tag{11.85}$$

Differentiating Eq. (11.85) leads to

$$I(t) = I_c \sec^2\left[\frac{\sqrt{I_c(I_0 - I_c)}}{S}(t_c - t)\right] \tag{11.86}$$

Let

$$I_0 - I_c = aS^2 \tag{11.87}$$

Then Eq. (11.86) becomes

$$I(t) = I_c \sec^2[\sqrt{aI_c}(t_c - t)] \tag{11.88}$$

Equation (11.88) is the Overton model.
   The entropy of the Overton model can be expressed as

$$H(I) = -0.5 \int_{I_c}^{I_0} \frac{(I - I_c)^{-0.5}}{(I_0 - I_c)^{0.5}} \ln \frac{0.5(I - I_c)^{-0.5}}{(I_0 - I_c)^{0.5}} dI = -1 - \ln 0.5 + \ln(I_0 - I_c) \tag{11.89}$$

Example 11.5.   The Overton equation has two parameters, a and I_c, and a time parameter t_c, as shown in Eq. (11.88). From observations given in Example 8.1, compute the Overton equation parameters, using the entropy theory and the least-squares method. Compute the infiltration rate, using parameters estimated in these two ways, and compare these rates with observed rates.

Solution.   For the Overton equation the value of S is 4.28 cm. Using the entropy method, the value of I_c is 3.1 cm/h and the value of a is 0.5; by the least-squares method and by minimizing the sum of squares of deviations between the computed infiltration rates and observed infiltration rates, I_c is 2.83 and a is 0.63. Figure 11.6 shows the observed and computed infiltration rates.

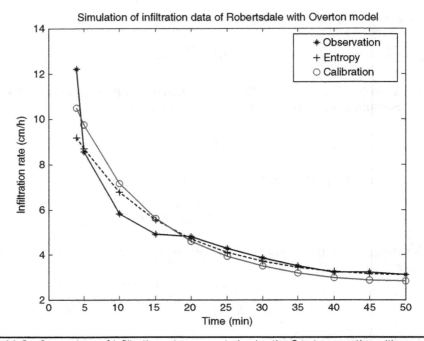

FIGURE 11.6   Comparison of infiltration rates computed using the Overton equation with parameters determined using the entropy theory and the least-squares method (calibration) with the observed infiltration rate values.

11.2.6   Holtan Equation

Let the constraints be defined by Eqs. (11.3) and (11.78). Using POME and the method of Lagrange multipliers, $f(I)$ is obtained as Eq. (11.79) and eventually Eq. (11.81):

$$f(I) = \frac{(-\lambda_1 + 1)(I - I_c)^{-\lambda_1}}{(I_0 - I_c)^{-\lambda_1 + 1}} = b(I - I_c)^{-\lambda_1} = bi^{-\lambda_1} \tag{11.90}$$

where

$$b = \frac{-\lambda_1 + 1}{(I_0 - I_c)^{-\lambda_1 + 1}} \qquad i = I - I_c \tag{11.91}$$

Let

$$-\lambda_1 = \frac{1 - m}{m} \tag{11.92}$$

Equation (11.90) is the probability density function of the corresponding infiltration rates by the Holtan equation and can be recast as

$$f(I) = bi^{-(1-m)/m} \tag{11.93}$$

and Eq. (11.91) as

$$b = \frac{1}{mi_0^{1/m}} \qquad i_0 = I_0 - I_c \tag{11.94}$$

Substituting Eq. (11.90) in Eq. (11.7), one obtains

$$dJ = -Sbmi^{(1-m)/m}di \tag{11.95}$$

Integration of Eq. (11.95) yields

$$S - J = j = Sbm \, i^{1/m} \qquad j = S - J \tag{11.96}$$

Equation (11.96) can be expressed as

$$\frac{1}{sbm} j^m = -\frac{dj}{dt} \tag{11.97}$$

Integrating Eq. (11.97), one obtains

$$j = \left(S^{1-m} - \frac{1-m}{sbm} t \right)^{\frac{1}{1-m}} \tag{11.98}$$

Differentiation of Eq. (11.98) with respect to t and simplification yield

$$I = I_c + \frac{1}{S(I_0 - I_c)^{1/m}} \left[S^{1-m} - \frac{1-m}{S(I_0 - I_c)^{1/m}} t \right]^{\frac{m}{1-m}} \tag{11.99}$$

Let

$$a = \frac{1}{S(I_0 - I_c)^{1/m}} \tag{11.100}$$

Equation (11.99) becomes

$$I = I_c + a[S^{1-m} - a(1-m)t]^{\frac{m}{1-m}} \tag{11.101}$$

Equation (11.101) is the Holtan equation with parameter a given by Eq. (11.100). The entropy of the Holtan model can be computed as follows:

$$H(I) = -\frac{1}{m} \int_{I_c}^{I_a} \frac{(I - I_c)^{(1-m)/m}}{(I_0 - I_c)^{1/m}} \ln \frac{(I - I_c)^{-(1-m)/m}}{(I_0 - I_c)^{1/m}} dI = -\frac{1}{2} - \frac{\ln 2}{3} + \ln(I_0 - I_c) \tag{11.102}$$

Example 11.6.   The Holtan equation has three parameters, a, m, and I_c. From observations given in Example 11.1, compute the Holtan equation parameters using entropy theory and the least-squares method. Compute the infiltration rate, using parameters estimated in these two ways, and compare these rates with the observed rates.

Solution.   For the Holtan equation the value of S is 4.28 cm. Using the entropy method, the value of I_c is 3.1 cm/h, the value of a is 1.03, and the value of $m = 1.5$; by the least-squares method and by minimizing the sum of squares of deviations between computed infiltration rates and observed infiltration rates, I_c is 2.51, a is 1.10, and m is 1.5. Figure 11.7 shows observed and computed infiltration rates.

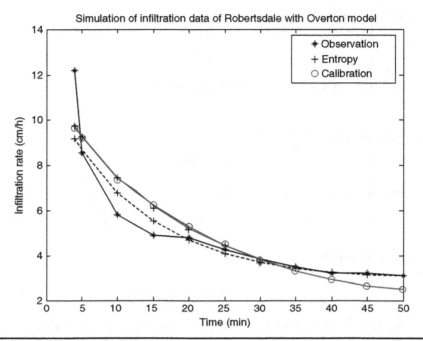

Simulation of infiltration data of Robertsdale with Overton model

Figure 11.7   Comparison of infiltration rates computed using the Holtan equation with parameters determined using the entropy theory and the least-squares method (calibration) with the observed infiltration rate values.

11.2.7   Singh-Yu Equation

The Singh-Yu infiltration equation (Singh and Yu, 1990) is expressed as

$$I(t) = I_c + a\frac{(S-J)^m}{J^n} \tag{11.103}$$

where a is a constant, S is a parameter, and m and n are exponents. For simplicity, it is assumed that $m = n = 1$. Equation (11.103) can then be simplified as

$$I(t) = I_c + a\frac{S-J}{J} \tag{11.104}$$

Substituting Eq. (11.104) in Eq. (11.1), one obtains

$$\frac{dJ(t)}{dt} = a\frac{S-J(t)}{J} \tag{11.105}$$

Integration of Eq. (11.105) yields

$$\int_0^{J(t)} \frac{J(t)}{S-J(t)}\,dJ = at \tag{11.106}$$

which yields

$$t = \frac{S}{a}\left(\frac{S-J}{S} - \ln\frac{S}{S-J} - 1\right)$$ (11.107)

Equation (11.107) can be considered as a parametric model of $I(t)$ with S as the characteristic parameter. This model is actually a combination of the Horton equation and the Green-Ampt equation (or the Philip two-term equation). When the denominator is removed from Eq. (11.104), it leads to the Horton equation. If the numerator is removed from Eq. (11.104), it leads to the Philip two-term equation. Equations (11.103) and (11.107) can be expressed in implicit form with t as function of $I(t)$. Equation (11.104) can be written as

$$J(t) = \frac{S[I(t) - I_c)]}{a + I(t) - I_c}$$ (11.108)

Substitution of Eq. (11.108) in Eq. (11.107) produces

$$J = \frac{S}{a}\left(\frac{I-I_c}{a+I-I_c} - \ln\frac{I-I_c}{a+I-I_c} - 1\right)$$ (11.109)

Following Singh (2011a,b), the constraints for the Singh-Yu equation can be defined as

$$C_0 = \int_b^c f(I)dl = 1$$ (11.110)

and

$$C_1 = \int_b^c \ln(a + I - I_c)f(I)dl = \overline{\ln(a + I - I_c)}$$ (11.111)

Equation (11.110) is the total probability law.

Using of the method of Lagrange multipliers for maximizing entropy subject to Eqs. (11.110) and (11.111), the probability density function of I is

$$f(I) = \exp[-\lambda_0 - \lambda_1 \ln(a + I - I_c)] = \exp(-\lambda_0)(a + I - I_c)^{-\lambda_1}$$ (11.112)

where the Lagrange multipliers $\lambda_i, i = 0,1$, can be determined with the use of Eqs. (11.110) and (11.111). Integration of Eq. (11.112) leads to the cumulative distribution function or simple probability distribution of I, $F(I)$:

$$F(I) = \frac{\exp(-\lambda_0)}{-\lambda_1 + 1}[(a + I - I_c)^{-\lambda_1+1} - (a + b - I_c)^{-\lambda_1+1}]$$ (11.113)

Through substitution of Eq. (11.112), the maximum entropy of $f(I)$ or I is

$$H = \lambda_0 + \lambda_1 \overline{\ln(a + I - I_c)}$$ (11.114)

Equation (11.114) shows that the entropy of the probability distribution of the infiltration rate or of the rate of infiltration itself depends only on the constraints, since the Lagrange multipliers themselves can be expressed in terms of the specified constraints.

Equation (11.112) contains two unknown Lagrange multipliers λ_0 and λ_1 which can be determined as follows. Substituting Eq. (11.112) in Eq. (11.110), one gets

$$\int_b^c (a + I - I_c)^{-\lambda_1} \exp(-\lambda_0) dl = 1 \Rightarrow \exp \lambda_0 = \frac{(a + c - I_c)^{-\lambda_1 + 1} - (a + b - I_c)^{-\lambda_1 + 1}}{\lambda_1 - 1} \tag{11.115}$$

Equation (11.115) can be expressed as

$$\lambda_0 = \ln[(a + c - I_c)^{-\lambda_1 + 1} - (a + b - I_c)^{-\lambda_1 + 1}] - \ln(\lambda_1 - 1) \tag{11.116}$$

Equation (11.116) expresses λ_0 as a function of λ_1. Differentiation of Eq. (11.116) with respect to λ_1 leads to

$$\frac{\partial \lambda_0}{\partial \lambda_1} = \frac{(a + c - I_c)^{-\lambda_1 + 1} \ln(a + c - I_c) - (a + c - I_c)^{-\lambda_1 + 1} \ln(b - I_c)}{(a + c - I_c)^{-\lambda_1 + 1} - (a + b - I_c)^{-\lambda_1 + 1}} - \frac{1}{\lambda_1 - 1} \tag{11.117}$$

On the other hand, Eq. (11.115) can also be written as

$$\lambda_0 = \ln \int_b^c (a + I - I_c)^{-\lambda_1} dl \tag{11.118}$$

Differentiating Eq. (11.118) and recalling the definition of the PDF of I, one finds

$$\frac{\partial \lambda_0}{\partial \lambda_1} = -\frac{\int_b^c (a + I - I_c)^{-\lambda_1} \exp(-\lambda_0) \ln(a + I - I_c) dl}{\int_b^c (a + I - I_c)^{-\lambda_1} \exp(-\lambda_0) dl} = -\overline{\ln(a + I - I_c)} \tag{11.119}$$

Equating Eq. (11.117) to Eq. (11.119), one obtains

$$\frac{(a + c - I_c)^{-\lambda_1 + 1} \ln(a + c - I_c) - (a + b - I_c)^{-\lambda_1 + 1} \ln(a + b - I_c)}{(a + c - I_c)^{-\lambda_1 + 1} - (a + c - I_c)^{-\lambda_1 + 1}} - \frac{1}{\lambda_1 - 1} = -\overline{\ln(a + I - I_c)} \tag{11.120}$$

Equation (11.120) expresses λ_1 in terms of the constraint $\overline{\ln(a + I - I_c)}$ and the limits of integration b and c. Unfortunately the relation is not explicit, but it can be easily solved numerically.

Inserting Eq. (11.116) in Eq. (11.112) and taking $c \to \infty$, $b \to I_c$, and $\lambda_1 = 2$, one obtains

$$f(I) = \frac{a}{(a + I - I_c)^2} \tag{11.121}$$

Equation (11.121) gives the PDF underlying the Singh-Yu (S-Y) infiltration equation. The choice of $\lambda_1 = 2$ leads to the desired S-Y equation. Then the CDF is given by integrating Eq. (11.121):

$$f(I) = 1 - \frac{a}{a + I - I_c} \tag{11.122}$$

The entropy of I can now be written as

$$H(I) = -\ln a + 2\overline{\ln(a + I - I_c)} \tag{11.123}$$

Equation (11.123) shows that higher infiltration rates exhibit higher entropy and hence higher uncertainty. Infiltration is higher in the beginning and decreases as time progresses, eventually attaining a steady-state value or tending to zero. This means that measurements of infiltration should be more closely spaced in time in the beginning, and the time interval of measurement can increase with the progression of time. Thus, it will be prudent to keep the interval of measurement variable.

Inserting Eq. (11.121) in Eq. (11.6a) gives

$$\frac{1}{S}\frac{dJ}{dI} = -\frac{a}{(a + I - I_c)^2} \tag{11.124}$$

Integrating Eq. (11.124) with the condition that $J = 0$ when $I \to \infty$, one gets

$$J = \frac{aS}{a + I - I_c} \tag{11.125}$$

Equation (11.125) can be written as

$$a = (a + I - I_c)\frac{J}{S} \tag{11.126}$$

Inserting Eq. (11.124) in Eq. (11.126), one obtains

$$\frac{dJ}{dt} - \frac{aS}{J} = -a \tag{11.127}$$

Integrating Eq. (11.127) with the condition that $J = 0$ at $t = 0$ gives

$$t = \frac{1}{a}\left[S\ln\left(\frac{S}{S - J}\right) - J\right] \tag{11.128}$$

Differentiating Eq. (11.128) and then equating to the continuity equation, Eq. (11.5), one obtains

$$\frac{a + I - I_c}{I - I_c} = \frac{S}{S - J} \tag{11.129}$$

Substituting Eq. (11.129) in Eq. (11.128) gives Eq. (11.109)—the Singh-Yu equation.

Example 11.7. For the data in Example 11.1, compute the parameters of the Singh-Yu equation, using the entropy theory and the least-squares method. Compare the computed infiltration rate with the observed rates.

Solution. The value of parameter a was found close to 1 cm/h. The agreement between the computed and observed infiltration rates was close for the data. Figure 11.8 shows the infiltration rates computed by the Singh-Yu infiltration equation along with the observed infiltration rates. However, note that there was no calibration of parameters. Even then the infiltration equation yielded quite close infiltration values. This is just to show that the equation does have potential. The value of entropy was 1.52 bits.

11.2.8   General Equation

Singh (2011b) has presented a general framework for deriving the Horton, Kostiakov, Green-Ampt, Philip two-term, Holtan, and Overton infiltration equations discussed in the preceding sections. The general framework leads to these equations and can obviate the need for deriving them separately. The key to the general framework is the derivation of (1) a general probability density function for infiltration and (2) a general relation between cumulative infiltration and infiltration capacity. The general probability density function can specialize into probability density functions associated with individual equations.

For the general framework, the constraints C_i, $i = 0, 1$, can be defined as Eq. (11.110) and

$$C_1 = \int_b^c \ln(I - I_c) f(I) dl = \overline{\ln(I - I_c)} \tag{11.130}$$

Equation (11.130) defines the mean of the logarithm of infiltration capacity in excess of the steady infiltration rate.

Maximization of entropy, subject to Eqs. (11.110) and (11.130), using the method of Lagrange multipliers, leads to the probability density function of I in terms of given constraints

$$f(I) = \exp[-\lambda_0 - \lambda_1 \ln(I - I_c)] = \exp(-\lambda_0)(I - I_c)^{-\lambda_1} \tag{11.131}$$

where the Lagrange multipliers $\lambda_r, r = 0, 1$, can be determined with the use of Eqs. (11.110) and (11.130). Integration of Eq. (11.131) leads to the cumulative distribution function or simply the probability distribution of I, $F(I)$:

$$F(I) = \frac{\exp(-\lambda_0)}{-\lambda_1 + 1}[(I - I_c)^{-\lambda_1+1} - (b - I_c)^{-\lambda_1+1}] \tag{11.132}$$

Substitution of Eq. (11.131) in Eq. (11.110) results in the maximum entropy of $f(I)$ or I:

$$H = \lambda_0 + \lambda_1 \overline{\ln(I - I_c)} \tag{11.133}$$

Equation (11.133) shows that the entropy of the probability distribution of infiltration capacity or of the infiltration capacity itself depends only on the constraints, since the Lagrange multipliers themselves can be expressed in terms of the specified constraints.

Equation (11.131) contains the Lagrange multipliers λ_0 and λ_1 which can be determined as follows. Substituting Eq. (11.131) in Eq. (11.110), one gets

$$\int_b^c (I - I_c)^{-\lambda_1} \exp(-\lambda_0) dl = 1 \Rightarrow \exp \lambda_0 = \frac{(c - I_c)^{-\lambda_1+1} - (b - I_c)^{-\lambda_1+1}}{-\lambda_1 + 1} \tag{11.134}$$

Equation (11.123) can be expressed as

$$\lambda_0 = \ln[(c - I_c)^{-\lambda_1+1} - (b - I_c)^{-\lambda_1+1}] - \ln(-\lambda_1 + 1) \tag{11.135}$$

Equation (11.135) expresses λ_0 as a function of λ_1, and its differentiation with respect to λ_1 leads to

$$\frac{\partial \lambda_0}{\partial \lambda_1} = \frac{(c-I_c)^{-\lambda_1+1}\ln(c-I_c) - (b-I_c)^{-\lambda_1+1}\ln(b-I_c)}{(c-I_c)^{-\lambda_1+1} - (b-I_c)^{-\lambda_1+1}} - \frac{1}{\lambda_1-1} \tag{11.136}$$

On the other hand, Eq. (11.134) can also be written as

$$\lambda_0 = \ln \int_b^c (I-I_c)^{-\lambda_1}\, dl \tag{11.137}$$

Differentiating Eq. (11.137), one finds

$$\frac{\partial \lambda_0}{\partial \lambda_1} = -\frac{\int_b^c (I-I_c)^{-\lambda_1}\ln(I-I_c)dl}{\int_b^c (I-I_c)^{-\lambda_1}\, dl} \tag{11.138}$$

Multiplying and dividing Eq. (11.138) by $\exp(-\lambda_0)$, one obtains

$$\frac{\partial \lambda_0}{\partial \lambda_1} = -\frac{\int_b^c (I-I_c)^{-\lambda_1}\exp(-\lambda_0)\ln(I-I_c)dl}{\int_b^c (I-I_c)^{-\lambda_1}\exp(-\lambda_0)dl} = -\overline{\ln(I-I_c)} \tag{11.139}$$

Equating Eq. (11.139) to Eq. (11.136), one obtains

$$\frac{(c-I_c)^{-\lambda_1+1}\ln(c-I_c) - (b-I_c)^{-\lambda_1+1}\ln(b-I_c)}{(c-I_c)^{-\lambda_1+1} - (b-I_c)^{-\lambda_1+1}} - \frac{1}{\lambda_1-1} = -\overline{\ln(I-I_c)} \tag{11.140}$$

Equation (11.140) expresses λ_1 in terms of the constraint $\overline{\ln(I-I_c)}$ and the limits of integration b and c. Unfortunately the relation is not explicit, but it can be easily solved numerically.

Inserting Eq. (11.135) in Eq. (11.131), one obtains

$$f(I)\frac{(-\lambda_1+1)(I-I_c)^{-\lambda_1}}{(c-I_c)^{-\lambda_1+1} - (b-I_c)^{-\lambda_1+1}} \tag{11.141}$$

Equation (11.141) gives the PDF corresponding to the constraints given by Eqs. (11.110) and (11.130). It may underlie a general infiltration equation which has yet to be determined. The general CDF is given by the integration of Eq. (11.141) as

$$F(I) = \frac{(I-I_c)^{-\lambda_1+1} - (b-I_c)^{-\lambda_1+1}}{(c-I_c)^{-\lambda_1+1} - (b-I_c)^{-\lambda_1+1}} \tag{11.142}$$

The entropy can now be written as

$$H(I) = \ln[(c-I_c)^{-\lambda_1+1} - (b-I_c)^{-\lambda_1+1}] - \ln(-\lambda_1+1) + \lambda_1\overline{\ln(I-I_c)} \tag{11.143}$$

Both Eqs. (11.141) and (11.142) as well as Eq. (11.143) contain only one unknown λ_1 which can be determined using Eq. (11.140).

Now for deriving a general infiltration equation, Eq. (11.141) is inserted in Eq. (11.7), and the result is

$$\frac{1}{S}\frac{dJ}{dI} = -\frac{(-\lambda_1+1)(I-I_c)^{-\lambda_1}}{(c-I_c)^{-\lambda_1+1}-(b-I_c)^{-\lambda_1+1}} \tag{11.144}$$

Integrating Eq. (11.144) with the condition that $J = 0$ when $I = c$, one gets

$$J = S\frac{(c-I_c)^{-\lambda_1+1}-(I-I_c)^{-\lambda_1+1}}{(c-I_c)^{-\lambda_1+1}-(b-I_c)^{-\lambda_1+1}} \tag{11.145}$$

Equation (11.145) expresses the relation between infiltration capacity and cumulative infiltration. Recalling the continuity equation, Eq. (11.1), Eq. (11.145) can be written as

$$\frac{dJ}{dt} = \left\{(c-I_c)^{-\lambda_1+1} - \frac{J}{S}[(c-I_c)^{-\lambda_1+1}-(b-I_c)^{-\lambda_1+1}]\right\}^{\frac{1}{-\lambda_1+1}} \tag{11.146}$$

By integration of Eq. (11.146) and then differentiation with respect to time and use of Eq. (11.144), the general infiltration equation can be expressed as

$$I(t) = I_c + \frac{-\lambda_1+1}{\lambda_1}\frac{S}{(c-I_c)^{-\lambda_1+1}-(b-I_c)^{-\lambda_1+1}}$$

$$\left(\frac{\lambda_1}{-\lambda_1+1}\frac{(c-I_c)^{-\lambda_1+1}-(b-I_c)^{-\lambda_1+1}}{S}\times\left[t+\frac{S(-\lambda_1+1)(c-I_c)^{-\lambda_1+1}}{\lambda_1[(c-I_c)^{-\lambda_1+1}-(b-I_c)^{-\lambda_1+1}]}\right]\right)^{1/\lambda_1} \tag{11.147}$$

Equation (11.147) is the entropy-based general infiltration equation corresponding to the constraints given by Eqs. (11.110) and (11.130), with the Lagrange multiplier λ_1 determined from Eq. (11.140). Equation (11.147) will be true if λ_1 is less than unity. Note that it is not possible to directly derive the aforementioned six infiltration equations from Eq. (11.147), for different equations have different limits of integration and different values of λ_1. Nevertheless, it is instructive to derive some special cases of Eq. (11.147).

Horton Equation
Let $c = I_0$ and $b = I_c$. Then Eq. (11.145) yields

$$J = S\frac{(I_0-I_c)^{-\lambda_1+1}-(I-I_c)^{-\lambda_1+1}}{(I_0-I_c)^{-\lambda_1+1}} \tag{11.148}$$

Let $\lambda_1 = 0$. Then Eq. (11.148) becomes

$$J = S\frac{(I_0-I_c)-(I-I_c)}{(I_0-I_c)} = S\left(1-\frac{I-I_c}{I_0-I_c}\right) \tag{11.149}$$

Taking advantage of the continuity Eq. (11.1), one can write Eq. (11.149) as

$$S \frac{dJ}{dt} = (I_0 - I_c)(S - J) \tag{11.150}$$

Integrating Eq. (11.150) with the condition that $J = 0$ at $t = 0$, one gets

$$J = S \left[1 - \exp \left(-\frac{I_0 - I_c}{S} t \right) \right] \tag{11.151}$$

Differentiating Eq. (11.151) with respect to time and equating it to Eq. (11.5), one obtains the Horton equation.

Kostiakov Equation
Let $b = 0$ and $\lambda_1 = 2$. Equation (8.145) becomes

$$J = S \frac{(c - I_c)^{-1} + (I - I_c)^{-1}}{(c - I_c)^{-1} + I_c^{-1}} \tag{11.152}$$

Now let $c \to \infty$. Then Eq. (11.152) becomes

$$\frac{J}{S} = \frac{I_c}{I - I_c} \tag{11.153}$$

With the use of Eq. (11.1), Eq. (11.153) can be written as

$$(I - I_c)J = SI_c \quad \text{or} \quad J \frac{dJ}{dt} = SI_c \tag{11.154}$$

Integration of Eq. (11.154), with the condition that $J = 0$ when $t = 0$, results in

$$J = (2I_c S)^{0.5} t^{0.5} \tag{11.155}$$

Differentiation of Eq. (11.155) leads to the Kostiakov equation.

Philip Two-Term Equation
Differentiating of Eq. (11.155) and making use of Eq. (11.1), one gets the Philip equation.

Green-Ampt Equation
Let $b = 0$ and $\lambda_1 = 2$ and $c \to \infty$. Following the same steps as above, Eq. (11.148) can be written as

$$\frac{J}{S} = \frac{I_c}{I - I_c} \tag{11.156}$$

In the continuity equation, Eq. (11.1), it is assumed that I_c tends to 0. Then Eq. (11.156) can be written as

$$\frac{dJ}{dt} = I_c + \frac{SI_c}{J}$$

(11.157)

Integration of Eq. (11.157) with the condition that $J = 0$ at $t = 0$ produces the Green-Ampt equation.

Holtan Equation

Let $-\lambda_1 + 1 = 1/m$, $b = I_c$, and $c = I_0$. Then Eq. (11.145) becomes

$$J = S\frac{(I_0 - I_c)^{1/m} + (I - I_c)^{1/m}}{(I_0 - I_c)^{1/m}}$$

(11.158)

where m is an exponent. Equation (11.158) can be written as

$$J = S\left[1 - \frac{(I - I_c)^{1/m}}{(I_0 - I_c)^{1/m}}\right]$$

(11.159)

Coupling Eq. (11.159) with Eq. (11.1), one obtains

$$\frac{dJ}{dt} = (I_0 - I_c)\frac{(S - J)^m}{S^m}$$

(11.160)

Integrating Eq. (11.160) with the condition that $J = 0$ at $t = 0$, one finds

$$J = S - \left[S^{-m+1} - \frac{(I_0 - I_c)(-m + 1)}{S^m}\right]^{\frac{1}{-m+1}}$$

(11.161)

Differentiating of Eq. (11.161) and using Eq. (11.1), one gets the Holtan equation.

Overton Equation

Let $\lambda_1 = 0.5$, $b = I_c$, and $c = I_0$. Then Eq. (11.145) becomes

$$J = S\frac{(I_0 - I_c)^{0.5} - (I - I_c)^{0.5}}{(I_0 - I_c)^{0.5}}$$

(11.162)

Coupling Eq. (11.162) with Eq. (11.11) with I_c tending to 0 leads to

$$\frac{dJ}{I_c + (I_0 - I_c)(1 - J/S)^2} = dt$$

(11.163)

Integration of Eq. (11.163) with the condition that $J = S$ when $t = t_c$ (t_c is a value of time) and a little algebraic manipulation yield

$$J = S - \sqrt{\frac{I_c S^2}{I_0 - I_c}} \tan\left[\sqrt{\frac{(I_0 - I_c)I_c}{S^2}}(t_c - t)\right]$$

(11.164)

Differentiating Eq. (11.164), one gets the Overton equation.

Questions

1. Obtain four data sets (labeled I, II, III, and IV) on infiltration in Robertsdale loamy sand, Stilson loamy sand, and Troupe sand in the Georgia Coastal Plain, collected by Rawls et al. (1976) and published in a report by the Agriculture Research Service of the U.S. Department of Agriculture. Characteristics of these infiltration observation data sets are given in Table 11.1. In the table, I_c is the constant (steady) rate of infiltration almost near the end of the infiltration experiment or the duration D; I_c' is the constant rate of infiltration before the end of the experiment; I_0 is the initial infiltration rate given a few minutes later than the start of infiltration ($t = 0$); I_0' is the adjusted initial infiltration rate for the Horton equation; t_c is the time to an approximately constant rate of infiltration where t_c may be less than the duration of the experiment D; and S is the maximum soil moisture retention until time t_c.

   For data set I, the infiltration rate reached a lower value at $t = 50$ min and thereafter fluctuated around 3.10 cm/h (1.22 in./h) [corresponding to the cumulative infiltration of 4.28 cm (1.686 in.)]. Thus in this case $t_c = 50$ min and $I_c = 1.22$ in./h. The initial infiltration rate at $t = 4$ min was 12.21 cm/h (4.807 in./h). It was therefore assumed that the constant infiltration rate I_c' was 3.10 cm/h (1.22 in./h). The actual initial infiltration rate (at $t = 0$) should be larger than the 12.21 (cm/h) (4.807 in./h) which is the value at $t = 4$ min. Extrapolation yielded the initial infiltration rate $I_c' = 14.0$ cm/h at $t = 0$. Since the connotation of parameter S may differ from one infiltration equation to another, it may have different values for different equations. Therefore, S_1 was used to denote parameter S for the Horton model, which is determined by the final soil moisture minus the initial soil moisture; and S_2 was used to denote parameter S for the Kostiakov, Philip, and Green-Ampt equations, and which was equal to the final accumulated infiltration. Likewise, S_3 was used to denote parameter S for the Overton and Holtan equations, which was equal to the accumulated infiltration until t_c. These parameter values were obtained from observations are given in Table 11.2. In a similar manner, values of I_c, I_c', I_0, I_c', S_1, S_2, S_3, and t_c were obtained for data sets II, III, and IV, as shown in Table 11.2. Observed values of infiltration are given in Table 11.3.
   Express the cumulative infiltration curve of the Horton equation. For data set I, construct this curve. Give the values of the curve used for its construction.

2. Compute the infiltration rate by the Horton equation for data set II, and compare with the observed values. Compute the relative error.

3. Compute the infiltration rate by the Horton equation for data set III, and compare with the observed values. Compute the relative error.

4. Compute the infiltration rate by the Horton equation for data set IV, and compare with the observed values. Compute the relative error.

5. Determine the sensitivity of the Horton equation to parameter S.

6. Compute the infiltration rate by the Kostiakov equation for data set I, and compare with the observed values. Compute the relative error.

7. Compute the infiltration rate by the Kostiakov equation for data set II, and compare with the observed values. Compute the relative error.

Soil Type	Code	ID	I_c at D (cm/h)	I_c at t_c (cm/h)	I_0 (cm/h)	I_0' (cm/h)	S_1 (cm)	S_2 (cm)	S_3 (cm)	t_c (min)	Duration of Observations D
Robertsdale loamy sand	I	09091D	2.42	3.10	12.21	14	7.61	4.17	4.28	50	120
Robertsdale loamy sand	II	09091W	2.25	1.93	8.24	14	4.90	0.76	2.40	50	120
Stilson loamy sand	III	10101W	2.89	2.96	12.81	8	7.04	1.68	4.99	50	91
Troupe sand	IV	12112D	4.38	4.38	11.60	14	12.14	2.59	11.21	90	90

TABLE 11.2 Parameters from Observations (after Rawls, 1976)

Data Sets I and II

Time from Start of Rain (min)	Infiltration Rate (in./h) Data Set I: 09091D	Infiltration Rate (in./h) Data Set II: 09091W
4	4.807	3.245
5	3.365	2.163
10	2.289	1.112
15	1.922	1.023
20	1.881	0.873
25	1.680	0.804
30	1.518	0.785
35	1.380	0.796
40	1.264	0.779
45	1.270	0.798
50	1220	0.758
55	1.253	0.775
60	1.265	0.789
65	1.321	0.802
70	1.252	0.839
75	1.303	0.869
80	1.243	0.871
85	1.226	0.822
90	1.136	0.857
95	0.989	0.956
100	1.025	0.936
105	1.055	0.913
110	0.910	0.972
115	0.961	0.919
120	0.953	0.886

Data Sets III and IV

Time from Start of Rain (min)	Infiltration Rate (in./h) Data Set III: 10101W	Infiltration Rate (in./h) Data Set IV: 12112D
5	5.047	6.249 (at 6 min)
10	4.447	1.690
15	2.747	1.653
20	2.048	1.879
25	1.643	1.888
30	1.432	1.724
35	1.366	1.734
40	1.231	1.751
45	1.270	1.755
50	1.165	1.785
55	1.225	1.763
60	1.254	1.756
65	1.204	1.785
70	1.254	1.845
75	1.284	1.744
80	1.184	1.843
85	1.181	1.720
90	1.167	1.630
91	1.168	

TABLE 11.3 Experimental Field Infiltration Observations (after Rawls et al., 1976)

8. Compute the infiltration rate by the Kostiakov equation for data set III, and compare with the observed values. Compute the relative error.

9. Compute the infiltration rate by the Kostiakov equation for data set IV, and compare with the observed values. Compute the relative error.

10. Compute the probability density function associated with the Kostiakov equation, and plot it.

11. Compute the entropy function of the Kostiakov equation, and plot it.

12. Compute the infiltration rate by the Philip equation for data set II, and compare with the observed values. Compute the relative error.

13. Compute the infiltration rate by the Philip equation for data set II, and compare with the observed values. Compute the relative error.

14. Compute the infiltration rate by the Philip equation for data set III, and compare with the observed values. Compute the relative error.

15. Compute the infiltration rate by the Philip equation for data set IV, and compare with the observed values. Compute the relative error.

16. Compute the probability density function associated with the Philip equation, and plot it.

17. Compute the entropy function of the Philip equation, and plot it.

18. Compute the infiltration rate by the G-A equation for data set I, and compare with the observed values. Compute the relative error.

19. Compute the infiltration rate by the G-A equation for data set II, and compare with the observed values. Compute the relative error.

20. Compute the infiltration rate by the G-A equation for data set III, and compare with the observed values. Compute the relative error.

21. Compute the infiltration rate by the G-A equation for data set IV, and compare with the observed values. Compute the relative error.

22. Compute the probability density function associated with the G-A equation, and plot it.

23. Compute the entropy function of the G-A equation, and plot it.

24. Compute the infiltration rate by the Overton equation for data set I, and compare with the observed values. Compute the relative error.

25. Compute the infiltration rate by the Overton equation for data set II, and compare with the observed values. Compute the relative error .

26. Compute the infiltration rate by the Overton equation for data set III, and compare with the observed values. Compute the relative error.

27. Compute the infiltration rate by the Overton equation for data set IV, and compare with the observed values. Compute the relative error.

28. Compute the probability density function associated with the Overton equation, and plot it.

29. Compute the entropy function of the Overton equation, and plot it.

30. Compute the infiltration rate by the Holtan equation for data set I, and compare with the observed values. Compute the relative error.

31. Compute the infiltration rate by the Holtan equation for data set II, and compare with the observed values. Compute the relative error.

32. Compute the infiltration rate by the Holtan equation for data set III, and compare with the observed values. Compute the relative error.

33. Compute the infiltration rate by the Holtan equation for data set IV, and compare with the observed values. Compute the relative error.

34. Compute the probability density function associated with the Holtan equation, and plot it.

35. Compute the entropy function of the Holtan equation, and plot it.

36. Compute the infiltration rate by the Singh-Yu equation for data set I, and compare with the observed values. Compute the relative error.

37. Compute the infiltration rate by the Singh-Yu equation for data set II, and compare with the observed values. Compute the relative error.

38. Compute the infiltration rate by the Singh-Yu equation for data set III, and compare with the observed values. Compute the relative error.

39. Compute the infiltration rate by the Singh-Yu equation for data set IV, and compare with the observed values. Compute the relative error.

40. Compute the probability density function associated with the Singh-Yu equation, and plot it.

41. Compute the entropy function of the Singh-Yu equation, and plot it.

42. Compare the infiltration rates by the seven equations for data set I, and compare with the observed values. Compute the relative error.

43. Compare the infiltration rates by the seven equations for data set II, and compare with the observed values. Compute the relative error .

44. Compare the infiltration rates by the seven equations for data set III, and compare with the observed values. Compute the relative error.

45. Compare the infiltration rates by the seven equations for data set IV, and compare with the observed values. Compute the relative error.

References

Crawford, N. H., and Linsley R. K. (1966). Digital simulation in hydrology: Stanford watershed model IV. *Technical Report No. 39*, Stanford University, Palo Alto, Calif.

Donigian, A. S., and Imhoff J. (2006). History and evolution of watershed modeling derived from the Stanford watershed model. In: *Watershed Models*, edited by V. P. Singh and D. K. Frevert, CRC Press, Boca Raton, Fla. pp. 21–45.

Green, W. H., and Ampt C. A. (1911). Studies on soil physics, I. Flow of air and water through soils. *Journal of Agricultural Sciences*, vol. 4, pp. 1–24.

Holtan, H. N. (1961). A concept of infiltration estimates in watershed engineering. ARS41-51, U.S. Department of Agriculture, Agricultural Research Service, Washington, D.C.

Horton, R. I. (1938). The interpretation and application of runoff plot experiments with reference to soil erosion problems. *Proceedings, Soil Science Society of America Proceedings*, vol. 3, pp. 340–349.

Kostiakov, A. N. (1932). On the dynamics of the coefficient of water percolations in soils. Sixth Commission, International Society of Soil Science, Part A, pp. 15–21.

Overton, D. E. (1964). Mathematical refinement of an infiltration equation for watershed engineering. ARS41-99, U.S. Department of Agriculture, Agricultural Research Service, Washington, D.C.

Philip, J. R. (1957). Theory of infiltration, Parts 1 and 4. *Soil Science*, vol. 85, no. 5, pp. 345–357.

Rawls, W., Yates, P. and Asmussen L. (1976). Calibration of selected infiltration equations for the Georgia plain. ARS-S-113, U.S. Department of Agriculture, Agricultural Research Service, New Orleans, La.

Singh, V. P., ed. (1995). *Computer Models of Watershed Hydrology*. Water Resources Publications, Littleton, Colo.

Singh, V. P. (1998). *Entropy-Based Parameter Estimation in Hydrology*. Kluwer Academic Publishers, Boston.

Singh, V. P. (2010a). Entropy theory for derivation of infiltration equations. *Water Resources Research*, vol. 46, pp. 1–20. W03527, doi: 10.1029/2009WR008193.

Singh, V. P. (2010b). Tsallis entropy theory for derivation of infiltration equations. *Transactions of the American Society of Agricultural and Biological Engineers*, vol. 53, no. 2, pp. 447–463.

Singh, V. P. (2011a). Derivation of the Singh-Yu infiltration equation using entropy theory. *Journal of Hydrologic Engineering, ASCE*, vol. 16, no. 2, pp. 187–191.

Singh, V. P. (2011b). A Shannon entropy-based general derivation of infiltration equations. *Transactions of the American Society of Agricultural and Biological Engineers*, vol. 54, no. 1, pp. 123–129.

Singh, V. P., and Frevert D. K. eds. (2002a). *Mathematical Models of Large Watershed Hydrology*. Water Resources Publications, Highlands Ranch, Colo.

Singh, V. P., and Frevert D. K. eds. (2002b). *Mathematical Models of Small Watershed Hydrology and Applications*. Water Resources Publications, Highlands Ranch, Colo.

Singh, V. P., and Frevert D. K. eds. (2006). *Watershed Models*. CRC Press, Boca Raton, Fla.

Singh, V. P., and Woolhiser D. A. (2002). Mathematical modeling of watershed hydrology. *Journal of Hydrologic Engineering*, vol. 7, no. 4, pp. 270–292.

Singh, V. P., and Yu F. X. (1990). Derivation of infiltration equation using systems approach. *Journal of Irrigation and Drainage Engineering*, vol. 116, no. 6, pp. 837–858.

Woodbury, A. D. (2012). Comment on Entropy theory for derivation of infiltration equations by V. P. Singh. Water Resources Research, vol. 48, no. 4, DOI: 10.1029/2012WR012157.

Woodbury, A. D., and Ulrych T. J. (1996). Minimum relative entropy inversion: Theory and application to recovering the release history of a groundwater contaminant. *Water Resources Research*, vol. 32, no. 9, pp. 2671–2681, DOI: 10.1029/95WR03818.

Further Reading

Brakensiek, D. L., and Onstad C. A. (1977). Parameter estimation of the Green-Ampt equation. *Water Resources Research*, vol. 13, no. 2, pp. 335–359.

Cahoon, J. (1988). Kostiakov infiltration parameters from kinematic wave model. *Journal of Irrigation and Drainage Engineering*, vol. 124, no. 2, pp. 127–130.

Collis-George, N. (1977). Infiltration equations for simple soil systems. *Water Resources Systems*, vol. 13, no. 2, pp. 395–403.

Corradini, C., Melone, F. and Smith R. E. (1994). Modeling infiltration during complex rainfall sequences. *Water Resources Research*, vol. 30, no. 1, pp. 2777–2784.

Corradini, C., Melone, F. and Smith R. E. (1994). A unified model for infiltration during complex rainfall patterns. *Journal of Hydrology*, vol. 192, pp. 104–124.

Fok, Y. S., and Chiang S. H. (1984). 2-D infiltration equations for furrow irrigation. *Journal of Irrigation and Drainage Engineering*, vol. 110, no. 2, pp. 208–217.

Fok, Y. S. (1975). A comparison of the Green-Ampt and Philip two term infiltration equations. *Transactions of the American Society of Agricultural and Biological Engineers*, vol. 18, pp. 1073–1075.

Gifford, G. F. (1976). Applicability of some infiltration formulae to rangeland infiltrometer data. *Journal of Hydrology*, vol. 28, pp. 1–11.

Mein, R. G., and Larson C. L. (1973). Modeling infiltration during steady rain. *Water Resources Research*, vol. 9, no. 2, pp. 384–394.

Mishra, S. K., Kumar, S. R. and Singh V. P. (1999). Calibration of a general infiltration model. *Hydrological Processes*, vol. 13, pp. 1691–1718.

Mishra, S. K., Tyagi, J. V. and Singh V. P. (2003). Comparison of infiltration models. *Hydrological Processes*, vol. 17, pp. 2629–2652.

Morel-Seytous, H. J., and Khanji J. (1974). Derivation of an equation of infiltration. *Water Resources Research*, vol. 10, no. 4, pp. 795–800.

Parlange, J. -Y. (1971). Theory of water movement in soils. 2. One-dimensional infiltration. *Soil Science*, vol. 111, no. 3, pp. 170–296.

Philip, J. R. (1954). An infiltration equation with physical significance. *Soil Science*, vol. 77, pp. 153–157.

Philip, J. R. (1969). Theory of infiltration. Advances in Hydroscience, vol. 5, edited by V. T. Chow. Academic Press, New York, pp. 215–296.

Philip, J. R. (1991). Hillslope infiltration: Planar slopes. *Water Resources Research*, vol. 27, pp. 109–117.

Rawls, W. J., and Asmussen L. E. (1973). Susburface flow in the Georgia coastal plain. *Journal of Irrigation and Drainage Engineering*, vol. 99, no. IR3, pp. 375–386.

Rawls, W. J., and Asmussen L. E. (1974). Neutron probe filed calibration for soils in the Georgia coastal plain. *Soil Science*, vol. 116, no. 4, pp. 262–265.

Rawls, W. J., Brakensiek, D. L. and Miller N. (1983). Green-Ampt infiltration parameters from soils data. *Journal of Hydraulic Engineering*, vol. 109, no. 1, pp. 62–70.

Skaggs, R. W., Huggins, L. F., Monke, E. J. and Foster G. R. (1969). Experimental evaluation of infiltration equations. *Transactions of the American Society of Agricultural and Biological Engineers*, vol. 16, no. 6, pp. 822–828.

Smith, R. E. (1972). Infiltration envelope: Results from a theoretical infiltrometer. *Journal of Hydrology*, vol. 17, pp. 1–21.

Smith, R. E., and Parlange J.-Y. (1978). A parameter-efficient hydrologic infiltration model. *Water Resources Research*, vol. 14, no. 3, pp. 533–538.

Snyder, W. M. (1971). A proposed watershed retention function. *Journal of Irrigation and Drainage Engineering*, vol. 97, no. IR1, pp. 193–201.

Swartzendruber, D., and Youngs E. G. (1974). A comparison of physically based infiltration equations. *Soil Science*, vol. 117, no. 3, pp. 165–167.

Van Genuchten, M. T. (1980). A closed from equation for predicting the hydraulic conductivity of unsaturated soils. *Soil Science Society of America Journal*, vol. 44, pp. 892–898.

Whisler, F., D., and Bower H. (1970). Comparison of methods for calculating vertical drainage and infiltration for soils. *Journal of Hydrology*, vol. 10, pp. 12–19.

Wilson, B. N., Slack, D. C. and Young R. A. (1982). A comparison of three infiltration models. *Transactions of the American Society of Agricultural and Biological Engineers*, vol. 25, pp. 349–356.

CHAPTER 12

Soil Moisture

Soil moisture occupies a central position in the hydrologic cycle, interfacing between land surface hydrologic processes and the atmosphere, on one hand, and between land surface processes and the lithosphere (groundwater zone), on the other hand. The zone of soil moisture (also called the vadose zone) is often called the gatekeeper in hydrology. Soil moisture is fundamental to analysis and evaluation of droughts, generation of runoff, irrigation scheduling and management, maintaining salt balance and reducing water logging, determination of evapotranspiration, sustaining ecological health, and spread of bacterial and viral activities. Because of its ubiquitous use, recent years have witnessed considerable emphasis on the measurement of soil moisture using, e.g., neutron probes, TDR probes, and remote sensing techniques. In the case of remote sensing, soil moisture estimates are obtained within a depth of no more than 5 cm (Ulaby et al., 1996), and modeling methods are needed to estimate the entire soil moisture profile. The objective of this chapter is to apply entropy theory to describe the one-dimensional movement of soil moisture in unsaturated soils and illustrate its application using experimental and field observations. Entropy theory permits a probabilistic description of soil moisture. Al-Hamdan and Cruise (2010) were the first to apply entropy theory to derive soil moisture profiles, and their work will be followed here.

12.1   Soil Moisture Zone

The porous medium below the land surface can be divided into two media: one between the water table and the land surface and the other below the water table. The *water table* is defined as the surface on which the fluid pressure in the pores of a medium is exactly atmospheric. This means that the hydraulic head at any point on the water table must equal the elevation of the water table at that point. The porous medium below the water table is saturated, i.e., the pores are filled with water, and can be referred to as groundwater or the saturated geologic zone. The porous medium above the water table is often divided into three zones. There exists a narrow zone immediately above the water table, called the capillary zone or fringe, where the porous medium is tension-saturated but the pressure head is negative. This zone is also called the tension-saturated zone. This pressure is the air entry pressure or bubbling pressure. The medium above the capillary fringe is called the unsaturated zone or vadose zone or zone of aeration. In this zone, pores are partially filled with water and partially filled with air. This means that water in the soil pores is under surface

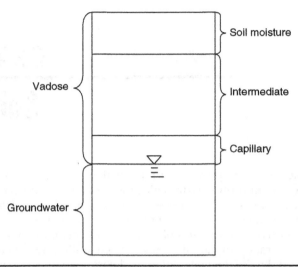

Figure 12.1   Zones below the land surface or subsurface zones: vadose zone, including soil moisture, intermediate, and capillary; and groundwater.

tension forces, and thus the pressure will be negative. In this zone both the moisture content θ and the hydraulic conductivity K are functions of the pressure head ψ. Furthermore, the θ-ψ relationship is hysteretic, and the same is true of the K-ψ relationship. Thus these relationships during wetting are somewhat different from those during drying. From an agricultural standpoint, the vadose zone can be further divided into two zones. The zone below the land surface is the zone in which agricultural crops grow, and it may thus be called the root zone. This is also referred to as the soil moisture zone. Below this zone is the intermediate zone or percolation zone or transmission zone. Figure 12.1 is a schematic of these zones.

12.2   Soil Moisture Profile

When water is applied to the land surface either artificially by irrigation or naturally by rainfall, part of this water infiltrates the soil, depending, of course, upon the soil characteristics, soil treatment, vegetation, and antecedent soil moisture conditions. The movement of moisture upon entry at the surface will depend on the duration for which water is applied at the surface and the moisture existing beforehand. The soil will first get saturated at the surface, and the saturation (or wetting) front will move downward until it reaches the water table. This is called the wetting phase. In this case the distribution of soil moisture monotonically decreases from the surface to the water table or up to a point of concern, as shown in Fig. 12.2.

When the supply of water is cut off, the downward movement of moisture will continue and the soil will start to drain and to dry. This is called the drying phase, in which the drying front moves downward. In this case the distribution of moisture is monotonically increasing downward, as shown in Fig. 12.3.

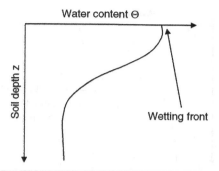

FIGURE 12.2 Soil moisture profile under wetting.

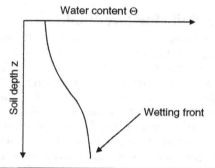

FIGURE 12.3 Soil moisture profile under drying.

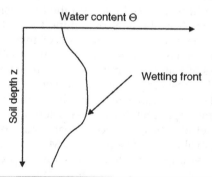

FIGURE 12.4 Soil moisture profile showing wetting and drying fronts.

Between these two phases there exists a situation where the distribution of moisture is monotonically increasing downward up to a point (zone 1) but then decreasing downward (zone 2). In this case one can divide the unsaturated zone into these two zones, as shown in Fig. 12.4. The three types of soil moisture profiles are shown in Fig. 12.5.

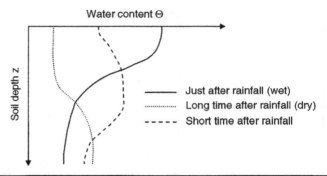

FIGURE **12.5** A schematic of three moisture distributions.

12.3 Estimation of Soil Moisture Profile

The approaches to estimating the soil moisture profile using near-surface soil moisture observations can be grouped into (1) regression, (2) inverse, (3) intelligence, and (4) water balance (Kostov and Jackson, 1993; Walker et al., 1997). Regression techniques relate near-surface soil moisture observations to wetting and drying separately at specific locations. For shallow depths Arya et al. (1983), Bruckler et al. (1988), and Srivastava et al. (1997), among others, found regression techniques to yield satisfactory estimates, but the development of regression relations needs sufficient observations at each location and these relations cannot be transferred to other locations.

In the inverse approaches, since the microwave brightness temperature profile is drawn from soil moisture estimates, with the use of inverse techniques, the remotely sensed brightness temperature can be employed for estimating the soil moisture (Kostov and Jackson, 1993). Intelligence techniques are based on artificial neural networks, genetic algorithms, fuzzy logic, and the like. Using a priori information on the hydrologic properties of soils, they determine the soil moisture content at different depths. Methods of determination include correlations between the surface soil moisture and that at lower layers (Kondratyev et al., 1977); energy-based methods with radiative properties of soil at different soil moisture states (Reutov and Shukto, 1986); and models using hydrostatic principles (Jackson et al., 1987). The radiative properties lead to a variety of soil moisture shapes, as shown in Fig. 12.6.

The water balance approach incorporates soil moisture as the output in the water balance (De Troch et al., 1996). In recent years soil moisture observations have been assimilated into hydrologic models (Das and Mohanty, 2006). Combining soil moisture observations with hydrologic models seems a more promising approach (Kostov and Jackson, 1993).

12.4 Derivation of Soil Moisture Profiles

It is assumed that θ is the *effective saturation*, defined as

$$\theta = \frac{\Theta - \Theta_0}{n - \Theta_0}$$

(12.1)

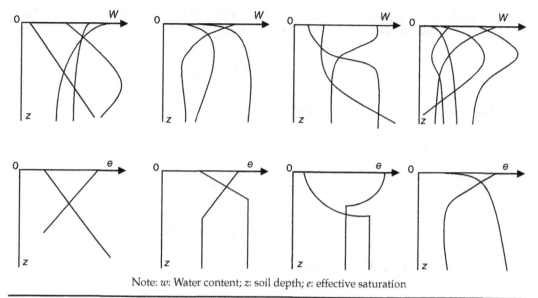

Note: *w*: Water content; *z*: soil depth; *e*: effective saturation

FIGURE **12.6** Moisture distributions (after Reutov and Shutko, 1986).

where Θ is the moisture content, Θ_0 is the initial moisture content or the moisture content (hygroscopic water) that cannot be extracted by plants, and n is the porosity. The objective is to derive θ as a function of distance z from the soil surface. The derivation involves two main steps: (1) deriving the cumulative probability distribution function (CDF) of θ and (2) hypothesizing a relation between the CDF of θ and then expressing θ as a function of z (Al-Hamdan and Cruise, 2010; Singh, 2010).

12.4.1 Case 1: Wetting Phase

Case 1, referred to as wetting phase, occurs during and immediately after rainfall. The moisture content is the highest near the surface and decreases downward. To derive the probability density function (PDF) of moisture content over the range $\theta = \theta_L$ at $z = L$ and $\theta = \theta_u$ at $z = 0$ using the principle of maximum entropy (POME), the following constraints are defined:

$$\int_{\theta_L}^{\theta_u} f(\theta)\, d\theta = 1 \tag{12.2}$$

$$\int_{\theta_L}^{\theta_u} \theta f(\theta)\, d\theta = \bar{\theta} \tag{12.3}$$

Applying POME and using the method of Lagrange multipliers, one gets

$$f(\theta) = \exp(-\lambda_0 - \lambda_1 \theta) \tag{12.4}$$

where λ_0 and λ_1 are the Lagrange multipliers which can be determined using Eqs. (12.2) and (12.3). Substituting Eq. (12.4) in Eq. (12.2), one gets

$$\int_{\theta_L}^{\theta_u} \exp(-\lambda_0 - \lambda_1\theta)\,d\theta = 1 \Rightarrow \lambda_0 = \ln\int_{\theta_L}^{\theta_u}\exp(-\lambda_1\theta)\,d\theta \tag{12.5}$$

Equation (12.5) can be solved explicitly as

$$\int_{\theta_L}^{\theta_u} \exp(-\lambda_0 - \lambda_1\theta)\,d\theta = 1 \Rightarrow \lambda_0 = -\ln\lambda_1 + \ln[\exp(-\lambda_1\theta_L) - \exp(-\lambda_1\theta_u)] \tag{12.6}$$

Differentiating Eq. (12.6) with respect to λ_1, one obtains

$$\frac{\partial\lambda_0}{\partial\lambda_1} = -\frac{1}{\lambda_1} - \frac{\theta_u\exp(-\lambda_1\theta_u) - \theta_L\exp(-\lambda_1\theta_L)}{\exp(-\lambda_1\theta_u) - \exp(-\lambda_1\theta_L)} \tag{12.7}$$

Differentiating Eq. (12.5), one obtains

$$d\theta\frac{\partial\lambda_0}{\partial\lambda_1} = \frac{\displaystyle\int_{\theta_L}^{\theta_u}\exp(-\lambda_0 - \lambda_1\theta)(-\theta)\,d\theta}{\displaystyle\int_{\theta_L}^{\theta_u}\exp(-\lambda_0 - \lambda_1\theta)} = -\bar{\theta} \tag{12.8}$$

Equating Eq. (12.8) to Eq. (12.7), one gets

$$\frac{1}{\lambda_1} + \frac{\theta_u\exp(-\lambda_1\theta_u) - \theta_L\exp(-\lambda_1\theta_L)}{\exp(-\lambda_1\theta_u) - \exp(-\lambda_1\theta_L)} = \bar{\theta} \tag{12.9}$$

Equation (12.9) gives the relation between λ_1 and $\bar{\theta}$.
    If θ_L is assumed to be zero, then Eq. (12.9) becomes

$$\frac{1}{\lambda_1} + \frac{\theta_u\exp(-\lambda_1\theta_u)}{\exp(-\lambda_1\theta_u) - 1} = \bar{\theta} \tag{12.10}$$

The probability density function of θ is now defined as

$$f(\theta) = \frac{\lambda_1\exp(-\lambda_1\theta)}{\exp(-\lambda_1\theta_L) - \exp(-\lambda_1\theta_u)} \tag{12.11a}$$

Integration of Eq. (12.11a) leads to the CDF of θ as

$$F(\theta) = \int_{\theta_L}^{\theta} f(\theta)\,d\theta = \int_{\theta_L}^{\theta}\frac{\lambda_1\exp(-\lambda_1\theta)}{\exp(-\lambda_1\theta_L) - \exp(-\lambda_1\theta_u)}\,d\theta = \frac{\exp(-\lambda_1\theta_L) - \exp(-\lambda_1\theta)}{\exp(-\lambda_1\theta_L) - \exp(-\lambda_1\theta_u)} \tag{12.11b}$$

To express the total soil moisture θ in terms of distance or the z coordinate, it is hypothesized that the cumulative probability distribution function of θ is a linear function of the distance from the soil:

$$F(\theta) = 1 - \frac{z}{L} \qquad f(\theta) = -\frac{1}{L}\frac{dz}{d\theta} \qquad (12.12)$$

where L is the water table depth or some distance to which moisture travels and z is the distance from the surface to any point along the depth. Equation (12.12) assumes that all values of z are equally likely between 0 and L, and all values of θ are equally likely. This equation may not be valid under all conditions but is attractive for its simplicity.

Substituting Eq. (12.11a) in Eq. (12.12) and integrating, one gets

$$\theta = \frac{1}{\lambda_1} \ln \frac{L}{L\exp(-\lambda_1\theta_u) + z[\exp(-\lambda_1\theta_L) - \exp(-\lambda_1\theta_u)]} \qquad (12.13)$$

where parameter λ_1 is obtained from Eq. (12.9).

For $\theta_L = 0$, Eq. (12.11a) reduces to

$$f(\theta) = \frac{\lambda_1 \exp(-\lambda_1\theta)}{1 - \exp(-\lambda_1\theta_u)} \qquad (12.14a)$$

Likewise, Eq. (12.11b) reduces to

$$F(\theta) = \int_{\theta_L}^{\theta} f(\theta)\,d\theta = \int_{\theta_L}^{\theta} \frac{\lambda_1 \exp(-\lambda_1\theta)}{1 - \exp(-\lambda_1\theta_u)}\,d\theta = \frac{1 - \exp(-\lambda_1\theta)}{1 - \exp(-\lambda_1\theta_u)} \qquad (12.14b)$$

and Eq. (12.13) to

$$\theta = \frac{1}{\lambda_1} \ln \frac{L}{L\exp(-\lambda_1\theta_u) + z[1 - \exp(-\lambda_1\theta_u)]} \qquad (12.14c)$$

Example 12.1. Examine the response of θ with variations in z, θ_L, and θ_u as well as λ_1.

Solution. The soil moisture profile for the wetting phase is given by Eq. (12.13). For computing the soil moisture profiles, let $L = 45$ cm, $\theta_u = 0.98$, and $\theta_L = 0.05$. For different values of the Lagrange multiplier $\lambda_1 = (1, 3, 5, 8, 12, 15)$, the soil moisture profile computed using Eq. (12.14c) is shown in Fig. 12.7.

Figure 12.7 shows that soil moisture profiles are sensitive to λ_1. When λ_1 is small (for example, $\lambda_1 = 1$), the profile is nearly like a straight line; when λ_1 increases, the profile becomes more curved. For computing soil moisture profiles for different values of θ_u, let $L = 45$ cm, $\lambda_1 = 3$, $\theta_L = 0.05$, surface effective saturation $\theta_u = [0.50, 0.60, 0.70, 0.80, 0.90, 1.00]$. Then the soil moisture profiles, based on different θ_u values, are computed as shown in Fig. 12.8.

Figure 12.8 shows that the 0- to 20-cm depth soil moisture profile is very sensitive to different surface effective saturation values, but the deeper soil moisture profile is not as sensitive. For computing soil moisture profiles for different values of θ_L, let $L = 45$ cm, $\lambda_1 = 3$, $\theta_u = 0.98$, and bottom effective saturation $\theta_L = [0.05, 0.10, 0.15, 0.20, 0.25, 0.30]$. Then the soil moisture profiles based on the different θ_L values are computed as shown in Fig. 12.9.

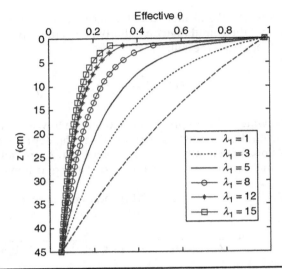

Figure 12.7 Soil moisture profiles for different values of the Lagrange parameter λ_1 (wet case).

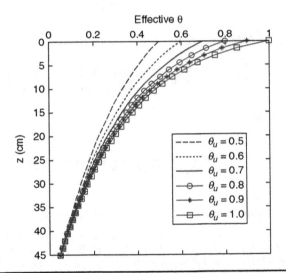

Figure 12.8 Soil moisture profiles for different values of θ_u (wet case).

From Fig. 12.9, it is seen that near the soil surface layer (that is, 0- to 3-cm depth) the soil moisture profile is not very sensitive to different bottom effective saturation values. However, the deeper the soil layer is, the more sensitive the soil moisture profile is to different bottom effective saturation values.

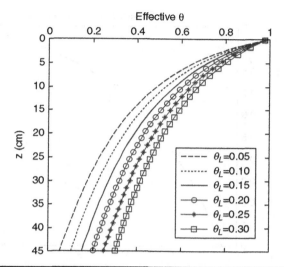

FIGURE 12.9   Soil moisture profiles for different values of θ_L (wet case).

Example 12.2.   Plot the probability density function $f(\theta)$ as well as cumulative probability distribution function $F(\theta)$ for the wetting case.

Solution.   The probability density function (PDF) of θ for the wet case is given by Eq. (9.11a), where $\theta_L \le \theta \le \theta_u$, and the cumulative distribution function (CDF) of θ is given by Eq. (12.13). Assuming that $\lambda_1 = 3$, $\theta_u = 0.98$, and $\theta_L = 0.05$, the PDF and CDF of soil moisture are computed as shown in Figs. 12.10 and 12.11, respectively.

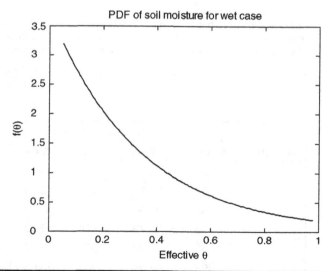

FIGURE 12.10   PDF of soil moisture for wet case.

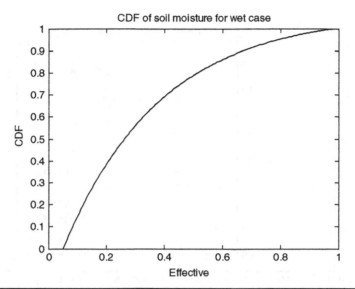

FIGURE **12.11** CDF of soil moisture for wet case.

Example 12.3. Compute the Lagrange multiplier values for the data set given in Table 12.1.

Solution. For the data of Melone et al. (2006), rainfall lasts for 8 h. Therefore, the effective saturation value is taken as that at the beginning of 8 h, as shown in Table 12.1. The Lagrange multiplier value λ_1 is calculated using Eq. (12.9). To that end, Eq. (12.9) is rewritten in MATLAB as

$$g(\lambda_1) = \frac{1}{\lambda_1} + \frac{\theta_u \exp(-\lambda_1\theta_u) - \theta_L \exp(-\lambda_1\theta_L)}{\exp(-\lambda_1\theta_u) - \exp(-\lambda_1\theta_L)} - \bar{\theta} \tag{12.15}$$

where $g(\lambda_1)$ is a function of λ_1. Then by using the **gsolve** function in MATLAB, the nonlinear function $f(\lambda_1) = 0$ is solved to get the Lagrange multiplier λ_1. For each hour observation, the Lagrange multiplier value λ_1 is obtained. After the λ_1 value is found, the λ_0 value is calculated using the following equation:

$$\lambda_0 = -\ln\lambda_1 + \ln[\exp(-\lambda_1\theta_L) - \exp(-\lambda_1\theta_u)] = \ln\frac{\exp(-\lambda_1\theta_L) - \exp(-\lambda_1\theta_u)}{\lambda_1} \tag{12.16}$$

Time (h)	Effective Saturation				
	5 cm	**15 cm**	**25 cm**	**35 cm**	**45 cm**
1	0.631	0.143	0.138	0.145	0.152
2	0.76	0.67	0.129	0.143	0.154
3	0.846	0.83	0.742	0.143	0.149
4	0.916	0.878	0.828	0.572	0.154
6	0.955	0.907	0.878	0.717	0.817
8	0.993	0.921	0.903	0.785	0.916

TABLE **12.1** Experimental Data for a Wet Case (Melone et al., 2006)

Time (h)	θ_u	θ_L	θ_{mean}	λ_1	λ_0
1	0.631	0.152	0.242	10.806	−4.028
2	0.76	0.154	0.371	2.949	−1.719
3	0.846	0.149	0.542	−1.110	0.216
4	0.916	0.154	0.670	−3.018	1.554
6	0.955	0.817	0.855	22.605	−21.631
8	0.993	0.916	0.904	153.563	−145.697

TABLE 12.2 Computed Lagrange Multiplier λ_0 and λ_1 Values for Different Hours for the Wet Case

Thus, the Lagrange multipliers are obtained as shown in Table 12.2.

Note that at time = 8 h, the values of λ_1 = 153.563 and λ_0 = −145.697 are very large compared to the values at other times. This may be partly due to experimental errors.

Example 12.4. Plot the effective saturation as a function of time for different depths for the wet case.

Solution. Based on the observations shown in Table 12.1, the effective saturation is plotted as a function for different values of depths, as shown in Fig. 12.12. This gives an indication of the soil moisture profile variability in space and time.

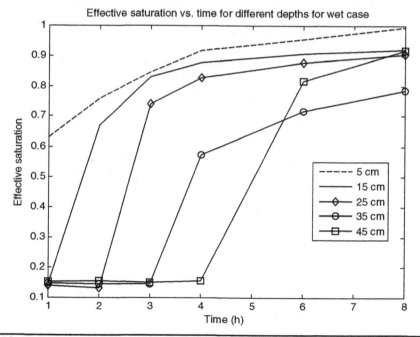

FIGURE 12.12 Effective saturation versus time for different depths for the wet case.

Example 12.5. Compute and plot the soil moisture profiles for a wet case shown in Table 12.1, and compare with the observed profiles. How well do entropy-based profiles compare? Compute the relative errors.

Solution. For the wet case, the Lagrange multiplier values have already been calculated, and the θ_u and θ_L values (see Table 12.2) are already known. Therefore, the effective saturation is calculated using Eq. (12.14c). For illustration, only the hour 1 data are used for calculation. The parameters needed to calculate the effective saturation are as follows (extracted from Table 12.2).

Time (h)	θ_u	θ_L	θ_{mean}	λ_1	λ_0
1	0.631	0.152	0.242	10.806	−4.028

Therefore, the computed and observed soil moisture profiles are computed as shown in Table 12.3, where the relative error is calculated as

$$\text{relative error} = \theta_{observed} \frac{-\theta_{computed}}{\theta_{observed}}.$$

The computed and observed moisture profiles are graphed in Fig. 12.13. It is seen from Table 12.3 and Fig. 12.13 that the entropy-based profiles are in reasonably good agreement with the observed profiles in the wet case.

Depth (cm)	Observed θ	Computed θ	Relative Error (%)
5	0.631	0.351	0.443
15	0.143	0.253	−0.767
25	0.138	0.206	−0.493
35	0.145	0.175	−0.208
45	0.152	0.152	0.000

TABLE 12.3   Computed and Observed Soil Moisture Profiles for the Wet Case

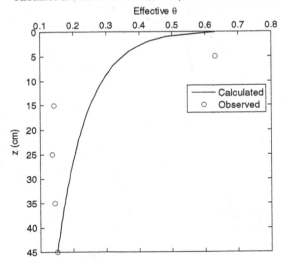

FIGURE 12.13   Calculated and observed soil moisture profiles for the wet case at hour 1.

12.4.2   Case 2: Drying Phase

This case, referred to as the drying phase, occurs long after rainfall. Consequently, the moisture at the surface ($z = 0$) is the lowest and highest at $z = L$. To derive the PDF of the moisture content profile using POME, the constraints are as defined by Eqs. (12.2) and (12.3) with the limits of integration interchanged:

$$\int_{\theta_u}^{\theta_L} f(\theta)\, d\theta = 1 \tag{12.17}$$

$$\int_{\theta_u}^{\theta_L} \theta f(\theta)\, d\theta = \bar{\theta} \tag{12.18}$$

Applying POME and the method of Lagrange multipliers, one gets

$$f(\theta) = \exp(-\lambda_0 - \lambda_1 \theta) \tag{12.19}$$

where λ_0 and λ_1 are Lagrange multipliers which can be determined using Eqs. (12.15) and (12.18). Substituting Eq. (12.19) in Eq. (12.17), one finds

$$\int_{\theta_u}^{\theta_L} \exp(-\lambda_0 - \lambda_1 \theta)\, d\theta = 1 \Rightarrow \lambda_0 = \ln \int_{\theta_u}^{\theta_L} \exp(-\lambda_1 \theta)\, d\theta \tag{12.20}$$

Equation (12.20) can be solved explicitly as

$$\int_{\theta_u}^{\theta_L} \exp(-\lambda_0 - \lambda_1 \theta)\, d\theta = 1 \Rightarrow \lambda_0 = -\ln \lambda_1 + \ln[\exp(-\lambda_1 \theta_u) - \exp(-\lambda_1 \theta_L)] \tag{12.21}$$

Differentiating Eq. (12.21) with respect to λ_1, one obtains

$$\frac{\partial \lambda_0}{\partial \lambda_1} = -\frac{1}{\lambda_1} + \frac{\theta_L \exp(-\lambda_1 \theta_L) - \theta_u \exp(-\lambda_1 \theta_u)}{\exp(-\lambda_1 \theta_u) - \exp(-\lambda_1 \theta_L)} \tag{12.22}$$

Differentiating Eq. (12.20) with respect to λ_1, one obtains

$$d\theta \frac{\partial \lambda_0}{\partial \lambda_1} = \frac{\int_{\theta_u}^{\theta_L} \exp(-\lambda_0 - \lambda_1 \theta)(-\theta)\, d\theta}{\int_{\theta_u}^{\theta_L} \exp(-\lambda_0 - \lambda_1 \theta)} = -\bar{\theta} \tag{12.23}$$

Equating Eq. (12.22) to Eq. (12.23), one gets

$$\frac{1}{\lambda_1} + \frac{\theta_u \exp(-\lambda_1 \theta_u) - \theta_L \exp(-\lambda_1 \theta_L)}{\exp(-\lambda_1 \theta_u) - \exp(-\lambda_1 \theta_L)} = \bar{\theta} \tag{12.24}$$

Equation (12.24) gives the relation between λ_1 and $\bar{\theta}$.

If θ_u is assumed to be zero, then Eq. (12.24) reduces to

$$\frac{1}{\lambda_1} + \frac{\theta_L \exp(-\lambda_1 \theta_L)}{\exp(-\lambda_1 \theta_L) - 1} = \bar{\theta} \tag{12.25}$$

The probability density function of θ is now defined as

$$f(\theta) = \frac{\lambda_1 \exp(-\lambda_1 \theta)}{\exp(-\lambda_1 \theta_u) - \exp(-\lambda_1 \theta_L)} = \frac{dz}{d\theta} \frac{1}{L} \tag{12.26a}$$

The cumulative distribution function of θ for the dry case can be obtained by integration of Eq. (12.26a)

$$F(\theta) = \int_{\theta_u}^{\theta} f(\theta) d\theta = \int_{\theta_u}^{\theta} \frac{\lambda_1 \exp(-\lambda_1 \theta)}{\exp(-\lambda_1 \theta_u) - \exp(-\lambda_1 \theta_L)} d\theta$$

$$= \frac{\exp(-\lambda_1 \theta_u) - \exp(-\lambda_1 \theta)}{\exp(-\lambda_1 \theta_u) - \exp(-\lambda_1 \theta_L)} \qquad \theta_u \le \theta \le \theta_L \tag{12.26b}$$

If $\theta_u = 0$,

$$f(\theta) = \frac{\lambda_1 \exp(-\lambda_1 \theta)}{1 - \exp(-\lambda_1 \theta_L)} = \frac{1}{L} \frac{dz}{d\theta} \tag{12.27a}$$

and

$$F(\theta) = \frac{1 - \exp(-\lambda_1 \theta)}{1 - \exp(-\lambda_1 \theta_L)} \tag{12.27b}$$

To derive the soil moisture content as a function of distance z, the hypothesis for the CDF of θ is formulated as

$$F(\theta) = \frac{z}{L} \qquad f(\theta) = \frac{1}{L} \frac{dz}{d\theta} \tag{12.28}$$

Substituting Eq. (12.26a) in Eq. (12.28) and integrating, one gets

$$\theta = \frac{1}{\lambda_1} \ln \frac{L}{L \exp(-\lambda_1 \theta_u) - z[\exp(-\lambda_1 \theta_u) - \exp(-\lambda_1 \theta_L)]} \tag{12.29}$$

where parameter λ_1 is obtained from Eq. (12.24).

If $\theta_u = 0$, Eq. (12.29) reduces to

$$\theta = \frac{1}{\lambda_1} \ln \frac{L \lambda_1^2}{L \lambda_1^2 - z[1 - \exp(-\lambda_1 \theta_L)]} \tag{12.30}$$

Equation (12.29) or (12.30) provides an explicit relation between θ and z.

Example 12.6. Examine the response of θ with variations in z, θ_L, and θ_u as well as λ_1.

Solution. The soil moisture profile is given by Eq. (12.29). For computing the soil moisture profile for different values of λ_1, let $L = 60$ cm, $\theta_u = 0.05$, $\theta_L = 0.98$, and the Lagrange parameter $\lambda_1 = [-10, -5, -3, -1, 1, 3]$. Then the soil moisture profiles are computed as shown in Fig. 12.14, which shows that the profiles are very sensitive to λ_1. When λ_1 is negative, the soil moisture profile form is convex; when λ_1 is positive, the form of the soil moisture profile becomes concave.

For computing the soil moisture profile for different values of θ_u, let $L = 60$ cm, $\lambda_1 = -1$, $\theta_L = 0.98$, and the surface effective saturation $\theta_u = [0.05, 0.10, 0.15, 0.20, 0.25, 0.30]$. Then the soil moisture profiles are computed as shown in Fig. 12.15, which shows that the 0- to 40-cm depth soil moisture profile is very sensitive to different surface effective saturation values. However, the deeper soil moisture profile is not very sensitive to different surface effective saturation values under the conditions that $L = 60$ cm, $\lambda_1 = -1$, and $\theta_L = 0.98$.

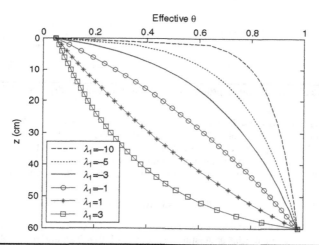

FIGURE 12.14 Soil moisture profiles with different λ_1 values (dry case).

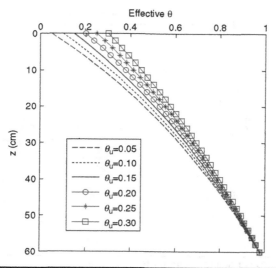

FIGURE 12.15 Soil moisture profiles with different θ_u values (dry case).

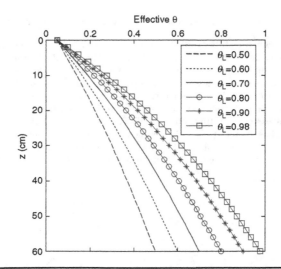

Figure 12.16   Soil moisture profiles with different θ_L values (dry case).

For computing the soil moisture profile for different values of θ_L, let $L = 60$ cm, $\lambda_1 = -1$, $\theta_u = 0.05$, and the bottom effective saturation $\theta_L = [0.50, 0.60, 0.70, 0.80, 0.90, 0.98]$. Then the soil moisture profiles are computed as in Fig. 9.16, which shows that near-surface soil layer (0- to 2-cm depth) of the soil moisture profile is not very sensitive to different bottom effective saturation values. However, the deeper the soil layer is, the more sensitive the soil moisture profile becomes to different bottom effective saturation values.

Example 12.7.   Plot the probability density function $f(\theta)$ as well as $F(\theta)$.

Solution.   The PDF and CDF of soil moisture for the dry case are given by Eqs. (12.26a) and (12.26b), respectively, where $\theta_u \le \theta \le \theta_L$. Assuming that $\lambda_1 = -1$, $\theta_u = 0.05$, and $\theta_L = 0.98$, the PDF and CDF of soil moisture for the dry case are computed as shown in Figs. 12.17 and 12.18, respectively.

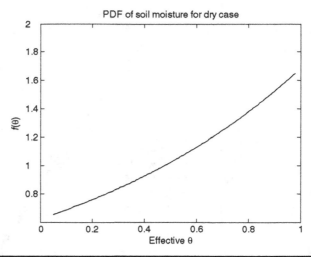

Figure 12.17   PDF of soil moisture for the dry case.

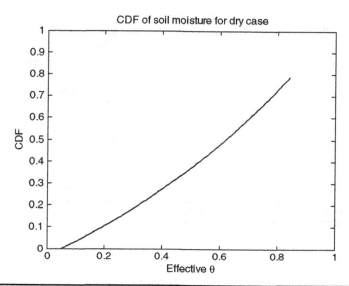

FIGURE **12.18** CDF of soil moisture for the dry case.

Example 12.8. Compute the Lagrange multiplier values for the data given in Table 12.4.

Solution. The mean moisture content for the dry case is given by Eq. (12.24) which is rewritten in MATLAB as

$$g(\lambda_1) = \frac{1}{\lambda_1} + \frac{\theta_u \exp(-\lambda_1\theta_u) - \theta_L \exp(-\lambda_1\theta_L)}{\exp(-\lambda_1\theta_u) - \exp(-\lambda_1\theta_L)} - \bar{\theta} \tag{12.31}$$

Using the **gsolve** function in MATLAB, the nonlinear function $g(\lambda_1) = 0$ is solved to get λ_1. Then for each observation the Lagrange multiplier value λ_1 is obtained. Thereafter, the Lagrange parameter λ_0 is calculated using the equation

$$\lambda_0 = -\ln \lambda_1 + \ln[\exp(-\lambda_1\theta_u) - \exp(-\lambda_1\theta_L)] = \ln \frac{\exp(-\lambda_1\theta_u) - \exp(-\lambda_1\theta_L)}{\lambda_1} \tag{12.32}$$

Date	\multicolumn{8}{Effective Saturation}
	5 cm	10 cm	15 cm	20 cm	25 cm	35 cm	45 cm	60 cm
July 21	0.45	0.5	0.625	0.75	0.85	0.875	0.875	0.95
July 22	0.4	0.45	0.55	0.7	0.825	0.875	0.875	0.95
July 23	0.35	0.425	0.5	0.625	0.825	0.875	0.875	0.95
July 24	0.325	0.375	0.475	0.575	0.8	0.875	0.875	0.95
July 25	0.3	0.35	0.45	0.55	0.8	0.875	0.875	0.95
July 26	0.275	0.325	0.425	0.5	0.75	0.875	0.875	0.95
July 28	0.25	0.275	0.35	0.45	0.575	0.875	0.875	0.95
July 29	0.225	0.275	0.35	0.45	0.6	0.875	0.875	0.925

TABLE **12.4** Experimental Data for the Dry Case in July (Menziani et al., 2006)

Date	θ_u	θ_L	θ_{mean}	λ_1	λ_0
July 21	0.45	0.95	0.734	−1.669	0.504
July 22	0.4	0.95	0.703	−1.123	0.176
July 23	0.35	0.95	0.678	−0.942	0.115
July 24	0.325	0.95	0.656	−0.577	−0.097
July 25	0.3	0.95	0.644	−0.534	−0.092
July 26	0.275	0.95	0.622	−0.247	−0.241
July 28	0.25	0.95	0.575	0.614	−0.717
July 29	0.225	0.925	0.572	0.077	−0.401

TABLE 12.5   The Computed Lagrange Multipliers λ_0 and λ_1 for Different Observations in July for the Dry Case

The computed Lagrange multipliers are given in Table 12.5.

Example 12.8.   Plot effective saturation as a function of time for different depths for the dry case. This will give an indication of the soil moisture profile variability in space and time.

Solution.   Based on the observations in Table 12.4, the soil moisture profile is computed, as shown in Fig. 12.19, which gives an indication of the soil moisture profile variability in space and time for the dry case in July. Note that the effective saturation values at depths of 35 and 45 cm are the same; therefore, they are in the same line in the figure.

Example 12.9.   Compute and plot the soil moisture profiles for a dry case, and compare with the observed profiles. How well do entropy-based profiles compare? Compute the relative errors.

Solution.   The Lagrange multiplier values have already been calculated, and the θ_u and θ_L values (see Table 12.5) are already known. Therefore, the effective saturation is calculated using Eq. (12.29). Here the data for July 21 are chosen. The parameters used to calculate the effective saturation are shown in the following table again (extracted from Table 12.5).

Date	θ_u	θ_L	θ_{mean}	λ_1	λ_0
July 21	0.45	0.95	0.734	−1.669	0.504

The computed and observed profiles and the relative error between them are shown in Table 12.6, where the relative error is calculated as before.

The computed and observed profiles are graphed in Fig. 12.20.

It is seen from Table 12.6 and Fig. 12.20 that the entropy-based profiles are in reasonably good agreement with the observed profiles in the dry case for the July data.

12.4.3   Case 3: Mixed Phase

Case 3, referred to as mixed phase, is a dynamic case that occurs sometime after rainfall has ceased. The soil moisture distribution can be divided into two parts. In the upper part the moisture increases from the ground surface down up to a maximum

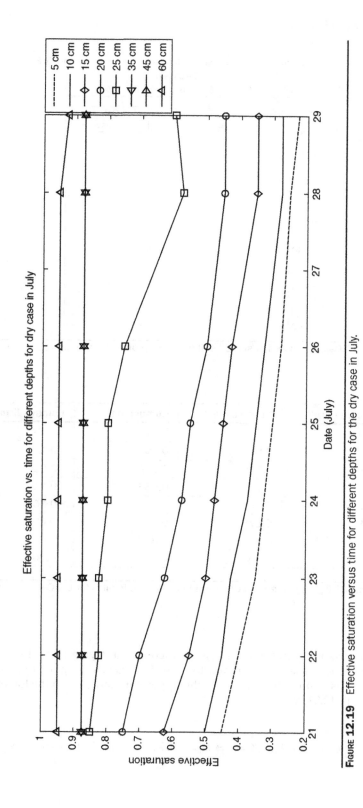

FIGURE 12.19 Effective saturation versus time for different depths for the dry case in July.

487

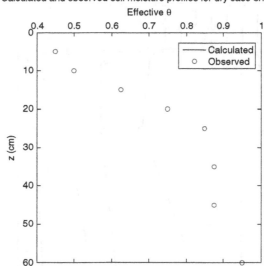

Calculated and observed soil moisture profiles for dry case on July 21

FIGURE **12.20** Calculated and observed soil moisture profiles for the dry case on July 21.

Depth (cm)	Observed θ	Computed θ	Relative Error
5	0.45	0.512	−0.137
10	0.5	0.568	−0.136
15	0.625	0.619	0.010
20	0.75	0.666	0.112
25	0.85	0.710	0.165
35	0.875	0.789	0.098
45	0.875	0.859	0.019
60	0.95	0.950	0.000

TABLE **12.6** Computed and Observed Soil Moisture Profiles for Dry Case on July 21

value at a certain point, say, d. This is analogous to a drying front. In the lower part, the moisture decreases from point d to point L. This is analogous to a wetting front. Thus, POME is applied twice, once to the upper part and once to the lower part. Thus, for $z < d$, Eq. (12.21) yields

$$\int_{\theta_u}^{\theta_d} \exp(-\lambda_0 - \lambda_1 \theta)\, d\theta = 1 \Rightarrow \lambda_0 = -\ln \lambda_1 + \ln[\exp(-\lambda_1 \theta_u) - \exp(-\lambda_1 \theta_d)] \qquad (12.33)$$

Equation (12.24) yields

$$\frac{1}{\lambda_1} + \frac{\theta_u \exp(-\lambda_1\theta_u) - \theta_d \exp(-\lambda_1\theta_d)}{\exp(-\lambda_1\theta_u) - \exp(-\lambda_1\theta_d)} = \overline{\theta}_{up} \tag{12.34}$$

where $\overline{\theta}_{up}$ is the average soil moisture in the upper part, and θ_d is the moisture at $z = d$. From Eq. (12.29), one can write

$$\theta = \frac{1}{\lambda_1} \ln \frac{d\lambda_1^2}{d\lambda_1^2 \exp(-\lambda_1\theta_u) - z[\exp(-\lambda_1\theta_u) - \exp(-\lambda_1\theta_d)]} \tag{12.35}$$

Likewise, for $z > d$, one obtains from Eqs. (12.6), (12.9), and (12.13), respectively,

$$\lambda_0 = -\ln\lambda_1 + \ln[\exp(-\lambda_1\theta_L) - \exp(-\lambda_1\theta_d)] \tag{12.36}$$

$$\frac{1}{\lambda_1} + \frac{\theta_0 \exp(-\lambda_1\theta_d) - \theta_d \exp(-\lambda_1\theta_L)}{\exp(-\lambda_1\theta_d) - \exp(-\lambda_1\theta_L)} = \overline{\theta}_{low} \tag{12.37}$$

$$\theta = \frac{1}{\lambda_1} \ln \frac{(L-d)\lambda_1^2}{(L-d)\lambda_1^2 \exp(-\lambda_1\theta_d) + (z-d)[\exp(-\lambda_1\theta_L) - \exp(-\lambda_1\theta_d)]} \tag{12.38}$$

where $\overline{\theta}_{low}$ is the average soil moisture in the lower part.

The average moisture content in the upper part and that in the lower part are defined as

$$\overline{\theta}_{up} = \frac{\int_0^d \theta(z)\,dz}{d} \tag{12.39}$$

$$\overline{\theta}_{low} = \frac{\int_d^L \theta(z)\,dz}{L-d} \tag{12.40}$$

Now the value of d needs to be determined. Using kinematic wave theory, Singh (1997) derived an expression for the depth of the wetting front as

$$d = \frac{K}{\theta_u} t \tag{12.41}$$

where K is the hydraulic conductivity and θ_u is the soil moisture at time t. It may be interesting to examine the response of θ with variations in z, θ_L, and θ_u as well as λ_1.

12.5 Mean Soil Moisture Content

The mean soil moisture content that has been employed in deriving the soil moisture profiles can be derived using mass balance. Consider a watershed element that can be represented as a two-dimensional grid in both horizontal and vertical dimensions.

In the vertical dimension, soil is represented as a column where each grid element represents a layer. In the soil column the net water W can be expressed using the mass balance (Al-Hamdan and Cruise, 2010) as

$$W = P + \Delta R - \mathrm{ET} - \Delta w_s \tag{12.42}$$

in which P is the precipitation, ET is the evapotranspiration, ΔR is the runoff entering the grid minus the runoff leaving the grid, and Δw_s is the change in water content of the surface layer. All variables in Eq. (12.42) are in depth units. The mean moisture content can be expressed in terms of W as

$$\bar{\theta} = \frac{W/(nL)}{n - \theta_0} \tag{12.43}$$

where L represents the root zone.

12.6   Soil Moisture Variation in Time

To map soil moisture at a given site as function of time, Al-Hamdan and Cruise (2010) noted that the total volume of water in the soil column, the total depth of the soil column that the water reaches at the time, and the hydraulic conductivity must be determined. The amount of water added to the soil profile controls the shape of the moisture profile. If the total amount of water exceeded the water deficit of the soil profile, then the entire soil column would be saturated at that time. In that model calculations would not be needed. On the other hand, if the amount of water was less than the soil water deficit, then POME-based equations derived previously can be used to determine how the water would be distributed in the soil profile.

The first step is to compute the amount of water using Eq. (12.42). Of course, the calculation will depend on the case at hand. For example, if it is a dry case, then evapotranspiration will be dominant and other terms, such as precipitation or lateral flow, may be either absent or negligible. In that case, the amount of water filling the soil profile can be calculated by starting from the initial value (e.g., saturation) and subtracting evapotranspiration during each time step from the previous value of the amount of water. Evapotranspiration can be computed using one of the standard methods, such as the Penman-Monteith equation (Penman, 1948; Monteith, 1965).

If it is a wet case, then precipitation will be dominant and evopotranspiration and lateral unsaturated flow will be negligible. The amount of water filling the soil profile can be calculated by starting from the initial value (e.g., field capacity) and adding the precipitation during each time step to the previous value of the amount of water.

The next step is to compute the depth of the soil column that the water reaches at a particular time. This depth is needed in the wet case, especially in the beginning of the rainfall event, where the water still needs time to reach the bottom of the root zone. A long time after the beginning of rainfall, the total depth will be the entire root zone depth.

Questions

1. Compute the soil moisture profiles for different Lagrange parameter values for both wet and dry cases. How sensitive are these profiles to the parameter? It may also be interesting to examine the response of the soil moisture profile to the initial, final, and mean soil moisture content. Also plot the PDF and CDF of soil moisture.

2. Compute the Lagrange multiplier values for the data set in Question 1.

3. Plot the effective saturation as a function of time for different depths for both wet and dry cases. This will give an indication of the soil moisture profile variability in space and time.

4. Compute and plot the soil moisture profiles for two wet cases and two dry cases, and compare with the observed profiles. How well do entropy-based profiles compare? Compute the errors.

5. Compute and plot the effective saturation as a function of time for at least two depths. Compare it with the observed effective saturation. Compute the errors.

6. Evaluate the sensitivity of θ with respect to parameters for the wet case.

7. Compute parameters for the wet case shown in Table 12.7. Plot the probability density function $f(\theta)$ as well as $F(\theta)$. Also evaluate the sensitivity of θ with respect to parameters.

Θ_u	Θ_m	Θ_L	Data #	Data Source
0.860	0.543	0.000	Data 1	Soil Experiment_2
0.453	0.427	0.416	Data 2	Soil Experiment_2
0.471	0.232	0.086	Data 3	Soil Experiment_1
0.432	0.263	0.080	Data 4	Soil Experiment_1

TABLE 12.7   Soil Moisture for the Wet Case

8. Compute parameters for the dry case shown in Table 12.8. Plot the probability density function $f(\theta)$ as well as $F(\theta)$. Also evaluate the sensitivity of θ with respect to parameters.

Θ_L	Θ_M	Θ_u	Data #	Data Source
0.250	0.210	0.100	1	USDA
0.260	0.230	0.170	2	USDA
0.775	0.374	0.130	3	Soil Experiment_2
0.465	0.394	0.322	4	Soil Experiment_1

TABLE 12.8   Soil Moisture for the Dry Case

9. Compute parameters for the mixed phase shown in Table 12.9. Plot the probability density function $f(\theta)$ as well as $F(\theta)$. Also evaluate the sensitivity of θ with respect to parameters.

Θ_U	Θ_{M_I}	Θ_d	Data #	Data Source
0.351	0.364	0.417		
Θ_d	Θ_{M_II}	Θ_L	1	Soil Experiment_2
0.417	0.340	0.337		
Θ_U	Θ_{M_I}	Θ_d	Data #	Data Source
0.379	0.404	0.428		
Θ_d	Θ_{M_II}	Θ_L	2	Soil Experiment_2
0.428	0.335	0.330		
Θ_U	Θ_{M_I}	Θ_d	Data #	Data Source
0.334	0.369	0.410		
Θ_d	Θ_{M_II}	Θ_L	3	Soil Experiment_2
0.410	0.278	0.080		

TABLE 12.9   Soil Moisture for the Mixed Phase

10. Using the soil moisture observations given in Table 12.10, determine the values of $\bar{\theta}$, θ_u, and θ_L. Specify L. Then compute the Lagrange multiplier λ_1, plot θ as a function of z, and compare it with the observed values.

Date	Effective Saturation							
	5 cm	10 cm	15 cm	20 cm	25 cm	35 cm	45 cm	60 cm
Sept. 15	0.15	0.175	0.225	0.275	0.375	0.85	0.925	1
Sept. 16	0.145	0.17	0.22	0.27	0.368	0.838	0.913	1
Sept. 17	0.14	0.165	0.215	0.265	0.36	0.825	0.9	1
Sept. 18	0.135	0.16	0.21	0.26	0.353	0.813	0.888	1
Sept. 19	0.13	0.155	0.205	0.255	0.345	0.8	0.875	1
Sept. 20	0.125	0.15	0.2	0.25	0.338	0.788	0.863	1
Sept. 22	0.115	0.145	0.195	0.24	0.323	0.75	0.85	1
Sept. 23	0.11	0.14	0.19	0.235	0.315	0.725	0.825	1
Sept. 24	0.105	0.135	0.185	0.23	0.308	0.7	0.8	1
Sept. 25	0.1	0.13	0.18	0.225	0.3	0.675	0.788	1
Sept. 26	0.095	0.13	0.175	0.22	0.293	0.65	0.775	1
Sept. 27	0.09	0.125	0.17	0.215	0.285	0.65	0.75	1

TABLE 12.10   Experimental Data for the Dry Case in September (Menziani et al., 2006)

References

Al-Hamdan, O. Z., and Cruise J. F. (2010). Soil moisture profile development from surface observations by principle of maximum entropy. *Journal of Hydrologic Engineering, ASCE*, vol. 15 no. 5, pp. 327–337.

Arya, L. M., Richter, J. C., and Paris J. F. (1983). Estimating profile water storage from surface zone soil moisture measurements under bare field conditions. *Water Resources Research*, vol. 19, no. 2, pp. 403–412.

Bruckler, L., Witono, H., and Stengel P. (1988). Near surface soil moisture estimation from microwave measurements. *Remote Sensing of Environment*, vol. 26, pp. 101–121.

Das, N. N., and Mohanty B. P. (2006). Rootzone soil moisture assessment using remote sensing and vadose zone modeling. *Vadose Zone Journal*, vol. 5, pp. 296–307.

De Troch, F. P., Troch, P. A. Su, Z. and Lin D. S. (1996). Application of remote sensing for hydrological modeling. In: *Distributed Hydrological Modeling*, edited by M. B. Abbott and J. Refsgaard. Kluwer Academic Publishers, Dordrecht, The Netherlands, pp. 165–191.

Jackson, T. J., Hawley, M. E. and O'Neill P. E. (1987). Preplanting soil moisture using passive microwave sensors. *Water Resources Bulletin*, vol. 23, no. 1, pp. 11–19.

Kondratyev, K. Y., Melentyev, V. V. Rabinovich, Y. I. and Shulgina E. M. (1977). Passive microwave remote sensing of soil moisture. *Proceedings, 11th Symposium on Remote Sensing of the Environment*, Environmental Research Institute of Michigan, Ann Arbor, pp. 1641–1661.

Kostov, K. G., and Jackson T. J. (1993). Estimating profile soil moisture from surface layer measurements—a review. *Proceedings, International Society for Optical Engineering*, vol. 1941, pp. 125–136.

Melone, F., Corradini, C. Morbidelli, R. and Saltalippi C. (2006). Laboratory experimental check of a conceptual model for infiltration under complex patterns. *Hydrological Processes*, vol. 20, pp. 439–452.

Menziani, M., Pugnaghi, S. de Leva, M. and Vincenzi S. (2008). Vertical water content distribution in the unsaturated soil by TDR measurements at shallow depth. *Conference on Water Observation and Information System for Decision support*, BALWOIS (2006). BAWOIS, 2008, http://balwoi-s.mpl.ird.fr/balwois/administration/full_paper/ffp-738.pdf

Monteith, J. L. (1965). Evaporation and the environment. *Proceedings, Symposium of Society of Exploration Biology*, vol. 19, pp. 265–234.

Penman, H. L. (1948). Natural vegetation from open water, bare soil and grass. *Proceedings, Royal Society London*, vol. A193, pp. 120–145.

Reutov, E.A., and Shutko A.M. (1986). Prior-knowledge based soil moisture determination of microwave radiometry. *Soviet Journal of Remote Sensing*, vol. 5, no. 1, pp. 100–125.

Singh, V. P. (1997). *Kinematic Wave Modeling in Water Resources: Environmental Hydrology*. Wiley, New York.

Singh, V. P. (2010). Entropy theory for movement of moisture in soils. *Water Resources Research*, vol. 46, pp. 1–12, 2010, W03516, doi:10.1029/2009WR008288.

Srivastava, S. K., Yograjan, N. Jayaraman, V. Nageswara, P. P. and Chandrsekhar M. G. (1997). On the relationship between ERS-1 SAR/backscatter and surface/subsurface soil moisture variations in vertisols. *Acta Astronautica*, vol. 40, no. 10, pp. 693–699.

Ulaby, F. T., Dubois, P. C. and Zyl J. V. (1996). Radar mapping of surface soil moisture. *Journal of Hydrology*, vol. 184, pp. 57–84.

Walker, J. P., Troch, P. A. Mancini, M. Willgoose, G. R. and Kalma J. D. (1997). Profile soil moisture estimation using the modified IEM. *Transactions of the IEEE*, vol. 26, pp. 1263–1265.

Further Reading

Anchal, K. J., and Murty V. V. N. (1985). Simulation of soil moisture profiles for scheduling of irrigations. *Agricultural Water Management*, vol. 10, no. 2, pp. 175–181.

Assouline, S., Tessier, D. and Bruand A. (1998). A conceptual model of the soil water retention curve. *Water Resources Research*, vol. 34, no. 2, pp. 223–231.

Bates, J. E. and Shepard, H. K. (1993). Measuring complexity using information fluctuation. *Physics Letters A*, vol. 172, pp. 416–425.

Bernard, R., Vauclin, M. and Vidal-Madjar D. (1981). Possible use of active microwave remote sensing data for prediction of regional evaporation by numerical simulation of soil water movement in the unsaturated zone. *Water Resources Research*, vol. 17, no. 6, pp. 1603–1610.

Biswas, B. C., and Dasgupta S. K. (1979). Estimation of soil moisture at deeper depth from surface layer data. *Mausam*, vol. 30, no. 4, pp. 511–516.

Bruckler, L., and Witono H. (1989). Use of remotely sensed soil moisture content as boundary conditions in soil-atmosphere water transport modeling: 2. Estimating soil water balance. *Water Resources Research*, vol. 25, no. 12, pp. 2437–2447.

Capehart, W. J., and Carlson T. N. (1994). Estimating near-surface soil moisture availability using a meteorologically driven soil-water profile model. *Journal of Hydrology*, vol. 160, no. 1-4, pp. 1–20.

Charbeneau, R. J. (1984). Kinematic models for soil moisture and solute transport. *Water Resources Research*, vol. 20, no. 6, pp. 699–706.

Crosson, W. L., Laymon, C. A. Inguva, R. and Schamschula M. (2002). Assimilating remote sensing data in a surface flux-soil moisture model. *Hydrological Processes*, vol. 16, pp. 1645–1662.

Ducharne, A., and Laval K. (2000). Influence of the realistic description of soil water-holding capacity on the global water cycle in a GCM. *American Meteorological Society*, vol. 13, pp. 4393–4413.

Engman, E. T. (1992). Soil moisture needs in Earth sciences. *Proceedings of the International Geoscience and Remote Sensing Symposium (IGARSS)*, pp. 477–479.

Georgakakos, K. P., and Baumer O. W. (1996). Measurement and utilization of on-site soil moisture data. *Journal of Hydrology*, vol. 184, pp. 131–152.

Germann, P., and Beven K. (1985). Kinematic wave approximation to infiltration into soils with sorbing macropores. *Water Resources Research*, vol. 21, pp. 990–996.

Heathman, G. C. (1992). *Data Report: Profile Soil Moisture*. U.S. Department of Agriculture, Washington, D.C.

Heathman, G. C. (1994). *Data Report: Profile Soil Moisture*. U.S. Department of Agriculture, Washington, D.C.

Heathman, G. C., Starks, P. J. Ahuja, L. R. and Jackson T. J. (2003). Assimilation of surface soil moisture to estimate profile soil water content. *Journal of Hydrology*, vol. 279, no. 1-4, pp. 1–17.

Hollenbeck, K. J., and Jensen K. H. (1998). Maximum-likelihood estimation of unsaturated hydraulic parameters. *Journal of Hydrology*, vol. 210, pp. 192–205.

Houser, P. R., Shuttleworth, W. J. Famiglietti, J. S. Gupta, H. V. Syed, K. H. and Goodrich, D. C. (1998). Integration of soil moisture remote sensing and hydrologic modeling using data assimilation. *Water Resources Research*, vol. 34, no. 12, pp. 3405–3420.

Jackson, T. J. (1980). Profile soil moisture from surface measurements. *Journal of Irrigation and Drainage Division, Proc. ASCE*, vol. 106, no. IR2, pp. 81–92.

Jackson, T. J., Schmugge, T. J. Nicks, A. D. Coleman, G. A. and Engman, E. T. (1981). Soil Moisture Updating and Microwave Remote Sensing for Hydrological Simulation. *Hydrological Sciences Bulletin*, vol. 26, no. 3, pp. 305–319.

Jackson, T. J., Hawley, M. E. and O'Neill P. E. (1987). Preplanting soil moisture using passive microwave sensors. *Water Resources Bulletin*, vol. 23, no. 1, pp. 11–19.

Jain, S. K., Singh, V. P. and van Genuchten M. T. (2004). Analysis of soil water retention data using artificial neural networks. *Journal of Hydrologic Engineering, ASCE*, vol. 9, no. 5, pp. 415–420.

Jaynes, E. T. (1957a). Information theory and statistical mechanics I. *Physical Review*, vol. 106, no. 4, pp. 620–630.

Jaynes, E. T. (1957b). Information theory and statistical mechanics II. *Physical Review*, vol. 108, no. 2, pp. 171–190.

Koekkoek, E. J. W., and Booltink H. (1999). Neural network models to predict soil water retention. *European Journal of Soil Science*, vol. 50, pp. 489–495.

Kolovos, A., Christakos, G. Serre, M. L. and Miller C. T. (2002). Computational Bayesian maximum entropy solution of a stochastic advection-reaction equation in the light of site-specific information. *Water Resources Research*, vol. 38, no. 12, 1318, doi: 10.1029/2001WR000743.

Li, J., and Islam S. (1999). On the estimation of soil moisture profile and surface fluxes partitioning from sequential assimilation of surface layer soil moisture. *Journal of Hydrology*, vol. 220, no. 1-2, pp. 86–103.

Li, J., and Islam S. (2002). Estimation of root zone soil moisture and surface fluxes partitioning using near surface soil moisture measurements. *Journal of Hydrology*, vol. 259, no.1-4, pp. 1–14.

Mays, D. C., Faybishenko, B. A. and Finsterle S. (2002). Information entropy to measure temporal and spatial complexity of unsaturated flow in heterogeneous media. *Water Resources Research*, vol. 38, no. 12, 1313, doi: 10.1029/2001WR001185.

Monteith, F. I. (1981). Evaporation and surface temperature. *Quarterly Journal of the Meterorological Society*, vol. 107, pp. 1–27.

Or, D., Leij, F.J., Snyder, V. and Ghezzehei A. (2000). Stochastic model for posttillage soil pore space evolution. *Water Resources Research*, vol. 36, no. 7, pp. 1641–1652.

Ottlé, C., and Vidal-Madjar D. (1994). Assimilation of soil moisture inferred from infrared remote sensing in a hydrological model over the Hapex-Mobilhy region. *Journal of Hydrology*, vol. 158, pp. 241–264.

Pachepsky, Y., Guber, A. Jacques, D. Simunek, J. Genuchten, M. T. V. Nicholson, T. and Cady R. (2006). Information content and complexity of simulated soil water fluxes. *Geoderma*, vol. 134, no. 3-4, pp. 253–266.

Penman, H. L. (1963). Vegetation and hydrology. *Technical Communication 53*, Commonwealth Bureau of Soils, Harpenden, England.

Penman, H. L. (1956). Evaporation: An introductory survey. *Netherlands Journal of Agricultural Sciences*, vol. 1, pp. 9–29, 151–153, 156.

Ragab, R. (1995). Towards a continuous operational system to estimate the root-zone soil moisture from intermittent remotely sensed surface moisture. *Journal of Hydrology*, vol. 173, pp. 1–25.

Russo, D. (1988). Determining soil hydraulic properties by parameter estimation: On the selection of a model for the hydraulic properties. *Water Resources Research*, vol. 24, no. 3, pp. 453–459.

Schmugge, T. J., Jackson, T. J. and McKim H. L. (1980). Survey of methods for soil moisture determination. *Water Resources Research*, vol. 16, no. 6, pp. 961–979.

Singh, V. P. (1989). *Hydrologic Systems*, vol. 2: *Watershed Modeling*. Prentice Hall, Englewood Cliffs, N.J.

Singh, V. P., and Joseph E. (1994). Kinematic wave model for soil moisture movement with plant root extraction. *Irrigation Science*, vol. 14, pp. 188–198.

Singh, V. P. (1997). The use of entropy in hydrology and water resources. *Hydrological Processes*, vol. 11, no. 6, pp. 587–626.

Singh, V. P. (2000). The entropy theory as a tool for modeling and decision-making in environmental and water resources. *Water SA*, vol. 26, no. 1, pp. 1–12, 20.

Singh, V. P. (2001). Kinematic wave modelling in water resources: A historical perspective. *Hydrological Processes*, vol. 15, no. 4, pp. 671–706.

Smith, R. E. (1983). Approximate soil water movement by kinematic characteristics. *Soil Science Society of America Journal*, vol. 47, pp. 3–8.

Walker, J. P., Willgoose, G. R. and Kalma J. D. (2001). One-dimensional soil moisture profile retrieval by assimilation of near-surface observations: A comparison of retrieval algorithms. *Advances in Water Resources*, vol. 24, no. 6, pp. 631–650.

Walker, J. P., Willgoose, G. R. and Kalma J. D. (2001). One-dimensional soil moisture profile retrieval by assimilation of near-surface measurements: A simplified soil moisture model and field application. *Journal of Hydrometeorology*, vol. 2, pp. 356–373.

Western, A. W., and Blöschl G. (1999). On the spatial scaling of soil moisture. *Journal of Hydrology*, vol. 217, no. 3-4, pp. 203–224.

Yamada, T., and Kobayashi M. (1988). Kinematic wave characteristics and new equations of unsaturated infiltration. *Journal of Hydrology*, vol. 102, pp. 257–266.

Zhang, S., Li, H. Zhang, W. Qiu, C. and Li X. (2006). Estimating the soil moisture profile by assimilating near-surface observations with the ensemble Kalman Filter (EnKF). *Advances in Atmospheric Sciences*, vol. 22, no. 6, pp. 936–945.

Zhao, R. J., Zhuang, Y. Fang, L. R. Lin, X. R. and Zhang Q. S. (1980). The Xinanjiang model. *IAHS Publications*, no. 129, pp. 351–356.

Zotova, E. N., and Geller A. G. (1985). Soil moisture content estimation by radar survey data during the sowing campaign. *International Journal of Remote Sensing*, vol. 6, no. 2, pp. 353–364.

CHAPTER 13

Groundwater Flow

In this chapter, entropy theory is applied to derive the distribution of hydraulic head for the case of steady flow in confined and unconfined aquifers.

13.1 One-Constraint Formulation

It is assumed that the aquifer is characterized as homogeneous, isotropic, and one-dimensional, and flow in the aquifer is steady, as shown in Fig. 13.1. The hydraulic head is assumed to monotonically increase from some value h_0 at the lower reservoir to a maximum value H at the higher reservoir. Let h be the elevation of the hydraulic head at a distance x from the lower reservoir, and let L be the distance between the two reservoirs. It is further assumed that the hydraulic head is a random variable with a probability density function (PDF) $f(h)$. Then the objective is to derive the PDF. To that end, the entropy $E(h)$ of $f(h)$ or the hydraulic head h or the aquifer characterized by the head can be expressed as (Barbe et al. 1994)

$$E(h) = -\int_{h_0}^{H} f(h)\ln f(h)\,dh \tag{13.1}$$

To derive the least-biased $f(h)$, Eq. (13.1) is maximized, subject to specified constraints.

For simplicity, it is assumed that nothing is known a priori about the hydraulic head. Therefore, the total probability of h can be expressed as

$$\int_{h_0}^{H} f(h)\,dh = 1 \tag{13.2}$$

Equation (13.1) can be maximized, subject to Eq. (13.2), using the method of Lagrange multipliers. The lagrangian function Φ can be written as

$$\Phi = -\int_{h_0}^{H} f(h)\ln f(h)\,dh - (\lambda_0 - 1)[\int_{h_0}^{H} f(h)\,dh - 1] \tag{13.3}$$

where λ_0 is the zeroth Lagrange multiplier. Differentiating Eq. (13.3) with respect to $f(h)$ and equating the derivative to zero, one gets

$$\frac{\partial \Phi}{\partial f(h)} = -\ln f(h) - 1 - (\lambda_0 - 1) = 0 \tag{13.4}$$

497

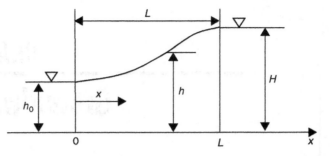

Figure 13.1 Flow in a homogenous aquifer.

Equation (13.4) leads to

$$f(h) = \exp(-\lambda_0) \tag{13.5}$$

Equation (13.5) is the PDF of h with the unknown Lagrange multiplier λ_0 which can be determined with the use of Eq. (13.2). Substitution of Eq. (13.5) in Eq. (13.2) gives

$$\int_{h_0}^{H} \exp(-\lambda_0) dh = 1 \tag{13.6}$$

Equation (13.6) yields

$$\lambda_0 = \ln(H - h_0) \tag{13.7}$$

Substitution of Eq. (13.7) in Eq. (13.5) results in

$$f(h) = \frac{1}{H - h_0} \tag{13.8}$$

It is hypothesized that the probability of the hydraulic head h being less than or equal to a given value can be expressed as a function of x/L, where x is the distance from the lower reservoir. The cumulative distribution function (CDF) of h is then cast as

$$F(h) = \frac{x}{L} \tag{13.9}$$

The PDF of h can be obtained by differentiating Eq. (13.9):

$$f(h) = \frac{1}{L}\frac{dx}{dh} \quad \text{or} \quad \frac{dh}{dx} = \frac{1}{Lf(h)} \tag{13.10}$$

Equation (13.10) expresses the hydraulic gradient in terms of the PDF of h.
  Equating Eq. (13.8) to Eq. (13.10), one finds

$$dh = \frac{dx}{L}(H - h_0) \tag{13.11}$$

Using the boundary condition $h = h_0$ at $x = 0$, the solution of Eq. (13.11) yields

$$h = h_0 + (H - h_0)\frac{x}{L} \tag{13.12}$$

Equation (13.12) describes the hydraulic head profile in an aquifer as a function of distance from the upstream end.

Substituting Eq. (13.12) in Eq. (13.9), one obtains the CDF of h in terms of the hydraulic head:

$$F(h) = \frac{h - h_0}{H - h_0} \tag{13.13}$$

Now the Shannon entropy can be expressed by inserting Eq. (13.8) in Eq. (13.1):

$$E(h) = \ln(H - h_0) \tag{13.14}$$

which is the primary entropy of the aquifer, since there is nothing known about the flow in the aquifer. Now the question arises as to the type of aquifer that Eq. (13.12) holds for.

13.1.1 Confined Aquifers

For a confined aquifer, Darcy's law can be expressed as

$$u = -K\frac{\partial h}{\partial x} \tag{13.15}$$

where u is the Darcy flux (volumetric rate of flow per unit cross-sectional area), $\partial h / \partial x$ is the head differential in the x direction, and K is the hydraulic conductivity. The negative sign in Eq. (13.15) implies that the flow is occurring in the direction of decreasing head. For one-dimensional steady flow in a confined aquifer, the continuity equation can be written as

$$\frac{\partial u}{\partial x} = 0 \tag{13.16}$$

Differentiating Eq. (13.15) with respect to x and then Eq. (13.16), one obtains

$$\frac{\partial^2 h}{\partial x^2} = 0 \tag{13.17}$$

Equation (13.17) is the Laplace equation. For solving the Laplace equation given by Eq. (13.17), the boundary conditions are $h = h_0$ at $x = 0$ and $h = H$ at $x = L$. The solution of Eq. (13.17) is actually the same as that given by Eq. (13.11). That is, the entropy-based solution shows that the Laplace equation does not assume any information about the flow and the head profile is a straight line in distance.

If A is the cross-sectional area of flow in the confined aquifer, then discharge Q through the aquifer can be written as

$$Q = -KA\frac{dh}{dx} \tag{13.18a}$$

In the case of steady confined flow, if the continuity equation and Darcy's equation are employed as another constraint, then the solution remains the same as before, provided

K and A are constant. This then suggests that Eq. (13.18a) does not add any new information. However, if either K or A or both are not constant, then new information will be obtained and the solution will be different. Substituting Eq. (13.11) in Eq. (13.18a), one gets

$$Q = \frac{KA}{L}(H - h_0) \tag{13.18b}$$

13.1.2   Unconfined Aquifer

For steady flow in an unconfined aquifer, Eqs. (13.1) to (13.14) will hold. The continuity equation can be expressed as

$$q = \int_0^h u\,dh \tag{13.19}$$

where q is the volumetric flow rate per unit width (transverse direction), h is the height of the water table, and u is the flow velocity. If it is assumed that the flux or velocity u is constant in the vertical direction, then Eq. (13.19) becomes

$$q = uh \tag{13.20}$$

By combining Eq. (13.20) with the Darcy equation, Eq. (13.15), the flow rate in the unconfined aquifer can be expressed as

$$q = -Kh\frac{dh}{dx} \tag{13.21}$$

Equation (13.21) can be written as

$$q\,dx = -Kh\,dh \tag{13.22}$$

Integrating Eq. (13.22), with the conditions $h = h_0$ at $x = 0$, and $h = H$ at $x = L$, one finds

$$q = \frac{K}{2L}(H^2 - h_0^2) \tag{13.23}$$

Equation (13.23) is the Dupuit-Forchheimer solution, which is not the same as Eq. (13.18b). Note that the hydraulic head profile is still the same as given by Eq. (13.12), which is linear.

13.2   Two-Constraint Formulation

It is assumed that the mean hydraulic head h_m is known from observations or theoretically. Then the mean head constraint can be defined as

$$\int_{h_0}^H hf(h)\,dh = h_m \tag{13.24}$$

To determine $f(h)$, Eq. (13.1) is maximized, subject to two constraints one given by the total probability expressed by Eq. (13.2) and the other given by Eq. (13.24). Again, entropy maximizing is done using the method of Lagrange multipliers where the lagrangian function becomes

$$\Phi = -\int_{h_0}^{H} f(h)\ln f(h)\,dh - (\lambda_0 - 1)\left[\int_{h_0}^{H} f(h)\,dh - 1\right] - \lambda_1\left[\int_{h_0}^{H} hf(h)\,dh - h_m\right] \tag{13.25}$$

where λ_0 and λ_1 are the Lagrange multipliers. Differentiating Eq. (13.25) with respect to $f(h)$ and equating the derivative to zero, one gets

$$\frac{\partial \Phi}{\partial f(h)} = -\ln f(h) - 1 - (\lambda_0 - 1) - \lambda_1 h = 0 \tag{13.26}$$

Equation (13.26) leads to

$$f(h) = \exp(-\lambda_0 - \lambda_1 h) \tag{13.27}$$

Equation (13.27) is the PDF of h with the unknown Lagrange multipliers λ_0 and λ_1, which can be determined with the use of Eqs. (13.2) and (13.24). Substitution of Eq. (13.27) in Eq. (13.2) gives

$$\int_{h_0}^{H} \exp(-\lambda_0 - \lambda_1 h)\,dh = 1 \tag{13.28}$$

Integration of Eq. (13.28) yields

$$\lambda_0 = \ln[\exp(-\lambda_1 h_0) - \exp(-\lambda_1 H)] - \ln \lambda_1 \tag{13.29}$$

Equation (13.28) can also be written as

$$\lambda_0 = \ln \int_{h_0}^{H} \exp(-\lambda_1 h)\,dh \tag{13.30}$$

Differentiating Eq. (13.30) with respect to λ_1, one obtains

$$\frac{\partial \lambda_0}{\partial \lambda_1} = -\frac{\int_{h_0}^{H} h\exp(-\lambda_1 h)\,dh}{\int_{h_0}^{H} \exp(-\lambda_1 h)\,dh} = \frac{\int_{h_0}^{H} h\exp(-\lambda_0 - \lambda_1 h)\,dh}{\int_{h_0}^{H} \exp(-\lambda_0 - \lambda_1 h)\,dh} = -h_m \tag{13.31}$$

Differentiation of Eq. (13.29) results in

$$\frac{\partial \lambda_0}{\partial \lambda_1} = \frac{H\exp(-\lambda_1 H) - h_0\exp(-\lambda_1 h_0)}{\exp(-\lambda_1 h_0) - \exp(-\lambda_1 H)} - \frac{1}{\lambda_1} \tag{13.32}$$

Equating Eq. (13.32) to Eq. (13.31), one finds

$$\frac{h_0\exp(-\lambda_1 h_0) - H\exp(-\lambda_1 H)}{\exp(-\lambda_1 h_0) - \exp(-\lambda_1 H)} + \frac{1}{\lambda_1} = h_m \tag{13.33}$$

Equation (13.33) expresses λ_1 in terms of the mean hydraulic head and hydraulic head values at the boundaries. Equations (13.29) and (13.33) can be used to determine the Lagrange multipliers λ_0 and λ_1.

By substituting Eq. (3.29) in Eq. (3.27), the PDF of h can be expressed as

$$f(h) = \frac{\lambda_1 \exp(-\lambda_1 h)}{\exp(-\lambda_1 h_0) - \exp(-\lambda_1 H)} \tag{13.34}$$

Note that equation has only one unknown Lagrange multiplier λ_1, which can be determined using Eq. (13.29).

Invoking the hypothesis on the CDF of h given by Eq. (13.9) and combining with Eq. (13.34), one gets

$$\frac{\lambda_1 \exp(-\lambda_1 h)}{\exp(-\lambda_1 h_0) - \exp(-\lambda_1 H)} dh = \frac{dx}{L} \tag{13.35}$$

Integration of Eq. (13.35) yields

$$\frac{\exp(-\lambda_1 h_0) - \exp(-\lambda_1 h)}{\exp(-\lambda_1 h_0) - \exp(-\lambda_1 H)} = \frac{x}{L} \tag{13.36}$$

Equation (13.36) can be expressed as

$$h = \frac{1}{\lambda_1}(\ln L - \ln\{\exp(-\lambda_1 h_0) - [\exp(-\lambda_1 h_0) - \exp(-\lambda_1 H)]x\}) \tag{13.37}$$

Interestingly, if Eq. (13.36) is approximated by retaining the first two terms in the expansion of the exponential terms, one gets

$$\frac{1 - \lambda_1 h_0 - 1 + \lambda_1 h}{1 - \lambda_1 h_0 - 1 + \lambda_1 H} = \frac{x}{L} \quad \text{or} \quad h = h_0 + (H - h_0)\frac{x}{L} \tag{13.38}$$

which is the same as Eq. (13.12).

It may be interesting to evaluate the derivative of the hydraulic head at the boundaries $x = 0$ and $x = L$. Differentiation of Eq. (13.27) yields

$$\frac{dh}{dx} = \frac{1}{\lambda_1}\frac{\exp(-\lambda_1 h_0) - \exp(-\lambda_1 H)}{\exp(-\lambda_1 h_0) - [\exp(-\lambda_1 h_0) - \exp(-\lambda_1 H)]x} \tag{13.39}$$

At the lower head boundary, at $x = 0$, $h = h_0$, and the derivative is

$$\frac{dh}{dx} = \frac{1}{\lambda_1}\frac{\exp(-\lambda_1 h_0) - \exp(-\lambda_1 H)}{\exp(-\lambda_1 h_0)} \tag{13.40}$$

At the greater head boundary, at $x = L$, $h = H$, and the derivative is

$$\frac{dh}{dx} = \frac{1}{\lambda_1}\frac{\exp(-\lambda_1 h_0) - \exp(-\lambda_1 H)}{\exp(-\lambda_1 h_0) - [\exp(-\lambda_1 h_0) - \exp(-\lambda_1 H)]L} \tag{13.41}$$

Equations (13.40) and (13.41) show that the derivatives are constant at the boundaries. If $\lambda_1 = 0$, then $dh/dx = 0$ at both boundaries, meaning that the hydraulic head surface becomes horizontal. Thus, near the vertical boundaries the flow lines are not perpendicular to the boundary as assumed in the flow net theory. Furthermore, Eq. (13.39) shows that in unconfined aquifers the piezometric head surface can never be horizontal and its gradient decreases with increasing x.

13.2.1 Unconfined Aquifers

Equation (13.21) can be expressed by deleting the negative sign due to the head increase with increasing x as

$$dx = \frac{K}{qL}h\,dh \tag{13.42}$$

Recall that the mean head can be expressed by taking advantage of Eq. (13.10) from the CDF hypothesis and Eq. (13.42) as

$$\int_{h_0}^{K} hf(h)\,dh = h_m = \int_{h_0}^{H} h\,dx = \frac{K}{2qL}\int_{h_0}^{H} h\,dh \tag{13.43}$$

Integrating Eq. (13.43), one finds

$$\int_{h_0}^{H} hf(h)\,dh = h_m = \frac{K}{2qL}(H^2 - h_0^2) \tag{13.44}$$

But the integration of Eq. (13.10) also yields the profile

$$\int_{h_0}^{h} f(h)\,dh = \frac{x}{L} + C \tag{13.45}$$

where C is the constant of integration evaluated at $h = H$ at $x = L$. Thus, Eq. (13.44) can also be used as a constraint, besides Eq. (13.2). In that case, the lagrangian function becomes

$$\Phi = -\int_{h_0}^{H} h(h)\ln f(h)\,dh - (\lambda_0 - 1)\left[\int_{h_0}^{H} f(h)\,dh - 1\right] - \lambda_1\left[\int_{h_0}^{H} hf(h)\,dh - \frac{K}{2qL}(H^2 - h_0^2)\right] \tag{13.46}$$

Differentiating Eq. (13.46) with respect to $f(h)$, one gets

$$\frac{\partial \Phi}{\partial f(h)} = -\ln f(h) - 1 - (\lambda_0 - 1) - \lambda_1 h = 0 \tag{13.47}$$

which yields

$$f(h) = \exp(-\lambda_0 - \lambda_1 h) \tag{13.48}$$

Equation (13.48) is the PDF of h with the unknown Lagrange multipliers λ_0 and λ_1 which can be determined with the use of Eqs. (13.2) and (13.44).

Substitution of Eq. (13.48) in Eq. (13.45) gives

$$\int_{H}^{h} \exp(-\lambda_0 - \lambda_1 h)\,dh = \frac{x - L}{L} \tag{13.49}$$

Integrating Eq. (13.49) with the condition $h = H$ at $x = L$ yields

$$h = -\frac{1}{\lambda_1}\left\{\lambda_0 + \ln\left[\frac{L-x}{L}\lambda_1 + \exp(-\lambda_0 - \lambda_1 H)\right]\right\} \tag{13.50}$$

Substitution of Eq. (13.48) in Eq. (13.2) for the first constraint yields

$$\exp(-\lambda_0 - \lambda_1 h_0) - \exp(-\lambda_0 - \lambda_1 H) = \lambda_1 \tag{13.51}$$

or

$$\exp(-\lambda_0) = \frac{\lambda_1}{\exp(-\lambda_1 h_0) - \exp(-\lambda_1 H)} \tag{13.52}$$

Inserting Eq. (13.48) in Eq. (13.44) for the second constraint, one gets

$$h_0 \exp(-\lambda_0 - \lambda_1 h_0) - H\exp(-\lambda_0 - \lambda_1 H) + \frac{1}{\lambda_1}[\exp(-\lambda_0 - \lambda_1 H) - \exp(-\lambda_0 - \lambda_1 h_0)]$$

$$= \frac{\lambda_1 K}{2Lq}(H^2 - h_0^2) \tag{13.53}$$

Equations (13.52) and (13.53) can be solved for the Lagrange multipliers λ_0 and λ_1. When these equations are combined, the result is

$$h_0 \exp(-\lambda_0 - \lambda_1 h_0) - H\exp(-\lambda_0 - \lambda_1 H) = \frac{K(H^2 - h_0^2) - 2Lq}{2Lq}[\exp(-\lambda_1 h_0) - \exp(-\lambda_1 H)] \tag{13.54}$$

Equation (13.54) has only one unknown parameter λ_1.

13.2.2   Confined Aquifer

The flow through a confined aquifer is given by Eq. (13.18a). Substituting Eq. (13.10) in Eq. (13.18a) and multiplying both sides by dh, one finds

$$Q\,dh\,Lf(h) = -KA\,dh \tag{13.55}$$

Substitution of Eq. (13.34) in Eq. (13.55) yields

$$QL\frac{\lambda_1 \exp(-\lambda_1 h)}{\exp(-\lambda_1 h_0) - \exp(\lambda_1 H)}dh = -KA\,dh \tag{13.56}$$

Integration of Eq. (13.56) leads to

$$Q = \frac{KA(H-h)}{[\exp(-\lambda_1 h) - \exp(-\lambda_1 H)][\exp(-h_0) - \exp(-\lambda_1 H)]} \tag{13.57}$$

13.3   Inverse Problems

A groundwater model may contain a finite number of parameters, such as hydraulic conductivity, transmissivity, or storativity. Each model parameter can be viewed as a random variable and will therefore have a PDF. The model parameters will have a multidimensional PDF. The objective is to determine the posterior PDF of the model, given a prior PDF and other information expressed as constraints. Thus, the posterior PDFs for groundwater and hydrogeologic applications can be estimated, as discussed by Woodbury and Ulrych (1993).

Consider a case where upper (U) and lower (L) bounds of all model parameters are known from data. This information suggests a uniform distribution or a boxcar PDF. Let k be a parameter. Then its PDF can be written as

$$q(k) = \begin{cases} \dfrac{1}{U-L} & L \le k \le U \\ 0 & \text{otherwise} \end{cases} \tag{13.58}$$

where L and U are the lower and upper bounds, respectively. If k is a random variable vector $k^T = (k_1, k_2, \cdots)$, then $q(k)$ is a multidimensional PDF. For simplicity we will deal with a univariate case.

The relative or cross entropy of k which has posterior probability density $p(k)$ and prior PDF $q(k)$ can be written as

$$H(p,q) = \int_L^U p(k) \ln \frac{p(k)}{q(k)} dk \tag{13.59}$$

Let the constraints be defined as expected value constraints that can be written as

$$\int_L^U g_j(k)p(k)dk = \overline{g_j(k)} = \overline{g}_j \qquad j = 1, 2, \ldots, m \tag{13.60}$$

To determine the PDF $p(k)$, $H(p, q)$ is minimized subject to Eq. (13.60) and

$$\int_L^U p(k)dk = 1 \tag{13.61}$$

Cross entropy minimizing can be done using the method of Lagrange multipliers where the lagrangian function ψ can be written as

$$\psi = \int_L^U p(k) \ln \frac{p(k)}{q(k)} dk + (\lambda_0 - 1)\left[\int_L^U p(k)dk - 1\right] + \sum_{j=1}^m \lambda_j \left[\int_L^U g_j(k)p(k)dk - \overline{g}_j\right] \tag{13.62}$$

where λ_j, $j = 0, 1, 2, \ldots, m$, are the Lagrange multipliers. Minimizing the variation in ψ with respect to $p(k)$, one obtains

$$0 = \ln \frac{p(k)}{q(k)} + \lambda_0 + \sum_{j=1}^m \lambda_j g_j(k) \tag{13.63}$$

Equation (13.63) yields

$$p(k) = q(k)\exp[-\lambda_0 - \sum_{j=1}^m \lambda_j g_j(k)] \tag{13.64}$$

Let $\exp(-\lambda_0) = c$. Then Eq. (13.64) becomes

$$p(k) = cq(k)\exp[-\sum_{j=1}^m \lambda_j g_j(k)] \tag{13.65}$$

Equation (13.65) can be written in multiplicative form as

$$p(k) = cq(k) \prod_{j=1}^{m} \exp[-\lambda_j g_j(k)] \tag{13.66}$$

In Eq. (13.66) the Lagrange multipliers can be determined with the use of Eqs. (13.60) and (13.61). First consider $L = 0$. Then $q(k) = 1/U$. Substituting Eq. (13.66) in Eq. (13.61), one gets

$$\frac{c}{U} \int_0^U \prod_{j=1}^{m} \exp[-\lambda_j g_j(k)] dk = 1 \tag{13.67}$$

Integrating each product term separately, one gets

$$\frac{c}{U} \prod_{j=1}^{m} \frac{1 - \exp(-\lambda_j U)}{\lambda_j} = 1 \tag{13.68}$$

Therefore,

$$c = \prod_{j=1}^{m} \frac{-\lambda_j U}{\exp(-\lambda_j U) - 1} \tag{13.69}$$

Substitution of Eq. (13.69) in Eq. (13.66) yields

$$p(k) = \prod_{j=1}^{m} \frac{-\lambda_j}{\exp(-\lambda_j U) - 1} \exp[-\lambda_j g_j(k)] \tag{13.70}$$

Substitution of Eq. (13.70) in Eq. (13.60) yields

$$\int_0^U \prod_{j=1}^{m} \frac{-\lambda_j U}{\exp(-\lambda_j U) - 1} \exp[-\lambda_j g_j(k)] g_j(k) dk = \overline{g_j} \qquad j = 1, 2, \dots, m \tag{13.71}$$

Equation (13.71) can be evaluated term by term

$$\int_0^U \frac{-\lambda_j U}{\exp(-\lambda_j U) - 1} \exp[-\lambda_j g_j(k)] g_j(k) dk = \overline{g_j} \tag{13.72}$$

Solution of Eq. (13.72) is

$$\frac{\exp(-\lambda_j U) \lambda_j U + \exp(-\lambda_j U) - 1}{\lambda_j [\exp(-\lambda_j U) - 1]} = \overline{g_j} \tag{13.73}$$

It is interesting that a first-order Taylor series expansion yields an explicit solution for λ_j:

$$\lambda_j \cong \frac{3}{2g_j^2} \left(U - 2g_j^2 \right) \tag{13.74}$$

This solution is less than satisfactory when the prior estimate g_j is close to the boundaries.

Finally, the posterior PDF is

$$p(k) = \prod_{j=1}^{m} \frac{-\lambda_j}{\exp(-\lambda_j U) - 1} \exp(-\lambda_j g_j) \qquad \overline{g}_j \neq \frac{U}{2} \qquad (13.75)$$

$$p(k) = \prod_{j=1}^{m} = \frac{1}{U} \qquad g_j = \frac{U}{2} \qquad (13.76)$$

These results can be extended to the prior ranging from L to U, using the transformation $u_* = U - L, k_* = k - L$, and $g_j = \overline{g} - L$.

Example 13.1. Consider hydraulic conductivity as a random variable having lower and upper bounds 4.32 meters per day (m/d) and 8.64 m/d, respectively (Woobdbury and Ulrych, 1993), for the Borden aquifer. The expected value is 6.05 m/d. Compute the posterior probability distribution of hydraulic conductivity.

Solution. It is known from given information that $L = 4.32$ m/d, $U = 8.64$ m/d, and there is one constraint

$$\int_L^u p(k)\,dk = \overline{g_1(k)} = 1 \qquad \text{and} \qquad \int_L^u kp(k)\,dk = \overline{g_2(k)} = 6.05$$

By transformation $k_* = k - L$, the new lower bound is 0, and the new upper and expected values are

$$u_* = U - L = 4.32 \qquad \text{and} \qquad g_* = \overline{g} - L = 1.67$$

From Eq. (13.74), the Lagrange multiplier can be computed as

$$\lambda_1 \cong \frac{3}{2g^2}(U - 2g^2) = \frac{3}{2 \times 1.67^2}(4.32 - 2 \times 1.67^2) = -0.677$$

Thus, the posterior distribution becomes

$$p(k) = \frac{-\lambda_1}{\exp(-\lambda_1 U) - 1} \exp(-\lambda_1 \overline{g}) = \frac{0.677}{\exp(0.677 \times 4.32) - 1} \exp(-0.677k) = 0.038 \exp(-0.677k)$$

13.4   Inverse Linear Problem

Here we discuss the estimation of model parameters, subject to constraints imposed by the observed data. It may be that unknown parameters are greater than measured data values. Then the inverse problem becomes undetermined, and an infinite number of parameters that satisfy the data may be found. Each model parameter is assumed as a random variable. The "true" model parameters are then regarded as the expected values of those random variables. In the context of entropy theory, an estimate of the prior PDF of the model parameter is given and some information on the parameters is known from observations as constraints. Then the PDF of parameters is determined that has minimum relative cross entropy. We can estimate model parameters by computing the expected values of the posterior PDF, together with appropriate confidence limits, as discussed by Woodbury and Ulrych (1998).

13.4.1   Generalized Linear Inversion

Following Woodbury and Ulrych (1996), consider a linear discrete problem

$$y = Gk \qquad (13.77)$$

where y represents a discrete set of known data, G denotes the known kernel functions, and k is the vector of model parameters that are unknown and are to be determined. Equation (13.77) can also be written in discrete form as

$$y_j = \sum_{n=0}^{N} g_{jn} k_n \tag{13.78}$$

The expected value of the random vector k can be written as

$$\bar{k} = \int_K k p(k) \, dk \tag{13.79}$$

in which the integration is over all the allowable values of K. Equation (13.78) can be written as

$$\bar{y_j} = \int_K p(k) \left[\sum_{n=0}^{N} g_{jn} k_n \right] \tag{13.80}$$

where $p(k)$ is the true PDF of k. Equation (13.80) is in the form of expected value constraint. The known expected values correspond to physical measurements, and if these measurements contain errors, the k values are allowed in the range $(0, U)$ where U is an upper bound.

Following Woodbury and Ulrych (1996), a prior PDF of k is

$$q(k) = \begin{cases} \dfrac{1}{U_n - L_n} & L_n \le k_n \le U_n \\ 0 & \text{otherwise} \end{cases} \tag{13.81}$$

where $n = 0, 1, 2, \ldots, N$, is the number of parameters and L_n and U_n are the lower and upper bounds for individual parameters, respectively. Then

$$q(k) = \prod_{n=0}^{N} \frac{-\beta_n}{\exp(-\beta_n U) - 1} \exp(-\beta_n k_n) \tag{13.82}$$

which is a multivariate truncated exponential, and the β_n are the Lagrange multipliers determined from the lower and upper bounds and the expected value constraints.

For K expected value constraints, one can write

$$p(k) = q(k) \exp\left(-\lambda_0 - \sum_{j=1}^{K} \lambda_j \sum_{n=0}^{N} g_{jn} k_n \right) \tag{13.83}$$

This can be written as

$$p(k) = c q(k) \exp\left(-\sum_{n=0}^{N} k_n \sum_{j=1}^{K} \lambda_j g_{jn} \right) \qquad c = \exp(-\lambda_0) \tag{13.84}$$

or in product form

$$p(k) = c q(k) \prod_{n=0}^{N} \exp\left(-k_n \sum_{j=1}^{K} \lambda_j g_{jn} \right) \tag{13.85}$$

Substituting for $q(k)$,

$$p(k) = c \prod_{n=0}^{N} \frac{-\beta_n}{\exp(-\beta_n U) - 1} \exp\left(-\beta_n k_n - k_n \sum_{j=1}^{K} \lambda_j g_{jn}\right)$$ (13.86)

or

$$p(k) = c \prod_{n=0}^{N} \frac{-\beta_n}{\exp(-\beta_n U) - 1} \exp(-k_n a_n)$$ (13.87)

where

$$a_n = \beta_n + \sum_{j=1}^{K} \lambda_j g_{jn}$$ (13.88)

Let

$$d_n = \frac{-\beta_n}{\exp(-\beta_n U) - 1}$$ (13.89)

Now applying Eq. (13.87) to the total probability constraint, one gets

$$p(k) = c \prod_{n=0}^{N} d_n \int_K \exp(-k_n a_n) dk$$ (13.90)

Integrating Eq. (13.90) term by term, one finds

$$p(k) = \prod_{n=0}^{N} \frac{-a_n}{\exp(-a_n U) - 1} \exp(-k_n a_n)$$ (13.91)

This is a multivariate truncated exponential PDF. The estimate $\bar{k}$ is the expected value of Eq. (13.91). Integrating it yields

$$\bar{k}_n = \frac{\exp(-a_n U) a_n U + \exp(-a_n U) - 1}{a_n [\exp(-a_n U) - 1]}$$ (13.92)

Equation (13.92) contains unknown Lagrange multipliers. Substituting in Eq. (13.80),

$$\bar{y}_j = \sum_{n=0}^{N} g_{jn} \bar{k}_n(\lambda)$$ (13.93)

Now we need to determine the Lagrange multipliers. This is done numerically, as discussed by Woodbury and Ulrych (1993).

Example 13.2. Consider the convolution representation of a plume's spatial distribution in one dimension as $C(x,t) = \int_0^t C_{in}(\tau) f_r(x, t-\tau) d\tau$, where the kernel function is defined as $f_r(x, t-\tau) = \frac{x}{2[\pi(t-\tau)^3]^{1/2}} \exp\left\{-\frac{[x-(t-\tau)]^{1/2}}{4(t-\tau)}\right\}$. The contamination release at $T = 300$ d is observed as shown in Fig. 13.2. Solve for the posterior $p(C_{in})$ and estimate C_{in}.

Solution. The above convolution can be written for the linear inverse model as

$$C(x,t) = \int_0^t C_{in}(\tau) f_r(x, t-\tau) d\tau = \sum_\tau \phi_{x,\tau} C_{in}(\tau)$$

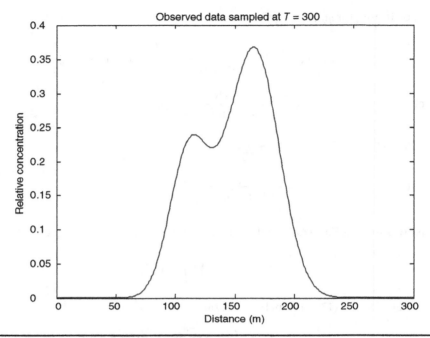

FIGURE **13.2**   Observed contaminant at $T = 300$ d.

where $C(x, t)$ is the given observation at distance x and time t and $C_{in}(t)$ is the release at the source, which is unknown. Thus, from Eq. (13.92) the expected $C_{in}(t)$ can be estimated from

$$\overline{C}_{in} = \frac{\exp(-a_n U)a_n U + \exp(-a_n U) - 1}{a_n[\exp(-a_n U) - 1]}$$

and the estimated output from Eq. (13.93) as

$$\overline{C}_{x,t} = \sum_{n=0}^{t} \phi_{xn}\overline{C}_{in}(n) = \sum_{n=0}^{t} \phi_{xn} \frac{\exp(-a_n U)a_n U + \exp(-a_n U) - 1}{a_n[\exp(-a_n U) - 1]}$$

Now the Lagrange multipliers a_n can be computed using the Newton-Ralphson method. The quantity $C_j(a)$ can be expressed from the first-order Taylor series as

$$C_t(a) \cong C(a^0) + (a - a^0)^T [\nabla C_t(a)]|_{a=a^0} = \overline{C}_t \qquad \text{for } t = 0, \ldots, 300$$

Define

$$\delta = a - a^0$$
$$\vec{v} = [\overline{C}_0 - C_0(a^0), \ldots, \overline{C}_{300} - C_{300}(a^0)]^T$$

and matrix M as

$$M_{kn} = \left(\frac{\partial C_n(a)}{\partial a_k} \right)\Bigg|_{(a=a^0)}$$

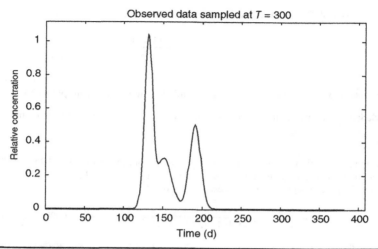

FIGURE 13.3 Estimated contaminant release.

Solve for $M\vec{\delta} = \vec{v}$, and the new value becomes $a = a^0 + \delta$. Compute iteratively until it converges. Starting from $a_n = [0, 0, 0,...]$, the a_n values converge to [375, 223, 117, 68.1, 21.7, 10.6, 5.53, 1.14, ...], which constitute the final solution.

Thus, the estimated

$$\overline{C_{in}} = \frac{\exp(-a_n U)a_n U + \exp(-a_n U) - 1}{a_n[\exp(-a_n U) - 1]}$$

is shown in Fig. 13.3.

13.4.2 Confidence Interval

The confidence interval about the mean $\overline{k}$ can be found by determining the cumulative distribution function for k. Let

$$\int_0^K p(k)\,dk = P(k) \tag{13.94}$$

Inserting Eq. (13.90) in Eq. (13.94) and then integrating term by term, one gets

$$P(k_n) = \frac{\exp(-a_n k_n) - 1}{\exp(-a_n U) - 1} \qquad 0 \le k_n \le U \tag{13.95}$$

If $k_n = U$, then $P(U) = 1$. If we determine $P^{-1}(k_n)$, then we can determine k_n corresponding to a given probability level, 95 percent. Let

$$P(k_n) = c\exp(-a_n k_n) - 1 \tag{13.96}$$

where

$$c = \frac{1}{\exp(-a_n U) - 1} \tag{13.97}$$

The inverse of P, denoted by P^{-1}, can be written as

$$k_n = -\frac{\log(p/c+1)}{a_n} \tag{13.98}$$

Example 13.3. Estimate the 95 percent confidence interval for Example 13.2.

Solution. For a two-sided confidence interval with $\alpha = 0.05$, one needs to compute the bounds for

$$P(L_{in} \le C_{in} \le U_{in}) = 95\%$$

which is equivalent to estimating the value corresponding to $P(C_{in}) = 0.025$ and $P(C_{in}) = 0.975$. For example, estimate the confidence intervals for C_{in} (173) of Fig. 13.3, where $U = 1.5$ and $a_n = 1.13$ from the previous example. From Eq. (13.97), one has

$$c = \frac{1}{\exp(-a_n U)-1} = \frac{1}{\exp(-1.13 \times 1.5)-1} = -1.225$$

Thus, from Eq. (13.98), the upper and lower bounds of C_{in} (173) of the 95 percent confidence interval are

$$U_{in} = -\frac{\log(P/c+1)}{a_n} = \frac{\log(-0.975/1.225+1)}{1.13} = 0.61$$

$$L_{in} = -\frac{\log(P/c+1)}{a_n} = \frac{\log(-0.025/1.225+1)}{1.13} = 0.01$$

Questions

1. Obtain a set of saturated hydraulic conductivity values corresponding to different locations for an aquifer. Plot these values as a histogram, and estimate the probability density function that will best fit the values. Then derive the density function using entropy theory and estimate its parameters by the theory. How well does it compare with the empirical (observed) density function?

2. Obtain a set of hydraulic head values at different times at a location of an unconfined aquifer. Plot the values as a histogram and determine the head density function. Now evaluate whether this density function can be estimated by entropy theory. What type of constraints will suffice for representing the empirical probability density function?

3. For the data in Question 2, determine the groundwater discharge at that location in the aquifer. Obtain additional data that will be needed.

4. Obtain a set of hydraulic head values at different times at a location of a confined aquifer. Plot the values as a histogram and determine the head density function. Now evaluate whether this density function can be estimated by entropy theory. What type of constraints will suffice for representing the empirical probability density function?

5. For the data in Question 4, determine the groundwater discharge at that location in the aquifer. Obtain additional data that will be needed.

6. Obtain a set of values of hydraulic conductivity and specify the lower and upper bounds. Compute the mean of the hydraulic conductivity values. Now assume

that the hydraulic conductivity values have a uniform probability distribution as a prior. Then compute the posterior probability distribution of hydraulic conductivity with constraint as mean hydraulic conductivity and compare it with the empirical probability density function.

7. Estimate the 95 percent confidence intervals for the posterior probability distribution derived in Question 6.

References

Barbe, D. E., Cruise, J. F. and Singh V. P. (1994). Derivation of a distribution for the hydraulic head in groundwater flow using entropy. In: *Stochastic and Statistical Methods in Hydrology and Environmental Engineering*, vol. 2, edited by K. W. Hipel. Kluwer Academic Publishers, Dordrecht, the Netherlands, pp. 151–161.

Woodbury, A. D., and Ulrych T. J. (1993). Minimum relative entropy: Forward probabilistic modeling. *Water Resources Research*, vol. 29, no. 8, pp. 2847–2860.

Woodbury, A. D., and Ulrych T. J. (1996). Minimum relative entropy inversion: Theory and application to recovering the release history of a groundwater contaminant. *Water Resources Research*, vol. 32, no. 9, pp. 2647–2681.

Woodbury, A. D., and Ulrych T. J. (1997). Minimum relative entropy and probabilistic inversion in groundwater hydrology. *Stochastic Hydrology and Hydraulics*, vol. 12, pp. 317–358.

Woodbury, A. D., and Ulrych T. J. (1998). Reply. *Water Resources Research*, vol. 34, no. 8, pp. 2081–2084.

Further Reading

Kabala, Z. J., and Skaggs T. H. (1998). Comment on "Minimum relative entropy inversion: Theory and application to recovering the release history of a groundwater contaminant" by A. D. Woodbury and T. D. Ulrych. *Water Resources Research*, vol. 34, no. 8, pp. 2077–2079.

Skaggs, T. H., and Kabala A. J. (1994). Recovering the release history of a groundwater contaminant. *Water Resources Research*, vol. 30, no. 1, pp. 71–79.

Tarantola, A. (1987). *Inverse Problem Theory*. Elsevier, New York.

Ulrych, T. J. (1992). Minimum relative entropy and inversion. *Proceedings of Geophysical Inversion Workshop, SIAM*, pp. 158–169.

Ulrych, T. J., and Bishop T. N. (1975). Maximum entropy spectral analysis and autoregressive deconvolution. *Reviews of Geophysics*, vol. 13, pp. 183–200.

Ulrych, T. J., Bassrei A. and Lane M. (1990). Minimum relative entropy inversion of 1D data with applications. *Geophysical Prospecting*, vol. 38, pp. 4965–4987.

Woodbury, A. D., and Ulrych T. J. (1995). Practical probabilistic groundwater modeling. *Groundwater*, vol. 33, no. 4, pp. 532–538.

PART IV

Surface Flow

CHAPTER **14**

Rainfall-Runoff Modeling

This chapter deals with the application of entropy theory to four aspects of rainfall-runoff modeling: (1) computation of runoff volume by the SCS-CN method, (2) peak discharge determination by the rational method, (3) runoff hydrograph determination by the unit hydrograph, and (4) computation of the response of a linear reservoir used as a conceptual model and the probability distribution of runoff.

14.1 Runoff Volume by SCS-CN Method

The Soil Conservation Service (SCS) (now called Natural Resources Conservation Service, or NRCS) *curve number* (CN) method was first published in 1956 in Section 4 of the *National Handbook of Soil Conservation Service*, U.S. Department of Agriculture. The SCS-CN method was originally developed for predicting runoff from small agricultural watersheds for individual rainfall events. It has since been revised several times and extended to rural, forest, and urban watersheds, and it is now applied to a range of environments, including erosion and water quality modeling. However, the core content and basic hypotheses of the method have remained essentially unchanged. In their treatise, Mishra and Singh (2003) presented a comprehensive discussion of the SCS-CN methodology. More recently, Hawkins et al. (2009) provided an excellent appraisal of the state of practice of curve number hydrology and discussed a number of philosophical questions, including: Is it science? Does it work? Can it be replaced? Can it be improved? They also made a number of recommendations on unfolding opportunities and needs for further investigations. Singh (2013) derived the SCS-CN method using entropy theory, and his derivation will be followed here.

14.1.1 Preliminaries

The SCS-CN method, as it is commonly called, is one of the most popular methods for computing runoff volume from small watersheds due to a rainstorm. Its popularity lies in its simplicity, on one hand, and its ability, on the other hand, to account for the effect of some key runoff-producing watershed characteristics, such as soil type, land use, hydrologic condition, and antecedent soil moisture condition. At the field level, it seems to work reasonably well, and that is also one of the reasons for its popularity. Despite several limitations, the method has been in frequent use, and it has received significant attention in recent years. Like any other method, the SCS-CN method is not without limitations which depend on the spatial and temporal scale at which the method is applied. For large watersheds the spatial distribution of rainfall intensity may become more important than the temporal distribution, and heterogeneity in infiltration characteristics may be large. The SCS-CN method may not be able to account for these heterogeneities.

Several investigators have discussed the theoretical basis of the SCS-CN method (Hawkins et al., 1985; Ponce and Hawkins, 1996; Bonta, 1997; Smith, 1997; Yu, 1998; Mishra and Singh, 1999). A significant amount of discussion has gravitated around the soil moisture retention capacity S, antecedent moisture condition (AMC), relation between AMCs, determination of curve numbers (CNs), storm rainfall P and size that are inherent in the method, and watershed size. AMC is divided into three categories, labeled I, II, and III, corresponding, respectively, to dry (lower extreme), medium, and wet (upper extreme) conditions. For CN between 50 and 95, Hawkins et al. (1985) showed that $S_I = 2.281 S_{II}$ and $S_{III} = 0.427 S_{II}$, and $S_I/S_{II} \cong S_{II}/S_{III} = 2.3$ (where subscripts refer to AMC categories). Likewise they defined CN_I and CN_{III} in terms of CN_{II}. They also developed guidelines for storm size as $P/S = 0.46$, $Q/P = 0.12$; $P(Q/S_{II} > 0) > 0.90$.

Although much of the discussion of the SCS-CN method has been deterministic, several investigators have also dealt with probabilistic aspects. Hjelmfelt et al. (1982) were probably the first to discuss the probabilistic aspects of the SCS-CN method by considering the CN and runoff Q variations. They found that the values of S determined from annual maximum events, when assumed to be log-normally distributed, led to the SCS-tabulated values for AMC I, AMC II, and AMC III, corresponding to 90, 50, and 10 percent cumulative probabilities of CN. Hawkins et al. (1985) showed that the AMC relationships and plots described the 90, 50, and 10 percent cumulative probabilities of runoff for a given P through CN. Defining the SCS-CN equation corresponding to AMC I, II, and III, for Q as Q_I, Q_{II}, and Q_{III}, they showed that at AMC I, there is a 10 percent chance of runoff depth Q exceeding that indicated by the Q_{II} equation. Similarly, for $CN_{II} = 80$ and $P = 2$ in., runoff expectation would be 90 percent for $Q > 0.11$ in., 50 percent for $Q > 0.56$ in., and 10 percent for $Q > 1.12$ in. for the Q_I, Q_{II}, and Q_{III} equations, respectively. From these bounds and plots one can define limits on P for which $P(Q) = 0$, that is, $P \le I_a$. When $P < 0.08S$ for condition III (C_{III}), when $P < 0.2S$ for condition II (C_{II}), and when $P < 0.465S$ for condition I (C_I), then $P(Q) = 0$.

Hawkins et al. (1985) noted that there is a conditional probability at $P > 0$ for which $Q = 0$. This probability denotes the fraction of occurrences when P is less than or equal to the initial abstraction. This happens when $P < 0.465S$ for AMC I, $P < 0.2S$ for AMC II, and $P < 0.085S$ for AMC III. Hawkins et al. (1985) derived a probability distribution of runoff using antecedent moisture condition–curve number relationships and found the distribution to be lognormal. In a recent study, Hawkins and Ward (2012) considered soil moisture retention S as a random variable, standardized it on the antecedent soil moisture condition II, found the standardized S to have a lognormal distribution, and then derived the expected value of runoff. These considerations lead to defining the storm size.

Smith and Montgomery (1979) argued that $P/S < 0.4$ poorly defines a CN. Hjelmfelt (1982) suggested a small storm as $P < 0.2S$ where S is obtained from the AMC I curve number. Hawkins et al. (1985) interpreted this threshold as $P < 0.456S$ or probability $(Q/S > 0) = 0.90$, where S is from AMC II; they went on to show that this small storm was indeed large from the runoff probability viewpoint. Defining relative storm size on the CN-based probabilities, they classified storm sizes as follows:

- $P/S_{II} \in (0, 0.085)$ or probability of $Q/S_{II} \in (0, 0.10)$ for very small
- $P/S_{II} \in (0.085, 0.20)$ or probability of $Q/S_{II} \in (0.10, 0.50)$ for small
- $P/S_{II} \in (0.20, 0.456)$ or probability of $Q/S_{II} \in (0.50, 0.90)$ for medium

- $P/S_{II} \in (0.456, 0.60)$ or probability of $Q/S_{II} \in (0.90, 0.95)$ for large
- $P/S_{II} > 0.60$ or probability of $Q/S_{II} > 0.95$ for very large. They noted that the definition of size based on P/S_{II} was also dependent on geologic and meteorological factors.

Fundamental to any calculation by the SCS-CN method is the determination of CN. The original CN determination used maximum annual events, and subsequent development considered P and Q as frequency distributions. Treating P and Q as separate frequency distributions, Bonta (1997) presented another method for determining the watershed CN. The methods differ in the use of measured P and Q data for 24-h period or event totals. He evaluated the effect of sample size on the accuracy and variability of CN estimates. The derived distribution method performed best when extreme 24-h totals were used and the actual sample size was greater than 25. The derived method was better than the median and average methods as well as the asymptotic method. Using Monte Carlo simulation, Bonta (1997) showed that the CN values estimated by his method were less variable for a wide range of sample sizes than by other methods. McCuen (2002) assumed CN to be a random variable and developed confidence intervals for it. He found the quantity $100 - CN$ to fit a gamma distribution. In another study, Schneider and McCuen (2005) discussed that the accuracy of CN is tied to the initial abstraction coefficient and went on to develop statistical guidelines for curve number generation which would lead to an improved SCS-CN method.

Assuming exponential distributions for the spatial variation of infiltration capacity and for the temporal variation of rainfall rate, Yu (1998) theoretically derived the SCS-CN method and showed that the potential retention or CN of a watershed is related to its spatially averaged infiltration capacity. Priestley (2000) questioned the use of the exponential distribution for the exponentially varying rainfall rate. Considering the spatial variation of infiltration storage with the SCS-CN method, he showed that the first derivative of the SCS equation would lead to the cumulative distribution of storage across a watershed, and this distribution was similar to the gamma distribution. The infiltration storage is the integral of the catchment infiltration rate. Yu (1998) showed that the cumulative distribution of the infiltration capacity can be estimated as the ratio of the derivative of the rainfall rate to the derivative of the infiltration rate. The potential retention is related to the spatially averaged infiltration capacity of the watershed.

Using the average rainfall rate, with the stipulation that the rainfall rate and average infiltration capacity across the watershed are both related to the rainfall duration, Istanbulluoglu (2000) expressed the contributing watershed fraction to runoff as $1 - \exp(P_e/S)$ and runoff in that fraction as $Q = P_e - S[1 - \exp(-P_e/S)]$, where P_e is the effective rainfall. Hawkins (1982) showed that the average loss rate was equal to $0.6142S$, and the cumulative probability distribution of infiltration rate I at a rainfall intensity equal to i yields the fraction of the watershed contributing to runoff. It is also the slope of the rainfall-runoff curve which can be computed by differentiating Q with respect to P. He showed that Yu's approach seems to be a ϕ index of the SCS infiltration. The slope of the SCS rainfall-runoff curve increases during continuing rainfall, and the rainfall-runoff relation defined by Yu (1998) has a constant slope of 54.55 percent, which leads to a fixed value of the contributing area.

Literature review shows that the spatial infiltration capacity, temporal rainfall rate, soil moisture retention, curve number, abstraction coefficient, and runoff have each been treated as random variables in different studies, and different distributions have been assumed or empirically derived. In this chapter the SCS-CN equation is revisited using entropy theory, and the implications arising therefrom are discussed. Also probability distributions of the variables of the SCS-CN method are derived. First, it will be pertinent to revisit the mathematical basis of the method.

14.1.2   Mathematical Basis

The SCS-CN method is based on mass conservation, expressed in volumetric form as

$$P = I_a + J - Q \tag{14.1}$$

and a hypothesis expressed as

$$\frac{Q}{P - I_a} = \frac{J}{S} \tag{14.2}$$

where

$$I_a = aS \tag{14.3}$$

in which Q is the direct or surface runoff; P is the precipitation or rainfall; J is the cumulative infiltration; I_a is the initial abstraction, including interception loss, surface storage, evaporation, and infiltration before time to ponding after runoff begins; S is the maximum soil moisture retention or the maximum amount of infiltration; and a is some fraction of S. The water that contributes to interception and surface storage is evaporated back to the atmosphere and contributes to neither runoff nor infiltration. The infiltrated water before the time to ponding may be interpreted as having satisfied the atmospheric demand of water absorption of the soil air column. Therefore, I_a can be deducted from the amount of total rainfall.

Equation (14.2) is referred to as proportional equality hypothesis because it equates the ratio of the actual amount of runoff Q to the maximum potential runoff (total rainfall minus initial abstraction, or $P - I_a$) to the ratio of the amount of actual infiltration to the amount of potential soil maximum retention S. Equation (14.3) relates the initial abstraction to the maximum soil moisture retention.

14.1.3   Derivation

Derivation of the SCS-CN method using entropy involves the following steps (Singh, 2013): (1) defining the Shannon entropy, (2) defining constraints, (3) maximizing the Shannon entropy, (4) determining parameters, (5) determining the probability density function, (6) hypothesizing a cumulative probability distribution function of runoff, and (7) deriving runoff as a function of rainfall. Each of these steps is now discussed.

Let Q be a random variable varying from 0 to $P - I_a$ and having a probability density function (PDF) of $f(Q)$ and cumulative distribution function (CDF) of $F(Q)$. Then the Shannon entropy of Q can be expressed as

$$H(Q) = -\int_0^{P-I_a} f(Q) \ln f(Q) dQ \tag{14.4}$$

Since it is not known what constraints Q must satisfy, only the total probability is invoked

$$\int_0^{P-I_a} f(Q) dQ = 1 \tag{14.5}$$

To determine the probability density function $f(Q)$, the Shannon entropy, given by Eq. (14.4), is maximized, subject to Eq. (14.5). Maximization can be done using the method of Lagrange multipliers as

$$L = -\int_0^{P-I_a} f(Q)\ln f(Q)\,dQ - (\lambda_0 - 1)\left[\int_0^{P-I_a} f(Q)dQ - 1\right] \quad (14.6)$$

where λ_0 is the Lagrange multiplier. Differentiating Eq. (14.6) with respect to $f(Q)$ and equating the derivative yield

$$f(Q) = \exp[-\lambda_0] \quad (14.7)$$

Equation (14.7) is the PDF of Q in terms of the Lagrange multiplier and has a constant value.

The Lagrange multiplier λ_0 can be determined by substituting Eq. (14.7) in Eq. (14.5) as

$$\int_0^{P-I_a} \exp(-\lambda_0)dQ = 1 \quad (14.8)$$

Equation (14.8) yields

$$P - I_a = \exp\lambda_0 \quad \text{or} \quad \lambda_0 = \ln(P - I_a) \quad (14.9)$$

Thus, the PDF of Q is obtained by inserting Eq. (14.9) in Eq. (14.7):

$$f(Q) = \frac{1}{P - I_a} \quad (14.10a)$$

Equation (14.10), with the use of Eq. (14.2), can also be written as

$$f(Q) = \frac{J}{SQ} \quad (14.10b)$$

Equation (14.10a) shows that the PDF is uniform over the interval 0 to $P - I_a$. The CDF of Q can be expressed as

$$F(Q) = \int_0^Q f(Q)dQ = \int_0^Q \frac{dQ}{P - I_a} = \frac{Q}{P - I_a} \quad (14.11)$$

Equation (14.11) shows that the CDF is a straight line over the space from 0 to $P - I_a$. Equation (14.11a) can also be written with the introduction of Eq. (14.2) as

$$F(Q) = \frac{J}{S} \quad (14.12)$$

To go from the probability domain to the real domain, it is hypothesized that the CDF of Q is equal to the ratio of cumulative infiltration J to the maximum potential soil moisture retention S as manifested by Eq. (14.12). Of course this hypothesis needs justification and empirical verification. However, a similar hypothesis has been employed by Chiu (1987) for deriving the velocity distribution, by Al-Hamdan and Cruise (2010) for deriving soil moisture profiles, by Singh (2011c) for deriving infiltration equations, among others.

Differentiating Eq. (14.12), one obtains

$$f(Q)dQ = \frac{dJ}{S} \tag{14.13a}$$

Substitution of Eq. (14.11) in Eq. (14.13) yields

$$\frac{dQ}{P-I_a} = \frac{dJ}{S} \tag{14.13b}$$

Integrating Eq. (14.13) with the initial condition that $Q = 0$ when $J = 0$, one finds

$$\frac{Q}{P-I_a} = \frac{J}{S} \tag{14.14}$$

Making use of Eq. (14.1), Eq. (14.14) can be written as

$$\frac{Q}{P-I_a} = \frac{P-I_a-Q}{S} \tag{14.15}$$

or

$$\frac{Q}{P-I_a} + \frac{Q}{S} = \frac{P-I_a}{S} \tag{14.16}$$

Equation (14.16) can be written as

$$\frac{Q(S+P-I_a)}{S(P-I_a)} = \frac{P-I_a}{S} \tag{14.17}$$

or

$$Q = \frac{(P-I_a)^2}{P+S-I_a} \tag{14.18}$$

Equation (14.18) can be written with the use of Eq. (14.3) as

$$Q = \frac{(P-aS)^2}{P+S(1-a)} \tag{14.19}$$

Equation (14.19) is the SCS-CN method for computing the runoff volume for a given rainfall event. Frequently, a is taken as 0.2; then Eq. (14.19) becomes

$$Q = \frac{(P-0.2S)^2}{P+0.8S} \tag{14.20}$$

Thus, Eq. (14.20) has only one parameter, S, and only one input.

Since the range of S is taken as 0 to ∞, it is mapped into a dimensionless curve number CN varying from 0 to 100 as

$$S = \frac{1000}{CN} - 10 \tag{14.21}$$

The value of CN depends on four factors: (1) soil type, (2) land use, (3) hydrologic condition, and (4) antecedent soil moisture condition. AMC has been classified into three groups: AMC I refers to the dry condition of a soil, AMC II corresponds to the normal or average condition, and AMC III corresponds to the wet condition. The values of CN have been extensively tabulated for AMC II, and tables are available for conversion from AMC I or AMC III to AMC II. For a given storm, an AMC value is obtained by computing the antecedent rainfall of 5-days (d).

Equation (14.19) can be written with the use of Eq. (14.21) as

$$Q = \frac{[P - a(1000/CN - 10)]^2}{P + (1-a)(1000/CN - 10)} \tag{14.22}$$

Now the PDF of Q can be cast in terms of CN by substituting Eq. (14.21) in Eq. (14.10b) as

$$f(Q) = \frac{J}{Q}\left(\frac{CN}{1000 - 10CN}\right) \tag{14.23}$$

Similarly, the CDF of Q can be written in terms of CN as

$$F(Q) = \frac{J \times CN}{1000 - 10CN} \tag{14.24}$$

Now the uncertainty of the SCS-CN method can be expressed by substituting Eq. (14.10a) in Eq. (14.4) as

$$H(Q) = -\int_0^{P-I_a} \frac{1}{P-I_a} \ln\left[\frac{1}{P-I_a}\right] dQ = \ln(P - I_a) \tag{14.25}$$

or with the use of Eq. (14.14),

$$H(Q) = \ln S - \ln J + \ln Q \tag{14.26}$$

Equation (14.25) can also be expressed in terms of the curve number as

$$H(Q) = \ln\left[P - a\left(\frac{1000}{CN} - 10\right)\right] \tag{14.27}$$

Likewise, Eq. (14.26) can be cast as

$$H(Q) = \ln\left(\frac{1000}{CN} - 10\right) - \ln J + \ln Q \tag{14.28}$$

An interesting implication of Eq. (14.25) is that the uncertainty in Q due to a given rainfall event computed by the SCS-CN method is directly related to the logarithm of the amount of rainfall minus the abstraction. This means that these quantities must be determined as accurately as possible. Furthermore, higher the value of $P - I_a$, the higher will be the uncertainty of Q. Since the initial abstraction is determined from the maximum soil moisture retention and in turn by the curve number, Eq. (14.27) states that the curve number should be determined as accurately as possible. Likewise, Eq. (14.26) states that the uncertainty of Q is directly related to the sum of the logarithm of the maximum soil moisture retention and the logarithm of runoff minus the logarithm of cumulative infiltration. This suggests that infiltration acts as a filter as regards uncertainty. This is also seen from Eq. (14.28), which shows that the uncertainty of runoff reduces with the increase in the value of curve number.

14.1.4   Derivation of Probability Distributions of SCS-CN Variables

In the SCS-CN method, Hjelmfelt et al. (1982) considered CN (and Q) as random variables. Likewise, Hawkins et al. (1985) used the antecedent soil moisture as a random variable.

McCuen (2002) and Schneider and McCuen (2005) also treated CN as a random variable. Yu (1998) assumed the spatially varying infiltration capacity as well as temporally varying rainfall rate as exponentially distributed. In a recent study, Hawkins and Ward (2012) used S as a random variable. If any of the variables in the SCS-CN method (P, J, S, CN) is a random variable, then Q will also be a random variable and will have a probability distribution. Other random variables can be Q/S and $C = Q/(P - I_a)$. Each of these variables can have a different probability distribution, and distributions of these variables have been derived empirically. The objective in this chapter is to develop a general method using entropy for deriving the probability distribution of any of these variables. The derivation depends on whether the random variable is considered as continuous or discrete. In this section the variables are assumed continuous, as has been assumed in the SCS-CN literature.

The methodology for deriving a probability distribution using entropy theory involves the same steps as outlined in the derivation of the SCS-CN equation in the preceding section. Here only the constraints and the PDF resulting from entropy maximizing are discussed, and other steps are detailed in App. A. For any continuous random variable $X \in (a,b)$ that is nonnegative, the Shannon entropy $H(X)$ can be written as

$$H(X) = -\int_a^b f(x)\log f(x)\,dx = -\int_a^b \log f(x)\,dF(x) = E[-\log f(x)] \qquad (14.29)$$

where a and b are the lower and upper limits of X, respectively. For deriving $f(x)$, $H(X)$ is maximized, subject to specified constraints.

The information on a random variable can be expressed in many different ways, but it is more convenient to express it in terms of moments, such as the mean, variance, covariance, cross covariance, and linear combinations of these statistics. Thus, constraints encode relevant information. If observations are available, then one way to express information on the random variable $X \in (a,b)$ is in terms of constraints $C_r, r = 0,$ $1, 2, ..., n$, defined as

$$C_0 = \int_a^b f(x)\,dx = 1 \qquad (14.30)$$

$$C_r = \int_a^b g_r(x)f(x)\,dx = \overline{g_r(x)} \qquad r = 1, 2, ..., n \qquad (14.31)$$

where $g_r(x), r = 1, 2, ..., n$, represent some functions of x; n denotes the number of constraints; and $\overline{g_r(x)}$ is the expectation of $g_r(x)$, with $g_0(x) = 1$. Equation (14.30) states that the PDF must satisfy the total probability. Other constraints, defined by Eq. (14.31), have physical meaning. For example, if $r = 1$ and $g_1(x) = x$, Eq. (14.31) would correspond to the mean $\overline{x}$; likewise, for $r = 2$ and $g_2(x) = (x - \overline{x})^2$, it would denote the variance of x. For most hydrologic engineering problems, more than two or three constraints are not needed.

To obtain the least-biased $f(x)$, the entropy given by Eq. (14.29) can be maximized, subject to Eqs. (14.30) and (14.31) using the method of Lagrange multipliers, leading to the PDF of X

$$f(x) = \exp\left[-\lambda_0 - \sum_{r=1}^n \lambda_r g_r(x)\right] \qquad (14.32)$$

where the Lagrange multipliers λ_r, $r = 0, 1, 2, \ldots, n$, can be determined with the use of Eqs. (14.30) and (14.31).

Integration of Eq. (14.32) leads to the cumulative distribution function or simply probability distribution of X, $F(x)$:

$$F(x) = \int_a^x \exp\left[-\lambda_0 - \sum_{r=1}^{n} \lambda_r g_r(x)\right] dx \qquad (14.33)$$

Substitution of Eq. (14.32) in Eq. (14.29) results in the maximum entropy of $f(x)$ or X:

$$H = \lambda_0 + \sum_{r=1}^{n} \lambda_r \overline{g_r(x)} = \lambda_0 + \sum_{r=1}^{n} \lambda_1 E[g_r(x)] = \lambda_0 + \sum_{r=1}^{n} \lambda_r C_r \qquad (14.34)$$

where $E[g_r(x)]$ is the expectation of $g_r(x)$. Equation (14.34) shows that the entropy of the probability distribution of X depends only on the specified constraints, since the Lagrange multipliers themselves can be expressed in terms of the specified constraints. Entropy maximization results in the probability distribution that is most conservative and hence most informative. If a distribution with lower entropy were chosen, then it would imply the assumption of information which is not available. On the other hand, a distribution with higher entropy would violate the known constraints. Thus, it can be stated that the maximum entropy leads to a probability distribution of a particular macrostate among all possible configurations of events under consideration.

Three distributions have been employed in the literature for fitting different SCS-CN variables: exponential, gamma, and lognormal. Here we only provide their equations but defer their derivations to App. 14B.

The exponential PDF of X can be written as

$$f(x) = \frac{1}{\overline{x}} \exp\left(-\frac{x}{\overline{x}}\right) \qquad (14.35)$$

where $\overline{x}$ is the mean of X. The gamma PDF can be written as

$$f(x) = \frac{\lambda_1^k}{\Gamma(k)} x^{k-1} \exp(-\lambda_1 x) \qquad (14.36)$$

where k and λ_1 are parameters. The two-parameter lognormal PDF is

$$f(x) = \frac{1}{x\sqrt{2\pi}\,\sigma_y} \exp\left[-\frac{1}{2}\left(\frac{\ln x - m_y}{\sigma_y^2}\right)^2\right] \qquad (14.37)$$

where $y = \ln x$, $m_y = $ mean of Y, and σ_y is the standard deviation of Y.

14.1.5 Probability Distributions of Discrete SCS-CN Variables Using Cross-Entropy

Let there be a discrete random variable X with a specific value denoted as x. The variable can be CN, S, $C = Q/(P - I_a)$, Q/S, or Q. The domain or region of X can be finite or infinite $R : [x_0, x_{r+1}]$ of real numbers which can be partitioned into $r + 1$ subintervals: $R_0 = [x_0, x_1)$, $R_1 = [x_1, x_2)$, ..., $R_r = [x_r, x_{r+1})$, shown in Fig. 14.1. Each sub-interval is like a class interval. The objective is to derive the PDF of X and this can be accomplished using cross entropy which involves prior and posterior distributions.

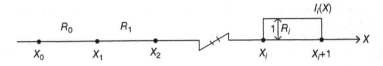

FIGURE 14.1   Indicator function

Let a prior PDF of X be given or assumed denoted as $g(x)$ and its CDF as $G(x)$. Noted that $g(x)$ is positive everywhere in domain R. Then, for the ith subinterval,

$$G_i = \int_{R_i} g(x)\,dx \tag{14.38}$$

The objective is to determine the posterior PDF of X, denoted as $w(x)$, and CDF $W(x)$. The objective is to seek a posterior PDF of X, $w(x)$, with CDF $W(x)$ that minimizes the cross-entropy function D, defined as

$$D(w,g) = \int_R w(x)\log\frac{w(x)}{g(x)}\,dx = \int_R w(x)\log w(x)\,dx - \int_R w(x)\log g(x)\,dx \tag{14.39}$$

Minimization of cross entropy can be achieved by satisfying the given or known information codified in the form of constraints on X. Since some of the variables, such as CN, in the SCS-CN method take on discrete values, it may be appropriate to specify the information in terms of fractile constraints, which can be specified as follows.

If observations on the random variable are available, then information can be obtained in the form of fractile pairs $(x_i, W(x_i))$, $i = 1, 2, ..., r$. That is, the observed values of the random variable X can be arranged as a sequence of $r+1$ elements, each of which is a realization of an independent identically distributed (IID) random variable with distribution $W(x)$ such that

$$W(x_i) = W_i = \int_{R_i} w(x)\,dx = \frac{i}{r+1} \qquad i = 1, 2, ..., r+1 \tag{14.40}$$

Equation (14.40) defines fractile constraints, which can be denoted as

$$C_i : W(x_i) = W_i \quad W_i \in (0,1) \qquad i \in n = \{0, 1, 2, ..., r\} \tag{14.41}$$

Now the constraints for individual intervals or interval constraints can be defined as

$$C_i = \int_{R_i} w(x)\,dx = W_i \qquad i \in n \qquad \sum_{i=1}^{n} W_i = 1 \tag{14.42}$$

Using the indictor function I, Eq. (14.42) can be written in terms of expectation over the whole domain as

$$C_i = \int_R I_i(x)w(x)\,dx - W_i = 0 \qquad i \in n \tag{14.43}$$

where, as shown in Fig. 14.1, I_i is the indicator function for the ith interval, defined as

$$I_i(x) = \begin{cases} 1 & \text{if } x \in R_i \text{ or } x_i \le x < x_{i+1} \\ 0 & \text{otherwise} \end{cases} \tag{14.44}$$

Now Eq. (14.39) is minimized to derive the posterior PDF $w(x)$ with CDF $W(x)$, subject to the constraints defined by Eq. (14.43), using the method of Lagrange multipliers. To that end, one can write the lagrangian function L as

$$L = D(w, g) + \sum_{i=0}^{n} \lambda_i C_i = \int_R w(x)[\log w(x) - \log g(x)] + \sum_{i=1}^{n} \lambda_i [I_i(x)w(x) - W_i] \qquad (14.45)$$

Recalling the Euler-Lagrange calculus of variation, differentiating Eq. (14.45) with respect to $w(x)$, and equating the derivative to zero, one gets

$$\frac{\partial L}{\partial w(x)} = 0 = \log w(x) - \log g(x) + 1 + \sum_{i=1}^{n} \lambda_i I_i(x) \qquad (14.46)$$

Therefore, Eq. (14.46) yields

$$w(x) = g(x)\exp[-1 - \sum_{i=1}^{n} \lambda_i I_i(x)] \qquad (14.47)$$

Equation (14.47) is the posterior PDF, $w(x)$, in terms of the prior PDF $g(x)$ and Lagrange multipliers λ_i, $i = 1, 2, \cdots, n$, depending on the fractile constraints.

Let $U(x) = \exp[-1 - \sum_{i=1}^{n} \lambda_i I_i(x)]$ which is a step function that has a constant u_i in interval I_i:

$$u_i = \exp(-1 - \lambda_i) \qquad i \in n \qquad (14.48)$$

as shown in Fig. 14.2. Therefore, Eq. (14.48) becomes

$$w(x) = g(x)U(x) \qquad (14.49)$$

Equation (14.49) is the posterior PDF of X in terms of the prior PDF $g(x)$ and the Lagrange multipliers embedded in the step function $U(x)$.

Substitution of Eq. (14.49) in Eq. (14.39) yields the minimum cross entropy value as

$$D_{min} = \int_R w(x)\log U(x)dx = \int_R g(x)U(x)\log U(x)dx \qquad (14.50)$$

In discrete form, Eq. (14.50) can be expressed as

$$D_{min} = \sum_{i=1}^{n} W_i \log U_i \qquad (14.51)$$

where U_i corresponds to the ith interval. The minimum cross entropy can be interpreted as the directional distance between the prior PDF and the sample PDF, suggesting that the prior PDF should be appropriately chosen to minimize the cross entropy.

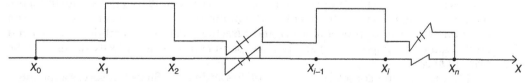

FIGURE 14.2 Stepwise function.

Now with the use of Eqs. (14.49) and (14.48), constraint Eq. (14.43) can be written as

$$W_i = \int_R I_i(x)g(x)U(x)dx = U_i \int_{R_i = I_i} g(x)dx = U_i G_i \qquad (14.52)$$

where G_i is the prior probability P of $x \in I_i$, or

$$G_i = P(x_{i+1}) - P(x_i) \qquad (14.53)$$

The constraints given by Eq. (14.41) or (14.42) are therefore satisfied when

$$U_i = \frac{W_i}{G_i}, \qquad i \in n \qquad (14.54)$$

Therefore, Eq. (14.49) becomes

$$w(x) = g(x)\frac{W_i}{G_i} \qquad i \in n \qquad (14.55)$$

Then with the use of Eqs. (14.54), (14.49), and (14.41) the minimum cross entropy becomes

$$D_{min} = \sum_{i=1}^{n} W_i \log \frac{W_i}{G_i} \qquad (14.56)$$

$$D_{min} = \frac{1}{n+1} H(x;w) - \log(n+1) \qquad (14.57)$$

where $H(x; w)$ is the Shannon entropy.

The minimum cross-entropy-based PDF $w(x)$, given by Eq. (14.47), retains the piecewise form of the prior PDF $g(x)$ scaled over each interval I_i by the constant factor $u_i = W_i/G_i$, given by Eq. (14.54). The minimum cross entropy is a function of the r prior distribution values G_r and the fractile constraints W_r. It is independent of the values assumed by the prior distribution at any other point, and it is not directly dependent on the prior distribution $g(x)$.

The minimum value of the cross-entropy given by Eq. (14.56) is equivalent to that for discrete distributions, corresponding to a set of $r + 1$ possible events with prior probabilities G_i and minimum cross entropy probabilities W_i. With the use of Eqs. (14.46), (14.47), and (14.53) in Eq. (14.48) and integrating,

$$W(x) = W_i + u_i[G(x) - G_i] \qquad x \in I_i \qquad i \in n \qquad (14.58)$$

Note that the posterior PDF $w(x)$ given by Eq. (14.47) is the prior PDF scaled over each interval; that is, $w(x)$ has the piecewise form of $g(x)$ scaled by $U(x)$ over each interval I_i as given by Eq. (14.54), as shown in Fig. 14.3. The prior PDF $g(x)$ does not influence the posterior PDF $w(x)$ at the fractiles. The minimum value of the cross entropy is a function of the r prior probabilities G_i and r fractile constraints W_i; it does not depend on the values assumed by $g(x)$ at any other point and does not directly depend on the prior PDF $g(x)$.

Now the question of selecting a prior PDF is addressed. Since $g(x)$ serves as an interpolation function between fractiles and more importantly an extrapolation function

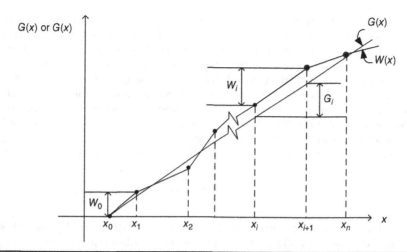

FIGURE 14.3 Reference distribution $G(x)$, posterior distribution $W(x)$, and interval probabilities G_i and P_i.

outside the fractile range $[x_1, x_r]$, it should be selected judiciously. The selection of a prior PDF can be made with the use of the Shannon entropy. Of course, the selection may vary from one problem to another.

14.2   Instantaneous Unit Hydrograph

The unit hydrograph, or more appropriately the instantaneous unit hydrograph (IUH), represents the response of a linear time-invariant system; i.e., the watershed is a linear and time-invariant system, and the IUH is the response of the watershed due to an instantaneous rainfall of unit volume. The IUH has dimensions of $1/T$ if surface runoff is expressed in the units of L/T, or has dimensions of L^2/T if surface runoff is expressed in the units of L^3/T. The surface runoff hydrograph or discharge $Q(t)$ as a function of time is obtained by convoluting IUH with the rainfall excess hyetograph $I(t)$ as

$$Q(t) = \int_0^t h(t-\tau)I(\tau)\,d\tau \qquad (14.59)$$

The discharge at the outlet of a watershed in response to a given rainfall event is a function of rainfall intensity and the duration, infiltration, and watershed characteristics. For a given rainfall event, $Q(t)$ can be expressed as a function of the area contributing to it. It takes time for the flow being contributed by any area to appear at the outlet; one may write $Q(t)$ as a function of the distance or length of travel, since length and area are related. But the time of travel depends on the length of travel, and hence one can also express $Q(t)$ simply as a function of time, wherein the parameters in the function will reflect the effect of different factors affecting runoff. In a similar manner, one can also state that IUH is a function of time, although realistically it also depends on rainfall-excess characteristics (space-time distribution), infiltration, and watershed characteristics.

It turns out that $h(t)$ has the characteristics of a probability distribution $f(t)$, wherein the time of travel is taken as a random variable. Once this premise is accepted, one can employ the principle of maximum entropy (POME) to derive the probability distribution of the time of travel, t, or $h(t)$, depending on the information given.

Many different distributions have been proposed to represent $h(t)$ (Rai et al., 2008; Singh, 1998). However, the basis for proposing most of these distributions depends entirely on fitting empirical data. The purpose of this section is to show, using an example, the derivation of $h(t)$, subject to given information, thus avoiding arbitrariness in the selection of an appropriate function.

Let $h(t)$ be represented by $f(t)$. The Shannon entropy of $f(t)$ can be written as

$$H(t) = -\int_0^\infty f(t) \ln[f(t)] dt \tag{14.60}$$

The objective is to determine $f(t)$ by maximizing $H(t)$ given by Eq. (14.60), subject to specified constraints. Thus, $f(t)$ can be derived following the methodology described in Chap. 1. Here we illustrate the derivation of a distribution derived by Lienhard (1964, 1972) using the Maxwell-Boltzmann statistics for the dimensionless unit hydrograph.

Clearly,

$$\int_0^\infty f(t) dt = 1 \tag{14.61}$$

Let $f(t)$ satisfy the following constraints:

$$\int_0^\infty \ln t \, f(t) dt = \overline{\ln t} \tag{14.62}$$

$$\int_0^\infty t^2 f(t) dt = \overline{t^2} \tag{14.63}$$

Following the POME methodology and using the method of Lagrange multipliers for maximizing $H(t)$, one obtains

$$f(t) = \exp\left(-\lambda_0 - \lambda_1 \ln t - \lambda_2 t^2\right) \tag{14.64}$$

where $\lambda_i, i = 0, 1, 2$, are Lagrange multipliers. Inserting Eq. (14.64) in Eq. (14.61), one gets

$$\exp \lambda_0 = \int_0^\infty \exp(-\lambda_1 \ln t - \lambda_2 t^2) dt = \int_0^\infty t^{-\lambda_1} \exp(-\lambda_2 t^2) dt \tag{14.65}$$

Equation (14.65) can be expressed in terms of gamma function as

$$\exp \lambda_0 = \lambda_2^{(\lambda_1 - 1)/2} \Gamma\left(\frac{1 - \lambda_1}{2}\right) \tag{14.66}$$

Therefore,

$$f(t) = \frac{\lambda_2^{(1-\lambda_1)/2}}{\Gamma((1-\lambda_1)/2)} \exp(-\lambda_1 \ln t - \lambda_2 t^2) \tag{14.67}$$

denoting $\lambda_2 = 3/(2k^2)$, where k is the mean travel or residence time. Then

$$f(t) = \frac{1}{k^{1+\lambda_1}\Gamma((1-\lambda_1)/2)}\left(\frac{t}{k}\right)^{-\lambda_1}\exp\left[-\frac{3}{2}\left(\frac{t}{k}\right)^2\right] \tag{14.68}$$

Let $\lambda_2 = -2$. Then

$$f(t) = \frac{1}{k\Gamma(3/2)}\left(\frac{3}{2}\right)^{3/2}\left(\frac{t}{k}\right)^2\exp\left[-\frac{3}{2}\left(\frac{t}{k}\right)^2\right] \tag{14.69}$$

which is what Lienhard (1964) obtained using the Maxwell-Boltzmann statistic.

Example 14.1. Consider a small watershed W-3, located 15 mi north of Stillwater, Oklahoma, Black Bear Creek, in the Arkansas River basin. It has an area of 92 acres. The watershed conditions were 100 percent of area in native grass: 32 percent in hay meadow in good condition, 46 percent in pasture in fair condition, and 22 percent in pasture in poor condition. A storm event occurred on this watershed for which rainfall-runoff data are available, given in Table 14.1. The total amount of rainfall was

Time of Day	Rainfall Intensity (in./h)	Time of Day	Runoff (cfs)	Time of Day	Rainfall Intensity (in./h)	Time of Day	Runoff (cfs)
2:45	0.0	2:59	0.00		0	4:05	38.20
3:00	1.4	3:00	0.00		0	4:10	38.67
3:30	1.4	3:06	0.13		0	4:18	36.76
4:00	0.34	3:10	0.31		0	4:28	31.41
4:45	0.15	3:12	0.26		0	4:38	26.56
5:50	0.13	3:19	0.45		0	4:50	24.44
6:20	0.08	3:22	1.37		0	5:08	18.70
	0	3:24	2.39		0	5:24	15.14
	0	3:26	3.24		0	5:50	11.79
	0	3:28	3.75		0	6:20	9.11
	0	3:30	4.25		0	6:56	7.32
	0	3:34	6.39		0	7:20	5.62
	0	3:35	7.25		0	7:42	4.43
	0	3:40	7.29		0	8:02	3.55
	0	3:42	9.79		0	8:28	2.71
	0	3:44	12.58		0	8:58	2.08
	0	3:45	14.60		0	9:32	1.41
	0	3:50	16.77		0	10:00	1.23
	0	3:53	20.85		0	10:30	0.88
	0	3:56	25.12		0	11:12	0.71
	0	3:58	29.42		0	12:10	0.53
	0	4:01	33.78		0	12:56	0.44

TABLE 14.1 Rainfall-Runoff Event of October 11, 1973 (To convert runoff in cubic feet per second to inches per hour, multiply by 0.01078.)

1.5 in., and the total amount of surface runoff was 0.78 in. Compute the unit hydrograph from observations, and compare it with the one derived using Eq. (14.69). Compute the discharge hydrograph, using the IUH, and compare it with the observed hydrograph.

Solution.   For given rainfall-runoff values, the mean travel time is estimated from the lag time between the center of rainfall 3:15 and the peak of runoff 4:10, which suggests the value of mean travel time $k = 0.917$ h. Thus, $\lambda_2 = 3/(2k^2) = 1.788$, and $\lambda_1 = -2$. The IUH can be obtained from Eq. (14.69) as

$$f(t) = \frac{1}{k\Gamma(3/2)}\left(\frac{t}{k}\right)^2 \exp\left[-\frac{3}{2}\left(\frac{t}{k}\right)^2\right]$$

$$= \frac{1}{0.917 \times 0.886}\left(\frac{t}{0.917}\right)^2 \exp\left[-\frac{3}{2}\left(\frac{t}{0.917}\right)^2\right] = 1.46t^2 \exp(-1.784t^2)$$

The unit hydrograph with $\Delta t = 0.25$ is 0, 0.08, 0.23, 0.30, 0.25, ... Then with the IUH so determined, the runoff can be obtained by convoluting the unit hydrograph with rainfall excess. In discrete form, this can be expressed as

$$Q(t) = \int_0^t h(t-\tau)I(\tau)d\tau = \sum_{\tau=0}^{t} f(t-\tau)I(\tau) = f(t)I(0) + f(t-1)I(1) + \cdots + f(0)I(t)$$

For example,

$$Q(0) = \sum_{\tau=0}^{t} f(t-\tau)I(\tau) = 0 \times 1.4 = 0 \text{ in./h}$$

$$Q(0.25) = \sum_{\tau=0}^{t} f(t-\tau)I(\tau) = 0.08 \times 1.4 + 0 \times 1.4 = 0.029 \text{ in./h}$$

$$Q(0.50) = \sum_{\tau=0}^{t} f(t-\tau)I(\tau) = 0.23 \times 1.4 + 0.08 \times 1.4 + 0 \times 1.4 = 0.11 \text{ in./h}$$

The computed runoff hydrograph is shown in Fig. 14.4.

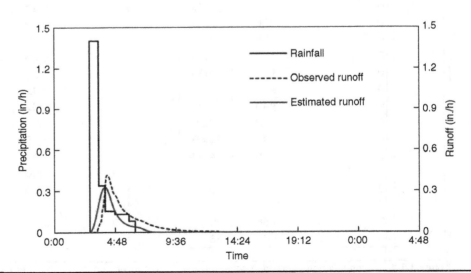

FIGURE 14.4   Computed and observed surface runoff hydrographs.

14.3   Volume-Peak Discharge Relation

The term *volume* here means the runoff volume under the hydrograph, and the hydrograph should be defined in a consistent manner so that all hydrographs are comparable. For example, the beginning and end of a hydrograph should be delineated using an objective criterion. Rogers and Zia (1982) reported an empirical equation that seems to have necessary consistency:

$$Q_T = Q_0 + Q_p^{0.6} \qquad (14.70)$$

where Q_0 is the base flow discharge (in cubic feet per second) immediately prior to the hydrograph rise, Q_p is the peak discharge of the hydrograph, and Q_T is the discharge identifying the termination discharge.

The volume-peak discharge relation was first investigated by Rogers (1980) and later by Rogers and Singh (1986, 1988) and Singh and Aminian (1986). This relation is also called the drainage basin–peak discharge rating curve or standardized peak discharge distribution. It is simply a peak discharge rating curve for a drainage basin in which peak discharge is plotted versus runoff (uniformly distributed over the basin) rather than stage. In short, it can also be called a peak discharge rating curve or drainage basin rating curve. In this case, no base flow separation is needed. It can also be applied to snowmelt hydrographs, and the basin can be of any size.

The peak discharge rating curve, as suggested by Rogers (1980), is a linear relation on logarithmic paper with the runoff volume on the abscissa and the peak discharge on the ordinate, as shown in Fig. 14.5. An investigation by Singh and Aminian (1986) of a large number of watersheds from the United States and other countries concluded that the linear relationship was satisfactory for a number of watersheds; the relationship improved if the storms were separated into winter and summer storms, excluding

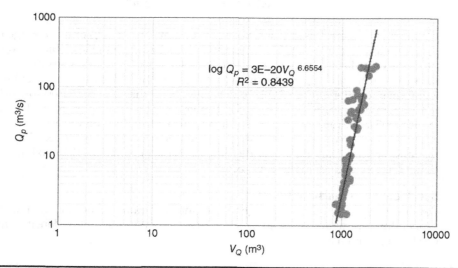

FIGURE 14.5   Linear relation between log of runoff volume and log of peak discharge for gauging station (USGS-08082000-Salt-FK-Brazos) at Brazos River.

snowstorms, and if rainfall excess and surface runoff peak discharge were used; and the intercept would reflect the individual watershed characteristics. The relation between peak discharge Q_p and volume of surface runoff V can be expressed as

$$Q_p = aV_Q{}^m \tag{14.71a}$$

or
$$\log Q_p = \log a + m \log V \tag{14.71b}$$

where a and m are parameters. The slope m of the linear line (in log domain) can be considered as an indicator of the linearity or nonlineariy of the drainage basin. If peak discharge is directly proportional to runoff volume, then $m = 1$ and the basin is linear. A value of m other than 1 is an indication of nonlinearity, and then peak discharge is a power function of the runoff volume. The drainage basin nonlinearity is caused by the nonuniform distribution of losses where the principal loss to runoff is infiltration. Drainage basins that have nonuniform infiltration are nonlinear unless this loss is compensated for in some manner by some other losses. The value of m can range from 1 to at least -0.83. For a majority of basins the value of m is positive, and a smaller value would correspond to greater nonlinearity. It is not uncommon to have values of m as low as 0.4. Negative values of m are an anomaly and may occur in basins with sinks or caverns.

14.4 Derivation of Peak Discharge Rating Curve

The procedure for deriving the peak discharge rating curve is similar to that described in Sec. 14.3. It is assumed that temporally peak discharge Q_p is a random variable with a probability density function denoted as $f(Q_p)$. The Shannon entropy of Q_p, or of $f(Q_p)$, $H(Q_p)$, can be expressed as

$$H(Q_p) = -\int_0^\infty f(Q_p) \ln f(Q_p) dQ_p \tag{14.72}$$

Equation (14.72) expresses a measure of uncertainty about $f(Q_p)$ or the average information content of sampled Q_p. The objective here is to derive the least-biased $f(Q_p)$ which can be accomplished by maximizing $H(Q_p)$, subject to specified constraints, in accordance with the principle of maximum entropy (POME). Maximizing $H(Q_p)$ is equivalent to maximizing $f(Q_p) \ln f(Q_p)$. To determine $f(Q_p)$ that is least biased toward what is not known as regards discharge, the principle of maximum entropy developed by Jaynes (1957a, b, 1982) is invoked, which requires specification of certain information, expressed in terms of what is called constraints, on peak discharge. According to POME, the most appropriate probability distribution is the one that has the maximum entropy or uncertainty, subject to these constraints.

Specification of Constraints

For deriving $f(Q_p)$, following Singh (1998), the constraints to be specified are the total probability law, which must always be satisfied by the probability density function of discharge, written as

$$C_1 = \int_0^{Q_{max}} f(Q_p) dQ_p = 1 \tag{14.73}$$

and
$$C_2 = \int_0^{Q_{max}} [\ln Q_p] f(Q_p) dQ_p = \overline{\ln Q_p} \tag{14.74}$$

Equation (14.73) is the first constraint defining the total probability law C_1, and Eq. (14.74) is the second constraint C_2 defining the mean of the logarithm of the discharge values.

Maximization of Entropy

To obtain the least-biased probability density function of Y, $f(Q_p)$, the Shannon entropy, given by Eq. (14.72), is maximized following POME, subject to Eqs. (14.73) and (14.74). To that end, the method of Lagrange multipliers is employed. The lagrangian function then becomes

$$L = -\int_0^{Q_{max}} f(Q_p) \ln f(Q_p) dQ_p - (\lambda_0 - 1)\left[\int_0^{Q_{max}} f(Q_p) dQ_p - C_1\right] - \lambda_1 \left[\int_0^{Q_{max}} \ln Q_p f(Q_p) dQ_p - C_2\right] \tag{14.75}$$

where λ_0 and λ_1 are the Lagrange multipliers. Recalling the Euler-Lagrange equation of calculus of variation and differentiating Eq. (14.75) with respect to f, noting that f is variable and Q_p is parameter, and equating the derivative to zero, one obtains

$$\frac{\partial L}{\partial f} = 0 = -[\ln f(Q_p) + 1] - (\lambda_0 - 1) - \lambda_1 \ln Q_p \tag{14.76}$$

Derivation of Probability Distribution

Equation (14.76) leads to the entropy-based probability density function (PDF) of discharge as

$$f(Q_p) = \exp[-\lambda_0 - \lambda_1 \ln Q_p] \text{ or } f(Q_p) = \exp(-\lambda_0) Q_p^{-\lambda_1} \tag{14.77}$$

The PDF of Q_p contains the Lagrange multipliers λ_0 and λ_1 which can be determined using Eqs. (14.73) and (14.74). The cumulative probability distribution function of Y can be obtained by integrating Eq. (16.46) as

$$F(Q_p) = \frac{\exp(-\lambda_0)}{-\lambda_1 + 1} Q_p^{-\lambda_1 + 1} \tag{14.78}$$

Maximum Entropy

Substitution of Eq. (14.77) in Eq. (14.72) yields the maximum entropy or uncertainty of discharge:

$$H(Q_p) = \lambda_0 + \lambda_1 \overline{\ln Q_p} \tag{14.79}$$

Determination of Lagrange Multipliers

Substitution of Eq. (14.77) in Eq. (14.73) leads to

$$\lambda_0 = -\ln(-\lambda_1 + 1) + (-\lambda_1 + 1)\ln[Q_{max}] \tag{14.80}$$

Differentiating Eq. (14.80) with respect to λ_1 produces

$$\frac{\partial \lambda_0}{\partial \lambda_1} = \frac{1}{-\lambda_1 + 1} - \ln Q_{max} \tag{14.81}$$

On the other hand, substitution of Eq. (14.77) in Eq. (14.73)) can also be written as

$$\lambda_0 = \ln \int_0^{Q_{max}} Q_p^{-\lambda_1} dQ_p \tag{14.82}$$

Differentiating Eq. (14.82) with respect to λ_1, one gets

$$\frac{\partial \lambda_0}{\partial \lambda_1} = -\frac{\int_0^{Q_{max}} (\ln Q_p) Q_p^{-\lambda_1} dQ_p}{\int_0^{Q_{max}} Q_p^{-\lambda_1} dQ_p} \tag{14.83}$$

Multiplying and dividing Eq. (14.83) by $\exp(-\lambda_0)$, and using Eq. (14.73), one finds

$$\frac{\partial \lambda_0}{\partial \lambda_1} = -\frac{\int_0^{Q_{max}} (\ln Q_p) \exp(-\lambda_0) Q_p^{-\lambda_1} dQ_p}{\int_0^{Q_{max}} \exp(-\lambda_0) Q_p^{-\lambda_1} dQ_p} = -\overline{\ln Q_p} \tag{14.84}$$

Equating Eq. (14.84) to Eq. (14.81), one obtains an expression for λ_1 in terms of the constraint and the limits of integration of Q_p:

$$\frac{1}{-\lambda_1 + 1} - \ln Q_{max} = -\overline{\ln Q_p} \tag{14.85}$$

Example 14.2. For data obtained from the gauging station (USGS-08082000-Salt-FK-Brazos) at Brazos River, one observed $Q_{max} = 199.8$ m³/s and $\overline{\ln Q_p} = 1.08$ m³/s. Compute the values of the Lagrange multipliers in Eq. (14.77).

Solution. From Eq. (14.85),

$$\lambda_1 = 1 - \frac{1}{\ln Q_{max} - \overline{\ln Q_p}} = 1 - \frac{1}{\ln 199.8 - 1.08} = 0.22$$

and from Eq. (14.80),

$$\lambda_0 = -\ln(-\lambda_1 + 1) + (-\lambda_1 + 1)\ln Q_{max} = -\ln(-0.22 + 1) + (-0.22 + 1)\ln 199.8 = 1.90$$

Peak Discharge Rating Curve

Let the maximum runoff volume be denoted by V_{max}. It is then assumed that all values of peak discharge Q_p measured at the watershed outlet for any storm between 0 and V_{max} are equally likely. In reality this is not highly unlikely because at different times different values of peak discharge do occur. This is also consistent with the laplacian principle of insufficient reason. Then the cumulative probability distribution of peak discharge can be expressed as the ratio of the runoff volume to the maximum runoff volume. The probability of peak discharge being equal to or less than a given value of Q_p is V/V_{max}; for any runoff volume (measured at the outlet) less than a given value V, the peak discharge is less than a given value, say, Q_p. Thus the cumulative distribution function of peak discharge $F(Q_p) = P$ (peak discharge $\leq$ a given value of Q_p), where P = probability, can be expressed as

$$F(Q_p) = \frac{V}{V_{max}} \tag{14.86}$$

Here $F(Q_p)$ denotes the CDF, and Q_p = peak discharge. Note that on the left side the argument of function F in Eq. (14.86) is variable Q_p, whereas on the right side the variable is V. The CDF of Q_p is not linear in terms of V, unless Q_p and V are linearly related. Of course, it is plausible that $F(Q_p)$ might have a different form.

Since Eq. (14.86) constitutes the fundamental hypothesis employed here for deriving the volume-peak discharge relation using entropy, it will be useful to evaluate its validity. This hypothesis (i.e., the relation between the cumulative probability $F(Q_p)$ and the ratio V/V_{max}) should be tested for a large number of watersheds. Also note that a similar hypothesis has been employed when using entropy theory for deriving infiltration equations by Singh (2010a, 2010b), soil moisture profiles by Singh (2010c), and velocity distributions by Chiu (1987).

The probability density function is obtained by differentiating Eq. (14.86) with respect to Q_p:

$$f(Q_p) = \frac{dF(Q_p)}{dQ_p} = \frac{1}{V_{max}} \frac{dV}{dQ_p} \quad \text{or} \quad f(Q_p) = \left(V_{max} \frac{dQ_p}{dV} \right)^{-1} \tag{14.87}$$

The term $f(Q_p) dQ_p = F(Q_p + dQ_p) - F(Q_p)$ denotes the probability of peak discharge being between Q_p and $Q_p + dQ_p$.

Substituting Eq. (14.77) in Eq. (14.87), one gets

$$\left. \frac{\exp(-\lambda_0)}{-\lambda_1 + 1} Q^{-\lambda_1+1} \right|_0^{Q_p} = \left. \frac{1}{V_{max}} V \right|_0^V \tag{14.88}$$

Equation (14.88) yields

$$Q_p = \left[\frac{\exp \lambda_0 (-\lambda_1 + 1)}{V_{max}} V \right]^{\frac{1}{-\lambda_1+1}} \tag{14.89}$$

Equation (14.89) is an expression of the peak discharge rating curve. Let

$$a = \left[\frac{\exp \lambda_0}{V_p} (-\lambda_1 + 1) \right]^{\frac{1}{-\lambda_1+1}} \tag{14.90}$$

and

$$m = \frac{1}{-\lambda_1 + 1} \tag{14.91}$$

When plotted on a log-log paper, Eq. (14.91) will be a straight line. It may now be interesting to evaluate the Lagrange multipliers and hence parameters a and m.

Determination of Lagrange Multipliers

Substitution of Eq. (14.77) in Eq. (14.73) yields Eq. (14.80), which can be written as

$$\exp\lambda_0 = \frac{Q_{max}^{-\lambda_1+1}}{-\lambda_1+1} \qquad \text{or} \qquad \lambda_0 = -\ln(-\lambda_1+1)+(-\lambda_1+1)\ln Q_{,max} \qquad (14.92)$$

where Q_{max} is the peak discharge when $V = V_{max}$.

Differentiating Eq. (14.92) with respect to λ_1, one obtains

$$\frac{\partial\lambda_0}{\partial\lambda_1} = -\ln Q_{max} + \frac{1}{-\lambda_1+1} \qquad (14.93)$$

One can also write, from the substitution of Eq. (14.77) in Eq. (14.73), which yields Eq. (14.82):

$$\lambda_0 = \ln \int_0^{Q_{max}} Q_p^{-\lambda_1}\, dQ_p \qquad (14.94)$$

Differentiating Eq. (14.94) with respect to λ_1 and simplifying, one obtains

$$\frac{\partial\lambda_0}{\partial\lambda_1} = -\overline{\ln Q_p} \qquad (14.95)$$

Equating Eq. (14.95) to Eq. (14.93) leads to an estimate of λ_1 given by Eq. (14.85). Therefore, exponent m of the power form relation becomes

$$m = \ln Q_{max} - \overline{\ln Q_p} \qquad (14.96)$$

Equation (14.96) shows that the exponent m of the power form rating curve can be estimated from the values of the logarithm of maximum peak discharge at the maximum runoff volume and the average of the logarithmic values of peak discharge. The greater the difference between these logarithm values, the higher will be the exponent.

The Lagrange multiplier λ_0 can now be expressed as

$$\lambda_0 = \frac{\ln Q_{max}}{\ln Q_{max} - \overline{\ln Q_p}} + \ln(\ln Q_{max} - \overline{\ln Q_p}) \qquad (14.97a)$$

Therefore,

$$a = \frac{Q_{max}}{V_{max}^{\ln Q_{max} - \overline{\ln Q_p}}} \qquad (14.97b)$$

The PDF of Q_p can be expressed as

$$f(y) = \exp(-\lambda_0)Q_p^{1/m-1} = \frac{(Q_{max})^{1/(\ln Q_{max} - \overline{\ln Q_p})}}{\ln Q_{max} - \overline{\ln Q_p}} Q_p^{1/m-1} \qquad (14.98)$$

and the CDF as

$$F(Q_p) = m\exp(-\lambda_0)Q_p^{1/m} = (Q_{max})^{1/(\ln Q_{max} - \overline{\ln Q_p})}Q_p^{1/m} \qquad (14.99)$$

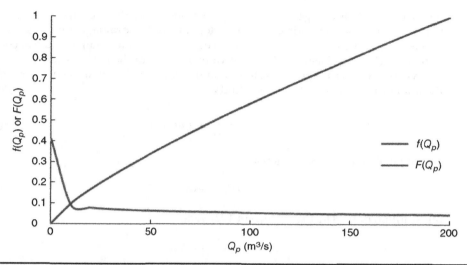

FIGURE 14.6 PDF and CDF of peak discharge values observed at a gauging station (USGS08082000-Salt-FK-Brazos) at Brazos River.

For $m < 1$, the probability density function monotonically increases from 0 to $\exp(-\lambda_0)Q_{max}^{1/m-1}$. Figure 14.6 shows a PDF of peak discharge values observed at a gauging station (USGS-08082000-Salt-FK-Brazos) at the Brazos River.

The entropy (in napiers) of the peak discharge distribution can be obtained by substituting Eq. (14.98) in Eq. (14.72):

$$H = \lambda_0 - \left(\frac{1}{m}-1\right)\overline{\ln Q_p} = -\frac{\ln Q_{max}}{\ln Q_{max} - \overline{\ln Q_p}} + \ln\left(\ln Q_{max} - \overline{\ln Q_p}\right) - \left(\frac{1}{m}-1\right)\overline{\ln Q_p} \quad (14.100)$$

Example 14.3. Use the previous data, where the Lagrange multipliers are computed as $\lambda_0 = 1.90$ and $\lambda_1 = 0.22$, to compute parameters a and m. Then plot the PDF and CDF of Q_p.

Solution. From Eqs. (14.90) and (14.91),

$$a = \left[\frac{(\exp\lambda_0)(-\lambda_1 + 1)}{V_{max}}\right]^{\frac{1}{-\lambda_1+1}} = \left[\frac{(\exp 1.90)(-0.22 + 1)}{2279}\right]^{\frac{1}{-0.22+1}} = 0.0004$$

$$m = \frac{1}{-\lambda_1 + 1} = \frac{1}{1 - 0.22} = 1.28$$

With these parameters, the PDF and CDF can be determined using Eqs. (14.98) and (14.99), as plotted in Fig. 14.6.

14.4 Rational Method

Following the procedure in Sec. 14.3, the rational method can be derived by using entropy theory. The rational method is normally written as

$$Q = CIA \quad (14.101)$$

where Q is peak discharge in cubic feet per second, C is a dimensionless constant whose value depends on the hydrologic characteristics of the drainage area, I is the rainfall intensity in inches per hour for a duration equal to or greater than the time of concentration of the drainage basin, and A is the drainage area of the basin in acres.

Assuming Q as a random variable and maximizing the Shannon entropy subject to the total probability constraint, one finds

$$C_1 = \int_0^{Q_{max}} f(Q)dQ = 1 \qquad (14.102)$$

and the PDF of Q can be written as

$$f(Q) = \exp(-\lambda_0) \qquad (14.103)$$

Substitution of Eq. (14.103) in Eq. (14.102) yields

$$\lambda_0 = \ln Q_{max} \qquad (14.104)$$

Therefore, the PDF of Q becomes

$$f(Q) = \frac{1}{Q_{max}} \qquad (14.105)$$

and the CDF of Q is

$$F(Q) = \frac{Q}{Q_{max}} \qquad (14.106)$$

and entropy is

$$H(Q) = \ln Q_{max} \qquad (14.107)$$

Let the maximum rainfall intensity for a given duration be denoted by I_{max}. It is then assumed that all values of peak discharge Q_p measured at the watershed outlet for any storm between 0 and Q_{max} are equally likely. In reality, this is not highly unlikely because at different times different values of peak discharge do occur. This is also consistent with the laplacian principle of insufficient reason. Then the cumulative probability distribution of peak discharge can be expressed as a ratio of rainfall intensity to maximum rainfall intensity. The probability of peak discharge being equal to or less than a given value of Q is I/I_{max}; for any rainfall intensity less than a given value, I, the peak discharge is less than a given value, say, Q; thus the cumulative distribution function of peak discharge $F(Q) = P$ (peak discharge $\leq$ a given value of Q), where P = probability, can be expressed as

$$F(Q) = \frac{I}{I_{max}} \qquad (14.108)$$

where $F(Q)$ denotes the cumulative distribution function, and Q = peak discharge. Note that on the left side the argument of function F in Eq. (14.108) is variable Q, whereas on the right side the variable is I. Of course, it is plausible that $F(Q)$ might have a different form.

The probability density function is obtained by differentiating Eq. (14.108) with respect to Q:

$$f(Q) = \frac{dF(Q)}{dQ} = \frac{1}{I_{max}} \frac{dI}{dQ} \quad\quad (14.109)$$

Substituting Eq. (14.107) in Eq. (14.109), one gets

$$Q = \left(\frac{Q_{max}}{I_{max}}\right) I \quad\quad (14.110)$$

Equation (14.110) is a form of the rational method in that

$$CA = Q_{max} / I_{max} \qu\quad (14.111)$$

Note that Q_{max} can be expressed as $Q_{max} = CAI_{max}$, where C is the coefficient accounting for the fraction of rainfall that becomes runoff and the remainder of rainfall is lost to runoff. Therefore, Eq. (14.110) can recast as

$$Q = \left(\frac{CAI_{max}}{I_{max}}\right) I = CIA \qu\quad (14.112)$$

which is the rational method. It is interesting that the value of C for a watershed is the ratio of the maximum runoff to the product of watershed area and maximum rainfall. Of course, this observation needs to be tested on natural watersheds.

14.5   Linear Reservoir Model

A watershed varies spatially and has a moisture storage capacity s, $0 \leq s < \infty$, that may vary from one place to another. Theoretically it may be possible to imagine a large number of storage elements in the watershed, each having a storage capacity s_i, $i = 1, 2, ..., N$, for N a large number. The moisture storage in an element is $0 \leq s_i < \infty$. Thus, the watershed can be represented by a population of these storage capacities. Let the storage capacity s be considered a random variable. In a way, one can consider s corresponding to the random property of the watershed at the microscopic level. Then the probability density function of s denoted by $f(s)$ describes the overall population of storage capacity. Jowitt (1991) employed entropy to derive a single-reservoir surface runoff model. The discussion in this section closely follows his work.

Surface runoff, denoted by q, is assumed to be a function of the watershed wetness or actual moisture or water storage, denoted by w in a manner similar to a linear reservoir. This represents the macroscopic behavior of the watershed. Clearly, $0 \leq w \leq s$. It is reasonable to assume w to be a random variable as well, with a probability density function defined by $f(w)$. Since the wetness is bounded by s, surface runoff q occurs when $w = s$ or when the storage is full. Therefore it is more appropriate to define w conditioned on s, or $w | s$, and the conditional probability density function $f(w | s)$.

14.5.1   Mass Conservation

The balance among rainfall, soil moisture, and surface runoff must be satisfied at all times. This balance can be expressed in terms of mean quantities as

$$\frac{dW}{dt} = p - q \qu\quad (14.113)$$

where p is rainfall intensity (assumed rainfall excess intensity), W is the mean wetness, and q is the mean surface runoff. It is further hypothesized that there is a linear relation between q and W

$$q = kW \qquad (14.114)$$

where k is a constant (approximately equal to the mean residence time or lag time). Combining Eqs. (14.113) and (14.114), one can write

$$\frac{dW}{dt} = p - kW \qquad (14.125)$$

Because W and q are watershed mean quantities, they are derived from watershed microscopic properties or their corresponding random variables.

14.5.2   Constraints

One can define the mean properties at the watershed scale. The probability density function of moisture storage capacity must satisfy

$$\int_0^\infty f(s)\,ds = 1 \qquad (14.116)$$

The mean storage capacity, denoted by S, can be expressed as

$$\int_0^\infty sf(s)\,ds = S \qquad (14.117)$$

The conditional probability density function $f(w\,|\,s)$ will satisfy

$$\int_0^\infty f(s)\int_0^s f(w\,|\,s)\,dw\,ds = 1 \qquad (14.118)$$

The mean wetness can now be defined as

$$\int_0^\infty f(s)\int_0^s wf(w\,|\,s)\,dw\,ds = W \qquad (14.119)$$

14.5.3   Discharge Hydrograph

Since W is expressed by Eq. (14.119), it may be recalled that in Eq. (14.115) q is proportional to the wetness but only in those storage elements that are full. The proportional relation must be weighted by $f(w\,|\,s)$, reflecting a measure of wetness evaluated at $w = s$. Thus, Eq. (14.115) can be expressed as

$$\frac{dW}{dt} = p - k\int_C sf(w\,|\,s)\,ds \qquad (14.120)$$

where the integral is evaluated along the contour $C \equiv w = s$, $0 \le s < \infty$. Integration over s takes into account the distribution of storage capacities throughout the watershed. For evaluation of Eq. (14.120), the probability density functions $f(s)$ and $f(w\,|\,s)$ must be expressed. These density functions are derived using POME.

14.5.4 Water Storage Capacity Density Function

The probability density function $f(s)$ is derived by maximizing entropy, subject to the constraints expressed by Eqs. (14.116) and (14.117). Entropy is expressed as

$$H(s) = -\int_0^\infty f(s) \ln f(s) ds \qquad (14.121)$$

Maximization of Eq. (14.121) is accomplished using the method of Lagrange multipliers. Therefore, the Lagrange function L is written as

$$L = -\int_0^\infty f(s) \ln f(s) ds - (\lambda_0 - 1)\left[\int_0^\infty f(s) ds - 1\right] - \lambda_1\left[\int_0^\infty sf(s) ds - S\right] \qquad (14.122)$$

where λ_0 and λ_1 are Lagrange multipliers.

Differentiating Eq. (14.122) with respect to $f(s)$ and equating the derivative to zero, one gets

$$\frac{\partial L}{\partial f(s)} = 0 \Rightarrow \ln f(s) = -\lambda_0 - \lambda_1 s \qquad (14.123)$$

Equation (14.123) results in

$$f(s) = \exp(-\lambda_0 - \lambda_1 s) = a \exp(-\lambda_1 s) \qquad a = \exp(-\lambda_0) \qquad (14.124)$$

Substituting Eq. (14.124) in Eq. (14.116), one obtains

$$\int_0^\infty a \exp(-\lambda_1 s) ds = 1 \Rightarrow \frac{a}{\lambda_1} \qquad \lambda_0 = -\ln \lambda_1 \qquad (14.125)$$

Substitution of Eq. (14.124) in Eq. (14.117) leads to

$$\int_0^\infty sa \exp(-\lambda_1 s) ds = 1 \Rightarrow \frac{1}{\lambda_1} = S \qquad (14.126)$$

Using Eqs. (14.125) and (14.126) in Eq. (14.124), one obtains

$$f(s) = \frac{1}{S} \exp\left(\frac{-s}{S}\right) \qquad (14.127)$$

which, as expected, is an exponential distribution with mean storage capacity as a parameter. Also note that

$$\lambda_0 = -\ln \lambda_1 = \ln S \qquad (14.128)$$

14.5.5 Probability Density Function for Water Storage Conditional on Moisture Storage

Following the above procedure, the entropy of water storage conditioned on moisture storage, $f(w \mid s)$, to be maximized is expressed as

$$H = -\int_0^\infty f(s) \int_0^s f(w \mid s) \ln f(w \mid s) dw \, ds \qquad (14.129)$$

Maximization of Eq. (14.129) is accomplished, subject to Eqs. (14.118) and (14.119). Using the method of Lagrange multipliers, the Lagrange function L can be written as

$$L = -\int_0^\infty f(s) \int_0^s f(w|s) \ln f(w|s) dw\, ds$$

$$-(\alpha_0 - 1)\left[\int_0^\infty f(s) \int_0^s f(w|s) dw\, ds - 1\right] - \alpha_1 \left[\int_0^\infty f(s) \int_0^s wf(w|s) dw\, ds - W\right] \quad (14.130)$$

where α_0 and α_1 are the Lagrange multipliers. Differentiating Eq. (14.140) with respect to $f(w|s)$ and equating the derivative to zero, one obtains

$$\frac{\partial L}{\partial f(w|s)} = 0 \Rightarrow f(w|s) = \exp(-\alpha_0 - \alpha_1 w) = b \exp(-\alpha_1 w), \qquad b = \exp(-\alpha_0) \quad (14.131)$$

Substitution of Eq. (14.131) in Eq. (14.118) yields

$$\int_0^\infty f(s) \int_0^s b \exp(-\alpha_1 w) dw\, ds = 1 \quad (14.132)$$

Equation (14.132) yields

$$b = \frac{\alpha_1}{1 - \exp(-\alpha_1 s)} \qquad \alpha_0 = -\ln\left[\frac{\alpha_1}{1 - \exp(-\alpha_1 s)}\right] \qquad \text{for all } s \quad (14.133)$$

Inserting Eq. (14.143) in Eq. (14.131), one gets

$$f(w|s) = \frac{\alpha_1 \exp(-\alpha_1 w)}{1 - \exp(-\alpha_1 s)} \qquad \text{for all } s \quad (14.134)$$

Parameter α_1 can be estimated by using Eq. (14.131) in Eq. (14.119) as

$$\int_0^\infty f(s) \frac{1 - \exp(-\alpha_1 s) - \alpha_1 s \exp(-\alpha_1 s)}{1 - \exp(-\alpha_1 s)} ds = W \quad (14.135)$$

Equations (14.133) and (14.135) can be used to estimate parameters α_0 and α_1. However, these equations are nonlinear, and a closed-form analytical solution does not seem tractable.

14.5.6   Runoff Hydrograph

Equation (14.120) prescribes the hydrograph of runoff for which $f(s)$ and $f(w|s)$ are specified by Eqs. (14.127) and (14.134), and W by Eq. (14.135). Therefore, for $C \equiv w = s$, $0 \le s < \infty$,

$$\frac{dW}{dt} = p(t) - k \int_0^\infty \lambda_1 \exp(-\lambda_1 s)\left[\frac{\alpha_1 s \exp(-\alpha_1 s)}{1 - \exp(-\alpha_1 s)}\right] ds \quad (14.136)$$

Parameterizing Eq. (14.136) yields

$$\frac{dW}{dt} = \frac{dW}{d\alpha_1}\frac{d\alpha_1}{dt}$$

(14.137)

Using Eq. (14.136), one writes Eq. (14.137) as

$$\frac{d\alpha_1}{dt}\frac{d}{d\alpha_1}\int_0^\infty \lambda_1 \exp(-\lambda_1 s)\left[\frac{1-\exp(-\alpha_1 s)-\alpha_1 s\exp(-\alpha_1 s)}{1-\exp(-\alpha_1 s)}\right]ds$$

$$= p(t) - k\int_0^\infty \frac{\lambda_1\alpha_1 s\exp[-(\lambda_1+\alpha_1)s]}{1-\exp(-\alpha_1 s)}ds$$

(14.138)

Equation (14.138) becomes

$$\frac{d\alpha_1}{dt}\int_0^\infty \lambda_1 \exp(-\lambda_1 s)\left[\frac{\alpha_1^2 s^2\exp(-\alpha_1 s)-[1-\exp(-\alpha_1 s)]^2}{\alpha_1^2[1-\exp(-\alpha_1 s)]^2}\right]ds$$

$$= p(t) - k\int_0^\infty \frac{\lambda_1\alpha_1 s\exp[-(\lambda_1+\alpha_1)s]}{1-\exp(-\alpha_1 s)}ds$$

(14.139)

Equation (14.139) does not seem amenable to analytical solution.
   Recall Riemann's zeta function $\zeta(z,\tau)$

$$\zeta(z,\tau) = \sum_{n=0}^\infty \frac{1}{(\tau+n)^z} \qquad \text{Re } z > 1$$

(14.140)

and the associated z-parameter phi function $\Phi(\upsilon,z,\tau)$

$$\Phi(\upsilon,z,\tau) = \sum_{n=0}^\infty \frac{\upsilon^n}{(\tau+n)^z} \qquad |\upsilon| < 1, \qquad \tau \neq 0, -1, -2, \dots$$

(14.141)

Note that

$$\Phi(1,z,\tau) = \zeta(z,\tau)$$

(14.142)

The integrals below have the same form as in Eq. (14.139):

$$\int_0^\infty \frac{x^{z-1}\exp(-cx)}{1-\exp(-bx)}dx = \frac{1}{b^z}\Gamma(z)\zeta(z,\frac{c}{b}) \qquad \text{Re } c > 0 \qquad \text{Re } z > 1$$

(14.143)

$$\int_0^\infty \frac{x^{z-1}\exp(-cx)}{1-\upsilon\exp(-x)^2}dx = \Gamma(z)[\Phi(\upsilon,z-1,c-1)-(c-1)\Phi(\upsilon,z,c-1)]$$

(14.144)

$$[\text{Re } z > 0 \qquad \text{Re } c > 0 \qquad |\arg(1-\upsilon)| < \pi]$$

Denote $\alpha_1/\lambda_1 = y$, a dimensionless parameter. Then Eq. (14.139) can be expressed as

$$\frac{dy}{dt} = \frac{\lambda_1\left[py^2 - ky\zeta\left(2,1+\frac{1}{y}\right)\right]}{\frac{2}{y}\left[\zeta\left(2,\frac{1}{y}\right) - \frac{1}{y}\zeta\left(3,\frac{1}{y}\right)\right] - 1} \qquad y > 0 \qquad (14.145)$$

$$\frac{dy}{dt} = \frac{\lambda_1\left[py^2 + ky\zeta\left(2,-\frac{1}{y}\right)\right]}{-\frac{2}{y}\left[\zeta\left(2,-\frac{1}{y}\right) + \frac{1}{y}\zeta\left(3,-\frac{1}{y}\right)\right] - 1} \qquad y < 0 \qquad (14.146)$$

When $y = 0$, expansions of terms in Eq. (14.146) lead to

$$\frac{dy}{dt} = \frac{12}{\lambda_1(p-k)} \qquad (14.147)$$

$$W = \frac{1}{\lambda_1}y - \zeta\left(2,1+\frac{1}{y}\right)\frac{1}{\lambda_1 y^2} \qquad y > 0$$

$$= \frac{1}{\lambda_1 y} + \zeta\left(2,-\frac{1}{y}\right)\frac{1}{\lambda_1 y^2} \qquad y < 0 \qquad (14.148)$$

$$= \frac{\lambda_1}{2} \qquad y = 0$$

$$q = -k\zeta(2,1+\frac{1}{y})\frac{1}{y} \qquad y > 0$$

$$= k\zeta(2,-\frac{1}{y})\frac{1}{y} \qquad y < 0 \qquad (14.149)$$

$$= k \qquad y = 0$$

The watershed storage and runoff are related as

$$q = k(1 - \lambda_1 yW) \qquad (14.150)$$

Note that

$$W(-y) = \frac{1}{\lambda_1 - W(y)} \qquad (14.151)$$

Example 14.4.   Plot the watershed wetness W versus the runoff q for various values of $k = \lambda_1$. What do you conclude from the plot? What are the upper and lower bounds for the wetness?

Solution.   For y ranging from -10 to 10, W and q are computed from Eqs. (14.148) and (14.149) respectively. Observe that for each y value, there is one W and related q. Thus, by relating the W and q values computed from the same value of y, the plot of W versus q can be made for various values of k, as shown in Fig. 14.7. It can be seen from the figure that the upper bound of wetness increases with k or λ_1, and the lower bound is zero. When $k = \lambda_1 = 1$, the wetness monotonically increases from 0 to 0.9.

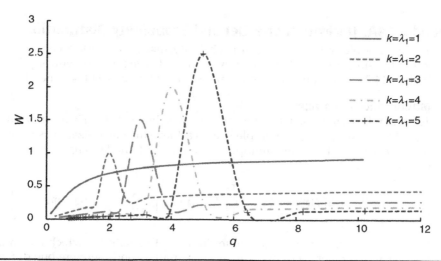

FIGURE **14.7** Variation of q with W. [$k = \lambda_1$]

Example 14.5. Plot the relation between W and q for various values of the dimensionless parameter y. It will be a coaxial diagram.

Solution. Consider $k = \lambda_1 = 1$, again W and q are computed from Eqs. (14.148) and (14.149), respectively, for y varies from -10 to 10. In the first quadrant, W is plotted versus q, which is the same as that in Fig. 14.7. In the second quadrant W is plotted versus y for both positive and negative values; these two curves meet at $W = 0.5$ for $y = 0$. In the fourth quadrant q is plotted versus y for both positive and negative values; these two curves meet at $q = 1$ for $y = 0$. Obviously, the third quadrant is y versus y, where it is a straight line. The relation between W and q is plotted for various values of y, as shown in Fig. 14.8.

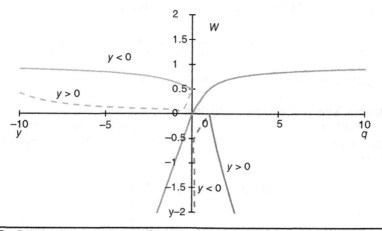

FIGURE **14.8** Relation between W and q for various values of y.

Appendix 14A: Derivation of a General Probability Distribution

The methodology for deriving a probability distribution for a continuous random variable using entropy theory is now discussed. The Shannon entropy $H(X)$ is given by Eq. (14.29), and constraints on X are expressed by Eqs. (14.30) and (14.31).

Maximization of Entropy

To obtain the least-biased $f(x)$, the entropy given by Eq. (14.29) is maximized, subject to Eqs. (14.30) and (14.31); one simple way to achieve the maximization is through the use of the method of Lagrange multipliers. To that end, the lagrangian function L can be constructed as

$$L = -\int_a^b f(x)\ln f(x)dx - (\lambda_0 - 1)\left[\int_a^b f(x)dx - C_0\right] - \sum_{r=1}^n \lambda_r\left[\int_a^b f(x)g_r(x)dx - C_r\right] \quad (14A.1)$$

where $\lambda_1, \lambda_2, ..., \lambda_n$ are the Lagrange multipliers. To obtain $f(x)$ which maximizes L, one may recall the Euler-Lagrange calculus of variation, and therefore one differentiates L with respect to $f(x)$ (noting X is a parameter and f is a variable), equates the derivative to zero, and obtains

$$\frac{\partial L}{\partial f} = 0 \Rightarrow -[1 + \ln f(x)] - (\lambda_0 - 1) - \sum_{r=1}^n \lambda_r g_r(x) = 0 \quad (14A.2)$$

Probability Distribution

Equation (14A.2) leads to the probability density function of X in terms of the given constraints

$$f(x) = \exp[-\lambda_0 - \sum_{r=1}^n \lambda_r g_r(x)] \quad (14A.3)$$

where the Lagrange multipliers λ_r, $r = 0,1, 2, ..., n$, can be determined with the use of Eqs. (14.30) and (14.31). Equation (14A.3) is also written as

$$f(x) = \frac{1}{Z(\lambda_1, \lambda_2, ..., \lambda_n)}\exp\left[-\sum_{r=1}^n \lambda_r g_r(x)\right] \quad (14A.4)$$

where $Z = \exp\lambda_0$ is called the partition function.

Determination of Lagrange Multipliers

The POME-based probability distribution contains Lagrange multipliers $\lambda_0, \lambda_1, \lambda_2, ..., \lambda_n$ as parameters which can be determined by inserting Eq. (14A.3) in Eqs. (14.30) and (14.31):

$$\exp(\lambda_0) = \int_a^b \exp\left[-\sum_{r=1}^n \lambda_r g_r(x)\right]dx \qquad r = 1,2, ..., n \quad (14A.5)$$

and

$$C_r = \int_a^b g_r(x)\exp\left[-\sum_{r=1}^n \lambda_r g_r(x)\right]dx \qquad r = 1, 2, ..., n \quad (14A.6)$$

Equation (14A.5) can be written for the zeroth Lagrange multiplier as

$$\lambda_0 = \ln \int_a^b \exp\left[-\sum_{r=1}^n \lambda_r g_r(x)\right] dx \qquad r = 1, 2, ..., n \qquad (14A.7)$$

Equation (14A.7) shows that λ_0 is a function of $\lambda_1, \lambda_2, ..., \lambda_n$ and expresses the partition function Z as

$$Z(\lambda_1, \lambda_2, ..., \lambda_m) = \int_a^b \exp[-\sum_{r=1}^m \lambda_r g_r(x)] dx \qquad (14A.8)$$

Differentiating Eq. (14A.7) with respect to λ_r, one gets

$$\frac{\partial \lambda_0}{\partial \lambda_r} = -\frac{\int_a^b g_r(x)\exp\left[-\sum_{r=1}^n \lambda_r g_r(x)\right] dx}{\int_a^b \exp\left[-\sum_{r=1}^n \lambda_r g_r(x)\right] dx} \qquad r = 1, 2, ..., n \qquad (14A.9)$$

Multiplying the numerator as well as the denominator of Eq. (14A.9) by $\exp(-\lambda_0)$, one obtains

$$\frac{\partial \lambda_0}{\partial \lambda_r} = -\frac{\int_a^b g_r(x)\exp\left[-\lambda_0 - \sum_{r=1}^n \lambda_r g_r(x)\right] dx}{\int_a^b \exp\left[-\lambda_0 - \sum_{r=1}^n \lambda_r g_r(x)\right] dx} \qquad r = 1, 2, ..., n \qquad (14A.10)$$

The denominator in Eq. (14A.10) equals 1 by virtue of Eq. (14.30). Therefore, Eq. (14A.10) becomes

$$\frac{\partial \lambda_0}{\partial \lambda_r} = -\int_a^b g_r(x)\exp\left[-\lambda_0 - \sum_{r=1}^n \lambda_r g_r(x)\right] dx \qquad r = 1, 2, ..., n \qquad (14A.11)$$

Note that substitution of Eq. (A.3) in Eq. (14.31) yields

$$C_r = \int_a^b g_r(x)\exp\left[-\lambda_0 - \sum_{r=1}^n \lambda_r g_r(x)\right] dx \qquad r = 1, 2, ..., n \qquad (14A.12)$$

Therefore, Eqs. (14A.12) and (14A.10) yield

$$\frac{\partial \lambda_0}{\partial \lambda_r} = -C_r \qquad r = 1, 2, ..., n \qquad (14A.13)$$

Likewise, Eq. (14A.7) can be written analytically as

$$\lambda_0 = \lambda_0(\lambda_1, \lambda_2, .., \lambda_n) \qquad (14A.14)$$

Equation (14A.14) can be differentiated with respect to λ_r, $r = 1, 2, ..., n$, and each derivative can be equated to the corresponding derivative in Eq. (14A.13). This would lead to a system of $n - 1$ equations with $n - 1$ unknowns, whose solution would lead to the expression of Lagrange multipliers in terms of constraints.

Appendix 14B: Derivation of Exponential, Gamma, and Lognormal Probability Distributions

The exponential, gamma, and lognormal probability distributions have been found appropriate for representing different SCS-CN variables. These distributions are therefore derived following the methodology discussed in App. 14A.

Exponential Distribution

Following Singh (1998), constraints for the exponential distribution can be expressed as Eq. (14.30) and

$$\int_a^b xf(x)dx = \bar{x} \tag{14B.1}$$

Then the entropy-based PDF of X from Eq. (14A.3) becomes

$$f(x) = \exp(-\lambda_0 - \lambda_1 x) = a\exp(-\lambda_1 x) \qquad c = \exp(-\lambda_0) \tag{14B.2}$$

which is the POME-based distribution with λ_0 and λ_1 as parameters. These parameters can be determined with the aid of Eqs. (14.30) and (14B.1).
Inserting Eq. (14B.2) in Eq. (14.30), one gets

$$c = \frac{\lambda_1}{\exp(-\lambda_1 a) - \exp(-\lambda_1 b)} \tag{14B.3}$$

Therefore, Eq. (14B.2) becomes

$$f(x) = \frac{\lambda_1}{\exp(-\lambda_1 a) - \exp(-\lambda_1 b)} \exp(-\lambda_1 x) \tag{14B.4}$$

Equation (14B.4) is the truncated exponential distribution. Inserting Eq. (14B.4) in Eq. (14B.1), one can determine λ_1 in terms of $\bar{x}$, but a closed form solution is not tractable. However, if $a = 0$ and $b \to \infty$, then the solution is straightforward:

$$c = \lambda_1 \tag{14B.5}$$

$$\lambda_1 = \frac{1}{\bar{x}} \tag{14B.6}$$

Therefore,

$$f(x) = \frac{1}{\bar{x}}\exp\left(-\frac{x}{\bar{x}}\right) \tag{14B.7}$$

Gamma Distribution

The constraint equations for the gamma distribution can be written (Singh, 1998) as Eq. (14.30), (14B.1) and

$$\int_a^b \ln x \, f(x) dx = \overline{\ln x} \tag{14B.8}$$

Then the entropy-based PDF from Eq. (14A.3) becomes

$$f(x) = ax^{-\lambda_2} \exp(-\lambda_1 x) \qquad a = \exp(-\lambda_0) \tag{14B.9}$$

where a, λ_0, and λ_1 are parameters which can be determined by substitution of Eq. (14B.9) in Eq. (14.30), (14B.1), and (14B.8). The zeroth Lagrange multiplier from Eq. (14A.7) becomes

$$\lambda_0 = \ln \int_a^b \exp(-\lambda_1 x - \lambda_2 \ln x) dx \tag{14B.10}$$

An explicit solution of Eq. (14B.10) is not tractable. However, if $a = 0$ and $b \to \infty$, then the solution is straightforward:

$$\lambda_0 = (\lambda_2 - 1)\ln \lambda_1 + \ln \Gamma(1 - \lambda_2) \tag{14B.11}$$

$$\frac{\lambda_2 - 1}{\lambda_1} = \bar{x} \tag{14B.12}$$

$$\ln \lambda_1 = \psi(k) - \overline{\ln x}, \; k = 1 - \lambda_2, \psi(k) = \frac{d\Gamma(k)}{dk} \tag{14B.13}$$

Then, Eq. (14B.9) becomes

$$f(x) = \frac{\lambda_1^{\,k}}{\Gamma(k)} x^{k-1} \exp(-\lambda_1 x) \tag{14B.14}$$

Lognormal distribution

The constraint equations for the log-normal distribution can be written as (Singh, 1998) as Eq. (14.30), (14B.8), and

$$\int_a^b (\ln x - c)^2 f(x) dx = \overline{(\ln x - c)^2} \qquad c = \overline{\ln x} \tag{14B.15}$$

Equation (14B.15) can be expressed as

$$\int_a^b (\ln x)^2 f(x) dx - c^2 = S_y^{\,2} + c^2 \tag{14B.16}$$

where S_y is the standard deviation of Y, $Y = \ln X$. Then one can write $E[y] = m_y$ or $E[x] = E[\exp y] = m_x$, $m_x = E[x]$, and $E[y^2] = \sigma_y^2 + m_y^2$. Then the entropy-based PDF from Eq. (14A.3) becomes

$$f(x) = \exp[-\lambda_0 - \lambda_1 \ln x - \lambda_2 (\ln x)^2] \tag{14B.17}$$

where λ_0, λ_1, and λ_2 are the Lagrange multipliers as parameters which can be determined by substitution of Eq. (B.17) in Eqs. (14.30), and (14B.8), and (14B.15). The zeroth Lagrange multiplier can be expressed from Eq. (14A.7) as

$$\lambda_0 = \ln \int_a^b \exp[-\lambda_1 \ln x - \lambda_2 (\ln x - c)^2] dx \qquad (14B.18)$$

However, an explicit solution of Eq. (14B.18) is not tractable for arbitrary values of a and b. However, for $a = 0$ and $b \to \infty$ the solution can be written as follows:

$$\lambda_0 = \frac{1}{2} \ln \pi - \frac{1}{2} \ln \lambda_2 + \frac{(\lambda_1 - 1)^2}{4\lambda_2} \qquad (14B.19)$$

$$\lambda_1 = 1 - \frac{m_y}{\sigma_y^2} \qquad (14B.20)$$

$$\lambda_2 = \frac{1}{2\sigma_y^2} \qquad (14B.21)$$

Finally, the resulting distribution is

$$f(x) = \frac{1}{x\sqrt{2\pi}\, \sigma_y} \exp\left[-\frac{1}{2} \left(\frac{\ln x - m_y}{\sigma_y^2} \right)^2 \right] \qquad (14B.22)$$

Questions

1. Consider any small USDA-ARS agricultural watershed. Obtain data of rainfall and runoff events and land use for several years for this watershed. Compute the curve number values for these events. Then compute the empirical probability distribution of discharge, and compare this distribution with the one obtained using entropy theory. How well do the two distributions compare?

2. For the watershed in Question 1, obtain the values of the rainfall amount and peak runoff for at least 15 rainfall-runoff events, and plot them on log-log paper. Fit a curve and specify the parameters of the curve. Estimate the parameters using the entropy method, and then compute the curve. Compare the empirical and entropy-based curves.

3. For the watershed in Question 1, select a storm event and obtain rainfall-runoff data for this event. Compute the unit hydrograph from observations, and compare it with the one derived using entropy theory. Then compute the discharge hydrograph using the IUH, and compare it with the observed hydrograph.

4. For Question 2, determine the probability density function and probability distribution function of peak discharge and plot them. Also obtain the distributions using empirical observations on the watershed and compare the two sets of distributions.

5. For the watershed in Question 1, compute the joint distribution of rainfall amount and peak discharge empirically. Then compute the joint distribution

with the entropy method. Compare the two distributions. (You can use the contingency table.)

6. For Question 5, compute the conditional distribution of peak discharge for given rainfall amount empirically, and compare it with the one estimated using entropy.

References

Al-Hamdan, O. Z., and Cruise J. F. (2010). Soil moisture profile development from surface observations by principle of maximum entropy. *Journal of Hydrologic Engineering, ASCE*, vol. 15, no. 5, pp. 327–337.

Chiu, C. L. (1987). Entropy and probability concepts in hydraulics. *Journal of Hydraulic Engineering*, vol. 113, no. 5, pp. 583–600.

Hawkins, R. H. (1982). Interpretation of source area variability in rainfall-runoff relations. In: *Rainfall-Runoff Relationship*, edited by V.P. Singh, pp. 303–324, Water Resources Publications, Littleton, Colorado.

Jaynes, E. T. (1957a). Information theory and statistical mechanics, I. *Physical Reviews*, vol. 106, pp. 620–630.

Jaynes, E. T. (1957b). Information theory and statistical mechanics, II. *Physical Reviews*, vol. 108, pp. 171–190.

Jaynes, E. T. 1982. On the rationale of maximum entropy methods. *Proceedings of the IEEE* vol. 70; pp. 939–952.

Lienhard, J. H. (1964). A statistical mechanical prediction of the dimensionless unit hydrograph. *Journal of Geophysical Research*, vol. 69, no. 24, pp. 5231–5238.

Lienhard, J. H. (1972). Prediction of the dimensionless unit hydrograph. *Nordic Hydrology*, vol. 3, pp. 107–109.

Jowitt, P. W. (1991). A maximum entropy view of probability-distributed catchment models. *Hydrological Sciences Journal*, vol. 36, no. 2, pp. 123–134.

Priestley, S. (2000). Discussion of "Theoretical justification of SCS method for runoff estimation" by B. Yu (1998). *Journal of Irrigation and Drainage Engineering*, ASCE, Vol. 126, No. 1, pp. 75.

Rai, R.K., Sarkar, S. and Gundekar H.G. (2008). Adequacy of two-parameter beta distribution functions for deriving unit hydrograph. *Water Resources Management*, vol. 39, No. 3, pp. 899-929.

Rogers, W. F. (1980). A practical model for linear and nonlinear runoff. *Journal of Hydrology*, vol. 46, pp. 51–78.

Rogers, W. F., and Singh V. P. (1986). Evaluating flood retarding structures. *Advances in Water Resources*, vol. 9, pp. 236–244.

Rogers, W. F., and Singh V. P. (1988). Drainage basin peak discharge rating curve. Hydrological Processes, vol. 2, pp. 245–253.

Rogers, W. F., and Zia H. A. (1982). Linear and nonlinear runoff from large drainage basins. *Journal of Hydrology*, vol. 55, pp. 267–278.

Singh, V. P. (1998). *Entropy-Based Parameter Estimation in Hydrology*. Kluwer Academic Publishers, Boston.

Singh, V. P. (2010a). Entropy theory for derivation of infiltration equations. *Water Resources Research*, vol. 46, pp. 1–20, W03527, doi:10-1029/2009WR008193.

Singh, V. P. (2010b). Tsallis entropy theory for derivation of infiltration equations. *Transactions of the ASABE*, vol. 53, no. 2, pp. 447–463.

Singh, V. P. (2010c). Entropy theory for movement of moisture in soils. *Water Resources Research*, vol. 46, pp. 1–12, W03516, doi:10-1029/2009WR008288.

Singh, V. P. (2013). SCS-CN method revisited using entropy theory. *Transactions of the ASABE*, vol. 56, no. 5, pp. 1805–1820.

Singh, V. P. and Aminian H. (1986). An empirical relation between volume and peak of direct runoff. *Water Resources Bulletin*, vol. 22, no. 5, pp. 735–750.

Smith, R. E. and Montgomery R. J. (1979). Discussion of "Runoff curve numbers for partial area watersheds" by R. H. Hawkins. *Journal of Irrigation and Drainage Engineering*, ASCE, vol. 106, no. IR4, pp. 379–381.

Further Reading

Arnold, J. G., Srinivasan, R., Muttiah, R. S. and Williams J. R. (1998). Large area hydrologic modeling and assessment, Part I: model develpment. *Journal of the American Water Resources Association*, vol. 34, pp. 73–89.

Auerswald, K., and Haider J. (1995). Runoff curve numbers for small grain under German cropping conditions. *Journal of Environmental Management*, vol. 47, pp. 223–228.

Beasley, D. B., and Huggins L. F. (1980). ANSWERS (*Area non-point source watershed environment simulation*). User's Manual. Department of Agricultural Engineering, Purdue University, West Lafayette, Ind.

Bonta, J. V. (1997). Determination of watershed curve number using derived distributions. *Journal of Irrigation and Drainage Engineering*, vol. 123, no. 1, pp. 28–36.

Carsel, R. F., Smith, C. N., Mulkey, L. A., Dean, J. D. and Jowise, J. D. (1984). *Users manual for the pesticide root zone model (PRZM)*. Release 1. U. S. Environmental Protection Agency (ed.), EPA-600/3-84-109, Athens, G.

Chen, C.-L. (1982). An evaluation of the mathematics and physical significance of the Soil Conservation Service curve number procedure for estimating runoff volume. In: *Rainfall-Runoff Modeling*, edited by V. P. Singh, Water Resources Publications, Littleton, Colo., pp. 387–418.

Choi, J. Y., Engle, B. A. and Chung H. W. (2002). Daily streamflow modeling and assessment based on the curve number technique. *Hydrological Processes*, vol. 16, no. 16, pp. 3131–3150.

Garen, D. C., and Moore D. S. (2005). Curve number hydrology in water quality modeling: Uses, abuses, and future directions. *Journal of the American Water Resources Association*, vol. 34, pp. 377–388.

Hawkins, R. H. (1973). Improved prediction of storm runoff in mountain watersheds. *Journal of Irrigation and Drainage Engineering*, vol. 99, no. 4, pp. 519–523.

Hawkins, R. H. (1978a). Effects of rainfall intensity on runoff curve numbers. *Utah Agric. Exp. Station Journal* 2288, Utah.

Hawkins, R. H. (1978b). Runoff curve numbers with varying site moisture. *Journal of Irrigation and Drainage Division*, vol. 104, no. 4, pp. 389–398.

Hawkins, R. H. (1993). Asymptotic determination of runoff curve numbers from data. *Journal of the irrigation and Drainage Division*, ASCE, vol. 119, no. 2, pp. 334–345.

Hawkins, R. H., Hjelmfelt, A. T., and Zevenbergen, A. W. (1985). Runoff probability, storm depth, and curve numbers. *Journal of Irrigation and Drainage Engineering*, vol. 111, no. 4, pp. 330–339.

Hawkins R. H. and Ward, T. J. (2012). Expected values of event runoff with curve number theory. Paper presented at ASCE-EWRI Congress held May, 2012, in Albuquerque, N. Mex.

Hawkins, R. H., Ward, T. J. Woodward, D. E. and van Mullen, J. A. (2009). *Curve Number Hydrology: State of the Practice.* Report of ASCE/EWRI Task Committee, Reston, V.

Hjelmfelt, A. T., Kramer, L. A. and Burwell R. E. (1982). Curve numbers as random variables. In: *Rainfall-Runoff Relationship,* edited by V. P. Singh. Water Resources Publications, Littleton, Colo. pp. 365–370.

Kapur, J. N. and Kesavan, H. K. (1992). *Entropy Maximization Principles with Applications.* Academic Press, New York.

Knisel, W. G. (1980). CREAMS: A fieldscale model for chemical, runoff, and erosion from agricultural management systems. USDA Science and Education Administration, Conservation Report no. 26, Washington, D.C.

Lahlou, N., Shoemaker, L. Choudhary, S. Elmer, R. Hu, A., Manguerra, H. and Parker, A. (1998). *BASINS V.2.0 User's manual.* U. S. Environmental Protection Agency Office of Water, EPA-823-B-98–006, Washington.

Leonard, R. A., Knisel, W. G. and Still, D. A. (1987). GLEAMS: Groundwater loading effects of agricultural managements systems. *Transactions of ASAE,* vol. 30, pp. 1403–1418.

Lienhard, J. H. and Meyer, P. L. (1967). A physical basis for the generalized gamma distribution. *Quarterly of Applied Mathematics,* vol. 25, pp. 330.

Lienhard, J. H., and Davis, L. B. (1964). An extension of statistical mechanics to the description of a broad class of macrosystems. *Journal of Applied Mathematics and Mechanics (ZAMP),* vol. 22, pp. 85–96.

McCuen, R. H. (2002). Approach to confidence interval estimation for curve numbers. *Journal of Hydrologic Engineering,* vol. 7, no. 1, pp. 43–48.

Mishra, S. K., Jain, M. K., Bhunya, P. K., and Singh, V. P. (2005). Field applicability of the SCS-CN-based Mishra-Singh general model and its variant. *Water Resources Management,* vol. 19, no. 1, pp. 37–62.

Mishra, S. K., Jain, M. K. and Singh V. P. (2004). Evaluation of SCS-CN based model incorporating antecedent moisture. *Water Resources Management,* vol. 18, no. 6, pp. 567–589.

Mishra, S. K., and Singh V. P. (1999). Another look at SCS-CN method, *Journal of Hydrologic Engineering,* Vol. 4, no. 3, pp. 257–264.

Mishra, S. K., and Singh V. P. (2003). *Soil Conservation Service Curve Number (SCS-CN) Methodology.* Kluwer Academic, Dordrecht, The Netherlands.

Moore, R. J. (1985). The probability distributed principle and runoff production at point and basin scales. *Hydrological Sciences Journal,* vol. 30, no. 2, pp. 273–297.

Moore, R. J., and Clarke R. J. (1981). A distributed function approach to rainfall runoff modeling. *Water Resources Research,* vol. 17, no. 5, pp. 1367–1382.

Ponce, V. M., and Hawkins R. H. (1996). Runoff curve number: Has it reached maturity? *Journal of Hydrologic Engineering, ASCE,* vol. 1, no. 1, pp. 11–19.

Rallison, R. E., and Miller N. (1982). Past, present and future SCS runoff procedure. In: *Rainfall-runoff Relationship,* edited by V. P. Singh. Water Resources Publications, Fort Collins, Colo., pp. 353–364.

Schneider, L. E., and McCuen R. H. (2005). Statistical guidelines for curve number generation. *Journal of Irrigation and Drainage Engineering,* vol. 131, no. 3, pp. 282–290.

Shannon, C. E. (1948). A mathematical theory of communications, I and II. *Bell System Technical Journal,* vol. 27, pp. 379–443.

Sharpley, A. N., and Williams, J. R., eds. (1990). EPIC-Erosion/productivity impact calculator: 1. Model documentation. *U. S. Department of Agriculture Bulletin 1768,* Washington.

Singh, V. P., ed. (1995). *Computer Models of Watershed Hydrology*. Water Resources Publications, Highlands Ranch, Colo.

Singh, V. P., and Frevert, D. K., eds. (2002a). *Mathematical Models of Small Watershed Hydrology and Applications*. Water Resources Publications, Highlands Ranch, Colo.

Singh, V. P. and Frevert, D. K., eds. (2002b). *Mathematical Models of Large Watershed Hydrology*. Water Resources Publications, Highlands Ranch, Colo.

Singh, V. P. and Frevert, D. K., eds. (2006). *Watershed Models*. CRC Press, Boca Raton, Floa.

Sonuga, J. O. (1976). Entropy principle applied to rainfall-runoff process. *Journal of Hydrology*, vol. 30, pp. 81–94.

Smith, R. E. (1978). A proposed infiltration model for use in simulation of field-scale watershed hydrology. Paper presented at the USDA-ARS Nonpoint Pollution Modeling Workshop, February 14–16, 10p.

Smith, R. E. (1997). Discussion of 'Runoff curve number: Has it reached maturity?' by V. M. Ponce and R. H. Hawkins. *Journal of Hydrologic Engineering*, vol. 2, no. 3, pp. 145–147.

Walker, W. R., Prestwich, C. and Spofford, T. (2006). Development of the revised USDA-NRCS intake families for surface irrigation. *Agricultural Water Management*, vol. 85, pp. 157–164.

Young, R. A., Onstad, C. A., Bosch, D. D. and Anderson W. P. (1989). AGNPS: A nonpoint source pollution model for evaluating agricultural watersheds. *Journal of Soil and Water Conservation*, vol. 44, pp. 168–173.

Yu, B. (1998). Theoretical justification of SCS method for runoff estimation. *Journal of Irrigation and Drainage Engineering*, vol. 124, no. 6, pp. 306–310.

CHAPTER 15

Stream Flow Simulation

Stream flow simulation plays an important role in water resources planning and management. The key requirement for stream flow simulation is that synthetic stream flow sequences preserve the required statistical properties, such as the mean, standard deviation, skewness, kurtosis, and lagone correlation of the historical record. This chapter discusses entropy-based simulation of monthly stream flow at a single site which hopefully is capable of preserving key statistics, such as the mean, standard deviation, skewness, and lag-one correlation.

15.1 Simulation Models

A number of models for stream flow simulation have been proposed, and these models can be classified into two groups: parametric and nonparametric. A commonly used parametric model for synthetic stream flow generation is the autoregressive moving average (ARMA) model, which is quite flexible and can be used for annual as well as seasonal stream flow simulation. The ARMA model with a large number of parameters is based on the gaussian assumption that is satisfied by transforming stream flow data to be normal, and then the model is fitted. An alternative to the ARMA model for simulating seasonal stream flow is the disaggregation model which has been widely applied. In the disaggregation model, annual or aggregated stream flow is generated with an appropriate model, and then the generated stream flow is disaggregated to obtain monthly or component stream flow. The disaggregation model ensures additive property between high and low stream flow levels, but has many parameters that need to be estimated. To reduce the number of parameters, several parsimonious models have been proposed, such as the condensed disaggregation model and stepwise disaggregation model. Koutsoyiannis and Manetas (1996) proposed another parametric disaggregation model, taking into account stream flow at both large scale (e.g., yearly) and small scale (e.g., monthly) and Koutsoyiannis (2001) generalized this model to couple stochastic models at different time scales.

One of the drawbacks of parametric models is the difficulty of parameter estimation, especially when the number of parameters is large. Moreover, the normal distribution assumption usually made about the marginal distribution of stream flow data may not hold in reality. Therefore, transformation techniques, such as logarithmic and power, to render the data normal are often applied. However, there are three potential drawbacks associated with such transformations. First, some bias of the stochastic properties in the original domain may be caused when data are simulated in the transformed domain. Second, negative values may be generated. Third, nongaussian features, such as skewness and bimodality, cannot be captured and reproduced efficiently.

Furthermore, it is hard for a usual parametric model to capture the nonlinear relationships that may be observed in the historical record.

Nonparametric models have provided an attractive alternative and can overcome the drawbacks of parametric models. Lall (1995) provided a review of the application of nonparametric models in hydrology. The advantages of these models are that they avoid model selection, minimize (or avoid) parameter estimation, and do not make any assumption about the underlying probability distribution. A nonparametric model generates a time series by resampling directly from the historical record and can thus preserve the empirical statistical structure.

Nonparametric stream flow simulation is often based on bootstrap techniques or kernel density estimation. Lall and Sharma (1996) proposed a nearest-neighbor bootstrap method for resampling monthly stream flow, while probabilistically preserving the dependence structure. To reproduce the serial correlation of historical data, Vogel and Shallcross (1996) suggested the moving block bootstrap (MBB) method by resampling the observed time series in approximately independent blocks, and they compared the method with parametric methods for generating annual stream flow series. Sharma et al. (1997) proposed a nonparametric method for monthly stream flow simulation, applying the conditional density function with gaussian kernel, and Sharma and O'Neill (2002) extended that method to impose a long-term dependence in the simulated stream flow by incorporating an aggregated variable (denoted as NPL model). Salas and Lee (2010) developed a nonparametric method using the K-nearest neighbor (KNN) resampling technique with gamma kernel perturbation that can generate data different from the historical record for single-site seasonal stream flow simulation. For this method, two approaches—one with the aggregate variable (denoted as the KGKA model) and another with the pilot variable (denoted as the KGKP model)—were developed to preserve annual variability.

Nonparametric methods have also been applied for seasonal stream flow simulation with the disaggregation approach. Tarboton et al. (1998) developed a nonparametric disaggregation model for simulation based on the conditional distribution obtained by kernel density estimation. To address the issue of inefficiency of the kernel density method in higher dimensions, Prairie et al. (2007) applied a fast KNN-based bootstrap approach instead of constructing and simulating from the conditional distribution. Lee et al. (2010) proposed a space-time disaggregation model based on the KNN coupled with a genetic algorithm that can overcome the shortcomings of the models proposed by Prairie et al. (2007) and Koutsoyiannis and Manetas (1996). Based on KNN resampling, Nowak et al. (2010) proposed another space-time disaggregation algorithm for disaggregating annual flow into daily flow at different sites.

To simulate stream flow, an assumption about the underlying data distribution is often made, especially for parametric models. However, many stream flow records cannot be characterized by commonly assumed probability distributions. The ability to preserve the cross boundary relation (e.g., the correlation between the last season of the previous year and the first season of the current year) and the generation of negative values are two issues that emerge for both parametric and nonparametric models (Lee, et al., 2010). To address the first issue, Mejia and Rousselle (1976) made a modification to link past and present values being disaggregated. A practical way to address this problem is to start the generation from a season in which the correlation is small. However, this does not work when all correlations between seasons are high. The issue of negative values arises due to the use of normal transformation in parametric models and application of the gaussian kernel in nonparametric models. Generally, negative values generated during simulation can be disregarded. However, this solution may not be appropriate when too many negative values are generated.

15.2   Stream Flow Simulation at a Single Site

Hao and Singh (2011) simulated monthly stream flow at a single site using entropy theory, wherein a probability density distribution is derived without the assumption of normality or the use of a normal transformation. The entropy-based simulation seems to preserve the cross-correlation and avoids the generation of negative values. It can also be extended to incorporate higher moments and more lag correlations, if needed. Using statistical properties, such as the mean, standard deviation, skewness, and lag-one correlation, as constraints, it can construct the joint probability density function of stream flow of two adjacent months and derive the conditional density function.

Entropy theory-based stream flow simulation entails seven steps: (1) defining the Shannon entropy, (2) formulating the principle of maximum entropy (POME), (3) specification of constraints, (4) derivation of a joint probability density function, (5) estimation of distribution parameters, (6) derivation of conditional distribution, and (7) sampling from the conditional distribution. Each of these steps is now discussed.

15.2.1   Shannon Entropy

For a continuous random variable $X \varepsilon$ $[a,b]$ with a probability density function (PDF) $f(x)$, the Shannon entropy H_1 is defined as (Shannon, 1948; Shannon and Weaver, 1949)

$$H_1 = -\int_a^b f(x)\ln f(x)dx \tag{15.1}$$

where x is a value of random variable X with lower limit a and upper limit b. For a bivariate case involving continuous random vector (X, Y) with joint probability density function $f(x, y)$ over the space $[a, b] \times [c, d]$, the Shannon entropy can be defined as

$$H_2 = -\int_c^d\int_a^b f(x,y)\ln f(x,y)dx\,dy \tag{15.2}$$

where y is a value of random variable Y with lower limit c and upper limit d. The objective is to determine the least-biased bivariate PDF $f(x, y)$, using the principle of maximum entropy (POME).

15.2.2   Principle of Maximum Entropy

POME, formulated by Jaynes (1982), states that for making inferences based on incomplete information, the probability distribution that has the maximum entropy permitted by the given information should be selected. The least-biased probability density function of stream flow can then be derived by the maximization of entropy with the information known from historical data, which can be expressed in the form of constraints. Accordingly, to derive the bivariate PDF, constraints must be specified for X and Y separately and jointly.

15.2.3   Specification of Constraints

For stream flow simulation, statistics such as the mean, standard deviation, skewness, and lag-one correlation must be preserved. These statistics can be regarded as constraints for deriving the least-biased distribution of stream flow. Then sampling from the distribution can be expected to preserve these required statistics. The mean, standard deviation, and skewness can be determined through the first three moments. To determine the correlation between stream flows of two adjacent months (say, January and February), the joint PDF is needed for which constraints in general form can be stated as

$$\int_c^d \int_a^b f(x, y)\, dx\, dy = 1 \tag{15.3}$$

$$\int_c^d \int_a^b g_i(x, y) f(x, y)\, dx\, dy = \overline{g_i} \qquad i = 1, 2, \ldots, m \tag{15.4}$$

Specifically,

$$\int_c^d \int_a^b x^i f(x, y)\, dx\, dy = \overline{x^i} \qquad i = 1, 2, 3 \tag{15.5}$$

$$\int_c^d \int_a^b y^{i-3} f(x, y)\, dx\, dy = \overline{y^{i-3}} \qquad i = 4, 5, 6 \tag{15.6}$$

$$\int_c^d \int_a^b xy f(x, y)\, dx\, dy = \overline{xy} \qquad i = 7 \tag{15.7}$$

where $g_i(x, y)$ or g_i is a known function of random vector (X, Y), which can be specified as $g_1 = x, g_2 = x^2, g_3 = x^3, g_4 = y, g_5 = y^2, g_6 = y^3$ and $g_7 = xy$ for the proposed method; $\overline{g_i}$ is the expected value of the function $g_i(x, y)$ (e.g., if $g_1(x, y) = x$, then $\overline{x}$ is the mean of X) estimated from the historical record; $\overline{x^i}$ and $\overline{y^{i-3}}$ are the first and third noncentral moments of random variables X and Y, respectively; $\overline{xy}$ is the expectation of XY, and m is the number of constraints ($m = 7$ in this case). The constraint in Eq. (15.3) ensures that the integration of the probability density function over the whole interval will be unity, which is often termed the *normalization condition* or the *total probability theorem*.

Note that to preserve the cross-correlation between stream flow in December of the previous year and that in January of the current year, the expected value $\overline{g_7}$ in Eq. (15.7) is specified through the cross product of stream flows of the two months with lag-one, or specifically

$$\overline{g_7} = \frac{1}{n} \sum_{t=2}^n x_{t-1,12} y_{t,1} \tag{15.8}$$

where t is the year index; n is the length of the year in the historical record; $x_{t-1,12}$ is the December stream flow of the year previous to the t year; and $y_{t,1}$ is the January stream flow of the t year.

15.2.4   Derivation of Joint Probability Density Function

To derive the joint PDF of stream flows of two adjacent months (say, January and February), the entropy given by Eq. (15.2) is maximized, subject to the constraints given by Eqs. (15.3) and (15.4). This can be accomplished using the calculus of variation. Denoting the parameters for the joint PDF of January and February stream flow as $\Phi_{1,2} = [\lambda_0, \lambda_1, \ldots, \lambda_7]$, where $\lambda_0, \lambda_1, \ldots, \lambda_7$ are the Lagrange multipliers, the lagrangian L can be expressed as

$$L = -\int_c^d \int_a^b f(x, y) \ln f(x, y)\, dx\, dy - (\lambda_0 - 1)\left[\int_c^d \int_a^b f(x, y)\, dx\, dy - 1 \right]$$

$$- \sum_{i=1}^m \lambda_i \left[\int_c^d \int_a^b g_i(x, y) f(x, y)\, dx\, dy - \overline{g_i} \right] \qquad i = 1, 2, \ldots, m \tag{15.9}$$

Differentiating L with respect to f and setting the derivative to zero, one obtains the least-biased probability density function with representation of $g_i(x, y)$ with their specific values as (Kesavan and Kapur, 1992)

$$f(x,y) = \exp\left[-\sum_{i=0}^{m} \lambda_i g_i(x,y)\right] = \exp(-\lambda_0 - \sum_{i=1}^{3} \lambda_i x^i - \sum_{i=4}^{6} \lambda_i y^{i-3} - \lambda_7 xy) \qquad (15.10)$$

The joint PDF given by Eq. (15.10) has unknown Lagrange multipliers λ_i ($i = 1, \ldots, m$) that need to be determined.

15.2.5 Parameter Estimation

For a stream flow series, it may be convenient to scale all values to [0, 1] without loss of generality. For the original data (OD) with the maximum (MX) and minimum (MN), the scaled data (SD) can be expressed as SD = (OD – MN)/(MX – MN). It is easy to check that this scaling method does not affect the skewness of the original data.

Substitution of Eq. (15.10) in Eqs. (15.3) and (15.4) results in a set of nonlinear equations whose solution results in the Lagrange multipliers. In the normalization condition in Eq. (15.3), one can obtain the zeroth Lagrange multiplier λ_0 as a function of other multipliers as

$$\exp \lambda_0 = \int_c^d \int_a^b \exp\left[-\sum_{i=1}^{m} \lambda_i g_i(x,y)\right] dx\, dy \qquad (15.11)$$

Generally the analytical solution of the Lagrange multipliers does not exist (for $m > 2$), and numerical solution is resorted to. One way of solving Lagrange multipliers is to find the minimum of the convex function Γ, expressed as

$$\Gamma = \lambda_0 + \sum_{i=1}^{m} \lambda_i \overline{g_i} = \ln \int_c^d \int_a^b \exp\left[-\sum_{i=1}^{m} \lambda_i g_i(x,y)\right] dx\, dy + \sum_{i=1}^{m} \lambda_i \overline{g_i} \qquad i = 1, 2, \ldots, m \qquad (15.12)$$

Newton's method can be applied to solve the optimization problem in Eq. (15.12) to obtain the Lagrange multipliers $\lambda = [\lambda_1, \ldots, \lambda_7]$. Starting from some initial value $\lambda_{(0)}$, one updates $\lambda_{(1)}$ through the equation

$$\lambda_{(1)} = \lambda_{(0)} - H^{-1} \frac{\partial \Gamma}{\partial \lambda_i} \qquad i = 1, 2, \ldots, m \qquad (15.13a)$$

where the gradient is expressed as

$$\frac{\partial \Gamma}{\partial \lambda_i} = \overline{g_i} - \int_c^d \int_a^b \exp\left[-\sum_{i=0}^{m} \lambda_i g_i(x,y)\right] g_i(x,y) dx\, dy \qquad i = 1, 2, \ldots, m \qquad (15.13b)$$

and the hessian matrix H is expressed as

$$H = \begin{bmatrix} \mathrm{var}(x,x) & \mathrm{cov}(x,x^2) & \cdots & \mathrm{cov}(x,xy) \\ \mathrm{cov}(x^2,x) & \mathrm{var}\,x^2 & \cdots & \mathrm{cov}(x^2,xy) \\ \cdots & \cdots & \cdots & \cdots \\ \mathrm{cov}(xy,x) & \mathrm{cov}(xy,x^2) & \cdots & \mathrm{var}\,xy \end{bmatrix} \qquad (15.13c)$$

in which elements of the Hessian matrix are expressed as

$$H_{i,j} = \text{cov}(g_i, g_j) = \int_c^d \int_a^b \exp\left[-\sum_{i=0}^m \lambda_i g_i(x, y)\right] g_i(x, y) g_j(x, y) dx\, dy$$

$$-\int_c^d \int_a^b \exp\left[-\sum_{i=0}^m \lambda_i g_i(x, y)\right] g_i(x, y) dx\, dy \qquad (15.14)$$

$$\bullet \int_c^d \int_a^b \exp\left[-\sum_{i=0}^m \lambda_i g_i(x, y)\right] g_j(x, y) dx\, dy \qquad i = 1, 2, \ldots, m$$

where $\text{cov}(g_i, g_j) = \text{var } g_i$ and H^{-1} is the inverse of the hessian matrix H. The MATLAB function *fminsearch* can be used to solve this optimization problem in Eq. (15.12) to obtain the Lagrange multipliers.

For the generation of monthly stream flow, 12 joint PDFs of two adjacent months are needed, and the corresponding parameters $(\Phi_{1,2}, \Phi_{2,3}, \ldots, \Phi_{12,1})$ of each joint PDF have to be estimated. Each Lagrange multiplier in the joint PDF in Eq. (15.7) is related to one statistical property that is to be preserved. For instance, in parameters $\Phi_{1,2} = [\lambda_0, \lambda_1, \ldots, \lambda_7]$ of the joint PDF of stream flow of January and February, λ_1, λ_2, and λ_3 relate to the mean, standard deviation, and skewness of the January streamlfow, λ_4, λ_5, and λ_6 relate to the mean, standard, deviation, and skewness of the February stream flow; and λ_7 is the mutual parameter relating to the lag-one correlation of stream flows of the two months. Likewise, parameter $\Phi_{2,3}$ of the joint PDF of stream flow of February and March relates to the required statistics for the February and March stream flow and so on. When more statistics (e.g., kurtosis) are required to be preserved, one can incorporate the corresponding Lagrange multipliers in the model. Thus, this model is quite flexible and can be extended to incorporate more statistics, if needed.

15.2.6   Derivation of Conditional Distribution

The marginal distribution for X can be obtained by integrating the joint PDF $f(x, y)$ given by Eq. (15.10) over Y as

$$f(x) = \int_0^1 f(x, y) dy = \exp(-\lambda_1 x - \lambda_2 x^2 - \lambda_3 x^3 - \lambda_0) \int_0^1 \exp(-\lambda_4 y - \lambda_5 y^2 - \lambda_6 y^3 - \lambda_7 xy) dy$$

$$(15.15)$$

Then one can obtain the conditional density function of Y, given $X = x$, as

$$f(y \mid x) = \frac{f(x, y)}{f(x)} = \frac{\exp(-\lambda_4 y - \lambda_5 y^2 - \lambda_6 y^3 - \lambda_7 xy)}{\int_0^1 \exp(-\lambda_4 y - \lambda_5 y^2 - \lambda_6 y^3 - \lambda_7 xy) dy} \qquad (15.16)$$

The conditional cumulative distribution function $F_{Y|X}(y|x)$ of Y given $X = x$ can be written as

$$F_{Y|X}(y \mid x) = \int_0^y f(t \mid x)\, dt \qquad (15.17)$$

The cumulative distribution function for variable X, $G(x)$, can be obtained from Eq. (15.15) as

$$G(x) = \int_0^x f(t)\, dt \qquad (15.18)$$

15.2.7 Sampling

Several techniques can be employed to generate random values from the bivariate distribution, such as the conditional distribution method, the transformation method, the acceptance/rejection method, and the composition method (Johnson, 1987; Balakrishnan and Lai, 2009). To sample from the continuous joint PDF $f(x, y)$ to obtain the random values of (X, Y), the conditional distribution method is discussed in this study. For the generation of stream flow while preserving the correlation between adjacent months, stream flow values of one months can be generated given the stream flow value of its previous month from the conditional distribution. When the method is applied to the generation of monthly data of each year, 12 monthly conditional density functions have to be used sequentially. To illustrate this method, assume that the simulation starts from January and n year data are to be generated. Let $x_{t,s}$ denote stream flow for the month (or period) s of year t ($s = 1, 2, ..., 12, t = 1, 2, ..., n$). The step-by-step simulation procedure for generating random values of each month can be summarized as follows:

1. Generate a uniform random value w_1 in [0, 1] which can be done with the use of random number generator function *rand* in MATLAB. This w_1 value can be considered the cumulative probability, which then corresponds to a specific value of x, say x_0. Derive the marginal density function $f_0(x)$ for stream flow of December, using Eq. (15.15) with parameters $\Phi_{12,1}$ estimated earlier. Then from Eq. (15.18) with the corresponding marginal density function $f_0(x)$ one can write the cumulative distribution function as

$$G(x_0) = \int_0^{x_0} f_0(t)\, dt = w_1$$

Then one can obtain x_0 from the inverse cumulative distribution function as

$$x_0 = G^{-1}(w_1)$$

where x_0 is considered as the initial value for the sequential generation in the following steps.

2. With the initial value x_0 generated in step 1, one can generate the value for the first month (January) of the first year $x_{1,1}$ from the conditional cumulative distribution function in Eq. (15.17) with parameters $\Phi_{12,1}$. To that end, generate another uniform distributed random value w_2 in [0, 1], which is considered to be the conditional cumulative probability corresponding to a specific value $x_{1,1}$, given the initial value x_0. This can be expressed with Eq. (15.17) as

$$F_{Y|X}(x_{1,1} | x_0) = w_2$$

Then the first value $x_{1,1}$ can be generated by solving the above equation.

3. Now the value $x_{1,2}$ for February of the first year can be generated, given the value of the previous month (January) $x_{1,1}$ generated in step 2 with parameters $\Phi_{1,2}$ following the same procedure as described above.

4. Repeat step 3 until $x_{1,12}$, the stream flow of December of the first year, is generated while parameters $\Phi_{2,3}, \Phi_{3,4}, ..., \Phi_{11,12}$ are used at each step. In this manner, the monthly stream flow of the first year is obtained.

5. Now one can start to generate steam flow of the second year. To that end, first generate value $x_{2,1}$, the stream flow of January of the second year, given the value of $x_{1,12}$, the stream flow of December of the first year, with the method described in step 2 with parameter $\Phi_{12,1}$.

6. Repeat step 5 until the value $x_{2,12}$, the stream flow of December of the second year, is generated while the parameters $\Phi_{1,2}, \Phi_{2,3}, \dots, \Phi_{11,12}$ are used at each step. In this manner, the monthly stream flow of the second year is obtained.

7. Repeat steps 5 and 6 until the monthly stream flow of the nth year is generated.

In the steps above, numerical integration is performed to generate random values from the inverse (conditional) cumulative distribution function. The generated data are then scaled back to their original domain, and one stream flow sequence of length n is generated.

Example 15.1.   Consider the observed stream flow for one gauging station. Use the first moment as the constraint to derive the entropy-based distribution. Generate 100 sequences of stream flow with 100 yr of stream flow in each sequence. Compare the mean, standard deviation, and skewness of observed and generated stream flow.

Solution.   The station selected is the USGS 08082500 at Brazos River in Seymour, Texas. The annual stream flow data from 1964 to 2011 are used for this example. The stream flow data are given in Table 15.1.

With the first moment as a constraint, the PDF can be expressed as $f(x) = \exp(-\lambda_0 - \lambda_1 x)$. It has been shown that this distribution is the exponential distribution when the lower bound a of stream flow is set to 0 and the upper bound b can be set to ∞.

Year	Flow (cfs)	Year	Flow (cfs)	Year	Flow (cfs)	Year	Flow (cfs)
1964	87.7	1976	106.8	1988	104.3	2000	154
1965	196.6	1977	202.4	1989	212.2	2001	175.5
1966	490.4	1978	167.4	1990	537.2	2002	133.9
1967	350.4	1979	161.3	1991	461.1	2003	117.3
1968	230.7	1980	295.3	1992	702.8	2004	268.6
1969	258	1981	161.9	1993	133.7	2005	516.5
1970	147.9	1982	606.9	1994	103.1	2006	106.2
1971	311	1983	113.7	1995	255	2007	404.2
1972	527.2	1984	278.8	1996	178.5	2008	98.3
1973	304.3	1985	340.7	1997	259.6	2009	99.6
1974	112.3	1986	434	1998	61.1	2010	439.2
1975	296.4	1987	742.1	1999	260.2	2011	36.6

TABLE 15.1   Annual Stream Flow (cfs) from the Station USGS 08082500 at Brazos River in Seymour, Texas, 1964–2011

The next step is to determine the Lagrange multipliers. To that end, Eq. (15.12) becomes

$$\Gamma = \lambda_0 + \lambda_1 \bar{x} \int_0^\infty \exp[-\lambda_1(x)] + \lambda_1 \bar{x}$$

The MATLAB function *fminsearch* can be used for maximization. The maximum of Γ is obtained as -0.3262 when the Lagrange multiplier λ_1 takes on the value of 3.7668. Then λ_0 can be obtained:

$$\lambda_0 = \ln \int_0^\infty \exp[-\lambda_1(x)] = -1.3262$$

Now a uniformly distributed random value w_1 is generated with the MATLAB function *rand*. For example, $w_1 = 0.1$ and the corresponding stream flow value is x_1. Note that

$$F(x) = \int_0^x \exp(1.3262 - 3.7668\,x)\,dx$$

Therefore,

$$F(x) = 0.1 \qquad \text{or} \qquad \int_0^{x_1} \exp(1.3262 - 3.7668x)\,dx = 0.1$$

Then one obtains

$$x_1 = F^{-1}(0.1) = 0.0280$$

In a similar manner, the stream flow for other years $x_2, \ldots, x_n$ can be generated. A total of 100 sequences of stream flow are generated with the above steps; and the mean, standard deviation, and skewness of each sequence can be computed. Sequences and statistics of each sequence can be denoted as follows:

Sequences	Mean m	Standard Deviation s	Skewness k
$x_1^1, x_2^1, \ldots, x_n^1$	m_1	s_1	k_1
$x_1^2, x_2^2, \ldots, x_n^2$	m_2	s_2	k_2
$\ldots$	$\ldots$	$\ldots$	$\ldots$
$x_1^{100}, x_2^{100}, \ldots, x_n^{100}$	m_{100}	s_{100}	k_{100}

One can compute the median of each statistic and then compare with the statistics of the observed stream flow. For example, the median of the simulated mean ($m_1, m_2, \ldots, m_{100}$) can be obtained as 0.2707 with the MATLAB function *median*, which is close to the observed mean 0.2655. Comparison of simulated and observed statistics is shown in Table 15.2 which reveals that the mean is preserved well, but the standard deviation and skewness are not preserved as well.

Statistics	Mean *m*	Standard Deviation s	Skewness k
Observed statistics	0.2655	0.1725	1.0183
Median of simulated statistics of Example 15.1	0.2656	0.2604	1.7333
Median of simulated statistics of Example 15.2	0.2622	0.1684	0.5863
Median of simulated statistics of Example 15.3	0.2651	0.1694	0.9301

TABLE 15.2   Comparison of Observed and Simulated Statistics

Example 15.2. Derive the entropy-based distribution, using the first and second moments as constraints. Determine the values of the constraints for the same data set as in Example 15.1. Generate 100 stream flow sequences with 100 yr of stream flow in each sequence. Compare the mean, standard deviation, and skewness of the observed and generated stream flows.

Solution. With the first and second moments as constraints, the PDF can be expressed as

$$f(x) = \exp(-\lambda_0 - \lambda_1 x - \lambda_2 x^2)$$

The lower bound a of the interval is set to 0, and the upper bound b can be set to some relatively large number (b is set to 40 times the maximum observed value in this example).

To determine the Lagrange multipliers, Eq. (15.12) becomes

$$\Gamma = \lambda_0 + \lambda_1 \bar{x} + \lambda_2 \overline{x^2} = \ln \int_0^b \exp(-\lambda_1 x - \lambda_2 x^2) + \lambda_1 \bar{x} + \lambda_2 \bar{x}$$

The maximum of Γ is found to be -0.4509, and the Lagrange multipliers are $\lambda_1 = -3.6845$ and $\lambda_2 = 9.9283$. Then λ_0 is obtained as

$$\lambda_0 = \ln \int_0^b \exp(-\lambda_1 x - \lambda_2 x^2) = -0.4619$$

Now a uniformly distributed random value w_1 is generated with the MATLAB function *rand.* For example, for $w_1 = 0.1$ the corresponding stream flow value is x_1. Note that

$$F(x) = \int_0^x \exp(0.4619 + 3.6845x - 9.9283x^2)\, dx$$

Therefore,

$$F(x) = 0.1 \quad \text{or} \quad \int_0^{x1} \exp(0.4619 + 3.6845x - 9.9283x^2) = 0.1$$

Then one obtains

$$x_1 = F^{-1}(0.1) = 0.0572$$

The simulated statistics are shown in Table 15.2, which reveals that both the mean and standard deviation are preserved well. However, skewness is not preserved well.

Example 15.3. Derive the entropy-based distribution, using the first three moments as constraints. Determine the constraints for the same data set as in Example 15.1. Generate 100 stream flow sequences with 100 yr of stream flow in each sequence. Compare the values of mean, standard deviation, and skewness of observed and generated stream flows.

Solution. To reduce the computational complexity, the original data are scaled to [0, 1]. For the original data (OD) with maximum value (MX) and minimum value (MN), the scaled data (SD) are expressed as SD = [OD − (1 − d)MN)]/[(1 + d)MX − (1 − d)MN], where d is a scale parameter and is selected as 0.05. The lower bound a for the integration is set to 0, and the upper bound b can be set to 2. With the first three moments as constraints, the PDF can be expressed as

$$f(x) = \exp(-\lambda_0 - \lambda_1 x - \lambda_2 x^2 - \lambda_3 x^3)$$

To determine the Lagrange multipliers, Eq. (15.12) becomes

$$\Gamma = \lambda_0 + \lambda_1 \bar{x} + \lambda_2 \bar{x}^2 + \lambda_3 \bar{x}^3 = \ln \int_0^b \exp(-\lambda_1 x - \lambda_2 x^2 - \lambda_3 x^3) + \lambda_1 \bar{x} + \lambda_2 \bar{x}^2 + \lambda_3 \bar{x}^3$$

The maximum of Γ is -0.2282, and the Lagrange multipliers are $\lambda_1 = -0.9810$, $\lambda_2 = 5.3319$, and $\lambda_3 = -1.0301$. Then λ_0 is obtained as

$$\lambda_0 = \ln \int_0^b \exp(-\lambda_1 x - \lambda_2 x^2 - \lambda_3 x^3) = -0.6230$$

Now a uniformly distributed random value w_1 is generated with the MATLAB function *rand*. For example, $w_1 = 0.1$ and the corresponding stream flow value is x_1. Note that

$$F(x) = \int_0^x \exp(0.6230 + 0.9810 x - 5.3319 x^2 + 1.0301 x^3)\, dx$$

Therefore, $F(x) = 0.1$ or

$$\int_0^{x1} \exp(0.6230 + 0.9810 x - 5.3319 x^2 + 1.0301 x^3) = 0.1$$

Then one obtains

$$x_1 = F^{-1}(0.1) = 0.0525$$

The simulated statistics are also found in Table 15.2, which shows that the mean, standard deviation, and skewness are preserved well.

15.3 A Case Study: Testing with Synthetic Data

To test the performance of the proposed method to recover the density function of the known model and reproduce the statistics of samples from it, Hao and Singh (2011) selected the bivariate normal distribution, which has several useful properties and is commonly employed. The joint PDF $f(x, y)$ for a continuous random vector (X, Y) can be expressed as

$$f(x, y) = \frac{1}{2\pi \sigma_1 \sigma_2 \sqrt{1-\rho^2}} \exp\left\{-\frac{1}{2(1-\rho^2)}\left[\left(\frac{x-u_x}{\sigma_x}\right)^2 + \left(\frac{y-u_y}{\sigma_y}\right)^2 - 2\rho\left(\frac{x-u_x}{\sigma_x}\right)\left(\frac{y-u_y}{\sigma_y}\right)\right]\right\}$$

(15.19)

where u_x and u_y are the means of random variables X and Y; σ_x and σ_y are the standard deviations of X and Y; and ρ is the correlation between X and Y. Parameters are specified as $[u_x, u_y] = [0.4, 0.6]$, $[\sigma_x, \sigma_y] = [0.1, 0.1]$, and $\rho = 0.8$. One sample record of length 3000 is drawn from the underlying bivariate normal distribution, which is considered representative of the observed values for the purpose of fitting and evaluating the proposed model. Values of random variables X and Y in the sample record are within the interval $[0, 1]$, and data scaling is not needed. The modeling and sampling steps in Sections 15.2 are then used.

To quantify the performance of the proposed model in recovering the marginal and joint PDFs, the mean square error (MSE) is computed as

$$\text{MSE} = \frac{1}{n}\sum_{i=1}^{n}(s_i - p_i)^2$$

(15.20)

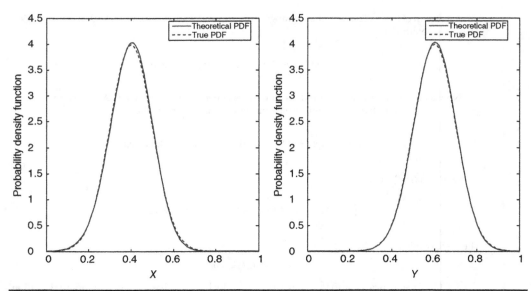

15.1   Theoretical marginal PDF and underlying true marginal PDF for variables X and Y.

where n is the length of the observed data, s_i and p_i ($i = 1, \dots, n$) are the theoretical densities from the estimated density function in Eq. (15.10) or (15.15) and the true density at the ith observed value.

Theoretical marginal PDFs of random variables X and Y in Eq. (15.15) estimated from the sample record together with the underlying true density functions are plotted as shown in Fig. 15.1. As can be seen, the theoretical density of X (or Y) is virtually indistinguishable from that of the true density of X (or Y). The MSE values between the theoretical marginal probability density and underlying true probability density of the observed values are 0.0032 for variable X and 0.0014 for variable Y, respectively. Thus, the marginal PDF estimated from the sample record can reproduce the true marginal PDF well.

Contours of the theoretical joint PDF in Eq. (15.10) estimated from the same sample record and the underlying true joint PDF are plotted as shown in Fig. 15.2. The contour lines of the theoretical PDF match those of the underlying true PDF well, although a little divergence exists for the contour lines with low density (<2) when random variables X and Y had extreme values. The MSE value between the theoretical joint probability density and the underlying true joint probability density of the observed values was 0.11. This shows that the proposed method can reproduce the underlying true joint PDF well.

To quantify the performance of the model, the bias and relative bias (RB) of the simulated statistics were computed as

$$\text{Bias} = S_m - X_O \qquad \text{RB} = \frac{S_m - X_O}{X_O} \qquad (15.21)$$

where S_m is the median of simulated statistic for the generated sequences and X_O is the statistic for the historical record.

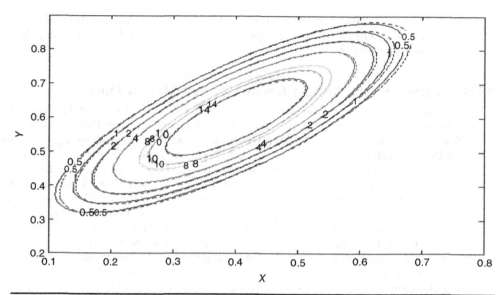

Figure 15.2 Contours of theoretical joint PDF and underlying true joint PDF. The continuous line is for the theoretical marginal PDF and the dotted line is for the underlying true marginal PDF.

A total of 100 sequences of random vector (X, Y) with length 3000 in each sequence were generated with the sampling steps discussed above (with period $s \leq 2$) based on the joint PDF in Eq. (15.19) estimated from the sample record. Statistics of the generated data and sample record (or observed values) with different percentiles, together with bias and relative bias for each statistic, are shown in Table 15.3. The relative bias for the mean, standard deviation, skewness, and lag-one correlation is 0.051, −0.28, 6.55, −0.26 percent for random variable X and −0.021, −0.17, −20.86, and −0.26 percent for random variable Y, respectively. The median values of statistics of generated data are close to those of the sample record. Among these statistics, skewness is reproduced with the largest relative bias. Generally, all statistics are preserved well, although skewness is not preserved as well as other statistics.

More sample records are drawn from the bivariate normal distribution to fit and evaluate the entropy-based model, as is done with the sample record. Results from

	Sim. X	Obs. X	Bias (10^{-3})	Relative Bias (%)	Sim. Y	Obs. Y	Bias (10^{-3})	Relative Bias (%)
Mean	0.400	0.399	0.21	0.051	0.600	0.601	−0.13	−0.021
Standard deviation	0.099	0.099	−0.28	−0.28	0.099	0.099	−0.17	−0.17
Skewness	−0.097	−0.091	−5.96	6.55	−0.024	−0.031	6.43	−20.86
Lag-one correlation	0.796	0.799	−2.10	−0.26	0.796	0.799	−2.10	−0.26

Table 15.3 Comparison of Simulated and Observed Statistic Values of Synthetic Data

these simulations show that the entropy-based model can recover the underlying true marginal and joint PDFs well and preserve the mean, standard deviation, skewness, and lag-one correlation well.

15.4   A Case Study: Simulation of Monthly Stream Flow

Hao and Singh (2011) applied the entropy-based method to the monthly stream flow of 10 sites (sites 11 through 20) in the Colorado River basin from 1906 to 2003 (Lee and Salas, 2006). These data can be found at this website: http://www.usbr.gov/lc/region/g4000/NaturalFlow/previous.html. For each station, stream flow data are scaled to the interval [0, 1] for each month. With the use of constraints in Eqs. (15.1) to (15.3), first parameters $(\Phi_{1,2}, \Phi_{2,3}, ..., \Phi_{12,1})$ of each joint PDF of two adjacent months in Eq. (15.10) are estimated. Then the conditional distribution is derived from the known joint PDF by using Eqs. (15.10) and (15.15). Thereafter, samples are drawn sequentially.

The proposed method is applied to 100 generated flow sequences, each of 100 yr length, to test the entropy method. This simulation is termed S_1. Statistics of generated data and historical record are compared using box plots. The central mark of the box is the median, and the end lines of the box represent the 25th and 75th percentiles. The whiskers are the extremes of simulated data points that are not considered to be outliers. The continuous line connecting the star marks represents the statistics of the historical record. A wide box plot signifies large variability. When a statistic falls in the box plot, the performance is considered to be good. The closer the median of the statistic of generated data is to that of the historical record, the better the performance of the model is.

The theoretical PDFs fit the observed histograms relatively well, except that the bimodality in the density of scaled stream flow is not resolved, as shown in Fig. 15.3. Contours of the theoretical joint densities together with the empirical density are shown

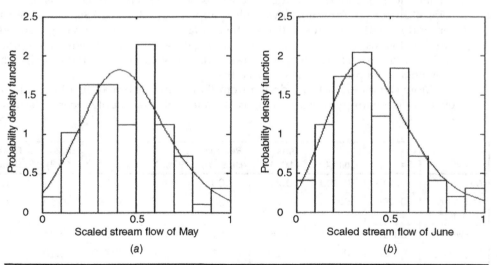

FIGURE **15.3**   Histogram of the scaled stream flow compared to the estimated probability density function for Colorado River at Lees Ferry. (a) Scaled stream flow of May; (b) scaled stream flow of June.

in Fig. 15.4*a* and *b* with the historical record plotted as stars. The historical record spread along the contours, as seen in Fig. 15.4*a*, and the theoretical values of the joint density also match the empirical values well for the most part in Fig. 15.4*b*.

Statistics of generated data and historical record for all months of simulation S_1 are shown in Fig. 15.5. The bias and relative bias for each required statistic are illustrated in Fig. 15.6 and Table 15.2. From Fig. 15.5, median values of the simulated mean, standard deviation, and lag-one correlation are close to those of the historical record, while little divergence exists between the median values of skewness of generated data and that of historical record. All statistics of the mean, standard deviation, skewness, and lag-one correlation fall in the box plots. The relative biases of the mean, standard deviation, and lag-one correlation are lower than 5 percent, and that of skewness is lower than 10 percent for most months. Specifically, the maximum relative bias values for the mean, standard deviation, skewness, and lag-one correlation are −1.47 percent in April, −4.07 percent in September, −10.95 percent in November, and −3.70 percent in October, respectively. The relative bias of the simulated skewness is relatively high and is not preserved as well as other statistics. The lag-one correlation between December stream flow of the previous year and January stream flow of the current year is also preserved well.

In Fig. 15.6, the bias of the simulated maximum value is negative and that of the simulated minimum value is positive for all months, which is also illustrated in Fig. 15.5. The entropy method underestimates maximum values and overestimates minimum values. The reason is that the theoretical joint PDF is estimated from the scaled data within the interval [0, 1]. Thus, the generated values from the resulting conditional

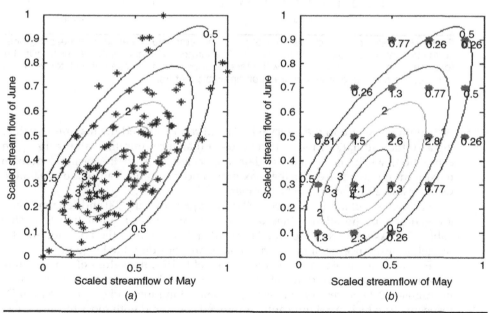

FIGURE 15.4 Contours of the theoretical and true joint PDF of scaled May and June stream flow with historical record for Colorado River at Lees Ferry. (*a*) Contours of the theoretical joint PDF and historical record plotted as stars; (*b*) contours of the theoretical joint PDF and empirical density plotted as circles.

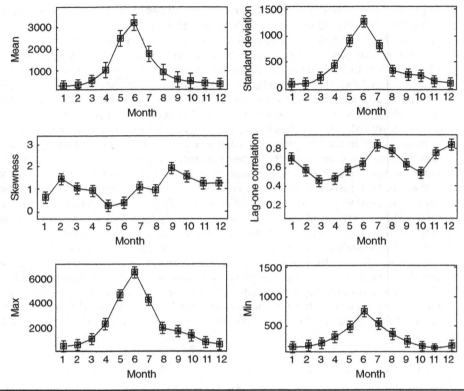

FIGURE **15.5** Comparison of statistics of the mean, standard deviation, skewness, lag-one correlation, maximum and minimum values of generated data, and historical record with box plot for simulation S_1. Continuous lines with stars for each month represent the historical record. The units for the mean and standard deviation are cubic meters.

distributions lie within the interval [0, 1] and cannot exceed the maximum value, causing an underestimation, and cannot be lower than the minimum value, causing an overestimation, when the generated data are scaled back to the original domain. The maximum relative biases for the maximum and minimum values are −5.60 percent in October and 28.12 percent in November.

For each simulation, S_2 and S_3, 100 flow sequences with 200 and 400 random values, are generated. The relative bias of simulated statistics of simulation S_3 is shown in Table 15.3. As seen from Fig. 15.7(a, b), all statistics fall in the box plot and thus the model performances in simulation S_2 and S_3 are considered to be good.

For the relative bias of simulation S_3 in Table 15.4, the values of maximum relative bias in the mean, standard deviation, skewness, and lag-one correlation are 0.47 percent in August, −2.29 percent in January, −3.28 percent in June, and −1.96 percent in October, respectively, which are significantly improved compared with those in simulation S_1. As can be seen, the absolute value of relative bias of simulated skewness is reduced from 10.95 percent in simulation S_1 to 3.28 percent in simulation S_3, and the skewness is preserved almost as well as other statistics in simulation S_3. The maximum and minimum

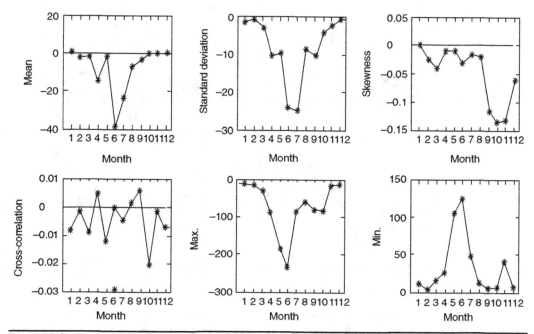

FIGURE 15.6 Bias in the mean, standard deviation, skewness, lag-one correlation, maximum values, and minimum values for simulation S_1. The units for the mean, standard deviation, maximum values, and minimum values are in cubic meters.

values are also preserved better, when the number of random sample values increases, although the overestimation and underestimation still exist. As shown in Table 15.4, the maximum relative bias values for the simulated maximum and minimum values are −1.29 percent in June and 11.39 percent in November.

Comparison of statistics between the generated data and historical record of selected months January, May, and September, in simulations S_1 and S_3 with different percentiles, is shown in Table 15.5. For the simulation of skewness of May stream flow, the 25th, 50th, and 75th percentiles of the simulated skewness are 0.120, 0.228, and 0.376 in simulation S_1 and 0.179, 0.239, and 0.286 in simulation S_3, respectively, while the skewness of the historical record is 0.238. The median value of simulated skewness 0.239 in simulation S_3 is closer to 0.238 than 0.228 for simulation S_1; and the interquartile range (difference between the values at 25th percentile and 75th percentile) is 0.107 in simulation S_3 which is smaller than 0.256 in simulation S_1.

Generally, the median values of statistics of simulation match those of the observations well when the number of generated random values is larger than 100; and these statistics can be improved when a longer record of random values is generated. This means that for a stream flow record with a length of around 100, the entropy-based model preserves the mean, standard deviation, skewness, and lag-one cross-correlation well.

The entropy-based method can be extended to incorporate higher moments and more lag correlations. For example, to preserve kurtosis in the simulation, two

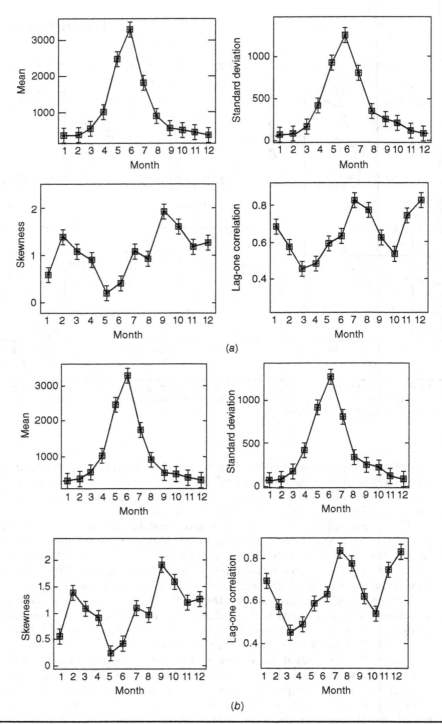

Figure 15.7 Comparison of statistics of the mean, standard deviation, skewness, and lag-one correlation of generated data and historical record with box plot for simulations S_2 and S_3. Continuous lines with stars for each month represent the historical record. The units for the mean and standard deviation are cubic meters. (a) Simulation S_2; (b) Simulation S_3.

Month Statistics	1	2	3	4	5	6	7	8	9	10	11	12
Mean	0.32	0.43	0.22	-0.06	-0.14	0.27	0.14	0.47	0.12	0.36	0.06	0.14
Standard deviation	-2.29	-0.51	0.13	-0.74	-0.86	-0.26	-1.26	0.26	-0.51	-0.26	-1.21	-1.39
Skewness	1.99	-2.25	-0.43	-1.10	0.39	-3.28	-1.67	-2.60	-1.22	-1.21	-2.17	-2.31
Lag-one correlation	-0.56	-0.85	-1.04	-0.27	-0.11	-0.60	-0.19	-0.60	0.37	-1.96	-0.02	-0.42
Maximum	-0.70	-0.51	-0.70	-0.76	-1.07	-1.29	-0.36	-0.76	-0.82	-0.84	-0.53	-0.48
Minimum	1.92	0.69	2.17	1.63	6.63	5.21	2.21	1.00	0.69	1.32	11.39	0.89

TABLE **15.4**   Relative Bias of Statistics for Each Month for Simulation S_3 (%)

		Simulation S_1	Simulation S_3	Percentile	Observation
January	Mean (m³/s)	286	288	25th	289
		290	290	50th	289
		295	293	75th	289
	Standard deviation (m³/s)	59	61	25th	64
		62	62	50th	64
		65	64	75th	64
	Skewness	0.492	0.526	25th	0.570
		0.572	0.582	50th	0.570
		0.675	0.639	75th	0.570
	Lag-one correlation	0.648	0.675	25th	0.695
		0.687	0.691	50th	0.695
		0.731	0.709	75th	0.695
May	Mean (m³/s)	2390	2446	25th	2476
		2475	2473	50th	2476
		2516	2512	75th	2476
	Standard deviation (m³/s)	870	905	25th	932
		922	924	50th	932
		962	939	75th	932
	Skewness	0.120	0.179	25th	0.238
		0.228	0.239	50th	0.238
		0.376	0.286	75th	0.238
	Lag-one correlation	0.541	0.571	25th	0.593
		0.581	0.592	50th	0.593
		0.623	0.613	75th	0.593
September	Mean (m³/s)	520	533	25th	542
		538	542	50th	542
		554	552	75th	542
	Standard deviation (m³/s)	220	232	25th	248
		238	246	50th	248
		266	259	75th	248
	Skewness	1.585	1.813	25th	1.956
		1.838	1.932	50th	1.956
		2.084	2.039	75th	1.956
	Lag-one correlation	0.565	0.591	25th	0.623
		0.629	0.625	50th	0.623
		0.670	0.650	75th	0.623

TABLE 15.5 Comparison Statistics of Simulated and Observed Stream Flows of January, May, and September in Simulation S_3

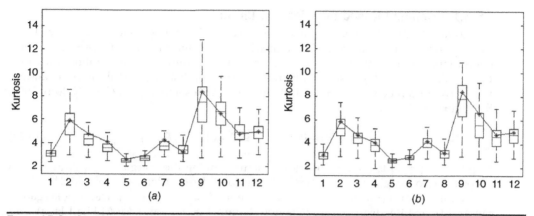

FIGURE 15.8 Comparison of values of kurtosis of generated data and the historical record with box plot for 100 sequences with 100 random values generated in each sequence. Continuous lines with stars for each month represent the historical record. (*a*) Proposed model; (*b*) extended model.

Lagrange multipliers associated with the fourth moments of variables X and Y would be added in Eq. (15.4). Then the stream flow can be generated based on the corresponding conditional distribution. Although the preservation of the fourth moment may not be essential and the sample instability problems with the estimation of higher moments may exist (Fiering, 1967), one simulation of this extension is shown to demonstrate the property of the proposed model. Comparison of the simulated kurtosis between the proposed model and the extended model is shown in Fig. 15.8*a* and *b*. It is seen that kurtosis was preserved better when the fourth moments were incorporated in the model.

15.5   Entropy-Copula Method for Single-Site Stream Flow Simulation

In the entropy-based stream flow simulation, the model requires specification of constraints for the statistics to be preserved, and that leads to as many Lagrange multipliers as statistics that need to be estimated, which may be tedious. Lee and Salas (2011) employed the copula method for annual stream flow simulation in which the copula-based joint distribution can be applied to model the dependence of the stream flow. Combining the entropy and copula methods, Hao and Singh (2012) developed an entropy-copula (EC) method for stream flow simulation, which is discussed here. In this method, the joint distribution is constructed using the copula method, where as the marginal distributions are constructed using the entropy method. The advantage of this method is that fewer parameters need to be estimated compared with the entropy method. The entropy-based marginal distributions do not rely on the gaussian assumption and are able to model the asymmetry property of stream flow without data transformation. The generation of negative values can be avoided. The entropy-based method also has the potential ability to resolve bimodality and is capable of modeling the nonlinear dependence of the stream flow.

15.5.1   Copula Method for Joint Distribution

The copula method has been applied extensively in hydrology, since it offers flexibility to construct multivariate probability distributions from marginal distributions. For a continuous random vector (X, Y) with marginal cumulative distribution functions (CDFs) $F_X(x)$ and $F_Y(y)$, respectively, the bivariate probability distribution of random vector (X, Y) can be expressed with its marginal CDFs and copula C as (Nelsen, 2006; Salvadori, 2007)

$$F(X, Y) = P(X \le x, Y \le y) = C[F_X(x), F_Y(y)] \tag{15.22}$$

Copula C represents the dependence structure linking the marginal distributions and maps the two marginal distributions into the joint distribution as $[0, 1]^2 \rightarrow [0, 1]$.

A number of copula families have been defined, such as the elliptical copula (gaussian and t), archimedean copula (Clayton, Gumbel, Frank, and Ali-Mikhail-Haq), and other copulas families (Joe, 1997; Nelsen, 2006; Bouyé and Salmon, 2009) and are also listed in App. 15A. For parameter estimation the exact maximum likelihood (EML) method and the inference functions for marginal probability distributions (IFM) can be used (Joe, 1997). For the EML method, the likelihood of including parameters of the marginal distributions and those of the copula function can be maximized to estimate the parameters simultaneously. For the IFM method, parameters of the marginal distributions and those of the copula function can be split, while the respective maximum likelihood functions can be estimated separately.

The conditional distribution can be derived from the copula function in Eq. (15.21). Let $U = F_X(x)$ with u denoting a specific value of random variable U and $V = F_Y(y)$ with v denoting a specific value of random variable V. From the joint distribution equation, Eq. (15.21), the conditional distribution can be defined as (Joe, 1997)

$$P(Y \le y \mid X = x) = C_{2|1}(F_Y(y) \mid F_X(x)) \tag{15.23}$$

where $C_{2|1}(v|u) = \partial C(u, v) / \partial u$.

For monthly stream flow simulation, it is desirable to preserve the correlation structure of the monthly stream flow. The joint distribution of stream flows of two adjacent months constructed from the copula method in Eq. (15.21) can be used to model the correlation. Moreover, an important feature of the copula-based joint distribution is that it can model the nonlinear dependence of stream flows.

15.5.2   Entropy-Copula method

The maximum entropy-based distribution can be used to preserve moment statistics (e.g., the mean, standard deviation, skewness, and kurtosis), and the copula-based joint distribution can be used to model the dependence structure (e.g., lag-one correlation) between stream flows of two adjacent months. In this way, samples from the copula-based joint distribution with the maximum entropy-based marginal distributions $F_X(x)$ and $F_Y(y)$ can be expected to preserve the mean, standard deviation, skewness, kurtosis, and lag-one correlation. The EC method is discussed for monthly stream flow simulation hereafter.

Now consider two adjacent months (say, January and February). Then the February stream flow can be generated from the conditional distribution given by Eq. (15.23), given the January stream flow. In this manner, the conditional distribution method is employed for the generation of random values from the bivariate distribution (Johnson, 1987; Cherubini et al., 2004). In total, 12 conditional distributions have to be used sequentially to generate the monthly stream flow. For a specific month j, the probability density function and the corresponding cumulative distribution F_j must be determined prior to stream flow generation. The cumulative distributions F_j and F_{j-1} (corresponding to v and u) are needed for generating the stream flow of month j. To generalize, let $x_{t,s}$ be the stream flow for month s of year t ($s = 1, 2, ..., 12, t = 1, 2, ..., n$). Then the simulation steps can be summarized as follows:

1. Generate a uniform random value $u_{1,1}$ in [0,1] that can be considered as the cumulative probability corresponding to a specific value of x, say, $x_{1,1}$ (January stream flow of the first year). If F_1 is the cumulative distribution of the January stream flow, then one can obtain $x_{1,1}$ from the inverse cumulative probability distribution function as $x_{1,1} = F_1^{-1}(u_{1,1})$.

2. Generate another uniform distributed random value w in [0,1] that is considered to be the conditional cumulative probability corresponding to a specific value $x_{2,1}$ (the February stream flow of the first year) with cumulative probability $u_{2,1}$, given the initial value $x_{1,1}$. Then one obtains $w = C_{2|1}(u_{2,1} | u_{1,1})$ from Eq. (15.23). And $u_{2,1}$ can be obtained as $u_{2,1} = C^{-1}_{2|1}(w)$ and then $x_{2,1}$ can be solved from $x_{2,1} = F_2^{-1}(u_{2,1})$ accordingly, where F_2 is the cumulative distribution of the February stream flow.

3. Repeat step 2 until $x_{12,1}$ (the December stream flow of the first year) is generated. Thus, monthly stream flows of the first year are generated.

4. With the generated $x_{12,1}$, $x_{1,2}$ (the January stream flow of the second year) can be generated with step 2. Similarly, stream flows of other months of the second year $x_{3,2}, x_{4,2}, ..., x_{12,2}$ are generated.

5. Repeat step 4 until $x_{12,n}$ is generated.

15.5.3   Extended Entropy-Copula method

Introducing an aggregate variable in the conditional distribution in a manner similar to that done by Sharma and O'Neill (2002), one can extend the EC method to preserve the interannual statistics (denoted the EEC method). For the monthly stream flow denoted by $X_1, X_2, ..., X_{12}, X_{13}, X_{14}, ..., X_n$, where $X_1, X_2, ..., X_{12}$ are the monthly stream flow values of the first year and so on, an aggregated stream flow can be defined as the summation of the previous m monthly stream flows:

$$Z_{t-1} = \sum_{i=1}^{m} X_{t-j} \tag{15.24}$$

Denoting the stream flows of two adjacent months as X_t and X_{t-1} with the corresponding aggregate variable Z_{t-1} defined as above, the joint distribution of random vector (Z_{t-1}, X_{t-1}, X_t) can be expressed by the copula method as

$$F_{z_{t-1}, x_{t-1}, x_t}(Z_{t-1}, X_{t-1}, X_t) = P(Z_{t-1} \leq z_{t-1}, X_{t-1} \leq x_{t-1}, X_t \leq x_t) = C(v_1, v_2, v_3) \tag{15.25}$$

where v_1, v_2, and v_3 are the corresponding cumulative probabilities of Z_{t-1}, X_{t-1}, and X_t, respectively. The conditional distribution of the monthly stream flow X_t given the previous monthly stream flow X_{t-1} and the aggregated stream flow Z_{t-1} can be expressed as (Embrechts et al., 2003)

$$F(X_t \mid Z_{t-1}, X_{t-1}) = C_{3|1,2}(v_3 \mid v_1, v_2) = \frac{\partial C^2(v_1, v_2, v_3)}{\partial v_1 \partial v_2} \cdot \left[\frac{\partial C^2(v_1, v_2)}{\partial v_1 \partial v_2} \right]^{-1} \qquad (15.26)$$

For the stream flow of month r ($r = 1, 2, ..., 12$) (with CDF F_r) and the corresponding aggregate stream flow (with CDF G_r), the probability density functions and the corresponding cumulative distribution functions must be estimated prior to stream flow generation. The cumulative distributions F_r, F_{r-1} and G_{r-1} [corresponding to v_3, v_2, and v_1 in Eq. (15.25)] are needed for the generation of the stream flow of month r. The procedure for generating monthly stream flow by the EEC method can be summarized as follows:

1. Assign random values to Z_{t-1} and X_{t-1}. Compute the corresponding cumulative probabilities $G_{t-1}(z_{t-1})$ (denoted as v_1) and $F_{t-1}(x_{t-1})$ (denoted as v_2).
2. Generate a uniform random value η in $[0,1]$ that is considered to be the conditional cumulative probability corresponding to a specific value x_t, given the initial values z_{t-1} and x_{t-1} (or v_1 and v_2). From Eq. (15.25), one obtains $C_{3|1,2}$ $(v_3 \mid v_1, v_2) = \eta$. The cumulative probability $F_t(x_t)$ (denoted as v_3) can be obtained as $v_3 = C_{3|1,2}^{-1}(\eta)$ and then x_t can be obtained as $X_t = F_t^{-1}(v_3)$.
3. Increase time step t and update the random values of X_{t-1} and Z_{t-1}.
4. Repeat steps 1 to 3 until the required length of the monthly stream flow is generated.

15.5.4 Case Study

Hao and Singh (2012) applied the entropy-copula method to monthly stream flow from the same 10 sites in the Colorado River basin as used in the case study described earlier (Hao and Singh, 2011) and compared with the results obtained from the entropy method. A total of 100 flow sequences with 100 yr of stream flow in each sequence were generated to assess the performance of the EC method. Moment statistics (including the mean, standard deviation, skewness, and kurtosis), dependence structure, and maximum and minimum values from the generated data were compared with those from the historical data. In addition, drought-related statistics, including the maximum drought length, maximum drought amount, maximum surplus length, maximum surplus amount, and storage capacity, were also used for comparison. Box plots were used for comparing the statistics of the historical and generated data. The relative error (RE) of the simulated statistics was used to further quantify the performance of the proposed method, which can be computed as $RE = (S_m - X_O)/X_O$, where S_m is the median of simulated statistics and X_O is the corresponding observed statistics from the historical data.

Joint Probability Distribution

The joint PDF of both generated and historic stream flow values was computed using the gaussian, Frank, and Clayton copulas. Both the gaussian copula and the Frank copula exhibit symmetric dependence with weaker dependence at both tails for the Frank copula, while the Clayton copula exhibits asymmetric dependence, strong in the left tail and weak in the right tail (Trivedi and Zimmer, 2007). A total of 500 pairs of the generated May and June stream flows are shown in Fig. 15.9, together with

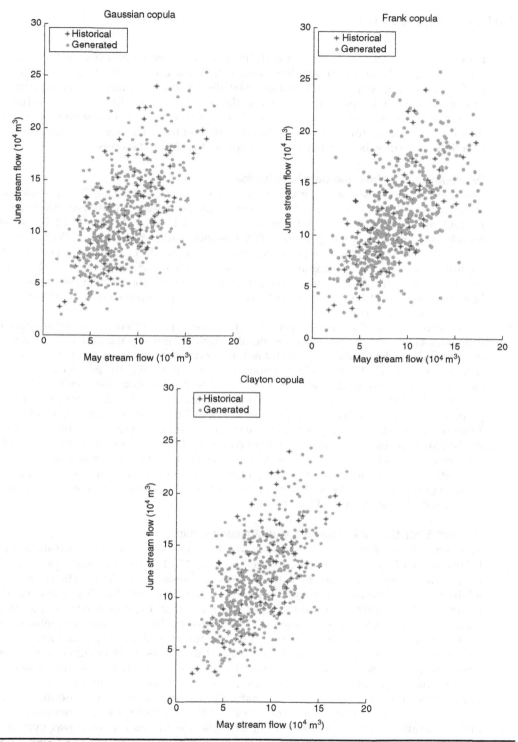

FIGURE 15.9 Comparison of the generated and historical May and June stream flow with different copula.

historical stream flow pairs. In general, the patterns of generated stream flow pairs matched those of historic stream flow pairs well. At the lower part the Clayton copula performed better, while for the middle part the gaussian copula performed better. Stream flow pairs for other months were also analyzed with different copula functions. Generally the patterns of generated stream flow pairs from different copulas matched those of historical pairs relatively well, and there was not one copula function that performed better than all other copulas for all the data sets.

Dependence Structure and Copula Selection

The Kendall and Spearman correlations were used for characterizing the dependence structure, and the preservation of these correlations was assessed. Different correlations for each month by different copulas are shown in Figs. 15.10, 15.11, and 15.12. For most months, the entropy-copula gaussian (ECG) method and entropy-copula Frank (ECF) method preserved the lag-one linear correlation well, while ECG method performed slightly better. The entropy-copula Clayton (ECC) method did not preserve the lag-one correlation well in most months. The gaussian and Frank copula also preserved the Spearman and Kendall correlation well, while the Clayton copula did not perform well in preserving these correlations.

Results showed that the dependence structure was affected by different copula functions. Generally the gaussian and Frank copulas preserved the dependence structure well, while the Clayton copula did not preserve the dependence structure well. To further compare the three copula functions, the AIC values of each copula function for each month were computed, and the number of each copula function with minimum Akaike Information Criterion (AIC) values was computed. Results showed that the gaussian, Frank, and Clayton copulas had 7, 3, and 2 cases of minimum values for all 12 months. This shows that generally there was not a unique copula function that fitted all the data pairs better, while the gaussian copula had the largest number of cases with minimum AIC values. The entropy-copula (gaussian) method preserved the lag-one correlation well for stream flows of most months. Furthermore, the incorporation of the copula concept enabled the modeling and preservation of nonlinear dependence. This is an attractive feature of the ECG method.

Moment Statistics and, Maximum and Minimum Values

Moment statistics, such as the mean, standard deviation, skewness, and kurtosis for the EC method with gaussian copulas, are shown in Fig. 15.13. The ECG method performed well in preserving these moment statistics, since all the statistics fell in the box plot. The relative error of mean and of standard deviation was lower than 5 percent for all months, and that of skewness and kurtosis was lower than 10 percent for most months, indicating a satisfactory preservation of these statistics. The maximum and minimum values from the ECG are shown in Fig. 15.14. The statistics of maximum and minimum values fell in the box plot for most months, and there was no clear difference in the performance of these three methods. The relative error for simulated maximum values was less than 10 percent for all months, and that for the simulated minimum values was under 15 percent for most months. Generally, the statistics of maximum and minimum values were preserved well for most months, although there was overestimation or underestimation for certain months. For example, the minimum value was overestimated for the November stream flow with a relative error of −36 percent. No negative values were generated by the ECG and EECG methods.

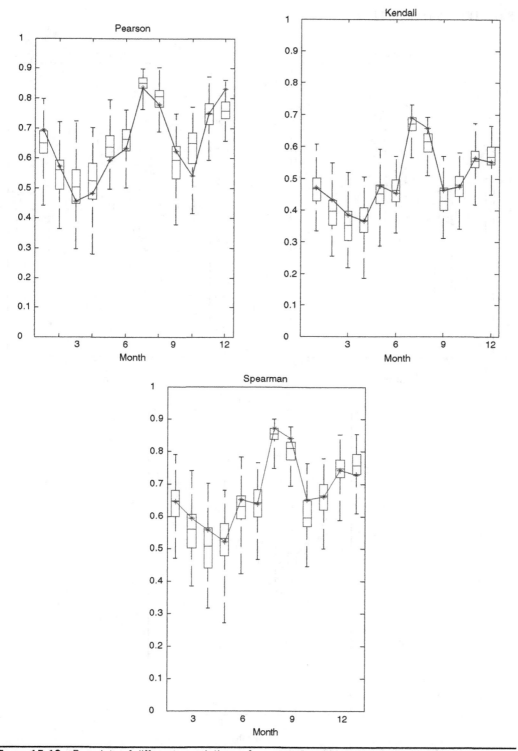

FIGURE 15.10 Box plots of different correlations of generated and historical data with EC method (with gaussian copula). Continuous lines with stars for each month represent statistics of historical data.

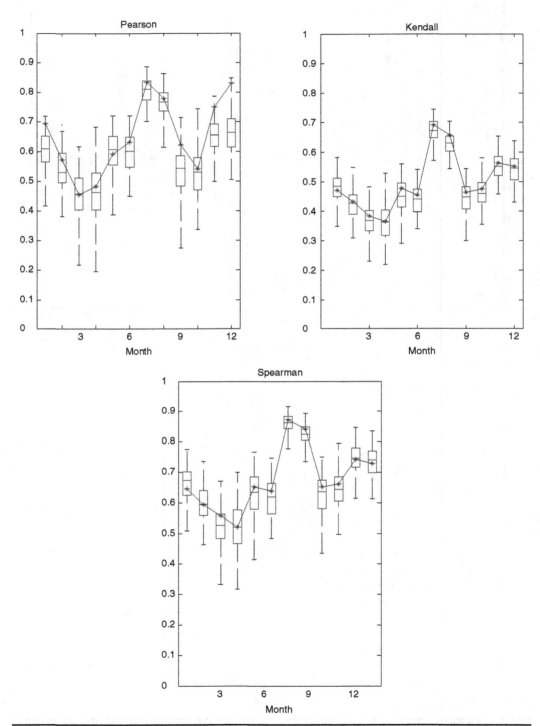

FIGURE 15.11 Box plots of different correlations of generated and historical data with EC method (with Frank copula). Continuous lines with stars for each month represent the statistics of historical data.

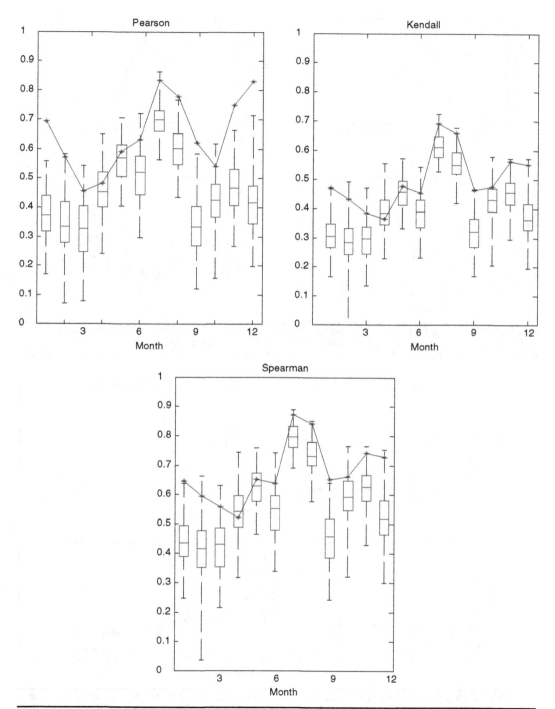

FIGURE 15.12   Box plots of different correlations of generated and historical data with EC method (with Clayton copula). Continuous lines with stars for each month represent the statistics of historical data.

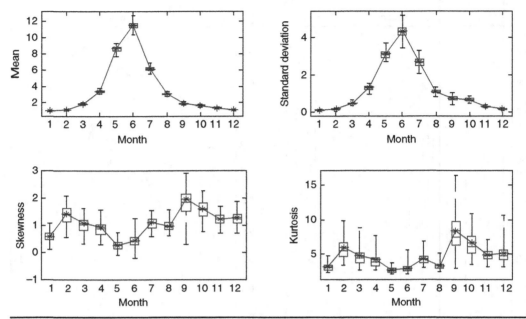

FIGURE 15.13   Box plots of the mean, standard deviation, skewness, and kurtosis of generated and historical data with EC method (with gaussian copula). Continuous lines with stars for each month represent the statistics of historical data. The units for the mean and standard deviation are cubic meters.

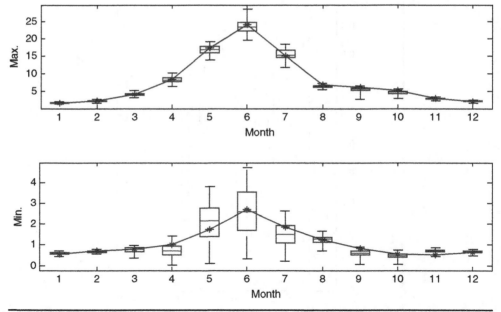

FIGURE 15.14   Box plots of maximum and minimum values of the historical and generated data with EC method (with gaussian copula). Continuous lines with stars for each month represent the statistics of historical data. The unit for maximum and minimum values is cubic meters.

High-Order Correlation

The preservation of the higher-order correlation of the proposed method was also assessed. Generally the ECG method did not preserve the lag-four correlation well, which is not unexpected, since only lag-one correlation is modeled by the ECG method. However, the lag-four correlation was preserved well for many months by the EECG method, although the lag-four correlation was not modeled directly.

Annual Statistics

Statistics of the annual stream flow were also assessed by adding the generated monthly stream flow of each year. Generated and historical statistics (including the mean, standard deviation, skewness, and lag-one correlation) of the annual stream flow from the ECG and EECG methods are shown in Table 15.6. The mean and standard variation were preserved well by both the ECG and extended ECG (EECG) methods. For example, the median of generated standard deviations ($10^4 \, m^3/s$) from the ECG and EECG methods were 11.72 and 12.19, while the observed value was 12.59. The EEC method performed slightly better in preserving the standard deviation. The two methods differed significantly in preserving the lag-one correlation. The median of the simulated lag-one correlation was 0.02 and 0.25 for the ECG and EECGs methods, respectively, while the observed value was 0.26. Thus, the EECG method preserved the lag-one correlation well, but the ECG method did not. However, neither the ECG nor the EECG method preserved the skewness well.

Inter-Annual Statistics

The generated monthly stream flow with the aggregate variable preserved the dependence between the current monthly stream flow and the annual stream flow of the previous year. Box plots of lag-one and lag-four correlation between the stream flow of a specific month (seasonal time scale) and the stream flow of the previous 12 months

Statistics	ECG method	EECG method	Percentile	Observation
Mean ($10^4 \, m^3/s$)	41.92	41.54	25th	42.71
	42.82	42.74	50th	42.71
	43.41	43.79	75th	42.71
Standard deviation ($10^4 \, m^3/s$)	11.21	11.88	25th	12.59
	11.72	12.62	50th	12.59
	12.48	13.28	75th	12.59
Skewness	0.33	0.37	25th	0.17
	0.53	0.58	50th	0.17
	0.71	0.75	75th	0.17
Lag-one correlation	−0.07	0.20	25th	0.26
	0.02	0.28	50th	0.26
	0.07	0.34	75th	0.26

TABLE 15.6   Comparison of Generated and Historical Annual Stream flows Statistics

(annual time scale) of the generated and historic data showed that the lag-one correlation was preserved well for all months, as expected from the structure of the EECG model. The lag-four correlation, though not directly modeled, was also preserved well. Results for the ECG model showed that these correlations were not preserved well, but the EECG method improved the preservation of interannual dependence.

Drought-Related Statistics
Box plots of the drought-related statistics (ratio of generated to historical) for the EECG method were constructed. The water demand level was expressed as a fraction of the historical mean, such as 0.7, 0.8, 0.9, and 1.0. For the ECG method, generally the deficit and storage properties were preserved relatively well, although some statistics were overestimated or underestimated. For example, the maximum surplus length was underestimated for the water demand level 0.8, and the maximum deficit length was overestimated for the water demand level 1.0. The result for the ECG method was also assessed, and it did not perform as well as the EECG method. This shows that the EECG model improved the preservation of drought-related statistics.

15.6   Entropy-Copula Method for Multisite Stream flow Simulation
For multi-site stream flow simulation in a river basin, it is desired that the statistical properties of stream flow at individual sites and the dependence structure among different sites be preserved. The copula method has been commonly used for modeling the dependence structure of multivariate random variables and for the multisite stochastic simulation (Nelsen, 2006; Bárdossy and Pegram, 2009). However, the ability of the commonly used parametric copulas to model dependencies in higher dimensions is rather restricted (Kao and Govindaraju, 2008; Chui and Wu, 2009). Hao and Singh (2013) applied the maximum entropy copula method for multisite monthly stream flow simulation in which the nonlinear dependence structure in higher dimensions among monthly stream flows at different sites can be modeled.

15.6.1   Maximum Entropy Copula
The maximum entropy copula has been developed based on entropy theory (Chui and Wu, 2009; Chu, 2011). Let U and V be the marginal probability distributions of random variables X and Y with u and v denoting realizations of U and V. For a copula density function $c(u, v)$, the entropy can be expressed as

$$W = -\int_0^1 \int_0^1 c(u,v)\log c(u,v)\,du\,dv \qquad (15.27)$$

For deriving the copula density function using entropy maximizing, constraints can be expressed as

$$\int_0^1 \int_0^1 c(u,v)g_i(u,v)\,du\,dv = g_i \qquad i = 1,2,\dots,n \qquad (15.28)$$

where g_i is the expectation of function $g_i(u, v)$, that is, $E[g_i(u, v)]$. To ensure that the integration of the copula density function over the entire space equates to 1, $g_1(u, v)$ can be specified as 1. To ensure that the marginal probability distribution of $c(u, v)$ is the uniform $[0, 1]$, moments of u and v can be specified as constraints (that is, u, u^2, u^3, v, v^2, and v^3 ...) to approximate the marginal properties numerically (Chu, 2011). To model the dependence structure, function $g(u, v)$ can be specified such that the expectation $E[g(u, v)]$ becomes some linear form of rank correlation. For example, when the pairwise product constraint $g(u, v) = uv$ is used, the commonly used Spearman rank correlation ρ can be linked to the constraint, i.e.,

$$\int_0^1 \int_0^1 uvc(u, v) du\, dv = \frac{\rho + 3}{12} \tag{15.29}$$

With the constraints specified by Eq. (15.28), the maximum entropy copula density function can be obtained as (Chui and Wu, 2009; Chu, 2011)

$$c(u, v) = \exp\left[-\lambda_0 - \sum_{r=1}^m (\lambda_r u^r + \lambda_{r+m} v^r - \lambda_{2m+1} uv) \right] \tag{15.30}$$

where m is the maximum order of moments and $\lambda_0, ..., \lambda_{2m+1}$ are the Lagrange parameters. Parameter λ_0 can be expressed as a function of other parameters as

$$\lambda_0 = \ln \int_0^1 \int_0^1 \exp\left[-\sum_{r=1}^m (\lambda_r u^r + \lambda_{r+m} v^r - \lambda_{2m+1} uv) \right] du\, dv \tag{15.31}$$

The dependence structure in terms of the Spearman rank correlation can be determined by using the joint probability density function in Eq. (15.30).

For modeling the multivariate dependence structure, the joint distribution can be defined by the copula density function which can be obtained by entropy maximization. Note that the derivation of the maximum entropy copula is separate from that of the marginal probability distributions.

15.6.2   Parameter Estimation

For the maximum entropy copula, the Lagrange multipliers λ_i ($i = 1, ..., 2m + 1$) in Eq. (15.30) must be determined. The Lagrange multipliers can be solved by finding the minimum of a convex function Γ expressed as (Kapur, 1989)

$$\Gamma = \lambda_0 + \sum_{i=1}^{2m+1} \lambda_i g_i \tag{15.32}$$

These parameters can be estimated using the Newton-Raphson iteration method (Wu, 2003; Hao and Singh, 2011). However, a high-dimension integration is involved in the parameter estimation for the multisite simulation to obtain the value of λ_0 in Eq. (15.31), which makes it more complicated than the single-site stream flow simulation. Hao and Singh (2013) employed an adaptive algorithm for numerical integration over hyper-rectangular region for the high-dimension integration (programmed as a MATLAB function ADAPT available from www.math.wsu.edu/faculty/genz/homepage) (Genz and Malik, 1980; Berntsen et al., 1991).

15.6.3 Simulation Methodology

For illustrative purposes, consider three sites from upstream to downstream, denoted as sites 1, 2, and 3. The marginal probability distribution of monthly stream flow at each site can be denoted by $(U_1, U_2, ...)$, $(V_1, V_2, ...)$ and $(W_1, W_2, ...)$ and their realizations as $(u_1, u_2, ...)$, $(v_1, v_2, ...)$, and $(w_1, w_2, ...)$, respectively. For site 1, the joint distribution $C(u_s, u_{s-1})$ of monthly stream flow for two adjacent months s and $s - 1$ must be determined, and the conditional distribution $C(u_s | u_{s-1})$ is used to generate the monthly stream flow (marginal) U_s, given the previous monthly stream flow (marginal) U_{s-1}. For the monthly stream flow of site 2, the joint distribution $C(u_s, v_{s-1}, v_s)$ has to be determined, and the conditional distribution $C(v_s | v_{s-1}, u_s)$ is used to generate the monthly stream flow V_s given stream flow V_{s-1} of site 2 and the monthly stream flow U_s of site 1. Similarly, for monthly stream flow W_s, the conditional distribution $C(w_s | w_{s-1}, u_s, v_s)$ can be used to generate the monthly stream flow W_s, given stream flow W_{s-1} of site 3, U_s of site 1, and V_s of site 2.

The methodology for generating the monthly stream flow (marginal) at each site can be summarized as follows:

1. Initialize the monthly stream flow at sites 1, 2, and 3 of the first month, that is, u_1, v_1 and w_1, by assigning random values from historical records.

2. With the initialized u_1, generate the monthly stream flow at site 1 for the second month u_2 from the conditional distribution $C(u_s | u_{s-1})$. With the generated u_2 and initialized value v_1, the monthly stream flow at site 2 for the second month v_2 can be generated from the distribution $C(v_s | v_{s-1}, u_s)$. With generated u_2, v_2, and initialized w_1, the monthly stream flow at site 3 for the second month w_2 can be generated from the distribution $C(w_s | w_{s-1}, u_s, v_s)$.

3. With generated u_2, v_2, and w_2, repeat step 2 to generate the monthly stream flow for the next month u_3, v_3, and w_3 for sites 1, 2, and 3, respectively.

4. Repeat step 3 to generate a sequence of monthly stream flows $u_4, ..., u_t, v_4, ..., v_t$ and $w_4, ..., w_t$ up to time t.

15.6.4 Case Study

Hao and Singh (2013) employed the entropy-copula method for generating monthly stream flow in the Colorado River basin. They considered three sites, namely, sites 21 (Paria River at Lees Ferry, Arizona), 22 (Little Colorado River near Cameron, Arizona), and 24 (Virgin River at Littlefield, Arizona), for illustrating the EC method. To illustrate the derivation of the joint probability density function for monthly stream flow at sites 1 and 2, for example, the marginal probabilities of monthly stream flow for month s at sites 1 and 2 can be denoted by u_s and v_s. From Eq. (15.30), the maximum entropy copula density function $c(u_s, v_{s-1}, v_s)$ with the moment constraints up to order 3 and pairwise product constraint can be expressed as

$$c(u_s, v_{s-1}, v_s) = \exp\left[-\lambda_0 - \sum_{i=1}^{3} \lambda_i u_s^{\ i} - \sum_{i=1}^{3} \lambda_{i+3} v_{s-1}^{\ i} - \sum_{i=1}^{3} \lambda_{i+6} v_s^{\ i} - \lambda_{10} u_s v_{s-1} - \lambda_{11} u_s v_s - \lambda_{12} v_{s-1} v_s \right]$$

(15.33)

The joint distribution $C(u_s, v_{s-1}, v_s)$ and conditional distribution $C(v_s | u_s, v_{s-1})$ can be obtained from the density function accordingly.

A total of 100 sequences of monthly stream flow (marginal) with the same length as the historical record (98 yr) were generated for each site by the simulation methodology. The scatter plots of the rank of observed stream flow pairs and one sequence of simulated stream flow pairs from the copula for March and April are shown in Fig. 15.15 (upper panel). The spread pattern of simulated stream flow pairs generally matched that of observed stream flow pairs of the two months well. For example, the monthly stream flow of March and April for site 3 showed strong dependence (Spearman correlation coefficient of 0.83), and most of the stream flow pairs spread along the diagonal. The simulated stream flow pairs spread near the diagonal with the Spearman correlation coefficient of 0.77. The scatter plots of the rank of observed monthly stream flow and one sequence of simulated stream flow pairs from the copula at different sites for the same month of March are also shown in Fig. 15.15 (lower panel). The Spearman correlation coefficients for simulated values were 0.59, 0.59, and 0.58, which were relatively close to those for observed stream flows (0.65, 0.68, and 0.67).

Box plots were constructed to show observed and simulated statistics, and the performance was judged to be good when a statistic fell within the box plot (Nowak et al., 2010; Salas and Lee, 2009). For most months, the medians of simulated statistics were within the box plot. The observed Spearman correlation feel within the box plots for most months. All simulations showed good results, and the dependence structure of monthly stream flow at each site and between different sites was preserved relatively well.

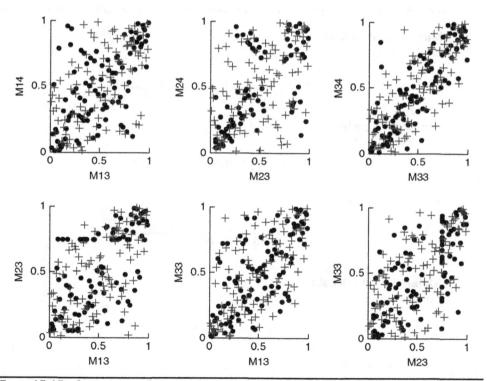

FIGURE 15.15 Scatter plots of observed (•) and simulated (+) monthly stream flow for different months at different sites (M_{pq} represents the monthly stream flow for month q at site p).

Appendix 15A: Bivariate Copulas and Their Corresponding Conditional Distributions

Bivariate copula density functions can be defined as $c(u,v)$ with the corresponding conditional density function defined as $c(v \mid u)$ and the parameter denoted by θ. Then v is solved for a specific conditional cumulative probability $w = c(v \mid u)$.

Bivariate Gaussian Copula
Copula function:

$$C(u, v) = \Phi_2(\Phi^{-1}(u), \Phi^{-1}(v); \theta)$$

Conditional distribution

$$C(v \mid u) = \Phi\left[\frac{\Phi_2(\Phi^{-1}(v) - \theta\Phi^{-1}(u))}{\sqrt{1-\theta^2}}\right]$$

Solving for v yields

$$v = \Phi(\rho\Phi^{-1}(u) + \sqrt{1-\theta^2}\,\Phi^{-1}(w))$$

where Φ and Φ_2 are the univariate and bivariate gaussian distributions, respectively.

Bivariate Frank Copula
Copula function:

$$C(u, v) = -\frac{1}{\theta}\ln\left[1 + \frac{(e^{-\theta u} - 1)(e^{-\theta v} - 1)}{e^{-\theta} - 1}\right]$$

Conditional distribution:

$$C(v \mid u) = \left[\frac{e^{-\theta u}}{(1 - e^{-\theta})(1 - e^{-\theta v})^{-1} - (1 - e^{-\theta u})}\right]$$

Solving for v gives

$$v = -\frac{1}{\theta}\ln\left[1 - \frac{1 - e^{-\theta}}{1 + e^{-\theta u}(w^{-1} - 1)}\right]$$

Bivariate Clayton Copula
Copula function:

$$C(u, v) = (u^{-\theta} + v^{-\theta} - 1)^{-1/\theta}$$

Conditional distribution:

$$C(v \mid u) = [1 + u^{\theta}(v^{-\theta} - 1)]^{-(1+\theta)/\theta}$$

Solving for v yields

$$v = [(w^{-(1+\theta)/\theta} - 1)u^\theta + 1]^{-1/\theta}$$

Appendix 15B: Trivariate Copulas and Their Corresponding Conditional Distributions

Trivariate Gaussian Copula

Trivariate copula functions are defined as $C(v_1, v_2, v_3)$ with the corresponding conditional distribution defined as $C(v_3 | v_1, v_2)$. The trivariate correlation matrix is denoted by $\Sigma = (\rho_{i,j}, i, j = 1, 2, 3)$ where ρ_{12}, ρ_{23}, and ρ_{13} are the correlation coefficients of cumulative probabilities v_1, v_2, and v_3, respectively. Then v_3 is solved for a specific conditional cumulative probability $w = C(v_3 | v_1, v_2)$.

Copula function:

$$C(v_1, v_2, v_3) = \Phi_\Sigma(\Phi^{-1}(v_1), \Phi^{-1}(v_2), \Phi^{-1}(v_3))$$

Conditional distribution:

$$C_{3|1,2}(v_3 | v_1, v_2) = \Phi\left[\frac{\Phi^{-1}(v_3) - a_1\Phi^{-1}(v_1) - a_2\Phi^{-1}(v_2)}{a_3}\right]$$

Solving for v_3 yields

$$v_3 = \Phi[a_1\Phi^{-1}(v_1) + a_2\Phi^{-1}(v_2) + a_3\Phi^{-1}(w)]$$

where Φ and Φ_Σ are the univariate and trivariate gaussian distributions, respectively, and a_1, a_2, and a_3 are the parameters defined as

$$a_1 = \frac{\rho_{13} - \rho_{12}\rho_{23}}{1 - \rho_{12}^2} \qquad a_2 = \frac{\rho_{23} - \rho_{12}\rho_{13}}{1 - \rho_{12}^2}$$

$$a_3 = \left(\frac{1 - \rho_{12}^2 - \rho_{13}^2 - \rho_{23}^2 + 2\rho_{12}\rho_{13}\rho_{23}}{1 - \rho_{12}^2}\right)^{0.5}$$

Questions

1. Download one data set of annual stream flow from the USGS website. Compute the mean u, and variance σ^2 of the data (or observations). With the mean and variance as constraints, derive the maximum entropy-based distribution $f(x)$.

2. Generate random values from the distribution $f(x)$ (simulation S_1, $n = 100$) derived in Question 1. Compute the statistics (i.e., mean and variance) of the generated data.

3. As in Question 2, generate $n = 500, 1000, 50{,}000$ (simulations S_2, S_3, and S_4) and compute the statistics of the generated data. Fill in the table.

Statistics Simulation and Observation	Mean u	Variance σ^2
$S_1 (n = 100)$		
$S_2 (n = 500)$		
$S_3 (n = 1000)$		
$S_4 (n = 5000)$		

Hints: The distribution $f(x)$ is normal. The method of generation of random values from the normal distribution with given mean and variance can be found in Excel, R, or MATLAB. The constraints are as follows:

$$\int_{-\infty}^{\infty}\int_{-\infty}^{\infty} f(x_1, x_2)\ln f(x_1, x_2)dx_1\, dx_2 = 1$$

$$\int_{-\infty}^{\infty}\int_{-\infty}^{\infty} x_1 f(x_1, x_2))\ln f(x_1, x_2)dx_1\, dx_2 = u_1$$

$$\int_{-\infty}^{\infty}\int_{-\infty}^{\infty} x_2 f(x_1, x_2)\ln f(x_1, x_2)dx_1\, dx_2 = u_2$$

$$\int_{-\infty}^{\infty}\int_{-\infty}^{\infty} (x_1 - m_1)^2 f(x_1, x_2)\ln f(x_1, x_2))\, dx_1\, dx_2 = \sigma^2{}_1$$

$$\int_{-\infty}^{\infty}\int_{-\infty}^{\infty} (x_2 - m_2)^2 f(x_1, x_2)\ln f(x_1, x_2)dx_1\, dx_2 = \sigma^2{}_2$$

$$\int_{-\infty}^{\infty}\int_{-\infty}^{\infty} (x_1 - m_1)(x_2 - m_2)f(x_1, x_2)\ln f(x_1, x_2)dx_1\, dx_2 = \rho\sigma_1\sigma_2$$

The lagrangian function L is

$$L = f(x_1, x_2) = \frac{1}{2\pi\sigma_1\sigma_1}\exp\left\{-\frac{1}{2(1-\rho^2)}\left[\frac{(x_1 - u_1)^2}{\sigma_1{}^2} + \frac{(x_2 - u_2)^2}{\sigma_2{}^2} - \frac{2\rho(x_1 - u_1)(x_2 - u_2)}{\sigma_1\sigma_2}\right]\right\}$$

4. Derive the conditional distribution of $f(x_t \mid x_{t-1})$. Generate the data from the conditional distribution. *Hints:* The resulting distribution is the bivariate normal distribution, and the conditional distribution is also a normal distribution.

5. Generate random values from the bivariate normal distribution.

References

Agmon, N., Alhassid, Y. and Levine R. (1979). An algorithm for finding the distribution of maximal entropy. *Journal of Computational Physics*, vol. 30, no. 2, pp. 250–258.

Balakrishnan, N., and Lai C. (2009). *Continuous Bivariate Distributions*. Springer, New York.

Bárdossy, A., and Pegram G. (2009). Copula based multisite model for daily precipitation simulation, *Hydrology and Earth System Sciences*, vol. 13, no. 12, p. 2299.

Berntsen, J., Espelid, T. O. and Genz A. (1991). An adaptive algorithm for the approximate calculation of multiple integrals. *ACM Transactions on Mathematical Software (TOMS)*, vol. 17, no. 4, pp. 437–451.

Bouyé, E., and M. Salmon (2009). Dynamic copula quantile regressions and tail area dynamic dependence in forex markets. *The European Journal of Finance*, vol. 15, no. 7, pp. 721–750.

Chu, B. (2011). Recovering copulas from limited information and an application to asset allocation. *Journal of Banking & Finance*, vol. 35, no. 7, pp. 1824–1842.

Chui, C., and Wu, X. (2009). Exponential series estimation of empirical copulas with application to financial returns. *Advances in Econometrics*, vol. 25, pp. 263–290.

Embrechts, P., Lindskog, F. and McNeil, A. (2003). *Modelling dependence with copulas and applications to risk management*, Elsevier, Amsterdam, The Netherlands.

Genz, A. C., and Malik, A. (1980). Remarks on algorithm 006: An adaptive algorithm for numerical integration over an N-dimensional rectangular region. *Journal of Computational and Applied Mathematics*, vol. 6, no. 4, pp. 295–302.

Fiering, M. (1967). *Stream flow Synthesis*, Harvard University Press, Cambridge, Mass.

Hao, Z., and Singh, V. P. (2011). Single-site monthly stream flow simulation using entropy theory. *Water Resources Research*, vol. 47, no. 9, W09528, doi:09510.01029/02010WR010208.

Hao, Z., and Singh, V. P. (2012). Entropy-copula method for single-site monthly stream flow simulation, *Water Resources Research*, vol. 48, no. 6, W06604, doi: 10.1029/2011WR011419.

Hao, Z., and Singh V. P. (2013). Modeling multisite stream flow dependence with maximum entropy copula. *Water Resources Research*, vol. 49, pp. 1–5, doi: 10.1002/wrcr.20523.

Jaynes, E. T. (1982). On the rationale of maximum-entropy methods. *Proceedings of the IEEE*, vol. 70, no. 9, pp. 939–952.

Johnson, M. (1987). *Multivariate Statistical Simulation*. Wiley, New York.

Joe, H. (1997). *Multivariate Models and Dependence Concepts*. Chapman & Hall, London.

Kao, S. C., and Govindaraju, R. S. (2008). Trivariate statistical analysis of extreme rainfall events via the Plackett family of copulas. *Water Resources Research*, vol. 44, no. 2, W02415, doi:02410.01029/02007WR006261.

Kapur, J. (1989). *Maximum-Entropy Models in Science and Engineering*. Wiley, New York.

Kesavan, H., and Kapur, J. (1992). *Entropy Optimization Principles with Applications*, Academic Press, New York.

Koutsoyiannis, D. (2001). Coupling stochastic models of different timescales. *Water Resources Research*, vol. 37, no. 2, pp. 379–391, doi:310.1029/2000WR900200.

Koutsoyiannis, D., and Manetas, A. (1996). Simple disaggregation by accurate adjusting procedures. *Water Resources Research*, vol. 32, no. 7, pp. 2105–2117, doi:2110.1029/2196WR00488.

Lall, U. (1995). Recent advances in nonparametric function estimation: Hydrologic application. *Reviews in Geophysics*, vol. 33, pp. 1093–1102.

Lall, U., and Sharma, A. (1996). A nearest neighbor bootstrap for resampling hydrologic time series. *Water Resources Research*, vol. 32, no. 3, pp. 679–693, 610.1029/1095WR02966.

Lee, T., and Salas, J. D. (2011). Copula-based stochastic simulation of hydrological data applied to Nile River flows. *Hydrology Research*, vol. 42, no. 4, pp. 318–330.

Lee, T., Salas, J. D. and Prairie J. (2010). An enhanced nonparametric stream flow disaggregation model with genetic algorithm. *Water Resources Research*, vol. 46, W08545, doi:10.1029/2009WR007761.

Mejia, J., and Rousselle, J. (1976). Disaggregation models in hydrology revisited. *Water Resources Research*, vol. 12, no. 2, pp. 185–186, doi:110.1029/WR012i1002p00185.

Nelsen, R. B. (2006). *An Introduction to Copulas*. Springer, New York.

Nowak, K., Prairie, J., Rajagopalan, B. and Lall, U. (2010). A non-parametric stochastic approach for multisite disaggregation of annual to daily stream flow. *Water Resources Research*, vol. 46, W08529, doi:10.1029/2009WR008530.

Prairie, J., Rajagopalan, B., Lall, U. and Fulp, T. (2007). A stochastic nonparametric technique for space-time disaggregation of stream flows. *Water Resources Research*, vol. 43, W03432, doi:03410.01029/02005WR004721.

Salas, J. D., and Lee, T. (2010). Nonparametric simulation of single site seasonal stream flows. *Journal of Hydrologic Engineering*, vol. 15, pp. 284–296.

Shannon, C., and Weaver, W. (1949). *The Mathematical Theory of Communication*. University of Illinois Press, Urbana.

Shannon, C. E. (1948). A mathematical theory of communications. *Bell System Technical Journal*, vol. 27, no. 7, pp. 379–423.

Sharma, A., and O'Neill, R. (2002). A nonparametric approach for representing inter-annual dependence in monthly stream flow sequences. *Water Resources Research*, vol. 38, no. 7, p. 1100, doi:1110.1029/2001WR000953.

Sharma, A., Tarboton, D. and Lall, U. (1997). Stream flow simulation: A non-parametric approach. *Water Resources Research*, vol. 33, no. 2, pp. 291–308, doi:210.1029/1096WR02839.

Tarboton, D., Sharma, A. and Lall, U. (1998). Disaggregation procedures for stochastic hydrology based on nonparametric density estimation. *Water Resources Research*, vol. 34, no. 1, pp. 107–119, doi:110.1029/1097WR02429.

Trivedi, P. K., and Zimmer, D. M. (2007). Copula modeling: An introduction for practitioners. *Foundations and Trends in Econometrics*, vol. 1, no. 1, pp. 1-111.

Vogel, R., and Shallcross, A. (1996). The moving blocks bootstrap versus parametric time series models. *Water Resources Research*, vol. 32, no. 6, pp. 1875–1882, doi:1810.1029/1896WR00928.

Wu, X. (2003). Calculation of maximum entropy densities with application to income distribution. *Journal of Econometrics*, vol. 115, no. 2, pp. 347–354.

Further Reading

AghaKouchak, A., Bardossy, A. and Habib, E. (2010). Copula-based uncertainty modeling: Application to multi-sensor precipitation estimates. *Hydrological Processes*, vol. 24, no. 15, pp. 2111–2124.

Chen, L., Singh, V. P. and Guo, S. (2013). Measure of correlation between river flows using the copula-entropy theory. *Journal of Hydrologic Engineering*, vol. 18, no. 12, pp. 1591–1606.

Chen, Y. D., Zhang, Q., Xiao, M. and Singh, V. P. (2013). Evaluation of risk of hydrological droughts by the trivariate Plackett copula in the East River basin (China). *Natural Hazards*, vol. 68, pp. 529–547.

Cherubini, U., Luciano, E. and Vecchiato, W. (2004). *Copula Methods in Finance*, Wiley, New York.

Chowdhary, H., and Singh, V. P. (2010). Reducing uncertainty in estimates of frequency distribution parameters using composite likelihood approach and copula-based bvariate distributions. *Water Resources Research*, vol. 46, p. 1–23, W11516, doi: 10.1029/2009WR008490.

Chowdhary, H., Escobar, L. A. and Singh, V. P. (2011). Identification of suitable copulas for bivariate frequency analysis of flood peak and flood volume data. *Hydrology Research*, vol. 42, no. 2–3, pp. 193–216.

Genest, C., Quessy, J. -F. and Rémillard, B. (2006). Goodness-of-fit procedures for copula models based on the probability integral transformation. *Scandinavian Journal of Statistics*, vol. 33, no. 2, pp. 337–366.

Genest, C., and Favre, A. -C. (2007). Everything you always wanted to know about copula modeling but were afraid to ask. *Journal of Hydrologic Engineering*, vol. 12, no. 4, pp. 347–368.

Genest, C., and Rémillard, B. (2005). Validity of the parametric bootstrap for goodness-of-fit testing in semiparametric models. Technical Rep. G-2005-51, Groupe d'Études et de Recherche en Analyse des Décisions, Montréal, Canada.

Grygier, J., and Stedinger, J. (1988). Condensed disaggregation procedures and conservation corrections for stochastic hydrology. *Water Resources Research*, vol. 24, no. 10, pp. 1574–1584, doi:1510.1029/WR1024i1010p01574.

Hao, Z., and Singh, V. P. (2013). Entropy-based method for bivariate drought analysis. *Journal of Hydrologic Engineering*, vol. 18, no. 7, pp. 780–786.

Hipel, K., and McLeod, A. (1978). Preservation of the rescaled adjusted range. 2: Simulation studies using Box-Jenkins models. *Water Resources Research*, vol. 14, no. 3, pp. 509–516, doi:10.1029/WR014i003p00509.

Hipel, K., McLeod, A. and McBean, E. (1979). Hydrologic generating model selection. *Journal of Water Resources Planning and Management Division, Proceedings, ASCE*, vol. 105, no. 2, pp. 223–242.

Lee, T., and Salas, J. D. (2006). Record extension of monthly flows for the Colorado River system. Bureau of Reclamation, U.S. Dept. of the Interior, Denver, Colo.

Lettenmaier, D., and Burges, S. (1977). An operational approach to preserving skew in hydrologic models of long-term persistence. *Water Resources Research*, vol. 13, no. 2, pp. 281–290, doi:210.1029/WR1013i1002p00281.

Loucks, D., Stedinger, J. and Haith, D. (1981). Water Resource Systems Planning and Analysis, Prentice-Hall, Englewood Cliffs, N.J.

Mead, L., and Papanicolaou, N. (1984). Maximum entropy in the problem of moments. *Journal of Mathematical Physics*, vol. 25, no. 8, pp. 2404–2417.

Prairie, J., Rajagopalan, B., Fulp, T. and Zagona, E. (2006). Modified *K*-NN model for stochastic stream flow simulation, *Journal of Hydrologic Engineering*, vol. 11, no. 4, pp. 371–378.

Renard, B. (2011). A Bayesian hierarchical approach to regional frequency analysis. *Water Resources Research*, vol. 47, W11513, doi: 10.1029/2010WR010089.

Salas, J., and Lee, T. (2010). Nonparametric simulation of single site seasonal stream flows. *Journal of Hydrologic Engineering*, vol. 15, no. 4, pp. 284–296.

Salas, J. D., and Delleur, J. (1980). *Applied Modeling of Hydrologic Time Series*, Water Resources Publication, Littleton, Colo.

Salvadori, G., and De Michele, C. (2004). Frequency analysis via copulas. Theoretical aspects and applications to hydrological events. *Water Resources Research*, vol. 40, W12511, doi: 10.1029/2004WR003133.

Salvadori, G., and De Michele, C. (2010). Mulivariaite multiparameter extreme value models and return periods: A copula approach. *Water Resources Research*, vol. 46, W10501, doi: 10.1029/2009WR009040.

Santos, E., and Salas, J. D. (1992). Stepwise disaggregation scheme for synthetic hydrology. *Journal of Hydraulic Engineering*, vol. 118, no. 5, pp. 765–784.

Savic, D., Burn, D. and Zrinji, Z. (1989). Comparison of stream flow generation models for reservoir capacity-yield analysis. *Water Resources Bulletin*, vol. 25, no. 5, pp. 977–983.

Serinaldi, F. (2009). A multisite daily rainfall generator driven by bivariate copula-based mixed distributions. *Journal of Geophysical Research*, vol. 114, D10103, doi: 10.1029/2008JD11258.

Schölzel, C., and Friederichs, P. (2008). Multivariate non-normally distributed random variables in climate research–introduction to the copula approach. *Nonlinear Processes in Geophysics*, vol. 15.

Singh, V. P., and Zhang, L. (2007). IDF curves using the Frank Archimedean copula. *Journal of Hydrologic Engineering, ASCE*, vol. 12, no. 6, pp. 651–662.

Sivakumar, B., and Berndtsson, R., eds. (2010). *Advances in Data-Based Approaches for Hydrologic Modeling and Forecasting*. World Science, Hackensack, N.J.

Song, S., and Singh, V. P. (2010). Meta-elliptical copulas for drought frequency analysis of periodic hydrologic data. *Stochastic Environmental Research and Risk Analysis*, vol. 24, no. 3, pp. 425–444.

Song, S. B., and Singh, V. P. (2010). Frequency analysis of droughts using the Plackett copula and parameter estimation by genetic algorithm. *Stochastic Environmental Research and Risk Analysis*, vol. 24, pp. 425–444.

Stedinger, J., Pei, D. and Cohn, T. (1985). A condensed disaggregation model for incorporating parameter uncertainty into monthly reservoir simulations. *Water Resources Research*, vol. 21, no. 5, pp. 665–675, doi:610.1029/WR1021i1005p00665.

Stedinger, J., and Vogel, R. (1984). Disaggregation procedures for generating serially correlated flow vectors. *Water Resources Research*, vol. 20, no. 1, pp. 47–56, doi:10.1029/WR1020i1001p00047.

Valencia, R. D., and Schaake, J. J. (1973). Disaggregation processes in stochastic hydrology. *Water Resources Research*, vol. 9, no. 3, pp. 580–585, doi:510.1029/WR1009i1003p00580.

Vogel, R., and Stedinger, J. (1988). The value of stochastic stream flow models in overyear reservoir design applications. *Water Resources Research*, vol. 24, no. 9, pp. 1483–1490, doi:1410.1029/WR1024i1009p01483.

Zhang, L., and Singh, V. P. (2006). Bivariate flood frequency analysis using the copula method. *Journal of Hydrologic Engineering, ASCE*, vol. 11, no. 2, pp. 150–164.

Zhang, L., and Singh, V. P. (2007). Gumbel-Hougaard copula for trivariate rainfall frequency analysis. *Journal of Hydrologic Engineering, ASCE*, vol. 12, no. 4, pp. 409–419.

Zhang, L., and Singh, V. P. (2007). Trivariate flood frequency analysis using the Gumbel-Hougaard copula. *Journal of Hydrologic Engineering, ASCE*, vol. 12, no. 4, pp. 431–439.

Zhang, L., and Singh, V. P. (2012). Bivariate rainfall and runoff analysis using entropy and copula theories. *Entropy*, vol. 14, pp. 1784–1812, doi: 10.3390/e140911784.

Zhang, Q., Li, J. and Singh, V. P. (2012). Application of Archimedean copulas in the analysis of the precipitation extremes: Effects of precipitation change. *Theoretical and Applied Climatology*, vol. 107, no. 1–2, pp. 255–264.

Zhang, Q., J. Li, V. P. Singh, and Xu, C. -Y. (2013). Copula-based spatio-temporal patterns of precipitation extremes in China. *International Journal of Climatology*, vol. 33, no. 5, pp. 1140–1152.

Zhang, Q., Xiao, M., Singh, V. P. and Chen X. (2013). Copula-based risk evaluation of droughts across the Pearl River Basin, China. *Theoretical and Applied Climatology*, vol. 111, no. 1, pp. 119–131, doi: 10.1007/s00704-012-0656-4, 2013.

CHAPTER 16

Hydrologic Frequency Analysis

ydrologic frequency analysis is fundamental to hydraulic design. Designs of hydraulic structures are based on either univariate or multivariate frequency analysis. The purpose of univariate frequency analysis is either to predict the frequency of occurrence (or return period) of a given magnitude of the design variable or to determine the quantity of the design variable for a given frequency (or return period) and determine the parameters of a given frequency distribution. For example, in the case of flood frequency analysis, instantaneous peak discharge is the design variable. From a practical standpoint it is desired that the distribution employed for frequency analysis be simple to use, have no more than three parameters, be consistent with the given information, and be minimally biased. Hydrologic variables are nonnegative. This means that the frequency distribution should have a lower bound at zero or a specified small value. Entropy theory enables to derive such distributions. Sonuga (1972) was among the first to employ the principle of maximum entropy (POME) in hydrologic frequency analysis. In the intervening four decades, many distributions have been derived, and their parameters have been estimated using entropy theory (Singh, 1998). Papalexiou and Koutsoyiannis (2012, 2013) provide an excellent discussion of generalized probability distributions for daily rainfall and annual maxima of daily rainfall. These distributions can also be employed for flood frequency analysis. In multivariate frequency analysis, two or more variates are analyzed. For example, frequency analysis of damage entails frequency of inundation and that of its duration. Similarly, risk assessment entails frequency of occurrence and frequency of ensuing consequences. This chapter discusses entropy theory–based flood frequency analysis.

16.1   General Derivation of a Probability Distribution

Derivation of a general probability distribution for frequency analysis entails (1) specification of the Shannon entropy, (2) specification of constraints, (3) maximization of entropy using the method of Lagrange multipliers, (4) derivation of probability distribution, and (5) determination of Lagrange multipliers. Each of these steps is now discussed.

16.1.1   Shannon Entropy

Let the hydrologic variable, say, flood discharge or annual maximum rainfall for a given duration, be represented as a continuous random variable $X \in [a, b]$ with a probability density function (PDF) $f(x)$. For $f(x)$, the Shannon entropy H can be defined as (Shannon, 1948; Shannon and Weaver, 1949)

$$H(f) = H(X) = H = -\int_a^b f(x) \ln f(x) dx \tag{16.1}$$

where x is a value of random variable X with lower limit a and upper limit b. And

$$\int_a^b f(x)\, dx = 1 \tag{16.2}$$

where a and b are the upper and lower limits of the integral or the range of variable X. The objective is to derive $f(x)$ which is achieved by maximizing entropy subject to given constraints.

16.1.2   Specification of Constraints

Often the maximization is done numerically. However, in a number of cases analytical solutions can be derived. To that end, m linearly independent constraints C_r $(r = 1, 2, ..., m)$ can be defined as

$$C_r = \int_a^b g_r(x) f(x)\, dx \qquad r = 1, 2, ..., m \tag{16.3}$$

where $g_r(x)$ are some functions over an interval (a, b).

16.1.3   Maximization of Entropy

To maximize $H(f)$ subject to Eqs. (16.2) and (16.3), one can employ the method of Lagrange multipliers and construct the lagrangian function L as

$$L = -\int_a^b f(x) \ln f(x)\, dx - (\lambda_0 - 1)\left[\int_a^b f(x)\, dx - 1\right] - \sum_{r=1}^m \lambda_r \left[\int_a^b f(x)\, g_r(x)\, dx - C_r\right] \tag{16.4}$$

where $\lambda_1, \lambda_2, ..., \lambda_m$ are the Lagrange multipliers. Differentiating L with respect to $f(x)$ only and equating the derivative to zero, one obtains

$$\frac{\partial L}{\partial f} = 0 \Rightarrow -[1 + \ln f(x)] - (\lambda_0 - 1) - \sum_{r=1}^m \lambda_r g_r(x) = 0 \tag{16.5}$$

16.1.4   Probability Distribution and Its Entropy

Equation (16.5) yields

$$f(x) = \exp[-\lambda_0 - \lambda_1 g_1(x) - \cdots - \lambda_m g_m(x)] \tag{16.6a}$$

Equation (16.6a) is the POME-based probability distribution containing Lagrange multipliers as parameters. Equation (16.6a) is also written as

$$f(x) = \frac{1}{Z(\lambda_1, \lambda_2, ..., \lambda_m)} \exp\left[-\sum_{r=1}^m \lambda_r g_r(x)\right] \tag{16.6b}$$

where $\exp \lambda_0 = Z$ is called the partition function, which is a function of other Lagrange multipliers: $Z = Z(\lambda_1, \lambda_2, ..., \lambda_m)$.

Integration of Eq. (16.6b) leads to the cumulative distribution function or simply probability distribution of X, $F(x)$, as

$$F(x) = \int_a^x \exp\left[-\lambda_0 - \sum_{r=1}^n \lambda_r g_r(x)\right] dx \qquad (16.7)$$

Substitution of Eq. (16.6a) in Eq. (16.1) yields the maximum entropy of $f(x)$ or X as

$$H = \lambda_0 + \sum_{r=1}^n \lambda_r \overline{g_r(x)} = \lambda_0 + \sum_{r=1}^n \lambda_r E[g_r(x)] = \lambda_0 + \sum_{r=1}^n \lambda_r C_r \qquad (16.8)$$

where $E[g(x)]$ is the expectation of $g(x)$. Equation (16.8) shows that the entropy of the probability distribution of X depends only on the specified constraints, since the Lagrange multipliers themselves can be expressed in terms of the specified constraints. Equations (16.1), (16.2), (16.3), (16.6a), and (16.8) constitute the building blocks for the application of entropy theory. Entropy maximization results in the probability distribution that is most conservative and hence most informative. If a distribution with lower entropy were chosen, then it would mean the assumption of information that is not available. On the other hand, a distribution with higher entropy would violate the known constraints. Thus, it can be stated that the maximum entropy leads to a probability distribution of a particular macrostate among all possible configurations of events under consideration.

16.1.5   Determination of Lagrange Multipliers

The next step is to determine the Lagrange multipliers $\lambda_1, \lambda_2, ..., \lambda_m$. There are three ways to determine these multipliers: the ordinary method, parameter space expansion method, and numerical method. Each is discussed here.

Ordinary Method

The Lagrange multipliers can be determined by inserting Eq. (16.6a) in Eqs. (16.2) and (16.3):

$$\exp \lambda_0 = \int_a^b \exp\left[-\sum_{r=1}^m \lambda_r g_r(x)\right] dx \qquad r = 1, 2, ..., m \qquad (16.9a)$$

and

$$C_r \exp \lambda_0 = \int_a^b g_r(x) \exp\left[-\sum_{r=1}^m \lambda_r g_r(x)\right] dx \qquad r = 1, 2, ..., m \qquad (16.9b)$$

Equation (16.7a) expresses the partition function Z as

$$Z(\lambda_1, \lambda_2, ..., \lambda_m) = \int_a^b \exp\left[-\sum_{r=1}^m \lambda_r g_r(x)\right] dx \qquad (16.10)$$

Differentiating Eq. (16.7a) with respect to λ_r, one gets

$$\frac{\partial \lambda_0}{\partial \lambda_r} = -\frac{\int_a^b g_r(x)\exp\left[-\sum_{r=1}^m \lambda_r g_r(x)\right]dx}{\int_a^b \exp\left[-\sum_{r=1}^m \lambda_r g_r(x)\right]dx}, r=1,2,\dots,m \qquad (16.11a)$$

Multiplying the numerator as well as the denominator of Eq. (16.11a) by $\exp(-\lambda_0)$, one obtains

$$\frac{\partial \lambda_0}{\partial \lambda_r} = -\frac{\int_a^b g_r(x)\exp\left[-\lambda_0-\sum_{r=1}^m \lambda_r g_r(x)\right]dx}{\int_a^b \exp\left[-\lambda_0-\sum_{r=1}^m \lambda_r g_r(x)\right]dx} \qquad r=1,2,\dots,m \qquad (16.11b)$$

The denominator in Eq. (16.11b) equals unity by virtue of Eq. (16.2). Therefore, Eq. (16.11b) becomes

$$\frac{\partial \lambda_0}{\partial \lambda_r} = -\int_a^b g_r(x)\exp\left[-\lambda_0-\sum_{r=1}^m \lambda_r g_r(x)\right]dx \qquad r=1,2,\dots,m \qquad (16.12)$$

Note that substitution of Eq. (16.6a) in Eq. (16.3) yields

$$C_r = \int_a^b g_r(x)\exp\left[-\lambda_0-\sum_{r=1}^m \lambda_r g_r(x)\right]dx \qquad r=1,2,\dots,m \qquad (16.13)$$

Therefore, Eqs. (16.12) and (16.13) yield

$$\frac{\partial \lambda_0}{\partial \lambda_r} = -C_r \qquad r=1,2,\dots,m \qquad (16.14)$$

Equation (16.14) expresses constraints as functions of the Lagrange multipliers.

Furthermore, differentiating Eq. (16.10) with respect to each Lagrange multiplier separately, one obtains

$$\frac{\partial \log[Z(\lambda_1,\lambda_2,\dots,\lambda_r)]}{\partial \lambda_r} = -C_r \qquad r=1,2,\dots,m \qquad (16.15)$$

Equating Eqs. (16.14) and (16.15), one obtains as many equations as the number of unknown Lagrange multipliers minus 1 that are solved to obtain the multipliers in terms of the given constraints.

Parameter Space Expansion Method
This method, developed by Singh and Rajagopal (1986), employs an enlarged parameter space and maximizes entropy subject to both the parameters and the Lagrange multipliers. An important implication of the enlarged parameter space is that the method is

applicable to virtually any distribution, expressed in direct form, having any number of parameters. For a continuous random variable X having a probability density function $f(x, \theta)$ with parameters θ, the entropy can be expressed as

$$H(X) = H(f) = -\int_{-\infty}^{+\infty} f(x; \theta) \ln f(x; \theta) \, dx \qquad (16.16)$$

The parameter θ, of this distribution, can be estimated by achieving the maximum of $H(f)$. To apply the method, the constraints are defined first. Next, the POME formulation of the distribution is obtained in terms of the parameters by using the method of Lagrange multipliers. This formulation is used to define H whose maximum is sought. If the probability distribution has n parameters, θ, $i = 1, 2, \dots n$, and the $n + 1$ Lagrange multipliers are λ, $i = 1, 2, \dots, n + 1$, then the point where $H[f]$ is a maximum is a solution of $2n - 1$ equations

$$\frac{\partial H[f]}{\partial \lambda_i} = 0 \qquad i = 1, 2, \dots, n + 1 \qquad (16.17a)$$

and

$$\frac{\partial H[f]}{\partial \theta_i} = 0 \qquad i = 1, 2, \dots, n \qquad (16.17b)$$

Solution of Eqs. (16.17a) and (16.17b) yields the estimates of the distribution parameters.

Numerical Method

Another way of determining the Lagrange multipliers is to employ the equation

$$\Gamma = \ln Z + \sum_{r=1}^{m} \lambda_r \bar{g}_r \qquad (16.18)$$

where Z is defined by Eq. (16.6b) which is determined using Eq. (16.2). The function given by Eq. (16.18) is a convex function of λ_1, λ_2, ..., λ_m everywhere (Mead and Papanicolaou, 1984). Thus, the Lagrange multipliers can be estimated by minimizing Eq. (16.18) using a numerical method, such as the Newton-Raphson method. Starting from some initial value $\lambda_{(0)}$, one can solve for the Lagrange parameters by updating $\lambda_{(1)}$ as follows:

$$\lambda_{(1)} = \lambda_{(0)} - H^{-1} \frac{\partial \Gamma}{\partial \lambda_i} \qquad i = 0, 1, 2, \dots, m \qquad (16.19a)$$

where Γ is the gradient expressed as

$$\frac{\partial \Gamma}{\partial \lambda_i} = \bar{g}_i - \int_a^b \exp\left[-\sum_{r=0}^{m} \lambda_r g_r(x) \right] g_i(x) \, dx \qquad i = 0, 1, 2, \dots, m \qquad (16.19b)$$

and H is the hessian matrix whose elements are expressed as

$$H_{i,j} = \int_a^b \exp\left(-\sum_{r=0}^{m} \lambda_r g_r(x) \right) g_i(x) g_j(x) \, dx \qquad i, j = 0, 1, 2, \dots, m \qquad (16.19c)$$

The elements of the hessian matrix can be expressed as

$$
H = \begin{bmatrix}
\text{var}(g_1(x), g_1(x)) & \text{cov}(g_1(x), g_2(x)) & \cdots & \text{cov}(x, g_m(x)) \\
\text{cov}(g_2(x), x) & \text{var}(g_1(x)) & \cdots & \text{cov}(g_2(x), g_m(x)) \\
\cdots & \cdots & \cdots & \cdots \\
\text{cov}(g_m(x), x) & \text{cov}(g_m(x), g_2(x)) & \cdots & \text{var}(g_m(x))
\end{bmatrix}
\tag{16.19d}
$$

Example 16.1.   Matz (1978) has shown that the first four moments would suffice in many cases for deriving $f(x)$. Therefore, express the first four moments as constraints: $g_i(x) = x^i$, $i = 1, 2, 3, 4$. Derive the maximum entropy-based probability density function. Use the data given in Table 16.1.

Year	Flow (cfs)	u	Year	Flow (cfs)	u	Year	Flow (cfs)	u
1900	135,000	2.16	1944	172,000	2.75	1969	35,500	0.57
1901	37,300	0.6	1945	139,000	2.22	1970	42,700	0.68
1902	65,900	1.05	1946	57,400	0.92	1971	12,100	0.19
1919	69,300	1.11	1947	46,400	0.74	1972	25,000	0.4
1920	70,500	1.13	1948	30,000	0.48	1973	41,500	0.66
1922	139,000	2.22	1949	53,200	0.85	1974	36,100	0.58
1923	56,300	0.9	1950	43,800	0.7	1975	74,300	1.19
1924	70,700	1.13	1951	15,600	0.25	1976	41,400	0.66
1925	58,000	0.93	1952	33,500	0.54	1977	61,800	0.99
1926	50,000	0.8	1953	65,200	1.04	1978	14,900	0.24
1927	43,800	0.7	1954	34,000	0.54	1979	58,700	0.94
1928	38,700	0.62	1955	30,700	0.49	1980	34,800	0.56
1931	66,300	1.06	1956	33,100	0.53	1981	59,900	0.96
1932	75,100	1.2	1957	137,000	2.19	1982	35,000	0.56
1933	40,900	0.65	1958	104,000	1.66	1983	31,500	0.5
1934	79,600	1.27	1959	14,200	0.23	1984	5,510	0.09
1935	139,000	2.22	1960	93,000	1.49	1985	34,400	0.55
1936	75,000	1.2	1961	88,000	1.41	1986	62,500	1
1937	133,000	2.12	1962	25,500	0.41	1987	60,400	0.96
1938	103,000	1.65	1963	18,100	0.29	1988	11,900	0.19
1939	54,200	0.87	1964	26,100	0.42	1989	50,300	0.8
1940	57,900	0.92	1965	137,000	2.19	1990	60,600	0.97
1941	150,000	2.4	1966	81,800	1.31	1991	43,700	0.7
1942	103,000	1.65	1967	16,100	0.26	1992	163,000	2.6
1943	53,800	0.86	1968	73,800	1.18			

TABLE 16.1   Annual Peak Flow Data for Brazos River Basin ($u = x/\bar{x}$)

Solution. The entropy-based PDF can be expressed as

$$f(x) = \exp(-\lambda_0 - \lambda_1 x - \lambda_2 x^2 - \lambda_3 x^3 - \lambda_4 x^4) \tag{16.20}$$

With the first four moments as constraints, the skewness, kurtosis, and multiple modes can be included in the resulting PDF (Zellner and Highfield, 1988). Each maximum of the polynomial inside the exponential corresponds to one mode, and thus multiple modes may exist in the maximum distribution (Smith, 1993). When the interval of random variable is from zero to infinity, the Lagrange parameter λ_4 should be nonnegative. Equation (16.20) has been applied by Eisenberger (1964) for fitting bimodal distributions.

Example 16.2. Estimate the four Lagrange multipliers in Eq. (16.20), using the numerical method.

Solution. The function in Eq. (16.18) becomes

$$\Gamma = \lambda_0 + \sum_{r=1}^{4} \lambda_r \overline{g}_r, \qquad r = 1, 2, 3, 4 \tag{16.21}$$

The Newton-Raphson method can be applied to solve the maximization problem by updating Lagrange parameters $\lambda_{(1)}$ from some initial value $\lambda_{(0)}$ through the equation

$$\lambda_{(1)} = \lambda_{(0)} - H^{-1} \frac{\partial \Gamma}{\partial \lambda_i} \qquad i = 1, 2, 3, 4 \tag{16.22a}$$

where Γ is the gradient expressed as

$$\frac{\partial \Gamma}{\partial \lambda_i} = \overline{g}_i - \int_a^b \exp\left[-\sum_{r=0}^{4} \lambda_r g_r(x)\right] g_i(x)\, dx \qquad i = 1, 2, 3, 4 \tag{16.22b}$$

and H is the hessian matrix with elements expressed as

$$H_{i,j} = \int_a^b \exp\left[-\sum_{r=0}^{4} \lambda_r g_r(x)\right] g_i(x) g_j(x)\, dx$$

$$-\int_a^b \exp\left[-\sum_{r=0}^{4} \lambda_r g_r(x)\right] g_i(x)\, dx \bullet \int_a^b \exp\left[-\sum_{r=0}^{4} \lambda_r g_r(x)\right] g_j(x)\, dx \qquad i, j = 1, 2, 3, 4 \tag{16.22c}$$

The elements of the hessian matrix H can be expressed as

$$H = \begin{bmatrix} \operatorname{var} x & \operatorname{cov}(x, x^2) & \cdots & \operatorname{cov}(x, x^4) \\ \operatorname{cov}(x^2, x) & \operatorname{var} x^2 & \cdots & \operatorname{cov}(x^2, x^4) \\ \cdots & \cdots & \cdots & \cdots \\ \operatorname{cov}(x^4, x) & \operatorname{cov}(x^4, x^2) & \cdots & \operatorname{var} x^4 \end{bmatrix} \tag{16.23}$$

Previously we stated that we can estimate the Lagrange multipliers by minimizing Eq. (16.21). To ensure the integration involved in the estimation is tractable, first we rescale the annual peak flow to [0, 1], using

$$x' = \frac{x - (1-d)\min(x)}{(1+d)\max(x) - (1-d)\min(x)}$$

where d is a small number; here we choose $d = 0.05$. The rescaled flood variable has the maximum entropy–based distribution as

$$f(x') = \exp\left(-\sum_{i=0}^{4}\lambda_i x''\right)$$

We have the maximum entropy–based distribution for the observed flood variable using the probability density transformation:

$$f(x) = f(x')\left|\frac{dx'}{dx}\right| = \frac{1}{(1+d)\max(x)-(1-d)\min(x)} = \frac{1}{C}f(x')$$

The first four noncentral moments for the rescaled flood variable are computed:

$$[E(x'), E(x'^2), E(x'^3), E(x'^4)] = [0.3270, 0.1572, 0.0959, 0.0674]$$

With the sample moments of the scaled variable as constraints, we can minimize Eq. (16.21) by using the Newton-Raphson method [i.e., Eq. (16.22)] starting from the initial estimate (that is, $\lambda_{(0)} = [\lambda_{1(0)}, \lambda_{2(0)}, \lambda_{3(0)}, \lambda_{4(0)}] = [0, 0, 0, 0]$). To illustrate the estimation procedure, we will use the initial estimate to show the first iteration.

1. Substituting the initial estimate into Eq. (16.9a), we have $\lambda_{0(0)} = 0$.

2. Substituting the estimated parameters into Eqs. (16.22b), and (16.23), we have

$$\left[\frac{\partial\Gamma}{\partial\lambda_1}, \frac{\partial\Gamma}{\partial\lambda_2}, \frac{\partial\Gamma}{\partial\lambda_3}, \frac{\partial\Gamma}{\partial\lambda_4}\right] = [-0.1730, -0.1762, -0.1541, -0.1326]$$

$$H = \begin{bmatrix} 0.3333 & 0.25 & 0.2 & 0.1667 \\ 0.25 & 0.2 & 0.1667 & 0.1429 \\ 0.2 & 0.1667 & 0.1429 & 0.1250 \\ 0.1667 & 0.1429 & 0.1250 & 0.1111 \end{bmatrix}$$

$$H^{-1} = \begin{bmatrix} 1.2E03 & -6.3E03 & 1.008E04 & -5.04E03 \\ -6.3E03 & 3.258E04 & -5.88E04 & 3.024E04 \\ 1.008E04 & -5.88E04 & 1.008E04 & -5.292E04 \\ -5.04E04 & 3.024E04 & -5.292E04 & 2.8224E04 \end{bmatrix}$$

3. Substituting the computed gradient and hessian matrix back into Eq. (16.22a), we have

$$\lambda_{(1)} = [-17.5090, 75.6008, -101.3483, 44.2724]$$

4. Repeat steps 1 to 3 until the objective function reaches the minimum. When the gradient reaches **0**, we will get the optimized Lagrange multipliers.

Based on the principle of maximum entropy discussed by Zellner and Highfield (1988), the lower and upper limits for the estimation of Lagrange multipliers are lower $= [-\infty, -\infty, -\infty, 0]$ and upper $= [+\infty, +\infty, +\infty, +\infty]$. Using the **fmincon** function in MATLAB, we have the following:

The estimated Lagrange multipliers are $\lambda = [-23.054, 91.9506, -123.4747, 55.8202]$ and $\lambda_0 = 0.8905$. The minimized objective function value is $\Gamma = -0.2796$. The gradient of the objective function is

$$\left[\frac{\partial\Gamma}{\partial\lambda_1}, \frac{\partial\Gamma}{\partial\lambda_2}, \frac{\partial\Gamma}{\partial\lambda_3}, \frac{\partial\Gamma}{\partial\lambda_4}\right] = [-1.18E-07, -1.74E-07, -4.63E-08, -4.22E-08]$$

The gradient of the objective function confirms that with the estimated Lagrange multipliers, the objective function reaches the minimum with the gradient close to **0**.

To illustrate graphically, Fig. 16.1 compares the empirical frequency with the entropy-based frequency distribution for both the scaled flood variable and the observed flood variable.

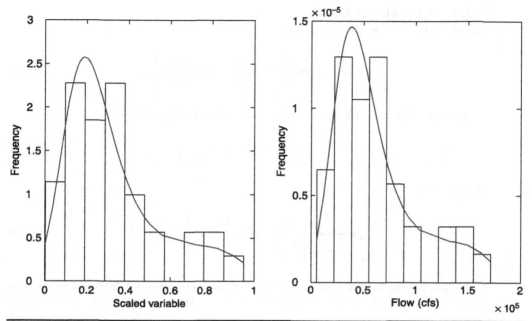

FIGURE 16.1   Comparison of empirical frequency with the PDF derived from entropy-based distribution.

16.2   General Probability Distribution

16.2.1   Specification of Constraints

A general probability distribution can be derived by specifying the following constraints:

$$\int_c^\infty \left(\frac{x-c}{b-c}\right)^a f(x)\,dx = E\left(\frac{x-c}{b-c}\right)^a = \overline{\left(\frac{x-c}{b-c}\right)^a} \tag{16.24}$$

$$\int_c^\infty \ln\frac{x-c}{b-c}\,f(x)\,dx = E\left[\ln\frac{x-c}{b-c}\right] = \overline{\ln\frac{x-c}{b-c}} \tag{16.25}$$

where a, b, and c are parameters.

16.2.2   Entropy-Based Distribution

The POME-based distribution can be written as

$$f(x) = \exp\left[-\lambda_0 - \lambda_1\left(\frac{x-c}{b-c}\right)^a - \lambda_2 \ln\frac{x-c}{b-c}\right] \qquad c \le x \le \infty \tag{16.26}$$

where λ_i, $i = 0, 1, 2$, are Lagrange multipliers.

16.2.3   Parameter Estimation

Inserting Eq. (16.26) in Eq. (16.2) yields

$$\exp \lambda_0 = \int_c^\infty \exp\left[-\lambda_1 \left(\frac{x-c}{b-c}\right)^a - \lambda_2 \ln\frac{x-c}{b-c}\right] dx \tag{16.27}$$

Equation (16.27) can be solved in terms of the gamma function:

$$\exp \lambda_0 = \frac{b-c}{a\lambda_1^{(1-\lambda_2)/a}}\Gamma\left(\frac{1-\lambda_2}{a}\right) \tag{16.28}$$

Equation (16.26) becomes

$$f(x) = \frac{a\lambda_1^{(1-\lambda_2)/a}}{(b-c)\Gamma\left(\dfrac{1-\lambda_2}{a}\right)}\left(\frac{x-c}{b-c}\right)^{-\lambda_2}\exp\left[-\lambda_1\left(\frac{x-c}{b-c}\right)^a\right] \tag{16.29}$$

For estimating the parameters of Eq. (16.29), one can use either the ordinary entropy method or the parameter space expansion method. The entropy of Eq. (16.29) can be written as

$$H(x) = \ln\frac{b-c}{a} + \frac{\lambda_2-1}{a}\ln\lambda_1 + \ln\Gamma\left(\frac{1-\lambda_2}{a}\right) + \lambda_2 E\left[\ln\frac{x-c}{b-c}\right] + \lambda_1 E\left[\left(\frac{x-c}{b-c}\right)^a\right] \tag{16.30}$$

Differentiating Eq. (16.30) with respect to λ_1, λ_2, a, b, and c and equating each derivative to zero, one finds

$$\frac{\partial H}{\partial \lambda_1} = 0 = \frac{\lambda_2-1}{a\lambda_1} + E\left[\left(\frac{x-c}{b-c}\right)^a\right] \tag{16.31a}$$

which leads to

$$\frac{1-\lambda_2}{a\lambda_1} = E\left[\left(\frac{x-c}{b-c}\right)^a\right] \Rightarrow \frac{\partial \lambda_0}{\partial \lambda_1} \tag{16.31b}$$

$$\frac{\partial H}{\partial \lambda_2} = 0 = \frac{1}{a}\ln\lambda_1 - \frac{1}{a}\psi\left(\frac{1-\lambda_2}{a}\right) + E\left[\ln\frac{x-c}{b-c}\right] \tag{16.32a}$$

This leads to

$$-\frac{1}{a}\ln\lambda_1 + \frac{1}{a}\psi\left(\frac{1-\lambda_2}{a}\right) = E\left[\ln\frac{x-c}{b-c}\right] \Rightarrow \frac{\partial \lambda_0}{\partial \lambda_2} \tag{16.32b}$$

$$\frac{\partial H}{\partial a} = 0 = -\frac{1}{a} - \frac{\lambda_2-1}{a^2}\ln\lambda_1 - \left(\frac{1-\lambda_2}{a^2}\right)\psi\left(\frac{1-\lambda_2}{a}\right) + \lambda_1 E\left[\left(\frac{x-c}{b-c}\right)^a \ln\frac{x-c}{b-c}\right] \tag{16.33a}$$

This results in

$$\frac{1}{a}+\frac{\lambda_2-1}{a^2}\ln\lambda_1+\frac{1-\lambda_2}{a^2}\psi\left(\frac{1-\lambda_2}{a}\right)=\lambda_1 E\left[\left(\frac{x-c}{b-c}\right)^a\ln\frac{x-c}{b-c}\right] \tag{16.33b}$$

$$\frac{\partial H}{\partial b}=0=\frac{1}{b-c}-\frac{\lambda_2}{b-c}-\frac{a\lambda_1}{b-c}E\left[\left(\frac{x-c}{b-c}\right)^a\right] \tag{16.34a}$$

This can be written as

$$\frac{1-\lambda_2}{a\lambda_1}=E\left[\left(\frac{x-c}{b-c}\right)^a\right] \tag{16.34b}$$

Equation (16.34b) is the same as Eq. (16.31b).

$$\frac{\partial H}{\partial c}=0=-\frac{1}{b-c}+\frac{\lambda_2}{b-c}+a\lambda_1 E\left\{\left(\frac{x-c}{b-c}\right)^{a-1}\left[-\frac{1}{b-c}+\frac{x-c}{(b-c)^2}\right]\right\}-\lambda_2 E\left[\frac{1}{x-c}\right] \tag{16.35a}$$

This can be simplified to

$$-\frac{a\lambda_1}{b-c}E\left[\left(\frac{x-c}{b-c}\right)^{a-1}\right]=\lambda_2 E\left[\frac{1}{x-c}\right] \tag{16.35b}$$

With the introduction of Eq. (16.31b),

$$(\lambda_2-1)\frac{E[(x-c)^{a-1}]}{E[(x-c)^a]}=\lambda_2 E\left[\frac{1}{x-c}\right] \tag{16.36}$$

We now have five equations: Eqs. (16.31b), (16.32b), (16.33b), (16.34b), and (16.35b). Equation (16.31b) is identical to Eq. (16.34b).

Using Eq. (16.32b) in Eq. (16.33b),

$$\frac{1}{a}+\frac{1-\lambda_2}{a}E\left[\ln\frac{x-c}{b-c}\right]=\lambda_1 E\left[\left(\frac{x-c}{b-c}\right)^a\ln\frac{x-c}{b-c}\right] \tag{16.37}$$

Using Eq. (16.31),

$$\frac{1}{a}+\lambda_1 E\left[\left(\frac{x-c}{b-c}\right)^a\right]E\left[\ln\frac{x-c}{b-c}\right]=\lambda_1 E\left[\left(\frac{x-c}{b-c}\right)^a\ln\frac{x-c}{b-c}\right]=\lambda_1\frac{\partial^2\lambda_0}{\partial\lambda_1\,\partial\lambda_2} \tag{16.38}$$

Now, multiplying Eqs. (16.31b) and (16.36), one finds

$$(\lambda_2-1)\frac{E[(x-c)^{a-1}]}{E[(x-c)^a]}=\lambda_2 E\left[\frac{1}{x-c}\right]\qquad \lambda_2\neq 0 \tag{16.39}$$

Equation (16.39) is the same as Eq. (16.36). Thus, four equations of interest are Eqs. (16.31b), (16.32b), (16.38), and (16.39).

16.2.3 Special Cases
Special cases that emanate from the general equation are briefly dealt with.

Exponential Distribution

If $a = 1$, $c = 0$, $\lambda_1 = 1$, and $\lambda_2 = 0$, Eq. (16.29) gives

$$f(x) = \frac{1}{b}\exp\left(-\frac{x}{b}\right) \tag{16.40}$$

This is an exponential distribution, with parameter b obtained from Eq. (16.31b) as

$$b = E(x) = \overline{x} \tag{16.41}$$

From Eq. (16.32b),

$$\psi(1) = E\left[\ln\frac{x}{b}\right] \tag{16.42}$$

From Eq. (16.38),

$$1 = E\left[\frac{x}{b}\right]E\left[\ln\frac{x}{b}\right] = E\left[\frac{x}{b}\ln\frac{x}{b}\right] \tag{16.43}$$

This is the identity for the exponential PDF. Equation (16.43) does not exist for this PDF.

Gamma Distribution and Pearson Type 3 Distribution

Here $a = 1$, $b = 1$, and $c = 0$ for the gamma PDF, and $a = 1$, $b > c \neq 0$ for the Pearson type 3 distribution. Equation (16.29) becomes

$$f(x) = \frac{1}{b-c}\frac{\lambda_1^{1-\lambda_2}}{\Gamma(1-\lambda_2)}\left(\frac{x-c}{b-c}\right)^{-\lambda_2}\exp\left[-\lambda_1\left(\frac{x-c}{b-c}\right)\right] \tag{16.44}$$

From Eq. (16.31b),

$$\frac{1-\lambda_2}{\lambda_1} = E\left[\frac{x-c}{b-c}\right] \tag{16.45}$$

From Eq. (16.32b),

$$-\ln\lambda_1 + \psi(1-\lambda_2) = E\left[\ln\frac{x-c}{b-c}\right] \tag{16.46}$$

From Eq. (16.38),

$$1 + \lambda_1 E\left[\frac{x-c}{b-c}\right]E\left[\ln\frac{x-c}{b-c}\right] = \lambda_1 E\left[\frac{x-c}{b-c}\ln\frac{x-c}{b-c}\right] \tag{16.47}$$

From Eq. (16.39),

$$\lambda_2 = \frac{1}{1 - E[1/(x-c)]E[x-c]} \tag{16.48}$$

Extreme Value Type 3 Distribution

Here $\lambda_1 = 1$, $\lambda_2 = 1 - a$, and $c \neq 0$. Equation (16.29) becomes

$$f(x) = \frac{a}{b-c}\left(\frac{x-c}{b-c}\right)^{a-1}\exp\left[-\left(\frac{x-c}{b-c}\right)^{a}\right] \tag{16.49}$$

From Eq. (16.31b),

$$1 = E\left[\left(\frac{x-c}{b-c}\right)^a\right] \tag{16.50}$$

From Eq. (16.32b),

$$\frac{1}{a}\psi(1) = E\left[\ln\frac{x-c}{b-c}\right] \tag{16.51}$$

From Eq. (16.38),

$$\frac{1}{a} + E\left[\left(\frac{x-c}{b-c}\right)^a\right]E\left[\ln\frac{x-c}{b-c}\right] = E\left[\left(\frac{x-c}{b-c}\right)^a\ln\frac{x-c}{b-c}\right] \tag{16.52}$$

This is an identity. From Eq. (16.39),

$$\frac{1-a}{a}E\left[\frac{1}{x-c}\right] + \frac{1}{b-c}E\left[\left(\frac{x-c}{b-c}\right)^{a-1}\right] = 0 \tag{16.53}$$

Normal Distribution

If $a = 2$, $\lambda_2 = 0$, and $c = 0$, Eq. (16.29) becomes

$$f(x) = \frac{2\lambda_1^{1/2}}{(b-c)\sqrt{\pi}}\exp\left[-\left(\frac{x-c}{b-c}\right)^2\right] \tag{16.54}$$

From Eq. (16.31b),

$$\frac{1}{2\lambda_1} = E\left[\left(\frac{x-c}{b-c}\right)^2\right] \tag{16.55}$$

From Eq. (16.32b),

$$-\frac{1}{2}\ln\lambda_1 + \frac{1}{2}\psi\left(\frac{1}{2}\right) = E\left[\ln\frac{x}{b}\right] \tag{16.56}$$

From Eq. (16.38),

$$\frac{1}{2} + \lambda_1 E\left[\left(\frac{x-c}{b-c}\right)^2\right]E\left[\ln\frac{x-c}{b-c}\right] = \lambda_1 E\left[\left(\frac{x-c}{b-c}\right)^2\ln\frac{x-c}{b-c}\right] \tag{16.57}$$

This is an identity. From Eq. (16.39),

$$E(x-c) = 0 \Rightarrow c = E(x) \tag{16.58}$$

Example 16.3. Fit a two-parameter gamma distribution and three-parameter Pearson type III distribution to the same data set as in Example 16.1. Compute quantiles for various values of the probability of exceedance.

Solution

Two-Parameter Gamma Distribution

For the gamma distribution, Eqs. (16.26), (16.28), (16.31b), and (16.32b) are rewritten as

$$f(x) = \exp(-\lambda_0 - \lambda_1 x - \lambda_2 \ln x) \qquad a = 1 \qquad b = 1 \qquad c = 0 \tag{16.59a}$$

$$\exp \lambda_0 = \frac{1}{\lambda_1^{1-\lambda_2}} \Gamma(1 - \lambda_2) \Rightarrow \lambda_0 = (\lambda_2 - 1) \ln \lambda_1 + \ln \Gamma(1 - \lambda_2) \tag{16.59b}$$

$$\frac{1 - \lambda_2}{\lambda_1} = E(x) \qquad \text{and} \qquad -\ln(\lambda_1) + \psi(1 - \lambda_2) = E[\ln x] \tag{16.59c}$$

Applying the annual peak discharge data from Example 16.1, we can compute the following sample moments:

$$E(x) = 62585.27 \qquad E[\ln x] = 10.84$$

Solving Eq. (16.59c) numerically, we have

$$\lambda_1 = 4.088 \times 10^{-5} \qquad \lambda_2 = -1.5584 \qquad \lambda_0 = 26.1793$$

$$f(x) = \exp(-26.1793 - 4.088 \times 10^{-5} x + 1.5584 \ln x)$$

$$= \frac{x^{1.5584}}{\exp(26.1793)} \exp(-4.088 \times 10^{-5} x) = \frac{(4.088 \times 10^{-5})^{2.5584} x^{2.5584-1}}{\Gamma(2.5584)} \exp(-4.088 \times 10^{-5} x)$$

This equation shows the entropy-based distribution can be expressed in the form of the gamma PDF. Before we compute the quantiles of the given exceedance probabilities, we compare the estimated entropy-based distribution with its empirical frequency shown in Fig. 6.2. Figure 6.2 indicates a reasonably good fit.

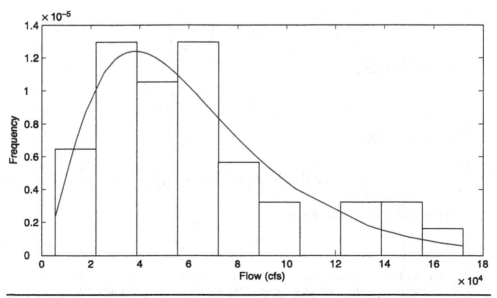

Figure 16.2   Comparison of entropy-based gamma PDF with the empirical frequency.

Setting the exceedance probability to 0.5, 0.1, 0.05, and 0.01, we can compute the corresponding quantiles (that is, 50, 90, 95, and 99 percent) as follows:

$$F(X \geq x_c) = \int_{x_c}^{\infty} f(x)\, dx = 1 - \int_{0}^{x_c} \frac{(4.088 \times 10^{-5})^{2.5584}\, x^{2.5584-1}}{\Gamma(2.5584)} \exp(-4.088 \times 10^{-5} x)\, dx$$

$$= 1 - GammaCDF(x_c; 2.5584, 4.088 \times 10^{-5})$$

Computing the inverse of the lower incomplete gamma function [we can directly use embedded computation routines in the computational software (i.e., MATLAB using **gamcdf**)], we have the quantiles for the given exceedance probability above:

$$Q_{50} = 5.4648 \times 10^4\, cfs \qquad Q_{90} = 1.1498 \times 10^5\, cfs \qquad Q_{95} = 1.3757 \times 10^5\, cfs \qquad Q_{99} = 1.8697 \times 10^5\, cfs$$

Three-Parameter Pearson Type III Distribution

For the Pearson type III distribution, we rewrite Eqs. (16.26), (16.28), (16.31b), (16.32b), and (16.35b) as:

$$f(x) = \frac{\lambda_1^{1-\lambda_2}}{(b-c)\Gamma(1-\lambda_2)} \left(\frac{x-c}{b-c}\right)^{-\lambda_2} \exp\left[-\lambda_1\left(\frac{x-c}{b-c}\right)\right] \qquad \text{where } a = 1 \qquad (16.60a)$$

$$\lambda_0 = \frac{(b-c)\Gamma(1-\lambda_2)}{\lambda_1^{1-\lambda_2}} \tag{16.60b}$$

$$\frac{1-\lambda_2}{\lambda_1} - E\left(\frac{x-c}{b-c}\right) = 0$$

$$-\ln\lambda_1 + \psi(1-\lambda_2) - E\left(\ln\frac{x-c}{b-c}\right) = 0 \tag{16.60c}$$

$$\frac{\lambda_1}{b-c} + \lambda_2 E\left(\frac{1}{x-c}\right) = 0$$

Using $\lambda_1 = 4.088 \times 10^{-5}$, $\lambda_2 = -1.5584$, $c = 0$ (from fitted gamma distribution) and $b = \min(Q)$ as the initial values, we can estimate the parameters numerically (i.e., via the Newton-Raphson method) as

$$\lambda_1 = 1.3351 \qquad \lambda_2 = -1.5608 \qquad b = 3.2627 \times 10^4 \qquad c = 0.5254 \qquad \lambda_0 = 9.9811$$

$$f(x) = \frac{1.3342^{2.5608}}{(3.2627 \times 10^4 - 0.5254)\Gamma(2.5608)} \left(\frac{x - 0.5254}{3.2627 \times 10^4 - 0.5254}\right)^{1.5608} \exp\left[-1.3351\left(\frac{x + 0.5254}{3.2627 \times 10^4 - 0.5254}\right)\right]$$

We can also rewrite this probability density function as a gamma density function

$$f(x) = \frac{1}{3.2627 \times 10^4 - 0.5254} \Gamma\left(\frac{x - 0.5254}{3.2627 \times 10^4 - 0.5254}, 2.5608, 1.3351\right) \quad \Gamma(x, a, b) = \frac{b^a}{\Gamma(a)} x^{a-1} \exp(-bx)$$

Figure 16.3 compares the entropy-based probability density function with the empirical frequency, which again indicates the fit is reasonably good.

Again, setting the exceedance probability to 0.5, 0.1, 0.05, 0.01, we can compute the corresponding quantiles as follows.

The CDF of the entropy-based Person type III distribution is given as

$$F(X \leq x) = \int_{0}^{x} \frac{1}{3.2627 \times 10^4 - 0.5254} \Gamma\left(\frac{x - 0.5254}{3.2627 \times 10^4 - 0.5254}; 2.5591, 1.3342\right) dx$$

$$= \int_{0}^{x'} \Gamma(x'; 2.5591, 1.3342)\, dx' = \Gamma CDF(x'; 2.5591, 1.3342)$$

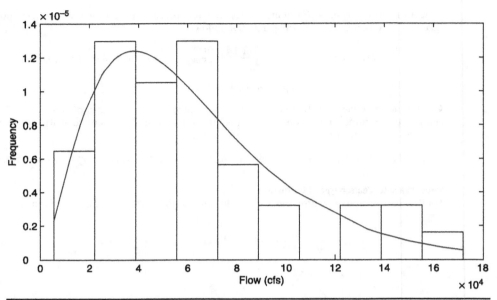

FIGURE 16.3   Comparison of entropy-based Pearson type III PDF with the empirical frequency.

where

$$x' = \frac{x - 0.5271}{3.2627 \times 10^4 - 0.5271}$$

Now applying the regular computation routine for the gamma distribution, we have

$$Q_{50} = 5.4649 \times 10^4 \, \text{cfs} \qquad Q_{90} = 1.1498 \times 10^5 \, \text{cfs} \qquad Q_{95} = 1.3756 \times 10^5 \, \text{cfs} \qquad Q_{99} = 1.8695 \times 10^5 \, \text{cfs}$$

Example 16.4.   For the same set of peak discharge data as in Example 16.1, compute the parameters of the generalized distribution. Compute quantiles for various values of the probability of exceedance.

Solution.   Using the parameters estimated for the Pearson type III distribution in Example 16.3 as the initial estimates, we can estimate the parameters for the generalized distribution using the Newton-Raphson method as follows:

$$\lambda_0 = 9.6778 \qquad \lambda_1 = 1.8585 \qquad \lambda_2 = -1.8380 \qquad a = 0.8789 \qquad b = 3.2627 \times 10^4 \qquad c = 0.5364$$

$$f(x) = \frac{0.8789\left(1.8585^{\frac{2.8380}{0.8789}}\right)}{(3.2627 \times 10^4 - 0.5364)\Gamma\left(\frac{2.8380}{0.8789}\right)}\left(\frac{x - 0.5364}{3.2627 \times 10^4 - 0.5364}\right)^{1.8380}$$

$$\times \exp\left[-1.8585\left(\frac{x - 0.5364}{3.2627 \times 10^4 - 0.5364}\right)^{0.8797}\right]$$

The cumulative probability distribution function can then be written as

$$F(X \le x) = \frac{1}{\Gamma\left(\frac{2.8380}{0.8789}\right)}\int_0^z z^{\frac{2.8380}{0.8797}-1}\exp(-z)\,dz = \Gamma CDF\left(z, \frac{2.8380}{0.8797}, 1\right) \qquad z = 1.8585\left(\frac{x - 0.5364}{3.2627 \times 10^4 - 0.5364}\right)$$

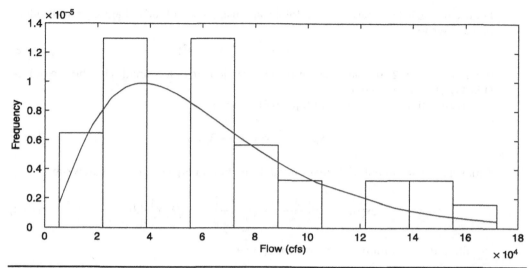

Figure 16.4 Comparison of entropy-based generalized probability distribution with empirical frequency.

Again, setting the exceedance probability to 0.5, 0.1, 0.05, and 0.01, we compute the corresponding quantiles:

$$Q_{50} = 5.0957 \times 10^4 \, \text{cfs} \qquad Q_{90} = 9.8990 \times 10^4 \, \text{cfs} \qquad Q_{95} = 1.1650 \times 10^5 \, \text{cfs} \qquad Q_{99} = 1.5432 \times 10^5 \, \text{cfs}$$

Figure 16.4 compares the entropy-based probability density function with the empirical frequency that indicates the fit is satisfactory.

16.3 Truncated Normal Distribution

It may be interesting to derive the truncated normal distribution for a continuous random variable X with nonnegative value. Let $f(x)$ be the desired distribution, which is obtained by maximizing the Shannon entropy:

$$H(x) = -\int_0^\infty f(x) \ln f(x) dx \tag{16.61a}$$

where

$$\int_0^\infty f(x) dx = 1 \tag{16.61b}$$

It is assumed that its mean $\bar{x}$ and variance S^2 are given and can be expressed as

$$\int_0^\infty x f(x) dx = E(x) = \bar{x} \tag{16.62a}$$

$$\int_0^\infty x^2 f(x) dx = S^2 + (\bar{x})^2 \tag{16.62b}$$

Following the POME-based methodology and using the method of Lagrange multipliers, one can write

$$f(x) = \exp(-\lambda_0 - \lambda_1 x - \lambda_2 x^2) \tag{16.63}$$

where λ_i, $i = 0, 1, 2$, are Lagrange multipliers which are determined with the use of Eqs. (16.61a), (16.62a) and (16.62b).

Substitution of Eq. (16.63) in Eq. (16.61a) yields

$$\exp \lambda_0 = \int_0^\infty \exp(-\lambda_1 x - \lambda_2 x^2)\, dx \tag{16.64}$$

Squaring the argument of the exponential function in Eq. (16.64), one can write

$$\exp \lambda_0 = \exp\left(\frac{\lambda_1^2}{4\lambda_2}\right) \int_0^\infty \exp\left[-\left(\frac{\lambda_1}{2\sqrt{\lambda_2}} + \sqrt{\lambda_2}\, x\right)^2\right] dx \tag{16.65}$$

Recalling the definition of the error function erf,

$$\text{erf}(z) = \frac{2}{\sqrt{\pi}} \int_0^z \exp(-u^2)\, du = 1 - \frac{2}{\sqrt{\pi}} \int_z^\infty \exp(-u^2)\, du \tag{16.66}$$

Letting $t = \sqrt{\lambda_2}\, x + \lambda_1/2\sqrt{\lambda_2}$, $dt = \sqrt{\lambda_2}\, dx$, one can write Eq. (16.65) as

$$\exp \lambda_0 = \frac{\exp(\lambda_1^2/4\lambda_2)}{\sqrt{\lambda_2}} \int_{\lambda_1/2\sqrt{\lambda_2}}^\infty \exp(-t^2)\, dt \tag{16.67}$$

Equation (16.67) can be expressed in terms of the error function as

$$\exp \lambda_0 = \frac{\exp(\lambda_1^2/4\lambda_2)}{\sqrt{\lambda_2}} \frac{\sqrt{\pi}}{2}\left[1 - \text{erf}\left(\frac{\lambda_1}{2\sqrt{\lambda_2}}\right)\right] \tag{16.68}$$

or

$$\lambda_0 = \frac{\lambda_1^2}{4\lambda_2} - \frac{1}{2}\ln \lambda_2 + \ln \frac{\sqrt{\pi}}{2} + \ln\left[1 - \text{erf}\left(\frac{\lambda_1}{2\sqrt{\lambda_2}}\right)\right] \tag{16.69}$$

Let $a = \lambda_1/(2\sqrt{\lambda_2})$. Equation (16.66) can be written as

$$\text{erf}(a) = \frac{2}{\sqrt{\pi}}\left(a - \frac{a^3}{1!3} + \frac{a^5}{2!5} - \frac{a^7}{3!7} + \cdots\right) \tag{16.70a}$$

Differentiating Eq. (16.69) with respect to λ_0, differentiating Eq. (16.67) with respect to λ_0, and equating the two derivatives to each other, one obtains

$$\frac{\partial \lambda_0}{\partial \lambda_1} = -\bar{x} = \frac{\lambda_1}{2\lambda_2} - \frac{2}{\sqrt{\pi}} \frac{(1/2\sqrt{\lambda_2})\exp(-\lambda_1^2/4\lambda_2)}{[1 - \text{erf}(\lambda_1/2\sqrt{\lambda_2})]} \tag{16.70b}$$

Multiplying both sides of Eq. (16.70b) by λ_1 and then using the definition of a, one gets

$$\lambda_1 \bar{x} = -2a^2 + \frac{2}{\sqrt{\pi}} \frac{a\exp(-a^2)}{[1 - \text{erf}(a)]} \tag{16.70c}$$

Equation (16.70c) can be expressed as

$$b = z - 2a^2 \qquad (16.71a)$$

where

$$b = \lambda_1 \bar{x} \qquad z = \frac{2}{\sqrt{\pi}} \frac{a\exp(-a^2)}{1 - \mathrm{erf}(a)} \qquad (16.71b)$$

Likewise, differentiating Eq. (16.68) with respect to λ_2 and noting that from Eq. (16.11a) $\partial\lambda_0/\partial\lambda_2 = -\overline{x^2} = -S^2 - (\bar{x})^2$, one gets

$$\frac{S^2}{(\bar{x})^2} = \frac{2a^2 + 2a^2 z - z^2}{4a^4 - 4a^2 z + z^2} = \frac{z(1-b) - b}{b^2} \qquad (16.72a)$$

Equation (16.72a) can be written in terms of the coefficient of variation (CV) as

$$\mathrm{CV} = \frac{S}{\bar{x}} = \sqrt{\frac{z(1-b) - b}{b^2}} \qquad (16.72b)$$

Note that $\lambda_2 = \lambda_1^2/(4a^2)$. Denoting $c = \lambda_2(\bar{x})^2$, one can write

$$c = \frac{a^2(\bar{x})^2}{4a^2} = \frac{b^2}{4a^2} \qquad (16.72c)$$

Thus, Eq. (16.63) can be expressed as

$$f(x) = \exp\left(-\lambda_0 - b\frac{x}{\bar{x}} - c\frac{x^2}{(\bar{x})^2}\right) \qquad x \ge 0 \qquad (16.73a)$$

where

$$\exp(-\lambda_0) = \frac{2\sqrt{c}}{\bar{x}d\sqrt{\pi}}\exp(-a^2) \qquad d = 1 - \mathrm{erf}(a) \qquad (16.73b)$$

Letting $K = 2\sqrt{c}/d\sqrt{\pi}$, then CDF can be written as

$$F(u) = K\int_0^u \exp(-a^2 - bu - cu^2)du \qquad u = x/\bar{x} \qquad u \ge 0 \qquad (16.74)$$

The return period T can be obtained from Eq. (16.74) written as

$$T(u) = \frac{1}{1 - F(u)} \qquad (16.75)$$

The values of parameters a, b, c, and K can be computed for a range of CV. One may compute the values of these parameters for a range of the CV values and tabulate them for future reference. Equation (16.74) can be evaluated for known values of parameters using a numerical method, such as gaussian quadrature, and then the corresponding values of T can be obtained from Eq. (16.75). One can also develop a table of u, $F(u)$, and T as well as develop graphs of u versus T for various values of CV. It may also be interesting to compare the variation of z with a and CV. For a less than particular value, z will approach zero; then for CV less than a particular value, parameter b would approach $-2a^2$ and c would approach a^2 such that the distribution would become a normal distribution.

Example 16.5. Using the annual instantaneous peak discharge from a station in the Brazos River basin given in Table 16.1, compute the frequency histogram for the peak discharge data. For a selected range of the CV tabulate the parameters of the distribution given by Eq. (16.74). Fit the truncated normal distribution to the observed data, and discuss how good the fit is.

Solution. From entropy theory, the cumulative probability function is determined to be

$$F(u) = K \int_0^u \exp(-a^2 - bu - cu^2)\, du \qquad u = \frac{x}{\bar{x}} \qquad u \geq 0$$

The parameter in the function can be defined as determined below:

$$CV = \frac{S}{\bar{x}} = \sqrt{\frac{z(1-b)-b}{b^2}} \qquad b = z - 2a^2 \qquad z = \frac{2}{\sqrt{\pi}} \frac{a\exp(-a^2)}{[1-\mathrm{erf}(a)]}$$

$$c = \frac{a^2(\bar{x})^2}{4a^2} = \frac{b^2}{4a^2} \qquad K = \frac{2\sqrt{c}}{d\sqrt{\pi}} \qquad d = 1 - \mathrm{erf}(a)$$

Parameters $a, b, c,$ and K of the distribution are computed for a range of values of CV and are presented in Table 16.2.

To evaluate this model, the annual peak flow data from the Brazos River for 74 yr, as in Table 16.1, are used.

$$\bar{x} = 6.26 \times 10^4 \qquad S = 3.96 \times 10^4 \qquad CV = \frac{S}{\bar{x}} = 0.63$$

The frequency and probability based on the observations are computed in Table 16.3 and presented in Figs. 16.5 and 16.6.

In this case $CV = S/\bar{x} = 0.63$, and we can estimate the parameter values of the CDF in Table 16.4. Table 16.5 lists the CDF values computed from the entropy-based truncated normal distribution and the empirical distribution for a set of observations. Figure 16.7 compares the truncated normal distribution and the empirical distribution. Table 16.5 and Fig. 16.7 both indicate reasonably good fit.

a	CV	b	c	K	a	CV	b	c	K
−0.05	0.747	−0.0583	0.3396	0.6225	−1.30	0.480	−3.5200	1.8329	0.7899
−0.10	0.738	−0.1204	0.3625	0.6107	−1.40	0.460	−3.7717	1.9514	0.8109
−0.20	0.720	−0.2573	0.4139	0.5937	−1.50	0.440	−4.5907	2.3417	0.8782
−0.30	0.700	−0.4129	0.4735	0.5844	−1.75	0.391	−6.1715	3.1092	1.0015
−0.40	0.680	−0.5893	0.5426	0.5819	−2.00	0.350	−8.0207	4.0207	1.1340
−0.50	0.660	−0.7889	0.6225	0.5855	−2.25	0.312	−10.1330	5.0705	1.2714
−0.60	0.640	−1.0145	0.7147	0.5948	−2.50	0.280	−12.5027	6.2527	1.4111
−0.70	0.620	−1.1378	0.7660	0.6014	−2.75	0.257	−15.1258	7.5633	1.5517
−0.80	0.600	−1.5532	0.9424	0.6288	−3.00	0.235	−18.0000	9.0002	1.6926
−0.90	0.572	−1.8714	1.0809	0.6529	−3.25	0.217	−21.1250	10.5625	1.8336
−1.00	0.560	−2.2252	1.2380	0.6813	−3.50	0.202	−24.5000	12.2500	1.9727
−1.10	0.520	−2.6169	1.4149	0.7138	−3.75	0.188	−28.1250	14.0625	2.1157
−1.20	0.500	−3.0479	1.6128	0.7501	−4.00	0.176	−32.0000	16.0000	2.2568

TABLE 16.2 Parameters for the CDF

Interval u	0–0.4	0.4–0.8	0.8–1.2	1.2–1.6	1.6–2.0	2.0–2.4	2.4–2.8
Frequency	9	24	24	4	3	8	2
Probability	0.12	0.32	0.32	0.05	0.04	0.11	0.03

TABLE 16.3 Frequency and Probability for u

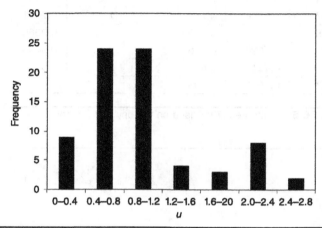

FIGURE 16.5 Frequency of u.

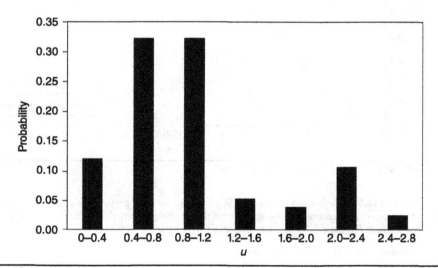

FIGURE 16.6 Probability of u.

CV	a	b	c	K
0.63	−0.6448	−1.1246	0.7605	0.5950

TABLE 16.4 Parameters for CDF

U	F(u)	Empirical Prob.	U	F(u)	Empirical Prob.
0.09	0.0371	0.0133	0.93	0.4982	0.5467
0.24	0.1064	0.0667	0.97	0.5213	0.6
0.4	0.1897	0.12	1.05	0.5663	0.6533
0.49	0.2398	0.1733	1.13	0.6096	0.7067
0.54	0.2684	0.2267	1.2	0.6459	0.76
0.57	0.2858	0.28	1.49	0.7761	0.8133
0.65	0.3328	0.3333	2.12	0.9369	0.8667
0.7	0.3624	0.3867	2.22	0.9495	0.92
0.8	0.4219	0.44	2.6	0.9774	0.9733
0.87	0.4632	0.4933			

TABLE 16.5   Computed $F(u)$ Based on Entropy and Observations

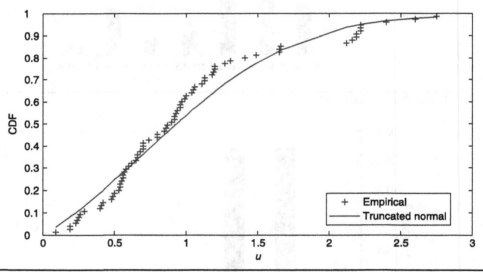

FIGURE 16.7   Computed and observed $F(u)$.

Example 16.6.   For various values of CV, plot u versus T on probability paper. For comparison you may also plot the same data for the normal distribution. Before doing the plotting, you may want to tabulate u, $F(u)$, and T for various values of CV.

Solution.   From the above calculations,

$$F(u) = K \int_0^u \exp(-a^2 - bu - cu^2)\, du \qquad u = \frac{x}{\bar{x}} \qquad u \geq 0$$

$$T(u) = \frac{1}{1 - F(u)}$$

The recurrence interval T and CDF $F(u)$ versus u are computed for various values of CV; the results are tabulated in Table 16.6 and presented in Fig. 16.8.

CV	u	F(u)	T(u)	CV	u	F(u)	T(u)
0.65	0.2	0.0098	1.01	0.35	0.2	0.0098	1.01
	0.4	0.0437	1.05		0.4	0.0437	1.05
	0.6	0.1293	1.15		0.6	0.1293	1.15
	0.8	0.2874	1.40		0.8	0.2874	1.40
	1	0.5020	2.01		1	0.5020	2.01
	1.2	0.7157	3.52		1.2	0.7157	3.52
	1.4	0.8719	7.81		1.4	0.8719	7.81
	1.6	0.9556	22.52		1.6	0.9556	22.52
	1.8	0.9883	85.47		1.8	0.9883	85.47
	2	0.9972	357.14		2	0.9972	357.14
0.6	0.2	0.0784	1.09	0.3	0.2	0.0035	1.00
	0.4	0.1760	1.21		0.4	0.0229	1.02
	0.6	0.2890	1.41		0.6	0.0918	1.10
	0.8	0.4107	1.70		0.8	0.2521	1.34
	1	0.5325	2.14		1	0.4963	1.99
	1.2	0.6459	2.82		1.2	0.7396	3.84
	1.4	0.7442	3.91		1.4	0.8985	9.85
	1.6	0.8234	5.66		1.6	0.9664	29.76
	1.8	0.8827	8.53		1.8	0.9844	64.10
	2	0.9241	13.18				
0.5	0.2	0.0466	1.05	0.25	0.2	0.0006	1.00
	0.4	0.1225	1.14		0.4	0.0082	1.01
	0.6	0.2309	1.30		0.6	0.0550	1.06
	0.8	0.3668	1.58		0.8	0.2122	1.27
	1	0.5163	2.07		1	0.5003	2.00
	1.2	0.6608	2.95		1.2	0.7883	4.72
	1.4	0.7832	4.61		1.4	0.9460	18.52
	1.6	0.8744	7.96		1.6	0.9854	68.49
	1.8	0.9339	15.13		1.8	0.9967	303.03
	2	0.9680	31.25				
0.45	0.2	0.0319	1.03	0.2	0.2	0.0000	1.00
	0.4	0.0948	1.10		0.4	0.0013	1.00
	0.6	0.1984	1.25		0.6	0.0228	1.02
	0.8	0.3423	1.52		0.8	0.1589	1.19
	1	0.5099	2.04		1	0.5003	2.00
	1.2	0.6741	3.07		1.2	0.8411	6.29
	1.4	0.8093	5.24		1.4	0.9797	49.26
	1.6	0.9029	10.30				
	1.8	0.9573	23.42				
	2	0.9838	61.73				

TABLE 16.6 Tabulation of the Distribution for Various Values of CV

CV	u	F(u)	T(u)	CV	u	F(u)	T(u)
0.4	0.2	0.0020	1.00	0.15	0.2	0.0000	1.00
	0.4	0.0682	1.07		0.4	0.0000	1.00
	0.6	0.1647	1.20		0.6	0.0039	1.00
	0.8	0.3161	1.46		0.8	0.0918	1.10
	1	0.5049	2.02		1	0.5006	2.00
	1.2	0.6916	3.24		1.2	0.9037	10.38
	1.4	0.8383	6.18		1.4	0.9878	81.97
	1.6	0.9297	14.22				
	1.8	0.9749	39.84				
	2	0.9905	105.26				

TABLE 16.6   Tabulation of the Distribution for Various Values of CV (*Continued*)

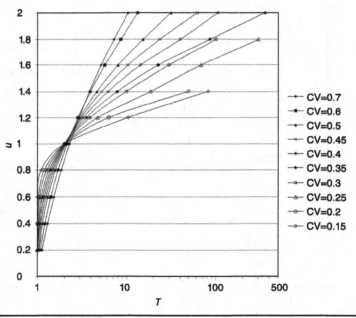

FIGURE 16.8   Return period versus *u* for different values of CV.

Example 16.7.   Compute the variation of z with a and CV. Then determine the conditions under which the truncated normal distribution becomes a normal distribution.

Solution

$$b = z - 2a^2$$

$$CV = \frac{S}{\bar{x}} = \sqrt{\frac{z(1-b)-b}{b^2}}$$

Table 16.7 shows the variation of z with a and CV. When a is less than -3.80 or CV is less than 0.186, then z becomes zero. Thus for CV less than 0.186, parameters b and c become $b = -2a^2$ and $c = a^2$, and the expression for the distribution is identical to that for a normal distribution.

Example 16.8. For the same set of peak discharge data as in Example 16.1, compute the parameters of the generalized distribution. Compute quantiles for various values of the probability of exceedance, and compare them with those obtained using the truncated normal distribution.

Solution. The parameters estimated for the truncated normal distribution are listed in Table 16.4. The parameters estimated for the generalized probability distribution are given in Example 16.4 as

$$f(x) = \frac{0.8789(1.8585^{\frac{2.8380}{0.8789}})}{(3.2627\times10^4 - 0.5364)\Gamma\left(\frac{2.8380}{0.8789}\right)}\left(\frac{x-0.5364}{3.2627\times10^4 - 0.5364}\right)^{1.8380}$$

$$\times \exp\left[-1.8585\left(\frac{x-0.5364}{3.2627\times10^4 - 0.5364}\right)^{0.8797}\right]$$

for the exceedance probability values of 0.5, 0.1, 0.05, and 0.01. The computed quantiles for the generalized distribution are computed in Example 16.4 as

$$Q_{50} = 5.0957\times10^4\,\text{cfs} \qquad Q_{90} = 9.8990\times10^4\,\text{cfs} \qquad Q_{95} = 1.1650\times10^5\,\text{cfs} \qquad Q_{99} = 1.5432\times10^5\,\text{cfs}$$

The truncated normal distribution for the transformed variable $Q/\bar{Q}$ is estimated in Example 16.5 as

$$f(u) = 0.595\exp[-(-0.6488)^2 + 1.1246u - 0.7605u^2] \qquad u = Q/\bar{Q}$$

$$Q_{50} = 5.8394\times10^4\,\text{cfs} \qquad Q_{90} = 1.1938\times10^5\,\text{cfs} \qquad Q_{95} = 1.3922\times10^5\,\text{cfs} \qquad Q_{99} = 2.1535\times10^5\,\text{cfs}$$

Figure 16.9 compares the generalized distribution, truncated normal distribution, and empirical distribution.

Example 16.9. Fit a two-parameter gamma distribution and three-parameter lognormal distribution to the same data set as in Example 16.1. Compute quantiles for various values of the probability of exceedance, and compare them with those obtained using the truncated normal distribution and the generalized distribution.

a	z	CV	A	z	CV
−0.25	−0.207629	0.720	−2.25	−0.008040	0.312
−0.50	−0.288978	0.661	−2.50	−0.002723	0.282
−0.75	−0.281796	0.606	−2.75	−0.000800	0.257
−1.00	−0.225271	0.550	−3.00	−0.000208	0.235
−1.25	−0.153752	0.493	−3.25	−0.000047	0.217
−1.50	−0.090735	0.440	−3.50	−0.000009	0.202
−1.75	−0.046487	0.391	−3.80	−0.000010	0.186
−2.00	−0.020715	0.348	−3.90	0.000000	0.182

TABLE 16.7 Variation of z with a and CV

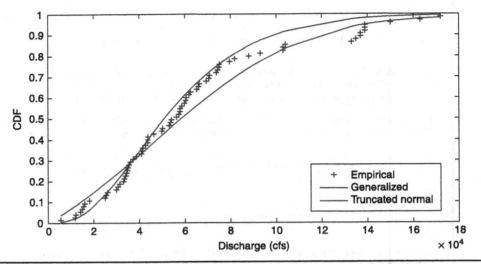

FIGURE 16.9   Comparison of empirical, generalized, and truncated normal distributions.

Solution

Two-Parameter Gamma Distribution

In Example 16.3, we fit the two-parameter gamma distribution to the discharge variable as

$$\frac{(4.088 \times 10^{-5})^{2.5584} \, x^{2.5584-1}}{\Gamma(2.5584)} \exp(-4.088 \times 10^{-5} x)$$

The quantiles with exceedance probability values of 0.5, 0.1, 0.05, and 0.01 were computed as

$$Q_{50} = 5.4648 \times 10^{4}\,\text{cfs} \qquad Q_{90} = 1.1498 \times 10^{5}\,\text{cfs} \qquad Q_{95} = 1.3757 \times 10^{5}\,\text{cfs} \qquad Q_{99} = 1.8697 \times 10^{5}\,\text{cfs}$$

Truncated Normal Distribution

The truncated normal distribution for the transformed variable $Q/\bar{Q}$ is estimated in Example 16.5 as

$$f(u) = 0.595 \exp[-(-0.6488)^2 + 1.1246u - 0.7605u^2] \qquad u = \frac{Q}{\bar{Q}}$$

The quantiles with exceedance probability values of 0.5, 0.1, 0.05, and 0.01 were computed as

$$Q_{50} = 5.8394 \times 10^{4}\,\text{cfs} \qquad Q_{90} = 1.1938 \times 10^{5}\,\text{cfs} \qquad Q_{95} = 1.3922 \times 10^{5}\,\text{cfs} \qquad Q_{99} = 2.1535 \times 10^{5}\,\text{cfs}$$

Three-Parameter Lognormal Distribution

Applying $E[\ln(Q-c)]$ and $E\{[\ln(Q-c)]^2\}$ as constraints, the entropy-based three-parameter lognormal distribution is written as

$$f(Q) = \exp\{-\lambda_0 - \lambda_1 \ln(Q-c) - \lambda_2 [\ln(Q-c)]^2\} \qquad c \le \min(Q) \qquad \lambda_2 > 0 \qquad (16.76)$$

Then we can express λ_0 as a function of $(\lambda_1, \lambda_2, c)$ as

$$\exp \lambda_0 = \int_c^\infty \exp\{-\lambda_1 \ln(Q-c) - \lambda_2 [\ln(Q-c)]^2\} dx \qquad (16.77a)$$

After applying some algebra, we have

$$\exp\lambda_0 = \frac{\sqrt{\pi}}{\sqrt{\lambda_2}}\exp\frac{(\lambda_1-1)^2}{4\lambda_2} \Rightarrow \lambda_0 = \frac{1}{2}\ln\pi - \frac{1}{2}\ln\lambda_2 + \frac{(\lambda_1-1)^2}{4\lambda_2} \tag{16.77b}$$

Taking the partial derivatives of λ_0 with respect to λ_1, λ_2, and c, we have

$$\frac{\partial\lambda_0}{\partial\lambda_1} = \frac{\lambda_1-1}{2\lambda_2} = -E[\ln(Q-c)]$$

$$\frac{\partial\lambda_0}{\partial\lambda_2} = -\frac{1}{2\lambda_2} - \frac{(\lambda_1-1)^2}{4\lambda_2^2} = E\{[\ln(Q-c)]^2\} \tag{16.77c}$$

$$\frac{\partial\lambda_0}{\partial c} = 0$$

The objective function to optimize the parameters is

$$\Gamma(\lambda_1,\lambda_2,c) = \frac{1}{2}\ln\pi - \frac{1}{2}\ln\lambda_2 + \frac{(\lambda_1-1)^2}{4\lambda_2} + \lambda_1\overline{\ln(Q-c)} + \lambda_2\overline{[\ln(Q-c)]^2} \tag{16.78}$$

Finally, by minimizing Eq. (16.78) with the constraints listed as Eq. (16.77c), we have

$$\lambda_0 = 238.7533 \qquad \lambda_1 = -41.9330 \qquad \lambda_2 = 1.9320 \qquad c = -1.3385\times10^4$$

We have the three-parameter lognormal probability density function

$$f(Q) = \exp\{-238.7533 + 41.9330\ln(Q+1.3385\times10^4) - 1.9320\,[\ln(Q+1.3385\times10^4)]^2\}$$

The quantiles with the exceedance probability of 0.5, 0.1, 0.05, and 0.01 computed using the bisection method are

$$Q_{50} = 5.3506\times10^4\,\text{cfs} \qquad Q_{90} = 1.15\times10^5\,\text{cfs} \qquad Q_{95} = 1.4106\times10^5\,\text{cfs} \qquad Q_{99} = 2.0506\times10^5\,\text{cfs}$$

Figure 16.10 compares the gamma, three-parameter lognormal, generalized, truncated normal, and empirical distributions. Figure 16.10 shows the gamma and three-parameter lognormal distributions fit the data better than the generalized and truncated normal distributions.

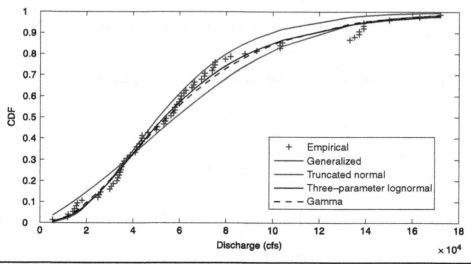

FIGURE 16.10 Comparison of three-parameter lognormal, gamma, generalized, truncated normal, and empirical distributions.

16.4   Bivariate Rainfall-Runoff Analysis

The discussion of univariate distributions can be extended to modeling a rainfall-runoff relation. Let rainfall be represented by R and runoff by Q. It is assumed that both R and Q are random variables. The objective is to determine the probability distribution of runoff Q given rainfall R. Runoff could be the amount generated by a rainfall event, or monthly runoff generated by monthly rainfall. One would determine the conditional probability distribution of runoff given rainfall, $f(q \mid r)$, where q is a specific value of Q and r is a specific value of R. This can be accomplished by maximizing the Shannon entropy. Note that the conditional Shannon entropy $H(Q \mid R)$ can be expressed as

$$H(Q \mid R) = -\iint f(r, q) \ln f(q \mid r) \, dq \, dr \qquad (16.79)$$

where $f(r, q)$ is the joint probability distribution of Q and R. Likewise, the joint entropy $H(R, Q)$ can be written as

$$H(R, Q) = -\iint f(r, q) \ln f(r, q) \, dq \, dr \qquad (16.80)$$

Note that

$$f(q \mid r) = \frac{f(r, q)}{g(r)} \qquad (16.81)$$

where $g(r)$ is the marginal probability distribution of R. With the introduction of Eq. (16.81) in Eq. (16.79), one gets

$$H(Q \mid R) = -\iint f(r, q) \ln \frac{f(r, q)}{g(r)} \, dq \, dr = -\iint f(r, q) \ln f(r, q) \, dr \, dq + \iint f(r, q) \ln g(r) \, dq \, dr$$
$$(16.82)$$

Since, $\iint f(r, q) \, dq = g(r)$, Eq. (16.82) can be written as

$$H(Q \mid R) = -\iint f(r, q) \ln f(r, q) \, dq \, dr + \int g(r) \ln g(r) \, dr$$

$$= -H(Q, R) + H(R) \qquad (16.83)$$

For maximizing $H(R, Q)$ the following constraints are specified:

$$\int_0^\infty \int_0^\infty f(r, q) \, dq \, dr = 1 \qquad (16.84a)$$

$$\int_0^\infty \int_0^\infty r^i f(r, q) \, dr \, dq = \overline{r^i} \qquad i = 1, \ldots, m \qquad (16.84b)$$

$$\int_0^\infty \int_0^\infty q^j f(r, q) \, dr \, dq = \overline{q^j} \qquad j = 1, \ldots, n \qquad (16.84c)$$

$$\int_0^\infty \int_0^\infty qr f(r, q) \, dq \, dr = \overline{qr} = S_{rq} + \overline{qr} \qquad (16.84d)$$

For $H(R)$, the constraints can be specified as

$$\int_0^\infty g(r)\, dr = 1 \tag{16.85a}$$

$$\int_0^\infty r^i g(r)\, dr = \overline{r^i} \qquad i = 1, \ldots, m \tag{16.85b}$$

Maximizing the joint entropy given by Eq. (16.80) using the method of Lagrange multipliers, one gets

$$f(r, q) = \exp\left(-\lambda_0 - \sum_{i=1}^m \lambda_i r^i - \sum_{j=1}^n \lambda_{j+m} q^j - \lambda_{m+n+1} rq\right) \tag{16.86}$$

where $\lambda = [\lambda_0, \lambda_1, \ldots, \lambda_{m+n+1}]$, are Lagrange multipliers which are determined using Eqs. (16.84a) to (16.84d). Substituting Eq. (16.86) in Eq. (16.84a), one gets

$$\exp \lambda_0 = \int_0^\infty \int_0^\infty \exp\left(-\sum_{i=1}^m \lambda_i r^i - \sum_{j=1}^n \lambda_{j+m} q^j - \lambda_{m+n+1} rq\right) dr\, dq \tag{16.87}$$

Substituting Eq. (16.86) in Eq. (16.84b), one gets

$$\int_0^\infty \int_0^\infty r^i f(r, q)\, dr\, dq = \overline{r^i} = \exp(-\lambda_0) \int_0^\infty \int_0^\infty r^i \exp\left(-\sum_{i=1}^m \lambda_i r^i - \sum_{j=1}^n \lambda_{j+m} q^j - \lambda_{m+n+1} rq\right) dr\, dq \tag{16.88}$$

Substituting Eq. (16.86) in Eq. (16.84c), one gets

$$\int_0^\infty \int_0^\infty q^j f(r, q)\, dr\, dq = \overline{q^j} = \exp(-\lambda_0) \int_0^\infty \int_0^\infty q^j \exp\left(-\sum_{i=1}^m \lambda_i r^i - \sum_{j=1}^n \lambda_{j+m} q^j - \lambda_{m+n+1} rq\right) dr\, dq \tag{16.89}$$

Substituting Eq. (16.86) in Eq. (16.84d), one gets

$$\overline{rq} = \exp(-\lambda_0) \int_0^\infty \int_0^\infty rq \exp\left(-\sum_{i=1}^m \lambda_i r^i - \sum_{j=1}^n \lambda_{j+m} q^j - \lambda_{m+n+1} rq\right) dr\, dq \tag{16.90}$$

Integrating Eq. (16.86) over q, one gets $g(r)$, the probability density function for r:

$$g(r) = \int_0^\infty \exp\left(-\lambda_0 - \sum_{i=1}^m \lambda_i r^i - \sum_{j=1}^n \lambda_{j+m} q^j - \lambda_{m+n+1} rq\right) dq \tag{16.91}$$

Similarly, integrating Eq. (16.86) over r, one gets $g(q)$, the probability density function for q:

$$g(q) = \int_0^\infty \exp\left(-\lambda_0 - \sum_{i=1}^m \lambda_i r^i - \sum_{j=1}^n \lambda_{j+m} q^j - \lambda_{m+n+1} rq\right) dr \tag{16.92}$$

Then the conditional distribution $f(q\,|\,r)$ can be written, using Eqs. (16.86) and (16.92),

$$f(q\,|\,r) = \frac{f(r,q)}{g(r)} = \frac{\exp\left(-\lambda_0 - \sum\limits_{i=1}^{m}\lambda_i r^i - \sum\limits_{j=1}^{n}\lambda_{j+m}q^j - \lambda_{m+n+1}rq\right)}{\int\limits_0^\infty \exp\left(-\lambda_0 - \sum\limits_{i=1}^{m}\lambda_i r^i - \sum\limits_{j=1}^{n}\lambda_{j+m}q^j - \lambda_{m+n+1}rq\right)dq} \tag{16.93}$$

Equation (16.93) denotes the conditional probability distribution, and its parameters need to be determined with the first m noncentral moments of rainfall, the first n noncentral moments of runoff, and the first moment of rainfall × runoff.

Consider a simple case, using the first and second noncentral moments as constraints for rainfall ($m = 2$) and first moment as the constraint for runoff ($n = 1$); Eqs. (16.84) and (16.92) can be written as

$$f(r,q) = \exp(-\lambda_0 - \lambda_1 r - \lambda_2 r^2 - \lambda_3 q - \lambda_4 rq) \tag{16.94a}$$

$$g(r) = \int_0^\infty \exp(-\lambda_0 - \lambda_1 r - \lambda_2 r^2 - \lambda_3 q - \lambda_4 rq)dq = \frac{\exp(-\lambda_0 - \lambda_1 r - \lambda_2 r^2)}{\lambda_3 + \lambda_4 r} \tag{16.94b}$$

$$g(q) = \frac{\sqrt{\pi}}{2\sqrt{\lambda_2}}\exp\left(-\lambda_0 - \lambda_3 q + \frac{(\lambda_1 + \lambda_4 q)^2}{4\lambda_2}\right)\operatorname{erfc}\left(\frac{\lambda_1 + \lambda_4 q}{2\sqrt{\lambda_2}}\right) \tag{16.94c}$$

Equation (16.84b) can be rewritten as the following two equations:

$$\bar{r} = \exp(-\lambda_0)\int_0^\infty \frac{r\exp(-\lambda_1 r - \lambda_2 r^2)}{\lambda_3 + \lambda_4 r}\,dr$$

$$= \frac{\exp(-\lambda_0)}{\lambda_4}\int_0^\infty \exp(-\lambda_1 r - \lambda_2 r^2)\,dr - \frac{\lambda_3 \exp(-\lambda_0)}{\lambda_4}\int_0^\infty \frac{\exp(-\lambda_1 r - \lambda_2 r^2)}{\lambda_3 + \lambda_4 r}\,dr$$

$$= \frac{\exp(-\lambda_0)}{\lambda_4}\int_0^\infty \exp(-\lambda_1 r - \lambda_2 r^2)\,dr - \frac{\lambda_3}{\lambda_4} \tag{16.95a}$$

$$\overline{r^2} = \exp(-\lambda_0)\int_0^\infty \frac{r^2\exp(-\lambda_1 r - \lambda_2 r^2)}{\lambda_3 + \lambda_4 r}\,dr$$

$$= \frac{\exp(-\lambda_0)}{\lambda_4}\int_0^\infty r\exp(-\lambda_1 r - \lambda_2 r^2)\,dr - \frac{\lambda_3 \exp(-\lambda_0)}{\lambda_4}\int_0^\infty \frac{r\exp(-\lambda_1 r - \lambda_2 r^2)}{\lambda_3 + \lambda_4 r}\,dr \tag{16.95b}$$

The last integral on the right side leads to $\bar{r}$, as given by Eq. (16.95a). Therefore, Eq. (16.95b) can be rewritten as

$$\overline{r^2} - \bar{r}\frac{\lambda_3}{\lambda_4} = \frac{\exp(-\lambda_0)}{\lambda_4}\int_0^\infty r\exp(-\lambda_1 r - \lambda_2 r^2)dr \tag{16.95c}$$

For $n = 1$, Eq. (16.84c) can be written as

$$\bar{q} = \int_0^\infty\int_0^\infty q\exp(-\lambda_0 - \lambda_1 r - \lambda_2 r^2 - \lambda_3 q - \lambda_4 rq)\,dr\,dq \tag{16.96}$$

Note that

$$\int_0^\infty q \exp[-(\lambda_3 + \lambda_4 r)q]\,dq = \frac{1}{(\lambda_3 + \lambda_4 r)^2} \tag{16.97}$$

Therefore, Eq. (16.96) can be written as

$$\bar{q} = \exp(-\lambda_0)\int_0^\infty \frac{\exp(-\lambda_1 - \lambda_2 r^2)}{(\lambda_3 + \lambda_4 r^2)}\,dr \tag{16.98}$$

Similarly,

$$\overline{rq} = \exp(-\lambda_0)\int_0^\infty r\,\frac{\exp(-\lambda_1 - \lambda_2 r^2)}{(\lambda_3 + \lambda_4 r)^2}\,dr \tag{16.99}$$

The conditional distribution, i.e., Eq. (16.93), can be rewritten as

$$f(q\,|\,r) = \frac{f(r,q)}{g(r)} = \frac{\exp(-\lambda_0 - \lambda_1 r - \lambda_2 r^2 - \lambda_3 q - \lambda_4 rq)}{\exp(-\lambda_0 - \lambda_1 r - \lambda_2 r^2)} \times (\lambda_3 + \lambda_4 r)$$

$$= (\lambda_3 + \lambda_4 r)\exp(-\lambda_3 q - \lambda_4 rq) \tag{16.100}$$

Equation (16.100) is the needed conditional probability distribution, and its parameters need to be determined from the specified constraints: the first and second noncentral moments of rainfall, the first moment of runoff, and the first moment of rainfall × runoff.

Example 16.10.   Using the individual rainfall-runoff events at the Waco experimental watershed (Y13) (Table 16.8), compute the marginal and joint distributions of rainfall and runoff. Then compute the conditional probability distribution of runoff for given rainfall. Also compute the conditional return periods.

Solution.   In this example, we will apply the simple case in which the first and second noncentral moments of rainfall, the first moment of runoff, and first moment of rainfall × runoff as the constraints to determine the joint distribution of rainfall and runoff [i.e., Eq. (16.94a)] with the constraints, Eqs. (16.95) to (16.99). Similar to Example 16.2, the Lagrange multipliers can be estimated by minimizing the following objective function:

$$\Gamma(\lambda_1,\lambda_2,\lambda_3,\lambda_4) = \ln\int_0^\infty\int_0^\infty \exp(-\lambda_1 r - \lambda_2 r^2 - \lambda_3 q - \lambda_4 rq)\,dr\,dq + \sum_{i=1}^{2}\lambda_i\,\overline{r^i} + \lambda_3\bar{q} + \lambda_4\,\overline{rq} \tag{16.101}$$

To ensure the integration is tractable, we choose the upper integration limit as 153.2 (that is, 20 times the maximum rainfall depth from the data set). From the sample rainfall and runoff, it is safe to assume the joint density function reaches 0 when r and q are greater than the upper limit. Applying the Newton-Raphson method as in Example 16.2, we estimate the Lagrange multipliers as

$$\lambda = [\lambda_0, \lambda_1, \lambda_2, \lambda_3, \lambda_4] = [3.6983, -1.4019, 0.2412, 0.8712, -0.0623]$$

Using the estimated Lagrange multipliers, the absolute values of gradient are computed as

$$\left[\left|\frac{\partial\Gamma}{\partial\lambda_1}\right|, \left|\frac{\partial\Gamma}{\partial\lambda_2}\right|, \left|\frac{\partial\Gamma}{\partial\lambda_3}\right|, \left|\frac{\partial\Gamma}{\partial\lambda_4}\right|\right] = \left[|\bar{r} - E(r)|, |\overline{r^2} - E(r^2)|, |\bar{q} - E(q)|, |\overline{rq} - E(rq)|\right]$$

$$= [7.76\times10^{-10}, 3.90\times10^{-9}, 7.83\times10^{-10}, 3.49\times10^{-7}]$$

Year	Rain Depth (in.)	Runoff (in.)	Year	Rain Depth (in.)	Runoff (in.)
1969	2.87	0.51	1989	3.52	2.20
1970	2.13	0.80	1990	3.43	2.20
1971	2.89	0.36	1991	2.55	1.62
1972	1.11	0.39	1992	3.25	0.52
1973	3.15	2.60	1993	1.55	0.14
1974	5.08	2.38	1994	3.14	0.25
1975	2.95	0.70	1995	2.4	1.03
1976	3.21	1.74	1995	2.49	0.75
1977	3.37	1.78	1996	2.24	0.30
1978	4.96	3.12	1997	2.15	1.25
1979	3.45	2.77	1997	3.87	3.56
1980	1.6	0.25	1998	2.32	1.91
1980	1.49	0.60	1998	4.48	0.46
1981	3.37	1.68	1999	1.44	0.33
1982	1.46	0.65	2000	4.04	2.74
1982	1.8	0.12	2001	4.05	2.76
1983	3.89	0.01	2002	3.67	2.46
1983	1.83	0.69	2003	3.18	1.43
1984	2.49	0.22	2004	4.47	3.07
1984	1.38	0.54	2005	7.65	3.53
1985	4.96	4.70	2006	2.64	0.92
1986	7.66	5.82	2007	5.88	4.43
1987	2.6	2.31	2008	2.57	0.95
1987	2.86	1.75	2009	5.19	1.79
1988	1.23	0.17	2010	2.67	1.67

TABLE 16.8   Rainfall-Runoff Events at Waco Experimental Watershed (Y13)

These absolute values of gradient indicate that the constraints are well preserved, and we can use the estimated Lagrange multiplier to build the marginal, joint, and conditional distributions. The joint distribution for rainfall and runoff is written as

$$f(r,q) = \exp(-3.6983 + 1.4019\,r - 0.2412\,r^2 - 0.8712\,q + 0.0623\,rq)$$

Substituting the estimated Lagrange multipliers into Eqs. (16.94b) and (16.94c), we have the marginal distribution for rainfall and runoff as

$$g(r) = \frac{\exp(-3.6983 + 1.4019\,r - 0.2412\,r^2)}{0.8712 - 0.0623\,r}$$

$$g(q) = \frac{\sqrt{\pi}}{2\sqrt{0.2412}} \exp\frac{(-1.4019 - 0.0623\,q)^2}{4(0.2412)} \operatorname{erfc}\frac{-1.4019 - 0.0623\,q}{2\sqrt{0.2412}}$$

$$\operatorname{erfc}(x) = 1 - \operatorname{erf}(x)$$

The conditional density function $f(q\,|\,r)$ is written as

$$f(q\,|\,r) = (0.8721 - 0.0623\,r)\exp(-0.8723\,q + 0.0623\,rq)$$

The marginal distribution for rainfall and runoff is plotted in Fig. 16.11 which shows a good-fit. Figure 16.12 plots the conditional distribution for different rainfall amounts.

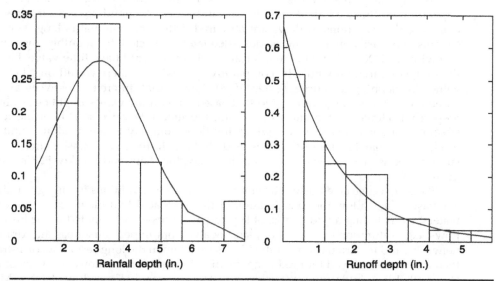

FIGURE 16.11   Marginal probability distributions for rainfall and runoff.

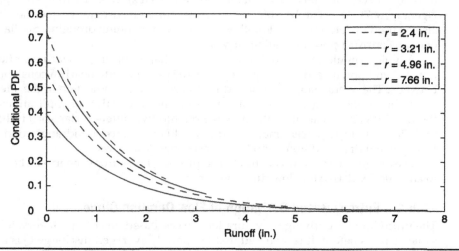

FIGURE 16.12   Conditional distribution of runoff for given rainfall depth.

16.5   Flow Duration Curve

A flow duration curve (FDC) at a gauging station is a plot of stream flow from high to low versus the percentage of time these values are either equaled or exceeded. The plot is constructed over a specified period of time scaled between 0 and 100 percent and considers the full range of flows, combining flow characteristics of a stream. An FDC is essentially a cumulative frequency curve of measured flow data over the specified time. It does not depict the chronological sequence of flows.

Usually, daily average discharge values are used for constructing an FDC, but depending on the need, the time interval can be a week, month, or season. However, details of the variations in flows are obscured if the time interval is long. For most streams the variation in flow is not adequately reflected by monthly discharges. Likewise, an FDC based on annual mean discharges would have little value, because the range of variation would be comparatively small and there would only be a few values for defining the curve. Foster (1934) showed that differences between an FDC based on daily discharge values and that based on monthly values might be as high as 35 percent. However, the effect of varying the time interval is not the same for all streams. For some large streams where the flow from day to day is almost uniform, weekly FDC may be almost the same as daily FDC. In the case of flashy streams with sudden floods lasting for a few hours in a day, the daily and weekly FDCs will be greatly different.

FDCs are constructed using the entire range of flow conditions for any given drainage basin. They allow for consideration of the integrated hydrologic condition of the watershed including a wide range of key runoff processes represented in the historical flows, e.g., saturation excess runoff, stormwater from impervious areas and drainage networks, and tile drain runoff. If the FDC is based upon the long-term flow of a stream, then it can be employed for predicting the distribution of future flows for water supply (Mitchell, 1957), hydropower (Hickox and Wessenauer, 1933), sediment load (Miller, 1951), and pollution (Searcy, 1959). An FDC can be used as a convenient way to compare one watershed with another. Flow duration curves constitute the foundation of load duration curves for total maximum daily loads (TMDLs) (U.S. Environmental protection Agency, 2007). They can be utilized to forecast future recurrence frequencies. They can also be used to determine the low-flow threshold for defining droughts. Similar to FDC, one can construct power duration curves.

A typical semilog FDC exhibits a sigmoidal shape, curving upward near flow duration of 0 and downward at a frequency near 100 percent, with nearly a constant slope in between. The overall slope of a flow duration curve is an indication of the stream flow variability at the gauge. For facilitating diagnostic and analytical applications of flow, the U.S. EPA (2007) classified the flow region into five different classes: 0 to 10 percent high flows, 10 to 40 percent moist conditions, 40 to 60 percent midrange conditions, 60 to 90 percent dry conditions, and 90 to 100 percent low flows.

The duration curves are constructed empirically, but entropy theory can be employed to analytically derive the flow duration curve.

16.5.1   Entropy-Based Derivation of Flow Duration Curve

The procedure for deriving flow duration curves, based on entropy theory, follows the same steps as before. It is assumed that temporally averaged discharge Q is a random variable with a probability density function denoted by $f(Q)$. The time interval for which the discharge is averaged can be any, but usually it is taken as 1 day(d). Each of these steps is now discussed.

Shannon Entropy

The Shannon entropy (Shannon, 1948; Shannon and Weaver, 1949) of discharge Q, or of $f(Q)$, denoted by $H(Q)$, can be expressed as

$$H = -\int_{Q_{min}}^{Q_{max}} f(Q)\ln f(Q)\, dQ \tag{16.102}$$

where Q_{min} and Q_{max} represent the minimum (or lower) and maximum (upper) limits of discharge for integration. Equation (16.102) expresses a measure of uncertainty about $f(Q)$ or the average information content of sampled Q. The objective here is to derive the least-biased $f(Q)$, which can be accomplished by maximizing $H(Q)$, subject to specified constraints, in accordance with the principle of maximum entropy (POME) (Jaynes, 1957, 1982). Maximizing $H(Q)$ is equivalent to maximizing $f(Q)\ln f(Q)$. To determine the $f(Q)$ that is least biased toward unknowns (in regard to discharge), the principle of maximum entropy developed by Jaynes (1957, 1982) is invoked. POME requires the specification of certain information, expressed in terms of constraints on discharge. According to POME, the most appropriate probability distribution is the one that has the maximum entropy or uncertainty, subject to these constraints.

Specification of Constraints

For deriving the flow duration curve, following Singh (1998) the constraints to be specified are the total probability law, which must always be satisfied by the probability density function of discharge, written as

$$C_1 = \int_{Q_{min}}^{Q_{max}} f(Q)\, dQ = 1 \tag{16.103}$$

and the mean discharge

$$C_2 = \int_{Q_{min}}^{Q_{max}} Q f(Q)\, dQ = \bar{Q} = Q_m = Q_{mean} \tag{16.104}$$

Equation (16.104) is the mean of the discharge values. Equation (16.103) is the first constraint C_1, defining the total probability law, and Eq. (16.104) is the second constraint C_2 defining the mean of discharge.

Maximization of Entropy

To obtain the least-biased probability density function of Q, $f(Q)$, the Shannon entropy, given by Eq. (16.102), is maximized following POME, subject to Eqs. (16.103) and (16.104). To that end, the method of Lagrange multipliers is employed. The Lagrangian function becomes

$$L = -\int_{Q_{min}}^{Q_{max}} f(Q)\ln = f(Q)\, dQ - (\lambda_0 - 1)\left(\int_{Q_{min}}^{Q_{max}} f(Q)\, dQ - C_1\right) - \lambda_1\left[\int_{Q_{min}}^{Q_{max}} Q f(Q)\, dQ - C_2\right] \tag{16.105}$$

where λ_0 and λ_1 are the Lagrange multipliers. Recalling the Euler-Lagrange equation from the calculus of variation and differentiating Eq. (16.105) with respect to f, noting that f is variable and Q is a parameter, and equating the derivative to zero, one obtains

$$\frac{\partial L}{\partial f} = 0 = -[\ln f(Q) + 1] - (\lambda_0 - 1) - \lambda_1 Q \qquad (16.106)$$

Derivation of Probability Distribution

Equation (16.106) leads to the entropy-based probability density function of discharge as

$$f(Q) = \exp(-\lambda_0 - \lambda_1 Q) \qquad (16.107)$$

The PDF of Q contains the Lagrange multipliers λ_0 and λ_1 which can be determined using Eqs. (16.103) and (16.104). The cumulative probability distribution function of Q can be obtained by integrating Eq. (16.107) as

$$F(Q) = \frac{\exp(-\lambda_0)}{\lambda_1}[\exp(-\lambda_1 Q_{min}) - \exp(\lambda_1 Q)] \qquad (16.108)$$

Note that when $Q_{min} = 0$, Eq. (16.108) reduces to

$$F(Q) = \frac{\exp(-\lambda_0)}{\lambda_1}[1 - \exp(\lambda_1 Q)] \qquad (16.109)$$

Maximum Entropy

Substitution of Eq. (16.107) in Eq. (16.102) yields the maximum entropy or uncertainty of discharge as:

$$H(Q) = \lambda_0 + \lambda_1 \bar{Q} \qquad (16.110)$$

Determination of Lagrange Multipliers

Substitution of Eq. (16.107) in Eq. (16.103) leads to

$$\lambda_0 = -\ln \lambda_1 + \ln[\exp(-\lambda_1 Q_{min}) - \exp(-\lambda_1 Q_{max})] \qquad (16.111)$$

Differentiating Eq. (16.111) with respect to λ_1 produces

$$\frac{\partial \lambda_0}{\partial \lambda_1} = -\frac{1}{\lambda_1} - \frac{Q_{min} \exp(-\lambda_1 Q_{min}) - Q_{max} \exp(-\lambda_1 Q_{max})}{\exp(-\lambda_1 Q_{min}) - \exp(-\lambda_1 Q_{max})} \qquad (16.112)$$

On the other hand, substitution of Eq. (16.107) in Eq. (16.103) can also be written as

$$\lambda_0 = \ln \int_{Q_{min}}^{Q_{max}} \exp(-\lambda_1 Q) \, dQ \qquad (16.113)$$

Differentiating Eq. (16.113) with respect to λ_1, one gets

$$\frac{\partial \lambda_0}{\partial \lambda_1} = -\frac{\displaystyle\int_{Q_{min}}^{Q_{max}} Q \exp(-\lambda_1 Q)\, dQ}{\displaystyle\int_{Q_{min}}^{Q_{max}} \exp(-\lambda_1 Q)\, dQ} \qquad (16.114)$$

Multiplying and dividing Eq. (16.114) by $\exp(-\lambda_0)$ and using Eq. (16.107), one finds

$$\frac{\partial \lambda_0}{\partial \lambda_1} = -\frac{\displaystyle\int_{Q_{min}}^{Q_{max}} Q \exp(-\lambda_0 - \lambda_1 Q)\, dQ}{\displaystyle\int_{Q_{min}}^{Q_{max}} \exp(-\lambda_0 - \lambda_1 Q)\, dQ} = -\bar{Q} = -Q_{mean} \qquad (16.115)$$

Equating Eq. (16.112) to Eq. (16.114) yields an expression for λ_1 obtained in terms of the constraint and the limits of integration of Q:

$$-\frac{1}{\lambda_1} - \frac{Q_{min} \exp(-\lambda_1 Q_{min}) - Q_{max} \exp(-\lambda_1 Q_{max})}{\exp(-\lambda_1 Q_{min}) - \exp(-\lambda_1 Q_{max})} = -\bar{Q} \qquad (16.116)$$

Equation (16.116) expresses a relation between the Lagrange multiplier λ_1 and Q_{max} and Q_{min} and can be simplified as follows:

Let $M = \lambda_1 Q_{max}$. Then $M(Q_{min}/Q_{max}) = \lambda_1 Q_{min}$. Dividing both sides of Eq. (16.116) by Q_{max} one gets

$$\frac{\bar{Q}}{Q_{max}} = \frac{1}{M} + \frac{Q_{min}/Q_{max} \exp(-M Q_{min}/Q_{max}) - \exp(-M)}{\exp(-M Q_{min}/Q_{max}) - \exp M} \qquad (16.117)$$

If $Q_{min} \cong 0$, then Eq. (16.117) becomes

$$\frac{\bar{Q}}{Q_{max}} = \frac{1}{M} - \frac{\exp(-M)}{1 - \exp M} \qquad (16.118)$$

Hypothesis on Cumulative Distribution of Flow

Since the flow duration curve is a relation between flow discharge and the percentage of time the flow is equaled or exceeded, it is assumed that all temporally averaged values of discharged Q measured at the gauging station between Q_{min} and Q_{max} are equally likely. In reality this is not highly unlikely because at different times different values of discharge do occur. This is also consistent with the laplacian principle of insufficient reason. Then the cumulative probability distribution of discharge can be expressed as 1 minus the percentage of time (or the ratio of time to the period of time under consideration, say, 365 days for daily discharge). The probability of discharge being equal to or less than a given value of Q, or the cumulative distribution function of discharge $F(Q) = P(\text{discharge} \le \text{a given value of } Q)$, $P = $ probability, can be expressed as

$$F(Q) = 1 - \frac{t}{T} = 1 - \tau \qquad \tau = \frac{t}{T} \qquad (16.119a)$$

where t is time (say, days) and T is the duration under consideration (say, 365 days). Note that on the left side the argument of function F in Eq. (16.119a) is variable Q, whereas the on the right side the variable is τ. The CDF of Q is not linear in terms of Q, unless Q and τ are linearly related. Also note that a similar hypothesis has been employed when using entropy theory for deriving infiltration equations by Singh (2010a, b), soil moisture profiles by Al-Hamdan and Cruise (2010) and Singh (2010c), and velocity distributions by Chiu (1987). Of course, it is plausible that $F(Q)$ might have a different form. A more general form can be expressed as

$$F(Q) = 1 - a\left(\frac{t}{T}\right)^b = 1 - a\tau^b \qquad \tau = \frac{t}{T} \tag{16.119b}$$

where a and b are constants.

Since Eq. (16.119b) constitutes the fundamental hypothesis employed here for deriving the flow duration curve using entropy, it will be useful to evaluate its validity. This hypothesis (i.e., the relation between the cumulative probability $F(Q)$ and the ratio t/T) should be tested for a large number of natural rivers. Also note that a similar hypothesis has been employed when using entropy theory for deriving hydrologic relations (Cui and Singh, 2012). It may also be emphasized that even if the above hypothesis is not strictly valid, it will not greatly influence the results, because it merely allows entropy theory to lead to the flow duration curves desired.

The probability density function is obtained by differentiating Eq. (16.119b) with respect to Q:

$$dF(Q) = f(Q)\,dQ = -ab\left(\frac{1}{T}\right)^b t^{b-1}\,dt \tag{16.120}$$

The term $f(Q)dQ = F(Q + dQ) - F(Q)$ denotes the probability of discharge being between Q and $Q + dQ$.

Substituting Eq. (16.107) in Eq. (16.120) and integrating from Q to Q_{max}, one obtains

$$\exp(-\lambda_1 Q) = \exp(-\lambda_1 Q_{max}) + \lambda_1 \exp\lambda_0\, a\left(\frac{t}{T}\right)^b \tag{16.121}$$

Replacing the term $\exp\lambda_0$ from Eq. (16.111), Eq. (16.121) leads to

$$Q = -\frac{1}{\lambda_1}\ln\left\{\exp(-\lambda_1 Q_{max}) - [\exp(-\lambda_1 Q_{max}) - \exp(-\lambda_1 Q_{min})]a\left(\frac{t}{T}\right)^b\right\} \tag{16.122}$$

Diving both sides by Q_{max} and using $M = \lambda_1 Q_{max}$, one can simplify Eq. (16.122) to

$$\frac{Q}{Q_{max}} = -\frac{1}{\lambda_1 Q_{max}}\ln\left\{\exp(-\lambda_1 Q_{max}) - [\exp(-\lambda_1 Q_{max}) - \exp(-\lambda_1 Q_{min})]a\left(\frac{t}{T}\right)^b\right\} \tag{16.123}$$

or

$$\frac{Q}{Q_{max}} = -\frac{1}{M}\ln\left\{\exp(-M) - \left[\exp(-M) - \exp\left(\frac{-MQ_{min}}{Q_{max}}\right)\right]a\left(\frac{t}{T}\right)^b\right\} \tag{16.124}$$

If $Q_{min} \cong 0$, Eq. (16.124) can be written as

$$\frac{Q}{Q_{max}} = -\frac{1}{M} \ln\left\{ \exp(-M) - [\exp(-M) - 1]a\left(\frac{t}{T}\right)^b \right\}$$ (16.125)

Equation (16.124) or (16.125) explicitly expresses the relation between discharge and time.

Example 16.11. Consider daily stream flow data from a gauging station 02131000 on the Pee Dee River, South Carolina, United States. The Pee Dee River is large and wild, having a length of about 232 mil (373 km). Most of the land bordering the river is floodplain forest, and the lower part of the river floodplain has been developed for rice. The climate of the Pee Dee River basin is typical for the southeastern United States, where the average temperature is about 69°F and the average annual precipitation is 47.04 in. (1195 mm). The gauging station is operated by the U.S. Geological Survey (USGS). The drainage area for this station is 8830 mi^2 (22,870 km^2). The flow record is from 1938 to 2011, and the number of days of record is 26,903, as shown in Table 16.9.

Estimate the values of parameters a, b, and M by fitting Eq. (16.124) to the observed FDC, using the least squares method. Plot the histogram of these parameters.

Solution. Parameters a, b, and M are estimated by fitting Eq. (16.124) to the observed FDC using the least squares method, and their histograms are plotted in Figs. 16.13 to 16.15.

Example 16.12. Assemble the minimum, maximum, and mean discharge values for each year for the gauging station. Estimate the values of parameter M.

Solution. The minimum, maximum, and mean discharge values for each year are assembled for gauging station 02131000, as shown in Table 16.10. Then parameter M is estimated, as shown in Table 16.10.

Example 16.13. Relate M to the ratio of mean discharge to the maximum discharge.

Solution. The frequency histogram of M is shown in Fig. 16.15. It is seen that the value of M most frequently has a narrow range between 2 and 4, and an average value of 3 might be a good approximation. Then M is related to the ratio of mean discharge to maximum discharge, as shown in Fig. 16.16. It seems that M has a one-to-one relation with the mean and peak discharge ratio; and if this relation is valid for other gauges, then this will have a lot of useful applications.

Example 16.14. Compute the entropy-based FDCs, using the parameters estimated as above, and compare with observed FDCs. Also construct the 95 percent confidence bands and comment.

Solution. The entropy-based FDCs were computed using the parameters estimated as above and were compared with observed FDCs, as shown in Fig. 16.17 for gauging station 02131000. Also, 95 percent confidence bands were constructed as shown for this gauging station in Fig. 16.18. Clearly, the two FDCs are in good agreement.
The entropy values reflect the relative goodness of fit. The entropy values are quite comparable, and agreement between observed and computed flow duration curves is also comparable. Since the differences between entropy values for different curves are small, it is difficult to judge the fit based on entropy values alone. The lower values reflect less uncertainty, meaning a better fit, and this was observed somewhat.

Example 16.15. Evaluate the sensitivity of Eq. (16.124) to changes in its parameters.

Solution. The flow duration curve has one dependent variable Q, one independent variable t, and three parameters M, a, and b. The sensitivity of the flow duration curve to these parameters is determined; i.e., one can write $\partial Q/\partial M$, $\partial Q/\partial a$, and $\partial Q/\partial b$. Each derivative of Q with respect to a given parameter

Year	Q_{max} (cfs)	Q_{min} (cfs)	Q_m (cfs)	Q_m/Q_{max}	M	Year	Q_{max} (cfs)	Q_{min} (cfs)	Q_m (cfs)	Q_m/Q_{max}	M
1938	65,500	1,660	10,967	0.1674	5.873	1976	36,200	1,280	12,363	0.3415	2.029
1939	34,600	1,000	5,610	0.1621	6.081	1977	69,600	2,610	14,504	0.2084	4.569
1940	17,100	1,540	6,099	0.3567	1.812	1978	99,900	1,670	14,207	0.1422	6.986
1941	33,700	1,040	7,565	0.2245	4.161	1979	42,500	2,450	15,533	0.3655	1.689
1942	34,400	1,360	10,637	0.3092	2.522	1980	18,800	1,200	5,279	0.2808	3.003
1943	50,700	1,040	10,945	0.2159	4.374	1981	33,800	900	11,548	0.3417	2.026
1945	44,300	1,620	11,960	0.2700	3.201	1982	45,800	1,600	15,277	0.3336	2.146
1946	30,100	1,520	8,226	0.2733	3.139	1983	63,500	1,220	17,288	0.2722	3.158
1947	68,200	1,560	14,179	0.2079	4.582	1984	29,500	1,530	6,836	0.2317	3.991
1948	47,800	2,760	13,542	0.2833	2.958	1985	30,700	1,090	7,684	0.2503	3.588
1949	25,600	2,380	8,408	0.3284	2.222	1986	95,200	968	13,590	0.1428	6.956
1950	21,100	958	6,282	0.2977	2.710	1987	20,100	1,270	7,260	0.3612	1.749
1951	61,500	880	10,245	0.1666	5.906	1988	46,500	1,620	11,467	0.2466	3.664
1952	39,700	1,300	9,920	0.2499	3.596	1989	49,900	1,070	16,475	0.3302	2.197
1953	59,500	720	8,441	0.1419	7.006	1990	51,200	921	17,593	0.3436	1.998
1954	31,600	826	7,305	0.2312	4.004	1991	35,700	1,460	8,733	0.2446	3.706
1955	21,500	1,260	6,263	0.2913	2.819	1992	46,700	1,860	18,297	0.3918	1.337
1956	30,000	1,700	8,128	0.2709	3.183	1993	32,000	1,680	10,692	0.3341	2.137
1957	51,700	2,800	15,636	0.3024	2.632	1994	37,000	1,300	10,220	0.2762	3.085

1958	28,800	2,210	8,632	0.2997	2.677	1995	29,100	1,380	12,198	0.4192	0.986
1959	69,300	3,450	20,090	0.2899	2.843	1996	32,800	1,660	11,642	0.3549	1.836
1960	45,500	2,800	10,597	0.2329	3.964	1997	43,100	1,180	15,363	0.3564	1.815
1961	38,600	1,640	11,822	0.3063	2.569	1998	22,400	743	6,155	0.2748	3.112
1962	53,400	1,830	11,258	0.2108	4.504	1999	21,600	1,100	7,548	0.3495	1.914
1963	47,000	1,160	11,153	0.2373	3.865	2000	17,800	767	4,033	0.2266	4.112
1964	61,500	3,340	16,132	0.2623	3.347	2001	15,600	664	3,232	0.2072	4.601
1965	56,300	1,770	9,482	0.1684	5.836	2002	97,500	1,200	16,646	0.1707	5.749
1966	25,900	1,780	5,577	0.2153	4.388	2003	37,400	1,260	7,120	0.1904	5.087
1967	36,700	1,000	8,873	0.2418	3.767	2004	31,400	854	9,975	0.3177	2.388
1968	32,200	1,230	10,027	0.3114	2.487	2005	19,200	977	6,721	0.3500	1.906
1969	31,300	1,830	7,847	0.2507	3.579	2006	36,700	653	10,087	0.2748	3.110
1970	47,300	1,970	12,525	0.2648	3.299	2007	19,200	1,030	4,678	0.2436	3.727
1971	46,000	2,080	14,567	0.3167	2.404	2008	24,900	1,390	8,007	0.3215	2.328
1972	72,200	2,460	17,832	0.2470	3.656	2009	52,100	1,350	12,212	0.2344	3.930
1973	35,100	2,190	11,801	0.3362	2.106	2010	17,200	1,160	5,658	0.3290	2.214
1974	83,500	2,280	16,677	0.1997	4.809	2011	21,000	1,630	6,513	0.3102	2.507
1975	22,800	1,490	9,016	0.3954	1.289						

TABLE 16.9 Yearly Flow Characteristics at Gauging Station 02131000

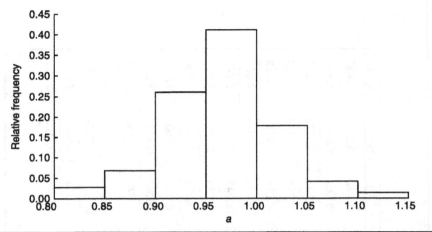

FIGURE 16.13   Histogram of values of *a* for gauging station 02131000.

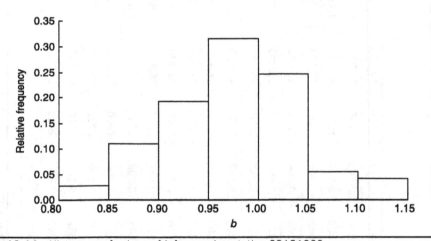

FIGURE 16.14   Histogram of values of *b* for gauging station 02131000.

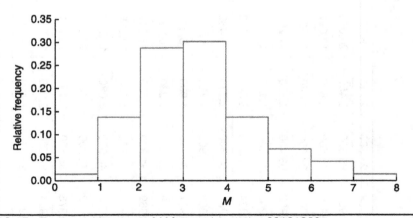

FIGURE 16.15   Frequency histogram of *M* for gauging station 02131000.

	Q_{max}	Q_{mean}	Q_{min}	Q_m/Q_{max}	M
Mean (cfs)	41,447	10,619	1,529	0.273	3.363
SD (cfs)	19,350	3,877	616	0.066	1.400
95% Confidence interval (CI) upper endpoint	45,961	11,517	1,672	0.288	3.690
95% CI lower endpoint	36,932	9,721	1,386	0.257	3.036

TABLE 16.10 Statistics of Flow Data for Gauging Station 02131000

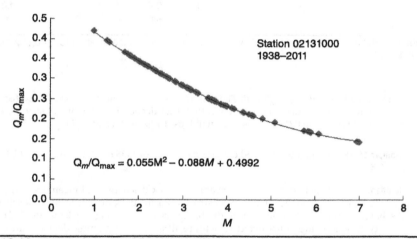

FIGURE 16.16 Relation between M and the ratio of mean discharge and peak discharge for gauging station 02131000.

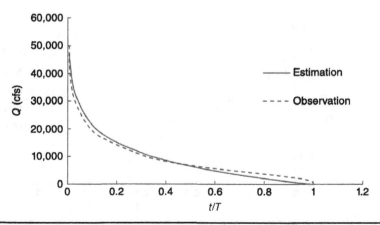

FIGURE 16.17 Entropy-based and observed FDCs for gauging station 02131000.

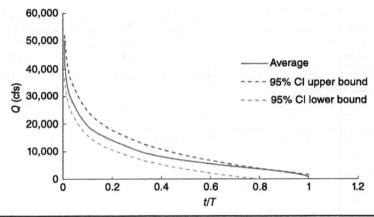

Figure 16.18   Entropy-based FDC with 95 percent confidence interval for gauging station 02131000.

can be called a *sensitivity coefficient*. The values of the coefficients help assess the effect of parameters on the flow duration curve and also reflect their significance. For economy of space, these coefficients are plotted for station 02131000, as shown in Figs. 16.19, 16.20, and 16.21.

Example 16.16.   Using the observed data for the period of 1938 to 2005, predict the FDC for 2011, 2009, and 2008.

Solution.   For prediction in future, the values of peak discharge and mean discharge are dealt with first. To that end, frequency plots are constructed for the peak discharge, mean discharge, minimum discharge, and ratio of mean to peak discharge, as shown in Figs. 16.22 to 16.25. Then probability density functions are constructed which are fitted by the gamma density function, as shown in Figs. 16.26 to 16.28. Since future values of peak, minimum, and mean discharge are subject to uncertainty, they can only be predicted for given probability values. Time series of annual peak, annual minimum, and mean annual discharge are plotted in Figs. 16.29 to 16.31. However, a simple autoregressive method cannot provide accurate prediction of the series due to low correlation of data.

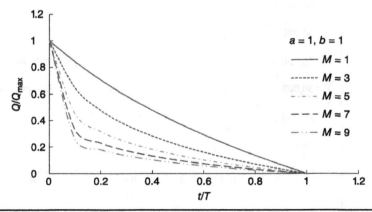

Figure 16.19   FDC for different *M* values for gauging station 02131000.

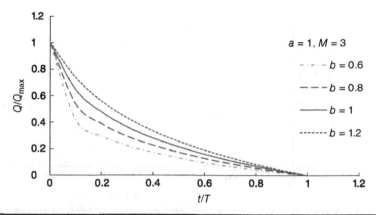

FIGURE **16.20** FDC for different *b* values for gauging station 02131000.

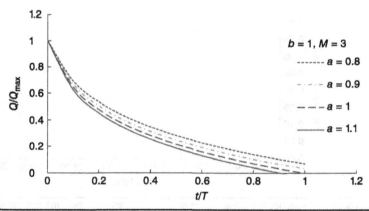

FIGURE **16.21** FDC for different *a* values for gauging station 02131000.

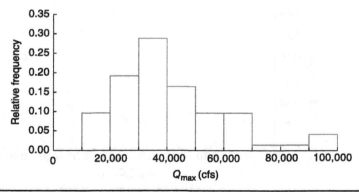

FIGURE **16.22** Frequency histogram of peak discharge for gauging station 02131000.

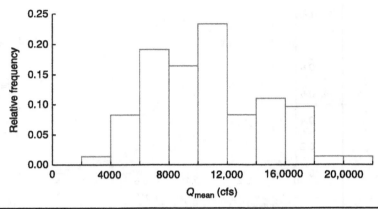

FIGURE 16.23   Frequency histogram of mean discharge for gauging station 02131000.

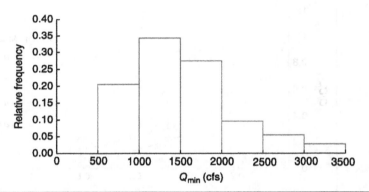

FIGURE 16.24   Frequency histogram of minimum discharge for gauging station 02131000.

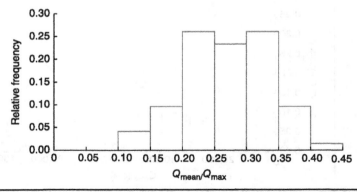

FIGURE 16.25   Frequency histogram of mean to peak discharge ratio for gauging station 02131000.

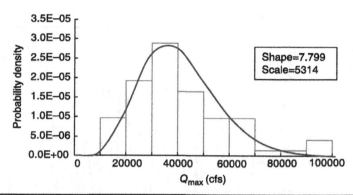

FIGURE 16.26   Observed and gamma-fitted probability density function of peak discharge ratio for gauging station 02131000.

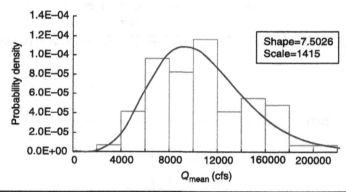

FIGURE 16.27   Observed and gamma-fitted probability density function of mean discharge for gauging station 02131000.

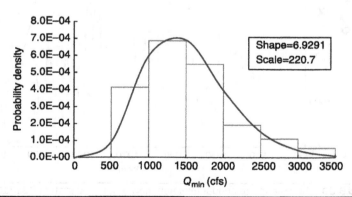

FIGURE 16.28   Observed and gamma-fitted probability density function of minimum discharge for gauging station 02131000.

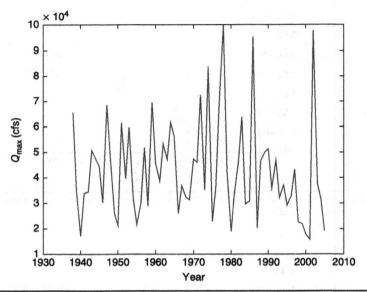

FIGURE **16.29** Time series of annual maximum flow at gauging station 02131000.

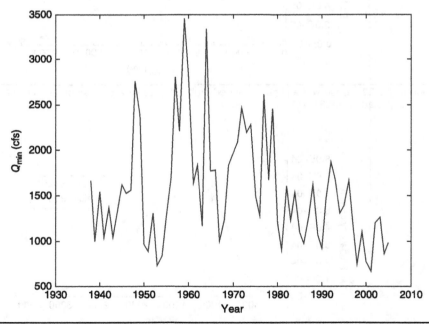

FIGURE **16.30** Time series of annual minimum flow at gauging station 02131000.

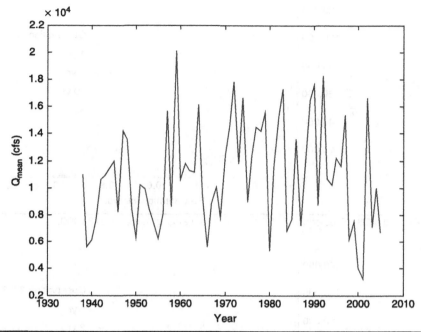

Figure 16.31 Time series of mean annual flow at gauging station 02131000.

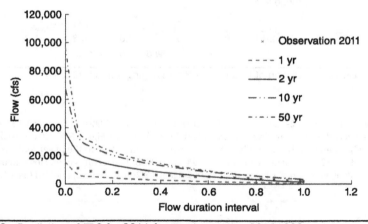

Figure 16.32 Prediction of FDC for 2011 for gauging station 02131000.

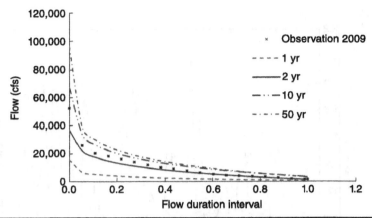

FIGURE 16.33 Prediction of FDC for 2009 for gauging station 02131000.

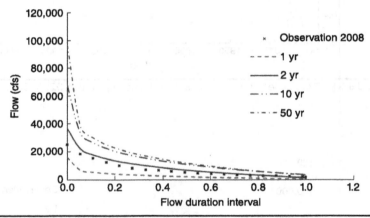

FIGURE 16.34 Prediction of FDC for 2008 for gauging station 02131000.

Therefore, for 1-, 2-, 10-, and 50-yr recurrence intervals, FDCs are constructed and compared with observed data for gauging station 02131000 as shown in Figs. 16.32 (2011), 16.33, (2009), and 16.34 (2008). Interestingly the predicted FDC for 2011 is very close to the observed 1-yr FDC, where as it is close to the 2-yr recurrence interval for 2009 and 2008. The result is even closer to the actual recurrence, since 2008 is roughly a 1.8-yr flow year, 2009 is a 2.2-yr flow year, and 2011 is a 1.3-yr flow year.

Questions

1. Obtain values of the annual maximum discharge (flood) for a gauging station on a river of your choice. Fit a two-parameter gamma distribution, two-parameter lognormal distribution, and three-parameter Pearson type III distribution to the flood data. Compute parameters using the maximum likelihood estimation method and entropy method. Then compute quantiles from the three distributions for various values of the probability of exceedance, and compare them with one

another as well as with the quantiles obtained directly from an appropriate plotting position formula.

2. Using the generalized extreme value distribution, do the same as in Question 1 and compare the quantiles.

3. Fit a two-parameter gamma distribution and three-parameter lognormal distribution to the same data set as in Question 1. Compute quantiles for various values of the probability of exceedance, and compare them with those obtained in Questions 1 and 2.

4. Obtain data for a set of individual rainfall-runoff events for a watershed. Compute the marginal and joint distributions of rainfall and runoff. Then compute the conditional probability distribution of runoff for given rainfall. Finally compute the conditional return periods.

5. Consider daily stream flow data for several years from a gauging station. Assemble the minimum, maximum, and mean discharge values for each year for the gauging station. Estimate the values of parameter M.

6. Construct the flow duration curve, using entropy for the gauging station in Questions 5, and determine how well it compares with the observed curve.

References

Al-Hamdan, O. Z., and Cruise, J. F. (2010). Soil moisture profile development from surface observations by principle of maximum entropy. *Journal of Hydrologic Engineering*, vol. 15, no. 5, pp. 327–337.

Chiu, C. L. (1987). Entropy and probability concepts in hydraulics. *Journal of Hydraulic Engineering*, vol. 113, no. 5, pp. 583–600.

Cui, H., and Singh, V. P. (2012). On the cumulative distribution function for entropy-based hydrologic modeling. *Transactions of the ASABE*, vol. 55, no. 2, pp. 429–438.

Eisenberger, I. (1964). Genesis of bimodal distributions. *Technometrics*, vol. 6, no. 4, pp. 357–363.

Foster, H. A. (1934). Duration curves. *Transactions of the ASCE*, vol. 99, pp. 1213–1267.

Hickox, G. H., and Wessenauer, G. O. (1933). Application of duration curves to hydroelectric studies. *Transactions of ASCE*, vol. 98, pp. 1276–1308.

Jaynes, E. T. (1957). Information theory and statistical mechanics, *I. Physical Review*, vol. 106, pp. 620–630.

Jaynes, E. T. (1982). On the rationale of maximum entropy methods. *Proceedings of the IEEE*, vol. 70, pp. 939–952.

Matz, A. (1978). Maximum likelihood parameter estimation for the quartic exponential distribution. *Technometrics*, vol. 20, no. 4, pp. 475–484.

Mead, L., and N. Papanicolaou (1984). Maximum entropy in the problem of moments. *Journal of Mathematical Physics*, vol. 25, no. 8, pp. 2404–2417.

Miller, C. R. (1951). Analysis of flow duration, sediment-rating curve method of computing sediment yield. U.S. Department of Interior, Bureau of Reclamation, Sedimentation Section, Hydrology Branch, Denver, Colo.

Mitchell, W. D. (1957). Flow-duration of Illinois streams. Illinois Department of Public Works and Buildings, Division of Waterways.

Papalexiou, S. M., and Koutsoyiannis, D. (2012). Entropy based derivation of probability distributions: A case study to daily rainfall. *Advances in Water Resources*, vol. 45, pp. 51–57, 2012.

Papalexiou, S. M., and Koutsoyiannis, D. (2013). Battle of extreme value distributions: A global survey on extreme daily rainfall. *Water Resources Research*, vol. 49, no. 1, pp. 187–201, doi:10.1029/2012WR012557, 2013.

Searcy, J. K. (1959). Flow-duration curves. USGS Water-Supply Paper 1542-A, U.S. Government Printing Office, Washington.

Shannon, C. E. (1948). A mathematical theory of communications, I and II. *Bell System Technical Journal*, vol. 27, pp. 379–443.

Shannon, C. E., and Weaver, W. (1949). *The Mathematical Theory of Communication*. University of Illinois Press, Urbana.

Singh, V. P. (1998). *Entropy-Based Parameter Estimation in Hydrology*. Kluwer Academic Publishers, Dirdrecht, The Netherlands.

Singh, V. P. (2010c). Entropy theory for movement of moisture in soils. *Water Resources Research*, vol. 46, pp. 1–12, W03516, doi:10-1029/2009WR008288.

Singh, V. P. and Rajagopal, A. K. (1986). A new method of parameter estimation for hydrologic frequency analysis. *Hydrological Science and Technology*, vol. 2, no. 3, pp. 33–40.

Smith, J. (1993). Moment methods for decision analysis. *Management Science*, vol. 39, no. 3, pp. 340–358.

Sonuga, J. O. (1972). Principle of maximum entropy in hydrologic frequency analysis. *Journal of Hydrology*, vol. 17, pp. 177–191.

U.S. Environmental Protection Agency (2007). An approach for using load duration curves in the development of TMDL. EPA-841-B-07-006, Watershed Branch, Office of Wetlands, Oceans and Watersheds, U.S. EPA, Washington.

Zellner, A., and R. Highfield (1988). Calculation of maximum entropy distributions and approximation of marginal posterior distributions. *Journal of Econometrics*, vol. 37, no. 2, pp. 195–209.

Further Reading

Arora, K., and Singh, V. P. (1987a). A comparative evaluation of the estimators of commonly used flood frequency models: 1. Monte Carlo simulation. Completion Report, Louisiana Water Resources Research Institute, Louisiana State University, Baton Rouge.

Arora, K., and Singh, V. P. (1987b). A comparative evaluation of the estimators of commonly used flood frequency models: 2. Computer programs. Completion Report, Louisiana Water Resources Research Institute, Louisiana State University, Baton Rouge.

Arora, K., and Singh, V. P. (1987c). An evaluation of seven methods for estimating parameters of the EVl distribution. In: *Hydrologic Frequency Modeling*, edited by V. P. Singh, D. Reidel Publishing Company, Boston, pp. 383–394.

Arora, K., and Singh, V. P. (1987d). On statistical intercomparison of EV 1 estimators by Monte Carlo simulation. *Advances in Water Resources*, vol. 10, no. 2, pp. 87–107.

Arora, K., and Singh, V. P. (1989a). A comparative evaluation of the estimators of log-Pearson type (LP) 3 distribution. *Journal of Hydrology*, vol. 105, pp. 19–37.

Arora, K., and Singh, V. P. (1989b). A note on the mixed moment estimation for the log-Pearson type 3 distribution. *Journal of the Institution of Engineers, Civil Engineering Division*, vol. 69, pt. CI5, pp. 298–301.

Arora, K., and Singh, V. P. (1990). A comparative evaluation of the estimators of the log Pearson type (LP) 3 distribution—A reply. *Journal of Hydrology*, vol. 117, pp. 375–376.

Fiorentino, M., Singh, V. P. and Arora, K. (1987b). On the two-component extreme value distribution and its point and regional estimators. In: *Regional Flood Frequency Analysis*, edited by V. P. Singh D. Reidel Publishing Company, Boston, pp. 257–272.

Fiorentino, M., Arora, K. and Singh, V. P. (1987a). The two-component extreme value distribution for flood frequency analysis: Another look and derivation of a new estimation method. *Stochastic Hydrology and Hydraulics*, vol. 1, pp. 199–208.

Guo, H., and Singh, V. P. (1992a). A comparative evaluation of estimators of extreme-value type III distribution by Monte Carlo simulation. Technical Report WRR24, Water Resources Program, Department of Civil Engineering, Louisiana State University, Baton Rouge.

Guo, H., and Singh, V. P. (1992b). A comparative evaluation of estimators of Pareto distribution by Monte Carlo simulation. Technical Report WRR25, Water Resources Program, Department of Civil Engineering, Louisiana State University, Baton Rouge.

Guo, H., and Singh, V. P. (1992c). A comparative evaluation of estimators of log-logistic distribution by Monte Carlo simulation. Technical Report WRR26, Water Resources Program, Department of Civil Engineering, Louisiana State University, Baton Rouge.

Guo, H., and Singh, V. P. (1992d). A comparative evaluation of estimators of two-component extreme-value distribution by Monte Carlo simulation. Technical Report WRR27, Water Resources Program, Department of Civil Engineering, Louisiana State University, Baton Rouge.

Hao, Z., and Singh, V. P. (2009a). Entropy-based parameter estimation for extended Burr XII distribution. *Stochastic Environmental Research and Risk Analysis*, vol. 23, pp. 1113–1122.

Jain, D., and Singh, V. P. (1986). Estimating parameters of EV1 distribution for flood frequency analysis. Water Resources Bulletin, vol. 23, no. 1, pp. 59–72.

Jain D., and Singh, V. P. (1987). Comparison of some flood distributions using empirical data. In: *Hydrologic Frequency Modeling*, edited by V. P. Singh D. Reidel Publishing Company, Boston, pp. 467–486.

Li, Y., Singh, V. P. and Cong, S. (1987). Entropy and its application in hydrology. In: *Hydrologic Frequency Modeling*, edited by V. P. Singh D. Reidel Publishing Company, Boston, pp. 367–382.

Singh, V. P. (1986). On the log-Gumbel (LG) distribution. *Hydrology*, vol. 8, no. 4, pp. 34–42.

Singh, V. P. (1987a). Reply to comments by H. N. Phien and V. T. V. Nguyen on derivation of the Pearson type III distribution by using the principle of maximum entropy (POME). *Journal of Hydrology*, vol. 90, pp. 355–357.

Singh, V. P. (1987b). On application of the Weibull distribution in hydrology. *Water Resources Management*, vol. 1, no. 1, pp. 33–43.

Singh, V. P. (1987c). On the extreme value (EV) type III distribution for low flows. *Hydrological Sciences Journal*, vol. 32, no. 4/12, pp. 521–533.

Singh, V. P. (1987d). Hydrologic modeling using entropy. The VII IHP Endowment Lecture, Centre of Water Resources, Anna University, Madras, India.

Singh, V. P. (1988b). *Hydrologic Systems*, vol. 1. *Rainfall-Runoff Modeling*. Prentice Hall, Englewood Cliffs, N. J.

Singh, V. P. (1989). Hydrologic modeling using entropy. *Journal of the Institution of Engineers, Civil Engineering Division*, vol. 70, Part CV2, pp. 55–60.

Singh, V. P. (1992). Entropy-based probability distributions for modeling of environmental and biological systems. In: *Structuring Biological Systems: A Computer Modeling Approach*, edited by S. S. Iyengar. CRC Press, Inc., Boca Raton, Fla., Chap. 6, pp. 167–208

Singh, V. P. (1996). Application of entropy in hydrology and water resources. Proceedings, *International Conference on From Flood to Drought*, IAHR-African Division, Sun City, South Africa, August 5–7.

Singh, V. P. (1997). The use of entropy in hydrology and water resources. *Hydrological Processes*, vol. 2, pp. 587–626.

Singh, V. P. (1997a). *Kinematic Wave Modeling in Water Resources: Environmental Hydrology*. Wiley, New York.

Singh, V. P. (1997b). The use of entropy in hydrology and water resources. *Hydrological Processes*, vol. 11, pp. 587–626.

Singh, V. P. (1997c). Effect of class interval size on entropy. *Stochastic Hydrology and Hydraulics*, vol. 11, pp. 423–431.

Singh, V. P. (1997b). The use of entropy in hydrology and water resources. *Hydrological Processes*, vol. 11, pp. 587–626.

Singh, V. P. (1997c). Effect of class interval size on entropy. *Stochastic Hydrology and Hydraulics*, vol. 11, pp. 423–431.

Singh, V. P. (1998a). *Entropy-Based Parameter Estimation in Hydrology*. Kluwer Academic Publishers, Boston.

Singh, V. P. (1998b). Entropy as a decision tool in environmental and water resources. Hydrology Journal, vol. 21, no. 1–4, pp. 1–12.

Singh, V. P. (2000a). The entropy theory as tool for modeling and decision making in environmental and water resources. *Water SA*, vol. 26, no. 1, pp. 1–11.

Singh, V. P. (2000b). Hierarchy of hydraulic geometry relations. *Proceedings, Eighth International Symposium on Stochastic Hydraulics*, edited by Z.-Y. Wang and S.-X. Hu. Beijing, China, pp. 51–62.

Singh, V. P. (2002a). Entropy theory in environmental and water resources modeling. In: *Advances in Civil Engineering: Water Resources and Environmental Engineering*, edited by J. N. Bandhopadhyay and D. Nagesh Kumar. Indian Institute of Technology, Kharagpur, India, pp. 1–11, January 3–6.

Singh, V. P. (2002b). Statistical analyses design. In: *Encyclopedia of Life Support Systems*, edited by A. Sydow. EOLSS Publishers Co., Ltd., Oxford, United Kingdom.

Singh, V. P. (2003). The entropy theory as a decision making tool in environmental and water resources. In *Entropy Measures, Maximum Entropy and Emerging Applications*, edited by Karmeshu. Springer-Verlag, Bonn, Germany, pp. 261–297.

Singh, V. P. (2003c). Toward a theory of bureaucracy for social development and its relation to thermodynamics. *New Global Development: International Journal of Comparative Social Welfare*, vol. 19, pp. 75–94.

Singh, V. P. (2005). Entropy theory for hydrologic modeling. In: *Water Encyclopedia: Oceanography; Meteorology; Physics and Chemistry; Water Law; and Water History, Art, and Culture*, edited by J. H. Lehr, Jack Keeley, Janet Lehr, and Thomas B. Kingery. Wiley, Hoboken, N.J., pp. 217–223.

Singh, V. P. (2010a). Entropy theory for derivation of infiltration equations. *Water Resources Research*, vol. 46, pp. 1–20, W03527, doi:10-1029/2009WR008193.

Singh, V. P. (2010b). Tsallis entropy theory for derivation of infiltration equations. *Transactions of the ASABE*, vol. 53, no. 2, pp. 447–463.

Singh, V. P. (2010c). Entropy theory for hydrologic modeling. *Beijing Normal University Journal of Research*, vol. 46, no. 3, pp. 229–240.

Singh, V. P., and Ahmad, M. (2004). A comparative evaluation of the estimators of the three-parameter generalized Pareto distribution. *Statistical Computation and Simulation*, vol. 74, no. 2, pp. 91–106.

Singh, V. P., and Chowdhury, P. K. (1985). On fitting gamma distribution to synthetic runoff hydrographs. *Nordic Hydrology*, vol. 16, pp. 177–192.

Singh, V. P., Cruise, J. F. and Ma, M. (1989). A comparative evaluation of the estimators of two distributions by Monte Carlo method. Technical Report WRR13, Water Resources Program, Department of Civil Engineering, Louisiana State University, Baton Rouge.

Singh, V. P., Cruise, J. F. and Ma, M. A. (1990a). A comparative evaluation of the estimators of the three-parameter lognormal distribution by Monte Carlo simulation. *Computational Statistics and Data Analysis*, vol. 10, pp. 71–85.

Singh, V. P., Cruise, J. F. and Ma, M. (1990b). A comparative evaluation of the estimators of the Weibull distribution by Monte Carlo simulation. *Journal of Statistical Computation and Simulation*, vol. 36, pp. 229–241.

Singh, V. P., and Deng, Z. Q. (2003). Entropy-based parameter estimation for kappa distribution. *Journal of Hydrologic Engineering*, ASCE, vol. 8, no. 2, pp. 81–92.

Singh, V. P., and Fiorentino, M., eds. (1992a). *Entropy and Energy Dissipation in Hydrology*. Kluwer Academic Publishers, Dordrecht, The Netherlands.

Singh, V. P., and Fiorentino, M. (1992b). A historical perspective of entropy applications in water resources. In: *Entropy and Energy Dissipation in Water Resources*, edited by V. P. Singh and M. Fiorentino. Kluwer Academic Publishers, Dordrecht, The Netherlands, pp. 21–61.

Singh, V. P., and Guo, H. (1995a). Parameter estimation for 2-parameter Pareto distribution by POME. *Water Resources Management*, vol. 9, pp. 81–93.

Singh, V. P., and Guo, H. (1995b). Parameter estimation for 3-parameter generalized Pareto distribution by POME. *Hydrological Science Journal*, vol. 40, no. 2, pp. 165–181.

Singh, V. P., and Guo, H. (1995c). Parameter estimation for 2-parameter log-logistic distribution (LLD) by POME. *Civil Engineering Systems*, vol. 12, pp. 343–357.

Singh, V. P., and Guo, H. (1997). Parameter estimation for 2 - parameter generalized Pareto distribution by POME. *Stochastic Hydrology and Hydraulics*, vol. 11, no. 3, pp. 211–228.

Singh, V. P., Guo, H. and Yu, F. X. (1993). Parameter estimation for 3-parameter log-logistic distribution. *Stochastic Hydrology and Hydraulics*, vol. 7, no. 3, pp. 163–178, 1993.

Singh, V. P., and Hao, L. (2009). Derivation of velocity distribution using entropy. 33rd IAHR Congress, August 9-14, 2009 Vancouver, Canada.

Singh, V. P., and Harmancioglu, N. B. (1997). Estimation of missing values with use of entropy. In: *Integrated Approach to Environmental Data Management Systems*, edited by N. B. Harmancioglu, N. Alpaslan, S. D. Ozkul, and V. P. Singh. Kluwer Academic Publishers, Dordrecht, The Netherlands, pp. 267–275.

Singh, V. P., Jain, S. K., and Tyagi, A. K. (2007). *Risk and Reliability Analysis in Civil and Environmental Engineering*. ASCE Press, Reston, Va.

Singh, V. P., Rajagopal, A. K., and Singh, K. (1986). Derivation of some frequency distributions using the principle of maximum entropy (POME). *Advances in Water Resources*, vol. 9, no. 2, pp. 91–106.

Singh, V. P., and Rajagopal, A. K. (1987a). Some recent advances in application of the principle of maximum entropy (POME) in hydrology. IAHS Publication no. 164, pp. 353–364.

Singh, V. P., and Rajagopal, A. K. (1987b). A new method of parameter estimation for hydrologic frequency analysis. *Hydrological Science and Technology*, vol. 2, no. 3, pp. 33–40.

Singh, V. P., and Singh, K. (1985a). Derivation of the gamma distribution by using the principle of maximum entropy (POME). *Water Resources Bulletin*, vol. 21, no. 6, pp. 941–952.

Singh, V. P., and Singh, K. (1985b). Derivation of the Pearson type (PT) III distribution by using the principle of maximum entropy. *Journal of Hydrology*, vol. 80, pp. 197–214.

Singh, V. P., and Singh, K. (1985c). Pearson type III Distribution and the principle of maximum entropy. *Proceedings of the 5th World Congress on Water Resources*, June 9–15, 1985, in Brussels, Belgium, vol. 3, pp. 1133–1146.

Singh, V. P., and Singh, K. (1987). Parameter estimation for TPLN distribution for flood frequency analysis. *Water Resources Bulletin*, vol. 23, no. 6, pp. 1185–1192.

Singh, V. P., and Singh, K. (1988). Parameter estimation for log-Pearson type III distribution by POME. *Journal of Hydraulic Engineering*, ASCE, vol. 114, no. 1, pp. 112–122.

Singh, K. and Singh, V. P. (1991). Derivation of bivariate probability density functions with exponential marginals. *Stochastic Hydrology and Hydraulics*, vol. 5, pp. 55–68.

Singh, V. P., Singh, K., and Rajagopal, A. K. (1985). Application of the principle of maximum entropy (POME) to hydrologic frequency analysis. Completion Report 06, Louisiana Water Resources Research Institute, Louisiana State University, Baton Rouge.

CHAPTER 17

Stream Flow Forecasting

Forecasting of stream flow is important for flood warning and flood protection, filling in and extension of hydrologic records, design of water resources projects, energy generation, navigation, irrigation, water supply, and sanitation and water pollution control. Depending on the purpose of the forecast, a forecast variable may be the surface water level, flood discharge, flood volume, time of flood occurrence, or duration of flood. This chapter discusses stream flow forecasting based on maximum entropy spectral analysis (MESA).

17.1 Preliminaries for Stream Flow Forecasting

For a forecast to be meaningful, it should be accurate, reliable, and timely, and one should be able to produce such forecasts under diverse hydrologic and climatic conditions. The accuracy of the forecast can be evaluated using performance measures, discussed in Chap. 3, such as the mean square error (MSE), forecast residual variance, and autocorrelation of forecast residuals. These measures provide an indication of the forecast accuracy if the effect of noise in the data is removed. Overestimating a flood forecast, say, by 10 cm may save thousands of lives, while its underestimation may lead to considerable loss of life and property. If human lives are endangered, one should minimize error in the forecast of flood peak and time to flood peak.

Flood forecasts must be sufficiently reliable that they are not easily disregarded. For example, flood hazard can be reduced by pre-emptying storage reservoirs to catch the forthcoming flood. If the prereleases are themselves damaging and if the flood peak is overestimated, this will cause greater harm than benefit, and few would be willing to take that risk. The forecast reliability can be evaluated using the coefficient of variation of residual error and the ratio of absolute error to the mean.

For a forecast to be useful, it must be timely. The longer the forecast time is, the longer the available time is for planning and mitigation measures. The *forecast lead time* (FLT) is defined as the maximum possible time in the future for which a meaningful forecast can be obtained. The forecast timeliness is evaluated using the error in forecasting the time to peak. The sampling time interval (STI) can be taken as any value, such as 1 month (mo), and FLT is always taken as a multiple of STI and may vary from 1 mo to several years.

17.1.1 Types of Forecasting

Stream flow is considered as a univariate stationary stochastic process, denoted by $Y(t)$ as a function of time t. It is assumed that measurements are available as a discrete time series x_t, $t \in (0, T)$, where T is the length of record. Let $\rho(k)$ be the autocorrelation function (acf)

655

of $Y(t)$, where k is the lag. In time series analysis, a common practice is to calculate $\rho(k)$ up to $T/4$ for nonperiodic series (Chatfield, 1984) and up to $T/2$ for periodic series (Box and Jenkins, 1976; Salas et al., 1980). For longer records there is usually no need to compute $\rho(k)$ in domain $t > T/4$ or $t > T/2$. For shorter records, however, correlations at $T/4$ or $T/2$ may still be significant. In such cases, $\rho(k)$ can be extended into domain $t > T/4$ or $t > T/2$. Three types of stream flow forecasting can be distinguished, as shown in Fig. 17.1.

Forward Forecasting

The values $y_t \in Y(t)$, $t > T$, are predicted from the past values and the acf. This type is designated as forward forecasting. In forecasting y_{T+2}, use is made of y_{T+1}; in forecasting y_{T+3}, use is made of y_{T+1} and y_{T+2}; and so on until y_{T+L}, where L is the maximum FLT. For forecasting the next unknown value of time series y_{T+1}, all known values of y_t and the extended acf are used.

Backward Forecasting or Reconstruction

Stream flow $Y(t)$ is available for time interval (T_1, T_2), and it is desired to fill in the missing record before time T_1. For that purpose, use is made of stream flow values during the record (T_1, T_2) and the associated acf which is computed from the known stream flow values and extended to the period $(0, T_1)$. This type is designated as backward forecasting or reconstruction.

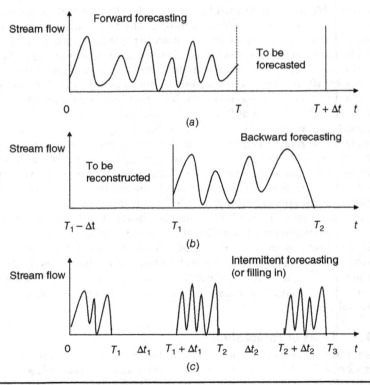

FIGURE 17.1 Types of stream flow forecasting.

Intermittent Forecasting or Filling of Records

Stream flow $X(t)$ is available on an intermittent basis, as shown in Fig. 17.1. Let Δt_1, Δt_2, and Δt_3 be the lengths of stream flow records 1, 2, and 3, respectively. The times of the first stream flow observation in each record are designated as T_1, T_2, and T_3. The times of last observation in each stream flow record are $T_1 + \Delta t_1$, $T_2 + \Delta t_1$, and $T_3 + \Delta t_1$, respectively. To fill in gap 1, use is made of stream flow record 1 (T_1, $T_1 + \Delta t_1$); to fill in gap 2, use is made of stream flow record 2 (0, $T_1 + \Delta t_1$); to fill in gap 3, use is made of record 3 (0, $T_2 + \Delta t_2$); and so on. For stream flow forecasting a complete reconstructed record (0, $T_3 + \Delta t_3$) is employed. Together with the stream flow record, the extended acf $\rho(k)$ for an appropriate number of lags is used. Thus, at first $\rho(k)$ is computed from record 1, then extended into domain ($\Delta t_1/2$, Δt_1) and into gaps 1 and 2. After filling in gaps 1 and 2, the acf is recomputed using record 2, and so on. After a certain point, $\rho(k)$ ceases to be significant. This is determined by the order of the model. This type is designated as intermittent forecasting or filling in of missing records.

17.1.2 Mean, Variance, and Normalization

Let there be a set of N stream flow observations y_t (y_1, y_2, ..., y_N), observed at time $t = 1$, ..., N. For a stationary stream flow process, the mean is constant and determined to be

$$\bar{y} = \frac{1}{N} \sum_{t=1}^{N} y_t \tag{17.1}$$

The variance is also constant and found to be

$$\sigma^2 = \frac{1}{N} \sum_{t=1}^{N} (y_t - \bar{y})^2 \quad \text{or} \quad \sigma^2 = \frac{1}{N-1} \sum_{t=1}^{N} (y_t - \bar{y})^2 \tag{17.2}$$

When calculations, are performed, data are often normalized as

$$y^* = \frac{y - \bar{y}}{\sigma} \tag{17.3}$$

so that the mean of the $y*$ values is 0 and the variance is 1. For simplicity, the normalized values $y*$ from now on will be denoted by y.

Example 17.1. Consider 4 yr of monthly stream flow (cfs) from station 09415000 at Colorado River, given in Table 17.1. Compute the mean and variance of the stream flow.

Solution. The mean and variance can be computed as follows:

$$\bar{y} = \frac{1}{N} \sum_{t=1}^{N} y_t = \frac{1}{48}(5279 + 6857 + 10{,}396 + \cdots + 6624 + 6428) = 9810.7 \text{cfs}$$

$$\sigma^2 = \frac{1}{N} \sum_{t=1}^{N} (y_t - \bar{y})^2 = \frac{1}{48}[(5279 - 9810.7)^2 + (6857 - 9810.7)^2 + \cdots + (66{,}428 - 9810.7)^2]$$

$$= 1.66 \times 10^7 \text{cfs}^2$$

The standard deviation $\sigma = \sqrt{1.66 \times 10^7} = 4079.5$ cfs. For normalized stream flow, $y^* = (y - 9810.7)/4079.5$. For example, the stream flow value of January 1966 after normalization becomes

$$y^* = \frac{5279 - 9810.7}{4079.5} = -1.11 \text{ cfs}$$

Month	Stream Flow (cfs) 1966	1967	1968	1969
1	5,279	5,188	5,654	5,151
2	6,857	6,739	7,253	6,691
3	10,396	10,217	11,478	10,147
4	11,961	11,754	18,786	11,706
5	15,361	15,078	16,172	15,434
6	14,885	14,116	15,608	13,971
7	12,841	12,400	13,332	12,182
8	14,616	13,895	14,986	13,519
9	5,827	5,507	5,947	5,338
10	5,432	4,978	5,409	5,013
11	6,792	6,317	7,146	6,624
12	6,725	6,899	6,878	6,428

TABLE 17.1 Monthly Stream Flow from Colorado River

17.1.3 Autocovariance Function and Autocorrelation Function

The autocovariance for a given lag n, denoted by R_n, can be defined as

$$R_n = \text{cov}(y_t, y_{t+n}) = \frac{1}{N-n}\sum_{t=1}^{N-n}(y_t - \bar{y})(y_{t+n} - \bar{y})$$

$$= \frac{1}{N-n}[(y_1 - \bar{y})(y_{1+n} - \bar{y}) + (y_2 - \bar{y})(y_{2+n} - \bar{y}) + \cdots + (y_{N-n} - \bar{y})(y_N - \bar{y})]$$

(17.4)

When $n = 0$, R_0 becomes

$$R_0 = \text{cov}(y_t, y_t) = \frac{1}{N}[(y_1 - \bar{y})(y_1 - \bar{y}) + (y_2 - \bar{y})(y_2 - \bar{y}) + \cdots + (y_N - \bar{y})(y_N - \bar{y})]$$

$$= \frac{1}{N}\sum_{t=1}^{N-n}(y_t - \bar{y})^2 = \sigma^2$$

(17.5)

The largest lag for the autocovariance is $N-1$, as the first and Nth values are the farthest apart. For $n = 1, 2, \ldots, N-1$, Eq. (17.4) can be expanded as

$$R_1 = \text{cov}(y_t, y_{t+1}) = \frac{1}{N-1}\sum_{t=1}^{N-1}(y_t - \bar{y})(y_{t+1} - \bar{y})$$

$$= \frac{1}{N-1}[(y_1 - \bar{y})(y_2 - \bar{y}) + (y_2 - \bar{y})(y_3 - \bar{y}) + \cdots + (y_{N-1} - \bar{y})(y_N - \bar{y})$$

(17.6)

$$R_2 = \text{cov}(y_t, y_{t+2}) = \frac{1}{N-2}\sum_{t=1}^{N-2}(y_t - \bar{y})(y_{t+2} - \bar{y})$$

$$= \frac{1}{N-2}[(y_1 - \bar{y})(y_3 - \bar{y}) + (y_2 - \bar{y})(y_4 - \bar{y}) + \cdots + (y_{N-2} - \bar{y})(y_N - \bar{y})$$

(17.7)

$$\vdots$$

$$R_{N-1} = \text{cov}(y_t, y_{t+N-1}) = (y_1 - \bar{y})(y_N - \bar{y})$$

(17.8)

Note that

$$R_{-1} = \text{cov}(y_t, y_{t-1}) = \frac{1}{N-1} \sum_{t=N}^{2} (y_t - \bar{y})(y_{t-1} - \bar{y})$$

$$= \frac{1}{N-1}[(y_N - \bar{y})(y_{N-1} - \bar{y}) + (y_{N-1} - \bar{y})(y_{N-2} - \bar{y}) + \cdots + (y_2 - \bar{y})(y_1 - \bar{y})] \qquad (17.9)$$

$$= \frac{1}{N-1} \sum_{t=1}^{N-1} (y_t - \bar{y})(y_{t+1} - \bar{y}) = \text{cov}(y_t, y_{t+1}) = R_1$$

Thus, for any negative lag n,

$$R_{-n} = \text{cov}(y_t, y_{t-n}) = \frac{1}{N-n} \sum_{t=N}^{n+1} (y_t - \bar{y})(y_{t-n} - \bar{y})$$

$$= \frac{1}{N-n}[(y_N - \bar{y})(y_{N-n} - \bar{y}) + (y_{N-1} - \bar{y})(y_{N-n-1} - \bar{y}) + \cdots + (y_{n+1} - \bar{y})(y_1 - \bar{y})] \qquad (17.10)$$

$$= \frac{1}{N-n} \sum_{t=1}^{N-n} (y_t - \bar{y})(y_{t+n} - \bar{y}) = \text{cov}(y_t, y_{t+n}) = R_n$$

The autocovariance function can be written as an nth-order matrix $\mathbf{R}$ whose element (i, j) is $R_{i,j} = R_{j-i}$, for $i, j = 1, 2, \ldots, n$. Thus,

$$\mathbf{R} = \begin{bmatrix} R_0 & R_1 & \cdots & R_{n-1} & R_n \\ R_{-1} & R_0 & \cdots & R_{n-2} & R_{n-1} \\ \cdots & \cdots & \cdots & \cdots & \cdots \\ R_{-n+1} & R_{-n+2} & \cdots & R_0 & R_1 \\ R_{-n} & R_{-n+1} & \cdots & R_{-1} & R_0 \end{bmatrix} \qquad (17.11a)$$

Following Eq. (17.10), matrix Eq. (17.11a) becomes a symmetric matrix

$$\mathbf{R} = \begin{bmatrix} R_0 & R_1 & \cdots & R_{n-1} & R_n \\ R_1 & R_0 & \cdots & R_{n-2} & R_{n-1} \\ \cdots & \cdots & \cdots & \cdots & \cdots \\ R_{n-1} & R_{n-2} & \cdots & R_0 & R_1 \\ R_n & R_{n-1} & \cdots & R_1 & R_0 \end{bmatrix} \qquad (17.11b)$$

Recall that a Toeplitz matrix is defined as a diagonal constant matrix. For example, the following matrix is a Toeplitz matrix:

$$\begin{bmatrix} a & b & c & d & e \\ f & a & b & \ddots & d \\ g & f & a & \ddots & c \\ h & \ddots & \ddots & \ddots & b \\ k & h & g & f & a \end{bmatrix}$$

It can be seen that the autocovariance matrix given by Eq. (17.11a) or Eq. (17.11b) is a Toeplitz matrix, since each descending diagonal from left to right is constant.

Autocorrelation is obtained by dividing the autocovariance by the variance, and it varies from -1 to 1. The larger absolute value of autocorrelation means the values of the variable are more correlated.

$$\rho_n = \frac{R_n}{R_0} = \frac{\sum_{t=1}^{N-n}(y_t - \overline{y})(y_{t+n} - \overline{y})}{\sum_{t=1}^{N-n}(y_t - \overline{y})^2} \tag{17.12}$$

Specifically,

$$\rho_0 = \frac{R_0}{R_0} = 1 \tag{17.13}$$

$$\rho_1 = \rho_{-1} = \frac{R_1}{R_0} \tag{17.14}$$

$$\vdots$$

$$\rho_{N-1} = \rho_{-N+1} = \frac{R_{N-1}}{R_0} \tag{17.15}$$

Analogously, the autocorrelation function of nth order can also be written as a Toeplitz matrix

$$\begin{bmatrix} \rho_0 & \rho_1 & \cdots & \rho_n \\ \rho_1 & & & \vdots \\ \vdots & & & \\ \rho_n & \cdots & & \rho_0 \end{bmatrix} \tag{17.16}$$

Example 17.2. Compute autocovaraince and autocorrelation of the stream flow data given in Table 17.1.

Solution. The autocovariance and autocorrelation for monthly stream flow are computed for different lags using Eqs. (17.4) and (17.12). For example,

$$R_0 = \frac{1}{N}\sum_{t=1}^{N}(y_t - \overline{y})^2 = \frac{1}{48}[(5279 - 9810.7)^2 + (6857 - 9810.7)^2 + \cdots + (66,428 - 9810.7)^2]$$

$$= 1.66 \times 10^7 \, \text{cfs}^2$$

$$R_1 = \frac{1}{N-1}\sum_{t=1}^{N-1}(y_t - \overline{y})(y_{t+1} - \overline{y})$$

$$= \frac{1}{47}[(5279 - 9810.7)(6857 - 9810.7) + (6857 - 9810.7)(10,396 - 9810.7) + \cdots$$

$$+ (6624 - 9810.7)(6428 - 9810.7)] = 1.08 \times 10^7 \, \text{cfs}^2$$

$$\rho_1 = R_1 / R_0 = 1.08 / 1.66 = 0.649$$

The results for other lags are shown in Table 17.2. Figure 17.2 plots the autocorrelation of monthly stream flow.

In Fig. 17.2 autocorrelations at lag 1, 6, and 12 are most significant, which means the monthly stream flow contains a periodic pattern at these frequencies.

Lag (month)	Autocovariance (cfs²)	Autocorrelation
0	16,642,320	1.000
1	10,804,714	0.649
2	4,325,852	0.260
3	−80,814	−0.005
4	−4,806,304	−0.289
5	−10,346,791	−0.622
6	−12,465,402	−0.749
7	−9,617,360	−0.578
8	−3,471,326	−0.209
9	135,579	0.008
10	3,532,104	0.212
11	7,914,707	0.476
12	11,081,145	0.666

TABLE 17.2   Autocovariance and Autocorrelation Computed for Lags to 12 Months

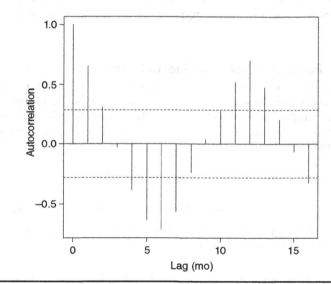

FIGURE 17.2   Autocorrelation of monthly stream flow.

17.1.4   Fourier Series and Fourier Transform

If the number of observations N is odd, then set $2q + 1 = N$. If the number of observations N is even, then let $2q = N$. The stream flow time series can be expressed by the Fourier series as

$$
\begin{aligned}
y_t &= \alpha_0 + \alpha_1 \cos(2\pi f_1 t) + \beta_1 \sin(2\pi f_1 t) + \alpha_2 \cos(2\pi f_2 t) + \beta_2 \sin(2\pi f_2 t) + \cdots \\
&\quad + \alpha_q \cos(2\pi f_2 t) + \beta_q \sin(2\pi f_2 t) \\
&= \alpha_0 + \sum_{k=1}^{q} \alpha_k \cos(2\pi f_k t) + \beta_k \sin(2\pi f_k t)
\end{aligned}
\tag{17.17}
$$

where frequency $f_k = k/N$ and α_k and β_k are the Fourier coefficients,

$$
\alpha_0 = \overline{y}
\tag{17.18a}
$$

$$
\alpha_k = \frac{2}{N} \sum_{t=1}^{N} y_t \cos(2\pi f_k t)
\tag{17.18b}
$$

$$
\beta_k = \frac{2}{N} \sum_{t=1}^{N} y_t \sin(2\pi f_k t)
\tag{17.18c}
$$

These coefficients can be computed by the least-squares method. When normalized data are considered, $\alpha_0 = 0$ hence it can be ignored. Frequencies $f_k = k/N$ are called the harmonics of the fundamental frequency $1/N$, and they allow f to vary continuously from 0 to 0.5.

17.1.5   Power Spectrum and Spectral Density

For each frequency f_k, there is one corresponding spectral power. The highest frequency is 0.5 cycle per unit time interval, since the smallest period is 2 intervals. For example, if the time interval is considered 1 mo in the case of monthly stream flow, the smallest period of the cycle T is 2 mo. Accordingly, $f = 1/T$, giving the largest frequency as 0.5. The *spectral power* at a given frequency is defined as

$$
x_k = \frac{N}{2}(\alpha_k^2 + \beta_k^2)
\tag{17.19}
$$

Example 17.3.   Compute the Fourier coefficients α_k and β_k and the spectral power for the data in Table 17.1.

Solution.   The Fourier coefficients α_k and β_k are computed using Eq. (17.18), and the spectral power is computed using Eq. (17.19). When $k = 1$,

$$
f_1 = \frac{k}{N} = \frac{1}{48} = 0.021
$$

$$
\alpha_1 = \frac{2}{N} \sum_{t=1}^{N} y_t \cos(2\pi f_1 t)
$$

$$
= \frac{2}{48}[5279\cos(2\pi \times 0.21) + 6857\cos(2\pi \times 0.21 \times 2) + \cdots + 6428\cos(2\pi \times 0.21 \times 48)]
$$

$$
= -422.111
$$

$$\beta_1 = \frac{2}{N}\sum_{t=1}^{N} y_t \sin(2\pi f_1 t)$$

$$= \frac{2}{48}[5279\sin(2\pi \times 0.21) + 6857\sin(2\pi \times 0.21 \times 2) + \cdots + 6428\sin(2\pi \times 0.21 \times 48)]$$

$$= -247.496$$

$$x_1 = \frac{N}{2}(\alpha_1^2 + \beta_1^2) = \frac{48}{2}(422.111^2 + 247.496^2) = 5.74 \times 10^6$$

Computation for other frequencies is similar, and results are shown in Table 17.3. In this example, the number of observations $N = 48$, the time interval is 1 mo, and the largest frequency is 0.5 and the corresponding $k = 24$.

The power spectrum is a function of frequency and is defined as Eq. (17.19) recast as

$$G(f) = \frac{N}{2}(a_f^2 + b_f^2) \tag{17.20}$$

k	f_k	α_k	β_k	x_k
1	0.021	−422.111	−247.496	5,746,365.885
2	0.042	83.346	668.167	10,881,443.640
3	0.063	26.054	−363.507	3,187,579.275
4	0.083	−4,960.966	1,076.547	618,483,225.076
5	0.104	295.895	−152.573	2,659,976.222
6	0.125	−236.156	5.278	1,339,139.537
7	0.146	273.242	118.463	2,128,675.931
8	0.167	888.792	−266.303	20,660,827.542
9	0.188	−20.290	192.538	899,583.947
10	0.208	56.459	−189.433	937,742.576
11	0.229	−241.027	143.309	1,887,156.099
12	0.250	1,300.042	−580.667	48,654,770.708
13	0.271	−254.582	−123.748	1,923,015.117
14	0.292	145.268	230.116	1,777,347.317
15	0.313	−15.475	−268.489	1,735,817.290
16	0.333	−412.750	−953.133	25,891,807.625
17	0.354	148.429	−140.655	1,003,561.015
18	0.375	−211.761	78.278	1,223,282.213
19	0.396	208.475	104.750	1,306,419.765
20	0.417	−295.326	795.161	17,267,967.591
21	0.438	0.878	224.800	1,212,856.654
22	0.458	93.843	−279.882	2,091,377.968
23	0.479	−234.988	94.841	1,541,137.298
24	0.500	804.042	0.000	15,515,592.042

TABLE 17.3 Computation of Fourier Coefficients and Spectral Power

where

$$\alpha_f = \frac{2}{N} \sum_{t=1}^{N} y_t \cos(2\pi ft) \qquad (17.21a)$$

$$\beta_f = \frac{2}{N} \sum_{t=1}^{N} y_t \sin(2\pi ft) \qquad (17.21b)$$

Now the power spectrum, also called the periodogram, can be plotted using $G(f)$ versus f continuously. The spectral density is defined by dividing the power spectrum by the variance

$$P(f) = G(f)/\sigma^2 \qquad (17.22)$$

The variance σ^2 is computed from Eq. (17.2). Physically the spectral density is the distribution of variance in the frequency domain.

Example 17.4 Compute the spectral density of the stream flow data in Table 17.1.

Solution. The spectral density is computed with coefficients α_f and β_f using Eqs. (17.20) and (17.22). First divide f from 0 to 0.5 with an interval of 0.001 so that the computation of the spectral density is as continuous as possible. Take frequency of 0.083 as an example:

$$\alpha_f = \frac{2}{N} \sum_{t=1}^{N} y_t \cos(2\pi ft)$$

$$= \frac{2}{48}[5279\cos(2\pi \times 0.083) + 6857\cos(2\pi \times 0.083 \times 2) + \cdots + 6428\cos(2\pi \times 0.083 \times 48)]$$

$$= -4961$$

$$\beta_f = \frac{2}{N} \sum_{t=1}^{N} y_t \sin(2\pi ft)$$

$$= \frac{2}{48}[5279\sin(2\pi \times 0.083) + 6857\sin(2\pi \times 0.083 \times 2) + \cdots + 6428\sin(2\pi \times 0.083 \times 48)]$$

$$= 1075.6$$

$$G(f) = \frac{N}{2}(\alpha_f^2 + \beta_f^2) = \frac{48}{2}(422.111^2 + 247.496^2) = 6.18 \times 10^8$$

$$P(f) = G(f)/\sigma^2 = 6.18 \times 10^8 / 1.66 \times 10^6 = 37.168$$

The spectral density for other frequencies is computed in a similar manner and is plotted in Fig. 17.3.

It can be seen from Fig. 17.3 that most significant spectral density occurs at frequency 1/12, and the second most significant spectral density is at frequency 1/4. The power spectrum or the spectral density is related to the autocovariance or autocorrelation function as follows:

$$G(f) = \sum_{k=-N+1}^{N-1} R_k \exp(-i2\pi fk) \qquad f \in [0, 0.5] \qquad (17.23a)$$

$$P(f) = \sum_{k=-N+1}^{N-1} \rho_k \exp(-i2\pi fk) \qquad f \in [0, 0.5] \qquad (17.23b)$$

Note here that integration of $G(f)$ or f between limits of 0 to 0.5 is

$$\int_0^{0.5} G(f)df = \int_0^{0.5} \sum_{k=-N+1}^{N-1} R_k \exp(-i2\pi fk)df = \sum_{k=-N+1}^{N-1} R_k \int_0^{0.5} \exp(-i2\pi fk)\, df = R_0 \qquad (17.24a)$$

and integration of $P(f)$ between limits of 0 to 0.5 yields

$$\int_0^{0.5} P(f)df = 1 \qquad (17.24b)$$

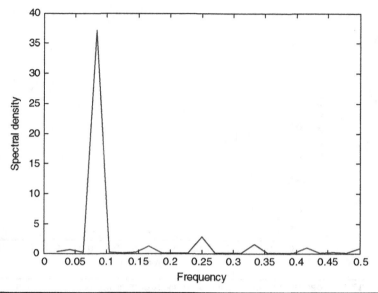

FIGURE 17.3 Spectral density of monthly stream flow.

Equations (17.23) show that the power spectrum or the spectral density is in the form of the Fourier transform of the autocovariance or autocorrelation function. Thus, inversely, autocovariance or autocorrelation can be written using the inverse Fourier transform as

$$R_k = \int_{-w}^{w} G(f)e^{i2\pi f k \Delta t}\, df \qquad -N \leq k \leq N \tag{17.25a}$$

$$\rho_k = \int_{-w}^{w} P(f)e^{i2\pi f k \Delta t}\, df \qquad -N \leq k \leq N \tag{17.25b}$$

Example 17.5. Compute the spectral density, using the stream flow data given in Table 17.1.

Solution. The spectral density is computed using Eq. (17.23b). First, f from 0 to 0.5 is divided with an interval of 0.001 so that the computation of the spectral density is as continuous as possible. Take a frequency of 0.083 as an example:

$$P(f) = \sum_{k=-N+1}^{N-1} \rho_k \exp(-i2\pi f k)$$

$$= 1 + 2[0.649\exp(-i2\pi \times 0.083 \times 1) + 0.26\exp(-i2\pi \times 0.083 \times 2) + \cdots + 0.666\exp(-i2\pi \times 0.083 \times 2)]$$

$$= 37.352$$

The spectral density for other frequencies is computed in the same manner and is plotted in Fig. 17.4. It can be seen from Fig. 17.4 that the spectral density is the same as in Fig. 17.3.

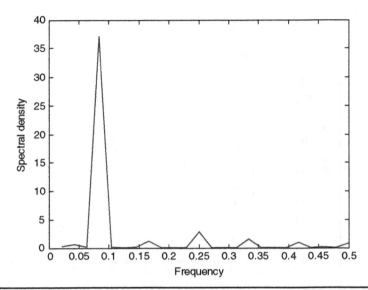

FIGURE **17.4** Spectral density of given monthly stream flow.

17.1.6 Autoregressive Linear Modeling

In the manner of autoregressive (AR) linear forecasting, the $(N + 1)$th value of y, denoted by y_{N+1}, can be forecasted using past observations y_N, y_{N-1}, ..., y_1. Thus, the value to be forecasted y_{N+1} is assumed to be a linear function of previous values and is defined as

$$\hat{y}_{N+1} = \sum_{j=1}^{N} a_j y_{N+1-j} \tag{17.26a}$$

$$y_{N+1} = \sum_{j=1}^{N} a_j y_{N+1-j} + \varepsilon_{N+1} = a_1 y_N + a_2 y_{N-1} + \cdots + a_N y_1 + \varepsilon_{N+1} \tag{17.26b}$$

where a_j is the prediction coefficient (it can also be written as $a_{n,n}$ to clarify its order as n), ε_{N-1} is the prediction error, and $\hat{y}_{N+1}$ represents the estimation of the true value of y_{N+1}. Equation (17.26b) is also called the AR model. The autocorrelation function can be obtained by multiplying Eq. (17.26b) throughout by $y(t - n)$ and taking the expectation as

$$\rho_k = E[y_N y_{N-k}] = a_1 \rho_{k-1} + a_2 \rho_{k-2} + \cdots + a_N \rho_{k-N} \tag{17.27}$$

Equation (17.27) can be rewritten for autocorrelation for lags from 1 to N as

$$\rho_1 = a_1 + a_2 \rho_1 + \cdots + a_N \rho_{N-1}$$
$$\rho_2 = a_1 \rho_1 + a_2 + \cdots + a_N \rho_{N-2} \tag{17.28}$$
$$\vdots$$
$$\rho_N = a_1 \rho_{N-1} + a_2 \rho_{N-2} + \cdots + a_N$$

which can be written in matrix form as

$$
\begin{pmatrix} \rho_1 \\ \rho_2 \\ \vdots \\ \rho_N \end{pmatrix} = \begin{pmatrix} 1 & \rho_1 & \cdots & \rho_{N-1} \\ \rho_1 & \rho_2 & \cdots & \rho_{N-2} \\ \vdots & \vdots & \cdots & \vdots \\ \rho_{N-1} & \rho_{N-2} & \cdots & 1 \end{pmatrix} \begin{pmatrix} a_1 \\ a_2 \\ \vdots \\ a_N \end{pmatrix}
\tag{17.29}
$$

On the other hand, the prediction error (bias) in Eq. (17.26b) can be defined as

$$
\varepsilon_{N+1} = y_{N+1} - \hat{y}_{N+1} = y_{N+1} - \sum_{j=1}^{N} a_j y_{N+1-j} = \sum_{j=0}^{N} -a_j y_{N+1-j}
\tag{17.30}
$$

where $a_0 = -1$, and the mean square error of the Nth-order prediction is defined as

$$
\upsilon_N = \overline{\varepsilon_{N+1}^2} = E(y_{N+1} - \hat{y}_{N+1})^2 = E\left(-\sum_{j=0}^{N} a_j y_{N+1-j}\right)^2 = E\left[\sum_{j=0}^{N} a_j^* y_{N+1-j}^* \sum_{j=0}^{N} a_j y_{N+1-j}\right]
\tag{17.31a}
$$

Since $\sum_{j=0}^{N} a_j y_{N+1-j}$ can be a complex polynomial, the conjugate is taken, where a_j^* and y_{N+1-j}^* are conjugates of a_j and y_{N+1-j}, respectively. Thus,

$$
\begin{aligned}
\upsilon_N &= E[(a_0^* y_{N+1}^* + a_1^* y_N^* + \cdots + a_N^* y_1^*)(a_0 y_{N+1} + a_1 y_N + \cdots + a_N y_1)] \\
&= E(a_0^* y_{N+1}^* a_0 y_{N+1} + a_0^* y_{N+1}^* a_1 y_N + \cdots + a_1^* y_N^* a_0 y_{N+1} + \cdots + a_N^* y_1^* a_0 y_{N+1} + \cdots) \\
&= E\left(\sum_{s=0}^{N} \sum_{j=0}^{N} a_s^* a_j y_{N+1-s}^* y_{N+1-j}\right) = \sum_{s=0}^{N} \sum_{j=0}^{N} a_s^* E(y_{N+1-s}^* y_{N+1-j}) a_j = \sum_{s=0}^{N} \sum_{j=0}^{N} a_s^* R_{s-j} a_j
\end{aligned}
\tag{17.31b}
$$

Note here that $\upsilon_0 = \overline{\varepsilon_1^2} = E(y_1 - \hat{y}_1)^2 = \sigma^2 = R_0$. Equation (17.31b) can be written in matrix form as

$$
\upsilon_N = \sum_{s=0}^{N} \sum_{j=0}^{N} a_s^* R_{s-j} a_j = \sum_{s=0}^{N} a_s^* \sum_{j=0}^{N} R_{s-j} a_j
$$

$$
= [a_0^* \quad a_1^* \quad \cdots \quad a_N^*] \begin{bmatrix} R_0 & R_1 & \cdots & R_N \\ R_1 & \cdots & & \vdots \\ \vdots & & & \\ R_N & \cdots & & R_0 \end{bmatrix} \begin{bmatrix} a_0 \\ a_1 \\ \vdots \\ a_N \end{bmatrix}
\tag{17.32}
$$

Coefficients $a_0, \ldots, a_N$ can be computed by solving for Eq. (17.29) using Newton's method or using the Levinson algorithm. The Levinson algorithm, also called the Durbin-Levinson algorithm, is a method for obtaining coefficients of the best linear predictor of y_{N+1} given $y_N, \ldots, y_1$. The Levinson algorithm has an advantage, as it involves the order of N^2 operations with memory storage on the order of N and is faster than the general gaussian elimination procedure. The Levinson algorithm computes the coefficient recursively from the previous order. Thus, one can denote the forecasting coefficient $a_0, \ldots, a_N$ as $a_{n,j}$ with an additional parameter n representing its order. If y_t is a zero-mean stationary process with autocovariance function R such that $R_0 > 0$ and $R_h \to 0$ as $h \to \infty$, then the forecasting coefficients $a_{n,j}$ and mean square errors v_n satisfy

$$a_{1,1} = \rho_1 \qquad v_0 = R_0 \qquad (17.33a)$$

$$a_{n,j} = a_{n-1,j} - a_{n,n}a_{n-1,n-j} \qquad \text{for } j < n \qquad (17.33b)$$

$$a_{n,n} = \frac{R_n - \sum_{j=1}^{n-1} a_{n-1,j}R_{n-j}}{v_{n-1}} \qquad (17.33c)$$

$$v_n = v_{n-1}(1 - a_{n,n}^2) \qquad (17.34)$$

According to the Durbin-Levinson algorithm, the forecasting coefficients can be computed recursively from the first order starting with $a_{1,1} = \rho_1$, $v_0 = 0$ to the second order, the third order, up to the Nth order. However, to be an efficient predictor, the prediction order should not be as high as the total number of observations N. The AR model order can be roughly found from the acf plot or can be determined from the Akaike information criterion (AIC) or bayesian information criterion (BIC) values. For example, it can be observed from the monthly stream flow autocorrelation plot, which has a periodic pattern with a period of 12 mo and most significant acf values showing at lags 1, 6, and 12. Thus, these predictor orders can be tried first.

Example 17.6.   Compute the AR coefficients using the monthly stream flow data given in Table 17.1. Forecast the monthly stream flow and plot the data.

Solution.   The coefficients for monthly stream flow are computed recursively as follows:
First order:

$$a_{1,1} = \rho_1 = 0.649 \qquad v_0 = R_0 = 1.66 \times 10^7$$

Second order:

$$v_1 = v_0(1 - a_{1,1}^2) = 1.66 \times 10^7(1 - 0.649^2) = 9.63 \times 10^6$$

$$a_{2,2} = \frac{R_2 - a_{1,1}R_1}{v_1} = \frac{4.33 \times 10^6 - 0.649 \times 1.08 \times 10^7}{9.63 \times 10^6} = -0.279$$

$$a_{2,1} = a_{1,1} - a_{2,2}a_{1,1} = 0.649 - (-0.279) \times 0.649 = 0.830$$

Third order:

$$v_2 = v_1(1 - a_{2,2}^2) = 9.63 \times 10^6(1 - 0.279^2) = 8.88 \times 10^6$$

$$a_{3,3} = \frac{R_3 - \sum_{j=1}^{2} a_{2,j}R_{3-j}}{v_2} = \frac{-80,814 - (0.83 \times 4.33 \times 10^6 - 0.279 \times 1.08 \times 10^7)}{8.89 \times 10^6} = -0.066$$

$$a_{3,2} = a_{2,2} - a_{3,3}a_{2,1} = -0.279 - (-0.066) \times 0.83 = 0.269$$

$$a_{3,1} = a_{2,2} - a_{3,3}a_{2,2} = -0.279 - (-0.066) \times (-0.279) = -0.297$$

The above steps are repeated for computing the coefficients until the 13th order where the AIC and BIC values are the smallest (see Table 17.4).

Thus, the monthly stream flow for 1970 is computed as an example.

a_1	0.5347	a_8	0.2048
a_2	−0.1435	a_9	−0.2499
a_3	0.0261	a_{10}	−0.1419
a_4	−0.3625	a_{11}	−0.1721
a_5	−0.1175	a_{12}	0.5485
a_6	−0.1904	a_{13}	−0.3532
a_7	−0.158		

TABLE 17.4   Forecasting Coefficients for AR (13)

For N increasing from 48 to 59:

$$\hat{y}_{N+1} = 0.535y_N - 0.1435y_{N-1} + \cdots + 0.549y_{N-11} - 0.353y_{N-12}$$

Or the forecasting parameters can be estimated using Newton's method by solving the matrix equation:

$$\begin{pmatrix} 0.649 \\ -0.260 \\ \vdots \\ 0.666 \end{pmatrix} = \begin{pmatrix} 1 & 0.649 & \cdots & 0.476 \\ 0.649 & -0.260 & & \\ \vdots & \vdots & & \vdots \\ 0.476 & & \cdots & 1 \end{pmatrix} \begin{pmatrix} a_1 \\ a_2 \\ \vdots \\ a_N \end{pmatrix}$$

Let

$$\mathbf{R} = \begin{pmatrix} 1 & 0.649 & \cdots & 0.476 \\ 0.649 & -0.260 & & \\ \vdots & \vdots & & \vdots \\ 0.476 & & \cdots & 1 \end{pmatrix} \quad \text{and} \quad \mathbf{r} = \begin{pmatrix} 0.649 \\ -0.260 \\ \vdots \\ 0.666 \end{pmatrix}$$

The forecasting coefficient can be solved from MATLAB command $\mathbf{R}\backslash\mathbf{r}$ with Newton's numerical method. The coefficient computed using Newton's method is listed in Table 17.5, which is comparable to that obtained from the Levinson algorithm.

The 95 percent prediction interval can be computed from Box and Jenkins (1970) as

$$b_m = 1.96s\sqrt{1 + \sum^{m-1} a_m^2}$$

a_1	0.543	a_8	0.205
a_2	−0.143	a_9	−0.248
a_3	0.025	a_{10}	−0.142
a_4	−0.351	a_{11}	−0.173
a_5	−0.119	a_{12}	0.549
a_6	−0.192	a_{13}	−0.352
a_7	−0.160		

TABLE 17.5   Forecasting Coefficients for AR (13)

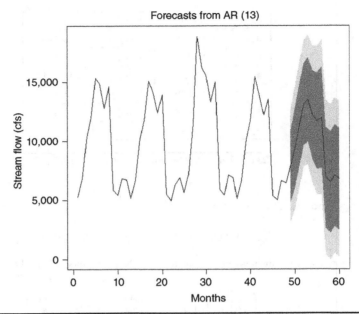

FIGURE **17.5**   Stream flow forecasted with coefficients computed with the Levinson algorithm.

For example, the interval for the first forecasted value with $s = 139.7$ cfs is

$$b_m = 1.96s\sqrt{1 + 0.534^2} = 310.6$$

Then the prediction interval for the first forecasted value is $y \pm b_m = [491.7 \text{ cfs}, 1112.9 \text{ cfs}]$. The forecasted stream flow and the confidence intervals are shown in Fig. 17.5.

17.2   Maximum Entropy Spectral Analysis for Stream flow Forecasting

The objective of maximum entropy spectral analysis (MESA) is to determine the power spectrum $P(f)$ with frequency f of a stationary process $X(t)$ by using the principle of maximum entropy (POME). This entails the following steps: (1) definition of entropy, (2) specification of constraints, (3) maximization of entropy, (4) determination of Lagrange multipliers, and (5) derivation of MESA. There are three ways in which entropy can be defined for doing entropy spectral analysis: (1) Burg entropy, (2) configuration entropy, and (3) cross entropy. Thus, entropy spectral analysis can be done in three ways.

17.3   Burg Entropy Spectral Analysis

Burg (1967, 1975) defined entropy $H(f)$ as

$$H(f) = \frac{1}{2}(\ln 2w) + \frac{1}{4w}\int_{-w}^{+w} \ln P(f)\, df \qquad (17.35a)$$

where w is the frequency band. Equation (17.35a) is known as the Burg entropy. Maximization of Eq. (17.35a) is equivalent to maximizing

$$H(f) = \int_{-W}^{W} \ln P(f)\, df \tag{17.35b}$$

since the first term of Eq. (17.35a) is constant. Here $W = 1/(2\Delta t)$ is the Nyquist foldover frequency; f is the frequency that varies from $-W$ to W; Δt is the sampling period; and $P(f)$ is the spectral density function. Here f is treated as a random variable. For example, for monthly stream flow, the sampling period is 1 mo, thus $\Delta t = 1$ mo, and $W = 1/(2\Delta t) = 0.5$, and the frequency f varies from -0.5 to 0.5.

17.3.1 Specification of Constraints
The Burg entropy is maximized, subject to the constraint given by autocorrelation function, as

$$\rho_n = \int_{-W}^{W} P(f)e^{i2\pi fn\Delta t}\, df \qquad -N \le n \le N \tag{17.36}$$

where N is usually taken from one-quarter up to one-half of the total record length, Δt is the sampling interval, and $i = \sqrt{-1}$. In practical applications, the theoretical autocorrelations are replaced by the sample values.

17.3.2 Maximization of Entropy
Equation (17.35b) is maximized subject to Eq. (17.36). The resulting spectrum is maximally consistent with the known acf, until lag n, and is minimally biased for the lags beyond. Entropy maximizing can be done with the use of the Lagrange multiplier method. Thus, the Lagrange function can be expressed as

$$\begin{aligned}
L(f) &= \int_{-W}^{W} \ln P(f)\, df - \sum_{n=-N}^{N} \lambda_n \left[\rho_n - \int_{-W}^{W} P(f)\exp(i2\pi fn\Delta t)\, df \right] \\
&= \int_{-W}^{W} \{\ln P(f) - \sum_{n=N}^{N} \lambda_n [P(f)e^{i2\pi fn\Delta t}\}\, df - \rho_n]
\end{aligned} \tag{17.37}$$

Taking the partial derivative of $L(f)$ with respect to $P(f)$ yields

$$\frac{\partial L(f)}{\partial P(f)} = \int_{-W}^{W} \left[\frac{1}{P(f)} - \sum_{n=N}^{N} \lambda_n e^{i2\pi fn\Delta t} \right] df = 0 \tag{17.38}$$

and one obtains

$$P(f) = \frac{1}{\sum\limits_{n=-N}^{N} \lambda_n e^{-i2\pi fn\Delta t}} \tag{17.39}$$

Equation (17.39) is the entropy-based spectral density function.

17.3.3 Determination of Lagrange Multipliers

Since computation of Eqs. (17.36) and (17.37) involves complex numbers and integration with $e^{-i2\pi f \Delta t}$, it is more convenient to do the analysis using the z transform, so that the integration can be done analytically around the unit circle in the complex z plane. Inputting Eq. (17.39) into the constraint equation, Eq. (17.36), one obtains

$$\rho_n = \int_{-W}^{W} \frac{1}{\sum_{s=-N}^{N} \lambda_s e^{-i2\pi f s \Delta t}} e^{i2\pi f n \Delta t} \, df \qquad -N \le n \le N \qquad (17.40)$$

Let

$$z = e^{-i2\pi f \Delta t} \qquad (17.41)$$

Then

$$dz = -i2\pi \Delta t \, e^{-i2\pi f \Delta t} df = -i2\pi \Delta t \, z \, df \qquad (17.42a)$$

Thus,

$$df = \frac{dz}{-i2\pi \Delta t \, z} \qquad (17.42b)$$

Replacing $e^{-i2\pi f \Delta t}$ and df by z and dz from Eqs. (17.42a) and (17.42b), respectively, one finds that Eq. (17.40) becomes

$$\rho_n = \int_{-W}^{W} \frac{z^{-n} \, df}{\sum_{s=-N}^{N} \lambda_s z^s} = \frac{1}{i2\pi \Delta t} \oint \frac{z^{-n-1} dz}{\sum_{s=-N}^{N} \lambda_s z^s} \qquad -N \le n \le N \qquad (17.43)$$

where integration is around the unit circle in the counterclockwise direction, as the integration on f from $-W$ to W is equivalent to that from $-\pi$ to π in the z plane; or the direction can also be determined from the following complex plane (see Fig. 17.6), where the horizontal axis represents the real part and the vertical axis is the imaginary part.

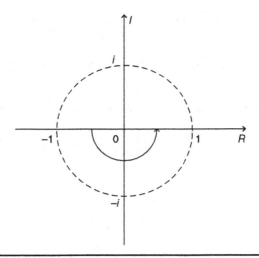

FIGURE 17.6 Complex z plane.

Equation (17.39) can be transformed as

$$P(f) = \frac{1}{\displaystyle\sum_{n=-N}^{N} \lambda_n z^n} \qquad (17.44)$$

which needs to be real and positive for $|z| = 1$. Recall that in time series analysis the AR process or the linear prediction process has the spectral density in the form of

$$P(f) = \frac{\sigma^2}{\left| 1 + \displaystyle\sum_{n=-N}^{N} a_n z^n \right|^2} \qquad (17.45)$$

where the denominator is the square of the characteristic function and σ^2 is the variance. Thus, equating the right sides of Eqs. (17.44) and (17.45), one obtains

$$\sum_{s=-N}^{N} \lambda_s z^s = \sigma^2 \left(1 + a_1 z + a_2 z^2 + \cdots + a_N z^Z \right) \left(a + a_1^* z^{-1} + a_2^* z^{-2} + \cdots + a_N^* z^{-N} \right)$$

$$= \sigma^2 \sum_{s=0}^{N} a_s z^s \sum_{s=0}^{N} a_s^* z^{-s} \qquad (17.46)$$

It can be found from Eq. (17.46) that for each z^s there is a unique Lagrange multiplier λ_s associated. Thus, by seeking the coefficient of the same order of z, the Lagrange multipliers can be expressed by the convolution of linear prediction coefficients as

$$\lambda_s = \sigma^2 \sum_{j=n}^{N} a_j a_{j-n}^* \qquad (17.47)$$

To be causal $\sum_{j=0}^{\infty} |a_j| \infty$, all the roots of the first polynomial in Eq. (17.47) are chosen to lie outside the unit circle, and thus all the roots of the second polynomial lie inside the unit circle. Otherwise, the process is noncausal, and the prediction will not converge.

Now to do linear predictions using MESA, inserting Eqs. (17.42a) and (17.42b) into Eq. (17.40), one obtains

$$\rho_n = \int_{-W}^{W} \frac{1}{\sigma^2 \displaystyle\sum_{s=0}^{N} a_s z^s \sum_{s=0}^{N} a_s^* z^{-s}} e^{i 2\pi f n \Delta t} \, df \qquad -N \le n \le N \qquad (17.48)$$

Let us assume that $\sigma^2 = 1/P_N$, and Eq. (17.48) can be recast as

$$\frac{P_N}{2\pi i} \oint \frac{z^{-n-1} dz}{\displaystyle\sum_{s=0}^{N} a_s z^s \sum_{s=0}^{N} a_s^* z^{-s}} = \rho_n \qquad -N < n < N \qquad (17.49)$$

where $P_N > 0$ is the prediction error defined later. Multiplying Eq. (17.49) by a_n^* and summing up for n from 0 to N, one obtains

$$\sum_{n=0}^{N} a_n^* \rho_{n-r} = \frac{P_N}{2\pi i} \oint \frac{z^{r-1} dz \sum_{n=0}^{N} a_n^* z^{-n}}{\sum_{s=0}^{N} a_s z^s \sum_{s=0}^{N} a_s^* z^{-s}} = \frac{P_N}{2\pi i} \oint \frac{z^{r-1} dz}{\sum_{s=0}^{N} a_s z^s} \qquad r \geq 0 \qquad (17.50)$$

Recall that Cauchy's integral formula gives

$$\oint \frac{z^{r-1} dz}{\sum_{s=0}^{N} a_s z^s} = \begin{cases} 2\pi i & r = 0 \\ 0 & r \geq 1 \end{cases} \qquad (17.51$$

Thus, integrating Eq. (17.50) using Cauchy's formula, one obtains

$$\sum_{n=0}^{N} a_n^* \rho_{n-r} = \frac{P_N}{2\pi i} \oint \frac{z^{r-1} dz}{\sum_{s=0}^{N} a_s z^s} = \frac{P_N}{2\pi i} 2\pi i = P_N \qquad r = 0 \qquad (17.52a)$$

$$\sum_{n=0}^{N} a_n^* \rho_{n-r} = \frac{P_N}{2\pi i} \oint \frac{z^{r-1} dz}{\sum_{s=0}^{N} a_s z^s} = \frac{P_N}{2\pi i} 0 = 0 \qquad r \geq 1 \qquad (17.52b)$$

Since P_N is real, taking the conjugate of Eqs. (17.52a) and (17.52b) yields

$$\sum_{n=0}^{N} \rho_{r-n} a_n = P_N \qquad r = 0 \qquad (17.53a)$$

$$\sum_{n=0}^{N} \rho_{r-n} a_n = 0 \qquad r \geq 1 \qquad (17.53b)$$

which can be written in matrix form as

$$\begin{bmatrix} \rho_0 & \rho_{-1} & \cdots & \rho_{-N} \\ \rho_1 & \rho_0 & & \rho_{1-N} \\ \vdots & \vdots & & \vdots \\ \rho_N & \rho_{N-1} & \cdots & \rho_0 \end{bmatrix} \begin{Bmatrix} 1 \\ a_1 \\ \vdots \\ a_N \end{Bmatrix} = \begin{Bmatrix} P_N \\ 0 \\ \vdots \\ 0 \end{Bmatrix} \qquad (17.54)$$

Equation (17.54) is the well-known equation for solving for the Nth-order prediction error filter, with error vector $[P_N, 0, \ldots, 0]^T$. Note that the autocorrelation matrix is positive definite only if P_N is positive; when $P_N = 0$, the matrix is singular. Both P_N and a_n can be solved for by using Newton's method, the Durbin-Levinson algorithm as shown in Example 17.6, or the Burg-Levinson algorithm, which is introduced here.

17.3.4   Burg-Levinson Algorithm

It is assumed that the $N \times N$ Toeplitz submatrix is positive definite and that the full $N + 1 \times N + 1$ Toeplitz matrix is at least nonnegative definite. Starting with the solution of the set of N equations, consider $(N-1)$th order with known P_{N-1} and $a_n{}^{N-1}$s, where $a_n{}^{N-1}$ represent the a_n coefficients in $(N-1)$th-order:

$$
\begin{bmatrix}
\rho_0 & \rho_{-1} & \cdots & \rho_{1-N} \\
\rho_1 & \rho_0 & \cdots & \rho_{2-N} \\
\vdots & \vdots & \cdots & \vdots \\
\rho_{N-1} & \rho_{N-2} & \cdots & \rho_0
\end{bmatrix}
\begin{Bmatrix}
1 \\
a_1^{N-1} \\
\vdots \\
a_{N-1}^{N-1}
\end{Bmatrix}
=
\begin{Bmatrix}
P_{N-1} \\
0 \\
\vdots \\
0
\end{Bmatrix}
\tag{17.55a}
$$

Taking the inverse conjugate of coefficient vector, one obtains

$$
\begin{bmatrix}
\rho_0 & \rho_{-1} & \cdots & \rho_{1-N} \\
\rho_1 & \rho_0 & \cdots & \rho_{2-N} \\
\vdots & \vdots & \cdots & \vdots \\
\rho_{N-1} & \rho_{N-2} & \cdots & \rho_0
\end{bmatrix}
\begin{Bmatrix}
a_{N-1}^{*N-1} \\
a_{N-2}^{*N-1} \\
\vdots \\
1
\end{Bmatrix}
=
\begin{Bmatrix}
0 \\
0 \\
\vdots \\
P_{N-1}^*
\end{Bmatrix}
\tag{17.55b}
$$

To compute P_N and a_n of the Nth order, Eq. (17.55) can be expanded into the Nth order by adding the last row and column to the autocorrelation matrix. Define $\sum_{n=0}^{N-1} \rho_{N-n} a_n^{N-1} = \Delta_N$, which is unknown. Thus, the expansion of Eq. (17.55a) will be

$$
\begin{bmatrix}
\rho_0 & \rho_{-1} & \cdots & \rho_{1-N} & \rho_{-N} \\
\rho_1 & \rho_0 & \cdots & \rho_{2-N} & \rho_{1-N} \\
\vdots & \vdots & \cdots & \vdots & \vdots \\
\rho_{N-1} & \rho_{N-2} & \cdots & \rho_0 & \rho_{-1} \\
\rho_N & \rho_{N-1} & \cdots & \rho_{-1} & \rho_0
\end{bmatrix}
\begin{Bmatrix}
1 \\
a_1^{N-1} \\
\vdots \\
a_{N-1}^{N-1} \\
0
\end{Bmatrix}
=
\begin{Bmatrix}
P_{N-1} \\
0 \\
\vdots \\
0 \\
\Delta_N
\end{Bmatrix}
\tag{17.56a}
$$

$$
\begin{bmatrix}
\rho_0 & \rho_{-1} & \cdots & \rho_{1-N} & \rho_{-N} \\
\rho_1 & \rho_0 & \cdots & \rho_{2-N} & \rho_{1-N} \\
\vdots & \vdots & \cdots & \vdots & \vdots \\
\rho_{N-1} & \rho_{N-2} & \cdots & \rho_0 & \rho_{-1} \\
\rho_N & \rho_{N-1} & \cdots & \rho_{-1} & \rho_0
\end{bmatrix}
\begin{Bmatrix}
0 \\
a_{N-1}^{*N-1} \\
\vdots \\
a_1^{*N-1} \\
1
\end{Bmatrix}
=
\begin{Bmatrix}
\Delta_N^* \\
0 \\
\vdots \\
0 \\
P_{N-1}^*
\end{Bmatrix}
\tag{17.56b}
$$

Now one introduces the reflection coefficient c_N with which one can minimize the forward and backward error together (see the explanation in the latter part). The reflection coefficient does not equal 1; otherwise estimated a_N^N would be 1 and not reversible. Multiplying Eq. (17.56b) by c_N and summing Eqs. (17.56a), one can obtain

$$
\begin{bmatrix}
\rho_0 & \rho_{-1} & \cdots & \rho_{1-N} & \rho_{-N} \\
\rho_1 & \rho_0 & \cdots & \rho_{2-N} & \rho_{1-N} \\
\vdots & \vdots & \cdots & \vdots & \vdots \\
\rho_{N-1} & \rho_{N-2} & \cdots & \rho_0 & \rho_{-1} \\
\rho_N & \rho_{N-1} & \cdots & \rho_{-1} & \rho_0
\end{bmatrix}
\left\{
\begin{Bmatrix}
1 \\
a_1^{N-1} \\
\vdots \\
a_{N-1}^{N-1} \\
0
\end{Bmatrix}
+ c_N
\begin{Bmatrix}
0 \\
a_{N-1}^{*N-1} \\
\vdots \\
a_1^{*N-1} \\
1
\end{Bmatrix}
\right\}
=
\left\{
\begin{Bmatrix}
P_{N-1} \\
0 \\
\vdots \\
0 \\
\Delta_N
\end{Bmatrix}
+ c_N
\begin{Bmatrix}
\Delta_N^* \\
0 \\
\vdots \\
0 \\
P_{N-1}^*
\end{Bmatrix}
\right\}
$$

$$\tag{17.57}$$

where $P_{N-1}^* = P_{N-1}$ (real). The second column vector on both sides of Eq. (17.57) is seen to be the complex conjugate reverse of the first column. The equation is valid for complex conjugate reverse vectors because the kth and the $(N-k)$th rows of the autocorrelation matrix are complex conjugate reverses. By definition Eq. (17.57) should also equal

$$\begin{bmatrix} R_0 & R_{-1} & \cdots & R_{1-N} & R_{-N} \\ R_1 & R_0 & \cdots & R_{2-N} & R_{1-N} \\ \vdots & \vdots & \cdots & \vdots & \vdots \\ R_{N-1} & R_{N-2} & \cdots & R_0 & R_{-1} \\ R_N & R_{N-1} & \cdots & R_{-1} & R_0 \end{bmatrix} \begin{Bmatrix} 1 \\ a_1^N \\ \vdots \\ a_{N-1}^N \\ a_N^N \end{Bmatrix} = \begin{Bmatrix} P_N \\ 0 \\ \vdots \\ 0 \\ 0 \end{Bmatrix} \qquad (17.58)$$

Equating the right terms of Eqs. (17.58) and (17.57), one gets

$$\begin{Bmatrix} P_N \\ 0 \\ \vdots \\ 0 \\ 0 \end{Bmatrix} = \begin{Bmatrix} P_{N-1} \\ 0 \\ \vdots \\ 0 \\ \Delta_N \end{Bmatrix} + c_N \begin{Bmatrix} \Delta_N^* \\ 0 \\ \vdots \\ 0 \\ P_{N-1} \end{Bmatrix} \qquad (17.59)$$

From the last row of Eq. (17.59),

$$0 = \Delta_N + c_N P_{N-1} \qquad (17.60a)$$

Thus, one obtains

$$c_N = \frac{-\Delta_N}{P_{N-1}} \qquad (17.60b)$$

where Δ_N can be computed from its definition

$$\Delta_N = \sum_{n=0}^{N-1} \rho_{N-n} a_n^{N-1} \qquad (17.61)$$

From the first row of Eq. (17.59), the prediction error of the next order can be determined:

$$P_N = P_{N-1} + c_N \Delta_N^* = P_{N-1} + c_N(-c_N P_{N-1})^* = P_{N-1}(1 - c_N c_N^*) = P_{N-1}(1 - |C_N|^2) \qquad (17.62)$$

The initial value of P_0 starts with R_0, and the 0th-order prediction error filter is simply the one-point filter with unit weight.

The prediction coefficient can also be determined by setting Eqs. (17.57) and (17.58) equal to each other:

$$\begin{Bmatrix} 1 \\ a_1^N \\ \vdots \\ a_{N-1}^N \\ a_N^N \end{Bmatrix} = \begin{Bmatrix} 1 \\ a_1^{N-1} \\ \vdots \\ a_{N-1}^{N-1} \\ 0 \end{Bmatrix} + c_N \begin{Bmatrix} 0 \\ a_{N-1}^{*N-1} \\ \vdots \\ a_1^{*N-1} \\ 1 \end{Bmatrix} \qquad (17.63)$$

As stated earlier, it is reversible only if $|C_n| \neq 1$. Equation (17.63) can also be written as follows:

$$
a_s^N = \begin{cases} 1 & s = 1 \\ a_s^{N-1} + c_N a_{N-s}^{*N-1} & 1 < s < N \\ c_N & s = N \end{cases}
\tag{17.64}
$$

Through the above procedure, a_n, P_N can be computed recursively from the previous order.

The reflection coefficient c_N is obtained by minimizing the backward and forward prediction error using the Burg entropy (Burg, 1975), which gives higher resolution than does the conventional method. Given $A(z)$ composed from the prediction coefficient of $(N-1)$th order, one can write

$$
A_{N-1}(z) = 1 + a_1^{N-1} z + a_2^{N-1} z^2 + \cdots + a_N^{N-1} z^{N-1}
\tag{17.65}
$$

From Eq. (17.60) the Nth order will be

$$
A_N(z) = 1 + (a_1^{N-1} + c_N a_{N-1}^{N-1})z + (a_2^{N-1} + c_N a_{N-2}^{N-1})z^2 + \cdots + (a_{N-1}^{N-1} + c_N a_1^{N-1})z^{N-1} + c_N z^N
\tag{17.66}
$$

The mean square error is given as

$$
MSE = \sum_{m=0}^{N} a_m^* y_{T-m}^* \sum_{n=0}^{N} a_n y_{T-n} = \overline{[A(z)y(z)]^2}
$$
$$
= \sum_{m=1}^{M} w_m \left| c_N y_1 + (a_1^{N-1} + c_N a_{N-1}^{N-1})y_2 + \cdots + (a_{N-1}^{N-1} + c_N a_1^{N-1})y_N + y_{N-1} \right|^2
\tag{17.67}
$$

where w_m are weights such that

$$
\sum_{m=1}^{M} w_m = 1
\tag{17.68}
$$

The forward prediction error is defined as

$$
f_m = a_{N-1} y_2 + \cdots + a_1 y_N + y_{N+1}
\tag{17.69}
$$

and backward prediction error as

$$
b_m = y_1 + a_1 y_2 + \cdots + a_{N-1} y_N
\tag{17.70}
$$

Thus, Eq. (17.67) can be written from forward and backward errors as

$$
MSE = \sum_{m=1}^{M} w_m \left| f_m + c_N b_m \right|^2 = \sum_{m=1}^{M} w_m \left| b_m + c_N f_m \right|^2
\tag{17.71}
$$

where c_N is obtained by minimizing Eq. (17.71), which is obtained as

$$c_N = \frac{-2 \sum_{m=1}^{M} w_m (b_m f_m)}{\sum_{m=1}^{M} w_m (b_m^2 + f_m^2)} \qquad (17.72)$$

Equation (17.72) is the key point that distinguishes the Burg MESA from other methods.

Example 17.7. Compute the prediction error and prediction coefficients up to seventh order for monthly flow from the Colorado River.

Solution. The autocorrelation up to $T/4$ is computed as given in Table 17.2 in Example 17.6.
Step 1: Starting from the zero order, compute P_0:

$$P_0 = \rho_0 = 1$$

Step 2: The first-order computation is done as follows:

$$\begin{bmatrix} \rho_0 & \rho_1 \\ \rho_1 & \rho_0 \end{bmatrix} \left\{ \begin{pmatrix} 1 \\ 0 \end{pmatrix} + c_1 \begin{pmatrix} 0 \\ 1 \end{pmatrix} \right\} = \begin{pmatrix} P_0 \\ \Delta_1 \end{pmatrix} + c_1 \begin{pmatrix} \Delta_1 \\ P_0 \end{pmatrix} = \begin{pmatrix} P_1 \\ 0 \end{pmatrix}$$

$$\begin{bmatrix} 1 & 0.649 \\ 0.649 & 1 \end{bmatrix} \left\{ \begin{pmatrix} 1 \\ 0 \end{pmatrix} + c_1 \begin{pmatrix} 0 \\ 1 \end{pmatrix} \right\} = \begin{pmatrix} 1 \\ \Delta_1 \end{pmatrix} + c_1 \begin{pmatrix} \Delta_1 \\ 1 \end{pmatrix}$$

$$\Delta_1 = \rho_1 = 0.649$$

$$c_1 = \frac{-\Delta_1}{P_0} = \frac{-0.649}{1} = -0.649$$

$$P_1 = P_0 (1 - |c_1|^2) = 1(1 - 0.649^2) = 0.579$$

$$a_1^1 = c_1 = -0.649$$

Step 3: The second-order computation is done as follows:

$$\begin{bmatrix} \rho_0 & \rho_1 & \rho_2 \\ \rho_1 & \rho_0 & \rho_1 \\ \rho_2 & \rho_1 & \rho_0 \end{bmatrix} \left\{ \begin{pmatrix} 1 \\ a_1^1 \\ 0 \end{pmatrix} + c_2 \begin{pmatrix} 0 \\ a_1^1 \\ 1 \end{pmatrix} \right\} = \begin{pmatrix} P_1 \\ 0 \\ \Delta_2 \end{pmatrix} + c_2 \begin{pmatrix} \Delta_2 \\ 0 \\ P_1 \end{pmatrix}$$

$$\begin{bmatrix} 1 & 0.649 & 0.260 \\ 0.649 & 1 & 0.649 \\ 0.260 & 0.649 & 1 \end{bmatrix} \left\{ \begin{pmatrix} 1 \\ -0.649 \\ 0 \end{pmatrix} + c_2 \begin{pmatrix} 0 \\ -0.649 \\ 1 \end{pmatrix} \right\} = \begin{pmatrix} 0.579 \\ 0 \\ \Delta_2 \end{pmatrix} + c_2 \begin{pmatrix} \Delta_2 \\ 0 \\ 0.579 \end{pmatrix}$$

$$\Delta_2 = \rho_2 + a_1^1 \rho_1 = 0.260 - 0.649 \times 0.649 = 0.421$$

$$c_2 = -\Delta_2 / P_1 = -0.421/0.579 = -0.727$$

$$P_2 = P_1 (1 - |c_2|^2) = 0.579(1 - 0.727^2) = 0.273$$

$$a_2^1 = a_1^1 + c_2 a_1^1 = -0.649 - 0.727 \times (-0.649) = -0.177$$

$$a_2^2 = c_2 = -0.727$$

Step 4: Higher-order computation is done by following the above steps. Expanding the matrix by one size each time, one can finally get the seventh-order prediction coefficients as given in Table 17.6.

a_0	1.000
a_1	−0.578
a_2	0.177
a_3	−0.004
a_4	0.434
a_5	−0.077
a_6	0.297
a_7	0.135

TABLE 17.6 Computation of Prediction Coefficients at 7th Order

17.3.5 Extension of Autocorrelation Function

For allowable values of the next lag autocorrelation, it is considered that one knows R_n (ρ_n) from $n = 0$ to $N − 1$ and one wants to estimate R_n (ρ_n). Given a set of complex numbers R_n, $|n| \leq N$, where $R_n = R^*_{-n}$, these numbers are the beginning of an autocovariance function if and only if matrix **R**, defined as Eq. (17.11), is nonnegative definite. Thus, the $N \times N$ Toeplitz submatrix will be nonnegative definite and ρ_N will have a permissible value only if the full $N + 1 \times N + 1$ matrix is also nonnegative definite. Consider when the $N \times N$ submatrix is singular, which has a rank of M. In this case (1, $a_1, \dots, a_M$) has the shortest prediction error, which satisfies

$$\begin{bmatrix} \rho_0 & \rho_1 & \cdots & \rho_M \\ \rho_1 & & & \vdots \\ \vdots & & & \\ \rho_M & \cdots & & \rho_0 \end{bmatrix} \begin{bmatrix} 1 \\ a_1 \\ \vdots \\ a_M \end{bmatrix} = \begin{bmatrix} 0 \\ 0 \\ \vdots \\ 0 \end{bmatrix} \tag{17.73}$$

and to have a nonnegative $N + 1 \times N + 1$ matrix, one must have

$$\sum_{j=0}^{M} \rho_{N-j} a_j = 0 \qquad \text{or} \qquad \rho_N = \sum_{j=1}^{M} \rho_{j-n} a_j \tag{17.74}$$

Therefore, if an autocorrelation matrix is singular, the next value of autocorrelation is uniquely specified by Eq. (17.74).

Now consider the $N \times N$ matrix is nonsingular. The $N + 1 \times N + 1$ matrix is nonnegative, if only P_N is nonnegative, since the determinant of the $N + 1 \times N + 1$ matrix equals P_N multiplied by the determinant of $N \times N$ matrix. If P_N needs to be nonnegative, it requires $|c_n| \leq 1$ from Eq. (17.62). Combining Eqs. (17.60a), (17.60b), and (17.61), one obtains

$$c_N P_{N-1} = \sum_{n=0}^{N-1} \rho_{N-n} a_n^{N-1} \tag{17.75a}$$

Thus,

$$\rho_N = -\sum_{n=1}^{N-1} \rho_{N-n} a_n^{N-1} - c_N P_{N-1} \qquad (17.75b)$$

It can be seen that when $|c_n| \le 1$, the value of ρ_N lies in the circle of radius P_{N-1} centered at $-\sum_{n=1}^{N-1} \rho_{N-n} a_n^{N-1}$ in the complex plane. Since $P_N = P_{N-1}(1-|c_N|^2)$ decreases, the radii of the sequence of circles will decrease unless $c = 0$.

Suppose one has the first $N + 1$ values of autocorrelation $\rho_0, \ldots, \rho_N$ and solves Eq. (17.54) for the Nth order prediction coefficient $(1, a_1, \ldots, a_N)^T$:

$$A(z) = 1 + a_1 z + a_2 z^2 + \cdots + a_N z^N \qquad (17.76)$$

If one considers zero extension of the reflection coefficient sequence starting with $c_{N+1} = 0$, from Eq. (17.75b), then one obtains

$$\rho_m = -\sum_{n=1}^{N} \rho_{m-n} a_n \qquad m > N \qquad (17.77)$$

It can be seen that the extension of autocorrelation is given by the convolution feedback operator. To show that the extension by Eq. (17.78) corresponds to MESA, one defines the function as

$$\Phi(z) = \frac{\rho_0}{2} + \rho_1 z + \rho_2 z^2 + \cdots + \rho_N z^N \qquad (17.78)$$

so that

$$\Phi(z) + \Phi^*(z^{-1}) = \frac{\rho_0}{2} + \rho_1 z + \rho_2 z^2 + \cdots + \frac{\rho_0}{2} + \rho_1 z^{-1} + \rho_2 z^{-2} + \cdots = \sum_{n=-N}^{N} \rho_n z^n \qquad (17.79)$$

which is the z transform of the autocorrelation.

Now convoluting $\Phi(z)$ and $A(z)$, one obtains

$$\Phi(z)A(z) = [1 + a_1 z + a_2 z^2 + \cdots + a_N z^N]\left[\frac{\rho_0}{2} + \rho_1 z + \rho_2 z^2 + \cdots + \rho_N z^N\right]$$

$$= \frac{\rho_0}{2}[1 + d_1 z + d_2 z^2 + \cdots + d_N z^N] = \frac{\rho_0}{2} D(z) \qquad (17.80)$$

where $D(z)$ is defined as $D(z) = [1 + d_1 z + d_2 z^2 + \cdots + d_N z^N]$ and $d_m = 0$ for $m > N$, according to Eq. (17.77).

Now from Eq. (17.80)

$$\Phi(z) = \frac{\rho_0}{2} \frac{D(z)}{A(z)} \qquad (17.81)$$

To derive the relationship between $D(z)$ and $A(z)$, one goes from $N = 2$ to $N = 3$. The first three terms of Eq. (17.77) in the convolution with $N = 2$ can be written as

$$\begin{bmatrix} \rho_0/2 & 0 & 0 \\ \rho_1 & \rho_0/2 & 0 \\ \rho_2 & \rho_1 & \rho_0/2 \end{bmatrix}\begin{bmatrix} 1 \\ a_1 \\ a_2 \end{bmatrix} = \frac{\rho_0}{2}\begin{bmatrix} 1 \\ d_1 \\ d_2 \end{bmatrix} \qquad (17.82)$$

The reminder of the second-order Toeplitz matrix must obey

$$
\begin{bmatrix}
\rho_0 & \rho_1 & \rho_2 \\
\rho_1 & \rho_0 & \rho_1 \\
\rho_2 & \rho_1 & \rho_0
\end{bmatrix}
\begin{bmatrix}
1 \\
a_1 \\
a_2
\end{bmatrix}
=
\begin{bmatrix}
P_2 \\
0 \\
0
\end{bmatrix}
\tag{17.83}
$$

Subtracting Eq. (17.82) from Eq. (17.83), one obtains

$$
\begin{bmatrix}
\rho_0/2 & \rho_1 & \rho_2 \\
0 & \rho_0/2 & \rho_1 \\
0 & 0 & \rho_0/2
\end{bmatrix}
\begin{bmatrix}
1 \\
a_1 \\
a_2
\end{bmatrix}
=
\begin{bmatrix}
P_2 \\
0 \\
0
\end{bmatrix}
-\frac{\rho_0}{2}
\begin{bmatrix}
1 \\
d_1 \\
d_2
\end{bmatrix}
\tag{17.84}
$$

Taking the complex conjugates of Eq. (17.84) and rearranging the order, one obtains

$$
\begin{bmatrix}
\rho_0/2 & 0 & 0 \\
\rho_1 & \rho_0/2 & 0 \\
\rho_2 & \rho_1 & \rho_0/2
\end{bmatrix}
\begin{bmatrix}
a_2^* \\
a_1^* \\
1
\end{bmatrix}
=
\begin{bmatrix}
0 \\
0 \\
P_2
\end{bmatrix}
-\frac{\rho_0}{2}
\begin{bmatrix}
d_2^* \\
d_1^* \\
1
\end{bmatrix}
\tag{17.85}
$$

Equation (17.82) is extended for third order as

$$
\begin{bmatrix}
\rho_0/2 & 0 & 0 & 0 \\
\rho_1 & \rho_0/2 & 0 & 0 \\
\rho_2 & \rho_1 & \rho_0/2 & 0 \\
\rho_3 & \rho_2 & \rho_1 & \rho_0/2
\end{bmatrix}
\begin{bmatrix}
1 \\
a_1 \\
a_2 \\
0
\end{bmatrix}
=\frac{\rho_0}{2}
\begin{bmatrix}
1 \\
d_1 \\
d_2 \\
0
\end{bmatrix}
+
\begin{bmatrix}
0 \\
0 \\
0 \\
\Delta_3
\end{bmatrix}
\tag{17.86}
$$

where $\Delta_3 = \rho_3 + \rho_2 a_1 + \rho_1 a_2$. Equation (17.85) is extended to

$$
\begin{bmatrix}
\rho_0/2 & 0 & 0 & 0 \\
\rho_1 & \rho_0/2 & 0 & 0 \\
\rho_2 & \rho_1 & \rho_0/2 & 0 \\
\rho_3 & \rho_2 & \rho_1 & \rho_0/2
\end{bmatrix}
\begin{bmatrix}
0 \\
a_2^* \\
a_1^* \\
1
\end{bmatrix}
=
\begin{bmatrix}
0 \\
0 \\
0 \\
P_2
\end{bmatrix}
-\frac{\rho_0}{2}
\begin{bmatrix}
0 \\
d_2^* \\
d_1^* \\
1
\end{bmatrix}
\tag{17.87}
$$

With Eqs. (17.86) and (17.87), the third order equation is generated with reflection coefficient c_3, which can be written as

$$
\begin{aligned}
&\begin{bmatrix}
\rho_0/2 & 0 & 0 & 0 \\
\rho_1 & \rho_0/2 & 0 & 0 \\
\rho_2 & \rho_1 & \rho_0/2 & 0 \\
\rho_3 & \rho_2 & \rho_1 & \rho_0/2
\end{bmatrix}
\left\{
\begin{bmatrix}
1 \\
a_1 \\
a_2 \\
0
\end{bmatrix}
+ c_3
\begin{bmatrix}
0 \\
a_2^* \\
a_1^* \\
1
\end{bmatrix}
\right\} \\[2mm]
&=\frac{\rho_0}{2}
\begin{bmatrix}
1 \\
d_1 \\
d_2 \\
0
\end{bmatrix}
+
\begin{bmatrix}
0 \\
0 \\
0 \\
\Delta_3
\end{bmatrix}
+ c_3
\left\{
\begin{bmatrix}
0 \\
0 \\
0 \\
P_2
\end{bmatrix}
-\frac{\rho_0}{2}
\begin{bmatrix}
0 \\
d_2^* \\
d_1^* \\
1
\end{bmatrix}
\right\} \\[2mm]
&=\frac{\rho_0}{2}
\left\{
\begin{bmatrix}
1 \\
d_1 \\
d_2 \\
0
\end{bmatrix}
- c_3
\begin{bmatrix}
0 \\
d_2^* \\
d_1^* \\
1
\end{bmatrix}
\right\}
+
\begin{bmatrix}
0 \\
0 \\
0 \\
\Delta_3
\end{bmatrix}
+
\begin{bmatrix}
0 \\
0 \\
0 \\
c_3 P_2
\end{bmatrix}
=\frac{\rho_0}{2}
\left\{
\begin{bmatrix}
1 \\
d_1 \\
d_2 \\
0
\end{bmatrix}
- c_3
\begin{bmatrix}
0 \\
d_2^* \\
d_1^* \\
1
\end{bmatrix}
\right\}
\end{aligned}
\tag{17.88}
$$

since $\Delta_3 + c_3 P_2 = 0$ from Eq. (17.80).

It can be shown by induction that $D(z)$ is built up from the sequence of reflection coefficients in exactly the same manner as the prediction error filters are. To visualize that the extension of Eq. (17.86) is by the maximum entropy spectrum, one needs to show that

$$\Phi(z) + \Phi^*(z^{-1}) = \sum_{n=-N}^{N} \rho_n z^n = \frac{\rho_0 \prod_{n=1}^{N}(1-|c_n|^2)}{A(z)A^*(z^{-1})}$$ (17.89)

which corresponds to the MESA assumption:

$$P(f) = \frac{1}{2W} \sum_{n=-N}^{N} \rho_n z^n = \frac{P_N \Delta t}{A(z)A^*(z^{-1})} = \frac{\rho_0 \prod_{n=1}^{N}(1-|c_n|^2)}{A(z)A^*(z^{-1})}$$ (17.90)

since $P_N = P_{N-1}(1-|c_N|^2) = \cdots = \rho_0 \prod_{n=1}^{N}(1-|c_N|^2)$ and $2W\Delta t = 1$. It can be proved by induction.

It is assumed that Eq. (17.89) is true for the $(N-1)$th order. To prove whether it is still true for the Nth order, let $A(z)$ and $D(z)$ be the $(N-1)$th order polynomials. Then if Eq. (17.89) is true for order $N-1$, then one has

$$\frac{\rho_0 \prod_{n=1}^{N-1}(1-|c_n|^2)}{A(z)A^*(z^{-1})} = H(z) + H^*(z^{-1})$$

$$= \frac{\rho_0}{2}\left[\frac{D(z)}{A(z)} + \frac{D^*(z^{-1})}{A^*(z^{-1})}\right] = \frac{\rho_0}{2}\frac{D(z)A^*(z^{-1}) + D^*(z^{-1})A(z)}{A(z)A^*(z^{-1})}$$ (17.91)

which can be reduced to

$$D(z)A^*(z^{-1}) + D^*(z^{-1})A(z) = 2\prod_{n=1}^{N-1}(1-|c_n|^2)$$ (17.92)

With the reflection coefficient c_N, $A'(z)$, and $D'(z)$ of the Nth order can be expressed with $A(z)$ and $D(z)$ of the $(N-1)$th order as

$$A'(z) = A(z) + c_N z^N A^*(z^{-1})$$ (17.93a)

$$D'(z) = D(z) - c_N z^N D^*(z^{-1})$$ (17.93b)

Thus, Nth order can be written as

$$\frac{\rho_0}{2}\left[\frac{D'(z)}{A'(z)} + \frac{D'^*(z^{-1})}{A'^*(z^{-1})}\right] = \frac{\rho_0}{2}\left[\frac{D(z) - c_N z^N D^*(z^{-1})}{A(z) + c_N z^N A^*(z^{-1})} + \frac{D^*(z^{-1}) - c_N^* z^{-N} D(z)}{A^*(z^{-1}) + c_N^* z^{-N} A(z)}\right]$$

$$= \frac{\rho_0}{2}\frac{[D(z)A^*(z^{-1}) + D^*(z^{-1})A(z)] - |c_N|^2[D(z)A^*(z^{-1}) + D^*(z^{-1})A(z)]}{[A(z) + c_N z^N A^*(z^{-1})][A^*(z^{-1}) + c_N^* z^{-N} A(z)]}$$ (17.94)

Lag (Mo)	Observed	Extended acf
13	0.7338	0.5884
14	0.4743	0.4203
15	0.1814	0.1365
16	−0.0430	−0.1459
17	−0.2858	−0.3698
18	−0.4926	−0.4915
19	−0.5560	−0.4668
20	−0.4243	−0.3044
21	−0.1379	−0.0656
22	0.0555	0.1707
23	0.1961	0.3447
24	0.3646	0.4164

TABLE 17.7   Extended Lag Computed by MESA

Using results from Eq. (17.91), Eq. (17.94) can be written as

$$\Phi(z) + \Phi^*(z^{-1}) = \frac{\rho_0 \prod_{n=1}^{N}(1-|c_N|^2)}{[A(z)+c_N z^N A^*(z^{-1})][A^*(z^{-1})+c_N^* z^{-N} A(z)]} = \frac{\rho_0 \prod_{n=1}^{N}(1-|c_N|^2)}{A'(z)A^*(z^{-1})} \tag{17.95}$$

which proves that Eq. (17.89) is also valid for the Nth order. Thus, the extension by Eq. (17.77) is proved by the maximum entropy spectrum.

Example 17.8.   Extend the autocorrelation function using Eq. (17.77) with the coefficients computed from Table 17.5 and plot it.

Solution.   From the given data, the autocorrelation function is extended using Eq. (17.77) with the coefficients from Table 17.7:

$$R_m = \sum_{n=1}^{m} R_{m-n} a_n$$

$$= -0.578 R_{m-1} + 0.177 R_{m-2} - 0.004 R_{m-3} + 0.434 R_{m-4} - 0.077 R_{m-5} + 0.297 R_{m-6} + 0.135 R_{m-7}$$

The extended autocorrelation function is shown in Fig. 17.7. Extension starts from lag 13th month, and it is seen from the figure that the extension generally fits the observations.

17.3.6   Error in Stream flow Forecasting

For the linear forecasting model (17.26), as defined in Eq. (17.28), the mean square error of the forecast is

$$\text{MSE} = \overline{\sum_{m=0}^{N} a_m^* y_{T-m}^* \sum_{n=0}^{N} a_n y_{T-n}} = \sum_{m=0}^{N}\sum_{n=0}^{N} a_m^* \overline{y_{T-m}^* y_{T-n}} a_n = \sum_{m=0}^{N}\sum_{n=0}^{N} a_m^* \rho_{m-n} a_n \tag{17.96}$$

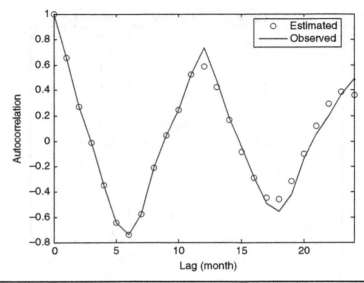

FIGURE 17.7   Extension of autocorrelation.

Let b_n, for n from 0 to N, be the solution to the autocorrelation matrix equation

$$
\begin{bmatrix}
\rho_0 & \rho_{-1} & \cdots & \rho_{1-N} & \rho_{-N} \\
\rho_1 & \rho_0 & \cdots & \rho_{2-N} & \rho_{1-N} \\
\vdots & \vdots & \cdots & \vdots & \vdots \\
\rho_{N-1} & \rho_{N-2} & \cdots & \rho_0 & \rho_{-1} \\
\rho_N & \rho_{N-1} & \cdots & \rho_{-1} & \rho_0
\end{bmatrix}
\begin{Bmatrix}
1 \\
b_1 \\
\vdots \\
b_{N-1} \\
b_N
\end{Bmatrix}
=
\begin{Bmatrix}
P_N \\
0 \\
\vdots \\
0 \\
0
\end{Bmatrix}
\tag{17.97}
$$

Using Eq. (17.97),

$$
\sum_{m=0}^{N}\sum_{n=0}^{N}(a_m^* - b_m^*)\rho_{m-n}(a_n - b_n) = \sum_{m=0}^{N}\sum_{n=0}^{N}(a_m^*\rho_{m-n}a_n - a_m^*\rho_{m-n}b_n - b_m^*\rho_{m-n}a_n + b_m^*\rho_{m-n}b_n)
$$

$$
= \sum_{m=0}^{N}\sum_{n=0}^{N}a_m^*\rho_{m-n}a_n - P_N
\tag{17.98}
$$

By rearranging, one finds

$$
\sum_{m=0}^{N}\sum_{n=0}^{N}a_m^*\rho_{m-n}a_n = P_N + \sum_{m=0}^{N}\sum_{n=0}^{N}(a_m^* - b_m^*)\rho_{m-n}(a_n - b_n)
\tag{17.99}
$$

It is clear that the minimum value occurs when $a_n = b_n$ and the MSE becomes P_N, which means that the forecasting coefficients need to be equal to the coefficients extending the autocorrelation function so that the prediction satisfies the least-squares prediction.

Example 17.9. Forecast the time series, using linear prediction.

Solution. Using the previous data, the prediction coefficients computed, shown in Table 17.5, are used to forecast the time series using linear prediction:

$$\hat{y}_T = \sum_{n=1}^{N} -a_n y_{T-n}$$

$$= -0.578 y_{T-1} + 0.177 y_{T-2} - 0.004 y_{T-3} + 0.434 y_{T-4} - 0.077 y_{T-5} + 0.297 y_{T-6} + 0.135 y_{T-7}$$

It can be seen from Fig. 17.8 that the prediction is similar to the one from the Durbin-Levinson algorithm; however, the confidence interval of the prediction is narrower than the previous one.

The spectral density function defined by Eq. (17.41) is plotted in the following by the maximum entropy spectral analysis:

$$P(f) = \frac{P_N \Delta t}{A(z) A^*(z^{-1})} = \frac{P_N \Delta t}{\left[1 + a_1 z + a_2 z^2 + \cdots + a_N z^Z\right]\left[a + a_1^* z^{-1} + a_2^* z^{-2} + \cdots + a_N^* z^{-N}\right]}$$

$$= \frac{0.3869}{(1 - 0.578 z + 0.177 z^2 + \cdots + 0.135 z^7)(1 - 0.578 z^{-1} + 0.177 z^{-2} + \cdots + 0.135 z^{-7})}$$

where $z = e^{-i2\pi f/\Delta t}$. In Fig. 17.9, $P(f)$ is plotted versus f.

It can be seen from Fig. 17.9 that the maximum value found at frequency $1/12$ and the second-largest value found at frequency $1/4$ match the observed spectral density plot, shown in Fig. 17.2.

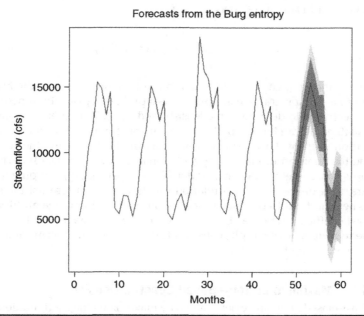

Forecasts from the Burg entropy

FIGURE 17.8   Forecast using MESA.

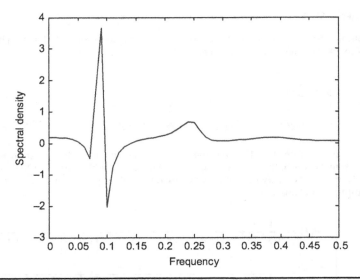

FIGURE 17.9   Spectral density from MESA.

17.4   Configuration Entropy Spectral Analysis

The configuration entropy $H(f)$ is an extension of the Burg entropy, defined (Frieden, 1972; Gull and Daniell, 1978) as

$$H(f) = -\int_{-W}^{W} P(f)\ln P(f)df \qquad (17.100)$$

where f is the frequency considered as a random variable, W is the Nyquist frequency, and $P(f)$ is the spectral density treated as a probability density function of frequency f. The objective is to determine the least-biased $P(f)$. To that end, whatever information is available on f can be encoded in terms of constraints. Thus, the development of configuration entropy–based spectral analysis method contains the following steps: (1) Define the configurational entropy, (2) construct constraints, (3) derive the maximum entropy–based spectral density, (4) define the cepstrum, (5) compute the Lagrange multipliers, (6) extend the autocorrelation function, and (7) forecast the stream flow time series beyond $t = T$. Cepstrum is used to compute the Lagrange multipliers and relate to forecast coefficients. Some of these steps are the same as for the Burg method and will not be repeated. For example, constraints can be defined in terms of the autocorrelation function given by Eq. (17.33).

17.4.1   Maximum Entropy–Based Spectral Density

The least-biased $P(f)$ can be determined by maximizing the configuration entropy in accordance with the principle of maximum entropy (Jaynes, 1957, 1982). Entropy maximizing can be done using the method of Lagrange multipliers. To that end, the lagrangian function can be formulated as

$$L(f) = -\int_{-W}^{W} P(f)\ln P(f)\,df - \sum_{n=-N}^{N} \lambda_n \left[\int_{-W}^{W} P(f)\exp(i2\pi fn\,\Delta t)\,df - \rho_n\right] \quad (17.101)$$

where λ_n, $n = 0, 1, 2, \ldots, N$, are the Lagrange multipliers. Taking the partial derivative of Eq. (17.101) with respect to $P(f)$ and equating the derivative to zero, one obtains:

$$\frac{\partial L(f)}{\partial P(f)} = 0 = -\int_{-W}^{W}\left[\ln P(f) + 1 + \sum_{n=-N}^{N}\lambda_n \exp(i2\pi fn\,\Delta t)\right]df \quad (17.102)$$

Equation (17.102) yields the spectral density:

$$P(f) = \exp\left(-1 - \sum_{n=-N}^{N}\lambda_n e^{i2\pi fn\,\Delta t}\right) \quad (17.103)$$

Equation (17.103) is the entropy-based spectral density with Lagrange multipliers λ_n that need to be determined. Note that unlike the spectral density in polynomial form derived from the Burg entropy, Eq. (17.103) is in exponential form. Therefore, the method for computing the Lagrange multipliers is different, and the cepstrum analysis is introduced to that end.

17.4.2 Definition of Cepstrum

Cepstrum is defined as the inverse Fourier transform of the log-magnitude of spectrum. Taking the Fourier transform of the original stream flow time series, one obtains

$$Y(f) = \sum_{n=-\infty}^{\infty} y(t)e^{-2\pi nif} \quad (17.104)$$

where $Y(f)$ is the Fourier transform of series $y(t)$. Taking the inverse Fourier transform of the log magnitude of Eq. (17.104), one obtains the cepstrum as

$$C(n) = \frac{1}{2\pi}\int_{-\pi}^{\pi} \log|Y(f)|\,e^{2\pi nif}\,df \quad (17.105)$$

Now, consider the cepstrum of the autocorrelation function as an autocorrelation series ρ_n instead of y_t. It is known that the Fourier transform of the autocorrelation function is related to the spectral density as

$$P(f) = \sum_{n=-\infty}^{\infty} \rho(n)e^{-2\pi nif} \quad (17.106)$$

Thus, similar to Eq. (17.105), the cepstrum of the autocorrelation function can be defined by the inverse Fourier transform of the log-magnitude of $P(f)$, which yields

$$e(n) = \frac{1}{2\pi}\int_{-\pi}^{\pi} \log|P(f)|\,e^{2\pi nif}\,df \quad (17.107)$$

The cepstrum of autocorrelation plays an important role in determining the Lagrange multipliers.

17.4.3   Computation of Lagrange Multipliers

Taking the logarithmic transform of Eq. (17.103) for computing the Lagrange multipliers, one obtains

$$1 + \log P(f) = -\sum_{n=-N}^{N} \lambda_n e^{i2\pi fn\Delta t} \tag{17.108}$$

Taking the inverse Fourier transform of Eq. (17.108), one obtains

$$\int_{-W}^{W} 1 + \log P(f) e^{i2\pi fn\Delta t}\, df = \int_{-W}^{W} \left(-\sum_{n=-N}^{N} \lambda_n e^{i2\pi fn\Delta t}\right) e^{i2\pi fn\Delta t}\, df \tag{17.109}$$

where the left side of Eq. (17.109) can be denoted as

$$\int_{-W}^{W} e^{i2\pi fn\Delta t}\, df = \int_{-W}^{W} \cos(2\pi fn\,\Delta t)\, df = \frac{\sin \pi n}{\pi n} \tag{17.110a}$$

Thus, when $n = 0$, Eq. (17.110a) is equivalent to 1; otherwise it equals 0. Therefore, it can be written by a delta function as

$$\int_{-W}^{W} e^{i2\pi fn\Delta t}\, df = \delta_n = \begin{cases} 1 & n = 0 \\ 0 & n \neq 0 \end{cases} \tag{17.110b}$$

The second part of the integral on the left side of Eq. (17.109) becomes

$$\int_{-W}^{W} \log P(f) e^{i2\pi fn\Delta t}\, df = e(n) \tag{17.110c}$$

and on the right side of Eq. (17.109), the Lagrange multipliers can be taken outside of the integrand as

$$\int_{-W}^{W} \left(-\sum_{n=-N}^{N} \lambda_n e^{i2\pi fn\Delta t}\right) e^{i2\pi fn\Delta t}\, df = -\sum_{s=-N}^{N} \lambda_s \int_{-W}^{W} e^{i2\pi f(n-s)\Delta t}\, df = \sum_{s=-N}^{N} \lambda_s \delta_{n-s} \tag{17.110d}$$

Substituting Eqs. (17.110a), (17.110c), and (17.110d) in Eq. (17.109) and integrating, one finds

$$\delta_n + e(n) = -\sum_{s=-N}^{N} \lambda_s \delta_{n-s} \tag{17.111}$$

Equation (17.111) can be expanded as a set of N linear equations:

$$\begin{aligned} \lambda_0 &= -1 - e(0) \\ \lambda_1 &= -e(1) \\ &\vdots \\ \lambda_N &= -e(N) \end{aligned} \tag{17.112}$$

Equation (17.112) shows that the Lagrange multipliers can be determined from the values of the cepstrum which entails the spectral density obtained from Eq. (17.103) by substituting the Lagrange multipliers. However, both Eqs. (17.104) and (17.107) involve integration of complex quantities and may cause computational difficulties. Nadeu (1992) developed a simple method for computing cepstrum for a finite length of data based on the use of the causal part of autocorrelation, where $\rho(n)$ is used only for $0 < n \leq N$ instead of $-N \leq n \leq N$. Using the onesided autocorrelation function, the relationship between the autocorrelation function and its cepstrum can be replaced by the recursive relation as

$$e(n) = \begin{cases} \dfrac{2}{\rho(0)}\left[\rho(n) - \displaystyle\sum_{k=1}^{n-1} \dfrac{k}{n} e(k)\, \rho(n-k)\right] & n > 0 \\[2mm] \log \rho(n) & n = 0 \\[2mm] 0 & n < 0 \end{cases} \tag{17.113}$$

Note that in order to compute $e(k)$, one needs autocorrelation from lag 0 to k. Thus, for given T lag autocorrelation, the cepstrum of autocorrelation can be computed up to lag T, and beyond this lag, the cepstrum is defined as 0. Then the calculated cepstrum from lag 0 to T can be used to compute the Lagrange multipliers, using Eq. (17.112).

Example 17.10.   For stream flow data from the Colorado River tabulated in Table 17.1, compute the cepstrum and Lagrange multipliers from Eq. (17.108).

Solution.   Using Eq. (17.113), $e(n)$ can be determined as

$$e(1) = \frac{2}{\rho(0)}\rho(1) = 2 \times 0.649 = 1.298$$

$$e(2) = \frac{2}{\rho(0)}\left[\rho(2) - \frac{1}{2}e(1)\rho(1)\right] = 2\left[0.26 - \frac{1}{2}1.298 \times 0.649\right] = -0.322$$

Thus, cepstrum is computed in this way up to lag 6 and is tabulated in Table 17.8. The Lagrange multipliers using Eq. (17.112) are

$$\lambda_0 = -1 - e(0) = -1$$

$$\lambda_1 = -e(1) = -1.298$$

and are given in Table 17.8.

Cepstrum		Lagrange Multipliers	
e(0)	0	λ_0	−1
e(1)	1.298	λ_1	−1.298
e(2)	−0.322	λ_2	0.322
e(3)	0.044	λ_3	−0.044
e(4)	−0.534	λ_4	0.534
e(5)	−0.555	λ_5	0.555
e(6)	−0.506	λ_6	0.506

TABLE 17.8   Estimated Cepstrum and the Lagrange Multipliers

17.4.4   Extension of Autocorrelation Function

The next issue is how to extend the autocorrelation function in such a manner that the configurational entropy is maximized. Let a_k be the extension coefficients, so that for $n > N$ the autocorrelation can be extended as

$$\rho_n = \sum_{k=1}^{m} \rho_{n-k} a_k \qquad n > N \tag{17.114}$$

where m is the model order and a_k, $k = 1, 2, \ldots, m$, are the extension coefficients. A proper model order needs to be determined so that the autocorrelation function is extended so as to be close to the observed function at the lowest possible order. The problem of extending the autocorrelation function now reduces to the computation of these coefficients. Note from Eq. (17.113) that the first n lags of cepstrum can be completely determined from the first n lags of autocorrelation and vice versa. Thus, it is possible to write the inverse function of Eq. (17.113)

$$\rho(n) = \begin{cases} \dfrac{\rho(0)}{2} e(n) + \sum_{k=1}^{n-1} \dfrac{k}{n} e(k)\, \rho(n-k)] & n > 0 \\ \exp e(n) & n = 0 \\ \rho(-n) & n < 0 \end{cases} \tag{17.115}$$

With determined model order m the autocorrelation function for $n > N$ can, therefore, be estimated as

$$\rho_n = \sum_{k=1}^{m-1} \dfrac{k}{m} e(k)\rho(n-k) \tag{17.116}$$

Equation (17.116) extends the autocorrelation function by maximizing the configuration entropy. Comparing Eq. (17.114) with Eq. (17.116), one finds the extension coefficients

$$a_k = \sum_{k=1}^{m-1} \dfrac{k}{m} e(k) \tag{17.117}$$

Coefficients a_k can be characterized as prediction coefficients and are used for stream flow series forecasting. Equation (17.115) is a linear combination of the previous values weighted with coefficients a_k. With these coefficients, stream flow forecasting would be exactly the same as with Burg entropy, which is done by the linear Eq. (17.26).

Example 17.11.   Extend the autocorrelation according to Eq. (17.116).

Solution.   Using Eq. (17.117), one finds the coefficient of extensions

$$a_1 = \sum_{k=1}^{m-1} \dfrac{k}{m} e(k) = \dfrac{1}{6}(-1.298) + \dfrac{2}{6}(0.322) + \cdots + \dfrac{5}{6}(0.555) = 0.687$$

Thus, the autocorrelation can be extended using

$$R_m = \sum_{n=1}^{m} R_{m-n} a_n$$

$$= 0.687 R_{m-1} - 0.217 R_{m-2} + 0.004 R_{m-3} - 0.543 R_{m-4} + 0.079 R_{m-5} - 0.278 R_{m-6} - 0.154 R_{m-7}$$

The extended acf is given in Table 17.9 and shown in Fig. 17.10.

Lag (month)	Observed	Extended acf
13	0.733	0.589
14	0.474	0.422
15	0.181	0.135
16	−0.043	−0.146
17	−0.285	−0.301
18	−0.492	−0.492
19	−0.556	−0.479
20	−0.424	−0.311
21	−0.137	−0.167
22	0.055	0.1707
23	0.196	0.345
24	0.364	0.421

TABLE 17.9   Extended Lag Computed by Configuration Entropy

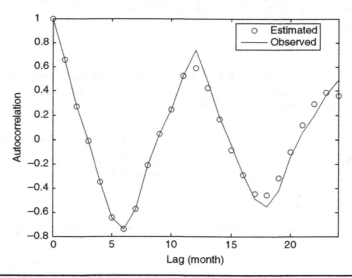

FIGURE 17.10   Extended autocorrelation using configuration entropy.

17.4.5   Stream Flow Forecasting

The extended autocorrelation in Eq. (17.115) is a linear combination of the previous values weighted with the coefficients a_k. Burg (1975) suggested weighing time series using the extension coefficients as

$$y_t = \sum_{k=1}^{m} y_{t-k} a_k \qquad t > T \tag{17.118}$$

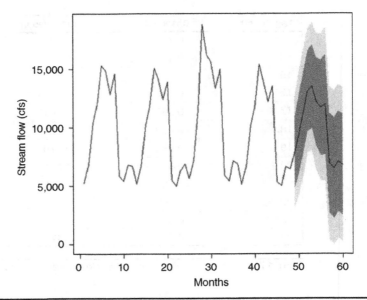

FIGURE **17.11**   Stream flow forecasting using configuration entropy.

Equation (17.118) represents the forecast using the entropy-based extended autocorrelation. It has been shown by Burg (1967) and Krstanovic and Singh (1991b) that Eq. (17.118) satisfies the least-squares prediction.

Example 17.12.   Forecast the stream flow time series using configuration entropy.

Solution.   From the previous data, the prediction coefficients, are computed and shown in Fig. 17.11. The linear prediction of stream flow can be computed as

$$\hat{y}_T = \sum_{n=1}^{N} -a_n y_{T-n}$$

$$= 0.687 y_{T-1} - 0.217 y_{T-2} + 0.004 y_{T-3} - 0.543 y_{T-4} + 0.079 y_{T-5} - 0.278 y_{T-6} - 0.154 y_{T-7}$$

17.5   Minimum Relative Entropy (Cross–Entropy) Spectral Analysis

Three different approaches to the minimum relative entropy-based spectral analysis consider (1) the spectral power as a random variable, (2) the frequency as a random variable and the normalized spectral as the probability distribution, and (3) an input-output filter.

17.5.1   Spectral Power as Random Variable

The spectral power defined by Eq. (17.19) is considered a random variable and applied to the minimum relative entropy. The constraints are given by the expected values of the spectral powers. As a result, the probability distribution of the spectral power can be obtained using the minimum relative entropy. From the definition of the spectral power, there is a spectral power or density value for each frequency. When all frequencies are taken together, the spectral power takes on a multi-dimensional aspect. Minimum relative entropy is employed first for a single dimension, then for two and more dimensions.

One Dimension

Considering the spectral power x as a random variable, one can define the minimum cross entropy as

$$H(q,p) = \int q(x) \ln \frac{q(x)}{p(x)} dx \qquad (17.119)$$

where $p(x)$ and $q(x)$ are prior and posterior distributions of the spectral power. In the case of the given total expected power, the constraints can be expressed as

$$1 = \int q(x) dx \qquad (17.120a)$$

and

$$P = \bar{x} = \int x q(x) dx \qquad (17.120b)$$

where $\bar{x} = P$ is the given expected spectral power. Equation (17.119) can be minimized using the method of Lagrange multipliers. The Lagrange function can be written as

$$L(q,p) = \int q(x) \ln \frac{q(x)}{p(x)} dx + \lambda_0 \left[\int q(x)\, dx - 1 \right] + \lambda_1 \left[\int x q(x)\, dx - \bar{x} \right] \qquad (17.121)$$

Taking the partial derivative with respect to $q(x)$, one obtains

$$q(x) = p(x) \exp(-1 - \lambda_0 - \lambda_1 x) \qquad (17.122)$$

Equation (17.122) is the posterior distribution obtained by minimizing the entropy based on prior information and given constraints.

The prior estimate of $p(x)$ is obtained from a prior guess of the process. For example, if the prior knowledge reveals that the spectral power of the underlying process follows an exponential distribution with expected mean S, then $p(x)$ can be written in the form

$$p(x) = \frac{1}{S} \exp\left(-\frac{x}{S}\right) \qquad (17.123a)$$

where the expected spectral power $S = \int x p(x)\, dx$ is the given prior information. Or if the prior knowledge reveals that underlying process is gaussian with mean S and standard deviation σ estimated from the prior guess, then $p(x)$ becomes

$$p(x) = \frac{1}{\sqrt{2\pi}\sigma} \exp\left[-\frac{(x-S)^2}{2\sigma^2} \right] \qquad (17.123b)$$

When the prior information is given in exponential form, the probability density using minimum relative entropy can be written as

$$q(x) = \frac{1}{S} \exp\left[-1 - \lambda_0 - \left(\lambda_1 + \frac{1}{S}\right) x \right] = \frac{1}{S} \exp(-1 - \lambda_0) \exp\left[-\left(\lambda_1 + \frac{1}{S}\right) x \right] \qquad (17.124)$$

It can be seen from Eq. (17.124) that if the posterior distribution is given by an exponential distribution, then the posterior distribution is also from the exponential family. Substituting in Eq. (17.120), one obtains

$$\frac{1}{P} = \frac{1}{S}\exp(-1-\lambda_0) = \lambda_1 + \frac{1}{S} \tag{17.125}$$

Thus, the Lagrange multipliers can be computed from Eq. (17.125) as

$$\lambda_1 = \frac{1}{P} - \frac{1}{S} \tag{17.126a}$$

$$\lambda_0 = \ln\left(\frac{1}{1+\lambda_1 S}\right) - 1 \tag{17.126b}$$

Example 17.13.   Compute the distribution of spectral power with an expected value of 2.57 and plot it.

Solution.   For given spectral power with an expected value of 2.57, the prior distribution is given as

$$p(x) = \frac{1}{3}\exp\left(-\frac{x}{3}\right)$$

One can have

$$\lambda_1 = \frac{1}{P} - \frac{1}{S} = \frac{1}{2.57} - \frac{1}{3} = 0.0558$$

$$\lambda_0 = \ln\left(\frac{1}{1+\lambda_1 S}\right) - 1 = \ln\left(\frac{1}{1+0.0558\times 3}\right) = -0.1548$$

Thus, $q(x) = 1/3\exp(-1+0.1548)\exp[-(0.0558+1/3)x]$, which is plotted in Fig. 17.12.

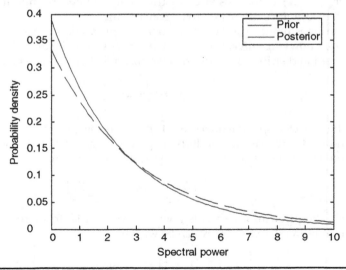

FIGURE **17.12**   Probability density function of spectral power.

Bivariate Case

Since the spectral power is expressed at each frequency, we can consider the spectral power at each frequency as a random variable, and the spectral power can be considered as a vector of random variables. Let x_{k1}, x_{k2} be spectral powers at frequency k_1 and k_2, respectively. The minimum cross entropy can be defined as

$$H(q,p) = \int_D q(x_{k1}, x_{k2}) \ln \frac{q(x_{k1}, x_{k2})}{p(x_{k1}, x_{k2})} \, dx_{k1} \, dx_{k2} \tag{17.127}$$

which is subject to given information in the form of

$$\overline{g}_r = \int_D q(x_{k1}, x_{k2}) g_r(x_{k1}, x_{k2}) \, dx_{k1} \, dx_{k2} \tag{17.128}$$

which can be taken as

$$\overline{x_{k1}} = \int_D x_{k1} q(x_{k1}, x_{k2}) \, dx_{k1} \, dx_{k2} = \overline{g_1} \tag{17.129a}$$

$$\overline{x_{k2}} = \int_D x_{k2} q(x_{k1}, x_{k2}) \, dx_{k1} \, dx_{k2} = \underline{g_2} \tag{17.129b}$$

$$\overline{x_{k1} x_{k2}} = \int_D x_{k1} x_{k2} q(x_{k1}, x_{k2}) \, dx_{k1} \, dx_{k2} = \overline{g_3} \tag{17.129c}$$

Thus, to minimize Eq. (17.127) using Lagrange multipliers, the Lagrange function can be expressed as

$$L(q,p) = \int_D q(x_{k1}, x_{k2}) \ln \frac{q(x_{k1}, x_{k2})}{p(x_{k1}, x_{k2})} \, dx_{k1} \, dx_{k2}$$

$$+ \sum_r \lambda_r \left[\int_D q(x_{k1}, x_{k2}) g_r(x_{k1}, x_{k2}) \, dx_{k1} dx_{k2} - \overline{g}_r \right] \tag{17.130}$$

Taking the partial derivative of Eq. (17.130) with respect to $q(\overline{x})$, $\overline{x} = \{x_{k1}, x_{k2}\}$, one obtains

$$q(x_{k1}, x_{k2}) = p(x_{k1}, x_{k2}) \exp\left[-1 - \sum_r \lambda_r g_r(x_{k1}, x_{k2}) \right] \tag{17.131}$$

Now, using the constraints shown explicitly in Eqs. (17.129a), (17.129b), and (17.129c), one finds

$$q(x_{k1}, x_{k2}) = p(x_{k1}, x_{k2}) \exp(-1 - \lambda_1 x_{k1} - \lambda_2 x_{k2} - \lambda_3 x_{k1} x_{k2}) \tag{17.132}$$

Equations (17.131) and (17.132) are the posterior distributions obtained by minimizing the entropy based on prior information and given constraints.

Example 17.14. Compute the joint distribution of two spectral powers with expected values of 2.57 and 3.45, respectively, and plot it.

Solution. Assume the prior distribution is given as

$$p(x_1, x_2) = \frac{1}{9} \exp\left(-\frac{x_1 + x_2}{3} \right)$$

Using the two constraints given as Eqs. (17.129a) and (17.129b),

$$2.57 = \int_D x_{k1} q(x_{k1}, x_{k2}) \, dx_{k1} \, dx_{k2}$$

$$3.45 = \int_D x_{k1} q(x_{k1}, x_{k2}) \, dx_{k1} \, dx_{k2}$$

The Lagrange multipliers solved by substituting Eq. (17.132) into constraints are

$$\lambda_1 = 20.02 \quad \text{and} \quad \lambda_2 = 16.99$$

Thus, the joint posterior distribution becomes

$$q(x_{k1}, x_{k2}) = 0.04 \exp(-20.02 x_{k1} - 16.99 x_{k2})$$

Multivariate Case

Consider that a single state $\bar{x} = (x_1, x_2, \ldots, x_k, \ldots, x_n)$, in domain D, has a probability density function $q(\bar{x})$ that is to be determined, which is assumed to be a posterior distribution. For any $x_k \in D$, $q(x_k) \geq 0$; consider $p(\bar{x})$ as the prior probability density function and $p(x_k) \geq 0$ for any $x_k \in D$. Each $q(x_k)$ is considered to be IID to the others. The *minimum cross entropy* can be defined as

$$H(q, p) = \int_D q(\bar{x}) \ln \frac{q(\bar{x})}{p(\bar{x})} \, d\bar{x} \tag{17.133}$$

which is subject to the given information in the form of constraints as

$$\bar{g}_r = \int_D q(\bar{x}) g_r(\bar{x}) \, d\bar{x} \tag{17.134}$$

where $g_r(\bar{x})$ is the rth g function of $\bar{x}$. Constraints can be given in terms of the total expected power or expected power at each frequency, which will lead to different solutions to the posterior distribution.

Thus, Eq. (17.133) can be minimized using the method of Lagrange multipliers:

$$L(q, p) = \int_D q(\bar{x}) \ln \frac{q(\bar{x})}{p(\bar{x})} \, d\bar{x} + \sum \lambda_r \left[\int_D q(\bar{x}) g_r(\bar{x}) \, d\bar{x} - \bar{g}_r \right] \tag{17.135}$$

Taking the partial derivative of Eq. (17.135) with respect to $q(\bar{x})$, one obtains

$$q(\bar{x}) = p(\bar{x}) \exp\left[-1 - \sum \lambda_r g_r(\bar{x})\right] \tag{17.136}$$

Equation (17.136) is the posterior distribution obtained by minimizing the entropy based on prior information and given constraints.

In the univariate case, $p(\bar{x})$ is obtained from the prior guess. If the prior information suggests that spectral powers identically follow exponential distributions, $p(\bar{x})$ can be written as a multivariate exponential distribution with the expected spectral powers S_k at each frequency, which is

$$p(x_k) = \frac{1}{S_k} \exp \frac{x_k}{S_k} \tag{17.137}$$

where the expected spectral power $S_k = \int x_k p(\bar{x}) d\bar{x}$ is the given prior information. According to prior information $p(\bar{x})$ can also be taken as a uniform or gaussian distribution. The probability densities are computed first by using the minimum cross entropy, which show that the powers are distributed as a multivariate exponential distribution. Now minimum cross-entropy is applied to the multivariate case with different constraints. In one case, the total expected spectral power is known rather than the expected values for each frequency being given. In the other case, the expected power at each frequency is given. However, for stream flow data the second case occurs more often.

17.5.2 Constraints in Terms of Total Expected Power per Discrete Frequency

The constraints can be written as

$$1 = \int_D q(\bar{x}) d\bar{x} \tag{17.138a}$$

$$P = \int_D \left[\frac{1}{n} \sum_{k=1}^{n} x_k \right] q(\bar{x}) \, d\bar{x} \tag{17.138b}$$

where P is the total expected power and D is the domain of the spectral power. The lagrangian function for minimizing the cross entropy with constraint Eqs. (17.138a) and (17.138b) can be written as

$$L(q,p) = \int q(\bar{x}) \ln \frac{q(\bar{x})}{p(\bar{x})} \, d\bar{x} + \lambda_0 \left[\int_D q(\bar{x}) d\bar{x} - 1 \right] + \lambda_1 \left[\int_D \sum_{k=1}^{n} x_k q(\bar{x}) \, d\bar{x} - P \right] \tag{17.139}$$

Taking the partial derivative of Eq. (17.139) with respect to $q(\bar{x})$, one gets

$$\frac{\partial L(q,p)}{\partial q(\bar{x})} = \ln \frac{q(\bar{x})}{p(\bar{x})} + 1 + \lambda_0 + \lambda_1 \sum_{k=1}^{n} x_k \tag{17.140}$$

Thus, the posterior distribution $q(\bar{x})$ will have the form

$$q(\bar{x}) = p(\bar{x}) \exp \left(-1 - \lambda_0 - \lambda_1 \sum_{k=1}^{n} x_k \right) \tag{17.141}$$

Consider a uniform prior distribution $p(\bar{x}) = C$. By defining a constant $A = p(\bar{x}) \exp(-1 - \lambda_0)$, Eq. (17.141) can be written as

$$q(\bar{x}) = A \exp \left(-\lambda_1 \sum_{k=1}^{n} x_k \right) \tag{17.142}$$

Since $\int_D q(\bar{x}) \, d\bar{x} = A \int_D \exp \left(-\lambda_1 \sum_{k=1}^{n} x_k \right) d\bar{x} = 1$ from Eq. (17.138a), by integration from 0 to infinity with respect to each x_k, one obtains

$$A = \lambda_1^n \tag{17.143}$$

To solve for the Lagrange multiplier, inserting Eqs. (17.142) and (17.143) into Eq. (17.137b), one obtains

$$
P = \frac{1}{n}\int_D \sum_k x_k \lambda_1^n \exp\left(-\lambda_1 \sum_k x_k\right) d\bar{x} = \frac{\lambda_1^n}{n}\sum_k \int dx_k \; x_k \exp(-\lambda_1 x_k) \prod_{m\neq k} \exp(-\lambda_1 x_m)\, dx_m
$$

$$
= \frac{\lambda_1^n}{n}\sum_k \frac{1}{\lambda_1^{n+1}} = \frac{1}{\lambda_1}
$$

(17.144)

since

$$
\int dx_k \; x_k \exp(-\lambda_1 x_k) \prod_{m\neq k} \exp(-\lambda_1 x_m)\, dx_m = \left[\int dx_k \; x_k \exp(-\lambda_1 x_k)\right]\left[\prod_{m\neq k}\int \exp(-\lambda_1 x_m)\, dx_m\right]
$$

$$
= \frac{1}{\lambda_1^2}\cdot\frac{1}{\lambda_1^{n-1}}
$$

Thus, the posterior distribution becomes

$$
q(\bar{x}) = A\exp\left(-\lambda_1 \sum_{k=1}^n x_k\right) = \lambda_1^n \exp\left(-\lambda_1 \sum_{k=1}^n x_k\right) = \prod_{k=1}^n \frac{1}{P}\exp\left(-\frac{x_k}{P}\right)
$$

(17.145)

The posterior distribution is a multivariate, independent, identically distributed (IID) exponential distribution with mean P. This means that the posterior probability distribution computed using the minimum cross entropy with uniform prior distribution maintains the same expected value with the given information.

Example 17.15.   Compute the distribution of spectral power with total expected power, given as 0.1956, and plot it.

Solution.   Given the total expected spectral power $P = 0.1956$, the posterior probability distribution of each frequency x_k is computed by using Eq. (17.144)

$$
q(x_k) = \frac{1}{P}\exp\left(-\frac{x_k}{P}\right) = \frac{1}{0.1956}\exp\left(-\frac{x_k}{0.1956}\right)
$$

which is plotted in Fig. 17.13.

17.5.3   Expected Spectral Power P_k at Each Frequency

Consider a uniform prior distribution again, $p(\bar{x}) = C$. The second constraint changes from Eq. (17.134a) to

$$
1 = \int_D q(\bar{x})\, d\bar{x}
$$

(17.146a)

and from Eq. (17.138b) to

$$
P_k = E[x_k] = \int x_k q(\bar{x})\, d\bar{x}
$$

(17.146b)

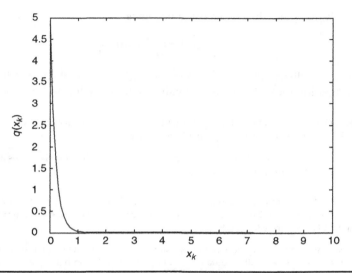

Figure 17.13 Posterior probability density given total expected power.

By minimizing cross entropy, subject to Eqs. (17.146*a*) and (17.146*b*), the posterior distribution becomes

$$q(\vec{x}) = p(\vec{x})\exp\left[-1 - \lambda_0 - \sum_{k=1}^{n}\lambda_k x_k\right] \qquad (17.147)$$

Let $A = p(\vec{x})\exp(-1-\lambda_0)$. To compute the Lagrange multipliers, inserting Eq. (17.147) into Eq. (17.146*a*), one obtains

$$1 = \int A\exp\left(-\sum_{k=1}^{n}\lambda_k x_k\right)dx_k = A\prod_{k=1}^{n}\int \exp(-\lambda_k x_k)\,dx_k \qquad (17.148)$$

Thus,

$$A = \prod_{k=1}^{n}\lambda_k \qquad (17.149)$$

Then the posterior distribution becomes

$$q(\vec{x}) = \prod_{k=1}^{n}\lambda_k \exp(-\lambda_k x_k) \qquad (17.150)$$

By inserting Eq. (17.150) into Eq. (17.146), λ_k can be solved as

$$P_k = \int x_k \lambda_k \exp(-\lambda_k x_k)\,dx_k \int \prod_{m\neq k}^{n}\lambda_m \exp(-\lambda_m x_m)\,dx_m = \frac{1}{\lambda_k} \qquad (17.151)$$

as the second part of the integration equals 1.

Finally, the posterior distribution becomes

$$q(\vec{x}) = \prod_{k=1}^{n} \frac{1}{P_k} \exp\left(-\frac{x_k}{P_k}\right) \tag{17.152}$$

The posterior distribution is still the multivariate exponential distribution; however, it is no longer identical. The expected value is the same for each frequency.

Example 17.16.   For Example 17.15, consider that spectral power at each frequency P_k varies from 0.076 to 0.295. Compute the posterior probability density function.

Solution.   From Eq. (17.152) the posterior probability density function becomes a family of exponential distribution functions, as shown in Fig. 17.14.

Example 17.17.   Obtain additional monthly stream flow data from the Colorado River at the same station as in Table 17.1. Compute the spectral powers and plot them. Also compute the spectral powers using fast Fourier transform (FFT).

Solution.   Monthly stream flow data from station 09415000 can be obtained from the USGS Web site from the early 1900s to 2013 (data are too large for tabulation here). For the 102-yr series, one can obtain 17 sample spectral power series. All the spectral powers have the same pattern, with peak obtained at 1/12 frequency, and second- and third-largest peaks at frequencies of 1/6 and 1/4, respectively. The spectral power for 2002 to 2007 is plotted as an example in Fig. 17.15.

The spectral powers computed using the FFT are shown in Fig. 17.16.

There are in total 17 samples obtained, which means for each frequency there are 17 possible values. Now let us consider the peak spectral power, which occurs at frequency 1/12, as shown in Fig. 17.16. It has an expected mean of 39.05 and standard deviation of 4.285.

Now consider the normal distribution of the spectral power as the prior distribution. The posterior spectral power is in the form

$$q(x) = p(x)\exp(-1 - \lambda_0 - \lambda_1 x)$$

With a given expected mean of 39.05 as the constraint, the Lagrange multipliers can be solved as $\lambda_0 = 0.074$ and $\lambda_1 = -0.15$. Thus, the posterior distribution becomes as shown in Fig. 17.17.

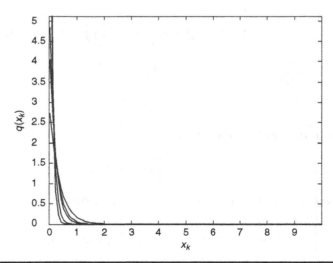

Figure 17.14   Computed posterior distributions.

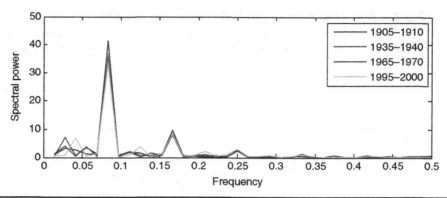

Figure 17.15   Spectral power as a function of frequency.

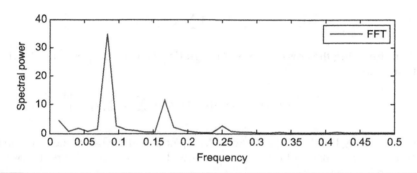

Figure 17.16   Spectral power for Colorado River from 2002 to 2007.

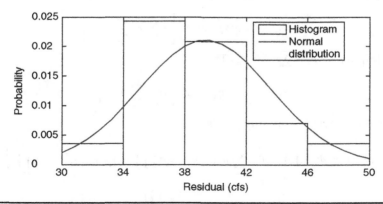

Figure 17.17   The posterior distribution of spectral power at 1/12 frequency.

17.5.4   Spectral Densities with Given Autocorrelation

The spectral density is computed from autocorrelations, and constraints are formed from the expected value of powers. Let T_k be actual power spectrum of $q(x)$, which is an unknown true value

$$T_k = \int_D x_k q(\bar{x})\, d\bar{x} \tag{17.153}$$

Then the autocovariance function can be written as a Fourier transform of the power spectrum

$$R_r = \sum_{k=-n}^{n} T_k \exp(2\pi i r\, \Delta t\, f_k) = \sum_{k=-n}^{n} T_k c_{rk} \qquad -N \le r \le N \tag{17.154}$$

where $c_{rk} = \exp(2\pi i r\, \Delta t\, f_k)$, $f_k = k/N$ the same definition as in Eq. (17.8), and N is the largest lag for the given autocorrelation.

Replacing T_k with Eq. (17.153), one finds the new constraint constructed from the autocovariance function is

$$R_r = \int_D \sum_{k=-n}^{n} x_k c_{rk} q(\bar{x})\, d\bar{x} \tag{17.155}$$

By minimizing the cross entropy through Eq. (17.131), the posterior distribution can be written as

$$q(\bar{x}) = p(\bar{x}) \exp\left(-1 - \lambda_0 - \sum_{r=-m}^{m} \lambda_r \sum_{k=-n}^{n} x_k c_{rk} \right) \tag{17.156}$$

In comparing Eq. (17.156) with Eq. (17.136), note that the Lagrange function has changed as a different constraint is used. Let us now discuss when a different type of prior information is given.

17.5.5   No Prior Information Given

If no prior information is given, one can consider a uniform distribution $p(\bar{x}) = C$. Let $A = p(\bar{x})\exp(-1-\lambda_0)$. Then Eq. (17.156) becomes

$$q(\bar{x}) = A \exp\left[-\sum_{r=-m}^{m} \lambda_r \sum_{k=-n}^{n} x_k c_{rk} \right] \tag{17.157}$$

To simplify, let $u_k = \sum_{r=-m}^{m} \lambda_r c_{rk}$. Then

$$q(\bar{x}) = A \exp\left(-\sum_{k=-n}^{n} u_k x_k \right) \tag{17.158}$$

Integration over 0 to infinity yields

$$1 = \int_0^\infty q(\bar{x})\, d\bar{x} = \int A \exp\left(-\sum_{k=-n}^{n} u_k x_k \right) dx_k = A \prod_{k=1}^{n} \int \exp(-u_k x_k)\, dx_k = A \prod_{k=1}^{n} \frac{1}{u_k} \tag{17.159a}$$

$$A = \prod_k u_k \tag{17.159b}$$

Thus, by inserting Eq. (17.156*b*) into Eq. (17.158), one obtains

$$q(\bar{x}) = \prod_{k=-n}^{n} u_k \exp\left(-\sum_{k=-n}^{n} u_k x_k\right)$$

(17.160)

Equation (17.160) is a multivariate exponential distribution. Thus, the expected value for each frequency can be obtained as

$$T_k = \frac{1}{u_k} = \frac{1}{\sum\limits_{r=-m}^{m} \lambda_r c_{rk}}$$

(17.161)

which is equivalent to that derived using the Burg MESA, Eq. (17.36). The Lagrange multiplier can be solved by substituting Eq. (17.161) into constraint (17.50) as

$$R_r = \sum_{k=-n}^{n} T_k c_{rk} = \sum_{k=-n}^{n} \left(\frac{c_{rk}}{\sum\limits_{r=-m}^{m} \lambda_r c_{rk}}\right)$$

(17.162)

Equation (17.162) can be solved using either Newton's method or the Burg-Levinson algorithm as previously.

17.5.6 Exponential Prior Distribution

If prior information is given as an exponential distribution with expected spectrum power S_k for each frequency, then the prior distribution becomes

$$p(\bar{x}) = \prod_{k} \frac{1}{S_k} \exp\left(-\frac{x_k}{S_k}\right)$$

(17.163)

The posterior distribution given by Eq. (17.156) then becomes

$$\begin{aligned}
q(\bar{x}) &= p(\bar{x}) \exp\left(-1 - \lambda_0 - \sum_{r=-m}^{m} \lambda_r \sum_{k=-n}^{n} x_k c_{rk}\right) \\
&= \exp(-1-\lambda_0) \prod_{k} \frac{1}{S_k} \exp\left(-\frac{x_k}{S_k}\right) \exp\left(-\sum_{r=-m}^{m} \lambda_r c_{rk} x_k\right) \\
&= \exp(-1-\lambda_0) \prod_{k} \frac{1}{S_k} \exp\left[-\left(u_k + \frac{1}{S_k}\right) x_k\right]
\end{aligned}$$

(17.164)

where $u_k = \sum_{r=-m}^{m} \lambda_r c_{rk}$, the same as previously defined.

To satisfy $\int_D q(\bar{x})\, d\bar{x} = 1$, Eq. (17.164) should have the following multivariate exponential distribution form:

$$q(\bar{x}) = \prod_{k=1}^{n} \left(u_k + \frac{1}{S_k}\right) \exp\left[-\left(u_k + \frac{1}{S_k}\right) x_k\right]$$

(17.165)

which means that $\exp(-1 - \lambda_0)$ must satisfy $\exp(-1 - \lambda_0) \prod_k 1/S_k = \prod_{k=-n}^{n} (u_k + 1/S_k)$, and because of the characteristic of the multivariate exponential distribution, the posterior power spectrum becomes

$$T_k = \frac{1}{1/S_k + \sum_{r=-m}^{m} \lambda_r c_{rk}} \tag{17.166}$$

The Lagrange multiplier λ_r can be computed by inserting Eq. (17.166) into Eq. (17.165), which yields

$$R_r = \sum_{k=-n}^{n} T_k c_{rk} = \sum_{k=-n}^{n} \left(\frac{c_{rk}}{1/S_k + \sum_{r=-m}^{m} \lambda_r c_{rk}} \right) \tag{17.167}$$

which can be solved by using Newton's method.

Example 17.18. Let the spectrum be given as shown in Fig. 17.18, whose autocovariance is known up to the fifth lag as [15.7511, 11.6149, 7.8699, 4.5411, 2.0145, 1.1413]. Consider the prior distribution as uniform. Compute the power spectrum for each frequency and plot it.

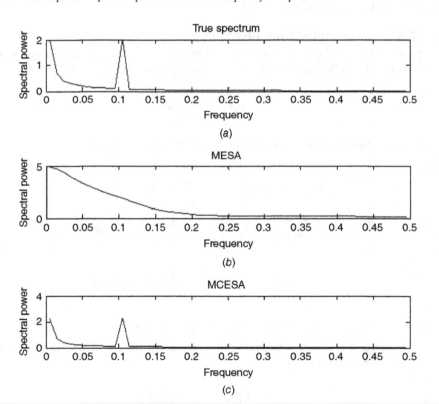

FIGURE **17.18** Computed spectral power.

Solution. Solving Eq. (17.166), one can compute the Lagrange multipliers as $\lambda_n = [1, -0.799, 0.0124, 0.0584, 0.121, -0.093]$.

Thus, the power spectrum for each frequency without prior information can be determined by

$$T_k = \frac{1}{\sum\limits_{r=-m}^{m} \lambda_r c_{rk}}$$

$$= \frac{1}{\exp(2\pi i f_k) - 0.799 \exp(4\pi i f_k) + \cdots + 0.121 \exp(10\pi i f_k) - 0.093 \exp(12\pi i f_k)}$$

It is plotted in Fig. 17.18b, which is equivalent to the result from Burg entropy. When prior information is given, consider the prior signal given as

$$S_k = \begin{cases} 2 & f_k = 0.105 \\ 0 & \text{otherwise} \end{cases}$$

The Lagrange multipliers are solved using Eq. (17.163), which are $\lambda_n = [-0.105, 0.113, -0.044, 0.010, 0.004, -0.000]$. Thus, the posterior spectral power is computed as

$$T_k = \frac{1}{1/S_k + \sum\limits_{r=-m}^{m} \lambda_r c_{rk}}$$

$$= \begin{cases} \dfrac{1}{\dfrac{1}{2} - 0.105 \exp(2\pi i f_k) + \cdots + 0.004 \exp(10\pi i f_k)} & f_k = 0.105 \\[3ex] \dfrac{1}{-0.105 \exp(2\pi i f_k) + \cdots + 0.004 \exp(10\pi i f_k)} & \text{otherwise} \end{cases}$$

The plot is shown in Fig. 17.18c. It can be seen from comparing Figs. 17.18b and c that both methods are capable of determine the decreasing trend of spectral power. But with prior information given at frequency $f_k = 0.105$, the spectral power estimated from the minimum relative entropy can determine the peak accurately as the true spectrum.

17.5.7 Frequency as a Random Variable

For applying the minimum relative entropy theory, frequency is considered as a random variable, and the normalized spectral density is treated as a probability density function. With constraints constructed from the autocorrelation function, the posterior spectral density function can be obtained by minimizing the relative entropy with a given prior distribution. The technique for computing the Lagrange multipliers is the same as that for the configuration entropy.

Development of Minimum Relative Entropy

With frequency f as a random variable, the minimum cross entropy is defined for the normalized spectral density as

$$H(q, p) = \int q(f) \ln \frac{q(f)}{p(f)} \, df \tag{17.168}$$

where $p(f)$ and $q(f)$ are normalized prior and posterior spectral density functions, respectively. In this case, constraints are given in the form of autocorrelation functions:

$$\rho_r = \int_{-W}^{W} q(f)\exp(2\pi i r \Delta t \, f)\, df \tag{17.169}$$

By minimizing the cross entropy, the posterior distribution can be written as

$$q(f) = p(f)\exp\left(-1 - \sum_{r=-m}^{m} \lambda_r e^{i2\pi f r \Delta t}\right) \tag{17.170}$$

Equation (17.170) is the posterior spectral density obtained by minimizing the entropy based on prior information and given constraints.

If the prior information is white noise and the prior spectral density is constant, then the posterior distribution becomes

$$q(f) = \exp\left(-1 - \sum_{r=-m}^{m} \lambda_r e^{i2\pi f r \Delta t}\right) \tag{17.171}$$

which is equivalent to the solution from the configuration entropy. The Lagrange multipliers can be computed as from the configuration entropy using cepstrum analysis, which is discussed in Sec. 17.4.2.

Computation of Lagrange Multipliers

Dividing both sides of Eq. (17.170) by $p(f)$, one obtains

$$\frac{q(f)}{p(f)} = \exp\left(-1 - \sum_{r=-m}^{m} \lambda_r e^{i2\pi f r \Delta t}\right) \tag{17.172}$$

Following the same steps stated in Sec. 17.4.3 and taking the inverse Fourier transform of the log magnitude of Eq. (17.172), one obtains

$$\int_{-W}^{W} [1 + \log q(f) - \log p(f)]e^{i2\pi f n \Delta t}\, df = \int_{-W}^{W} \left(-\sum_{n=-N}^{N} \lambda_n e^{i2\pi f n \Delta t}\right)e^{i2\pi f n \Delta t}\, df \tag{17.173a}$$

With the integrated result of Eq. (17.110), Eq. (17.173) can be integrated as

$$\delta_n + e(n) - e_p(n) = -\sum_{s=-N}^{N} \lambda_s \delta_{n-s} \tag{17.173b}$$

where $e_p(n)$ and $e(n)$ represent the cepstrum of prior and posterior distributions and δ_n is the delta function as shown in Eq. (17.110b).

Equation (17.173) can be expanded as a set of N linear equations:

$$\begin{aligned}
\lambda_0 &= -1 - e(0) + e_p(0) \\
\lambda_1 &= -e(1) + e_p(1) \\
&\vdots \\
\lambda_k &= -e(k) + e_p(k)
\end{aligned} \tag{17.174}$$

When a uniform prior distribution is given, the cepstrum $e_p = 0$ and the solution of Eq. (17.174) becomes the one derived using the configuration entropy.

Example 17.19. Using the Colorado River data in Table 17.1, assume the prior spectral density given as

$$p(f) = \begin{cases} 24f & 0 \leq f \leq 1/12 \\ 2.4 - 4.8f & 1/12 < f \leq 0.5 \end{cases}$$

which ensures having peak at frequency 1/12. Compute the cepstrum, Lagrange multipliers, and posterior distribution.

Solution. From Eq. (17.174), the Lagrange multipliers can be computed as shown in Table 17.10. Thus,

$$q(f) = p(f)\exp\left(-1 - \sum_{n=-N}^{N} \lambda_n e^{i2\pi fn\Delta t}\right)$$

$$= p(f)\exp\{-1 - 2[-1 - 1.268\cos(2\pi f) + 0.644\cos(4\pi f) + \cdots - 1.999\cos(12\pi f)]\}$$

Extension of Autocorrelation Function

Similar to the analysis from the configuration entropy in Sec. 17.4.3, the cepstrum can be calculated as

$$e(n) = e_p(n) + \begin{cases} \dfrac{2}{\rho(0)}\left\{\rho(n) - \sum_{k=1}^{n-1}\dfrac{k}{n}[e(k) - e_p(k)]\,\rho(n-k)\right\} & n > 0 \\ \log\rho(n) & n = 0 \\ 0 & n < 0 \end{cases} \qquad (17.175)$$

and vice versa

$$\rho(n) = \begin{cases} \dfrac{\rho(0)}{2}[e(n) - e_p(n)] + \sum_{k=1}^{m-1}\dfrac{k}{m}[e(k) - e_p(k)]\,\rho(n-k) & n > 0 \\ \exp[e(n) - e_p(n)] & n = 0 \\ \rho(-n) & n < 0 \end{cases} \qquad (17.176)$$

Posterior Cepstrum		Prior Cepstrum		Lagrange Multipliers	
$e(0)$	0	$e_p(0)$	0	λ_0	−1
$e(1)$	−0.319	$e_p(1)$	0.949	λ_1	1.268
$e(2)$	−0.156	$e_p(2)$	0.488	λ_2	0.644
$e(3)$	−0.106	$e_p(3)$	0.346	λ_3	0.452
$e(4)$	−0.532	$e_p(4)$	0.270	λ_4	0.802
$e(5)$	0.151	$e_p(5)$	0.217	λ_5	0.066
$e(6)$	0.683	$e_p(6)$	−1.316	λ_6	−1.999

TABLE 17.10 Computation of Cepstrum and Lagrange Multipliers

As can be seen by comparing Eq. (17.175) to Eq. (17.113), the only difference is in adding the cepstrum of the prior spectral estimation.

Let

$$a_n = \sum_{k=1}^{m-1} \frac{k}{m}[e(k) - e_p(k)] \tag{17.177}$$

such that

$$\rho_n = -\sum_{k=1}^{m} \rho_{n-k} a_k \qquad n > N \tag{17.178}$$

Example 17.20.   Using the Colorado River stream flow data, compute the prediction coefficient; extend the autocorrelation function, and plot it.

Solution.   Using the coefficient computed from Example 17.19, the prediction coefficient computed through Eq. (17.177) can be determined as

$$a_n = [1, 0.132, -0.334, 0.025, -0.231, -0.174]^T$$

Thus, the autocorrelation can be extended by using Eq. (17.178) with prediction coefficient a_n and is plotted in Fig. 17.19.

Furthermore, the stream flow can be forecasted, as shown in Figure 17.20, using the above coefficient with linear prediction, which becomes

$$\hat{y}_T = \sum_{n=1}^{N} -a_n y_{T-n}$$

$$= 0.132 y_{T-1} - 0.334 y_{T-2} + 0.025 y_{T-3} - 0.231 y_{T-4} - 0.174$$

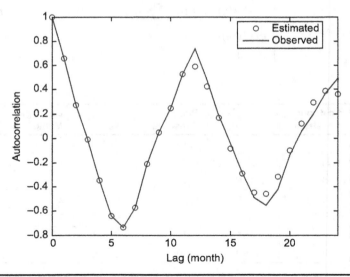

FIGURE **17.19**   Extended autocorrelation function.

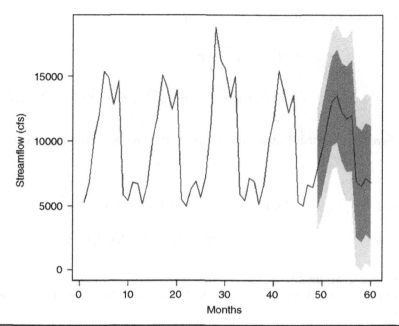

Figure 17.20   Stream flow forecast using minimum relative entropy.

17.5.8 Linear Filter Method

The linear filter method is another method to derive the spectral density by the minimum cross entropy spectral analysis. Consider that a signal with probability density function $p(\bar{x})$ is passed through a filter with characteristic function or transfer function $Y(f)$. By passing through the filter, the original spectral powers S_k will be multiplied by the factor with magnitudes A_k as

$$T_k = S_k A_k \tag{17.179}$$

where T_k is the true spectral power to be determined, or posterior spectral power as defined in Sec. 17.5.4. The multiplying factor is defined from the transfer function $Y(f)$ as

$$A_k = |Y(f_k)|^2 \tag{17.180}$$

where $f_k = k/N = \omega_k/2\pi$ as defined previously. Then the output signals $q(\bar{y})$ are related to the input $p(\bar{x})$ by using the magnitude

$$q(\bar{y}) = \frac{p(x_1, x_2, \ldots, x_n)}{A_1 A_2 \cdots A_n} \tag{17.181a}$$

where $A_k x_k = y_k$. Thus one can obtain

$$q(\bar{y}) = \frac{p(y_1/A_1, y_2/A_2, \ldots, y_n/A_n)}{A_1 A_2 \cdots A_n} \tag{17.182b}$$

The constraints shown in Eq. (17.154) can be recast using Eq. (17.179) as

$$R_r = \sum_{k=-n}^{n} T_k c_{rk} = \sum_{k=-n}^{n} S_k A_k c_{rk} \tag{17.183}$$

Then the minimum cross entropy becomes

$$H(q,p) = \int p(\bar{y}) \log \frac{p(\bar{y})}{q(\bar{y})} \, d\bar{y}$$

$$= \int p(\bar{y}) \log \left(\frac{p(y_1/A_1, y_2/A_2, \ldots, y_n/A_n)}{A_1 A_2 \cdots A_n} \right) d\bar{y} - \sum_k \log A_k \tag{17.184}$$

When the prior distribution is uniform, the filter becomes zero, and Eq. (17.184) becomes

$$H(q,p) = -\sum_k \log A_k \tag{17.185}$$

When the prior distribution is exponential, Eq. (17.184) reduces to

$$H(q,p) = -\sum_k (1 - A_k + \log A_k) \tag{17.186}$$

No Prior Information
Considering a uniform prior spectrum, the output is the same with the magnitude factor, thus

$$A_k = |Y(f_k)|^2 = T_k \tag{17.187}$$

and the entropy derived in Eq. (17.185) becomes

$$H(q,p) = -\sum_k \log T_k \tag{17.188}$$

For minimizing the relative entropy, subject to the constraint given by Eq. (17.183), one can write

$$\frac{\partial L(q,p)}{\partial T_k} = \frac{\partial \left[H(q,p) + \sum_{r=-m}^{m} \lambda_r T_k c_{rk} \right]}{\partial T_k} = \frac{\partial \left(-\sum_k \log T_k + \sum_{r=-m}^{m} \lambda_r T_k c_{rk} \right)}{\partial T_k}$$

$$= -\frac{1}{T_k} + \sum_{r=-m}^{m} \lambda_r c_{rk} = 0 \tag{17.189}$$

Thus, one gets $T_k = 1/\sum_{r=-m}^{m} \lambda_r c_{rk}$ which is the same as Eq. (17.161).

Exponential Prior Distribution
Considering an exponential prior distribution with the expected value S_k, passing through the filter, the output and input follow the relationship defined in Eq. (17.179). Thus, the entropy from Eq. (17.186) becomes

$$H(q,p) = -\sum_k (1 - A_k + \log A_k) = -\sum_k \left(1 - \frac{T_k}{S_k} + \log \frac{T_k}{S_k} \right) \tag{17.190}$$

Minimizing the relative entropy with no constraint, one can write

$$\frac{\partial L(q,p)}{\partial T_k} = \frac{\partial\left(H(q,p) + \sum\limits_{r=-m}^{m} \lambda_r T_k c_{rk}\right)}{\partial T_k}$$

$$= \frac{\partial\left[-\sum\limits_{k}\left(1 - \frac{T_k}{S_k} + \log\frac{T_k}{S_k}\right) + \sum\limits_{r=-m}^{m} \lambda_r T_k c_{rk}\right]}{\partial T_k} \qquad (17.191)$$

$$= \frac{1}{S_k} - \frac{1}{T_k} + \sum\limits_{r=-m}^{m} \lambda_r c_{rk} = 0$$

Solving for T_k, one obtains

$$T_k = \frac{1}{1/S_k + \sum\limits_{r=-m}^{m} \lambda_r c_{rk}} \qquad (17.192)$$

which is the same as Eq. (17.167).

Questions

1. Consider at least 10 yr of monthly stream flow (cfs) data from a USGS gauging station on any river. Compute the autocovariance and autocorrelation of these stream flow data for different lags, and plot them.

2. Compute the spectrum of stream flow from the data in Question 1. Also specify the Fourier coefficients and the spectral power for the data and tabulate them. Now compute the spectral density of the stream flow.

3. Compute the spectral density using the autocorrelation function of stream flow data given in Question 1.

4. Fit an AR model to the monthly stream flow given in Question 1. Then forecast the monthly stream flow and plot it.

5. Compute the prediction error and prediction coefficients up to an appropriate order for monthly flow in Question 1.

6. Extend the autocorrelation function and compute the data in Question 1.

7. Forecast the stream flow of Question 1, using linear prediction.

8. For the stream flow data in Question 1, compute the cepstrum and the Lagrange multipliers for the configuration entropy formulation.

9. Extend the autocorrelation using the configuration formulation.

10. Forecast the stream flow of Question 1, using the configuration entropy.

11. Compute the distribution of spectral power with an expected value of 2.25 and 3.25 and plot it.

12. Compute the joint distribution of two spectral powers with an expected value of 2.25 and 3.25, respectively, and plot it.

13. Compute the distribution of spectral power with total expected power given as 0.175 and plot it.

14. Consider that spectral power at each frequency P_k varies from 0.05 to 0.28. Compute the posterior probability density function.

15. Using the stream flow data, compute the prediction coefficient, extend the autocorrelation function, and plot it.

References

Box, G. E. P., and Jenkins G. M. (1970). *Time Series Analysis; Forecasting and Control.* Holden-Day, San Francisco, Calif.

Box, G. E. P., and Jenkins G. M. (1976). *Time Series Analysis: Forecasting and Control.* Holden-Day, San Francisco, Calif.

Burg, J. P. (1967). Maximum entropy spectral analysis. Paper presented at 37th Annual Meeting of Society of Exploration Geophysicists, Oklahoma City, Okl.

Chatfield, C. (1984). *The Analysis of Time Series: An Introduction.* Chapman and Hall, New York.

Frieden, B. R. (1972). Restoring with maximum likelihood and maximum entropy. *Journal of Optical Society of America*, vol. 62, pp. 511–518.

Jaynes, E. T. (1982). On the rationale of maximum-entropy methods. *Proceedings of the IRRR*, vol. 70, no. 9, pp. 939–952.

Krstanovic, P. F., and Singh, V. P. (1991b). A univariate model for long-term stream flow forecasting: 2. Application. *Stochastic Hydrology and Hydraulics*, vol. 5, no. 3, pp. 189–205.

Nadeu, C. (1992). Finite length cepstrum modeling—A simple spectrum estimation technique. *Signal Process*, vol. 26, no. 1, pp. 49–59.

Wood, A. W., and Lettenmaier, D. P. (2008). An ensemble approach for attribution of hydrologic prediction uncertainty. *Geophysical Research Letters*, vol. 35, L14401, doi:10.1029/2008GL034648.

Wood, A. W., Kumar, A. and Lettenmaier, D. P. (2005). A retrospective assessment of National Centers for Environmental Prediction climate model-based ensemble hydrologic forecasting in the western United States. *Journal of Geophysical Research—Atmospheres*, vol. 110, D04105, doi:10.1029/2004JD004508.

Woodbury, A. D., and Ulrych, T. J. (1993). Minimum relative entropy-forward probabilistic modeling. *Water Resources Research*, vol. 29, no. 8, pp. 2847–2860.

Woodbury, A. D., and Ulrych, T. J. (1996). Minimum relative entropy inversion: Theory and application to recovering the release history of a groundwater contaminant. *Water Resources Research*, vol. 32, no. 9, pp. 2671–2681.

Woodbury, A. D., and Ulrych, T. J. (1998). Minimum relative entropy and probabilistic inversion in groundwater hydrology. *Stochastic Hydrology and Hydraulics*, vol. 12, no. 5, pp. 317–358.

Wu, C. L., Chau, K. W. and Li, Y. S. (2009). Predicting monthly stream flow using data-driven models coupled with data-preprocessing techniques. *Water Resources Research*, vol. 45, W08432, doi:10.1029/2007WR006737.

Yakowitz, S. (1987). Nearest neighbor method for time series analysis. *Journal of Time Series Analysis*, vol. 8, no. 2, pp. 235–247.

Zealand, C. M., Burn, D. H. and Simonovic, S. P. (1999). Short term stream flow forecasting using artificial neural networks. *Journal of Hydrology*, vol. 214, no. 1–4, pp. 32–48.

Further Reading

Abbott, M. B., Bathurst, J. C. Cunge, J. A. O'Connell, P. E. and Rasmussen, J. (1986). An introduction to the European Hydrological System—Systeme Hydrologique Europeen, SHE 2. Structure of a physically-based, distributed modeling system. *Journal of Hydrology*, vol. 87, no. 1–2, pp. 61–77.

Bath, M. (1974). *Spectral Analysis in Geophysics*. vol. 7, *Developments in Solid Earth Geophysics*. Elsevier, Amsterdam.

Bendat, J. S., and Piersol, A. G. (1971). *Random Data: Analysis and Measurement Procedures*. Wiley-Interscience, New York.

Beven, K., and Freer, J. (2001). A dynamic TOPMODEL. *Hydrological Processes*, vol. 15, no. 10, pp. 1993–2011.

Beven, K. J., Kirkby, M., Schofield, J. N. and Tagg, A. F. (1984). Testing a physically-based flood forecasting-model (TOPMODEL) for 3 UK catchments. *Journal of Hydrology*, vol. 69, no. 1–4, pp. 119–143.

Box, G. E. P., and Cox, D. R. (1964). An analysis of transformations. *Journal of Royal Statistical Society*, B, vol. 26, no. 2, pp. 211–252.

Burg, J. P. (1969). A new analysis technique for time series data. Paper presented at NATO Advanced Institute on Signal Processing with Emphasis on Underwater Acoustics.

Burg, J. P. (1972a). The relationship between maximum entropy spectra and maximum likelihood spectra. *Geophysics*, vol. 37, no. 2, pp. 375-376.

Burg, J. P. (1972). The relationship between maximum entropy spectra and maximum likelihood spectra. In: *Modern Spectral Analysis*, edited by D. G. *Childers*. pp. 130–131. IEEE Press, New York.

Burg, J. P. (1975). Maximum entropy spectral analysis. Unpublished Ph.D. dissertation, Stanford University, Palo Alto, Calif.

Chatfield, C. (1975). *The Analysis of Time Series: Theory and Practice*. Chapman and Hall, London.

Chiew, F. H. S., Peel, M. C. and Western, A. W. (2002). Application and testing of the simple rainfall-runoff model SIMHYD. In: *Mathematical Models of Small Watershed Hydrology and Applications*, edited by V. P. Singh and D. K. Frevert. Water Resources Publications, Littleton, Colo.

Cloke, H. L., and Pappenberger, F. (2009). Ensemble flood forecasting: A review. *Journal of Hydrology*, vol. 375, no. 3–4, pp. 613–626.

Dalezios, N. R., and Tyraskis, P. A. (1999). Maximum entropy spectra for regional precipitation analysis and forecasting. *Journal of Hydrology*, vol. 109, pp. 25–42.

Edward, J. A., and Fitelson, M. M. (1973). Notes on maximum-entropy processing. *IEEE Transactions, Information Theory*, vol. 19, no. 2., pp. 232–234.

Edwards, J. A., and Fitelson, M. M. (1973). Notes on maximum-entropy processing. *IEEE Transactions on Information Theory*, vol. IT-19, pp. 232–234.

Elshorbagy, A., Simonovic, S. P. and Panu, U. S. (2002). Estimation of missing stream flow data using principles of chaos theory. *Journal of Hydrology*, vol. 255, no. 1–4, pp. 123–133.

Fougere, P. F., Zawalick, E. J. and Radoski, H. R. (1976). Spontaneous line splitting in maximum entropy power spectrum analysis. *Physics of the Earth and Planetary Interiors*, vol. 12, pp. 201–207.

Frausto-Solis, J., Pita, E. and Lagunas, J. (2008). Short-term stream flow forecasting: ARIMA versus neural networks. *American Conference on Applied Mathematics (MATH'08)*, Harvard University, Cambridge, Mass., pp. 402–407.

Galeati, G. (1990). A comparison of parametric and non-parametric methods for runoff forecasting. *Hydrological Sciences Journal*, vol. 35, no. 1, pp. 79–94.

Gull, S. F., and Daniell G. J. (1978). Image reconstruction from incomplete and noisy data. *Nature*, vol. 272, pp. 686–690, doi: 10.1038/272686a0.

Haltiner, J. P., and Salas, J. D. (1988). Development and testing of a multivariate, seasonal ARMA (1,1) model. *Journal of Hydrology*, vol. 104, no. 1–4, pp. 247–272.

Hannan, E. J., and Kavalieris, L. (1984). Multivariate linear time-series models. *Advanced Applied Probability*, vol. 16, no. 3, pp. 492–561.

Harris, F. J. (1978). On the use of windows for harmonic analysis with the discrete Fourier transform. *IEEE Proceedings*, vol. 66, no. 1, pp. 51–83.

Hedgge, B. J., and Masselink, G. (1996). Spectral analysis of geomorphic time series: Auto-spectrum. *Earth Surface Processes and Landforms*, vol. 21, pp. 1021–1040.

Hipel, K. W., and McLeod, A. I. (1994). *Time Series Modelling of Water Resources and Environmental Systems. Developments in Water Science*, vol. 45. Elsevier, Burlington.

Jain, D., and Singh, V. P. (1986). A comparison of transformation-methods for flood frequency-analysis. *Water Resources Bulletin*, vol. 22, no. 6, pp. 903–912.

Jaynes, E. T. (1957). Information theory and statistical mechanics. *Physical Review*, vol. 106, no. 4, pp. 620–630.

Jenkins, G. M., and Watts, D. G. (1968). *Spectral Analysis and Its Applications*. Holden-Day, San Francisco, Calif.

Jimenez, C., Mcleod, A. I. and Hipel, K. W. (1989). Kalman filter estimation for periodic autoregressive-moving average models. *Stochastic Hydrology and Hydraulics*, vol. 3, no. 3, pp. 227–240.

Johnson, R. W., and Shore, J. E. (1983). Which is better entropy expression for speech processing: -SLogS or LogS? Report, Naval Research Laboratory, Washington.

Jones, R. H., and Brelsfor, W. M. (1967). Time series with periodic structure. *Technometrics*, vol. 9, no. 1, pp. 403–408.

Kalman, P. E. (1960). A new approach to linear filtering and prediction problems. *ASME Transactions, Journal of Basic Engineering*, vol. 83D, pp. 95–108.

Kapur, J. N. (1999). *Maximum Entropy Models in Science and Engineering*. Wiley Eastern, New Delhi, India.

Kapur, J. N., and Kesavan, H. K. (1997). *Generalized Maximum Entropy Principle (with Applications)*. Sandford Educational Press, University of Waterloo, Waterloo, Canada.

Kapur, J. N., and Kesavan, H. K. (1992). *Entropy Optimization Principles with Applications*. Academic Press, New York.

Karlsson, M., and S. Yakowitz (1987). Nearest-neighbor methods for nonparametric rainfall runoff forecasting. *Water Resources Research*, vol. 23, no. 7, pp. 1300–1308.

Katsakos-Mavromichalis, N.A., Tzannes, M.A. and Tzannes, N. S. (1985). Frequency resolution: A comparative study of four entropy methods. *Kybernetes*, vol. 15, no. 1, pp. 25–32.

Krause, P., Boyle, D. P. and Base F. (2005). Comparison of different efficiency criteria for hydrological model asssessment. *Advances in Geosciences*, vol. 5, pp. 89–97.

Kaveh, M., and Cooper G. R. (1976). An empirical investigation of the properties of the autoregressive spectral estimator. *IEEE Transactions on Information Theory*, vol. IT-22, no. 3, pp. 313–323.

Kerpez, K. J. (1989). The power spectral density of maximum entropy charge constrained sequences. *IEEE Transactions*, vol. 35, no. 3, pp. 692–695.

Krstanovic, P. F., and Singh, V. P. (1991a). A univariate model for long-term stream flow forecasting .1. Development. *Stochastic Hydrology and Hydraulics*, vol. 5, no. 3, pp.173–188.

Krastanovic, P. F., and Singh, V. P. (1991a). A univariate model for long-term stream flow forecasting: 2. Application. *Stochastic Hydrology and Hydraulics*, vol. 5, pp. 189–205.

Krastanovic, P. F., and Singh, V. P. (1993a). A real-time flood forecasting model based on maximum entropy spectral analysis: I. Development. *Water Resources Management*, vol. 7, pp. 109–129.

Krastanovic, P. F., and Singh, V. P. (1993b). A real-time flood forecasting model based on maximum entropy spectral analysis: II. Application. Water Resources Management, vol. 7, no. 2, pp 131–157.

Lacoss, R. T. (1971). Data adaptive spectral analysis methods. *Geophysics*, vol. 36, no. 4, pp. 661–675.

Lall, U., and Sharma, A. (1996). A nearest neighbor bootstrap for resampling hydrologic time series. *Water Resources Research*, vol. 32, no. 3, pp. 679–693, 610.1029/1095WR02966.

Legates, D. R., and Davis, R. E. (1997). The continuing search for an anthropogenic climate change signal: Limitations of correlation-based approaches. *Geophysical Research Letters*, vol. 24, no. 18, pp. 2319–2322.

Legates, D. R., and McCabe, G. J. (1999). Evaluating the use of "goodness-of-fit" measures in hydrologic and hydroclimatic model validation. *Water Resources Research*, vol. 35, no. 1, pp. 233–241.

Mack, Y. P., and Rosenblatt, M. (1979). Multivariate K-nearest neighbor density estimates. *Journal of Multivariate Analysis*, vol. 9, no. 1, pp. 1–15.

Nadeu, C., Sanvicente, E. and Bertranm, M. (1981). A new algorithm for spectral estimation. Paper presented at International Conference on Digital Signal Processing, Florence, Italy.

Nash, J. E., and Sutcliffe J. V. (1970). River flow forecasting through conceptual models: Part1: A discussion of principles. *Journal of Hydrology*, vol. 10, pp. 282–290.

Newman, W. I. (1977). Extension to the maximum entropy method. *IEEE Transactions on Information Theory*, vol. IT-23, no. 1, pp. 89–93.

Noonan, J. P., Tzannes, N. S. and Costfello, T. (1976). On the inverse problem of entropy maximizations. *IEEE Transactions on Information Theory*, vol. IT-22, pp. 120–123.

Nowak, K., Prairie, J., Rajagopalan, B. and Lall, U. (2010). A non-parametric stochastic approach for multisite disaggregation of annual to daily stream flow. *Water Resources Research*, vol. 46, W08529, doi:10.1029/2009WR008530.

Oppenheim, A. V., and Schafer, R. W. (1975). *Digital Signal Processing*, Prentice Hall, Englewood Cliffs, N. J.

Padmanabhan, G., and Rao, A. R. (1986). Maximum entropy spectra of some rainfall and river flow time series from southern and central India. *Theoretical and Applied Climatology*, vol. 37, pp. 63–73.

Padmanabhan, G., and Rao, A. R. (1999). Maximum entropy spectral analysis of hydrologic data. *Water Resources Research*, vol. 24, no. 9, pp. 1519–1533.

Papademetriou, R. C. (1998). Experimental comparison of two information-theoretic spectral estimators. Paper presented at *Signal Processing Proceedings*, 1998. ICSP '98.

Papoulois, A. (1973). Minimum-bias windows for high-resolution spectral estimates. *IEEE Transactions on Information Theory*, vol. IT-19, no. 1, pp. 9–12.

Papoulis, A. (1981). Maximum-entropy and spectral estimation—A review. *IEEE Transactions, Acoustic Speech*, vol. 29, no. 6, pp. 1176–1186.

Pardo-Iguzquiza, E., and Rodriguez-Tovar F. J. (2005). MAXENPER: A program for maximum spectral estimation with assessment of statistical significance by the permutation test. *Computers and Geosciences*, vol. 31, pp. 555–567.

Pardo-Iguzquiza, E., and Rodriguez-Tovar F. J. (2005). Maximum entropy spectral analysis of climatic time series revisited: Assessing the statistical significance of estimated spectral peaks. *Journal of Geophysical Research*, vol. 111, pp. 1–8, D010102, doi:10.1029/2005JD0006293.

Prairie, J., Rajagopalan, B., Fulp, T., and Zagona, E. (2006). Modified K-NN model for stochastic stream flow simulation. *Journal of Hydrologic Engineering*, vol. 11, no. 4, pp. 371–378.

Salas, J. D., and Lee, T. (2010). Nonparametric simulation of single site seasonal stream flows. *Journal of Hydrologic Engineering*, vol. 15, no. 4, pp. 284–296.

Salas, J. D., and Obeysekera, J. T. B. (1982). ARMA model identification of hydrologic time-series. *Water Resources Research*, vol. 18, no. 4, pp. 1011–1021.

Salas, J. D., Boes, D. C. and Smith, R. A. (1982). Estimation of ARMA models with seasonal parameters. *Water Resources Research*, vol. 18, no. 4, pp. 1006–1010.

Schaefli, B., and Gupta H. V. (2007). Do Nash values have value? *Hydrological Process*, vol. 21, no. 15, pp. 2075–2080.

Sharma, A., and O'Neill R. (2002). A nonparametric approach for representing interannual dependence in monthly stream flow sequences. *Water Resources Research*, vol. 38, no. 7, p. 1100, doi:1110.1029/2001WR000953.

Sharma, A., Tarboton, D., and Lall, U. (1997). Stream flow simulation: A nonparametric approach. *Water Resources Research*, vol. 33, no. 2, pp. 291–308, doi:210.1029/1096WR02839.

Shore, J. E. (1979). Minimum cross-entropy spectral analysis. NRL Memorandum Report 3921, Naval Research Laboratory, Washington.

Shore, J. E. (1981). Minimum cross-entropy spectral-analysis. IEEE Transactions, *Acoustic Speech*, vol. 29, no. 2, pp. 230–237.

Singh, V. P. (1988). *Hydrologic Systems*. vol. 1: *Rainfall-Runoff Modeling*. Prentice Hall, Englewood Cliffs, N. J.

Singh, V. P. (1995). *Computer Models of Watershed Hydrology*, Water Resources Publications, Littleton, Colo.

Singh, V. P., and Frevert, D. K. (2002a). *Mathmatical Modeling of Small Watershed Hydrology and Applications*. Water Resources Publication, Littleton, Colo.

Singh, V. P., and Frevert, D. K. (2002b). *Mathmatical Modeling of Large Watershed Hydrology*. Water Resources Publication, Littleton, Colo.

Singh, V. P., and Frevert, D. K. (2006). Watershed Models. CRC Press, Boca Raton, Fla.

Toth, E., Brath, A. and Montanari, A. (2000). Comparison of short-term rainfall prediction models for real-time flood forecasting. *Journal of Hydrologic Engineering*, vol. 239, no. 1–4, pp. 132–147.

Tzannes, N. S., and Noonan, J. P. (1973). The mutual information principle and applications. *Information and Control*, vol. 22, pp. 1–12.

Tzannes, N. S., and Avgeris, T. G. (1981). A new approach to the estimation of continuous spectra. *Kybernetes*, vol. 10, pp. 123–133.

Tzannes, M. A., Politis, D. and Tzannes, N. S. (1985). A general method of minimum cross-entropy spectral estimation. *IEEE Transactions on Acoustics, Speech, and Signal Processing*, vol. ASSP-33, no. 3, pp. 748–752.

Ulrych, T. J., and Clayton, R. W. (1976). Time series modeling and maximum entropy. *Physics of the Earth and Planetary Interiors*, vol. 12, no. 2–3, pp. 188–200.

Van Den Bo, A. (1971). Alternative interpretation of maximum entropy spectral analysis. *IEEE Transactions, Information Theory*, vol. 17, no. 4, pp. 493–494.

Wu, N. -L. (1983). An explicit solution and data extension in the maximum entropy method. *IEEE Transactions on Acoustics, Speech, and Signal Processing*, vol. ASSP-31, no. 2, pp. 486–491.

Wang, D. C., and Salas, J. D. (1991). Forecasting stream flow for Colorado River systems. Report, Colorado Water Resources Research Institute. Colorado State University, Fort Collins.

Wang, E. L., Zhang, Y. Q. Luo, J. M. Chiew, F. H. S., and Wang, Q. J. (2011). Monthly and seasonal stream flow forecasts using rainfall-runoff modeling and historical weather data. *Water Resources Research*, vol. 47, W05516, doi:10.1029/2010WR009922.

Wang, Q. J., Robertson, D. E. and Chiew, F. H. S. (2009). A Bayesian joint probability modeling approach for seasonal forecasting of stream flows at multiple sites, *Water Resources Research*, vol. 45, W05407, doi:10.1029/2008WR007355.

Wragg, A., and Dowson, D. C. (1970) Fitting continuous probability density functions over $[0, \infty)$ using information theory ideas, *IEEE Transactions on Information Theory*, vol. 16, pp. 226–230.

Yevjevich, V. (1972). *Stochastic Processes in Hydrology*. Water Resources Publications, Fort Collins, Colo.

PART V

Environmental Conditions

CHAPTER 18

River Flow Regime Classification

Afluvial hydrosystem comprises not only the low flow channel delimited by the bed and banks but also secondary channels, exposed riverine sediments, islands, riparian zones, and floodplains (Petts and Amoros, 1997; Gurnell et al., 2000). These systems are sustained by lateral transfers of material, energy, and biota (Ward and Stanford, 1995). River flow and temperature, which vary in space and time, are the principal determinants driving fluvial hydrosystems (Hynes, 1970). The patterns of flow, including intra-annual and inter-annual variation, are necessary for many species to successfully complete their life-cycle (Poff and Ward, 1989; Richter et al., 1996). The relation between these flow patterns and temperature variations is the key to the sustainability of riverine ecosystems. Flow patterns and temperature variations can be characterized by their regimes.

Variations in the flow of a river help define its regime. A river flow regime represents the average seasonal behavior based on mean monthly values. The river regime depends on the interaction between climate, vegetation, soil, geology, basin morphometry, hydraulic geometry, land use, and man-made changes. Parde (1955) characterized seasonal variations of flow by their 'stable rhythm or irregularity from year to year.' There are several methods for classifying flow regimes, both local and global (Parde, 1955; Haines et al., 1988). This chapter discusses the employment of entropy to classify and evaluate flow regimes.

18.1 Flow Regime Classification Using Clustering

Regime classes can be established using cluster analysis of mean monthly flows (Haines et al. 1988), which can be expressed as percentages of mean annual flows. Data grouping or clusters can be identified using a statistical method. Cluster analysis finds a unique partition of data only if groups in the data are distinct and isolated from one another. However, in reality it is difficult to attain this distinction, because climate, vegetation, and soil, which influence stream flow, exhibit geographical commonalities and overlap transitions from one area to another. Statistical packages are available to do hierarchical clustering in which all individual values (or data points) are joined in the manner of a branching tree. In hierarchical clustering, the data are initially treated as individuals, then progressively joined or grouped, and thereafter as groups of individuals together. Using a stopping criterion, one can determine the hierarchical tree at an appropriate level.

Let there be two variables X and Y with values as x_i and y_i, $i = 1, 2, ..., n$. Then, a similarity measure can be defined in the form of correlation coefficient as:

$$S(x, y) = \frac{\sum_i (x_i \, y_i)}{\sqrt{\left(\sum_i x_i^2\right)\left(\sum_i y_i^2\right)}}$$

(18.1)

The stopping criterion due to Ratkowsky (1984) can be applied. This involves maximization of an optimization parameter computed from the data similarity matrix suggesting a two-group optimal partition. Another way is to seek a number of groups each with a significant number of individuals. Haines et al. (1988) found 8 groups for Australian rivers, and 15 groups for global rivers.

Example 18.1. Consider a matrix of similarity coefficients between eight individuals given in Table 18.1. Choose the optimum number of groups with the Ratkowsky stopping criterion (Ratkowsky, 1984).

Solution. There are three groups of individuals: 1-4, 5-6 and 7-8. Within each group the individuals are completely similar to each other, but between the members of different groups, the similarity of each pair of individuals is a. The criterion proposed for the optimum number of groups is based on finding n such that $C_n = (S_n - S_1)/n^{1/2}$ is a maximum. S_n is the overall average similarity when the N individuals are divided into n groups. S_1 is the overall average similarity when the N individuals form a single group. The procedure for clustering considers two main steps. The first step is the ungrouped one which computes the value of S_1 by averaging all the individual cell indices in the matrix. Note that the Jaccard index is 1 when A = B, otherwise it takes a value $\alpha < 1$. The second step is to determine the S_n value for each n group; here individuals are clustered in n groups ($n = 2, ..., n$). In each group, each individual (the whole column) is compared to each of the other members including itself. Let say, the Jaccard index is computed for each pair of individuals (here pair of column) including the targeted individual. S_n is finally the average of all individual similarity indices. This second step is actually performed gradually toward having n total clusters (each individual represents a cluster).

Now consider cases with different values of n (number of groups). For the ungrouped case, $n = 1$. Then, the average similarity of each of the individuals 1-4 with all other individuals, including itself is $(1 + a)/2$. For individuals 5-8, it will be $(1 + 3a)/4$. The overall average similarity S_1 is obtained by averaging all the eight individual average similarities and is equal to $(3 + 5a)/8$. For $n = 2$, consider two groups 1-4 and 5-8. The average similarity of each of individuals 1-4 with other group members, including itself is 1. The average of each of the individuals 5-8 with the other members of its group

Individual	1	2	3	4	5	6	7	8
1	1	1	1	1	A	A	A	A
2	1	1	1	1	A	A	A	A
3	1	1	1	1	A	A	A	A
4	1	1	1	1	A	A	A	A
5	A	A	A	A	1	1	A	A
6	A	A	A	A	1	1	A	A
7	A	A	A	A	A	A	1	1
8	A	A	A	A	A	A	1	1

TABLE 18.1 Eight Individuals

is $(1 + a)/2$. The overall average similarity is thus $(3 + a)/4$. For $n = 3$, consider three groups 1-4, 5-6 and 7-8. The overall average similarity is 1. This continues for any n more than 3. The values of C_n for various values of n are shown below:

N	S_n	$S_n - S_1$	C_n
1	$(3 + 5a)/8$	0	0
2	$(3 + a)/4$	$3(1 - a)/8$	$0.265(1 - a)$
3	1	$5(1 - a)/8$	$0.361(1 - a)$
4	1	$5(1 - a)/8$	$0.313(1 - a)$

The maximum value of C_n occurs for $n = 3$ which will be chosen as the optimum number of groups (except in the case of $a = 1$, where in all C_n values will be zero).

18.2   Regime Classification Using the Uncertainty Coefficient

Data on variables can be arranged in contingency tables for purposes of determining degrees of association of variables. Let X and Y denote two random variables whose data are arranged in a contingency table, where $X : \{x_i, i = 1, 2, ..., N\}$ with probability distribution $P : \{p_i, i = 1, 2, ..., N\}$ and $Y : \{y_i, i = 1, 2, ..., M\}$ with probability distribution $Q : \{q_i, i = 1, 2, ..., M\}$. A contingency table is composed of rows and columns whose intersections form cells, and frequencies of data points in the cells provide the data for computing entropy. The entropy of X can be computed as:

$$H(X) = -\sum_{i=1}^{N} p_i \ln p_i \tag{18.2}$$

Likewise, entropy of Y can be computed as

$$H(Y) = -\sum_{j=1}^{M} q_j \ln q_j \tag{18.3}$$

Here p_i is the proportion of frequency of each cell relative to the frequency of its host column in the array, p_j is the proportion of frequency of each cell relative to the frequency of its host row in the array, N is the number of columns, and M is the number of rows.

In a similar manner, entropy can be used to assess the information content between pairs of variables. Entropy characteristics of two-way contingency tables can be evaluated. For a contingency table the entropy can be expressed as

$$H = -\sum_{i=1}^{N} \sum_{j=1}^{M} p_{ij} \ln p_{ij} \tag{18.4}$$

where H is the total entropy of the contingency table, p_{ij} is the proportion of total points in cell (i, j), N is the number of columns, and M is the number of rows. H is highest when all cells have the same number of points, pointing to the highest disorder in the data and least association or dependence between variables.

Now conditional entropies can be expressed. Entropy of X given Y is computed as

$$H(X|Y) = -\sum_{j=1}^{M} \sum_{i=1}^{N} \frac{p_{ij} \ln p_{ij}}{p_j} \tag{18.5}$$

and entropy of Y given X as

$$H(Y|X) = -\sum_{j=1}^{M}\sum_{i=1}^{N}\frac{p_{ij}\ln p_{ij}}{p_i} \tag{18.6}$$

To quantify the degree of association between two variables in a pair, one can compute the coefficient of uncertainty U for each variable, considering the effect of the other variable of the pair. The uncertainty of X given Y, $U(X|Y)$, can be expressed as

$$U(X|Y) = \frac{H(X) - H(X|Y)}{H(X)} \tag{18.7}$$

Similarly, the uncertainty of Y given X, $U(Y|X)$, can be expressed as

$$U(Y|X) = \frac{H(Y) - H(Y|X)}{H(Y)} \tag{18.8}$$

Equations (18.7) and (18.8) show that U varies from 0 to 1, where 0 indicates no relation between X and Y, and 1 indicates a perfect correlation between X and Y. Note that $U(X|Y) \neq U(Y|X)$. These two coefficients of uncertainty can be combined into a single symmetrical uncertainty coefficient, $U(X, Y)$ by taking a weighted average of the two uncertainty measures:

$$U(X,Y) = \frac{H(X)U(X|Y) + H(Y)U(Y|X)}{H(X) + H(Y)} \tag{18.9}$$

Using these uncertainty coefficients or similarity measures, variables can be clustered.

Example 18.2.   Consider values of two random variables X and Y as given in Table 18.2. Calculate the uncertainty measures for these values.

Solution.   The first step is to calculate the relative frequencies for X and Y. This is done as shown in Table 18.3.
From Table 18.3, the marginal entropies $H(X) = 2.11$ bits and $H(Y) = 2.06$ bits.
Now to calculate the conditional entropies, we need to construct the contingency table and the relative frequency table for X and Y. Tables 18.4 and 18.5 give the contingency table and relative frequency tables for X and Y, respectively.

Class Interval for X	Number of Values (n)	Class Interval for Y	Number of Values (n)
0–50	12	0–30	26
50–100	20	30–60	17
100–150	19	60–90	8
150–200	6	90–120	4
200–250	1	120–150	3
250–300	0	150–180	2
300–350	2	—	—

TABLE 18.2   Class Interval Table for Random Variables X and Y

Class Interval for X	Relative Frequency p_i	$\log p_i$	$p_i \log p_i$	Class Interval for Y	Relative Frequency p_i	$\log p_i$	$p_i \log p_i$
0–50	0.200	−2.321	−0.464	0–30	0.433	−1.206	−0.522
50–100	0.333	−1.584	−0.528	30–60	0.283	−1.819	−0.515
100–150	0.316	−1.659	−0.525	60–90	0.133	−2.906	−0.387
150–200	0.100	−3.322	−0.332	90–120	0.066	−3.907	−0.260
200–250	0.016	−5.906	−0.098	120–150	0.050	−4.322	−0.216
250–300	0	0	0	150–180	0.033	−4.907	−0.164
300–350	0.033	−4.907	−0.164				

TABLE 18.3 Relative Frequencies of X and Y

| Y | X | | | | | | | |
	0–50	50–100	100–150	150–200	200–250	250–300	300–350	Sum
0–30	9	7	8	1	0	0	1	26
30–60	2	6	5	4	0	0	0	17
60–90	0	4	3	0	0	0	1	8
90–120	1	3	0	0	0	0	0	4
120–150	0	0	3	0	0	0	0	3
150–180	0	0	0	1	1	0	0	2
Sum	12	20	19	6	1	0	2	60

TABLE 18.4 Contingency Table for X and Y

| Y | X | | | | | | | |
	0–50	50–100	100–150	150–200	200–250	250–300	300–350	$p(y_j)$
0–30	0.150	0.116	0.133	0.016	0	0	0.0167	0.433
30–60	0.033	0.100	0.083	0.066	0	0	0	0.283
60–90	0	0.067	0.05	0	0	0	0.0167	0.133
90–120	0.016	0.050	0	0	0	0	0	0.066
120–150	0	0	0.05	0	0	0	0	0.05
150–180	0	0	0	0.016	0.016	0	0	0.033
$p(x_j)$	0.200	0.333	0.316	0.100	0.016	0	0.033	—

TABLE 18.5 Relative Frequency Table for X and Y

From Table 18.5, $H(X|Y) = 1.648$ bits and $H(Y|X) = 1.601$ bits.

The coefficient of uncertainty U for each variable, considering the effect of the other variable, can be determined:

$$U(X|Y) = \frac{H(X) - H(X|Y)}{H(X)} = \frac{2.11 - 1.648}{2.11} = 0.219$$

$$U(Y|X) = \frac{H(Y) - H(Y|X)}{H(Y)} = \frac{2.06 - 1.601}{2.06} = 0.223$$

Let us recall that the symmetrical uncertainty coefficient $U(X,Y)$ is the weighted average of the uncertainty coefficient of each variable X and Y. Given the marginal entropies $H(X) = 2.11$ bits and $H(Y) = 2.06$ bits considered, the weighted average is obtained:

$$U(X,Y) = \frac{H(X)U(X|Y) + H(Y)U(Y|X)}{H(X) + H(Y)} = 0.221$$

18.3   Regime Classification and Measure of Flow Stability

In a flow regime classification, the times of occurrence of high and low flows can be identified. The question arises: How probable is it for an identified flow pattern to occur each year? The associated question is: How rhythmic or irregular is this flow pattern? In other words, how stable or unstable is the flow regime? Using a flow series, Krasovskaia and Gottschalk (1992) and Krasovskaia et al. (1994) have shown that different flow regimes can occur from year to year in response to climatic variability; and flow regimes classified, based on monthly means, can deviate from the flow regimes that occur most frequently when each year is classified separately. Indeed the deviation can occur in up to 45 percent of cases. Certain regimes are more stable than others, depending on the regime type.

Krasovskaia (1995, 1997) developed a quantitative measure of the stability of river flow regimes using entropy. The measure can be obtained from observed flow series and can be used for regime classification. The discussion in what follows draws from her work. The development of an entropy-based measure involves (1) classification of mean monthly flows into different types, (2) identification of discriminating periods for different classes, (3) specification of the instability index, (4) computation of the instability index value for each regime type, and (5) computation of the instability index for all flow series.

The first step is to classify mean monthly flows into different regime types. For example, Krasovskaia et al. (1994) classified Scandinavian flow regimes based on high water into three types: H1 (dominant snowmelt high water), H2 (transition to secondary high water), and H3 (dominant rain high water); and based on low water into three types: L1 (dominant low in winter), L2 (transition zone, two low-water periods in different seasons), and L3 (dominant summer low water). For regions with no snow, regime type H1 will vanish. Under high water, they used three flow types: Max1 (highest observed flow), Max2 (second-highest observed flow), and Max3 (third-highest observed flow). Likewise they used two flow types for low water: Min1 (the lowest observed flow) and Min2 (second-lowest observed flow). Then there can be different combinations of types based on high water and low water, such as H1L1, H2L1, H2L2, H2L3, H3L3, etc. Each combination corresponds to a certain instability index.

The second step is to identify the discriminating period for each classification. Krasovskaia et al. (1994) used three discriminating periods for high water: April–August,

September–November, and December–March; and two for low water: January–April and May–December.

The third step is to compute entropy as an index of flow regime instability:

$$H = -\sum_{i=1}^{N} p_i \ln p_i \qquad (18.10)$$

where N denotes the number of periods used for the discrimination of a particular regime type and p_i is the probability that the particular regime type will occur in the ith period. For example, under high water in the above discussion there are three discriminating periods, and under low water there are two discriminating periods. The probability of observing the assigned regime type during each individual year in a flow series with a certain instability index is obtained as follows. First, one determines p_i in Eq. (18.10), the values of the percentage of the years in the series with the assigned regime type and those with a different regime. Second, this computation is repeated for all maxima and minima involved in the regime definition. The values so obtained will yield the corresponding values of instability index for one of many possible combinations, such as when all maxima and minima involved in the defined flow regime exhibit the same probability of occurrence within a discriminating period.

The fourth step is to compute the instability index value for the regime type by summing the entropy values for the maxima and the minima obtained as above. The fifth step is to compute the instability index for all flow series by averaging values across samples. In this manner, the percentages of cases with assigned regime type can be determined.

Example 18.3. It is known that the instability from Eq. (18.10) will attain a maximum if all p_i are equal. Determine the instability index for the regime type involving three maxima and two minima. Also compute the instability index for the regime type with two maxima and two minima as well as the regime type with three maxima and one minimum.

Solution. The probability p_i that the maxima during high water will occur during one of the three periods will be 1/3, and the probability that the minima will occur during one of the two periods will be 1/2. Table 18.6 shows the calculation of the instability index for a regime with three maxima and two minima.

The instability index from Eq. (18.10) for high water will be 3.2958, and under low water it will be 1.386. The maximum instability index value for a flow regime entailing three maxima and two minima will therefore be the sum of these two values, or 4.682. The instability index for the regime type with two maxima and two minima will be: 2.196 + 1.386 = 3.582, and with three maxima and one minimum it will be 3.2958 + 0.693 = 3.989.

Example 18.4. Select mean monthly discharge data for the Brazos River in Texas. Define its flow regime. Calculate the instability index for each flow regime, and comment on the river flow stability.

Solution. A total of 10 stream gauge stations within the Brazos River basin are selected for the purpose of flow regime classification and calculation of the instability index. The details of the stations are given in Table 18.7.

The monthly flow series for the mentioned stations for the period from 1964 to 2004 is downloaded and used for further analysis. For each station, the long-term monthly mean for individual months is determined, and then they are used for flow regime classification. A total of four discriminating periods are chosen for the flows. The four seasons are winter (December to March), spring (March to June), summer (June to September) and fall (September to December). The genetic origins of river flow such as snowmelt, rain, or groundwater are utilized to describe the flow regime (Krasovkaia, 1997).

Discriminating Period under Consideration	Flow Type	Instability Index
Discriminating period 1 for high water: April–August	Maxima 1	1/3 ln 1/3 = 0.366
	Maxima 2	1/3 ln 1/3 = 0.366
	Maxima 3	1/3 ln 1/3 = 0.366
Discriminating period 2 for high water: September–November	Maxima 1	1/3 ln 1/3 = 0.366
	Maxima 2	1/3 ln 1/3 = 0.366
	Maxima 3	1/3 ln 1/3 = 0.366
Discriminating period 3 for high water: December–March	Maxima 1	1/3 ln 1/3 = 0.366
	Maxima 2	1/3 ln 1/3 = 0.366
	Maxima 3	1/3 ln 1/3 = 0.366
Discriminating period 1 for low water: January–April	Minima 1	1/2 ln 1/2 = 0.346
	Minima 2	1/2 ln 1/2 = 0.346
Discriminating period 2 for low water: May–December	Minima 1	1/2 ln 1/2 = 0.346
	Minima 2	1/2 ln 1/2 = 0.346

TABLE 18.6   Calculation of Instability Index for Regime with Three Maxima and Two Minima

Station Name	Location
Brazos River near Seymour	Latitude 33°34′51″, longitude 99°16′02″
Clear Fork at Fort Griffin	Latitude 32°56′04″, longitude 99°13′27″
Brazos near Glenrose	Latitude 32°15′32″, longitude 97°42′08″
North Bosque River near Clifton	Latitude 31°47′09″, longitude 97°34′04″
Brazos River near Waco	Latitude 31°32′09″, longitude 97°04′23″
Cowhouse Creek at Pidcock	Latitude 31°17′05″, longitude 97°53′05″
Little River near Little River	Latitude 30°57′59″, longitude 97°20′45″
Little River at Cameron	Latitude 30°50′06″, longitude 96°56′47″
Davidson Creek near Lyons	Latitude 30°25′10″, longitude 96°32′24″
Brazos River near Hempstead	Latitude 30°07′44″, longitude 96°11′15″

TABLE 18.7   Details of Chosen Stations within Brazos River Basin

A total of three groups are formed. We consider two maxima and two minima for each flow series, and we check whether the times of occurrence of the two maxima and two minima are consistent for the flow series belonging to the same regime. Details are given in Table 18.8.

The time of occurrence of the highest flow indicates its main source of origin. The time for the second-highest maximum flow either strengthens the indicated flow source (when it occurs in the same period as the highest flow) or shows that there is another source (when it occurs during a different discriminating period). The time of occurrence of the first lowest water indicates whether the minimum is due to low temperature or high temperature and lack of precipitation.

Now we calculate the probability of occurrence of each maximum and minimum in each discriminating period for all the years in a flow series. The probability of occurrence for all maxima and minima within each discriminating period is given in Table 18.9.

Regime Type	Station	Maximum 1	Maximum 2	Minimum 1	Minimum 2
Group 1	1	May	June	August	September
Group 1	6	May	June	September	August
Group 1	8	May	June	September	October
Group 1	9	May	June	September	August
Group 2	2	June	May	January	December
Group 2	3	May	June	January	December
Group 2	4	May	June	January	July
Group 3	5	May	March	July	September
Group 3	7	May	February	August	July
Group 3	10	February	June	August	July

TABLE 18.8   Details of Flow Series and Corresponding Regime Types

Flow Series	Discriminating Period	Flow Type	Probability of Occurrence p_i	Instability Index $p_i \ln p_i$	Total Instability Index for the Flow Series
1	Winter	Maximum 1	0.268	0.353	
	Spring		0.439	0.361	
	Summer		0	0	
	Fall		0.292	0.359	
	Winter	Maximum 2	0.391	0.367	
	Spring		0.391	0.367	
	Summer		0.122	0.256	
	Fall		0.098	0.227	4.681
	Winter	Minimum 1	0.171	0.302	
	Spring		0.024	0.091	
	Summer		0.391	0.367	
	Fall		0.414	0.365	
	Winter	Minimum 2	0.195	0.319	
	Spring		0.122	0.256	
	Summer		0.463	0.356	
	Fall		0.219	0.332	

TABLE 18.9   Calculation of Total Instability Index for the Flow Series

Flow Series	Discriminating Period	Flow Type	Probability of Occurrence p_i	Instability Index p_i ln p_i	Total Instability Index for the Flow Series
2	Winter	Maximum 1	0.146	0.281	
	Spring		0.415	0.365	
	Summer		0.268	0.353	
	Fall		0.171	0.302	
	Winter	Maximum 2	0.171	0.302	
	Spring		0.146	0.281	
	Summer		0.366	0.368	
	Fall		0.317	0.364	
	Winter	Minimum 1	0.463	0.356	4.962
	Spring		0.195	0.319	
	Summer		0.220	0.333	
	Fall		0.122	0.257	
	Winter	Minimum 2	0.195	0.319	
	Spring		0.585	0.313	
	Summer		0.171	0.302	
	Fall		0.049	0.147	
3	Winter	Maximum 1	0.146	0.281	
	Spring		0.439	0.361	
	Summer		0.220	0.333	
	Fall		0.195	0.319	
	Winter	Maximum 2	0.073	0.191	
	Spring		0.098	0.227	
	Summer		0.610	0.302	
	Fall		0.220	0.333	
	Winter	Minimum 1	0.244	0.344	4.756
	Spring		0.220	0.333	
	Summer		0.488	0.350	
	Fall		0.049	0.147	
	Winter	Minima 2	0.146	0.281	
	Spring		0.195	0.319	
	Summer		0.268	0.353	
	Fall		0.146	0.281	

TABLE 18.9   Calculation of Total Instability Index for the Flow Series (*Continued*)

Flow Series	Discriminating Period	Flow Type	Probability of Occurrence p_i	Instability Index $p_i \ln p_i$	Total Instability Index for the Flow Series
4	Winter	Maximum 1	0.268	0.353	
	Spring		0.341	0.367	
	Summer		0.146	0.281	
	Fall		0.244	0.344	
	Winter	Maximum 2	0.268	0.353	
	Spring		0.146	0.281	
	Summer		0.220	0.333	
	Fall		0.366	0.368	
	Winter	Minimum 1	0.366	0.368	5.168
	Spring		0.098	0.227	
	Summer		0.390	0.367	
	Fall		0.146	0.281	
	Winter	Minimum 2	0.415	0.365	
	Spring		0.341	0.367	
	Summer		0.122	0.257	
	Fall		0.122	0.257	
5	Winter	Maximum 1	0.220	0.333	
	Spring		0.537	0.334	
	Summer		0.024	0.091	
	Fall		0.220	0.333	
	Winter	Maximum 2	0.049	0.147	
	Spring		0.024	0.091	
	Summer		0.463	0.356	
	Fall		0.463	0.356	
	Winter	Minimum 1	0.171	0.302	4.229
	Spring		0.098	0.227	
	Summer		0.610	0.302	
	Fall		0.122	0.257	
	Winter	Minimum 2	0.488	0.350	
	Spring		0.366	0.368	
	Summer		0.073	0.191	
	Fall		0.073	0.191	

TABLE **18.9** Calculation of Total Instability Index for the Flow Series (*Continued*)

Flow Series	Discriminating Period	Flow Type	Probability of Occurrence p_i	Instability Index $p_i \ln p_i$	Total Instability Index for the Flow Series
6	Winter	Maximum 1	0.268	0.353	
	Spring		0.512	0.343	
	Summer		0.049	0.147	
	Fall		0.171	0.302	
	Winter	Maximum 2	0.341	0.367	
	Spring		0.098	0.227	
	Summer		0.220	0.333	
	Fall		0.341	0.367	
	Winter	Minimum 1	0.268	0.353	
	Spring		0.195	0.319	5.013
	Summer		0.317	0.364	
	Fall		0.220	0.333	
	Winter	Minimum 2	0.366	0.368	
	Spring		0.415	0.365	
	Summer		0.146	0.281	
	Fall		0.073	0.191	
7	Winter	Maximum 1	0.317	0.364	
	Spring		0.463	0.356	
	Summer		0.073	0.191	
	Fall		0.146	0.281	
	Winter	Maximum 2	0.098	0.227	
	Spring		0.049	0.147	
	Summer		0.585	0.313	
	Fall		0.268	0.353	
	Winter	Minimum 1	0.220	0.333	
	Spring		0.098	0.227	4.467
	Summer		0.561	0.324	
	Fall		0.122	0.257	
	Winter	Minimum 2	0.512	0.343	
	Spring		0.341	0.367	
	Summer		0.073	0.191	
	Fall		0.073	0.191	

TABLE 18.9 Calculation of Total Instability Index for the Flow Series (*Continued*)

Flow Series	Discriminating Period	Flow Type	Probability of Occurrence p_i	Instability Index $p_i \ln p_i$	Total Instability Index for the Flow Series
8	Winter	Maximum 1	0.195	0.319	
	Spring		0.439	0.361	
	Summer		0.195	0.319	
	Fall		0.171	0.302	
	Winter	Maximum 2	0.195	0.319	
	Spring		0.098	0.227	
	Summer		0.268	0.353	
	Fall		0.439	0.361	
	Winter	Minimum 1	0.244	0.344	4.865
	Spring		0.122	0.257	
	Summer		0.390	0.367	
	Fall		0.244	0.344	
	Winter	Minimum 2	0.390	0.367	
	Spring		0.512	0.343	
	Summer		0.024	0.091	
	Fall		0.073	0.191	
9	Winter	Maximum 1	0.171	0.302	
	Spring		0.415	0.365	
	Summer		0.122	0.257	
	Fall		0.293	0.360	
	Winter	Maximum 2	0.195	0.319	
	Spring		0.098	0.227	
	Summer		0.268	0.353	
	Fall		0.439	0.361	
	Winter	Minimum 1	0.244	0.344	4.783
	Spring		0.049	0.147	
	Summer		0.561	0.324	
	Fall		0.146	0.281	
	Winter	Minimum 2	0.463	0.356	
	Spring		0.366	0.368	
	Summer		0.073	0.191	
	Fall		0.098	0.227	

TABLE 18.9 Calculation of Total Instability Index for the Flow Series (*Continued*)

Flow Series	Discriminating Period	Flow Type	Probability of Occurrence p_i	Instability Index $p_i \ln p_i$	Total Instability Index for the Flow Series
10	Winter	Maximum 1	0.244	0.344	
	Spring		0.488	0.350	
	Summer		0.049	0.147	
	Fall		0.220	0.333	
	Winter	Maximum 2	0.000	0.000	
	Spring		0.024	0.091	
	Summer		0.829	0.155	
	Fall		0.146	0.281	
	Winter	Minimum 1	0.098	0.227	3.921
	Spring		0.268	0.353	
	Summer		0.537	0.334	
	Fall		0.098	0.227	
	Winter	Minimum 2	0.512	0.343	
	Spring		0.317	0.364	
	Summer		0.024	0.091	
	Fall		0.146	0.281	

TABLE 18.9   Calculation of Total Instability Index for the Flow Series (*Continued*)

By taking the average across all the flow series belonging to the same regime, the instability index for the particular regime can be determined:

Regime 1: 4.835
Regime 2: 4.962
Regime 3: 4.206

To interpret the instability index values obtained, let us first calculate the maximum possible values of instability index for two maxima, two minima with four discriminating periods. The instability index comes out to be 5.536. Determining the percentage of the maximum value of the index for each regime type, we can identify how stable or unstable the particular regime is:

Regime 1: 87.3 percent
Regime 2: 89.6 percent
Regime 3: 75.8 percent

Now we determine how stable the maxima and minima are within each regime type:

Maxima: Regime 1: 2.458 (88.8 percent)   Minima: Regime 1: 2.376 (85.84 percent)
Regime 2: 2.547 (92 percent)   Regime 2: 2.414 (87.21 percent)
Regime 3: 1.992 (71.9 percent)   Regime 3: 2.213 (79.95 percent)

18.4   Interpretation of Instability Index

A higher index value means greater instability. For example, the maximum value of the instability index means that the maxima and minima occur with equal likelihood in any of the discriminating periods used for regime classification. This will correspond to the flow regime being very unstable. On the other hand, a lower value of the instability index will correspond to a greater stability; i.e., the identified flow regime will be most probable in the flow series. In this manner, one can investigate the changes in flow regimes in time by comparing the instability indices for different time periods and can quantify the impact of climate change.

Example 18.5.   Consider Example 18.4 where we calculated the instability indices for all three regime types and for the maxima and minima within each regime type.

Solution.   The overall instability index for the regime types are 87.3, 89.6, and 75.8 percent, respectively. The higher the instability index, more unstable the regime type will be. To assess the level of instability, the index will be represented as the percentage of the maximum possible value of instability index possible for the particular case (depending on the number of discriminating periods and number of maxima and minima considered). Thus, it can be concluded that regime types 1 and 2 are unstable and regime type 3 is relatively unstable. If we look at the instability indices for the maxima and minima of each regime type separately, we find for maxima 88.8 percent and for minima 85.84 percent for regime type 1. Hence, both the maxima and minima for regime 1 are unstable. For regime 2, for the maximum it is 92 percent, and for the minimum is 87.21 percent. Thus both maxima and minima are unstable, by a larger degree than that for regime type 1. In the case of regime type 3, for the maximum it is 71.9 percent and for the minimum it is 79.95 percent. Thus the maxima and minima are relatively unstable, but much less so than the other two regime types.

18.5   Grouping of River Flow Regimes by Cross Entropy

River flow series can be aggregated into regime types using certain discriminating criteria. The method described above can be employed for grouping river flow regimes. Another method, developed by Krasovskaia (1997), employs minimization of an entropy-based objective function. This function uses a concept of information loss resulting from such an aggregation and determining the difference between the series aggregated into one group, i.e., inaccuracy of aggregation. This method can be described as follows.

To start, determine the number of discriminating periods N. This number is important, because the number of values (minima/maxima) is related to the N discriminating periods of the year. Let there be M different series. Then the probability of having the maximum (or minimum) monthly flow during a year i for a series j can be denoted by p_{ij}. It is assumed that a priori one can define probabilities for possible flow regimes for an ensemble of M flow series as q_{ik}, $i = 1, 2, ..., N$; $k = 1, 2, ..., G$, where G is the number of regime types. These are prior probabilities assumed known and are used to determine the flow regime for a series j. Then p_{ij}, $i = 1, 2, ..., N$; $j = 1, 2, ..., M$, can be seen as posterior probabilities. The distance between the two sets of probabilities, often called divergence or cross entropy (Kullback and Leibler, 1951), can be expressed as

$$I_j(p|q) = \sum_{i=1}^{N} p_{ij} \log \frac{p_{ij}}{q_{ik}} \tag{18.11}$$

Equation (18.11) can be interpreted as a measure of the information inaccuracy of the prediction of regime types q_{ik}, $i = 1, 2, ..., N$, by p_{ij}, that is, it measures whether the regime pattern for a series j belongs to a regime type k. If $I_j(p \mid q)$ is close to zero, then there is hardly any information loss and the regime pattern probably belongs; on the other hand, if it is large, it does not belong.

Consider a regime of type k applicable to a set of flow series S_k. Then a joint measure of information inaccuracy for this set can be expressed as

$$I_k(p \mid q) = \sum_{j \in S_k} \sum_{i=1}^{N} p_{ij} \log \frac{p_{ij}}{q_{ik}} \tag{18.12}$$

If one of the G regime types is assigned to each of the M series, then the inaccuracy of the classification can be written as

$$I_G(p \mid q) = \sum_{k=1}^{G} \sum_{j \in S_k} \sum_{i=1}^{N} p_{ij} \log \frac{p_{ij}}{q_{ik}} \tag{18.13}$$

Equation (18.13) sums the inaccuracies due to the difference between a prior regime type and a posterior regime type of the flow series belonging to this latter regime of total G types. When the number of groups decreases, the inaccuracy given by Eq. (18.13) will increase. This should be an inherent characteristic of the sample.

The total grouping inaccuracy I_{total} consists of two parts: within-set inaccuracy I_w and between-set inaccuracy I_b:

$$I_{\text{total}} = I_w + I_b \tag{18.14}$$

Considering Eq. (18.13), the total inaccuracy can be written as

$$I_{\text{total}} = \sum_{j \in S_k} \left(\sum_{i=1}^{N} p_{ij} \log \frac{p_{ij}}{q_{iG}} \right) = \left(\sum_{k=1}^{G} \sum_{j \in S_k} \sum_{i=1}^{N} p_{ij} \log \frac{p_{ij}}{q_{iG}} \right) \tag{18.15}$$

If the quantity within parentheses can be multiplied and divided by q_{ik}, then Eq. (18.15) can be cast as

$$I_{\text{total}} = \sum_{k=1}^{G} \sum_{j \in S_k} \left(\sum_{i=1}^{N} p_{ij} \log \frac{p_{ij} q_{ik}}{q_{iG} q_{ik}} \right) = \sum_{k=1}^{G} \sum_{j \in S_k} \sum_{i=1}^{N} p_{ij} \left(\log \frac{p_{ij}}{q_{ik}} + \log \frac{q_{ik}}{q_{iG}} \right) \tag{18.16}$$

Denoting the number of series aggregated to a regime k by k_j, Eq. (18.16) can be expressed as

$$I_{\text{total}} = \sum_{k=1}^{G} \sum_{j \in S_k} \sum_{i=1}^{N} p_{ij} \log \frac{p_{ij}}{q_{ik}} + \sum_{k=1}^{G} k_j \sum_{j \in S_k} \sum_{i=1}^{N} \frac{p_{ij}}{k_g} \log \frac{q_{ik}}{q_{iG}} \tag{18.17}$$

Noting that

$$\sum_{j \in S_k} \sum_{i=1}^{N} \frac{p_{ij}}{k_j} = \sum_{i=1}^{N} q_{ik}$$ (18.18)

Equation (18.18) can be written as

$$I_{\text{total}} = \sum_{k=1}^{G} \sum_{j \in S_k} \sum_{i=1}^{N} p_{ij} \log \frac{p_{ij}}{q_{ik}} + \sum_{k=1}^{G} k_j \sum_{i=1}^{N} q_{ik} \log \frac{q_{ik}}{q_{iG}}$$ (18.19)

Example 18.6. Consider Example 18.4, where 10 flow series within the Brazos River basin are considered. Calculate the information inaccuracies associated with each flow series, the within-group inaccuracy, and the total grouping inaccuracy.

Solution.

Number of flow series M, $j = 1, 2, \ldots, 10$.
Number of discriminating periods N, $i = 1, 2, \ldots, 4$ (spring, summer, fall, winter).
Number of regime types $G = 3$.
Maxima = 2 and minima = 2.

It is assumed that the prior probabilities q_{ik} define the probabilities of possible river flow regimes for the M flow series considered. Here, the prior probabilities are assumed to follow a uniform distribution, which means that the considered flow series may fall under any of the regime types with equal probability. Since we have defined three regime types for this case, the prior probability values will be $1/3$, since the considered flow series have an equal chance of belonging to any of the three regime types. The next step is to calculate the p_{ij} values which denote the probability of having maximum/minimum monthly flow during a period i for a series j. An alternative method that can be used to calculate the p_{ij} values is to consider the runoff volumes during a year. If the total volume for a particular year is V and the volume during a certain period i is v_i, then the ratio v_i/V gives the probability values. Table 18.10

Discriminating Period	Winter	Spring	Summer	Fall	Information Inaccuracy for the Flow Series
p_{ij}	0.2744	0.3603	0.2163	0.1489	0.058
	0.0793	0.2586	0.3846	0.2774	0.0304
	0.1733	0.2703	0.3065	0.2499	0.0705
	0.1800	0.3523	0.2496	0.2180	0.0655
	0.2716	0.4574	0.1700	0.1088	0.0176
	0.2279	0.3889	0.2379	0.1452	0.0531
	0.2774	0.4271	0.1733	0.1222	0.0329
	0.2231	0.3942	0.2747	0.1080	0.0404
	0.2504	0.3827	0.2428	0.1240	0.0491
	0.3536	0.3223	0.1595	0.1645	0.0514

TABLE 18.10 p_{ij} Values for the Flow Series

Regime Type	Inaccuracy (bits)
1	0.2005
2	0.0706
3	0.1018

TABLE 18.11 Within-Group Inaccuracies for the Three Regime Types

gives the p_{ij} values for all 10 flow series during each of the four discriminating periods. The information inaccuracy associated with each flow series can then be calculated as

$$I_j(p \mid q) = \sum_{i=1}^{N} p_{ij} \log \frac{p_{ij}}{q_{ik}}$$

The within-group information inaccuracy can be determined using Eq. (18.13). The total grouping inaccuracy can be calculated using Eq. (18.15). Table 18.11 gives the within-group inaccuracies and the total grouping inaccuracy for this example.

Total group inaccuracy: 0.3728 bits

Example 18.7.   Basically this example is a comment and proposition of an entropy-based stopping rule for clustering. Recall that the statistical stopping rule described by Mojena (1977) was designed particularly for hierarchical clustering. Later, Ratkowsky (1984) improved the approach of the stopping rule toward hierarchical and nonhierarchical clustering. He employed the Jaccard similarity coefficient which measures the similarity between A and B by

$$J(A, B) = \frac{|A \cap B|}{|A \cup B|}$$

Although Krasovskaia (1997) seemed to improve the stopping rule by using entropy, her method was typically designed for river flow regimes, and so to extend its application may be difficult. Here, we improve Ratkowsky's method by using entropy. We define a similarity index $\rho_T(A, B)$ as the ratio of the transentropy by the joint entropy:

$$\rho_T(A, B) = \frac{H(A) + H(B) - H(A, B)}{H(A, B)} \tag{18.20}$$

where $H(A)$ and $H(B)$ are the marginal entropy values for A and B, respectively, while $H(A, B)$ represents the joint entropy. This index is an entropy version of the Jaccard similarity coefficient. Two levels of computation are proposed: within-group and intergroup. The first step considers all individuals in a unique group and averages the indices to get the value of ρ_1. The second step consists of clustering the individuals into n groups. Within groups, individuals are assigned a unit similarity index for the intergroup similarity calculation; otherwise, the initial index is conserved. At each step of n groups, a corresponding similarity index ρ_n is computed. Just as Ratkowsky (1984) did, the stopping rule criterion $(\rho_n - \rho_1)/n^{1/2}$ is computed at each stage. We apply this approach to discriminate the year's seasons in the Texas High Plain region. Indeed, the Texas High Plain is well known for its agricultural activities but also for its limited availability. We intend to discriminate the year's season base on precipitation. Hence we use monthly precipitation data over a 119-yr period (1895 to 2013). The data are released by the Texas A&M Atmospheric Science Department (http://climatexas.tamu .edu) through the Full Network Estimated Precipitation project. From Table 18.12, we note that the approach discriminates the year in five seasons based on precipitation patterns. These five seasons can be identified with the common season known in general for the temperate climate (summer, autumn, winter, and spring).

	Jan	Feb	Mar	Apr	May	Jun	Jul	Aug	Sep	Oct	Nov	Dec
Jan	1.00	0.46	0.48	0.48	0.51	0.53	0.53	0.53	0.55	0.47	0.45	0.48
Feb	0.46	1.00	0.48	0.50	0.52	0.54	0.54	0.54	0.56	0.48	0.45	0.50
Mar	0.48	0.48	1.00	0.49	0.53	0.54	0.54	0.54	0.56	0.49	0.46	0.49
Apr	0.48	0.50	0.49	1.00	0.54	0.55	0.55	0.56	0.58	0.50	0.47	0.50
May	0.51	0.52	0.53	0.54	1.00	0.58	0.58	0.58	0.60	0.53	0.50	0.54
Jun	0.53	0.54	0.54	0.55	0.58	1.00	0.60	0.60	0.62	0.55	0.52	0.55
Jul	0.53	0.54	0.54	0.55	0.58	0.60	1.00	0.60	0.62	0.54	0.52	0.55
Aug	0.53	0.54	0.54	0.56	0.58	0.60	0.60	1.00	0.63	0.55	0.52	0.57
Sep	0.55	0.56	0.56	0.58	0.60	0.62	0.62	0.63	1.00	0.57	0.54	0.58
Oct	0.47	0.48	0.49	0.50	0.53	0.55	0.54	0.55	0.57	1.00	0.46	0.49
Nov	0.45	0.45	0.46	0.47	0.50	0.52	0.52	0.52	0.54	0.46	1.00	0.46
Dec	0.48	0.50	0.49	0.50	0.54	0.55	0.55	0.57	0.58	0.49	0.46	1.00
$\bar{\rho}$	0.54	0.55	0.55	0.56	0.58	0.60	0.60	0.60	0.62	0.55	0.53	0.56

n	ρ_n	$\rho_n - \bar{\rho}$	$\rho_n - \rho_1$	Clusters
1	0.57	0.00	0.00	Jun-Jul-Aug-Sep-Oct-Nov-Dec-Jan-Feb-Mar-Apr-May
2	0.78	0.21	0.15	Jun-Jul-Aug/ Sep-Oct-Nov-Dec-Jan-Feb-Mar-Apr-May
3	0.82	0.25	0.14	Jun-Jul-Aug/ Sep-Oct-Nov-Dec-Jan/ Feb-Mar-Apr-May
4	0.89	0.32	0.16	Jun-Jul-Aug/ Sep-Oct-Nov-Dec-Jan/ Feb-Mar/ Apr-May
5	0.97	0.40	0.18*	Jun-Jul-Aug/ Sep-Oct-Nov/ Dec-Jan/ Feb-Mar/ Apr-May
6	1.00	0.43	0.18	Jun-Jul-Aug/ Sep/ Oct-Nov/ Dec-Jan/ Feb-Mar/ Apr-May
7	1.00	0.43	0.16	Jun-Jul-Aug/ Sep/ Oct/ Nov/ Dec-Jan/ Feb-Mar/ Apr-May
8	1.00	0.43	0.15	Jun-Jul-Aug/ Sep/ Oct/ Nov/ Dec-Jan/ Feb-Mar/ Apr/ May
9	1.00	0.43	0.14	Jun-Jul-Aug/ Sep/ Oct/ Nov/ Dec/ Jan/ Feb-Mar/ Apr/ May
10	1.00	0.43	0.14	Jun-Jul-Aug/ Sep/ Oct/ Nov/ Dec/ Jan/ Feb-Mar/ Apr/ May
11	1.00	0.43	0.13	Jun-Jul/ Aug/ Sep/ Oct/ Nov/ Dec/ Jan/ Feb/Mar/ Apr/ May
12	1.00	0.43	0.12	Jun/Jul/ Aug/ Sep/ Oct/ Nov/ Dec/ Jan/ Feb/Mar/ Apr/ May

TABLE 18.12 Entropy-Based Clustering for Discrimination of Seasons in Texas High Plain

Questions

1. How good is Ratkowsky stopping criterion when choosing optimum number of groups?

2. Consider measurements of rainfall from a rain gauge for any two years. From these measurements, select the amount and duration for each rainfall event. The number of events may be 50 or more. Calculate uncertainty measures of rainfall amount and duration, assuming they are random variables.

3. Consider precipitation measurements in a given area. Determine the instability index of the precipitation regime.

4. Select mean monthly discharge data for a river and define the river flow regime. Calculate the instability index for the flow regime and comment on the flow stability.

5. Consider several (say 12) 10-flow series within a river basin. Calculate the information inaccuracy associated with each flow series, the within group inaccuracy, and the total grouping inaccuracy.

References

Krasovskaia, I., and Gottschalk, L. (1992). Stability of river flow regimes. *Nordic Hydrology*, vol. 23, pp. 137–154.

Krasovskaia, I., Arnell, N. W. and Gottschalk, L. (1994). Flow regimes in northern and western Europe: development and application of procedures for classifying flow regimes. In: *FRIEND: Flow Regimes from International Experiment and Network Data*, edited by P. Seuna, A. Gustard, N. W. Arnell, and G. A. Cole. (Proceedings of 2nd International FRIEND Conference, UNESCO, Braunschweig, Germany, October 1993), pp. 185–193, IAHS Publication no. 221.

Krasovskaia, I. (1995). Quantification of the stability of river flow regimes, *Journal of Hydrological Sciences*, vol. 40, pp. 587–598.

Krasovskaia, I. (1997). Entropy based grouping of river flow regimes. *Journal of Hydrology*, vol. 202, pp. 173–191.

Kullback, S., and Leibler, R. A. (1951). On information and sufficiency. *Annals of Mathematical Statistics*, vol. 22, pp. 79–86.

Parde, M. (1955). *Fleuves et Rivieres*. Colin, Paris.

Petts, G. E., and Amoros, C. (1997). *Fluvial Hydrosystems*. Chapman and Hall, London.

Poff, N. L., and Ward, J. V. (1989). Implications of streamflow variability and predictability for lotic community structure: A regional analysis of streamflow patterns. *Canadian Journal of Aquatic Sciences*, vol. 46, pp. 1805–1818.

Ratkowsky, D. A. (1984). A stopping rule and clustering method of wide applicability. Botanical Gazette, vol. 145, no. 4, pp. 518–523.

Richter, B., Baumgartner, J. V., Powell, J. and Braun, B. P. (1996). A method for assessing hydrologic alteration within ecosystems. *Conservation Biology*, vol. 10, pp. 1163–1174.

Ward, J. V., and Stanford, J. A. (1995). Ecological connectivity in alluvial ecosystems and its disruption by flow regulation. *Regulated Rivers*, vol. 11, pp. 105–119.

Further Reading

Gurnell, A. M., Petts, G. E., Harris, N., Ward, J. V., Tockner, K., Edwards, P. J. and Kollmann, J. (2000). Large wood retention in river channels: The case of the Fiume Tagliamento, Italy. *Earth Surface Processes and Landforms*, vol. 25, pp. 255–275.

Haines, A. T., Finlayson, B. L. and McMahon, T. A. (1988). A global classification of river regimes. *Applied Geography*, vol. 8, pp. 255–272.

Hynes, H. B. N. (1970). *The Ecology of Running Waters*. University of Toronto Press, Toronto, Canada.

Krasovskaia, I., and Gottschalk, L. (2002). River flow regimes in a changing climate. *Hydrological Sciences Journal*, vol. 47, no. 4, pp. 597–609.

Mojena, E. (1977). Hierarchical grouping methods and stopping rules: A evaluation. *The Computer Journal*, vol. 20, no. 4, pp. 359–363.

Ward, J. H. (1963). Hierarchical groupings to optimize an objective function. *Journal of the American Statistical Association*, vol. 58, pp. 236–244.

CHAPTER 19

Sediment Yield

Sediment yield from a watershed depends on the rainfall characteristics (amount, intensity, duration, and spatiotemporal distribution), soil characteristics (texture, structure, porosity, and spatial variability), land use, slope, and anthropogenic factors. Perhaps the most popular lumped model for computing sediment yield from small watersheds (agricultural, forest, and urban) is the universal soil loss equation (USLE) developed by Wischmeier and Smith (1965, 1978). Williams (1975) and Williams and Berndt (1977) modified the USLE by explicitly including the effect of runoff and designated it as MUSLE. Later Renard et al. (1993) further revised it and called it RUSLE. To apply USLE to large watersheds, the concept of delivery ratio (ratio of sediment generated to the amount of erosion) has been incorporated.

Another group of conceptual sediment yield models based on the hydrologic systems approach that became popular in the 1970s and 1980s includes the sediment graph and unit sediment graph. The sediment graph model was proposed by Rendon-Herrero (1974) and further investigated by Rendon-Herrero et al. (1980) and Singh and Chen (1982). Rendon-Herrero (1974, 1978) extended the unit hydrograph method to directly derive a unit sediment graph for a small watershed. Williams (1978) and Chen and Kuo (1974) extended the concept of the instantaneous unit hydrograph (IUH) to derive the instantaneous unit sediment graph (IUSG) and then determined the sediment discharge and yield.

Tyagi et al. (2008) extended the SCS-CN method (Soil Conservation Service, 1972) for determining sediment from agricultural watersheds. The empirical sediment yield models, including USLE and its amended versions MUSLE and RUSLE, and conceptual models yield satisfactory results for watersheds where needed data are available, but have limited capability for ungauged watersheds. To overcome the limitations of such lumped empirical models, a number of physically based models have been developed, incorporating the hydraulics of flow of water and sediment (detachment, suspension, deposition, and transport) (Foster and Lane, 1983; Lane and Nearing, 1989; Singh, 1983, 1989; Singh and Regl, 1983; Hairsine and Rose, 1992; Jin et al., 1999). These process-based models enhance our understanding of the mechanics of erosion and sediment transport, but it is difficult to estimate their parameters for ungauged or large watersheds. To overcome this limitation, the process-based models have been coupled with watershed hydrology models, such as SHETRAN (Figueiredo and Bathurst, 2007), EUROSEM (Morgan et al., 1998), MEDRUSH (Kirkby et al., 2002), and MODFSRS (Nunes et al., 2006). These models are deterministic. Although erosion and sediment transport are subject to considerable uncertainty, these models have not been modified to account for this uncertainty. de Araujo (2007) is perhaps the first to develop a hillslope sediment production equation accounting for this uncertainty by applying the entropy theory. The objective of this chapter is to discuss sediment yield estimation by using entropy following the work of de Araujo (2007).

19.1   Erosion Process: Preliminaries

Consider a small watershed with a stream network. When it rains, soil particles are dislodged due to the impact of raindrops as well as the force of flow exerted on the soil. The dislodged particles are transported by runoff generated by rainfall. The length to which these particles are transported by water depends on the rainfall, soil, slope, vegetation, and consequently generated runoff. Runoff is generated on hillslopes and moves downslope to the nearby stream. Small streams join bigger streams which may join the main stream or even bigger streams. In the discussion here, it is assumed that runoff generated on a hillslope directly enters the mainstream, and smaller streams are considered part of the overland flow plane, as shown in Fig. 19.1.

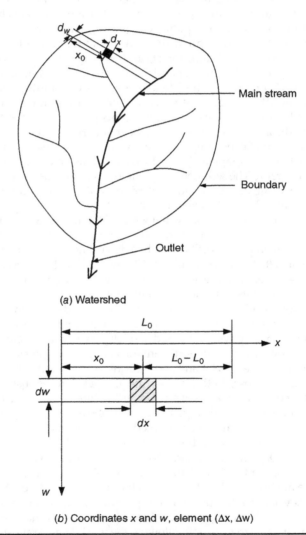

(a) Watershed

(b) Coordinates x and w, element (Δx, Δw)

FIGURE 19.1   (a) Representation of a watershed and (b) coordinates x and w.

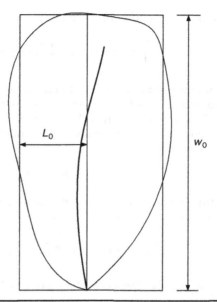

FIGURE 19.2   Rectangular representation of a watershed or a portion thereof.

Consider a stream dividing the hillslope into two parts: 1 and 2, as shown in Fig. 19.2. Now consider a small strip on which sediment production occurs as a result of rainfall. Let this strip be of width dw, as shown in Fig. 19.1. Consider a small element of thickness dx located on this strip, and let the distance be measured along x away from the boundary; that is, $x = 0$ at the boundary. This element has a cross-sectional area $dx\, dw$. Let this element be located at a distance x_0 from the watershed boundary; x_0 indicates the position of the initial effective hillslope erosion. If L_0 is the length of the strip between the divide and stream, then $L_0 - x_0$ is the distance from the element to the stream; L_0 is the slope length along which water moves. The length L_0 depends on where the strip is located on the w axis; in other words, it is a function of w and hence can be denoted as $L_0(w)$. Let E be the gross erosion per unit area (kg/m^2). Then the erosion on the element (also called elemental erosion) would be $E\, dx\, dw$. The objective is to compute the sediment yield of the strip denoted as $dQ_s(x,w)$ (kg/m^2) that feeds into the stream. This can be computed by integrating $E\, dx\, dw$ as

$$dQ_s(x, w) = \left[\int_{x_0}^{L_0(w)} E(x, w)dx \right] dw \tag{19.1}$$

For simplicity, let the watershed be represented by a symmetric rectangular shape with total drainage length w_0 and one-half the width as L_0. In this case $L_0(w) = L_0$ is independent of w and is constant. The area of the watershed A (km^2) is $2L_0 w_0$, or L_0 and can be defined as

$$L_0 = \frac{A}{2w_0} \tag{19.2}$$

Equation (19.1) can be integrated to provide sediment yield Q_s from the watershed:

$$Q_s = 2\int_0^{w_0} dQ_s(x, w)\, dx = \int_0^{w_0}\int_{x_0}^{L_0} E(x, w)\, dw\, dx \tag{19.3}$$

Here it is implied that the entire gross erosion from the strip becomes sediment yield; $E(x,w)$ is a function of both x and w and defines the spatially distributed gross erosion. For simplicity, let $E(x,w)$ be represented by an average value ξ. This means that the watershed is assumed to be homogeneous and the rainfall event is uniformly distributed. Then Eq. (19.3) becomes

$$Q_s = 2\xi w_0 (L_0 - x_0) \tag{19.4}$$

Equation (19.4) defines the sediment yield for a homogeneous watershed.

The gross erosion from the entire plain will be

$$G_e = 2\xi w_0 L_0 = \xi A = 2\int_0^{w_0}\int_0^{L_0(w)} \xi\, dx\, dw \tag{19.5}$$

The delivery ratio can now be defined as the ratio of sediment yield to gross erosion:

$$D_r = \frac{2\xi w_0 (L_0 - x_0)}{\xi A} = \frac{2w_0 (L_0 - x_0)}{A} \tag{19.6}$$

19.2   Consideration of Probability

In general, the shape of a watershed is complicated, and how L_0 varies as a function of w is difficult to express analytically. The distance x of the eroded sediment reaching the stream or drainage system varies with a number of factors. If the watershed is comprised of a large number of elements, then there will correspondingly a large number of x values which can be construed as a collection of random values. In other words, the distance of the eroded material can be considered a random variable X with a probability density function (PDF) $f(x)$ and cumulative distribution function (CDF) $F(x)$. The random variable X will vary from x_0 to a maximum value (potential travel distance) denoted by L_m. For a weak rainfall event, the sediment eroded only at $x \geq x_0 > 0$ is unlikely to reach the stream. In this case the maximum expected travel distance L_m will be less than L_0. Thus, $x \geq x_0 = L_0 - L_m$. On the other hand, if the rainfall event is strong, then $L_m \geq L_0$ and the eroded sediment may travel from the boundary to the stream. In this case, $x_0 = 0$. The term L_m is a parameter whose value depends on the rainfall and hillslope characteristics. Now the CDF of X can be defined as

$$F(x) = \int_0^{L_0 - x} f(x)\, dx \tag{19.7}$$

and

$$F(x, w) = \int_{L_0(w) - x}^{L_m} f(x)\, dx \tag{19.8}$$

Equation (19.8) defines the probability of eroded sediment reaching the stream.

With the above probabilistic considerations or Eqs. (19.7) and (19.8), the sediment yield equations can now be modified, reflecting the uncertainty associated with erosion and sediment yield prediction. Equation (19.1) becomes

$$dQ_s(x, w) = \left[\int_{x_0}^{L_0(w)} E(x, w)F(x,w)dx \right] dw \tag{19.9}$$

Note the probability $F(x, w)$ acts as a weighting factor.
Equation (19.3) becomes

$$Q_s = 2 \int_0^{w_0} \left\{ \int_{x_0}^{L_0} E(x,w) \left[\int_{L_0-x}^{L_m} f(x)dx \right] dx \right\} dw \tag{19.10}$$

Equation (19.10) simplifies to

$$Q_s = 2\xi w_0 \int_{x_0}^{L_0} \left[\int_{L_0-x}^{L_m} f(x)dx \right] dx \tag{19.11}$$

Now the delivery ratio D_r can be interpreted as the average probability F of a particle reaching the stream:

$$D_r = \frac{\int_0^{L_0} F(x)\,dx}{\int_0^{L_0} dx} \tag{19.12}$$

Inserting Eq. (19.8) in Eq. (19.12), one obtains

$$D_r = \frac{\int_0^{L_0} \left[\int_{L_0-x}^{L_m} f(x)\,dx \right] dx}{\int_0^{L_0} dx} \tag{19.13}$$

Combining Eqs. (19.11) and (19.12), one can write the sediment equation as

$$Q_s = \xi AD_r \tag{19.14}$$

19.3   Application of Entropy Theory

Equation (19.10) or (19.11) can be solved only if $f(x)$ is specified, and this can be obtained using entropy theory. Recalling the definition of the Shannon entropy (Shannon, 1948), one can write for the case of soil erosion

$$H(x) = - \int_0^{L_0} f(x) \ln f(x)\,dx \tag{19.15}$$

Equation (19.15) represents the uncertainty associated with the water erosion process. Maximization of H, implying the elimination of what is not known or specified, following the principle of maximum entropy (POME), yields $f(x)$. To that end, the first step is the specification of constraints.

19.3.1   Specification of Constraints

The PDF $f(x)$ must satisfy

$$\int_0^{L_0} f(x)\,dx = 1 \tag{19.16}$$

Equation (19.16) is the total probability constraint. The second constraint can be defined as the mean travel distance x_m

$$\int_0^{L_0} x f(x)\,dx = x_m \tag{19.17}$$

The physical meaning of Eq. (19.17) can be derived from hydraulic principles as follows.

Let the sediment discharge (kg/s) on a hillslope at a distance x from the boundary be denoted by $q_s(x)$ and the corresponding flow discharge (m³/s) by $Q(x)$. The average sediment concentration $\overline{C}(x)$ (kg/m³) will be $q_s(x)/Q(x)$. If $q_e(x)$ (kg/m²) is the cumulative gross erosion at a distance x, then one can express

$$q_s(x) = w_0 \int_0^x q_e(x) f(x)\,dx \tag{19.18}$$

where $f(x)\,dx$ represents the probability of a particle traveling the infinitesimal distance dx. The cumulative gross erosion can be represented as

$$q_e = \frac{E(x)}{T} x \tag{19.19}$$

where T denotes the duration of the event. For homogeneous watersheds, $E(x) = \xi$ if $0 \le x \le L_0$, but $E = 0$ if $x > L_0$, which occurs if $L_m > L_0$. For a homogeneous watershed, Eq. (19.17), in conjunction with Eq. (19.19), leads to the average sediment concentration c_m as

$$c_m = \frac{\xi w_0}{TQ} \int_0^{L_0} x f(x)\,dx \tag{19.20}$$

where the flow discharge Q (m³/s), on each side of the cell or element can be written as

$$Q = \frac{D}{T} w_0 L_0 \tag{19.21}$$

in which D is the runoff depth or amount of runoff. Equation (19.21) assumes uniform discharge during the duration of the event.

The maximum sediment concentration c_{max} of a hillslope discharge can be expressed, following Siepel et al. (2002), as

$$c_{max} = k_1 (1 - k_2)^{k_3} \frac{\rho S_0 u_m}{v_s} \left(\frac{\rho_s}{\rho_s - \rho} \right) \left(\frac{w_f}{w_f + 2D} \right) \tag{19.22}$$

where k_1, k_2, and k_3 are dimensionless parameters; ρ (kg/m³) is the density of water; ρ_s (kg/m³) is the density of wet sediment; S_0 is the bed slope; v_s is the average settling velocity (m/s) (or depositability); u_m is the average flow velocity; and w_f (m) is the width of flow.

Now recall the definition of stream power (Bagnold, 1977)

$$S_p = \gamma Q S_0 = \rho g S_0 R_h u_m \tag{19.23}$$

where S_p is the stream power (J/m²s), g is the acceleration due to gravity (m/s²), and R_h is the hydraulic radius (m). For flow between rills, w_f is much greater than D, meaning the last term in Eq. (19.22) equals unity, and R_h can be approximated as D. The average concentration c_m can be assumed as a function of c_{max} (Ni et al., 1999). Substitution of Eqs. (19.21), (19.22), and (19.23) into Eq. (19.20) leads to the average travel distance x_m of the sediment for a specified event

$$\int_0^{L_0} x f(x)\, dx = K_v \left(\frac{\rho_s}{\rho_s - \rho} \right) \frac{S_p L_0}{g \xi v_s} = x_m \tag{19.24}$$

in which K_v is the delivery parameter defined as

$$K_v = \left(\frac{c_m}{c_{max}} \right) k_1 (1 - k_2)^{k_3} \tag{19.25}$$

Note that the average travel distance of sediment particles depends on the delivery parameter, which inversely depends on the hillslope trap efficiency. Vegetative cover and morphology of the hillslope strongly influence the delivery parameter. A higher value of K_v means easier sediment transport and vice versa (Siepel et al., 2002). The average travel distance, given by Eq. (19.24), depends inversely on ξ. If an event is highly erosive, more sediment particles will be dislodged and more will be trapped, leading to a smaller value of x_m.

19.3.2 Maximization of Entropy

Equation (19.15) is maximized, subject to Eqs. (19.16) and (19.24), using the method of Lagrange multipliers. To that end, the lagrangian function L becomes

$$L = -\int_0^{L_0} f(x) \ln f(x)\, dx - (\lambda_0 - 1)\left[\int_0^{L_0} f(x)\, dx - 1 \right] - \lambda_1 \left[\int_0^{L_0} x f(x)\, dx - x_m \right] \tag{19.26}$$

Differentiating Eq. (19.26) with respect to f, while recalling the Euler-Lagrange calculus of variation, noting f as variable and x as parameter, and equating the derivative to zero, one obtains

$$\frac{\partial L}{\partial f} = 0 \Rightarrow -[\ln f(x) + 1] - (\lambda_0 - 1) - \lambda_1 x \tag{19.27a}$$

Equation (19.27a) gives

$$f(x) = \exp(-\lambda_0 - \lambda_1 x) \tag{19.27b}$$

where λ_0 and λ_1 are the Lagrange multipliers. Note that $f(x) = \exp(-\lambda_0)$ at $x = 0$.

19.3.3   Determination of Lagrange Multipliers

Substituting Eq. (19.27*b*) in Eq. (19.16) yields from the water divide up to the final position $x = x_0 + L_m$:

$$\exp \lambda_0 = \frac{1}{\lambda_1} \{1 - \exp[-\lambda_1 (x_0 + L_m)]\} \tag{19.28}$$

For a weak event, $L_m < L_0$ and $x_0 = L_0 - L_m$. Then the limits of integration will be from 0 to L_0, that is, on the entire hillslope length. On the other hand, for a strong event $L_m \geq L_0$ and $x_0 = 0$. Then the limits of integration will be from 0 to L_m, considering the potential distance.
   Substituting Eq. (19.28) in Eq. (19.27*b*), one gets the PDF of X

$$f(x) = \frac{\lambda_1}{1 - \exp[-\lambda_1 (x_0 + L_m)]} \exp(-\lambda_1 x) \tag{19.29}$$

Substituting Eq. (19.29) in Eq. (19.8), one gets the entropy-based CDF of X

$$F(x) = \frac{\lambda_1}{1 - \exp[-\lambda_1 (x_0 + L_m)]} \{\exp[-\lambda_1 (L_0 - x)] - \exp(-\lambda_1 L_m)\} \tag{19.30}$$

19.3.4   Sediment Production Equation

Substituting Eq. (19.27*b*) in Eq. (19.11), one gets

$$Q_s = 2\xi w_0 \int_{x_0}^{L_0} \left[\int_{L_0 - x}^{L_m} \exp(-\lambda_0 - \lambda_1 x)\, dx \right] dx = 2\xi w_0 \int_{x_0}^{L_0} F(x)\, dx \tag{19.31}$$

Inserting Eq. (19.30) in Eq. (19.31), one gets

$$Q_s = 2\xi w_0 \frac{1}{1 - \exp[-\lambda_1 (x_0 + L_m)]} \left\{ \frac{\exp[-\lambda_1 (L_0 - x_0)] - 1}{\lambda_1} - \exp(-\lambda_1 L_m)(L_0 - x_0) \right\} \tag{19.32}$$

Using the definition of D_r from Eq. (19.13), Eq. (19.32) can be expressed as

$$Q_s = 2\xi A \frac{1}{1 - \exp[-\lambda_1 (x_0 + L_m)]} \left\{ \frac{\exp[-\lambda_1 (L_0 - x_0)] - 1}{\lambda_1} - \exp(-\lambda_1 L_m)(L_0 - x_0) \right\} \tag{19.33}$$

Equation (19.33) is the entropy-based hillslope production equation for a watershed of area A, hillslope length L_0, and maximum sediment travel distance L_m for an event that generates gross erosion ξ.
   Equation (19.33) contains two parameters λ_1 and L_m which can be determined as follows. When a particle is detached from its position x from the watershed boundary (on a relatively linear hillslope), it may be trapped before reaching the stream. This means that this particle has not traveled a distance $L_0 - x$. The potential travel distance for this particle is L_m. Thus, the ratio of the remaining distance to the stream $L_0 - x$ to the potential travel distance L_m can be assumed to represent the probability of a sediment particle being trapped before reaching the stream. Then the probability of the particle reaching the stream can be expressed as

$$F(x) = 1 - \frac{L_0 - x}{L_m} \tag{19.34}$$

Differentiating Eq. (19.34), one finds

$$f(x) = \frac{1}{L_m}$$ (19.35)

Considering Eq. (19.29) for the average distance $x_m = (x_0 + L_0)/2$,

$$f(x_m) = \frac{1}{L_m} = \frac{\lambda_1}{1 - \exp[-\lambda_1(x_0 + L_m)]} \exp(-\lambda_1 x_m)$$ (19.36)

Substituting Eq. (19.29) in Eq. (19.23) yields

$$x_m = \frac{\lambda_1}{1 - \exp[-\lambda_1(x_0 + L_m)]} \left[-\frac{1}{\lambda_1} \exp(-\lambda_1 L_0) L_0 - \frac{1}{\lambda_1^2} \exp(-\lambda_1 L_0) - \frac{1}{\lambda_1^2} \right]$$ (19.37)

Equations (19.36) and (19.37) can be employed to determine λ_1 and L_m for each event in an iterative manner as follows: (1) Assume an initial value of L_m. Take L_0 as a starting value. (2) Compute λ_1 from Eq. (19.36). (3) Compute L_m from Eq. (19.37). (4) Compare this value of L_m with that in step 1. If the difference between the two values is large, use this most recent value as a starting value and repeat the steps until a satisfactory value of L_m is obtained. de Araujo (2007) noted that L_m varied from 0 to 20 times the slope length L_0, where λ_1 varied from 10^{-2} to 10^{-1} for 600 events.

19.4   Sediment Graph

A sediment graph, suggested by Rendon-Herrero (1974), is a linear relation on logarithmic paper between sediment yield (wash load) and surface runoff volume due to rainfall from small upland watersheds. He hypothesized that the slope of this straight line would remain approximately constant from one watershed to another. An investigation by Singh and Chen (1982) of 39 watersheds from 14 states in the United States concluded that the linear relationship was satisfactory for a number of watersheds; the relationship improved if the storms were separated into winter and summer storms, excluding snowstorms; and the intercept would reflect the individual watershed characteristics. The relation between sediment yield y and volume of surface runoff V_Q can be expressed as

$$y = a V_Q^b$$ (19.38)

or $$\log y = \log a + b \log V_Q$$ (19.39)

where a and b are parameters.

Example 19.1.   For relatively isolated storms on a watershed, obtain data on runoff and sediment yield, and plot graphically and fit a regression relation. How good is the regression relation?

Solution.   Runoff and sediment yield data are obtained from the Southwest Watershed Research Center, U.S. Department of Agriculture (http://tucson.ars.ag.gov/dap/), Tucson, Arizona. Table 19.1 lists isolated storms observed at Station 102 on the Walnut Gulch watershed, Arizona.
    Sediment yield is regressed on runoff volume, as shown in Fig. 19.3, where parameters $a = 184.57$ and $b = 0.9435$ are obtained by the least-squares method. The coefficient of determination is found to be 0.822, which suggests a good fit by regression.

Storm Start Time	Sediment Yield (g)	Runoff Volume (mm)	Storm Start Time	Sediment Yield (g)	Runoff Volume (mm)
7/28/1998 17:35	254	1.52	9/11/2002 12:55	637	3.11
8/3/1998 17:09	338	1.17	7/25/2003 18:15	1211	6.06
8/12/1998 19:05	1596	9.97	8/23/2003 14:41	1113	5.08
7/14/1999 16:20	441	6.41	9/8/2005 12:18	2678	13.32
7/14/1999 20:44	1449	6.41	7/23/2007 13:20	712	2.51
8/2/1999 18:54	1238	6.83	7/31/2007 15:34	2183	12.70
8/28/1999 18:14	1336	7.88	8/3/2007 13:27	625	2.41
6/29/2000 12:11	1183	6.39	8/4/2007 11:15	650	3.26
7/30/2000 16:11	2273	12.52	7/19/2008 21:28	2278	12.67
8/10/2000 15:40	1294	11.49	7/22/2008 14:51	1027	7.22
8/11/2000 12:41	152	0.96	7/25/2008 14:37	1812	11.54
8/28/2000 19:26	264	0.73	6/27/2009 17:13	1185	6.409
10/19/2000 14:00	819	3.53	7/25/2010 13:07	685	3.76
10/21/2000 11:33	257	1.44	7/26/2010 18:39	2972	17.09
10/22/2000 18:46	90	2.24	8/28/2010 14:20	346	1.83
10/22/2000 20:57	307	2.24	9/9/2011 12:48	388	1.70
11/6/2000 19:19	139	1.33	9/14/2011 11:28	246	1.57
7/26/2002 21:21	462	3.29	7/3/2012 20:25	660	4.39
8/4/2002 12:54	1189	7.72	7/4/2012 16:42	321	2.03
			9/3/2012 18:13	586	3.76

TABLE 19.1 Runoff Volume and Sediment Yield Data

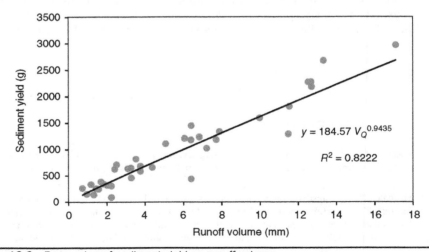

$$y = 184.57\, V_Q^{0.9435}$$

$$R^2 = 0.8222$$

FIGURE 19.3 Regression of sediment yield on runoff volume.

19.4.1   Derivation of Sediment Graph Using Entropy Theory

The procedure for deriving the sediment graph is similar to that described in the preceding section. It is assumed that temporally averaged sediment yield Y is a random variable with a probability density function denoted by $f(y)$. The Shannon entropy of Y, or of $f(y)$, $H(Y)$, can be expressed as

$$H(Y) = -\int_0^\infty f(y) \ln f(y)\, dy \tag{19.40}$$

Equation (19.40) expresses a measure of uncertainty about $f(y)$ or the average information content of sampled Y. The objective here is to derive the least-biased $f(y)$, which can be accomplished by maximizing $H(Y)$, subject to specified constraints, in accordance with the principle of maximum entropy. Maximizing $H(Y)$ is equivalent to maximizing $f(y) \ln f(y)$. To determine $f(y)$ that is least biased toward what is not known regarding sediment yield, the principle of maximum entropy developed by Jaynes (1957a, 1957b, 1982) is invoked, which requires specification of certain information, expressed in terms of constraints on sediment yield. According to POME, the most appropriate probability distribution is the one that has the maximum entropy or uncertainty, subject to these constraints.

Specification of Constraints

For deriving the sediment graph, following Singh (1998) the constraints to be specified are the total probability law which must always be satisfied by the probability density function of sediment yield, written as

$$C_1 = \int_0^\infty f(y)\, dy = 1 \tag{19.41}$$

and

$$C_2 = \int_0^\infty (\ln y)\, f(y)\, dy = \overline{\ln y} \tag{19.42}$$

Equation (19.41) is the first constraint C_1 defining the total probability law, and Eq. (19.42) is the second constraint C_2 defining the mean of the logarithm of sediment yield values.

Maximization of Entropy

To obtain the least-biased probability density function of Y, $f(y)$, the Shannon entropy, given by Eq. (19.40), is maximized following POME, subject to Eqs. (19.41) and (19.42). To that end, the method of Lagrange multipliers is employed. The lagrangian function then becomes

$$L = -\int_0^\infty f(y) \ln f(y)\, dy - (\lambda_0 - 1)\left[\int_0^\infty f(y)\, dy - C_1\right] - \lambda_1 \left[\int_0^\infty \ln y\, f(y)\, dy - C_2\right] \tag{19.43}$$

where λ_0 and λ_1 are the Lagrange multipliers. Recalling the Euler-Lagrange equation of calculus of variation and differentiating Eq. (19.43) with respect to f, noting that f is variable and y is parameter, and equating the derivative to zero, one obtains

$$\frac{\partial L}{\partial f} = 0 = -[\ln f(y) + 1] - (\lambda_0 - 1) - \lambda_1 \ln y \tag{19.44}$$

Derivation of Probability Distribution

Equation (19.44) leads to the entropy-based probability density function of sediment yield

$$f(y) = \exp(-\lambda_0 - \lambda_1 \ln y) \qquad \text{or} \qquad f(y) = \exp(-\lambda_0)\, y^{-\lambda_1} \tag{19.45}$$

The PDF of Y contains the Lagrange multipliers λ_0 and λ_1 which can be determined using Eqs. (19.41) and (19.42). The cumulative probability distribution function of Y can be obtained by integrating Eq. (19.45):

$$F(y) = \frac{\exp(-\lambda_0)}{-\lambda_1 + 1}\, y^{-\lambda_1 + 1} \tag{19.46}$$

Maximum Entropy

Substitution of Eq. (19.45) in Eq. (19.40) yields the maximum entropy or uncertainty of sediment yield:

$$H(Y) = \lambda_0 + \lambda_1 \overline{\ln y} \tag{19.47}$$

Determination of Lagrange Multipliers

Substitution of Eq. (19.45) in Eq. (19.41) leads to

$$\lambda_0 = -\ln(-\lambda_1 + 1) + (-\lambda_1 + 1)\ln y_D \tag{19.48}$$

where y_D is the maximum value of y. Differentiating Eq. (19.48) with respect to λ_1 produces

$$\frac{\partial \lambda_0}{\partial \lambda_1} = \frac{1}{-\lambda_1 + 1} - \ln y_D \tag{19.49}$$

On the other hand, substitution of Eq. (19.45) in Eq. (19.41)) can also be written as

$$\lambda_0 = \ln \int_0^{y_D} y^{-\lambda_1} dy \tag{19.50}$$

Differentiating Eq. (19.50) with respect to λ_1, one gets

$$\frac{\partial \lambda_0}{\partial \lambda_1} = -\frac{\displaystyle\int_0^{y_D} (\ln y) y^{-\lambda_1}\, dy}{\displaystyle\int_0^{y_D} y^{-\lambda_1}\, dy} \tag{19.51}$$

Multiplying and dividing Eq. (19.51) by $\exp(-\lambda_0)$, and using Eq. (19.41), one finds

$$\frac{\partial \lambda_0}{\partial \lambda_1} = -\frac{\displaystyle\int_0^{y_D} (\ln y)\exp(-\lambda_0) y^{-\lambda_1}\, dy}{\displaystyle\int_0^{y_D} \exp(-\lambda_0) y^{-\lambda_1}\, dy} = -\overline{\ln y} \tag{19.52}$$

Setting Eq. (19.52) to Eq. (19.49), one obtains an expression for λ_1 in terms of the constraint and the limits of integration of Q:

$$\lambda_1 = 1 - \frac{1}{\ln y_D - \overline{\ln y}} \tag{19.53}$$

Derivation of Sediment Graph

Let the maximum runoff volume be denoted by V_p. It is then assumed that all values of sediment yield y measured at the watershed outlet for any storm between 0 and V_p are equally likely. In reality, this is not highly unlikely because at different times different values of sediment yield do occur. This is also consistent with the laplacian principle of insufficient reason. Then the cumulative probability distribution of sediment yield can be expressed as the ratio of runoff volume to the maximum surface runoff volume. The probability of sediment yield being equal to or less than a given value of y is V_Q/V_p; for any surface runoff volume (measured at the outlet) less than a given value V_Q, the sediment yield is less than a given value, say y; thus the cumulative distribution function of sediment yield $F(y) = P(\text{sediment yield} \leq \text{a given value of } y)$, P = probability, can be expressed as

$$F(y) = \frac{V_Q}{V_p} \tag{19.54}$$

where $F(y)$ denotes the cumulative distribution function and y = sediment yield. Note that on the left side the argument of function F in Eq. (19.54) is variable y, whereas on the right side the variable is V_Q. The CDF of y is not linear in terms of V, unless y and V are linearly related. Of course, it is plausible that $F(y)$ might have a different form. Since Eq. (19.54) constitutes the fundamental hypothesis employed here for deriving the sediment graph using entropy, it will be useful to evaluate its validity. This hypothesis (i.e., the relation between the cumulative probability $F(y)$ and the ratio V_Q/V_p) should be tested for a large number of natural watersheds. Also note that a similar hypothesis has been employed when using the entropy theory for deriving infiltration equations by Singh (2010a, 2010b), soil moisture profiles by Singh (2010c), and velocity distributions by Chiu (1987).

The probability density function is obtained by differentiating Eq. (19.54) with respect to y:

$$f(y) = \frac{dF(y)}{dy} = \frac{1}{V_p} \frac{dV_Q}{dy} \quad \text{or} \quad f(y) = \left(V_p \frac{dy}{dV_Q} \right)^{-1} \tag{19.55}$$

The term $f(y)\, dy = F(y + dy) - F(y)$ denotes the probability of sediment yield being between y and $y + dy$.

Substituting Eq. (19.45) in Eq. (19.55), one gets

$$\frac{\exp(-\lambda_0)}{-\lambda_1 + 1} y^{-\lambda_1 + 1} \Big|_0^y = \frac{1}{V_p} V_Q \Big|_0^{V_Q} \tag{19.56}$$

Equation (19.56) yields

$$y = \left[\frac{\exp \lambda_0 (-\lambda_1 + 1)}{V_P} V_Q \right]^{\frac{1}{-\lambda_1 + 1}} \tag{19.57}$$

$$a = \left[\frac{\exp \lambda_0}{V_P} (-\lambda_1 + 1) \right]^{\frac{1}{-\lambda_1 + 1}} \tag{19.58}$$

and

$$b = \frac{1}{-\lambda_1 + 1} \tag{19.59}$$

Then Eq. (19.57) is an expression of the sediment graph, and is the same as Eq. (19.38). When plotted on a log-log paper, Eq. (19.57) will be a straight line. It may now be interesting to evaluate the Lagrange multipliers and hence parameters a and b.

Determination of Lagrange Multipliers

Substitution of Eq. (19.45) in Eq. (19.40) yields

$$\exp \lambda_0 = \frac{y_D^{-\lambda_1 + 1}}{-\lambda_1 + 1} \quad \text{or} \quad \lambda_0 = -\ln(-\lambda_1 + 1) + (-\lambda_1 + 1) \ln y_D \tag{19.60}$$

where y_D is the sediment yield when $V_Q = V_P$.
    Differentiating Eq. (19.60) with respect to λ_1, one obtains

$$\frac{\partial \lambda_0}{\partial \lambda_1} = -\ln y_D + \frac{1}{-\lambda_1 + 1} \tag{19.61}$$

One can also write from Eqs. (19.45) and (19.41)

$$\lambda_0 = \ln \int_0^{y_D} y^{-\lambda_1} \, dy \tag{19.62}$$

Differentiating Eq. (19.62) with respect to λ_1 and simplifying, one obtains

$$\frac{\partial \lambda_0}{\partial \lambda_1} = -\overline{\ln y} \tag{19.63}$$

Setting Eq. (19.63) equal to Eq. (19.61) leads to an estimate of λ_1:

$$\lambda_1 = 1 - \frac{1}{\ln y_D - \overline{\ln y}} \tag{19.64}$$

Therefore, exponent b of Eq. (19.38) becomes

$$b = \ln y_D - \overline{\ln y} \tag{19.65}$$

Equation (19.65) shows that the exponent b of the power form sediment graph can be estimated from the values of the logarithm of maximum sediment yield at the maximum surface runoff volume and the average of the logarithmic values of sediment yield. The larger the difference between these logarithm values, the larger the exponent will be.

The Lagrange multiplier λ_0 can now be expressed as

$$\lambda_0 = \frac{\ln y_D}{\ln y_D - \overline{\ln y}} + \ln(\ln y_D - \overline{\ln y}) \tag{19.66a}$$

Therefore,

$$a = \left\{ \frac{\exp[\ln y_D/(\ln y_D - \overline{\ln y})]}{V_P} \right\}^{\ln y_D - \overline{\ln y}} \tag{19.66b}$$

The PDF of Y can be expressed as

$$f(y) = \exp(-\lambda_0)y^{1/b-1} = \frac{\exp[-\ln y_D/(\ln y_D - \overline{\ln y})]}{\ln y_D - \overline{\ln y}} y^{1/b-1} \tag{19.67a}$$

and the CDF as

$$F(y) = b\exp(-\lambda_0)y^{1/b} = \exp[-\ln y_D/(\ln y_D - \overline{\ln y})]y^{1/b} \tag{19.67b}$$

For $b < 1$, the probability density function monotonically increases from 0 to $\exp(-\lambda_0)y_D^{1/b-1}$. Figure 19.4 shows a PDF of sediment yield values observed at gauging station 102 at Walnut Gulch watershed.

The entropy (in napiers) of the sediment yield distribution can be obtained by substituting Eq. (19.67) in Eq. (19.40):

$$H = \lambda_0 - \left(\frac{1}{b} - 1\right)\overline{\ln y} = -\frac{\ln y_D}{\ln y_D - \overline{\ln y}} + \ln(\ln y_D - \overline{\ln y}) - \left(\frac{1}{b} - 1\right)\overline{\ln y} \tag{19.68}$$

Example 19.2.   For the same data as in Example 19.1, determine the parameters of the relation between sediment yield and runoff using the entropy theory, and compare these parameter values with those obtained by regression. Using the entropy-based parameter values, plot sediment yield on the same graph, compare with the regression relation, and comment.

Solution.   The maximum runoff volume V_P is given as 17.09 mm and corresponding $y_D = 2972$ g. From Eq. (19.66b),

$$a = \left\{ \frac{\exp[\ln y_D/(\ln y_D - \overline{\ln y})]}{V_P} \right\}^{\ln y_D - \overline{\ln y}} = \left\{ \frac{\exp[\ln 2972/(\ln 2972 - 6.54)]}{17.09} \right\}^{\ln 2972 - 6.54} = 47.25$$

$$b = \ln y_D - \overline{\ln y} = \ln 2972 - 6.54 = 1.46$$

The estimated b value from the entropy method is greater than that from regression, thus there is more curve fitting using the entropy method. But the entropy method fit the large value better than regression. See Fig. 19.5.

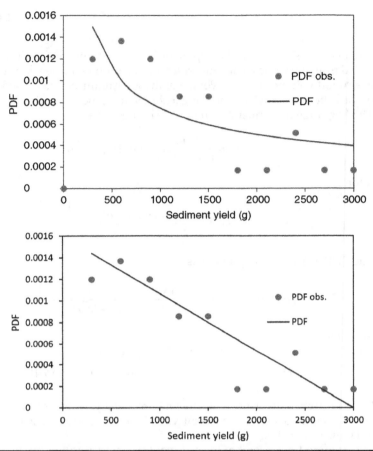

FIGURE 19.4 PDF of sediment yield values observed at gauging station 102 at Walnut Gulch watershed.

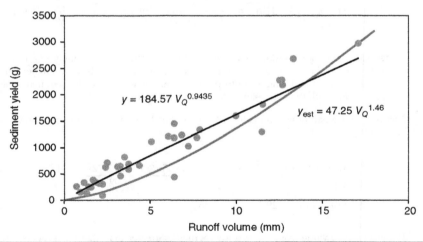

FIGURE 19.5 Relation between sediment yield and runoff using entropy. (Red line represents the values estimated from entropy method.)

19.5   Unit Sediment Graph

Hydrologic systems or conceptual approaches have been applied to determine the sediment discharge from small watersheds. For example, Williams (1978) extended the concept of unit hydrograph (UH) or instantaneous unit hydrograph (IUH), discussed in Chap. 14, to derive the unit sediment graph (USG) or instantaneous unit sediment graph (IUSG) for determining the sediment discharge from agricultural watersheds. Singh et al. (1982), Chen and Kuo (1984), and Srivastava et al. (1984) among others employed USG or IUSG. Sharma and Dickinson (1979a, 1979b) developed a conceptual model for daily and monthly sediment yield. The purpose of this section is to show how entropy theory can be applied to derive some of these conceptual models.

Sediment discharge $Q_s(t)$ at the watershed outlet at time t is the product of spatially average sediment concentration $c(t)$ and water discharge $Q(t)$ and can be expressed as

$$Q_s(t) = c(t)Q(t) \tag{19.69}$$

Equation (19.69) suggests two possibilities for application of entropy theory: (1) Sediment concentration is determined using an empirical or a conceptual approach, and discharge is determined using entropy theory. (2) Sediment discharge is directly determined by entropy theory. Both possibilities are briefly discussed here.

Since our focus here is on the use of unit hydrograph type of approaches, it may be useful to recall the UH and USG. The unit hydrograph is the distribution of runoff due to a unit amount of rainfall excess of one unit duration occurring uniformly over the watershed. Clearly the amount of runoff generated is unity, and its time distribution is UH. Here the input to the watershed is rainfall, but abstractions are subtracted and rainfall excess of uniform intensity is determined. The runoff thus generated is surface runoff. Analogously, USG can be defined as the distribution of sediment due to the unit amount of effective sediment of one unit duration that is generated uniformly over the watershed. The term *effective sediment* implies that the sediment that gets trapped during the course of its travel from the point of generation to the watershed outlet is subtracted, and the effective sediment is what constitutes the sediment yield. This sediment is essentially wash load. The effective sediment amount is the integral of effective erosion intensity; and analogous to rainfall excess intensity, the effective erosion intensity is uniform over its unit duration. The unit amount of effective sediment is generated by the unit amount of rainfall excess, and they both are assumed to occur uniformly over the watershed and the amount of sediment yield denoted by Y_s. If the unit duration tends to be infinitesimally small, then the resulting UH and USG will become IUH and IUSG, respectively.

Let $I(t)$ denote the rainfall excess intensity and $h(t)$ denote IUH. Then the unit hydrograph theory yields

$$Q(t) = \int_0^t h(t-\tau)I(\tau)\,d\tau \tag{19.70}$$

In a similar manner, if the effective sediment erosion intensity (ESEI) $E(t)$ denotes the amount of sediment erosion per unit of time and $h_s(t)$ denotes IUSG, then sediment discharge at any time t, denoted as $Q_s(t)$, can be written as

$$Q_s(t) = \int_0^t h_s(t-\tau)E(\tau)\,d\tau \tag{19.71}$$

19.5.1   Possibility 1: Discharge by Entropy Theory

In this case, the unit hydrograph is determined by entropy theory as discussed in Chap. 14. Depending on the constraints, entropy maximizing yields different expressions for IUH. Let the constraints be expressed as

$$\int_0^\infty (\ln t)\, h(t)\, dt = \overline{\ln t} \tag{19.72a}$$

$$\int_0^\infty t^c\, h(t)\, dt = \overline{t^c} \tag{19.72b}$$

where c is an exponent—an empirical parameter—but can be related to the hydraulics of flow. Equation (19.72a) expresses the expectation of the log values of times of travel or the mean travel time in the logarithmic domain, whereas Eq. (19.72b) expresses the moment of time of travel raised to power c. If, for example, $c = 1$, then Eq. (19.72b) expresses the mean travel time or the first moment about the origin; if $c = 2$, then it expresses the second moment about the origin (equal to the variance of time of travel minus the square of mean travel time); and so on.

Entropy maximizing, subject to Eqs. (19.72a) and (19.72b) as well as the total probability, leads to

$$h(t) = \frac{c\lambda_2^{(1-\lambda_1)/c}}{\Gamma\!\left(\dfrac{1-\lambda_1}{c}\right)} \exp(-\lambda_1 \ln t - \lambda_2 t^c) = \frac{c\lambda_2^{(1-\lambda_1)/c}}{\Gamma((1-\lambda_1)/c)} t^{-\lambda_1} \exp(-\lambda_2 t^c) \tag{19.73}$$

Equation (19.73) has three parameters: λ_1, λ_2, and c. Exponent c can be specified or determined by trial and error or can be estimated using the entropy method. Parameters λ_1, λ_2, and c are determined using entropy theory (Singh, 2013):

$$\frac{\ln \lambda_2}{c} - \frac{1}{\Gamma[(1-\lambda_1)/c]} \frac{\partial \Gamma[(1-\lambda_1)/c]}{\partial \lambda_1} = -\overline{\ln t} \tag{19.74a}$$

$$\frac{\lambda_1 - 1}{c\lambda_2} = -\overline{t^c} \tag{19.74b}$$

$$\frac{\lambda_1 - 1}{c\lambda_2^2} = -\overline{(t^c)^2} \tag{19.74c}$$

If $c = 2$, the IUH derived by Lienhard (1964) is

$$h(t) = \frac{1}{k\Gamma(3/2)}\left(\frac{t}{k}\right)^2 \exp\left[-\frac{3}{2}\left(\frac{t}{k}\right)^2\right] \tag{19.75}$$

By using Eq. (19.73) in Eq. (19.70), the runoff hydrograph can be determined. Using an expression for $c(t)$, Eq. (19.69) can be utilized to determine sediment discharge. Entropy-based approaches to sediment concentration in open channels were developed by Chiu et al. (2000), Choo (2000), and Cui and Singh (2014). These authors showed that the entropy-based equations predicted sediment concentration better than did empirical or hydraulics-based equations. Cui and Singh (2014) surveyed different formulations for sediment concentration.

19.5.2   Possibility 2: IUSG by Entropy Theory

In this case, IUSG is derived using entropy theory. First, a word about ESEI. For a given storm, the amount of sediment generated, or sediment yield, can be determined using the sediment yield–runoff relation discussed in Sec. 19.5.1. By dividing the sediment yield by the duration of rainfall excess, ESEI is estimated. This suggests that ESEI is a function of rainfall excess. Chen and Kuo (1984) expressed this function as a power function—ESEI $= aI^b$, where a and b are parameters—but it has not been adequately verified. Note that IUSG has the same dimension as IUH. Therefore, for deriving IUSG using entropy theory, the same framework can be employed and will not be repeated. Once $h_s(t)$ is determined, Eq. (19.71) is employed to determine $Q_s(t)$.

Example 19.3.   Obtain data on the rainfall, runoff, and sediment discharge for several storms on a given watershed. Determine the unit hydrograph and unit sediment graph for the storm.

Solution.   Rainfall, runoff, and sediment discharge data are obtained from the USDA Web site (http://tucson.ars.ag.gov/dap/). The data are given in Tables 19.2 and 19.3.

Time	Rainfall Rate (mm/h)	Depth (mm)
6/29/2000 12:29	15.24	11.938
6/29/2000 12:30	15.24	12.192
6/29/2000 12:31	11.43	12.446
6/29/2000 12:33	11.43	12.827
6/29/2000 12:35	22.86	13.208
6/29/2000 12:36	45.72	13.589
6/29/2000 12:37	38.10	14.351
6/29/2000 12:38	38.10	14.986
6/29/2000 12:39	45.72	15.621
6/29/2000 12:40	83.82	16.383
6/29/2000 12:41	99.06	17.78
6/29/2000 12:42	99.06	19.431
6/29/2000 12:43	99.06	21.082
6/29/2000 12:44	83.82	22.733
6/29/2000 12:45	45.72	24.13
6/29/2000 12:46	45.72	24.892
6/29/2000 12:47	15.24	25.654
6/29/2000 12:48	7.62	25.908
6/29/2000 12:51	1.09	26.289

Table 19.2   Rainfall Event of 6/29/2000 on Walnut Gulch Watershed

Time	Runoff Rate (mm/h)	Accumulated Runoff (mm)	Sediment Discharge (g/mm)
6/29/2000 12:30	1.28	0	128.00
6/29/2000 12:39	2.30	0.22	347.90
6/29/2000 12:42	16.64	0.67	4076.80
6/29/2000 12:45	49.92	3.18	6639.36
6/29/2000 12:48	29.18	6.65	1926.14
6/29/2000 12:52	5.63	7.88	478.72

TABLE 19.3 Runoff and Sediment Discharge for the 6/29/2000 Event on Walnut Gulch Watershed

For this storm event, the mean travel time is estimated from the lag time between the center of rainfall at 12:40 P.M. and peak runoff at 12:45 P.M., which is 5 min. Thus, k is estimated to be 0.083 h. From Eq. (19.75),

$$h(t) = \frac{1}{k\Gamma(3/2)}\left(\frac{t}{k}\right)^2 \exp\left[-\frac{3}{2}\left(\frac{t}{k}\right)^2\right]$$

$$= \frac{1}{0.083 \times 0.886}\left(\frac{t}{0.083}\right)^2 \exp\left[-\frac{3}{2}\left(\frac{t}{0.083}\right)^2\right] = 1974t^2 \exp(-218t^2)$$

Similarly, the mean travel time for sediment is estimated as 0.133 h, since the peak sediment discharge is observed at 12:48 P.M. Thus, one has

$$h_s(t) = \frac{1}{k\Gamma(3/2)}\left(\frac{t}{k}\right)^2 \exp\left[-\frac{3}{2}\left(\frac{t}{k}\right)^2\right]$$

$$= \frac{1}{0.133 \times 0.886}\left(\frac{t}{0.133}\right)^2 \exp\left[-\frac{3}{2}\left(\frac{t}{0.133}\right)^2\right] = 480t^2 \exp(-84.8t^2)$$

The IUH and IUSG are constructed, as plotted in Fig. 19.6.

Example 19.4. For the data in Example 19.3, determine the sediment concentration graph by dividing the sediment discharge by the water discharge. Plot concentration values on a graph.

Solution. The sediment concentration is calculated by dividing the sediment discharge by flow discharge, and it is plotted in Fig. 19.7.

Example 19.5. For the sediment yield data from a watershed, construct the cumulative probability distribution and probability density function of sediment yield. Determine the best-fit probability distribution.

Solution. The observed probability values of sediment yield are plotted in Fig. 19.8. The PDF and CDF are determined from Eqs. (19.67a) and (19.67b), respectively, with parameters determined in Example 19.3.

Example 19.6. Determine the sediment discharge for the rainfall-runoff event on the same watershed as in Example 19.1 by two methods based on entropy. These events should be different from the ones used in deriving the IUH and IUSG.

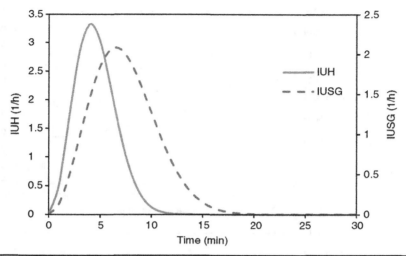

FIGURE 19.6   Estimated IUH and IUSG.

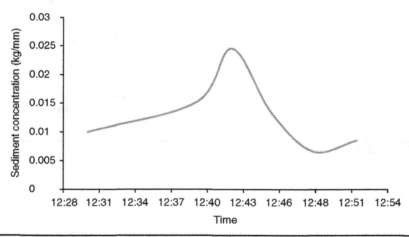

FIGURE 19.7   Sediment concentration as function of time.

Solution.   Storm event data are obtained from the same Web site as mentioned before and are shown in Table 19.4. For the rainfall-runoff event on 7/3/2012, rainfall started at 20:21 and runoff started at 20:25.

The first method computes runoff discharge from the IUH derived using entropy. Then multiplying runoff discharge by sediment concentration, the sediment discharge is obtained. Using the IUH estimated from Example 19.3 and interpolating the sediment concentration from Example 19.4, one computes the sediment discharge

$$Q_s(t) = cQ(t) = c \int_0^t h(t - \tau) I(\tau) \, d\tau$$

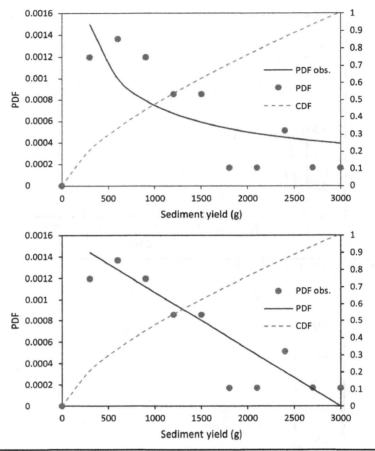

Figure 19.8 Probability distribution of sediment yield.

For example,

$$Q_c(0.1) = \sum_{\tau=0}^{t} h_c(t-\tau)I(\tau) = 0.01(0.461 \times 0.315 + 1.23 \times 1.27 + 0.98 \times 2.03 + \cdots) = 0.481 \text{ g/mm}$$

The second method uses IUSG derived using entropy and then uses Eq. (19.71) to obtain the sediment discharge

$$Q_s(t) = \int_0^t h_s(t-\tau)E(\tau)\,d\tau$$

where $E = aI^b$. First, parameters a and b are determined by plotting the effective soil erosion intensity versus effective rainfall intensity. Since effective soil erosion intensity data are not known, it is assumed that effective soil erosion intensity is uniform and is plotted versus uniform excess rainfall intensity, as shown in Fig. 19. 9. Uniform effective erosion intensity is obtained by dividing the sediment yield amount by rainfall excess duration. This is done for several storms. Thus, approximate first-order

Elapsed Time (min)	Rainfall Intensity (mm/h)	Cumulative Rainfall
0	0.311	11.43
2	0.315	15.24
3	0.637	38.1
4	1.27	45.72
5	2.032	45.72
6	2.794	38.1
7	3.429	45.72
8	4.191	53.34
9	5.08	60.96
10	6.096	60.96
11	7.112	60.96
12	8.128	68.58
13	9.271	68.58
14	10.414	60.96
15	11.43	60.96
16	12.446	76.2

TABLE 19.4 Rainfall Event for 7/3/2012

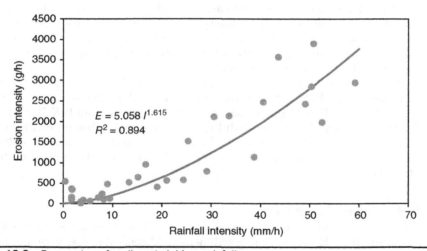

FIGURE 19.9 Regression of sediment yield on rainfall excess.

values of parameters are obtained by regression: $a = 5.058$ g/mm and $b = 1.615$. Then the IUSG obtained in Example 19.3 is convoluted. For example,

$$Q_s(0.1) = \sum_{\tau=0}^{t} h_s(t - \tau)E(\tau) = 0.461(5.058 \times 0.315^{1.615}) + 1.23(5.058 \times 1.27^{1.615}) + \cdots = 0.05 \text{ g/mm}$$

Sediment discharge thus computed is plotted in Fig. 19.10, which shows that sediment discharge values computed in two ways are comparable.

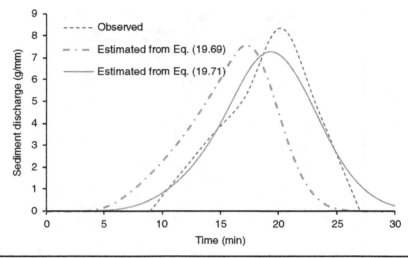

FIGURE 19.10 Estimated sediment discharge for event 7/3/2012.

19.6 Conditional Probability Distribution of Sediment Yield

Another method is to derive the bivariate runoff and sediment yield distribution using entropy theory. This will require the specification of constraints. Then the conditional probability distribution of sediment yield for given runoff can be derived. It is hypothesized that the derived joint distribution would be able to capture the full dependence structure between runoff and sediment yield.

Let runoff amount (V) be represented by X and sediment yield by Y, and let it be assumed that both X and Y are random variables. The objective is to determine the probability distribution of sediment yield Y, given runoff X. The objective is to determine the conditional probability distribution of sediment yield given rainfall, $F_{Y|X}(Y \leq y \mid X = x)$, where x is a specific value of X and y is a specific value of Y. The procedure is the same as for a single variable or the univariate case discussed in Chap. 1.

19.6.1 Shannon Entropy

Since a runoff-sediment yield distribution entails two dimensions, one representing runoff and the other representing sediment yield generated due to runoff, the Shannon entropy must be defined in both one and two dimensions. For univariate runoff or sediment yield, the Shannon entropy is

$$H(X) = -\int f_X(x) \ln f_X(x) \, dx \tag{19.76a}$$

where X here is a random variable V (or it can be Y), x is a specific value of X and y is a specific value of Y, and $f_X(x)$ is the probability density function of X, that is, of V or Y. Note that the upper limit for V is the value of the rainfall amount.

Likewise, the joint entropy $H(X, Y)$ can be written (Kapur and Kesavan, 1992)

$$H(X,Y) = -\iint f_{X,Y}(x,y) \ln f_{X,Y}(x,y)\, dx\, dy \qquad (19.76b)$$

where $f_{X,Y}(x,y)$ is the joint probability density function of X and Y. Similarly, the conditional Shannon entropy $H(Y \mid X)$ can be expressed as

$$H(Y \mid X) = -\iint f_{X,Y}(x,y) \ln f_{Y\mid X}(y \mid x)\, dy\, dx \qquad (19.76c)$$

where $f_{Y\mid X}(y \mid x)$ is the conditional probability density function of sediment yield, given runoff.

Note that

$$f_{Y\mid X}(y \mid x) = \frac{f_{X,Y}(x,y)}{f_X(x)} \qquad (19.77)$$

where $f_X(x)$ is the marginal probability density function of runoff. Substituting Eq. (19.77) in Eq. (19.76c), one can write the conditional entropy $H(Y \mid X)$ as

$$H(Y \mid X) = -\iint f_{X,Y}(x,y) \ln \frac{f_{X,Y}(x,y)}{f_X(x)}\, dy\, dx$$

$$= -\iint f_{X,Y}(x,y) \ln f_{X,Y}(x,y)\, dy\, dx + \iint f_{X,Y}(x,y) \ln f_X(x)\, dy\, dx \qquad (19.78a)$$

Since $\int f_{X,Y}(x, y)\, dy = f_X(x)$, Eq. (19.74) can be written as

$$H(Y \mid X) = -\iint f_{X,Y}(x,y) \ln f_{X,Y}(x,y)\, dy\, dx + \int f_X(x) \ln f_X(x)\, dx$$

$$= H(X,Y) - H(X) \qquad (19.78b)$$

19.6.2   Specification of Constraints

For maximizing the Shannon entropy (univariate as well as bivariate), constraints need to be specified first for maximizing the runoff entropy and then for maximizing the joint entropy of runoff and sediment yield. The gamma marginal distribution has been applied for a variety of hydrological analyses (e.g., Clarke, 1980; Yue, 2001). Preliminary data analysis shows that both runoff and sediment yield distributions can be represented by a gamma distribution. Therefore, following Singh (1998), the constraints on runoff for maximizing $H(X)$ can be expressed as

$$\int_0^\infty f_X(x)\, dx = 1 \qquad (19.79a)$$

$$\int_0^\infty x f_X(x)\, dx = \bar{x} \qquad (19.79b)$$

$$\int_0^\infty \ln(x) f_X(x)\, dx = \overline{\ln(x)} \qquad (19.79c)$$

Likewise, for maximizing $H(X, Y)$ the following constraints are specified:

$$\int_0^\infty \int_0^\infty f_{X,Y}(x,y)\, dx\, dy = 1 \tag{19.80}$$

$$\int_0^\infty \int_0^\infty x f_{X,Y}(x,y)\, dy\, dx = \overline{x} \tag{19.81a}$$

$$\int_0^\infty \int_0^\infty y f_{X,Y}(x,y)\, dy\, dx = \overline{y} \tag{19.81b}$$

$$\int_0^\infty \int_0^\infty \ln x f_{X,Y}(x,y)\, dy\, dx = \overline{\ln x} \tag{19.82}$$

$$\int_0^\infty \int_0^\infty \ln y f_{X,Y}(x,y)\, dy\, dx = \overline{\ln y} \tag{19.83}$$

$$\int_0^\infty \int_0^\infty xy f_{X,Y}(x,y)\, dy\, dx = \overline{xy} = S_{XY} + \overline{x} \cdot \overline{y} \tag{19.84}$$

Now note that the marginal distribution of runoff can be derived in two ways. First, it can be derived independently using the constraints given by Eqs. (19.79a) to (19.79c). Second, it can be derived from the joint distribution of runoff and sediment yield. However, both methods must lead to the same distribution, and this consistency was overlooked by Agrawal et al. (2005) who used the empirical distribution as the marginal distribution; hence their derivation is in error.

19.6.3   Maximization of Entropy and Probability Distributions

In accordance with the principle of maximum entropy, both the runoff entropy and the joint runoff–sediment yield entropy must be maximized. Since the procedure for entropy maximizing is the same, maximization of the joint entropy given by Eq. (19.76b) is illustrated here. Maximization can be achieved by using the method of Lagrange multipliers. To that end, the lagrangian function L is constructed as

$$L = \int_0^\infty \int_0^\infty f_{X,Y}(x,y) \ln f_{X,Y}(x,y)\, dx\, dy - (\lambda_0 - 1)\left[\int_0^\infty \int_0^\infty f_{X,Y}(x,y)\, dx\, dy - 1 \right]$$

$$-\lambda_1 \left[\int_0^\infty \int_0^\infty x f_{X,Y}(x,y)\, dx\, dy - \overline{x} \right] - \lambda_2 \left[\int_0^\infty \int_0^\infty \ln(x) f_{X,Y}(x,y)\, dx\, dy - \overline{\ln x} \right]$$

$$-\lambda_3 \left[\int_0^\infty \int_0^\infty y f_{X,Y}(x,y)\, dx\, dy - \overline{y} \right] - \lambda_4 \left[\int_0^\infty \int_0^\infty \ln(y) f_{X,Y}(x,y)\, dx\, dy - \overline{\ln y} \right]$$

$$-\lambda_5 \left[\int_0^\infty \int_0^\infty xy f_{X,Y}(x,y)\, dx\, dy - \overline{xy} \right] \tag{19.85}$$

where λ_i, $i = 0, 1, 2, 3, 4$, and 5, are the Lagrange multipliers. Differentiating Eq. (19.85) with respect to $f_{X,Y}(x,y)$, recalling the calculus of variation, and equating the derivative to zero, one obtains the joint distribution of rainfall and runoff as

$$f_{X,Y}(x,y) = \exp(-\lambda_0 - \lambda_1 x - \lambda_2 \ln x - \lambda_3 y - \lambda_4 \ln y - \lambda_5 xy) \qquad (19.86)$$

Equation (19.86) has six unknown Lagrange multipliers which are determined using Eqs. (19.79a) to (19.84).

The probability density function $f_X(x)$ can be obtained by integrating $f_{X,Y}(x, y)$ over y:

$$f_X(x) = \int_0^\infty f_{X,Y}(x,y)\, dy = \frac{\exp(-\lambda_0 - \lambda_1 x - \lambda_2 \ln x)}{(\lambda_3 + \lambda_5 x)^{1-\lambda_4}} \Gamma(1 - \lambda_4) \qquad (19.87)$$

where $\lambda_3 + \lambda_5 x > 0$. Similarly, the probability density function $f_Y(y)$ can be determined by integrating $f_{X,Y}(x, y)$ over x:

$$f_Y(y) = \int_0^\infty f_{X,Y}(x,y)\, dx = \frac{\exp(-\lambda_0 - \lambda_3 y - \lambda_4 \ln y)}{(\lambda_1 + \lambda_5 y)^{1-\lambda_2}} \Gamma(1 - \lambda_2) \qquad (19.88)$$

where $\lambda_1 + \lambda_5 y > 0$.

Now the conditional distribution of sediment yield, given runoff $f_{Y|X}(y \mid x)$, can be written, using Eqs. (19.86) and (19.87), as

$$f_{Y|X}(y \mid x) = \frac{f_{X,Y}(x,y)}{f_X(x)} = (\lambda_3 + \lambda_5 x)^{1-\lambda_4} \frac{\exp(-\lambda_3 y - \lambda_4 \ln y - \lambda_5 xy)}{\Gamma(1 - \lambda_4)} \qquad (19.89)$$

Equation (19.89) is the needed conditional probability distribution of sediment yield and its parameters need to be determined from the specified constraints.

19.6.4   Determination of Lagrange Parameters

Substituting Eq. (19.86) in Eq. (19.79), one gets

$$\int_0^\infty \int_0^\infty f_{X,Y}(x,y)\, dx\, dy = \exp(-\lambda_0) \int_0^\infty \frac{\exp(-\lambda_1 x - \lambda_2 \ln x) \Gamma(1 - \lambda_4)}{(\lambda_3 + \lambda_5 x)^{1-\lambda_4}}\, dx = 1 \qquad (19.90)$$

Equation (19.90) can be written for the zeroth Lagrange multiplier as

$$\lambda_0 = \ln \int_0^\infty \frac{\exp(-\lambda_1 x - \lambda_2 \ln x) \Gamma(1 - \lambda_4)}{(\lambda_3 + \lambda_5 x)^{1-\lambda_4}}\, dx \qquad (19.91)$$

Equation (19.91) can be rewritten as

$$\lambda_0 = \frac{1}{(\lambda_3)^{1-\lambda_4}} \ln \int_0^\infty \frac{\exp(-\lambda_1 x - \lambda_2 \ln x)}{\left(1 + (\lambda_5/\lambda_3)x\right)^{1-\lambda_4}}\, dx \qquad (19.92)$$

Equations (19.91) and (19.92) show that λ_0 is a function of $\lambda_1, \lambda_2, \lambda_3, \lambda_4$, and λ_5, but this function is not known. Nevertheless, one can write

$$\lambda_0 = \lambda_0(\lambda_1, \lambda_2, \lambda_3, \lambda_4, \lambda_5) \qquad (19.93)$$

Based on the constraints defined by Eqs. (19.80) to (19.84), the Lagrange multipliers must satisfy the following constraints:

$$\lambda_1 > 0 \qquad \lambda_2 < 1 \qquad \lambda_4 < 1 \qquad \lambda_3 + \lambda_5 x > 0 \qquad \lambda_1 + \lambda_5 y > 0 \qquad (19.94)$$

Now, the Lagrange multipliers can be estimated analytically or numerically. But in the entropy-based joint distribution, there is no analytical solution. Even though Taylor's series expansion seems a viable approach to obtain the Lagrange multipliers, it is not guaranteed that $|\lambda_5 x / \lambda_3| < 1$ unless $\lambda_5 \to 0$ for the Taylor series expansion to converge quickly enough to be applicable. Thus the Lagrange multipliers for the proposed distribution can be estimated numerically by using Newton's method as follows.

Let the constraints of the entropy-based bivariate distribution, i.e., Eqs. (19.79a) to (19.84), be $g_r(\overline{\lambda})$. Then, using the first-order Taylor series expansion, the constraints can be approximated as

$$a_r = g_r(\overline{\lambda}) \approx g_r(\overline{\lambda_0}) + (\overline{\lambda} - \overline{\lambda_0})^T \, |G|_{\overline{\lambda} = \overline{\lambda_0}} \qquad r = 0, 1, \ldots, 5 \qquad (19.95)$$

in which a_r is the sample moment pertaining to each constraint in Eqs. (19.79a) to (19.84); $g_r(\overline{\lambda})$ is the population moments calculated from Eqs. (19.79a) to (19.84); $\lambda\overline{\lambda} = [\lambda_0 \; \lambda_1 \; \lambda_2 \; \lambda_3 \; \lambda_4 \; \lambda_5]$, the Lagrange multipliers; $\overline{\lambda^0}$ is the initial Lagrange multiplier set; $\overline{\lambda^0} = [\ln[(x_{max} - x_{min})(y_{max} - y_{min})] \, 0 \, 0 \, 0 \, 0 \, 0]^T$; $x_{max}, x_{min}, y_{max},$ and y_{min} are the maxima and minima of observed runoff and sediment yield data; $|G|$ is the 6×6 symmetric matrix; and $G_{rj} = \partial g_r(\overline{\lambda})/\partial \lambda_j|_{\overline{\lambda} = \lambda^0}$, $r, j = 0, 1, \ldots, 5$. Now with the additional inequality constraints for the Lagrange multipliers [Eq. (19.94)], the Lagrange multipliers in Eq. (19.95) can be optimized using Newton's method by minimizing the difference between sample moments and the population moments obtained from Eqs. (19.80) to (19.84), that is, $a_r - g_r(\overline{\lambda})$. The parameters thus obtained are optimized, which then results in the maximum entropy-based bivariate rainfall–runoff distribution corresponding to the defined constraints.

19.6.5   Determination of Empirical Bivariate Joint Entropy

The empirical joint entropy may be considered one way to test whether the parametric joint distribution may be used to model the bivariate random variable appropriately. Similar to univariate analysis, nonparametric statistical methods, i.e., the frequency histogram and kernel density function, are generally applied to calculate the empirical joint entropy. Scott (1992) and Wand and Jones (1995) extensively discussed the importance of bin width for frequency histograms and bandwidth of the kernel density function during the process of estimation. They stated that in either the frequency histogram or the kernel density estimation approach, different bin widths or bandwidths play a significant role in preserving the underlying probability density function. In the kernel density estimation approach, different choices of kernel functions (e.g., the commonly used normal, box, triangular, Epanechnikov, etc.) should not result in significant differences for the estimated probability density function.

Here the product normal kernel density function is chosen for discussion. The appropriate bandwidth can be determined using the minimum integrated square error (MISE) criterion following the algorithm proposed by Botev et al. (2010), i.e., using the diffusion approach to optimize the kernel bandwidth. The bivariate joint density function for bivariate random variable $f_{X,Y}(x,y)$, using the product kernel function, can be written

$$\hat{f}_{X|Y}(\bar{x},\bar{y}) = \frac{1}{nh_x h_y} \sum_{i=1}^{n} \left[K\left(\frac{x_i - X_i}{h_x}\right) K\left(\frac{y_i - Y_i}{h_y}\right) \right]$$ (19.96a)

in which h_x and h_y are the kernel bandwidth for random variables X and Y, respectively; n is the sample size; and X_i, Y_i is the location where the density function is being estimated. Equation (19.96a) can be rewritten as:

$$\hat{f}_{X,Y}(\bar{x},\bar{y}) = \frac{1}{n} \sum_{i=1}^{n} \left[\frac{1}{h_x} K\left(\frac{x_i - X_i}{h_x}\right) \right] \left[\frac{1}{h_y} K\left(\frac{y_i - Y_i}{h_y}\right) \right]$$

$$= \frac{1}{n} \sum_{i=1}^{n} \left[K_{w_x}(x_i - X_i) \cdot K_{w_y}(y_i - Y_i) \right]$$ (19.96b)

in which

$$K_{w_x}(x_i - X_i) = \frac{1}{\sqrt{2\pi h_x}} \exp\left[-\frac{(x_i - X_i)^2}{2h_x} \right] \sim N(X_i, h_x)$$ (19.96c)

and

$$K_{w_y}(y_i - Y_i) = \frac{1}{\sqrt{2\pi h_y}} \exp\left[-\frac{(y_i - Y_i)^2}{2h_y} \right] \sim N(Y_i, h_y)$$ (19.96d)

Now MISE for the estimated joint probability density function can be written as

$$\text{MISE}\left\{\hat{f}_{X,Y}(x,y)\right\}(h_x, h_y) = \iint E\left[\hat{f}_{X,Y}(x,y;h_x,h_y) - f_{X,Y}(x,y) \right]^2 dx\, dy$$

$$+ \iint \text{var}_f \left[\hat{f}_{X,Y}(x,y;h_x,h_y) \right] dx\, dy$$ (19.97)

Following the algorithms proposed by Botev et al. (2010), the integrated square error can be minimized and the kernel bandwidth of X and Y can be optimized to preserve the underlying joint probability density function.

Then the empirical joint entropy can be calculated using the optimized bandwidth:

$$\text{Entropy}_{\text{emp}} = -\sum_{i=1}^{n}\sum_{i=1}^{n} \log\left[\hat{f}_{X,Y}\left(x_i, y_j; h_x^{\text{opt}}, h_y^{\text{opt}}\right) \right] \cdot \hat{f}_{X,Y}\left(x_i, y_j; h_x^{\text{opt}}, h_y^{\text{opt}}\right)$$ (19.98a)

or simply

$$\text{Entropy}_{\text{emp}} = E\left[\log\left[\hat{f}_{X,Y}\left(x_i, y_j; h_x^{\text{opt}}, h_y^{\text{opt}}\right) \right] \right]$$ (19.98b)

in which h_x^{opt} and h_y^{opt} are the optimized bandwidth for random variables X and Y, respectively.

19.6.6   Derivation of Return Periods

Return Period of Univariate Annual Series

The return period or recurrence interval T defines the expected time period on average for the occurrence of event $x \geq x_T$, where x_T is the design event, which can be expressed as

$$T = \frac{1}{P_X(x \geq X_T)} = \frac{1}{1 - P_X(x \leq X_T)} \tag{19.99}$$

Runoff amounts for certain return periods (that is, 5-, 25-, 50-, 100-yr) can be calculated using the marginal rainfall distribution.

Return Period of Bivariate Annual Series

For a joint probability density function of two random variables X and Y denoted by $f_{X,Y}(x, y)$ and the corresponding joint probability distribution function denoted by $F_{X,Y}(x, y)$, the joint return period of $X > X_T$ or $Y > Y_T$ can be written as

$$T_{X,Y}(x, y) = \frac{1}{1 - F_{X,Y}(x \leq X_T, y \leq Y_T)} \tag{19.100}$$

In runoff-sediment analysis, one may be interested in the return period of runoff given a particular rainfall amount, say, $T_{Y|X}(y \mid X = x)$ which is called the conditional return period. The conditional return period needs to be derived from the conditional probability density function $f_{X|Y=y}(x \mid Y = y)$. The conditional return period of sediment yield given runoff can be written as

$$T_{Y|X}(Y > y \mid X = x) = \frac{1}{1 - F_{Y|X}(Y \leq y \mid X = x)} = \frac{1}{1 - \int_0^y f_{Y|X}(t \mid X = x) \, dt}$$

$$= \frac{1}{1 - \int_0^y f_{X,Y}(x, t) / f_X(x) \, dt} \tag{19.101}$$

in which the upper integration limit is $y \leq x$ for runoff and sediment analysis.

Questions

1. For relatively isolated storms on a watershed, obtain data on runoff and sediment yield. Plot graphically and fit a regression relation. How good is the regression relation?

2. For the same data as in Question 1, determine the parameters of the relation between sediment yield and runoff, using entropy theory, and compare these parameter values with those obtained by regression. Using the entropy-based parameter values, plot sediment yield on the same graph, compare with the regression relation, and comment.

3. Obtain data on rainfall, runoff, and sediment discharge for several storms on a given watershed. Determine the unit hydrographs and unit sediment graphs for these storms.

4. For the data in Question 3, determine the sediment concentration graph by dividing the sediment discharge by the water discharge. Plot concentration values on a graph, and determine a relation for the concentration as a function of time.

5. Determine sediment discharge for two rainfall-runoff events on the same watershed as in Question 1 by two methods based on entropy. These events should be different from the ones used in deriving the IUH and IUSG.

6. For the sediment yield data from a watershed, construct the CPD and PDF of sediment yield. Determine the best-fit probability distribution.

References

Agrawal, D., Singh, J. K., and Kumar, A. (2005). Maximum Entropy-based Conditional Probability Distribution Runoff Model. *Biosystem Engineering*, vol. 90, no. 1, doi:10.1016/j.biosystemseng.2004.08.003.

de Araujo, J. C. (2007). Entropy-based equation to assess hillslope sediment production. *Earth Surface Processes and Landforms*, vol. 32, pp. 2005–2018.

Bagnold, R. A. (1977). Bed load transport in natural rivers. *Water Resources Research*, vol. 13, pp. 303–311.

Botev, Z. I. (2010). Kernel density estimation via diffusion. *The Annals of Statistics*, vol. 38, pp. 2916–2957.

Chen, V. J. and Kuo, C. Y. (1984). A study of synthetic sediment graphs for ungauged watersheds. *Journal of Hydrology*, vol. 84, pp. 35–54.

Chiu, C. L. (1987). Entropy and probability concepts in hydraulics. *Journal of Hydraulic Engineering*, vol. 113, no. 5, pp. 583–600.

Chiu, C. L., Jin, W. and Chen, Y. C. (2000). Mathematical models of distribution of sediment concentration. *Journal of Hydraulic Engineering*, vol. 126, no. 1, pp. 18328–18335.

Choo, T. H. (2000). An efficient method of the suspended sediment-discharge measurement using entropy concept. *Water Engineering Research*, vol. 1, no. 2, pp. 95–105.

Clarke, R. T. (1980). Bivariate gamma distributions for extending annual streamflow records from precipitation: Some large sample results. *Water Resources Research*, vol. 16, no. 5, pp. 863–870.

Cui, H., and Singh V. P. (2014). Suspended sediment concentration in open channels using tsallis entropy. *Journal of Hydrologic Engineering*, vol. 19, no. 5, pp. 966–977, in press.

Figueiredo, E. E., and Bathurst J. C. (2007). Runoff and sediment yield predictions in a semiarid region of Brazil using SHETRAN. In IAHS Decade on: Prediction in Ungaged Basins, edited by D. Schertzer, P. Hubert, S. Koide, and K. Takeuchi. IAHS Publication 309, Wallingford, United Kingdom, pp. 258–266.

Foster, G. R., and Lane L. J. (1983). Erosion by concentrated flow in farm fields. *Proceedings, D. B. Simons Symposium on Erosion and Sedimentation*, Colorado State University, Fort Collins, pp. 9.56–9.82.

Hairsine, P. B., and Rose, C. W. (1992). Modeling water erosion due to overland flow using physical principles: 1. Sheet flow. *Water Resources Research*, vol. 28, pp. 237–243.

Jaynes, E. T. (1957a). Information theory and statistical mechanics, I. *Physical Reviews*, vol. 106, pp. 620–630.

Jaynes, E. T. (1957b). Information theory and statistical mechanics, II. *Physical Reviews*, vol. 108, pp. 171–190.

Jaynes, E. T. (1982). On the rationale of maximum entropy methods. *Proceedings of the IEEE*, vol. 70, pp. 939–952.

Jin, D., Chen, H. and Guo, Q. (1999). A preliminary experimental study on non-linear relationship between sediment yield and drainage network development. *International Journal of Sediment Research*, IRTCES, vol. 14, no. 2, pp. 9–18.

Kirby, M. J., Abrahardt, R. J., Bathurst, J. C., Kilsby, .G., McMahon, M. L., Osborne, C. P., and J. B. Thornes, et al. (2002). MEDRUSH—A basin scale physically based model for forecasting runoff and sediment yield. In: *Mediterranean Desertification: A Mosaic of Processes and Responses*, edited by N. S. Geeson, C. J. Brandt and J. B. Thornes, Wiley, Chichester, United Kingdom, pp. 203–207.

Lane, L. J., and Nearing M. A., eds. (1989). Water erosion prediction project: Hillslope profile model documentation. NSERL Report. USDA-ARS, West Lafayette, Ind.

Lienhard, J. H. (1964). A statistical mechanical prediction of the dimensionless unit hydrograph. *Journal of Geophysical Research*, vol. 69. no. 24, pp. 5231–5238.

Morgan, R. P. C., Quinton, J. N., Smith, R. E., Govers, G., Poesen, J. W. A., Chisi, G. and Tori, D. (1998). The European soil erosion model (EUROEM). A process based approach for predicting soil loss from fields and small catchments. *Earth Surface Processes and Landforms*, vol. 23, pp. 527–544.

Ni, J., Wang, G. and Liao, Q., (1999). A correction model for hyper-concentrated flow. *International Journal of Sediment Research*, vol. 14, no. 2, pp. 145–148.

Nunes, de Lima, J. L. M. P., Singh, V. P., de Lima, M. I. P. and Vieira, G. N. (2006). Numerical modeling of surface runoff and erosion due to moving storms at the drainage basin scale. *Journal of Hydrology*, vol. 330, pp. 709–720.

Rendon-Herrero, O. (1974). Estimation of washload produced by certain small watersheds. *Journal of the Hydraulics Division, ASCE*, vol. 109, no. HY7, pp. 835–848.

Rendon-Herrero, O. (1978). Unit sediment graph. *Water Resources Research*, vol. 14, no. 5, pp. 889–901.

Rendon-Herrero, O., Singh, V. P. and Chen, V. J. (1980). ER-ES watershed relationship. *Proceedings, International Symposium on Water Resources Systems*, held December 1980 at University of Roorkee, Roorkee, India, vol. 1, pp. II-8–41-7.

Renard, K. G., Foster, G. R., Weesies, G. A., McCool, D. K. and Yoder, D. S. (1993). Predicting soil erosion by water—A guide to conservation planning with the revised Universal Soil Loss Equation-RUSLE. Agricultural Research Service, *Agriculture Handbook No. 703*, Publication, U.S. Department of Agriculture, Washington.

Scott, D. W. (1992), *Multivariate Density Estimation: Theory, Practice, and Visualization*, New York: John Wiley.

Shannon, C. E. (1948). A mathematical theory of communication. *Bell System Technical Journal*, vol. 27, no. 3, pp. 379–423.

Sharma, T. C., and Dickinson W. T. (1979a). Unit step and frequency response functions applied to the watershed fluvial system. *Journal of Hydrology*, vol. 40, pp. 323–335.

Sharma, T. C., and Dickinson W. T. (1979b). Discrete dynamic model of watershed sediment yield. *Journal of the Hydraulics Division, Proceedings of the ASCE*, vol. 105, no. HY5, pp. 555–571.

Siepel, A. C., Steenhuis, T. S., Rose, C. W., Parlange, J. Y. and McIsaac, G. F. (2002). A simplified hillslope erosion model with vegetation elements for practical applications. *Journal of Hydrology*, vol. 258, pp. 111–121.

Singh, V. P. (1983). Analytical solutions of kinematic equations for erosion on a plane: 2. Rainfall of finite duration. *Advances in Water Resources*, vol. 6, pp. 88–95.

Singh, V. P. (1989). *Hydrologic Systems*, vol. 2: Watershed Modeling. Prentice Hall, Englewood Cliffs, New Jersey.

Singh, V. P. (1998). *Entropy-Based Parameter Estimation in Hydrology*. Kluwer Academic Publishers, Boston.

Singh, V. P. (2010a). Entropy theory for derivation of infiltration equations. *Water Resources Research*, vol. 46, pp. 1–20, W03527, doi:10-1029/2009WR008193.

Singh, V. P. (2010b). Tsallis entropy theory for derivation of infiltration equations. *Transactions of the ASABE*, vol. 53, no. 2, pp. 447–463.

Singh, V. P. (2010c). Entropy theory for movement of moisture in soils. *Water Resources Research*, vol. 46, pp. 1–12, W03516, doi:10-1029/2009WR008288.

Singh, V. P. (2011). An IUH equation based on entropy theory. *Transactions of the ASABE*, vol. 54, no. 1, pp. 1–10.

Singh, V. P. (2013). *Entropy Theory in Environmental and Water Engineering*. John Wiley, New York.

Singh., V. P., Baniukiewicz, A. and Chen, V. J. (1982). An instantaneous unit sediment graph study for small upland watersheds. In: *Modeling Components of Hydrologic Cycle*, edited by V. P. Singh. Water Resources Publications, Littleton, Colo. pp. 539–554.

Singh, V. P., and Chen V. J. (1982). On the relation between sediment yield and runoff volume. In: *Modeling Components of Hydrologic Cycle*, edited by V. P. Singh. Water Resources Publications, Littleton, Colo. pp. 555–570.

Singh, V. P., and Regl R. R. (1983). Analytical solutions of kinematic equations for erosion on a plane: l. Rainfall of infinite duration. *Advances in Water Resources*, vol. 6, pp. 2–10.

Soil Conservation Service (1972). National Engineering Handbook, Section 4, Hydrology. U.S. Department of Agriculture, Washington, D.C.

Srivastava, P. K., Rastogi, R. A. and Chauhan, H. S. (1984). Prediction of storm sediment yield from a small watershed. *Journal of Agricultural Engineering*, vol. 21, no. 1–2, pp. 121–126.

Tyagi, J. V., Mishra, S. K., Singh, R. and Singh, V. P. (2008). SCS-CN based time-distributed sediment yield model. *Journal of Hydrology*, vol. 352, pp. 388–403.

Wand, M. P., and Jones M. C. (1995). *Kernel Smoothing*, London: Chapman & Hall.

Williams, J. R. (1975). Sediment routing for agricultural watersheds. *Water Resources Bulletin, AWRA*, vol. 11, no. 5, pp. 965–974.

Williams, J. R. (1978). Unit sediment graph. *Water Resources Research*, vol. 14, no. 5, pp. 889–901.

Williams, J. R. and Berndt, H. D. (1977). Sediment yield prediction based on watershed hydrology. *Transactions of the ASABE*, vol. 20, pp. 1100–1104.

Wischmeier, W. H., and Smith, D. D. (1965). Predicting rainfall—erosion losses-a guide to conservation planning. *Agricultural Handbook 537*, U.S. Department of Agriculture, Washington.

Wischmeier, W. H., and Smith, D. D. (1978). *Predicting rainfall erosion losses—a guide to conservation planning*. U.S. Department of Agriculture, Agricultural Research Service, Washington, D.C.

Yue, S. (2001). A bivariate gamma distribution for use in multivariate flood frequency analysis. *Hydrological Processes*, vol. 15, no. 6, pp. 1033–1045.

Further Reading

Anselmo, V., Galmacci, G., Singh, V. P. and Ubertini, L. (1981). Rainfall-runoff-sediment yield modeling by stochastic models: Preliminary results. *Annali Della Facolta Di Agravia, Universita Delgi Studi Di Perugia, Italy*, vol. 35, pp. 507–516.

Caroni, E., Singh, V. P. and Ubertini, L. (1984). Rainfall-runoff-sediment yield relation by stochastic modeling. *Hydrological Sciences Journal*, vol. 29, no. 2/6, pp. 203–218.

Cui, H., and Singh, V. P. (2012). On the cumulative distribution function for entropy-based hydrologic modeling. *Transactions of the ASABE*, vol. 55, no. 2, pp. 429–438.

Cui, H., and Singh, V. P. (2013). Two-dimensional velocity distribution in open channels using Tsallis Entropy. *Journal of Hydrologic Engineering*, vol. 18, no. 3, pp. 331–339.

Cui, H., and Singh, V. P. (2014). One-dimensional velocity distribution in open channels using Tsallis Entropy. *Journal of Hydrologic Engineering*, vol. 19, no. 2, pp. 290–299.

Lane, L. J., Shirley, E. D. and Singh, V. P. (1988). Modeling erosion on hillslopes. Chapter 10 In: *Modeling Geomorphological Systems*, edited by M. Anderson, Wiley, New York, pp. 287–308.

Mishra, S. K., and Singh, V. P. (2002). SCS-CN based hydrologic simulation package. In: *Mathematical Models of Small Watershed Hydrology and Applications,* edited by V. P. Singh and D. K. Frevert, Water Resources Publications, Littleton, Colo. pp. 391–464.

Mishra, S. K., Suresh Babu, P. and Singh, V. P. (2008). SCS-CN method. In: *Hydrology and Hydraulics,* edited by V. P. Singh, Water Resources Publications, Highlands Ranch, Colo. pp. 277–330.

Prasad, S. N., and Singh, V. P. (1982). A hydrodynamic model of sediment transport in rill furrows. International Association of Hydrological Science Publication 137, pp. 293–301.

Pruski, F. F., and Nearing M. A. (2002). Climate-induced changes in erosion during the 21st century for eight U.S. locations. *Water Resources Research*, vol. 38, no. 12, pp. 34–1 to 34–11, doi: 10.1029/2001WR000493.

Singh, V. P. (1973). Predicting sediment yield in Western United States, discussion. *Journal of the Hydraulics Division, Proceedings of ASCE*, vol. 99, no. HY10, pp. 1891–1894.

Singh, V. P. (1983). A mathematical study of erosion from upland areas. Technical Report WRRl, Water Resources Program, Department of Civil Engineering, Louisiana State University, Baton Rouge.

Singh, V. P., and Chen, V. J. (1983). The relationship between storm runoff and sediment yield. Technical Report WRR3, Water Resources Program, Department of Civil Engineering, Louisiana State University, Baton Rouge.

Singh, V. P., and Krstanovic, P. F. (1986). A stochastic model for sediment yield. In: *Multivariate Analysis of Hydrologic Processes,* edited by H. W. Shen, J. T. B. Obeysekera, V. Yevjevich, and D. G. DeCoursey, Colorado State University, Fort Collins, pp. 755–767.

Singh, V. P., Krstanovic, P. F. and Lane, L. J. (1988). Stochastic models of sediment yield. In: *Modeling Geomorphological Systems*, edited by M. Anderson, Wiley, New York, pp. 259–285.

Singh, V. P., and Scarlatos, P. D. (1985). Sediment transport in vertically two-dimensional manmade canals. *Proceedings of the 21st Biennial International Association for Hydraulic Research Congress,* held August 19–23, 1985, in Melbourne, Australia, vol. 3, pp. 577–582.

Soil Conservation Society of America (1977). Soil erosion: Prediction and control. *Proceedings of National Conference on Soil Erosion*, held May 24–26, 1976, Purdue University, West Lafayette, Ind.

Vanoni, V. A., ed. (1975). *Sedimentation Engineering. ASCE Manuals and Reports on Engineering Practice,* No. 54, ASCE Press, Reston, Va.

Yang, C. T. (1996). *Sediment Transport: Theory and Practice.* McGraw-Hill, New York.

Zhang, Q., Singh, V. P. and Chen, X. (2011). Influence of Three Gorges dam on streamflow and sediment load of the middle Yangtze River, China. *Stochastic Environmental Research and Risk Analysis*, vol. 26, pp. 569–579.

CHAPTER 20

Eco-Index

Freshwater resources are becoming increasingly limited as the population multiplies. Over one-half of the world's accessible runoff presently is appropriated for human use, and that fraction is projected to grow to 70 percent by 2025. A recent projection of water demand through 2025 indicated that ensuring a sustainable water supply would become increasingly challenging for large areas of the globe (Vorosmarty et al., 2000). In 31 countries, 450 million people already face serious shortages of water. These shortages occur almost exclusively in developing countries, some of which are ill equipped to address water shortages. By the year 2025, one-third of the world's population is expected to face severe to chronic water shortages. Allocating water for diverse and often competing traditional uses, such as industry, agriculture, urban, energy, waste disposal, and recreation, is now even more complex because of society's expectation, that for its health and integrity, ecosystems should receive adequate attention and accommodation. Freshwater and freshwater-dependent ecosystems provide a range of services for humans, including fish, flood protection, and wildlife. To maintain these services, water needs to be allocated to ecosystems, as it is allocated to other users. In the face of limited freshwater supply, there are multiple competing and conflicting demands.

With increasing concern for eco-needs, the scientific field of "eco-flows" has prospered in recent years with the result that there are more than 200 methods for their computation. These methods can be grouped into four categories: hydrological rules, hydraulic rating methods, habitat simulation methods, and holistic methodologies (Dyson et al., 2003; Tharme, 2003; Naiman et al., 2002; Postel and Richter, 2003). Past studies include those based on percentages of the natural mean or median annual flow, percentages of total divertible annual flow allocated to wet and dry seasons, and eco-flow prescriptions based on a percentage of total annual base flow plus a high-flow component derived as a percentage of mean annual runoff (Smakhtin et al., 2004). However, such guidelines have no documented empirical basis, and the temptation to adopt them may represent a risk to the future integrity and biodiversity of riverine ecosystems (Arthington, 1998). In the literature it is now recognized that the structure and function of a riverine ecosystem and many adaptations of its biota are dictated by the patterns of temporal variation in river flow or the "natural flow-regime paradigm reflected by Indicators of Hydrologic Alteration" (Richter et al., 1996; Poff et al., 1997; Lytle and Poff, 2004).

The objective of this chapter is to outline an entropy-based hydrologic alteration assessment of biologically relevant flow regimes using gauged flow data. The maximum entropy ordered weighted averaging method is used to aggregate noncommensurable biologically relevant flow regimes to fit an eco-index such that the harnessed level of the ecosystem is reflected. The methodology can serve as a guide for eco-managers when allocating water resources among potential users and to where they should concentrate their attention while mitigating the human-induced effects on natural flow regimes to have a sustainable development.

20.1 Indicators of Hydrologic Alteration

Indicators of hydrologic alteration (IHA) aim to protect a range of flows in a river. Richter et al. (1996) proposed 32 biologically relevant parameters, shown in Table 20.1, which jointly reflect different aspects of flow variability (magnitude, frequency, duration, and timing of flows). These are estimated from a natural daily flow time series at a site of interest (often at a gauging site). These parameters consider intra- and interannual variation of hydrologic regime which is necessary to sustain the ecosystem. In other words, a range of flow regime is considered to define the state of the ecosystem such that hydrologic requirements for all aquatic species are met. The ecosystem alteration is then assessed by comparing it with the natural system which is relatively unharnessed (Richter et al., 1996).

Group	Regime Characteristics	32 Parameters	Number of Parameters
Group 1: Magnitude of monthly water conditions	Magnitude Timing	Mean value for each calendar month	12
Group 2: Magnitude and duration of annual extreme water conditions	Magnitude Duration	Annual minimum/maximum of 1-d means Annual minimum/maximum of 3-d means Annual minimum/maximum of 7-d means Annual minimum/maximum of 30-d means Annual minimum/maximum of 90-d means	10
Group 3: Timing of annual extreme water conditions	Timing	Julian date of each annual 1-d minimum and maximum	2
Group 4: Frequency and duration of high and low pulses	Frequency Duration	Number of high and low pulses each year Mean duration of high and low pulses	2 + 2
Group 5: Rate and frequency of consecutive water condition changes	Rate of change	Means of all positive differences between daily values Means of all negative differences between daily values Number of rises Number of falls	1 + 1 + 1 + 1

Source: Richter et al. (1996).

TABLE 20.1 Hydrological Parameters Used in IHA

Although Richter et al. (1996), Poff et al. (1997), and Lytle and Poff (2004), among others, have emphasized why these 32 biologically relevant parameters are required to represent an ecosystem, one needs a tool that allows for the transmission of technical information in a summarized format, preserving the original meaning of data, using only the variables that best reflect the desired objective. Information on the 32 biologically relevant parameters and their values may not show the dependence among parameters and how important each of these parameters is. Furthermore, it is difficult to visualize 32 parameters spatially. In addition, the increasing tendency of public participation in water-related issues requires that results of technical analyses be presented in a way that can be understood and shared by all stakeholders, including those with little technical background. Therefore, narrowing the result to a single value that represents the ecosystem may be more desirable.

20.2   Probability Distributions of IHA Parameters

It is hypothesized that each of the 32 biologically relevant hydrologic parameters, proposed by Richter et al. (1996), can be considered as a random variable. Then for each variable the least-biased probability distribution can be obtained by maximizing entropy (Singh, 1998):

$$H = -\sum_{i=1}^{N} p_i \log p_i \qquad (20.1)$$

in accordance with the principle of maximum entropy (POME), subject to known constraints. In Eq. (20.1), H is the Shannon entropy, $p_1, p_2, \ldots, p_N$ are the values of the probabilities corresponding to the specific values x_i, $i = 1, 2, \ldots N$, of the hydrologic parameter X, and N is the number of values. These probabilities constitute the probability distribution $P = \{p_1, p_2, \ldots, p_N\}$ of parameter X: $\{x_i, i, = 1, 2, \ldots, N\}$ in question. For maximization, the constraints on X can be expressed in terms of averages or expected values of the parameter reflecting the state of the ecosystem as

$$\sum_{i=1}^{N} p_i g_j(x_i) = C_j \qquad j = 1, 2, \ldots, m \qquad (20.2)$$

$$\sum_{i=1}^{N} p_i = 1 \qquad p_i \geq 0 \qquad i = 1, 2, \ldots, N \qquad (20.3)$$

where C_j is the jth constraint, m is the number of constraints, and $g_j(x_i)$ is the jth function of X. Using the method of Lagrange multipliers, the maximization of H would lead to the least-biased P expressed as (Singh, 1998)

$$p_i = \exp\left[-\lambda_0 - \sum_{j=1}^{m} \lambda_j g_j(x_i)\right] \qquad i = 1, 2, \ldots, N \qquad (20.4)$$

In practical applications, functions $g_j(x_i)$ are expressed as simple moments, and the number of constraints is kept to 2 or 3. Thus, the first constraint would be the average, and the second constraint the second moment or variance. Once the least-biased probability distribution is determined using Eq. (20.4) in this manner, it is inserted in Eq. (20.1) to obtain the maximum entropy for each IHA parameter under consideration.

20.3   Computation of Nonsatisfaction Eco-Level

The nonsatisfaction level (NSL) (in absolute difference terms) for an nth parameter can be defined as

$$\text{NSL}_n = |(H_{\text{pre}} - H_{\text{post}})|_n \qquad n = 1, 2, \ldots, 32 \tag{20.5a}$$

or the nonsatisfaction level (NSL) (in relative difference terms) as

$$\text{NSL}_n = \frac{H_{\text{pre}} - H_{\text{post}}}{H_{\text{pre}}} \qquad n = 1, 2, \ldots, 32 \tag{20.5b}$$

where H_{pre} and H_{post} are the Shannon entropies for parameter n for pre- and postchange conditions, respectively. The change may be represented by a dam or levee or even a land use change such as urbanization. Equation (20.5a) or Eq. (20.5b) relates the lack of information about the ecosystem to the level of nonsatisfaction. The satisfaction level can be seen as to how much the system is unharnessed. The values of NSL are computed for all IHA parameters.

20.4   Computation of Eco-Index

The eco-index can be computed using the steps shown in Fig. 20.1. The values of the nonsatisfaction level of biological parameters are aggregated based on Yager's (1999) finding such that the final aggregation maximizes the information associated with each NSL. The ordered weighted averaging (OWA) operator introduced by Yager (1999) is a

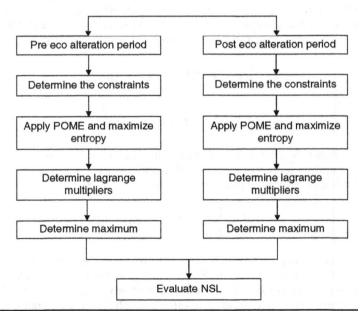

FIGURE **20.1**   Evaluation of NSL.

general type of operator that provides flexibility in the aggregation process such that the aggregated value is bounded between minimum and maximum values of input parameters. The OWA operator is defined as

$$F(a_1, a_2, \ldots, a_n) = \sum_{j=1}^{n} w_j b_j \qquad (20.6)$$

where the computed value of NSL for each of the 32 parameters is the argument a_i; b_j is the jth largest of a_i; and w_j is a collection of weights such that $w_j \in [0, 1]$ and $\sum w_j = 1$. Equation (20.6) can also be written as

$$\text{Eco-index} = F(a_1, \ldots, a_{32}) = F(\text{NSL}_1, \text{NSL}_2, \ldots, \text{NSL}_{32}) = \sum_{j=1}^{32} w_j b_j \qquad (20.7)$$

The methodology used for obtaining the OWA vector is based on work by Lamata (2004). This approach, which only requires the specification of just the orness value $(1 - \text{andness})$, generates a class of OWA weights that are called maximum entropy operator weighted averaging (ME-OWA) weights. The determination of weights $w_1, \ldots, w_{32}$, from a degree of optimism orness given by the decision maker requires the solution of an optimization problem formulated below. The objective function used for optimization tries to maximize the dispersion or entropy of weights, which calculates the weights to be the ones that use as much information as possible about the values of NSL in the aggregation. It is assumed here that weights are values of probability having the probability distribution $W = \{w_i, i = 1, 2, \ldots, n\}$.

Maximize

$$H(W) = -\sum_{i=1}^{n} w_i \log w_i \qquad (20.8)$$

subject to

$$\text{Orness}(W) = \frac{1}{n-1} \sum_{i=1}^{n} (n - i) w_i \qquad (20.9)$$

where $n = 32$ and $w_i \in [0, 1]$, $\sum w_i = 1$.

O'Hagan (1988) suggested an approach to obtain the OWA operators with maximal entropy of the OWA weights for a given level of orness:

Maximize

$$H = -\sum_{j=1}^{n} w_j \ln(w_j) \qquad (20.10)$$

subject to

$$\alpha = \frac{1}{n-1} \sum_{i=1}^{n} (n - i) w \qquad 0 \le \alpha \le 1 \qquad (20.11)$$

$$\sum_{i=1}^{n} w_i = 1 \qquad 0 \le w_i \le 1 \qquad i = 1, \ldots, n \qquad (20.12)$$

Fulle´r (2001) determined the optimal weighting vector using the method of Lagrange multipliers:

$$w_1[(n-1)\alpha + 1 - nw_1]^n = [(n-1)\alpha]^{n-1}\{[(n-1)\alpha - n]w_1 + 1\} \tag{20.13}$$

$$w_n = -\frac{[(n-1)\alpha - n]w_1 + 1}{(n-1)\alpha + 1 - nw_1} \tag{20.14}$$

Once w_1 and w_n are computed using Eqs. (20.13) and (20.14), the other weights w_i, $i = 1$, $2, \ldots, n-1$, are obtained:

$$w_j = \sqrt[n-1]{w_1^{n-j}w_n^{j-1}} \tag{20.15}$$

To calculate the OWA operator using the method of Fulle´r and Majlender (2001), orness can be set at 0.75 in view of a moderately optimistic strategy. It can be assumed that orness is 1 in the case of optimistic condition and orness is 0 in the case of pessimistic condition.

The orness characterizes the degree to which the aggregation is like an OR operator. For analysis an orness value of 0.75 can be assumed to make sure that the impact of NSL of all the IHA parameters is considered in the index development and to avoid assigning equal weights, as some of the parameters may have greatest influence in defining the underlying ecosystem. Then an array of weights w_j is generated using Eqs. (20.8) and (20.9).

Example 20.1.   Obtain data for a river and then summarize the hydrological parameters used in IHA for the river.

Solution.   The river chosen is Geum River in South Korea which is 398 km long and has a drainage area of 9912 km^2, and there are two multipurpose dams, shown in Fig. 20.2. The reason for choosing the river is that there is easy access to data. To examine the effect of dams, the daily flow data of four stations are used, as shown in Table 20.2.

The 33 biologically relevant parameters proposed by Richter et al. (1998) are shown in Table 20.3. These data can be obtained using the IHA software. In the IHA software, parameters can be calculated using parametric (mean and standard deviation) or nonparametric (percentile) statistics. For most situations nonparametric statistics are a better choice, because of the skewed nature of many hydrologic data sets (The Nature Conservancy, 2009). Only 12 parameters of monthly inflow were obtained from parametric statistics, because the mean is a better representation of the hydrologic characteristics than the median. Annual minimum 1-d mean values at Gongju station are shown in Table 20.4.

Example 20.2.   Using the entropy theory, derive and compute the PDFs of four most important IHA parameters, using the data in Example 20.1. Plot them.

Solution.   First, parameters 13 to 16, which are the annual minimum means of 1, 3, 7, and 30-d, have large NSL values, because inflow was improved by the dam, as shown in Tables 20.4 to 20.7. For analyzing the data obtained using the IHA software, histograms are constructed to find the probability density function that best represents the hydrologic characteristics using two goodness-of-fit tests: the Kolmogorov-Smirnov (K-S) test and the chi-square test. Results are shown in Table 20.8.

Second, the least-biased probability distribution is obtained by maximizing the Shannon entropy (Singh, 1998) given by Eq. (20.1). In Eq. (20.10), H is the Shannon entropy; $p_1, p_2, \ldots, p_N$ are the values of probabilities corresponding to the specific values x_i, $i = 1, 2, \ldots, N$, of the hydrologic parameter X; and N is the number of values. Using the method of Lagrange multipliers, the

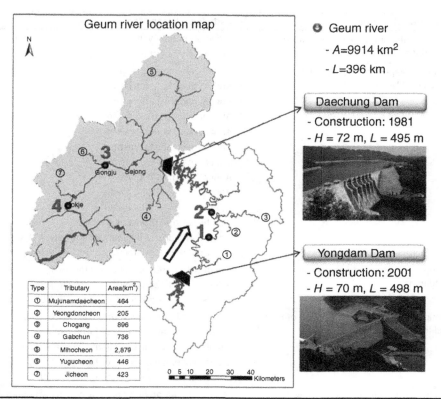

FIGURE 20.2 Study area: Geum River basin.

Type	Name	Predam period	Postdam period	Remarks
1	Hotan	1995–2001 (7 yr)	2002–2010 (9 yr)	Yongdam dam
2	Ogcheon			
3	Gongju	1966–1981 (16 yr)	1982–2001 (20 yr)	Daechung dam
4	Gyuam			

TABLE 20.2 Water Level Stations in Geum River Basin

maximization of H would lead to the least-biased P expressed by Eq. (20.4) (Singh, 1998). For maximization, constraints on X can be expressed in terms of averages or expected values of the parameter reflecting the state of the ecosystem as Eqs. (20.2) and (20.1). In practical applications, functions $g_j(x_i)$ are expressed as simple moments, and the number of constraints is kept to 2 or 3. Thus, the first constraint would be the average, and the second constraint would be the second moment or variance or some other type.

In a case of the gamma PDF, the least-biased PDF is expressed as

$$f(x) = \exp(-\lambda_0 - \lambda_1 x - \lambda_2 \ln x) \qquad (20.16)$$

IHA Parameter Group	Hydrologic Parameters
Group 1: Magnitude of monthly water conditions	Mean or median value for each calendar month Subtotal: 12 parameters (no. 1–12)
Group 2: Magnitude and duration of annual extreme water conditions	Annual minima, 1-d mean Annual minima, 3-d mean Annual minima, 7-d mean Annual minima, 30-d mean Annual minima, 90-d mean Annual maxima, 1-d mean Annual maxima, 3-d mean Annual maxima, 7-d mean Annual maxima, 30-d mean Annual maxima, 90-d mean Number of zero-flow days Base flow index: 7-d minimum flow/mean flow for year Subtotal: 12 parameters (nos. 13–24)
Group 3: Timing of annual extreme water conditions	Julian date of each annual 1-d minimum Julian date of each annual 1-d maximum Subtotal: 2 parameters (nos. 25, 26)
Group 4: Frequency and duration of high and low pulses	Number of low pulses within each water year Mean or median duration of low pulses (d) Number of high pulses within each water year Mean or median duration of high pulses (d) Subtotal: 4 parameters (nos. 27–30)
Group 5: Rate and frequency of water condition changes	Rise rates: Mean or median of all positive differences between consecutive daily values Fall rates: Mean or median of all negative differences between consecutive daily values Number of hydrologic reversals Subtotal: 3 parameters (nos. 31–33)

Source: The Nature Conservancy (2009).

TABLE 20.3   Summary of IHA Parameters

Using Eqs. (20.2) and (20.3), one concludes that

$$f(x) = \frac{1}{a\Gamma(b)}\left(\frac{x}{\alpha}\right)^{b-1} e^{-x/a} \tag{20.17}$$

$$a = \frac{1}{\lambda_1} \tag{20.18}$$

$$E[\ln x] - \ln \mu = \psi(b) - \ln b \tag{20.19}$$

In this manner, parameters in each case are obtained, as shown in Table 20.9.

	Preperiod						Postperiod				
Type	Year	X (m³/s)	Type	Year	X (m³/s)	Type	Year	X (m³/s)	Type	Year	X (m³/s)
1	1966	10.93	9	1974	21.45	1	1982	10.23	11	1992	38.73
2	1967	6.37	10	1975	20.68	2	1983	37.53	12	1993	8.56
3	1968	2.09	11	1976	8.81	3	1984	7.27	13	1994	24.53
4	1969	14.48	12	1977	1.36	4	1985	0.03	14	1995	61.01
5	1970	22.53	13	1978	1.11	5	1986	26.39	15	1996	20.58
6	1971	20.09	14	1979	34.56	6	1987	25.43	16	1997	10.18
7	1972	11.83	15	1980	31.98	7	1988	1.08	17	1998	68.48
8	1973	13.29	16	1981	20.41	8	1989	6.45	18	1999	58.85
						9	1990	28.36	19	2000	48.61
						10	1991	19.99	20	2001	24.38

TABLE 20.4   Annual Minimum one-d Means X at Gongju Station

	Preperiod						Postperiod				
Type	Year	X (m³/s)	Type	Year	X (m³/s)	Type	Year	X (m³/s)	Type	Year	X (m³/s)
1	1966	11.42	9	1974	24.02	1	1982	11.28	11	1992	40.26
2	1967	6.50	10	1975	17.88	2	1983	39.89	12	1993	9.963
3	1968	2.18	11	1976	11.97	3	1984	9.37	13	1994	24.53
4	1969	15.71	12	1977	1.45	4	1985	0.3833	14	1995	61.98
5	1970	34.43	13	1978	1.28	5	1986	38.35	15	1996	29.13
6	1971	18.11	14	1979	34.86	6	1987	30.53	16	1997	10.3
7	1972	12.48	15	1980	32.83	7	1988	2.737	17	1998	70.03
8	1973	15.63	16	1981	22.88	8	1989	7.013	18	1999	62.5
						9	1990	34.54	19	2000	51.25
						10	1991	22.29	20	2001	25.31

TABLE 20.5   Annual Minimum 3-d Means X at Gongju Station

On the other hand, parameters can also be estimated by the method of moments. In case of the gamma PDF, a and b can be calculated as

$$a = \frac{\sigma^2}{\mu} \qquad b = \left(\frac{\mu}{\sigma}\right)^2 \tag{20.20}$$

The PDF obtained by the moment method is compared with that obtained by the entropy method, as shown in Table 20.10 and Figs. 20.3–20.6. In the case of the normal PDF, results by the moment method and entropy method are same, so one line is drawn.

In the case of normal PDF, the least-biased PDF is expressed as

$$f(x) = \exp(-\lambda_0 - \lambda_1 x - \lambda_2 x^2) \tag{20.21}$$

Preperiod						Postperiod					
Type	Year	X	Type	Year	X	Type	Year	X	Type	Year	X
1	1966	12.29	9	1974	25.04	1	1982	15.02	11	1992	45.42
2	1967	7.17	10	1975	19.86	2	1983	41.47	12	1993	12.48
3	1968	2.82	11	1976	12.45	3	1984	11.09	13	1994	25.11
4	1969	18.17	12	1977	1.62	4	1985	0.8814	14	1995	62.53
5	1970	35.38	13	1978	1.97	5	1986	47.32	15	1996	31.83
6	1971	20.16	14	1979	35.45	6	1987	32.68	16	1997	10.73
7	1972	14.50	15	1980	35.38	7	1988	3.207	17	1998	72.05
8	1973	20.52	16	1981	25.36	8	1989	7.263	18	1999	68.62
						9	1990	35.39	19	2000	52.03
						10	1991	27.13	20	2001	25.77

TABLE 20.6   Annual Minimum 7-d Means in Gongju Station

Preperiod						Postperiod					
Type	Year	X (m³/s)	Type	Year	X (m³/s)	Type	Year	X (m³/s)	Type	Year	X (m³/s)
1	1966	21.12	9	1974	27.46	1	1982	28.36	11	1992	46.81
2	1967	10.99	10	1975	24.96	2	1983	57.56	12	1993	16.64
3	1968	7.48	11	1976	18.90	3	1984	23.79	13	1994	28.19
4	1969	36.25	12	1977	3.16	4	1985	3.875	14	1995	65.88
5	1970	39.99	13	1978	2.97	5	1986	62.28	15	1996	36.96
6	1971	24.91	14	1979	37.52	6	1987	45.81	16	1997	15.35
7	1972	34.00	15	1980	43.03	7	1988	6.44	17	1998	80.11
8	1973	26.70	16	1981	35.04	8	1989	8.868	18	1999	71.7
						9	1990	45.41	19	2000	53.64
						10	1991	35.08	20	2001	27.26

TABLE 20.7   Annual Minimum 30-d Means (X) in Gongju Station

Type	Parameter No.	Preperiod	Postperiod
1	13	Gamma	Gamma
2	14	Normal	Gamma
3	15	Gamma	Gamma
4	16	Normal	Gamma

TABLE 20.8   Chosen PDF of Each Parameter

Type	Parameter No.	Preperiod		Postperiod	
		A	**b**	**A**	**b**
1	13	1/0.1016	1.5	1/0.0359	0.94
2	14	—	—	1/0.0439	1.28
3	15	1/0.0923	1.66	1/0.0465	1.46
4	16	—	—	1/0.0567	2.15

TABLE 20.9   Parameter Estimation by Entropy Theory (Gamma PDF)

Gamma PDF (Moment)				Gamma PDF (Entropy)			
X (m³/s)	**Preperiod**	**X (m³/s)**	**Postperiod**	**X (m³/s)**	**Preperiod**	**X (m³/s)**	**Postperiod**
1.11	0.01	0.03	0.00	1.11	0.03	0.03	0.05
1.36	0.02	1.08	0.01	1.36	0.04	1.08	0.04
2.09	0.02	6.45	0.03	2.09	0.04	6.45	0.03
6.37	0.05	7.27	0.03	6.37	0.05	7.27	0.03
8.81	0.05	8.56	0.03	8.81	0.04	8.56	0.03
10.93	0.05	10.18	0.03	10.93	0.04	10.18	0.03
11.83	0.05	10.23	0.03	11.83	0.04	10.23	0.03
13.29	0.04	17.58	0.02	13.29	0.03	17.58	0.02
14.48	0.04	19.99	0.02	14.48	0.03	19.99	0.02
17.09	0.03	24.38	0.02	17.09	0.03	24.38	0.01
20.68	0.03	24.53	0.02	17.68	0.03	24.53	0.01
20.41	0.02	25.43	0.02	20.41	0.02	25.43	0.01
21.45	0.02	26.39	0.02	21.45	0.02	26.39	0.01
22.53	0.02	28.36	0.02	22.53	0.02	28.36	0.01
31.98	0.01	37.53	0.01	31.98	0.01	37.53	0.01
34.56	0.01	38.73	0.01	34.56	0.01	38.73	0.01
		48.61	0.01			48.61	0.01
		58.85	0.00			58.85	0.00
		61.01	0.00			61.01	0.00
		68.48	0.00			68.48	0.00

TABLE 20.10   Parameter Estimation of Annual Minimum 1-d Means at Gongju Station

Using Eqs. (20.12) and (20.13), the PDF is expressed as

$$f(x) = \frac{1}{b\sqrt{2\pi}} \exp\left[-\frac{(x-a)^2}{2b^2}\right] \tag{20.22}$$

$$a = \mu \qquad b = \sigma \tag{20.23}$$

Results for the preperiod are shown in Table 20.11.

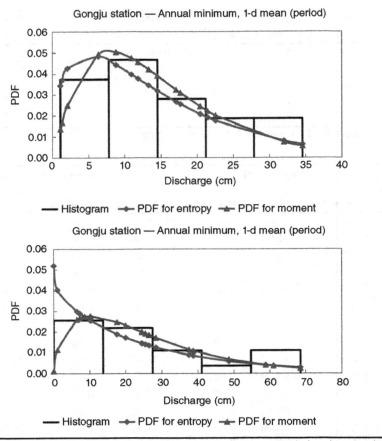

FIGURE 20.3   PDF of annual minimum 1-d mean corresponding to preperiod and postperiod.

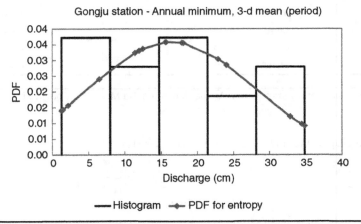

FIGURE 20.4   PDF of annual minimum 3-d mean corresponding to preperiod and postperiod.

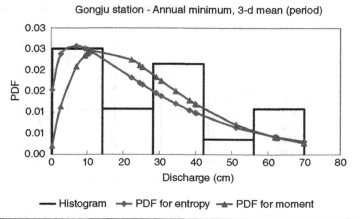

FIGURE 20.4   (*Continued*).

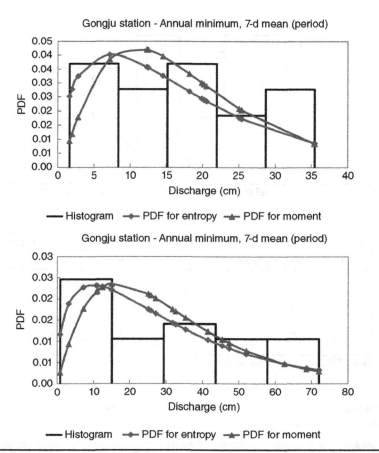

FIGURE 20.5   PDF of annual minimum 7-d mean corresponding to preperiod and postperiod.

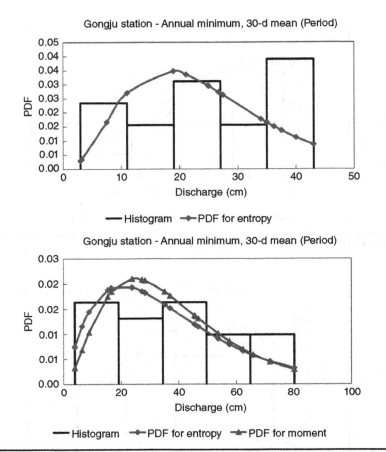

FIGURE **20.6**   PDF of annual minimum 30-d mean corresponding to preperiod and postperiod.

		Preperiod		Postperiod	
Type	Parameter No.	a	b	a	b
1	13	—	—	—	—
2	14	16.48	11.12	—	—
3	15	—	—	—	—
4	16	24.66	13.01	—	—

TABLE **20.11**   Parameter Estimation by Entropy Method (Normal PDF)

Example 20.3.   Compute the entropy of each of the four IHA parameters selected in Example 20.2 for the data in Example 20.1.

Solution.   The Lagrange multipliers and maximum entropy using POME are summarized in Table 20.12 (Singh, 1998), and the results are shown in Table 20.13.

Type	Normal	Log-Normal	Gamma
λ_0	$\frac{1}{2}\ln\pi - \frac{1}{2}\ln\lambda_2 + \frac{\lambda_1^2}{4\lambda_2}$	$\ln\sqrt{\pi} + \ln(\sqrt{2}\,\sigma_y) + \frac{\mu_y^2}{2\sigma_y^2}$	$(\lambda_2 - 1)\ln\lambda_1 + \ln\Gamma(1-\lambda_2)$
λ_1	$-\dfrac{\mu}{\sigma^2}$	$1 - \dfrac{\mu_y}{\sigma_y^2}$	$-\dfrac{1}{\mu}(\lambda_2 - 1)$
λ_2	$\dfrac{1}{2\sigma^2}$	$\dfrac{1}{2\sigma_y^2}$	$E[\ln x] - \ln\mu = \psi(1-\lambda_2) - \ln(1-\lambda_2)$
H_{max}	$\ln[\sigma(2\pi e)^{0.5}]$	$\ln[\sigma_y(2\pi e)^{0.5}] + \mu_y$	$\ln\left[\Gamma(1-\lambda_2)\left(\frac{1}{\lambda_1}\right)^{1-\lambda_2}\right] + \lambda_1\mu + \lambda_2 E[\ln x]$

TABLE 20.12 Summary of Lagrange Multipliers and Maximum Entropy

	Type	Parameter 13 Pre	Parameter 13 Post	Parameter 14 Pre	Parameter 14 Post	Parameter 15 Pre	Parameter 15 Post	Parameter 16 Pre	Parameter 16 Post
1	PDF	Gamma	Gamma	Normal	Gamma	Gamma	Gamma	Normal	Gamma
2	λ_0	3.3058	3.1638	4.4256	3.8868	3.8567	4.3592	5.2799	6.2523
3	λ_1	0.1016	0.0359	−0.1333	0.0439	0.0923	0.0465	−0.1456	0.0567
4	λ_2	−0.4986	0.0608	0.0040	−0.2767	−0.6617	−0.4604	0.0030	−1.1528
5	H_{max}	3.64	4.26	3.83	4.35	3.83	4.41	3.98	4.50

TABLE 20.13 Summary of Lagrange Multipliers and Maximum Entropy

Type	Parameter 13 Pre	Parameter 13 Post	Parameter 14 Pre	Parameter 14 Post	Parameter 15 Pre	Parameter 15 Post	Parameter 16 Pre	Parameter 16 Post
H_{max}	3.64	4.26	3.83	4.35	3.83	4.41	3.98	4.50
NSL	0.62		0.52		0.58		0.52	

TABLE 20.14 Summary of Lagrange Multipliers and Maximum Entropy

Example 20.4. Compute the nonsatisfaction level for the four IHA parameters selected in Example 20.2.

Solution. The nonsatisfaction level (NSL) for an nth parameter can be defined by Eq. (20.5a) where H_{pre} and H_{post} are the Shannon entropies for parameter n for pre- and postchange (say, dam) conditions, respectively. The results are given in Table 20.14.

Example 20.5. Compute the OWA operator for the data in Example 20.1.

Solution. The values of the nonsatisfaction level of biological parameters are aggregated based on Yager's (1999) method in which the ordered weighted averaging (OWA) operator is computed using Eq. (20.6). The OWA operators are computed by maximizing entropy, as shown in Table 20.15 and Fig. 20.7; the maximum entropy was 3.11 nats.

j	w_j	Entropy (nats)	Accumulative Entropy (nats)	j	w_j	Entropy (nats)	Accumulative Entropy (nats)
1	0.1024	0.2334	0.2334	18	0.0174	0.0704	2.5283
2	0.0922	0.2199	0.4532	19	0.0156	0.0650	2.5934
3	0.0831	0.2067	0.6599	20	0.0141	0.0600	2.6534
4	0.0749	0.1941	0.8540	21	0.0127	0.0554	2.7088
5	0.0674	0.1819	1.0359	22	0.0114	0.0511	2.7600
6	0.0608	0.1702	1.2060	23	0.0103	0.0471	2.8071
7	0.0547	0.1590	1.3651	24	0.0093	0.0434	2.8505
8	0.0493	0.1484	1.5135	25	0.0084	0.0400	2.8905
9	0.0444	0.1383	1.6518	26	0.0075	0.0368	2.9273
10	0.0400	0.1288	1.7806	27	0.0068	0.0339	2.9612
11	0.0360	0.1198	1.9004	28	0.0061	0.0312	2.9923
12	0.0325	0.1113	2.0117	29	0.0055	0.0286	3.0210
13	0.0293	0.1033	2.1150	30	0.0050	0.0263	3.0473
14	0.0264	0.0958	2.2108	31	0.0045	0.0242	3.0715
15	0.0237	0.0888	2.2996	32	0.0040	0.0222	3.0937
16	0.0214	0.0822	2.3819	33	0.0036	0.0204	3.1141
17	0.0193	0.0761	2.4580				
				Sum	1.0000	3.1141	

TABLE 20.15 OWA Operator and Entropy

Example 20.6. Compute the eco-index for the data from Example 20.1.

Solution. The eco-index can be computed with the computed value of NSL for each of the 33 parameters represented by the argument a_i and can be written as

$$\text{Eco-index} = F(a_1, a_2, \ldots, a_{33}) = F(\text{NSL}_1, \text{NSL}_2, \ldots, \text{NSL}_{33}) = \sum_{j=1}^{33} w_j b_j \qquad (20.24)$$

Results of eco-index at Gongju station are shown in Table 20.16, and the eco-index of four gauge stations is shown in Fig. 20.8.

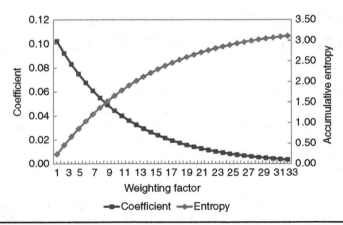

FIGURE 20.7 OWA operator and entropy.

Type		H_{pre}	H_{post}	NSL	NSL_{sort}	OWA	Eco-index
1	January	4.38	4.99	0.61	0.66	0.1024	0.0679
2	February	4.94	4.91	0.03	0.62	0.0922	0.0571
3	March	4.89	4.95	0.06	0.62	0.0831	0.0512
4	April	5.62	5.69	0.07	0.61	0.0749	0.0460
5	May	5.55	5.46	0.09	0.58	0.0674	0.0393
6	June	5.58	5.88	0.30	0.54	0.0608	0.0330
7	July	6.75	6.64	0.11	0.52	0.0547	0.0287
8	August	6.46	6.86	0.39	0.51	0.0493	0.0254
9	September	6.24	6.59	0.35	0.50	0.0444	0.0222
10	October	5.22	5.66	0.44	0.50	0.0400	0.0200
11	November	4.91	5.04	0.13	0.44	0.0360	0.0158
12	December	4.52	5.02	0.50	0.43	0.0325	0.0140
13	1-d min	3.64	4.26	0.62	0.42	0.0293	0.0123
14	3-d min	3.83	4.35	0.52	0.40	0.0264	0.0105
15	7-d min	3.83	4.41	0.58	0.39	0.0237	0.0094
16	30-d min	3.98	4.50	0.52	0.35	0.0214	0.0076
17	90-d min	4.59	4.72	0.14	0.30	0.0193	0.0059
18	1-d max	8.50	8.61	0.11	0.30	0.0174	0.0052
19	3-d max	8.10	8.26	0.16	0.27	0.0156	0.0043
20	7-d max	7.53	7.93	0.40	0.27	0.0141	0.0038
21	30-d max	6.76	7.06	0.30	0.27	0.0127	0.0034
22	90-d max	6.33	6.59	0.27	0.27	0.0114	0.0030
23	0-d	0.00	0	0.00	0.16	0.0103	0.0017
24	Base flow	1.58	1.52	0.06	0.14	0.0093	0.0013
25	Date min	5.88	6.31	0.43	0.13	0.0084	0.0011
26	Date max	4.44	4.71	0.27	0.11	0.0075	0.0008
27	Low pulse #	2.50	3.12	0.62	0.11	0.0068	0.0007
28	Low pulse L	3.17	2.62	0.54	0.09	0.0061	0.0005
29	High pulse #	2.74	3.16	0.42	0.07	0.0055	0.0004
30	High pulse L	2.33	2.83	0.50	0.06	0.0050	0.0003
31	Rise rate	3.12	2.85	0.27	0.06	0.0045	0.0003
32	Fall rate	2.39	3.06	0.66	0.03	0.0040	0.0001
33	Reversals	4.56	4.29	0.27	0.00	0.0036	0.0000
Sum							0.4932

TABLE 20.16 OWA Operator for Gongju Station

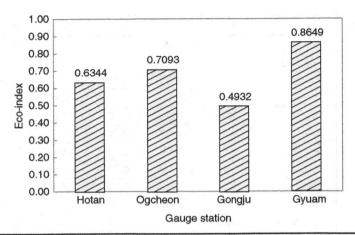

FIGURE **20.8** Eco-index.

20.5 Case Study: Application to Mekong Ecosystem

For application of the eco-index, the Mekong River ecosystem is considered. In the Mekong River system, a cascade of dams have been constructed during the 1980–1994 period. Thus, the prealteration period can be defined as the period from the beginning to the construction of the first dam of the dam cascade (1980 to 1994). For this system, the postalteration period covers the most available data set and the start of the dam construction period in the Upper Mekong (1995 to present, say, 2006).

The Mekong River originates in the Tibetan Plateau, China, at an elevation of over 5000 m, and flows through Laos, Thailand, Cambodia, and Vietnam (Lower Mekong countries) into the South China Sea. The Mekong River is 4,800 km long and has a drainage area of 795,000 km^2; it supports one of the world's most diverse fisheries, second only to Brazil's Amazon River. Over 60 million people depend on the Mekong River and its tributaries for food, water, transport, and many other aspects of their daily lives. By 2010 the population grew to between 75 and 90 million (Ringler, 2000). A cascade of eight dams have been proposed and/or are in development on the mainstream of the Upper Mekong River to meet the demands for energy, as shown in Fig. 20.9. Planned Manwan hydropower (HP) station, with an installed capacity of 1500 MW, was completed in 1996 and is in operation. Dachaoshan HP station (1350 MW) was completed in 2003. Xiaowan HP station (4200 MW) is under construction. Two others, Jinghong (1500 MW) and Nuzadu HP stations (5500 MW), are in progress.

On the other hand, a few demand-based dams have been proposed on the tributaries of the Lower Mekong, specifically in Laos' People's Democratic Republic (PDR). Therefore, the Mekong River Commission (MRC), regional governance of the Lower Mekong, is challenged to meet multiple water demands within the constraint of limited freshwater supply. The necessity to integrate ecosystem needs is also pronounced in Mekong's water management. Proper ecosystem management is essential to protect native biodiversity and ecological processes. This proposed and ongoing reservoir ecosystem may override the existing riverine ecosystems. Even though the completed Manwans storage capacity is not great, it plays a role one way or another in altering the natural flow regime. On the

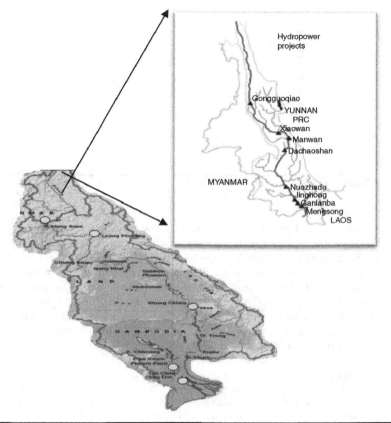

FIGURE 20.9 Location of stream gauges in the Lower Mekong.

other hand, the completion of the reservoir at Xiaowan before the year 2010 may adversely affect the sustained flow regime. Thus, the Lower Mekong River, which contributes 82 percent of the annual discharge (about 15,000 m³/s), is selected for the analysis of hydrologic alteration associated with the dam construction in the Upper Mekong basin.

The Mekong River stream is sustained mainly by the intraregional rainfall and river head ice melting. Annual precipitation and flow in the Lower Mekong are strongly seasonal. About 85 percent of the precipitation falls during the rainy season. Annual precipitation averages about 1680 mm across the basin (Ringler, 2000). One of the striking characteristics of the Lower Mekong's hydrologic regime is the flow regulation by the Great Lake in Cambodia, the largest permanent freshwater body in Southeast Asia. The Great Lake plays an important role in regulating the flow downstream of Phnom Penh. During high season, the Great Lake stores inflow from the Mekong River; this is reversed to the river during the low season. Due to this situation, the flow during high season can be reduced and the flow during low season can be increased from the downstream part of Phnom Penh.

Based on the availability of stream gauge records, five stream gauge sites, namely, Chiang Saen, Luang Prabang, Pakse, Tan Chau and Chao Doc, along the Lower Mekong

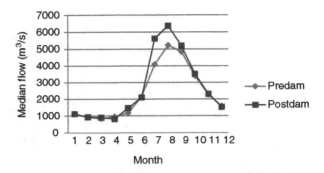

FIGURE 20.10   Monthly median flow at Chiang Saen.

River, as shown in Fig. 20.9, are selected for the analysis of hydrologic alteration associated with dam construction. However, one also must make certain to select the locations where most of the livelihoods of rural masses in the region depend on fish and fishery products. The period of analysis covered is 1980 to 2006, with the predam period being from 1980 to 1994. Figure 20.10 shows the magnitude of dam-induced alteration on natural flow regime at Chiang Saen gauging station which is the uppermost gauging station in the Lower Mekong. Figure 20.11 justifies the alteration of the natural flow regime through the measure of dispersion, which is the spread between the 25th and 75th percentiles, divided by the median.

Example 20.6.   Consider the case of the Lower Mekong, as shown in Fig. 20.9. One of the IHA parameters is the predam mean monthly flow for January at Chiang Saen gauging station, as shown in Table 20.17. The mean value of this parameter is 1163.73 m³/s, and its standard deviation is 154.43 m³/s. Compute the probability distribution for this IHA parameter.

Solution.   The given values of the mean and standard deviation are considered as constraints. Therefore, the probability distribution will be normal. Using these values, Eqs. (20.2) and (20.3) yield values of Lagrange multipliers as λ_0, λ_1, and λ_2 [34.3515, −0.0488, 2.096 × 10⁻⁵]. Substitution of these values in Eq. (20.4) yields the normal probability distribution given by Eq. (20.22), where $a = 1163.73$ m³/s and $b = 154.43$ m³/s. Insertion of this probability distribution in Eq. (20.1) gives the maximum entropy H of 6.46 nats for the mean flow for the month of January.

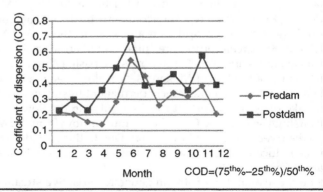

FIGURE 20.11   Coefficient of dispersion at Chiang Saen.

Year	Mean Flow (m³/s)	Year	Mean Flow (m³/s)	Year	Mean Flow (m³/s)
1980	1040	1985	1040	1990	1190
1981	1100	1986	1300	1991	1110
1982	1230	1987	1280	1992	1450
1983	946	1988	1200	1993	1030
1984	1450	1989	1030	1994	1060

TABLE 20.17   Predam Mean Flow (m³/s) for January at Chiang Saen

In this manner, the least-biased probability distributions can be determined for each IHA parameter for both predam and postdam conditions. Table 20.18 shows the values of constraints, Lagrange multipliers, and the form of the distribution for the mean flow for the month of January at Chiang Saene gauging station for both predam and postdam conditions.

Note that constraints may be different for different IHA parameters, depending on the shapes of their empirical distributions. For most parameters, however, the first two moments and hence the normal distribution suffice, providing the maximum entropy values. For a group 2 parameter that defines the magnitude and duration of the ecosystem, constraints are specified that lead to a lognormal distribution, giving the maximum entropy value associated with the information contained in the flow regimes of pre- and postalteration periods.

Figure 20.12 shows normally distributed mean monthly flow at Chiang Saen which is the uppermost gauge station in the Lower Mekong. It reveals the alteration of flow regime during the wet season (July to November). The ongoing cascade dam series in the Upper Mekong plays a role in altering the flow regime, which leads to the loss of information about the biologically relevant ecosystem.

The maximum entropy values are computed from Eq. (20.1) using the least biased probability distribution derived for each of the biologically relevant parameters for the post- and predam periods. Table 20.18 shows the values of maximum entropy for the mean flow for the month of January at Chiang Saen gauging station. The computed values of entropy for parameter group 1 and group 2, shown in Fig. 20.13 and 20.14, respectively, also justify this observation.

An interesting feature of Figs. 20.13 and 20.14 is that entropy values during the postdam period are on the high side compared to the predam condition at Chiang Saen, Luang

IHA Parameter Group	Parameter	Predam	Postdam
Parameter group 1	January	1. Constraints: Mean = 1163.73 m³/s SD = 154.43 m³/s 2. Lagrange multipliers: λ_0, λ_1, and λ_2 [34.3515, −0.0488, 2.096E-05] 3. Distribution: Normal 4. Maximum entropy: 6.46 nats	1. Constraints: Mean = 1126.11 m³/s SD = 147.17 m³/s 2. Lagrange multipliers: λ_0, λ_1, and λ_2 [35.1847, −0.0520, 2.308E-05] 3. Distribution: Normal 4. Maximum entropy: 6.41 nats

TABLE 20.18   Maximum Entropy Values for Mean Flow for January at Chiang Saen

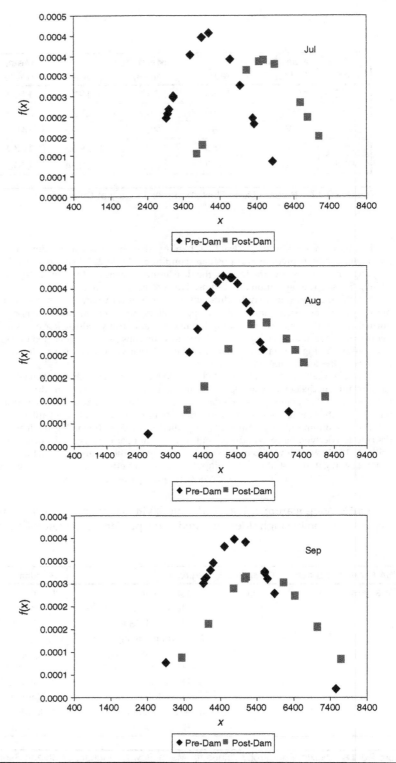

Figure 20.12 Distribution of mean monthly flow at Chiang Saen.

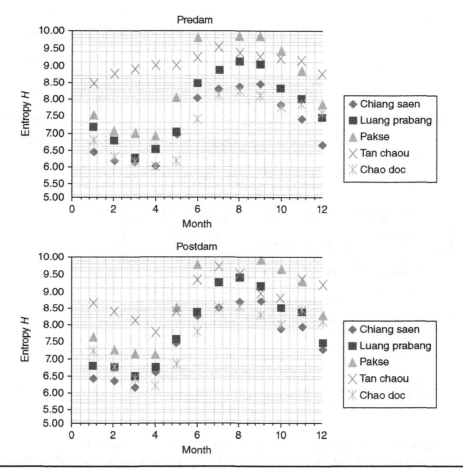

FIGURE 20.13   Entropy of group 1 parameters.

Prabang, and Pakse. However, the entropy value during the postdam period at Tan Chau in the Mekong Delta is lower than during the predam period. For parameter group 1 (Fig. 20.13), the value of entropy at Tan Chau also drops significantly, especially during the dry season which may harbor the habitat availability and food (Richter et al., 1996). These findings highlight that constraints on the flow regime are relaxed in the upper basin during the postdam period, whereas more constraints are added to biological flow regimes during the postdam period at Tan Chau. This is in accord with the increase in the distribution entropy as constraints are removed and drop with more constraints. This shows that the tributaries between the two gauges (Pakse and Tan Chau) will have to bear the burden of a further deteriorating ecosystem, even though dams have played a role.

If the claim is that dams have alone altered the flow regime, then a similar pattern of alteration should be observed at Tan Chau as well. Thus, one can reject this claim as the alteration at Tan Chau is totally opposite to what one can see at Pakse and above. Tonle Sap Lake, the largest permanent freshwater body in Southeast Asia, also plays an important role in regulating the flow downstream of Pakse. However, most of the

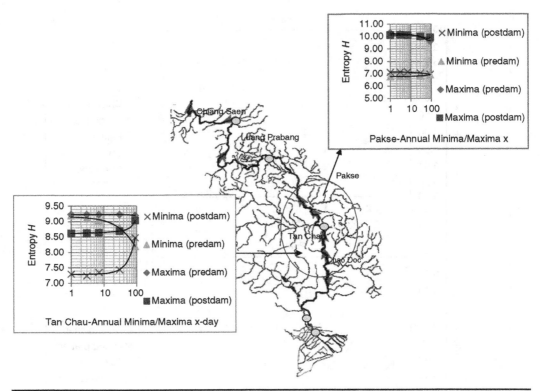

FIGURE 20.14 Entropy of group 2 parameter.

spawning sites are below Pakse. Movements of species are bidirectional based on their seasonal needs, as shown in Fig. 20.14. The Mekong region is vital for fish migration and its spawning below Pakse. As water flows downstream of Pakse, the alteration in the group 2 parameter accumulates and registers a high value at Tan Chau (Fig. 20.14). This may destroy the structuring of river channel morphology, physical habitat conditions, and aquatic ecosystems by abiotic versus biotic factors. Further the duration of stressful conditions, such as low oxygen, concentrated chemicals in aquatic environments, and aeration of spawning beds in channel sediments, may drop due to the alteration in parameter group 2 (Richter et al., 1996).

Table 20.19 shows the computed NSL for mean flow for the month of January at Chiang Saen gauging station. The steps required to determine a nonsatisfied level are outlined in Figs. 20.15 and 20.16, which show the total NSL for all the gauge sites and

| IHA Parameter group | Parameter | Entropy at Chiang Saen | | NSL |
		Predam	Postdam	
Group 1 parameter	January	6.46	6.41	0.05

TABLE 20.19 NSL Values for Mean Flow for January at Chiang Saen

for all IHA groups. The total NSL is significantly below the Pakse gauging location. Besides this, as explained earlier, IHA group 2 has gone through much alteration below Pakse. The total NSL for IHA group 3–group 5 is not that significant compared to the rest of the groups.

This procedure is followed for each gauging site.

The computed values of eco-index, presented in Figure 20.17, show that the Mekong ecosystem is vulnerable, especially below Pakse and thus in the Mekong Delta. The differences in the eco-index values between the Chiang Saen, Luang Prabang, and Pakse

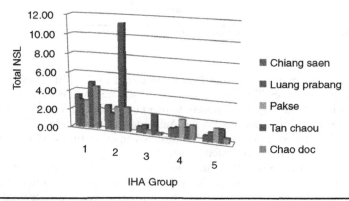

FIGURE 20.15   Total NSL for IHA groups.

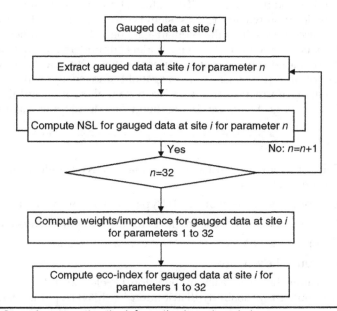

FIGURE 20.16   Steps for computing the information-based eco-index.

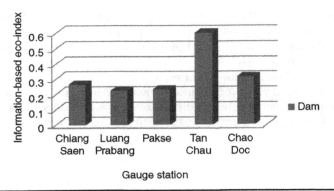

FIGURE 20.17   Eco-index along Lower Mekong-Dam Impact.

gauge stations may not be significant compared to the differences between Pakse and Tan Chau or between Tan Chau and Chao Doc. The point is that tributaries below Pakse may not have been regulated wisely to meet the need of the ecosystem, or eco-managers may have paid less attention to the message of cascade dams in the Upper Mekong. Considering the eco-needs of the Mekong River, allocating water resources in these tributaries may enrich the ecosystem below Pakse.

This exercise shows that an information-based eco-index that reflects the non-satisfaction flow regime can guide Lower Mekong eco-managers in a proper direction in allocating water resources among potential users, and concentrating their attention on mitigating the dam-induced effects and future alteration on sustained flow regime. It also gives an indication of the role of Lower Mekong tributaries in negating the effect of proposed dams on flow regimes. The values of eco-index at gauging locations indirectly portray the location where the ecosystem is severely altered. The relative values of this eco-index also show where the ecomanagers need to focus immediate attention. Many times the paucity of hydrological data hinders our understanding of the state of a system. The entropy-based eco-index uses observed flow. Often water resource development activities within a river basin rely on spatial homogeneity or heterogeneity. Therefore, having spatial information about the ecosystem alterations and their influences at the subbasin level or grid scale needs attention when developing water resources.

Questions

1. Select a river reach, and give a summary of hydrological parameters used in IHA for the reach.

2. Compute the entropy of two IHA parameters for the data in Question 1.

3. Compute the nonsatisfaction level for the two parameters used in Question 2.

4. Compute the OWA operator for the data in Question 1

5. Compute the eco-index.

6. Consider the river reach in Question 1. Select an IHA parameter for the reach. Compute the probability distribution for this IHA parameter.

References

Arthington, A. H. (1998). Comparative evaluation of environmental flow assessment techniques: Review of holistic methodologies. Land and Water Resources Research and Development Corporation Occasional Paper No. 26/98, Canberra, Australia.

Dyson, M., Berkamp, G. and Scanlon J. (2003). *Flow: The Essentials of Environmental Flows.* IUCN, Gland, Switzerland, and Cambridge, United Kingdom.

Fulle'r, R., and Majlender P. (2001). An analytic approach for obtaining maximal entropy OWA operator weights. *Fuzzy Sets and Systems,* vol. 124, no. 1, pp. 53–67.

Lamata, T. (2004). Ranking of alternatives with ordered weighted averaging operators. *International Journal of Intelligent Systems,* vol. 19, pp. 473–482.

Lytle, D. H., and Poff, N. L. (2004). Adaptation to natural flow regimes, *Trends in Ecology and Evolution,* vol. 19, pp. 94–100.

Naiman, R. J., Bunn, S. E., Nilsson, C., Petts, G. E., Pinay, G., and Thompson, L. C. (2002). Legitimizing fluvial ecosystems as users of water: An overview. *Environmental Management,* vol. 30, pp. 455–467.

O'Hagan, M. (1988). Aggregating template or rule antecedents in real-time expert systems with fuzzy set logic. Proceedings of the 22nd IEEE Asilomar Conf. on Signals, Systems, Computers, Pacific Grove, CA, 1988, pp. 81–689.

Poff, N. L., Allan, J. D., Bain, M. B., Karr, J. R., Prestegaard, K. L., Richter, B. D., Sparks, R. E. and Stromberg, J. C. (1997). The natural flow regime: A paradigm for river conservation and restoration. *BioScience,* vol. 47, pp. 769–784.

Postel, S., and Richter, B. D. (2003). *Rivers for Life: Managing Water for People and Nature.* Island Press, Washington.

Richter, B. D., Baumgartner, J. V., Braun, D. P. and Powell, J. (1998). A spatial assessment of hydrologic alteration within a river network. *Regulated Rivers: Research and Management,* vol. 14, pp. 329–340.

Richter, B. D., Baumgartner, J. V., Powel, J., and Braun, D. P. (1996). A method for assessing hydrologic alteration within ecosystems. *Conservation Biology,* vol. 10, pp. 1163–1174.

Ringler, C. (2000). Optimal allocation and use of water resources in the Mekong River Basin: Multi-country and intersectoral analyses, Doctoral dissertation, Bonn, Rheinischen Friedrich-Wilhelm's Universität.

Singh, V. P. (1998). The use of entropy in hydrology and water resources. *Hydrological Processes,* vol. 11, pp. 587–626.

Smakhtin, V., Revenga, C. and Doll, P. (2004). A pilot global assessment of environmental water requirements and scarcity. *Water International,* vol. 29, pp. 307–320.

Tharme, R. E. (2003). A global perspective on environmental flow assessment: Emerging trends in the development and application of environmental flow methodologies for rivers. *River Research and Applications,* vol. 19, pp. 397–442.

The Nature Conservancy (2009). *Indicators of Hydrologic Alteration, Version 7.1, User's Manual.*

Vorosmarty, C. J., P. Green, J., Salisbury, and Lammers, R. B. (2000). Global water resources vulnerability from climate change and population growth. *Science,* vol. 289, pp. 284–288.

Yager, R. R. (1999). Induced ordered weighted averaging operators. *IEEE Transactions on Systems, Man and Cybernetics,* vol. 29, pp. 141–150.

Further Reading

Amadore, L., W. C. Bolhofer, R. V. Cruz, R. B. Feir, C. A. Freysinger, S. Guill S, K. F. Jalal, et al. (1996). Climate change vulnerability and adaptation in Asia and the Pacific: Workshop summary. *Water, Air, and Soil Pollution*, vol. 92, pp. 1–12.

Arnell, N. W. (2006). Climate change and water resources: A global perspective. In: *Avoiding Dangerous Climate Change. Proceedings of the Exeter Conference*. Edited by H. J. Schellnhuber, W. Cramer, N. Nakicenovic, T. M. L. Wigley, and G. Yohe. Cambridge University Press, Cambridge, United Kingdom, pp. 167–175.

Bunn, S. E., and Arthington, A. H. (2002). Basic principles and ecological consequences of altered flow regimes for aquatic biodiversity. Environmental Management, vol. 30, pp. 492–507.

Chapman, E. C., and He, D. (1996). Downstream implications of China's dams on the Lancang Jiang (Upper Mekong) and their potential significance for greater regional cooperation, basinwide. In: *Development Dilemmas in the Mekong Region: Workshop Proceedings*, edited by B. Stensholt. Monash Asia Institute, Melbourne, Australia, pp. 16–24.

King, J. M., and Louw, D. (1998). Instream flow assessments for regulated rivers in South Africa using the building block methodology. *Aquatic Ecosystem Health and Management*, vol. 1, pp. 109–124.

Malmqvist, B., and Rundle, S. (2002). Threats to the running water ecosystems of the world. *Environmental Conservation*, vol. 29, pp. 134–153.

Mathews, R., and Richter, B. D. (2007). Application of the indicators of hydrologic alteration software in environmental flow setting. *Journal of American Water Resources Association*, vol. 43, no. 6, pp. 1400–1413.

Poff, N. L., Allan, J. D., Bain, M. B., Karr, J. R., Prestegaard, K. L., Richter, B. D., and Sparks, et al. (1997). The natural flow regime: A paradigm for river conservation and restoration. *BioScience*, vol. 47, pp. 769–784.

Richter, B. D., Baumgartner, J. V., Wigington, R. and Braun, D. P. (1997). How much water does a river need? *Freshwater Biology*, vol. 37, pp. 231–249.

Richter, B. D., Mathews, R., Harrison, D. L. and Wigington, R. (2003). Ecologically sustainable water management: Managing river flows for ecological integrity. *Ecological Applications*, vol. 13, pp. 206–224

Richter, B. D., and Richter, H. E. (2000). Prescribing flood regimes to sustain riparian ecosystems along meandering rivers. *Conservation Biology*, vol. 14, pp. 1467–1478.

Richter, B. D., Warner, A. T., Meyer, J. L. and Lutz, K. (2006). A collaborative and adaptive process for developing environmental flow recommendations. *River Research and Applications*, vol. 22, pp. 297–318.

Sparks, R. E. (1995). Need for ecosystem management of large rivers and their floodplains. BioScience, vol. 45, pp. 169–182.

Walker, K. F., Sheldon, F. and Puckridge, J. T. (1995). A perspective on dryland river ecosystems. *Regulated Rivers*, vol. 11, pp. 85–104.

Ward, J. V., and Stanford, J. A. (1995). The serial discontinuity concept: Extending the model to floodplain rivers. *Regulated Rivers: Research and Management*, vol. 10, pp. 159–168.

Yager, R. R. (1988). On ordered weighted averaging aggregation operators in multicriteria decision making. *IEEE Transactions on Systems, Man and Cybernetics*, vol. 18, pp. 183–190.

Index

Note: Page numbers followed by *f* denote figures; page numbers followed by *t* denote tables.